K. MACDONALD

Nelson

Biology 11

College Preparation

Authors

Maurice Di Giuseppe
Toronto Catholic District School Board

Douglas Fraser
District School Board Ontario North East

Barry LeDrew
Formerly of Newfoundland Department of Education

Jill Roberts
Formerly of Ottawa-Carleton District School Board

Program Consultant

Maurice Di Giuseppe
Toronto Catholic District School Board

Australia Canada Mexico Singapore Spain United Kingdom United States

Nelson Biology 11: College Preparation
By Maurice Di Giuseppe, Douglas Fraser,
Barry LeDrew, and Jill Roberts

Director of Publishing
David Steele

Publisher
Kevin Martindale

Executive Managing Editor, Development and Testing
Cheryl Turner

Program Managers
Lee Geller
John Yip-Chuck

Developmental Editors
Jackie Dulson
Lee Ensor
Lee Geller

Editorial Assistant
Mathew Roberts

Executive Managing Editor, Production
Nicola Balfour

Senior Production Editor
Debbie Davies-Wright

Copy Editor
Margaret Allen

Proofreader
Margaret Hoogeveen

Senior Production Coordinator
Sharon Latta Paterson

Creative Director
Angela Cluer

Art Director
Ken Phipps

Art Management
Kyle Gell
Allan Moon

Illustrators
AMID Studios
Greg Banning
Andrew Breithaupt
Steven Corrigan
Deborah Crowle
Kyle Gell design
Imagineering
Irma Ikonen
Kathy Karakasidis
Margo Davies LeClair
Dave Mazierski
Dave McKay
Marie Price
Dino Pulera
Bart Vallecoccia
Cynthia Watada

Design and Composition Team
AMID Studios
Kyle Gell design

Interior Design
Kyle Gell
Allan Moon

Cover Design
Peter Papayanakis

Cover Image
SPL/Photo Researchers

Photo Research and Permissions
Robyn Craig

Printer
Transcontinental Printing Inc.

Printed and bound in Canada
1 2 3 4 06 05 04 03

For more information contact Nelson, 1120 Birchmount Road, Toronto, Ontario, M1K 5G4.
Or you can visit our Internet site at http://www.nelson.com

National Library of Canada Cataloguing in Publication Data

Main entry under title:
Biology 11: college preparation / Maurice Di Giuseppe ... [et al.].

Includes index.
ISBN 0-17-626525-2

1. Biology–Textbooks. I. Di Giuseppe, Maurice II. Title: Biology eleven.

QH308.7.B58 2003 570
C2003-902215-3

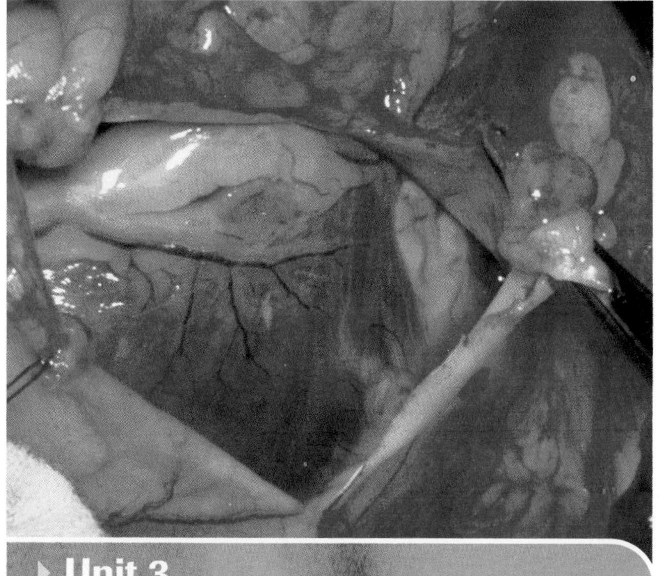

Unit 3 Animal Anatomy and Physiology

Unit 4 Plant Structure and Physiology

Unit 5 Environmental Science

Appendices

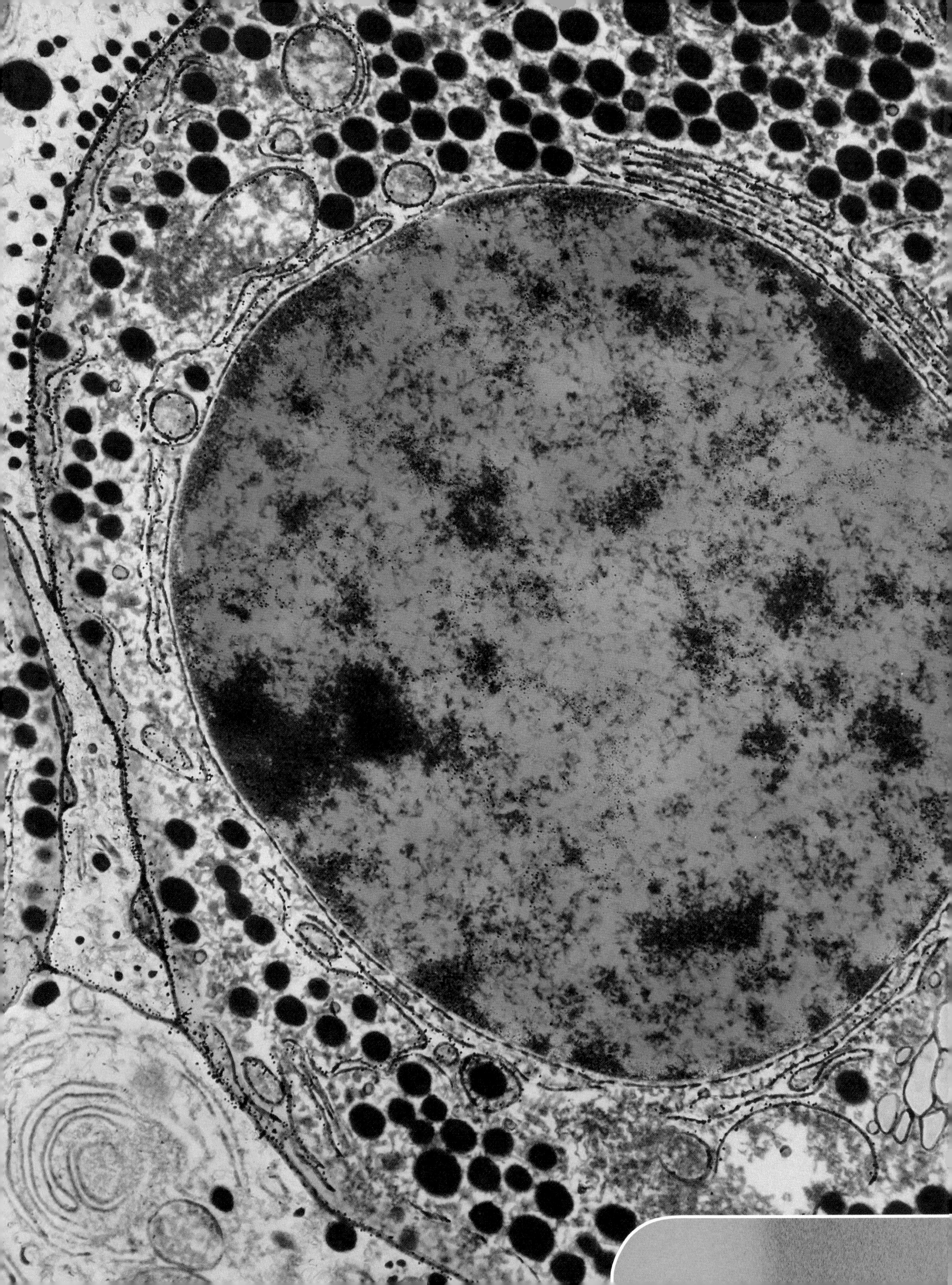

Reviewers

Advisory Panel

Christine Adam-Carr
Ottawa-Carleton Catholic District School Board, ON

Anu Arora
Peel District School Board, ON

Terry Brown
Thames Valley District School Board, ON

Thomas Joseph
Thunder Bay Catholic District School Board

Ted Laxton
Wellington Catholic District School Board, ON

Beth Lisser
Peel District School Board, ON

Stephanie Lobsinger
St. Clair Catholic District School Board, ON

Kimberley Walther
Durham Catholic District School Board, ON

Meredith White-McMahon
St. James-Assiniboia School Division, ON

Accuracy Reviewers

David Arthur
Ontario Society for Environmental Education

T.J. Beveridge PhD, FAAM, OAW, CRC, FRSC
Department of Microbiology, College of Biological Science, University of Guelph

Dr. John S. Greenwood
Associate Professor, Department of Botany, University of Guelph

Steve Harvey
Department of Physiology, University of Alberta

John Hoddinott
Biological Sciences, University of Alberta

Safety Reviewer

Ray Bowers
York/Seneca Institute for Science, Technology and Education

CONTENTS

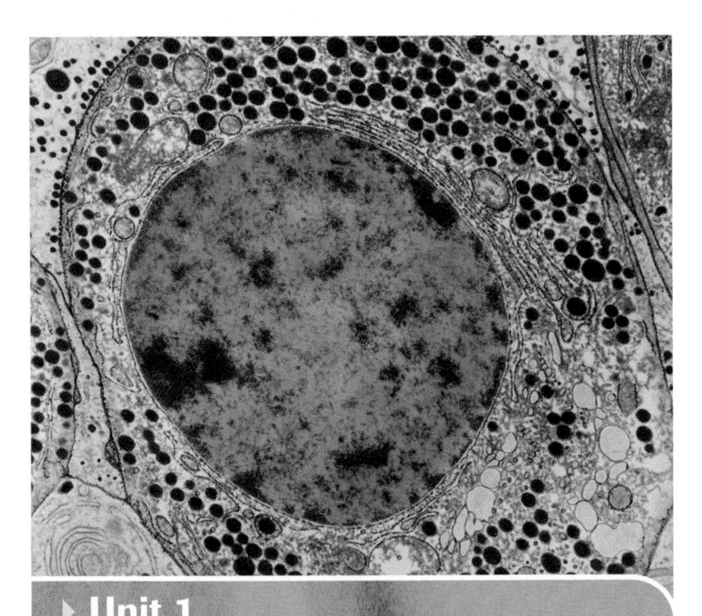

Unit 1 Cellular Biology

Unit 2 Microbiology

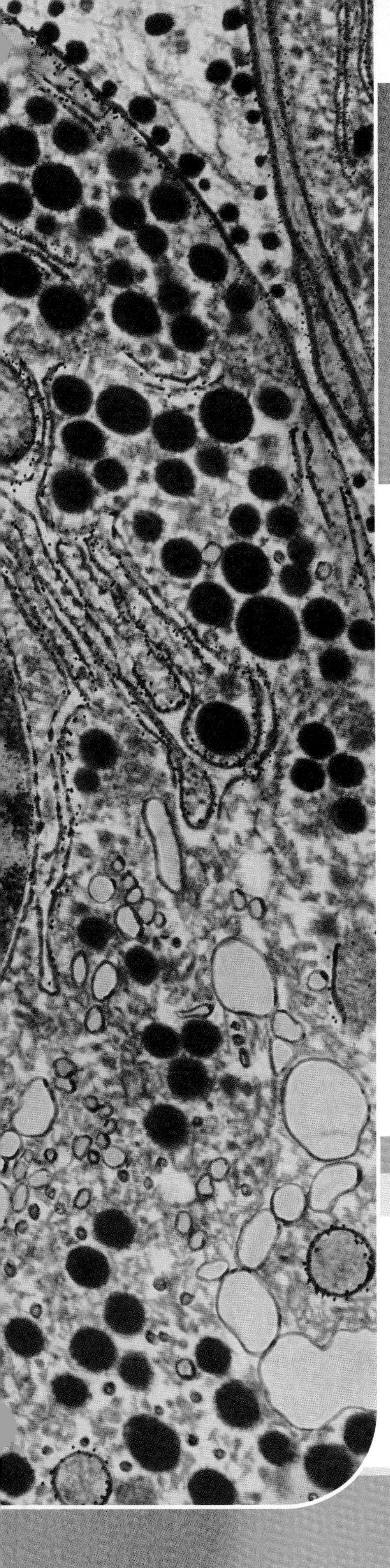

Nearly 300 years of biological research has confirmed that cells are the fundamental units of life. Your life depends on the health and well-being of the 100 trillion cells that make up your body. Virtually everything you do is aimed at ensuring the health and safety of your cells. Your nervous system makes your body avoid hot objects to prevent being burned. Eating and drinking supplies your cells with essential nutrients and water. Breathing brings vital oxygen into your cells and removes poisonous carbon dioxide. When you become ill, a doctor focuses on various groups of cells to determine the cause of your illness. Using a stethoscope, a nurse may examine the action of your heart muscle cells, and with a microscope, a lab technician may observe the condition of your blood cells. If your doctor suspects that your cells have experienced an abnormal change in structure or function, he or she may order an X-ray image to determine if the cells have changed their patterns of growth and reproduction. When the cause of your illness is determined, drugs may help your cells regain normal function.

Cells of other organisms are very useful. Yeast cells turn fruit juices into wine and cause bread dough to rise. Bacteria in your digestive system produce essential vitamins, moulds produce life-saving antibiotics, and stem cells may someday produce replacements for damaged organs. Learning about cells is a fundamental component of all the biological sciences.

Overall Expectations

In this unit, you will be able to

- demonstrate an understanding of the basic processes of cellular biology, including membrane transport, cellular respiration, photosynthesis, and enzyme activity;
- investigate the factors that influence cellular activity using appropriate laboratory equipment and techniques;
- demonstrate an understanding of the importance of cellular processes in your personal life as well as in the development and application of biotechnology.

ARE YOU READY?

Prerequisites

Concepts

- major cell organelles and their key functions
- differences between plant cells and animal cells
- elements, compounds, ions, and chemical reactions

Skills

- use a compound light microscope
- apply appropriate safety precautions when carrying out laboratory investigations
- select and use chemicals, glassware, and heating apparatuses in an appropriate and safe manner
- analyze social issues related to cellular biology
- formulate an opinion and support it with evidence
- communicate verbally and in writing

Knowledge and Understanding

1. Match the following names and functions to the labelled components of the cells in **Figure 1** (one name and one function per label).

   Names: *nucleus, cell wall, vacuole, flagellum, endoplasmic reticulum, Golgi apparatus, mitochondrion, chloroplast, cell membrane*

   Functions: *energy production, protein storage, protection, material transport within the cell, overall control, food production, locomotion, water and nutrient storage, gatekeeping*

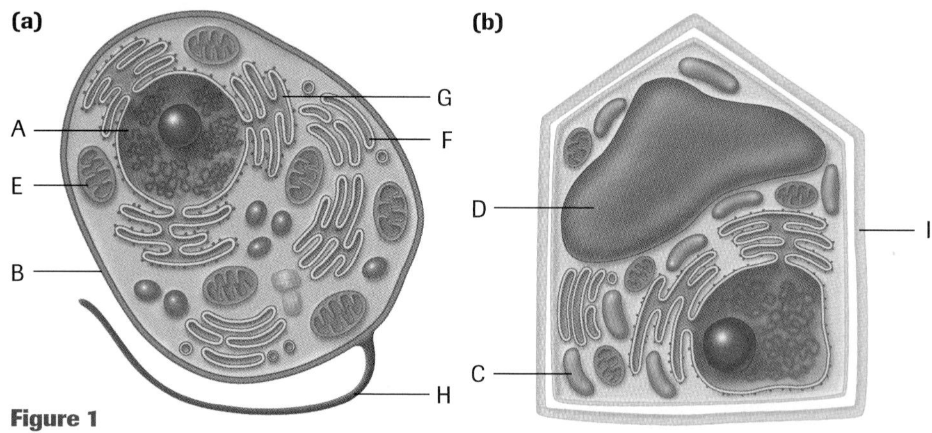

Figure 1
Generalized animal and plant cells

2. **Figures 2** and **3** illustrate two basic processes that move materials into and out of cells.
   (a) What process is represented by **Figure 2**?
   (b) What process is represented by **Figure 3**?
   (c) Name the structure labelled "A" in **Figure 3**.
   (d) What is the substance that passes through structure A in **Figure 3**?

Figure 2

A
before
after

Figure 3

3. Examine the following chemical equation:

   $$\underset{\text{methane}}{CH_4} + \underset{\text{oxygen}}{2O_2} \rightarrow \underset{\text{carbon dioxide}}{CO_2} + \underset{\text{water}}{H_2O}$$

   (a) Is the equation balanced? Explain.
   (b) Which of the reactants is a compound? Provide reasons for your answer.

4. Examine the following chemical formulas:

   $NaCl_{(s)}$ $NaOH_{(s)}$ $HCl_{(g)}$

   (a) Which compound forms an acidic solution when dissolved in water?
   (b) Which compound forms a basic solution when dissolved in water?
   (c) Which compound forms a neutral solution when dissolved in water?

Inquiry and Communication

5. Match the following parts of the compound light microscope to the labels in **Figure 4.**

   Microscope parts: *arm, ocular lens, coarse-adjustment knob, stage, fine-adjustment knob, base, condenser, objective lens, revolving nosepiece, stage clips, light source*

6. Place the following steps in the correct order for viewing a slide through a compound light microscope.
   (a) Rotate the nosepiece to the medium-power objective lens.
   (b) Use the fine-adjustment knob to bring the image into focus.
   (c) Place the slide on the stage and hold it in place with clips.
   (d) Use the coarse-adjustment knob to bring the low-power objective lens close to the slide.
   (e) Make sure the low-power objective lens is in place.

7. **Figure 5** is a wet mount of onion cells viewed under medium power in a compound light microscope.
   (a) What might be the cause of the dark circles in **Figure 5**?
   (b) How could the circles be avoided?

8. In your notebook, complete the following chart of measurements.

Decimal notation	Scientific notation
0.010 m	
	3.4×10^3 mL
385.5 g	

9. Complete the following metric conversions in your notebook.

10 cm	______mm
4.2 mm	______μm
3.90×10^{-9} m	______nm

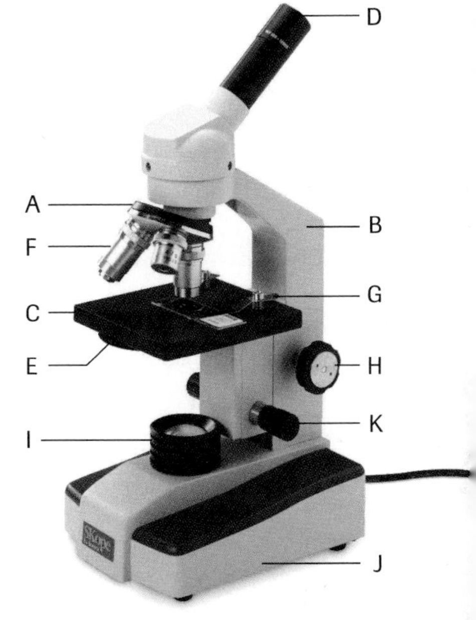

Figure 4
A compound light microscope

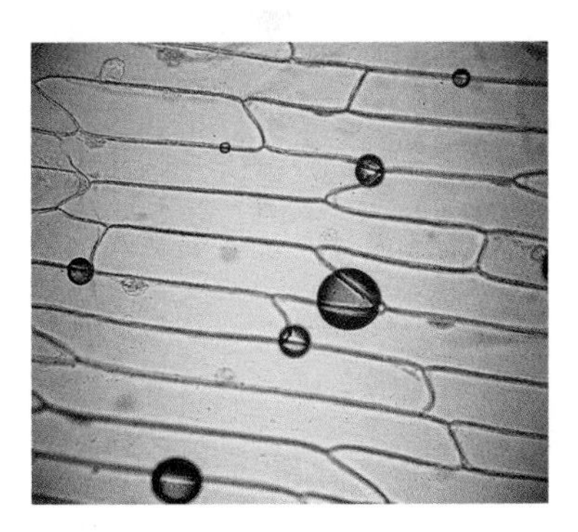

Figure 5
A wet mount of onion cells

GETTING STARTED

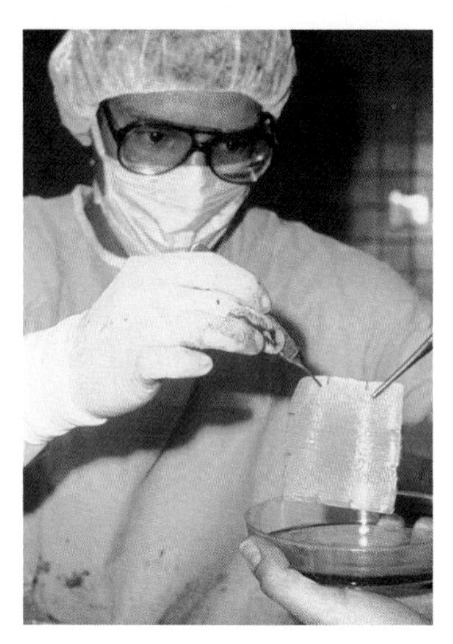

Figure 1
The first use of artificial skin to treat a third-degree burn occurred in 1981.

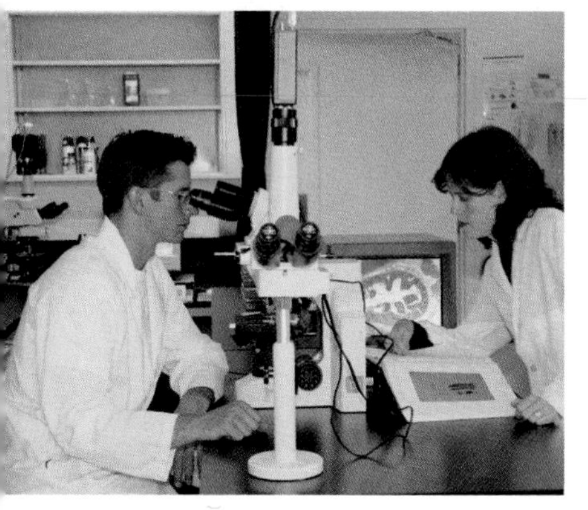

Figure 2
Biologists, technicians, and technologists conduct research and perform tests in laboratories.

Life on Earth began almost four billion years ago with the formation of the first cells. Since then, cells have multiplied and changed to form the huge number of different organisms we see today. Over the past several hundred years, we have learned so much about the structure and function of cells that we are able to put cells to work in beneficial ways and fix or prevent many of the problems that affect cells. In Canada, scientists have modified cells so that they produce chemicals called antibodies that fight diseases such as cancer, leprosy, and tetanus (lockjaw). Other Canadian scientists have fused white blood cells (that produce antibodies) to cancer cells (that multiply rapidly) to create new, mixed cells. These cells produce larger quantities of beneficial antibodies than the white blood cells would produce on their own.

When a person is burned, the skin in the affected area is destroyed. In the past, surgeons would replace burned skin with pieces of skin, called skin grafts, taken from unburned parts of the body. The grafts grow in the new location, but the area that the skin was taken from must also grow new skin cells. This was a painful process that can now be avoided with the use of artificial skin. Biologists grow artificial skin from skin cells incubated in petri dishes outside of the body, as seen in **Figure 1**. As a result, new skin grown in the laboratory can now be used to replace burned skin.

Scientists have also been able to change one type of cell into another. For example, skin cells have been forced to change into bone cells and, in more recent advances, fat cells have been changed into nerve cells. These are important advances in medical technology as they could provide a treatment for diseases such as osteoporosis and Parkinson's disease.

Modern cell research is conducted in laboratories all over the world (**Figure 2**). In addition to cell research, many laboratories, called diagnostic laboratories, conduct routine tests on samples of cells, tissues, and body fluids to determine the presence of disease or the type of disease a person or animal may have. A common medical laboratory procedure (called a biopsy) involves looking for cancer cells in a small sample of tissue that medical doctors remove from a patient's body. Other laboratories conduct studies on cells that are important in food and beverage industries. In these facilities, biologists, technicians, and technologists work as a team to discover more about how cells work and how they affect food processing.

REFLECT on your learning

1. How has the microscope benefited the study of biology?
2. There are four major classes of biological compounds: carbohydrates, lipids, proteins, and nucleic acids. Which class includes compounds that cells use as a ready source of energy? to store energy? to store genetic information?
3. What will happen to the size of a cell that is placed in distilled water? Why?
4. The chemical reactions of life must occur at a relatively fast rate. Most chemical reactions double in speed with every 10°C increase in temperature.
   (a) Why can't living cells use large amounts of heat to speed up reactions?
   (b) How do cells increase the speed of chemical reactions without using large amounts of heat?

▸ TRY THIS activity The Great Baking Powder–Yeast Challenge

Yeast are tiny one-celled fungi commonly used to brew beer and leaven bread (make it rise) (**Figure 3**).

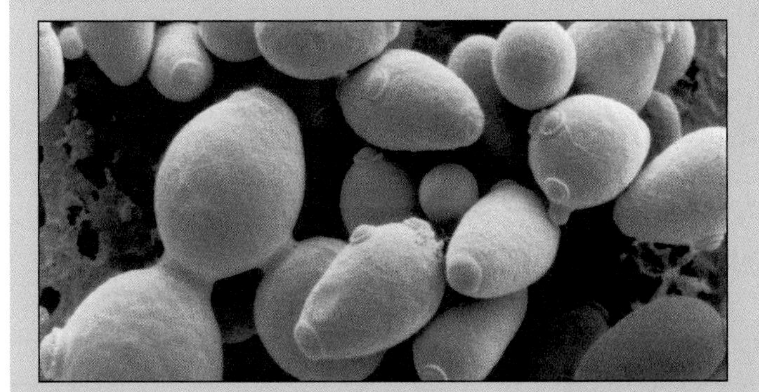

Figure 3
Yeast cells

Yeast cells make bread rise because they eat the sugar molecules in flour and release carbon dioxide gas as a waste product. However, you can make bread dough rise without yeast cells by adding a simple mixture of sodium bicarbonate (baking soda) and sodium aluminum sulfate (baking powder) to the bread batter.

Question
Which produces carbon dioxide gas more efficiently, yeast or baking powder?

Materials: safety goggles, 7 g active dry yeast, 7 g baking powder, 500 mL very warm water (40°C–45°C), 6 level tablespoons (each 14 mL) of sugar, 2 large rubber balloons, 2 identical empty 1-L water bottles, tape measure

Wear eye protection at all times.

1. Stretch the balloons by blowing each one up two or three times.
2. Add 250 mL of warm water and 3 level tablespoons (each 14 mL) of sugar to each of the 2 water bottles. Swirl until sugar is dissolved.
3. Simultaneously add 7 g dry yeast to one bottle and 7 g baking powder to the other bottle, and swirl until yeast and baking powder dissolve.
4. Attach a completely deflated balloon to the mouth of each bottle and set aside (**Figure 4**).

Figure 4

5. Measure the circumference of both balloons every minute for 15 min.

(a) Record your values in suitable chart form.
(b) Sketch a graph of balloon circumference versus time for yeast and baking powder on the same set of axes.
(c) Analyze your chart and graph to determine which substance produces carbon dioxide more efficiently.
(d) Describe similarities and differences in the rate of gas production in the first 5 min, the second 5 min, and the last 5 min.
(e) Does the baking powder experiment require sugar? Explain.
(f) Describe two possible sources of error in this procedure and make one suggestion for improvement.

1.1 Cell Theory and the Microscope

Figure 1
An animatronic chimpanzee

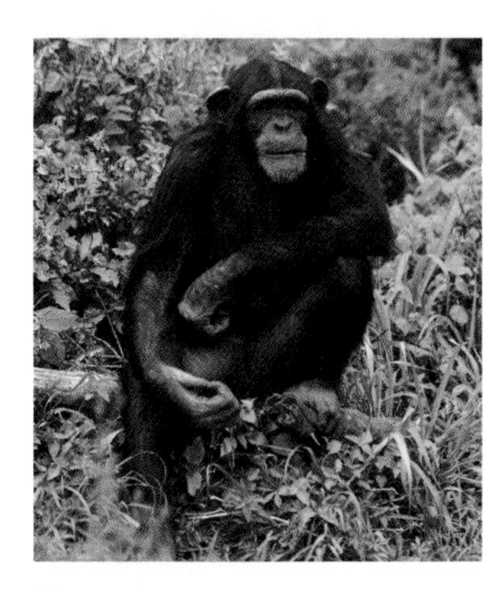

Figure 2
A real chimpanzee

Animatronics is the art and science of building robotic models of animals. Animatronic animals are programmed to be as realistic as possible and can be used in motion pictures instead of real animals, which might be injured. The animatronic chimpanzee in **Figure 1** shares many structural features with the real chimpanzee in **Figure 2**. However, no matter how lifelike an animatronic device may be, it is not alive. What characteristics distinguish a nonliving thing, such as a mechanical chimpanzee, from a living organism, such as a real chimpanzee? **Table 1** describes four basic characteristics that distinguish living things from nonliving things.

Table 1 Characteristics of Living Things

Characteristic	Example
• Living organisms obtain and use energy to power activities such as movement and growth.	Animals obtain energy by eating other organisms; plants obtain energy from light.
• Living organisms try to maintain a constant internal environment.	Humans keep a constant internal temperature by shivering when it's cold and sweating when it's hot.
• Living organisms reproduce.	Organisms produce new organisms like themselves.
• Living organisms are made of cells.	Cells are the smallest living things. Organisms can be composed of a single cell (unicellular organisms, such as amoeba) or many cells (multicellular organisms, such as mammals).

Cell Theory

After Robert Hooke discovered cells in 1655, scientists all over the world began to study their structures and functions. The study of cells became synonymous with biology.

In 1838, botanist Matthias Schleiden concluded that all plants are made of cells. In the following year, zoologist Theodore Schwann came to a similar conclusion about animal tissues. After concluding that plants, animals, and microorganisms were composed of cells, these two scientists stated the following scientific law:

- All living things are composed of one or more cells.
- Cells are the basic structural and functional units of life.

About twenty years later, in 1858, Rudolf Virchow, a medical doctor who conducted extensive studies on the causes of human disease, stated a further law:

- All cells arise from the division of other cells.

These three statements describe the basic nature of living things, and together are called cell theory.

Observing Cells

When you look at the skin on your hand with unaided eyes, you cannot see that it is composed of millions of tiny living cells. Observing cells is not easy. Much of what we know about cells could not have been discovered without the invention and development of the microscope. Early microscopes allowed scientists to discover the cell, and modern microscopes are so powerful that they help reveal how cells work.

Figure 3
A van Leeuwenhoek microscope

The Simple Light Microscope

The simple light microscope, like the one used by 17th-century biologist Anton van Leeuwenhoek, was hand held and composed of a single lens (**Figure 3**). Light from the sun, a candle, or a lamp was used to illuminate the specimen viewed with the simple light microscope.

The Compound Light Microscope

The compound light microscope (**Figure 4**) contains two lenses, an objective lens close to the specimen, and an ocular lens close to the observer's eye. The specimen is illuminated with sunlight or an electric lamp. The quality of a microscope may be expressed in terms of magnification and resolution.

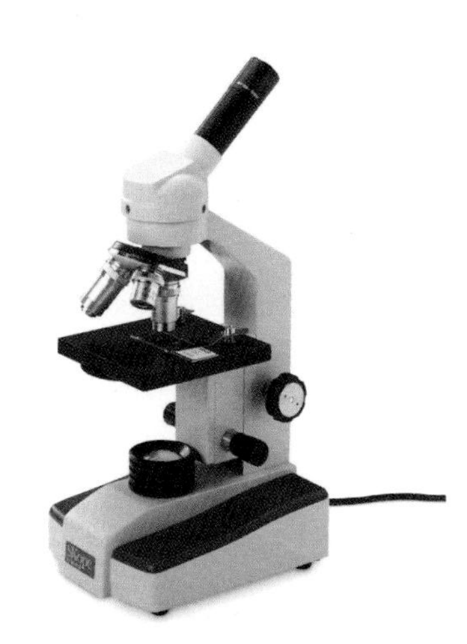

Figure 4
A compound light microscope

Magnification is the ability of lenses to enlarge the image of a specimen, and depends on the power of each lens. An image magnified 10× (10 times) by the ocular lens and 10× by the objective lens will be seen as 100× larger than the specimen looks to unaided eyes. The total magnification of a microscope is equal to the product of the ocular lens magnification and the objective lens magnification. The best compound light microscopes can magnify up to about 2000×.

Resolution refers to the amount of detail seen when viewing a specimen. A microscope with poor resolution produces a blurred image; one with good resolution produces a clear, sharp image in which features that are close together in the specimen appear separate and distinct.

magnification the ability of lenses to enlarge the image of a specimen

resolution the ability to show objects that are close together as separate and distinct

TRY THIS activity — Resolution

A microscope is most useful when it produces highly magnified images with good resolution. This means that a microscope produces a large image in which components of the specimen that are close together appear separate and distinct. In this activity, you will resolve components of a colour picture with a microscope.

Materials: compound light microscope, microscope slide and cover slip, scissors, colour picture from a magazine or a colour dot-matrix printer

1. Cut a small (1 cm × 1 cm) section of a colour image from a magazine or colour dot-matrix printout. Include parts of the picture containing red, green, or blue colour.
2. Place the section in the centre of a microscope slide and cover with a cover slip.
3. Examine the image with a microscope under low, medium, and high power.

(a) Describe the picture's qualities (e.g., graininess, etc.) when observed with unaided eyes.

(b) Describe the magnified images of the picture when examined under low, medium, and high power. How do the qualities of the magnified images compare with the image viewed with unaided eyes?

(c) Comment on the resolving abilities of your microscope.

Resolution depends on the type of light used to illuminate the specimen. The best compound microscopes, using visible light from the sun or electric lamps, cannot resolve (distinguish) parts of a specimen that are closer together than about 200 nm (2×10^{-7} m).

The Transmission Electron Microscope (TEM)

A transmission electron microscope (TEM) (**Figure 5**) uses an invisible beam of electrons. Since the electrons must pass through the specimen, the specimen must be sliced very thin. To accomplish this, the specimen is encased in plastic and shaved into very thin layers that are mounted onto a fine metal grid containing holes that allow the electrons to pass through. Light microscopes can be used to view living cells, but transmission electron microscopes can only be used to view dead cells (**Figure 6**).

The first useful transmission electron microscope was designed and built in 1937 by University of Toronto researchers James Hillier and Albert Prebus. This microscope could magnify up to 7000×. Modern transmission electron microscopes magnify up to 5 000 000×, and may resolve parts of a specimen that are up to 0.2 nm apart (the average distance between two atoms in a solid).

Figure 5
Transmission electron microscope

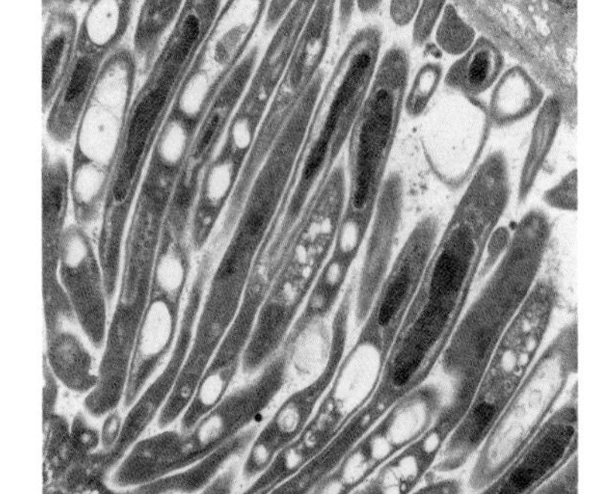

Figure 6
Transmission electron micrograph of a mosquito (*Anopheles*) gut

The Scanning Electron Microscope (SEM)

Unlike the transmission electron microscope, the scanning electron microscope (SEM) (**Figure 7**) reflects electrons from the surface of the specimen, allowing thicker specimens to be used. In fact, the surface features of whole animals such as small insects may be viewed using a scanning electron microscope (**Figure 8**). Scanning electron microscopes use magnets to focus a beam of electrons to a fine point. This thin stream of electrons scans the surface of the specimen and produces a three-dimensional image of the specimen on a television monitor. Scanning electron microscopes cannot magnify or resolve as well as transmission electron microscopes; however, they provide greater contrast and depth. The best SEMs magnify up to 300 000×, and have a maximum resolution of approximately 10 nm.

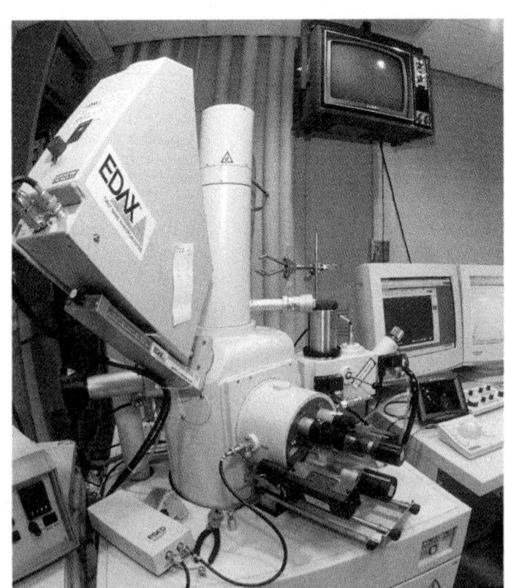

Figure 7
Scanning electron microscope

Figure 8
A fruit fly (*Drosophila*) as seen using a scanning electron microscope (magnification 60×)

DID YOU KNOW?

Powerful Microscopes
The most powerful transmission electron microscope, called a field-emission TEM, can resolve rows of atoms 0.01 nm apart!

Section 1.1 Questions

Understanding Concepts

1. State the cell theory in your own words.
2. Which had to come first, invention of the microscope or development of our present cell theory? Explain.
3. A virus contains a DNA molecule surrounded by a protein shell (**Figure 9**). Viruses can only reproduce when inside a living cell. Are viruses alive?

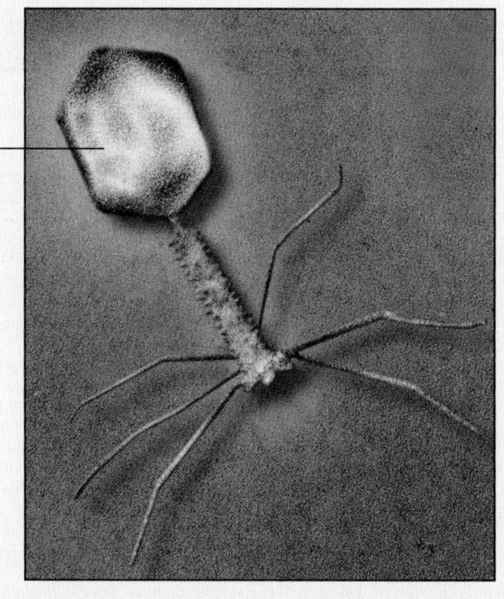

Figure 9

4. Distinguish between magnification and resolution.
5. State one advantage and one disadvantage of using a transmission electron microscope as compared with a scanning electron microscope.

Making Connections

6. The world's smallest microscope, called a microlens microscope, is approximately the size of the period at the end of this sentence.

   Conduct library and/or Internet research to answer the following questions about microlens microscopes.
   (a) What type of microscope is a microlens microscope?
   (b) What are some useful applications of a microlens microscope?

www.science.nelson.com

7. The following claims are seen on the package labels of two microscopes sold in the hobbies department of a toy store:

Claim	Price ($)
"Magnifies up to 5000×!"	69.99
"Magnifies up to 400×, and resolves to a maximum of 300 nm."	249.95

Which microscope is most likely the better buy? Explain.

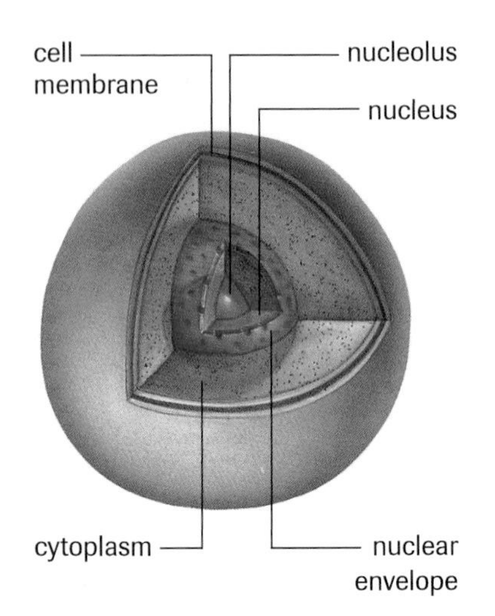

Figure 1
The basic structure of an animal cell

extracellular fluid the watery environment outside a cell

cytosol the aqueous solution inside a cell in which cell organelles are suspended

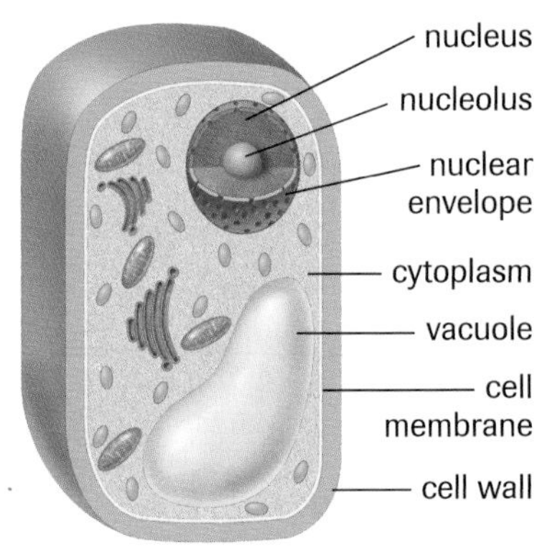

Figure 2
The basic structure of a plant cell

Cells are the basic structural and functional units of life. There are many different kinds of cells, which are specialized to carry out particular functions. In animals, muscle cells move limbs by contracting, bone cells produce the hard material that makes bones rigid, brain cells conduct electrical signals, and skin cells form waterproof sheets that cover your body. In plants, root cells absorb water from the environment, and green cells in leaves produce food. Although they perform different functions, cells have many common features.

General Structure of a Cell

A cell can be viewed as a factory with different departments, each carrying out specialized tasks that help keep the cell alive. Whether they are nerve cells in the brain of an animal, or epidermal cells in the leaf of a plant, cells are bathed in an aqueous (water-based) solution called **extracellular fluid**. On the inside, a cell contains an aqueous solution called **cytosol** in which a number of structures called cell organelles (little organs) are suspended (**Figure 1**). A special organelle called the nucleus acts as the control centre of the cell, issuing orders that determine how the other organelles do their work. Cytoplasm is the name given to the cytosol and cell organelles outside of the nucleus. The contents of an animal cell (cytosol and cell organelles) are separated from the extracellular fluid by a thin, flexible film called the cell membrane or plasma membrane. In addition to separating the interior of the cell from the external environment, the cell membrane holds the contents of the cell together and controls the movement of materials into and out of the cell. Plant cells have an additional outer covering called the cell wall (**Figure 2**). Cell walls are firm yet porous structures that give plants their rigidity while allowing water and dissolved materials to pass through easily. Wood is composed of the cell walls of dead plant cells.

Figure 3 shows a typical plant cell and a typical animal cell. Many of the smaller organelles identified in these diagrams cannot be seen with a compound light microscope, but instead, require the magnification and resolving power of a transmission electron microscope.

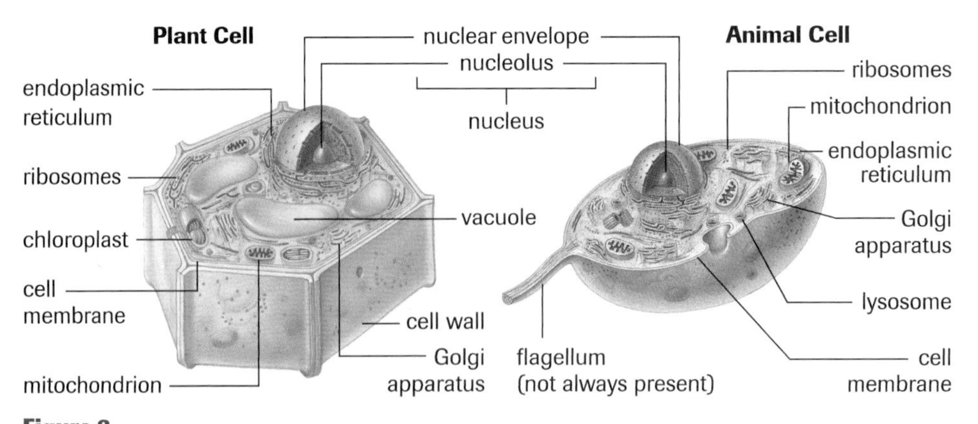

Figure 3
A typical plant cell and a typical animal cell

DID YOU KNOW ?

Cell Size
The smallest cell produced by humans is the sperm cell, and the largest cell is the egg cell. The largest cell produced by any organism is the ostrich egg!

Membranes

In addition to the membrane that surrounds the cell (the cell membrane), plant and animal cells have a number of organelles that are surrounded by membranes such as the nucleus, mitochondria, and chloroplasts. A membrane is composed of a bilayer (double layer) of fat (lipid) molecules, called **phospholipids**, and other compounds, called proteins and carbohydrates (**Figure 4**).

phospholipid a fat molecule in a cell membrane

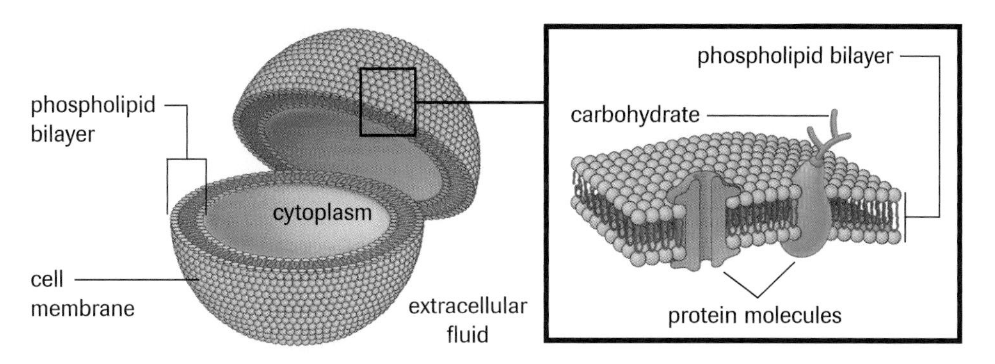

Figure 4
Biological membranes are composed of a phospholipid bilayer, proteins, and carbohydrates.

The parts of a membrane are always moving, and the membrane is said to be "fluid." That is why if a membrane is punctured with an ultra-fine needle it will not leak. Instead, the phospholipids and proteins will move in to fill the gap as the needle is removed. The membrane is more like a strong soap bubble than a rubber balloon.

Nucleus

As mentioned, the nucleus is the control centre of the cell and directs all of the cell's activities. The nucleus is separated from the cytoplasm by a porous double membrane called the **nuclear envelope**, and is filled with nucleoplasm. **Nucleoplasm** is a mixture of chemicals that stores information used by other cell organelles in carrying out their functions. Nucleoplasm is rich in compounds called nucleic acids. These include ribonucleic acid (RNA) and deoxyribonucleic acid (DNA).

nuclear envelope the double membrane surrounding the nucleus

nucleoplasm the fluid material inside the nucleus

Ribosomes

Ribosomes are organelles used by the cell to produce proteins (protein synthesis). Ribosomes are either floating in the cytoplasm or attached to membranes. In general, free-floating ribosomes produce proteins that are used inside the cell, and membrane-attached ribosomes manufacture proteins for use outside the cell. Ribosomes are so small (approximately 30 nm in diameter) that even when viewed with a transmission electron microscope they appear as small fuzzy dots (**Figure 5**).

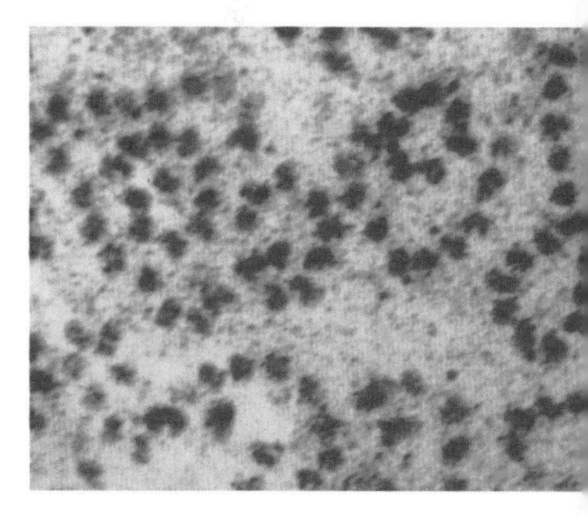

Figure 5
Ribosomes seen using a TEM (magnification 185 000×)

Endoplasmic Reticulum

The endoplasmic reticulum (ER) is a complicated system of membranous tubes and canals that connect with the nuclear envelope (**Figure 6**, on the next page). The endoplasmic reticulum is like a complex subway system with criss-crossing tunnels and stations. As you can see in **Figure 6**, there are two types of endoplasmic reticulum: **rough endoplasmic reticulum (RER)**, containing attached ribosomes, and **smooth endoplasmic reticulum (SER)**, with no ribosomes.

rough endoplasmic reticulum (RER) endoplasmic reticulum with attached ribosomes

smooth endoplasmic reticulum (SER) endoplasmic reticulum without ribosomes

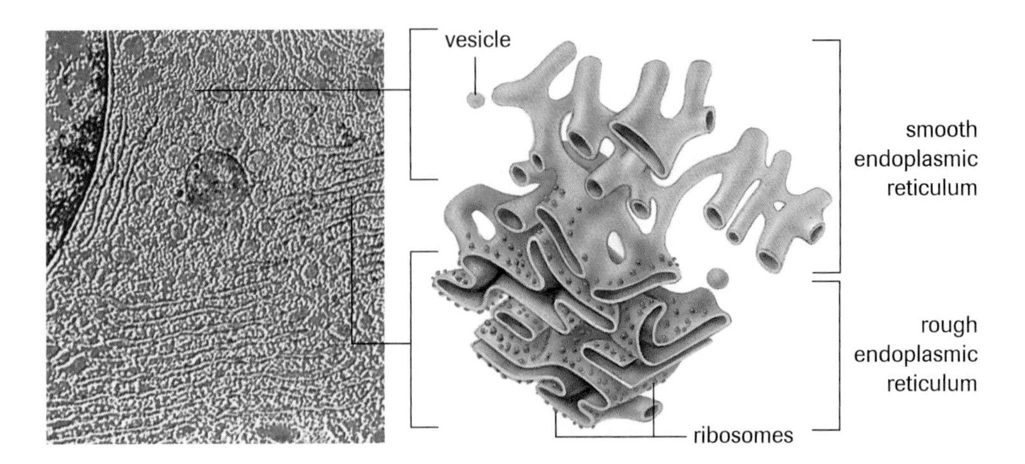

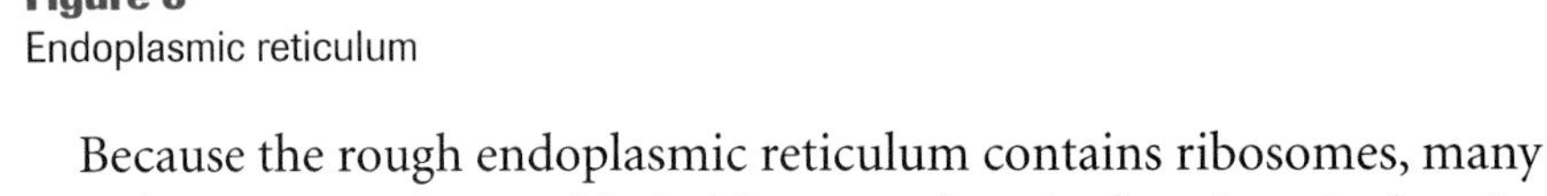

Figure 6
Endoplasmic reticulum

vesicle a membranous sac containing proteins and lipids that may be formed by the endoplasmic reticulum and Golgi apparatus

Because the rough endoplasmic reticulum contains ribosomes, many proteins are manufactured in it. The smooth endoplasmic reticulum is thought to be a place where molecules of fat are produced. The products of the endoplasmic reticulum become enclosed in membrane-bound structures called **vesicles**. These vesicles form when the membrane of the endoplasmic reticulum pinches off at the ends. These vesicles then travel toward another membranous organelle called a Golgi apparatus.

Golgi Apparatus

The Golgi apparatus is composed of several membranous tubes that, under the microscope, usually look like a stack of flattened balloons (**Figure 7**). It is named after Camillo Golgi, the medical doctor who first identified it in 1898. The Golgi apparatus chemically changes the fats and proteins produced in the endoplasmic reticulum and then packages them in vesicles. In many cases, the vesicles move through the cytoplasm, attach to the cell membrane, and release their contents into the extracellular fluid.

vesicle

Figure 7
The Golgi apparatus

Lysosomes (found only in animal cells)

Some of the vesicles formed by the Golgi apparatus are called lysosomes. Lysosomes are cell organelles containing proteins that can break down the molecules that cells are made of into their individual chemical components. Lysosomes are found only in animal cells, and may be used to digest food particles brought into the cell from the extracellular fluid. They are also used to destroy potentially dangerous microorganisms, such as bacteria and viruses, that may force themselves into the cell (**Figure 8**). When an animal cell gets old, lysosomes break open and decompose the entire cell. This process, in which the cell could be said to "commit suicide," is called **apoptosis**. The organism then uses the resulting compounds to build new cells. Hence, lysosomes are sometimes referred to as "suicide sacs."

apoptosis cell suicide carried out by lysosomes

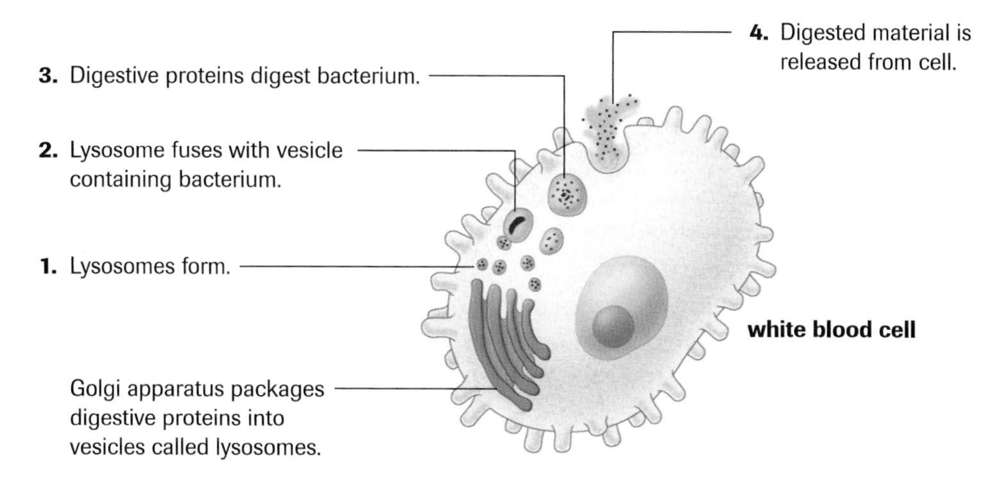

Figure 8
White blood cells use lysosomes to digest invading bacteria.

Mitochondria

Mitochondria (singular: mitochondrion) are circular or rod-shaped membranous organelles that float freely in the cytosol (**Figure 9**). Many important chemical reactions occur in mitochondria. These reactions contribute to **cellular respiration**, a series of chemical changes that produces compounds that cells use as a source of energy. Like the nucleus, the mitochondrion contains two membranes: a smooth outer membrane and a highly folded inner membrane. The folds of the inner membrane are called **cristae** (singular: crista). Cristae contain compounds that help carry out the reactions of cellular respiration, and provide a large surface area on which these reactions occur. Cells that require large amounts of energy, such as muscle cells in animals and root tip cells in plants, usually contain large numbers of mitochondria, each with many cristae. Cells that do not require large amounts of energy, such as most fat cells in animals and leaf cells in plants, have smaller numbers of mitochondria with fewer cristae. The interior space of a mitochondrion is filled with a protein-rich fluid called **matrix**. Many chemical reactions of cellular respiration also occur in the matrix.

cellular respiration a set of reactions partially occurring in mitochondria that produces compounds that cells use as a source of energy

crista an infolding of the inner membrane of mitochondria

matrix the protein-rich fluid that fills the interior space of a mitochondrion

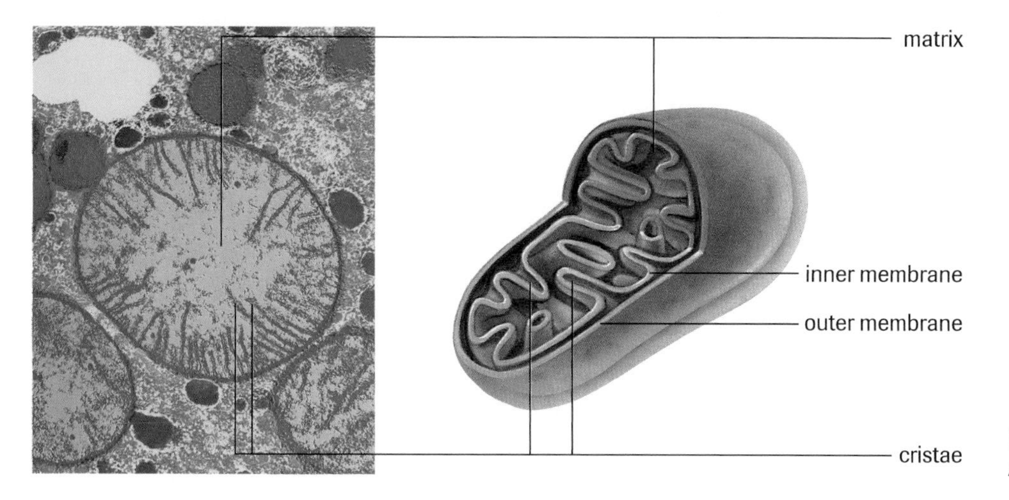

Figure 9
A mitochondrion

Plastids (found only in plant cells)

plastid an organelle in plant cells that stores useful materials or carries out vital functions such as photosynthesis

chloroplast a cell organelle found only in plant cells. It contains chlorophyll and carries out the reactions of photosynthesis

photosynthesis a set of chemical reactions that converts carbon dioxide and water into molecules that plant cells use for food

thylakoid a membrane-bound compartment found in chloroplasts

granum a stack of thylakoids found in chloroplasts

chlorophyll the green pigment that begins the process of photosynthesis. It is found in the thylakoid membranes of plant cell chloroplasts.

stroma the protein-rich fluid that fills the space between thylakoid membranes and the inner membrane of a chloroplast

Plastids are free-floating membranous organelles found only in the cytosol of plant cells. They are usually large enough to be seen with a compound light microscope. Plastids either store materials that are useful to the plant cell or perform important functions that keep the cell alive. One of the most important plastids is the **chloroplast**. Chloroplasts contain all of the chemicals necessary to perform **photosynthesis**, a set of chemical reactions that converts carbon dioxide and water (from the extracellular fluid) into molecules that plant cells use as food. Like mitochondria, chloroplasts possess two membranes, an outer and an inner membrane. However, in chloroplasts, other membranes are arranged into a system of interconnected compartments called **thylakoids**, and thylakoids are usually stacked on top of one another forming structures called **grana** (singular: granum) (**Figure 10**).

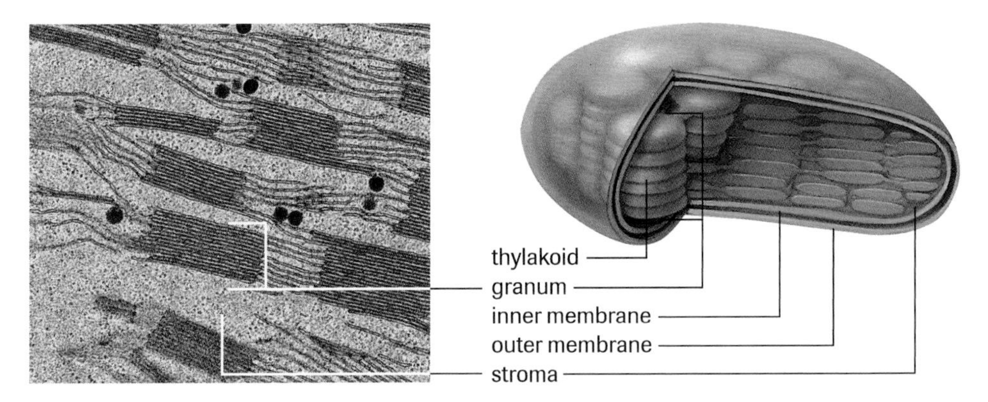

Figure 10
A chloroplast

The membranes of thylakoids contain molecules of green pigment called **chlorophyll**. Chlorophyll is responsible for the green colour of plants and is also responsible for the start of photosynthesis. The space between thylakoid membranes and the inner chloroplast membrane is filled with a protein-rich fluid called **stroma**. Stroma contains compounds needed for photosynthesis and DNA that allows chloroplasts to reproduce.

Other plastids in plant cells include amyloplasts and chromoplasts. Amyloplasts are white or colourless plastids that store the energy-rich products of photosynthesis in the form of starch. Chromoplasts are colourful plastids containing the red, orange, and yellow pigments commonly found in flowers, fruits, and vegetables.

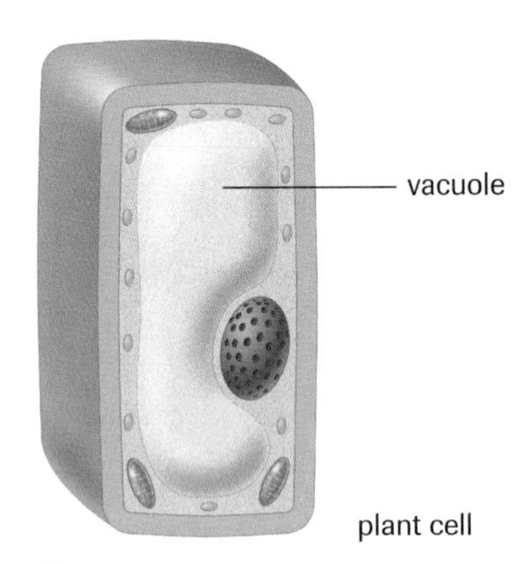

Figure 11
A vacuole in a plant cell

Vacuoles

turgor pressure pressure, caused by water in the cytoplasm and vacuoles, that helps keep the cell membrane pressed firmly against the cell wall of a plant cell

Vacuoles are usually large, membrane-bound sacs filled with a watery solution containing dissolved sugars, minerals, and proteins. They are common in plant cells, where the pressure inside one or two large vacuoles helps keep the cell membrane pressed firmly against the cell wall (**Figure 11**). This pressure, called **turgor pressure**, is responsible for the firm texture of fresh vegetables such as celery stalks and carrot roots, and of the stems and leaves of all plants. When plant cells lose water, the vacuoles shrink, and the decrease in turgor pressure causes structures such as stems and roots to become limp and wilted.

Section 1.2 Questions

Understanding Concepts

1. In your notebook, complete the following chart for **Figure 12**:

Cell part	Plant, animal, or both	Function	Figure 12 label
nucleus	both		
	plant	photosynthesis	
		stores water and nutrients	
rough endoplasmic reticulum			
		packages proteins	
mitochondria			

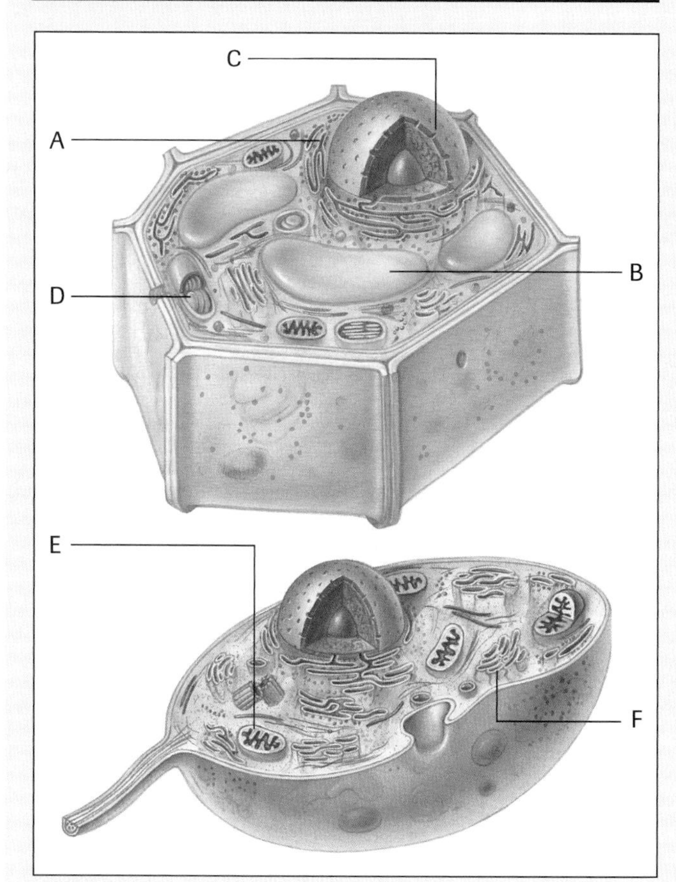

Figure 12

2. Why are plant cells generally firmer than animal cells?
3. Distinguish between nucleoplasm and cytoplasm.
4. Why do muscle cells contain more mitochondria than fat cells do?
5. (a) What is the main difference between the structure of smooth endoplasmic reticulum and the structure of rough endoplasmic reticulum?
   (b) Distinguish between the functions of smooth endoplasmic reticulum and rough endoplasmic reticulum.
6. Why does a cell membrane not leak when it is punctured with a needle?
7. Describe apoptosis.
8. Name the components of the labelled mitochondrion in **Figure 13**.

Figure 13

9. Describe a plant cell plastid other than the chloroplast.

Making Connections

10. Scientists in Canada have invented artificial cells that carry out some of the functions of real cells. Conduct library and/or Internet research to answer the following questions about artificial cells:
    (a) Describe the size and structure of a typical artificial cell.
    (b) Describe two uses of artificial cells.

www.science.nelson.com

11. Some scientists believe that the smallest cells on Earth are nanobes. Nanobes were discovered in 1998 by a group of Australian researchers. Conduct research to find out more about nanobes and the possibility that they are, in fact, the smallest organisms known.

www.science.nelson.com

1.3 Activity

Observing Cells

In this activity, you will prepare and examine wet mounts of animal cells and plant cells in order to observe the structural differences between the two types of cells. A wet mount is used to observe living cells that are thin and transparent enough for light to pass through. A wet mount is prepared by placing the specimen on a microscope slide, adding a drop of water, and covering with a cover slip, as in **Figure 1**.

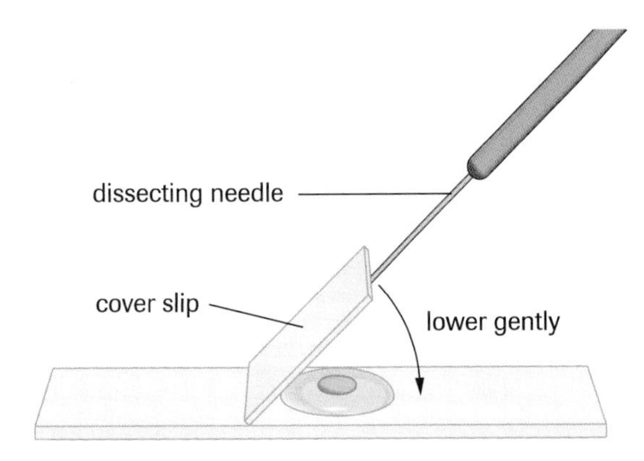

Figure 1
Preparing a wet mount

You will examine human cheek cells (your own) and liver cells as typical animal cells, and cells of potato, tomato, and a green aquatic plant as plant cells. Since some cells tend to be colourless, you will stain them with a dye to enhance their visibility.

Materials

- laboratory apron and gloves
- 2 microscope slides
- 2 cover slips
- 2 dissecting needles
- forceps
- dropper
- scalpel or single-edge razor blade
- transparent millimetre ruler or slide micrometer
- paper towels or filter paper
- distilled water at room temperature
- compound light microscope
- toothpicks
- petri dish
- saline solution (salt solution)
- methylene blue solution
- dilute iodine solution
- cotton swab
- potato
- fresh liver
- red tomato
- sprig of aquatic plant (*Ludwigia* or *Echinodorus*)

Wear a laboratory apron throughout this activity. Always cut away from yourself and others sitting near you in case the scalpel slips.

Procedure

There are six parts to this procedure. In Part 1, you will determine the diameter of the fields of view of your microscope under low, medium, and high power. The field of view is the bright white circle you see when looking through a microscope. In Parts 2, 3, 4, 5, and 6, you will prepare wet mounts of various plant and animal cells and view them under the microscope. You will examine the cells under low, medium, and high power, and use the field-of-view diameters determined in Part 1 to estimate cell sizes.

Part 1: Estimating the Size of Cells Viewed under a Microscope

Carry out the following procedure before proceeding to Parts 2, 3, 4, 5, and 6. Perform steps 1 to 4 if using a millimetre ruler. Perform steps 5 and 6 if using a slide micrometer.

1. With the low-power objective lens in place, lay a microscope slide on the stage and place a transparent millimetre ruler on the slide.
2. Viewing the microscope stage from the side, position the millimetre marks on the ruler immediately below the objective lens.
3. Looking through the eyepiece, focus the marks on the ruler using the coarse- and fine-adjustment knobs.
4. Move the ruler so that one of the millimetre marks is at the edge of the field of view.

(a) Measure and record the diameter of the low-power field of view in millimetres.

(b) Use the following equations to calculate the diameter of the medium- and high-power fields of view. Record these values for future use. (*Note*: Magnification is shortened to mag. in the equations.)

For medium power:

$$\text{diameter} = \text{total mag.}_{\text{low power}} \times \frac{\text{diameter}_{\text{low power}}}{\text{total mag.}_{\text{medium power}}}$$

For high power:

$$\text{diameter} = \text{total mag.}_{\text{low power}} \times \frac{\text{diameter}_{\text{low power}}}{\text{total mag.}_{\text{high power}}}$$

Go to (e).

5. With the medium-power objective lens in place, place the slide micrometer on the stage, and focus.

(c) Determine the diameter of the medium-power field of view by direct measurement, and record the value.

6. Switch to high power.

(d) Focus and determine the diameter of the high-power field of view by direct measurement. Record this value.

(e) Use the values you obtained for the diameters of the fields of view, and the following equation, to calculate the size of the cells you will be viewing in Parts 2, 3, 4, 5, and 6, when asked to do so.

Cell Diameter Equation:

$$\text{estimated cell diameter} = \frac{\text{diameter of field of view}}{\text{number of cells across field of view}}$$

Part 2: Examining Cheek Cells

7. To swab cheek cells, take a sterile cotton swab and move the swab over the inside of your cheek on one side of your mouth and along the outer lower side of your gums.

Wear gloves when swabbing cheek cells.

8. Prepare a wet mount by lightly spreading the swabbed material in a small drop of saline solution placed at the centre of a microscope slide, and gently covering with a cover slip, as in **Figure 1**. IMMEDIATELY place the used cotton swab in a container of absolute alcohol, as instructed by your teacher.

9. Examine the cheek cells under the microscope. *Remember*: Use the low-power lens first.

(f) If anything is observed, draw labelled diagrams of your observations.

10. Remove the slide from the microscope and place it on a piece of paper towel. Add one drop of methylene blue solution to one side of the cover slip. Place a small piece of paper towel or filter paper at the opposite side of the cover slip, as in **Figure 2**. The absorbent paper will draw the dye to the other edge of the cover slip, washing over and staining the cells.

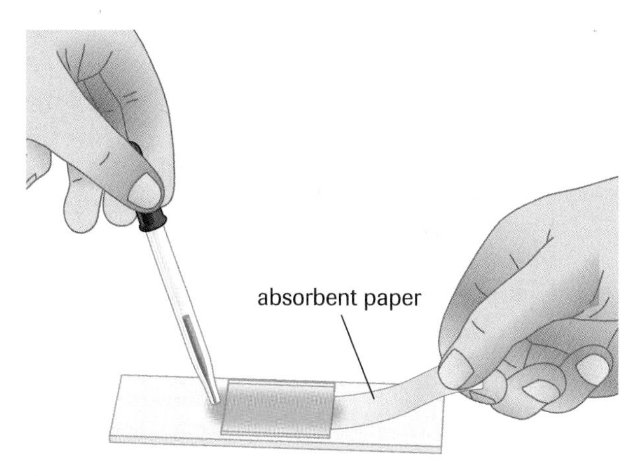

Figure 2
Staining a wet mount

11. Examine the stained specimen under the microscope.

(g) Using the high-power objective lens, estimate the number of times a typical cheek cell fits across the field of view, and use the Cell Diameter Equation given in Part 1, step 6(e) to calculate the size of the cell.

(h) Sketch and label a diagram of one or two separate cheek cells. Calculate the total magnification of the image, and include this value in your report.

12. After the cells have been observed, immerse the slide and the cover slip in a beaker of laboratory disinfectant, as directed by your teacher.
13. Slides and cover slips should be washed thoroughly, dried, and reused according to your teacher's instructions.

Part 3: Examining Aquatic Plant Leaf Cells

14. Obtain a young leaf from a sprig of a live aquatic plant and place it in a petri dish containing room-temperature water for 10 to 15 min.
15. Prepare a wet mount of the leaflet (the edges of leaflets are thin enough for viewing under a light microscope).
16. Examine the edges of the leaflet under the microscope. Note the arrangement of cells and veins at the leaf's edges. The plant cells are arranged in layers, one on top of the other. As you focus, you are going through different layers of cells. Find a thin area (one that does not have many layers) to study.
17. Pick a cell in which the nucleus (dark-coloured, round or oval structure) and chloroplasts (green discs) are visible. Note that the chloroplasts may be moving around in the cytoplasm. This movement is called cytoplasmic streaming. Cytoplasmic streaming helps circulate nutrients and other substances around the cell.

(i) Using the medium-power objective lens, estimate the number of times a typical leaf cell fits across the field of view, and use the Cell Diameter Equation given in Part 1, step 6(e) to calculate the size of the cell.

(j) Sketch and label a diagram of a leaf cell (you may include some of the neighbouring cells). Count the number of chloroplasts in the cell, and calculate the total magnification of the image. Include these values in your report.

18. Separate the cover slip from the slide. Discard the specimen according to your teacher's instructions, and wash and dry the slide and cover slip for use in Part 4.

Part 4: Examining Potato Cell Amyloplasts

19. Carefully use a single-sided razor blade or scalpel to make thin slices (sections) of potato. Cut the potato as thinly as possible.
20. Prepare a wet mount of a small piece of a potato section.
21. Examine the potato wet mount under the microscope and locate the amyloplasts.
22. Stain the specimen with iodine solution, using the same technique you used to stain your cheek cells. Iodine reacts with starch to produce a very dark blue colour. Wait 1 min and examine the amyloplasts again under the microscope.

(k) Using the medium-power objective lens, estimate the number of times a typical potato cell fits across the field of view, and use the Cell Diameter Equation given in Part 1, step 6(e) to calculate the size of the cell.

(l) Sketch and label a diagram of a potato cell. Calculate the total magnification of the image, and include this value in your report.

23. Separate the cover slip from the slide. Discard the potato specimen according to your teacher's instructions, and wash and dry the slide and cover slip for use in Part 5.

Part 5: Examining Red Tomato Cell Chromoplasts

24. Using a toothpick, carefully scrape some tomato pericarp (the tissue just under the peel).
25. Place a drop of water and a drop of methylene blue at the centre of a microscope slide, and prepare a wet mount by lightly spreading the pericarp through the liquid. Gently cover with a cover slip.
26. Examine the tomato wet mount under the microscope and locate the chromoplasts.

(m) Using the medium-power objective lens, estimate the number of times a typical tomato cell fits across the field of view, and use the Cell Diameter Equation given in Part 1, step 6(e) to calculate the size of the cell.

(n) Sketch and label a diagram of a tomato cell. Calculate the total magnification of the image, and include this value in your report.

27. Separate the cover slip from the slide. Discard the tomato specimen according to your teacher's instructions, and wash and dry the slide and cover slip for use in Part 6.

Part 6: Examining Liver Cells

28. Place one drop of crushed liver tissue on a clean, dry microscope slide. Add a drop of methylene blue.

29. Cover the sample with a cover slip, and examine the liver cells under the microscope.

(o) Using the high-power objective lens, estimate the number of times a typical liver cell fits across the field of view, and use the Cell Diameter Equation given in Part 1, step 6(e) to calculate the size of the cell.

(p) Sketch and label a diagram of a liver cell. Calculate the total magnification of the image, and include this value in your report.

30. Return laboratory equipment to its designated location, discard the slide and cover slip according to your teacher's instructions, and wash your hands with soap and water.

Analysis

Part 2: Examining Cheek Cells

(q) Describe similarities and differences between cheek cells and liver cells.

(r) What cell organelles were stained by methylene blue in cheek cells?

Part 3: Examining Aquatic Plant Leaf Cells

(s) Describe the shape of plant cells.

(t) What is the average number of chloroplasts per cell? What shape are the chloroplasts?

(u) Which are larger, the chloroplasts or the nuclei of leaf cells?

(v) Did you detect cytoplasmic streaming? If so, describe the process.

Part 4: Examining Potato Cell Amyloplasts

(w) Describe the change in the potato cells after they were treated with iodine solution.

(x) Where is starch stored in potato plants? How do you know?

Part 5: Examining Red Tomato Cell Chromoplasts

(y) Describe the colour and shape of tomato chromoplasts, and their size relative to the nucleus.

Synthesis

(z) Summarize similarities and differences between the plant and animal cells you examined in this activity. Include differences in cell sizes in your summary.

CAREER CONNECTION

The study of cells is called cytology, and a professional who studies cells is a cytologist. A cytologist prepares and observes cells with a microscope to detect abnormalities. Cytology is used extensively to diagnose cancer, and is also used to check for fetal abnormalities and infectious organisms in tissue samples. Describe this career: duties, education, job opportunities, and strengths and skills required.

www.science.nelson.com

1.4 Chemistry Basics

It is amazing to realize that millions of chemical reactions occur in your body at every moment. What is even more remarkable is that they happen without your even knowing it. We know that chemicals are critical to life because humans and other animals can live only a few weeks without food, several days without water, and a few minutes without oxygen. In most cases, the difference between health and illness depends on the proper functioning of chemical reactions. When reactions occur as they should, an organism is healthy. Medical experts identify faulty reactions and use drugs and other treatments to correct problems.

Table 1 Legend of atoms

Atom	Symbol
hydrogen	
oxygen	
nitrogen	
carbon	
chlorine	

Elements, Molecules, and Compounds

Chemistry is the study of matter. Matter is composed of tiny particles called atoms that contain protons, neutrons, and electrons. Elements are pure substances that are made up of a single kind of atom. The number of protons in the nucleus, called the *atomic number*, determines the identity of an atom. Atoms are represented by symbols, such as C for carbon and Cl for chlorine. The periodic table lists the elements in order of increasing atomic number (**Figure 1**). **Table 1** summarizes how common atoms are illustrated.

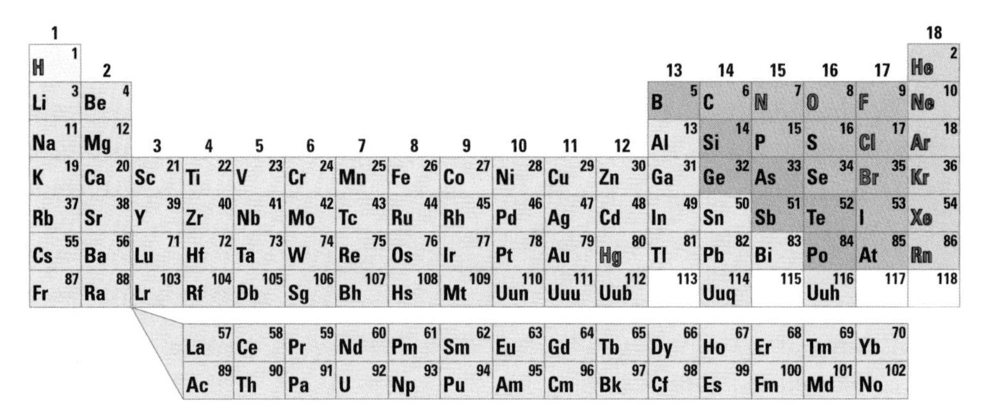

Figure 1
Periodic table of the elements

Molecules are combinations of two or more atoms held together by forces of attraction called chemical bonds. Molecules that contain atoms of one element only are called molecular elements. Oxygen in the air, $O_{2(g)}$, is a molecular element. Molecules that contain atoms of different elements are called compounds. Water, $H_2O_{(l)}$, and carbon dioxide, $CO_{2(g)}$, are examples of compounds (**Figure 2**). The subscripts $_{(l)}$, $_{(s)}$, $_{(g)}$, and $_{(aq)}$ refer to the state of the substance—liquid, solid, gas, or aqueous solution. The $_{(aq)}$ symbol means that the solute is dissolved in water.

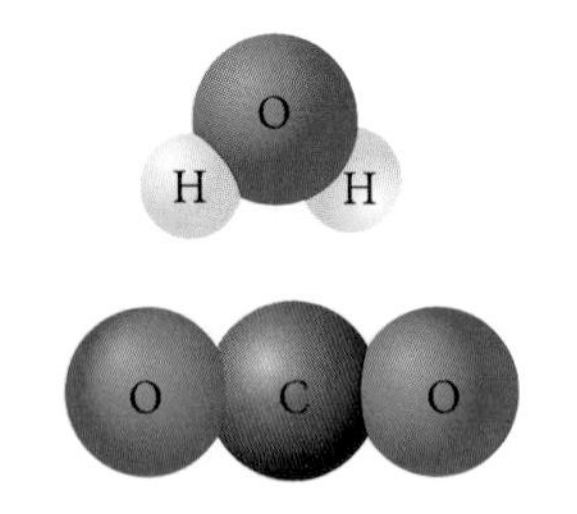

Figure 2
Examples of compounds

Chemical Bonds

Atoms form chemical bonds by gaining and losing electrons or by sharing electrons. An ion is an atom that becomes electrically charged by gaining or losing electrons. Bonds formed when oppositely charged ions come together are called **ionic bonds**. Sodium chloride (table salt) is a common ionic compound (**Figure 3**).

ionic bond the attraction between positive and negative ions in an ionic compound

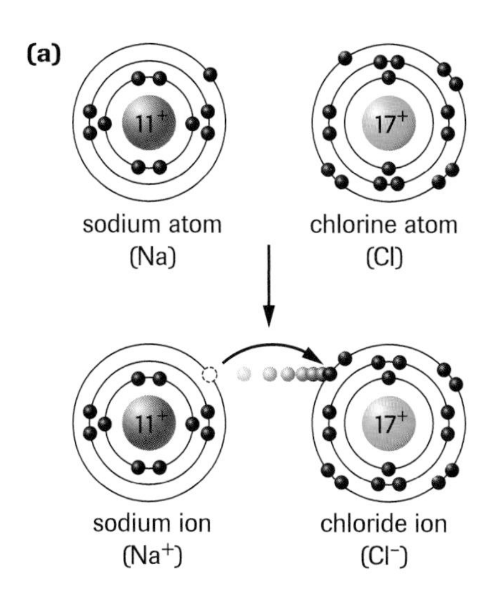

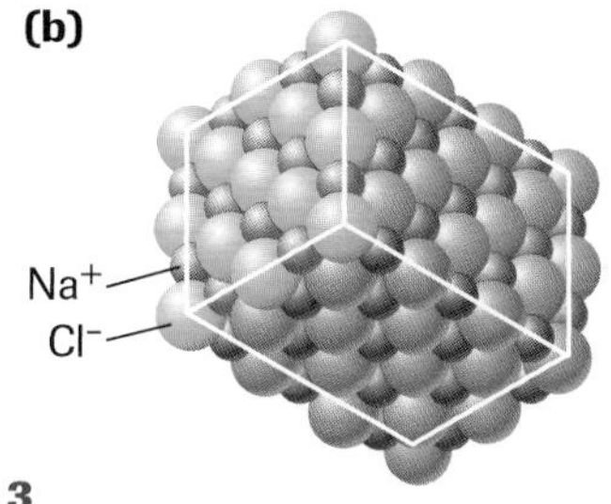

Figure 3
Formation of sodium chloride
(a) A sodium ion (Na^+) and a chloride ion (Cl^-) form when a sodium atom transfers an electron to a chlorine atom.
(b) Sodium chloride, $NaCl_{(s)}$, is composed of many sodium ions and chloride ions held together by ionic bonds in a crystalline lattice.

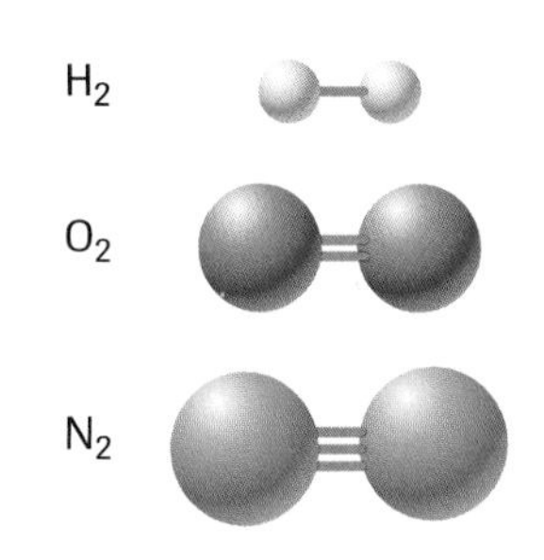

Figure 4
Molecules can have single (H_2), double (O_2), or triple (N_2) covalent bonds.

Covalent bonds form when two atoms share electrons. Molecules may have single, double, or triple covalent bonds. Hydrogen (H_2) has a single covalent bond, oxygen (O_2) has a double covalent bond, and nitrogen (N_2) has a triple covalent bond (**Figure 4**). The shared electrons in a covalent bond may be equally or unequally shared. In a hydrogen chloride molecule, HCl, the electron pair is unequally shared because chlorine has a larger, more positive nucleus (17 protons) than hydrogen (1 proton). This causes the hydrogen side of the molecule to have a partial positive charge (δ^+) and the chlorine side to have a partial negative charge (δ^-) (**Figure 5**). A covalent bond in which positive and negative charges are separated because of unequally shared electron pairs is called a **polar covalent bond**. Equally shared electrons form a **nonpolar covalent bond**, as in hydrogen, $H_{2(g)}$. Polar molecules are molecules that have a positively charged end and a negatively charged end. Nonpolar molecules do not have charged ends.

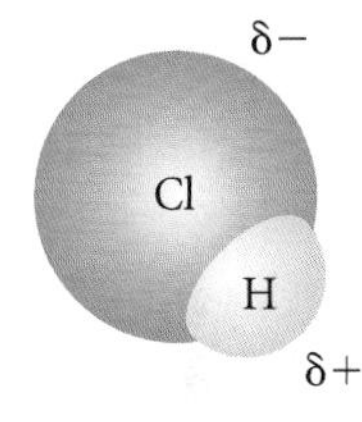

Figure 5
The hydrogen chloride molecule has a polar covalent bond in which the chlorine side of the bond has a partial negative charge ($\delta-$), and the hydrogen side of the bond has a partial positive charge ($\delta+$).

covalent bond a bond formed by the sharing of electrons between atoms in a molecule

polar covalent bond a covalent bond in which positive and negative charges are separated because of unequally shared electrons

nonpolar covalent bond a covalent bond in which charges are not separated because electrons are equally shared

Hydrogen Bonds

Covalent bonds are the strongest forces of attraction that hold the atoms of a molecule together. They are known as intramolecular bonds (bonds within a molecule). However, molecules are attracted to other molecules by a group of much weaker forces called intermolecular bonds (bonds between molecules). A single intermolecular bond is weak, but when they form between the atoms of very large molecules, they add up to a large total force of attraction (**Figure 6**).

The strongest intermolecular bond is the hydrogen bond. **Hydrogen bonds** form between highly polar molecules containing various combinations of hydrogen, oxygen, and nitrogen atoms. Water molecules form hydrogen bonds.

hydrogen bond the force of attraction between highly polar molecules containing combinations of oxygen, hydrogen, and nitrogen atoms

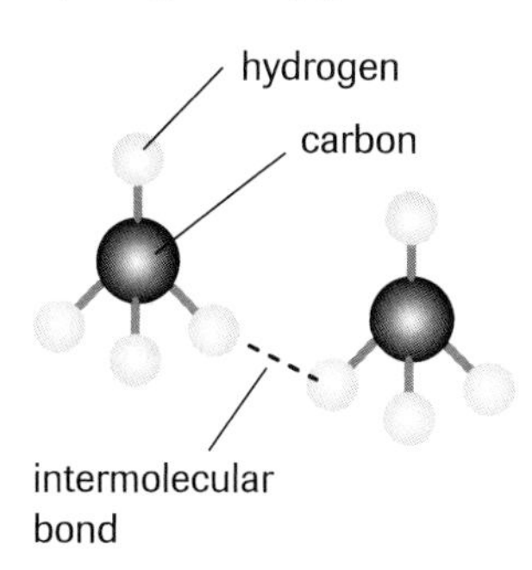

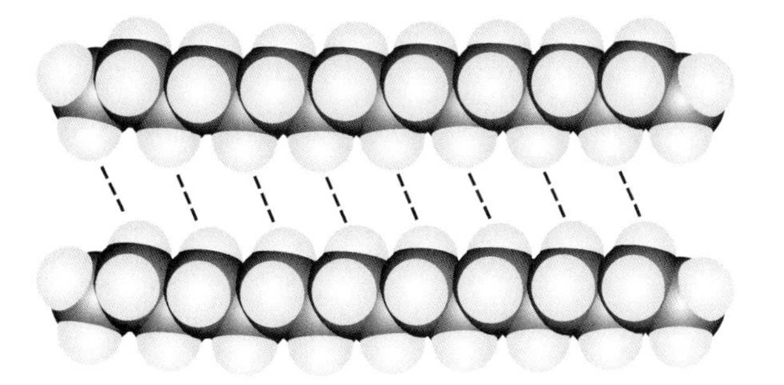

Figure 6
A single intermolecular bond (left side) is weak, but many bonds (right side) create a large force of attraction.

Water: A Very Important Substance

Water is probably the most important molecule on Earth. Water covers three-quarters of Earth's surface and makes up about two-thirds of your body's mass. Water is a polar molecule composed of one oxygen atom bonded to two hydrogen atoms by single polar covalent bonds. Water molecules are polar and form hydrogen bonds with other water molecules (**Figure 7**). Hydrogen bonds between water molecules are difficult to break and are responsible for water's high boiling point and strong surface tension (**Figure 8**).

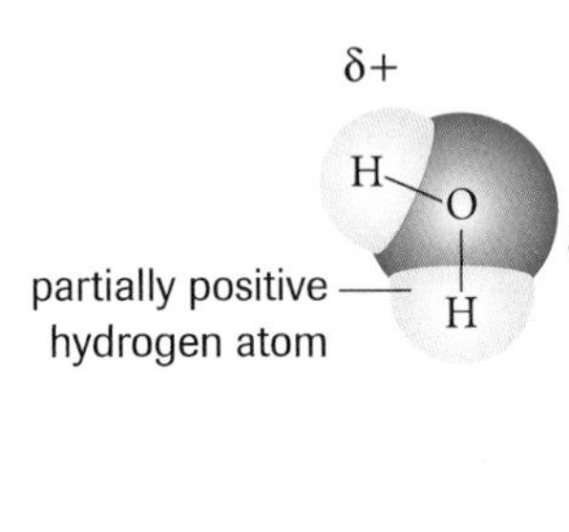

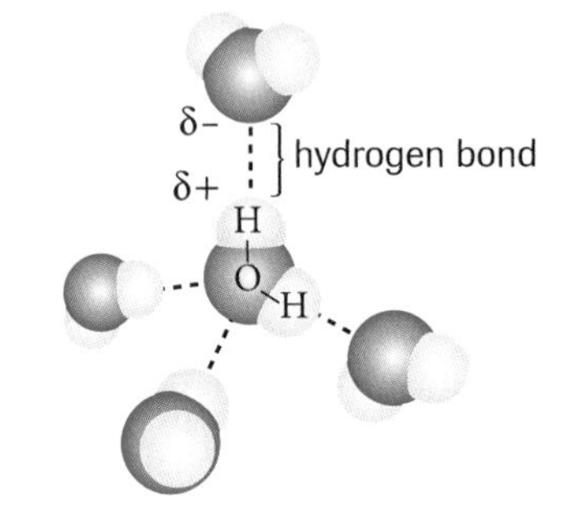

Figure 7
Water molecules form hydrogen bonds with each other.

Figure 8
The strong surface tension caused by hydrogen bonds between water molecules allows organisms like this water strider to walk on the water's surface.

Aqueous Solutions

A solution is a homogeneous (uniform) mixture of two or more pure substances. The pure substance in greater quantity is called the solvent, and the substance in lesser quantity is the solute. Thus, if 10 mL of alcohol is dissolved in 100 mL of water, the alcohol is the solute and the water is the solvent.

When predicting whether or not a substance will dissolve in water, scientists use the saying "like dissolves like." This means that polar solutes dissolve only in polar solvents and nonpolar solutes dissolve only in nonpolar solvents. Oil molecules (nonpolar) do not dissolve in water molecules (polar) because they cannot form hydrogen bonds with water. Instead, the water molecules form hydrogen bonds with one another, causing the nonpolar oil molecules to be excluded. For this reason, nonpolar molecules are described as **hydrophobic** (literally, "water-fearing"). Oil floats on water because it is hydrophobic and less dense than water (**Figure 9**). Substances that are attracted to water are described as **hydrophilic** (literally, "water-loving"). Hydrophilic substances include polar molecules such as alcohols and sugars, and ions such as sodium ions (Na^+) and chloride ions (Cl^-).

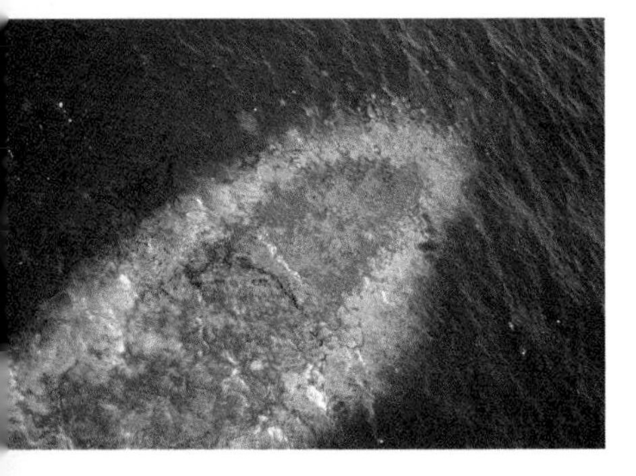

Figure 9
Oil floats on water.

hydrophobic a substance that does not dissolve in water because it is nonpolar

hydrophilic a substance that dissolves in water because it is polar

Concentration

The concentration of a solution is a ratio of the quantity of solute to the quantity of solution. This value may be expressed as a percentage. A vinegar label indicating 5% acetic acid means that there are 5 mL of pure acetic acid (a liquid) dissolved in every 100 mL of vinegar solution. When solids are dissolved in liquids, the concentration is usually given in units of grams of solute per litre of solution (g/L), or grams of solute per hundred millilitres of solution (g/100 mL). For example, if 10 g of sugar is dissolved in enough water to form 100 mL of solution, the concentration may be given as 0.1 g/mL.

Frequently, you may need several concentrations of the same solution. If you begin with a solution of known concentration (a stock solution), you may dilute the solution (decrease its concentration) by adding more solvent (usually water). Calculating the concentration of the diluted solution is straightforward if you know the concentration of the original solution and the volume of the final solution. For example, if water is added to 6% hydrogen peroxide until the total volume is doubled, the concentration becomes one-half the original value, or 3%.

Acids, Bases, and Buffers

Aqueous solutions may be classified as acidic, basic, or neutral, depending on their physical and chemical properties. **Table 2** lists the properties of acids, bases, and neutral solutions, and provides some examples.

Table 2 Acids, Bases, and Neutral Solutions

	Acids	Bases	Neutral solutions
Taste	sour	bitter	not sour or bitter
Feel	not slippery	slippery	not slippery
Litmus test	turns blue litmus red	turns red litmus blue	no change to colour of litmus
Examples	hydrochloric acid ($HCl_{(aq)}$) sulfuric acid ($H_2SO_{4(aq)}$)	sodium hydroxide ($NaOH_{(aq)}$) ammonia ($NH_{3(aq)}$)	pure water ($H_2O_{(l)}$) sodium chloride solution ($NaCl_{(aq)}$)

Acidic solutions contain high concentrations of hydrogen ions (H^+). Hydrogen ions are responsible for the properties of acids. Basic solutions contain high concentrations of hydroxide ions (OH^-). Hydroxide ions are responsible for the properties of bases. Neutral solutions contain equal numbers of hydrogen ions and hydroxide ions, and do not display the properties of acids or bases.

When an acid reacts with a base, water and a salt are produced. This is called a neutralization reaction. Neutralization results in a neutral solution.

$$\underset{\text{acid}}{HCl_{(aq)}} + \underset{\text{base}}{NaOH_{(aq)}} \rightarrow \underbrace{\underset{\text{water}}{H_2O_{(l)}} + \underset{\text{salt}}{NaCl_{(aq)}}}_{\text{neutral salt solution}}$$

The pH Scale

The acidity of a solution may be expressed in terms of pH. The pH scale was devised as a convenient way to distinguish between an acidic solution, a basic solution, and a neutral solution (**Figure 10**).

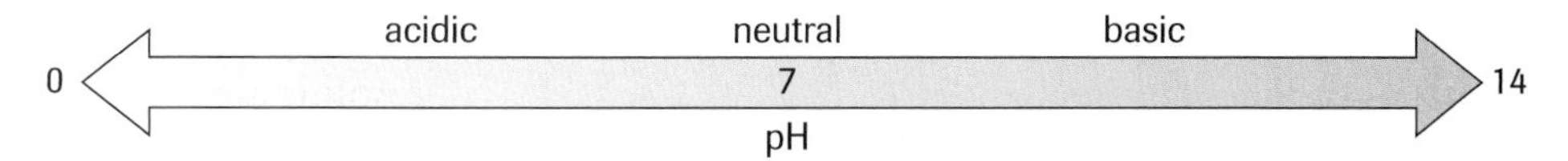

Figure 10
A solution whose pH is 7 will be neutral. Solutions with pH $>$ 7 are basic. Solutions with pH $<$ 7 are acidic.

The pH of pure water is 7. Any aqueous solution with pH = 7 is neutral. Solutions with pH greater than 7 are basic, indicating that there are more hydroxide ions than hydrogen ions in the solution. A solution with pH less than 7 is acidic. This indicates that there are more hydrogen ions than hydroxide ions in the solution.

buffer a solution containing chemicals that can neutralize small amounts of acid and base

DID YOU KNOW ?

Sprinters and Buffers

Sprinters take advantage of the carbonic acid–bicarbonate buffer system in blood by hyperventilating shortly before a race. This increases the pH of the blood slightly, which allows better control of the short-term buildup of lactic acid during the sprint.

Acid-Base Buffers

The components of living cells are sensitive to pH levels. Most cellular processes operate best in a neutral environment where pH = 7. However, cells have to contend with acids and bases all the time. Many of the reactions that occur within cells produce acids or bases. Many of the foods you eat are acidic. When your digestive system absorbs acidic nutrients, such as the acids in fruits, many vegetables, and salad dressings, the pH of your blood may be lowered. Living cells use buffers to resist such changes in pH. A **buffer** is a mixture of chemicals that can neutralize small amounts of acid (hydrogen ions) or base (hydroxide ions) that may be accidentally added to the mixture. In living organisms, buffers usually consist of equal concentrations of an acid and a base. The most important buffer in human extracellular fluid and blood is the carbonic acid–bicarbonate ion buffer. In this buffer, the acid is carbonic acid (H_2CO_3) and the base is bicarbonate ions (HCO_3^-). The carbonic acid component will neutralize small amounts of base that may enter the solution, and bicarbonate ions will neutralize small amounts of acid (**Figure 11**).

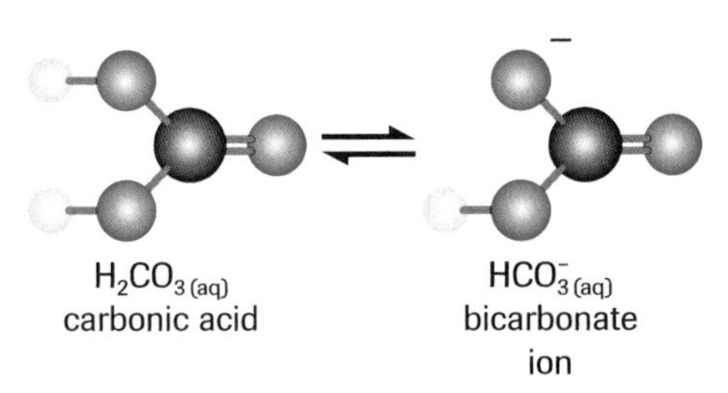

Figure 11
A carbonic acid–bicarbonate buffer contains equal concentrations of carbonic acid molecules and bicarbonate ions.

Section 1.4 Questions

Understanding Concepts

1. List the names of the elements and number of atoms of each element in each of the following compounds.
   (a) $C_2H_4O_2$ (b) H_2SO_4 (c) $C_6H_{12}O_6$
2. KBr is an ionic compound.
   (a) Name and describe the bonds that hold the ions in KBr together.
   (b) Sketch diagrams that show the formation of KBr from its elements.
3. A glass can be filled with water above the rim without spilling (**Figure 12**). Why is this possible?

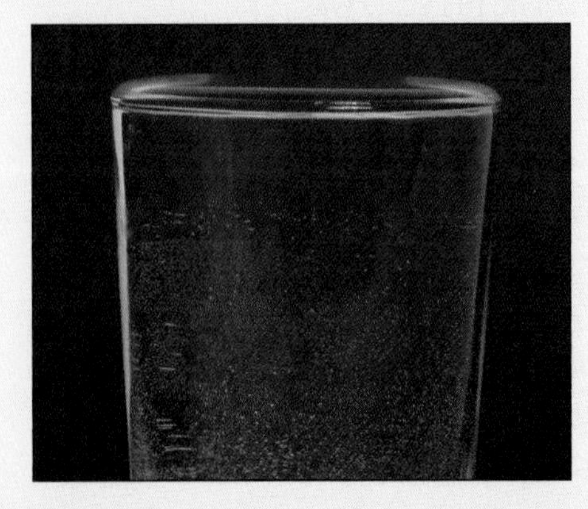

Figure 12

4. Molecule A in **Figure 13** is soluble in water, and molecule B is not soluble in water.

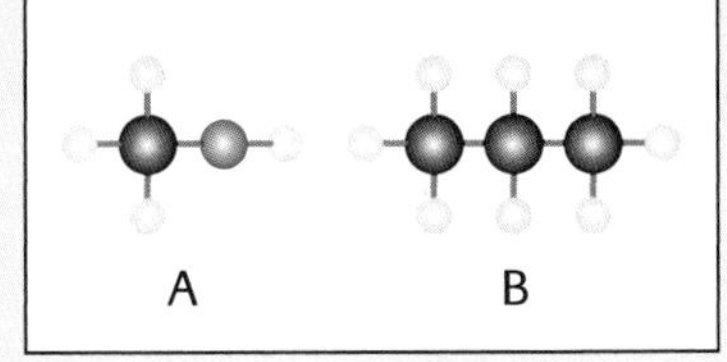

Figure 13

   (a) What type of chemical bonds are responsible for the solubility of molecule A in water?
   (b) Why will molecule B not dissolve in water?
5. The stomach produces hydrochloric acid and has a pH approximately equal to 2.
   (a) Are the contents of the stomach acidic, neutral, or basic?
   (b) Antacids are medications used to neutralize stomach acids. Should antacids be acidic, neutral, or basic compounds? Explain.

Applying Inquiry Skills

6. Water and varsol (a paint thinner) do not mix. Describe an experiment using these two materials that would determine whether Styrofoam is polar or nonpolar.

Biological Chemistry 1.5

With the exception of water, most of the molecules in living organisms contain the element carbon. Carbon is the lightest element that can form four stable covalent bonds with other elements. This allows it to form the enormous number of different compounds needed to support life.

Organic Chemistry

Compounds containing carbon atoms bound to other elements such as hydrogen, oxygen, and nitrogen are called organic compounds, and the study of organic compounds is called organic chemistry. Much of the human body is composed of organic compounds. While there are many different atoms found in the bodies of living organisms, the vast majority (over 96%) are carbon (C), hydrogen (H), oxygen (O), and nitrogen (N).

Many organic molecules contain rings or chains of carbon atoms with additional atoms attached (**Figure 1**). In many cases, the additional atoms form clusters called **functional groups** that give the organic molecule specific chemical properties. Chemical reactions between organic molecules usually involve the molecules' functional groups. **Table 1** lists the five most important functional groups in organic molecules. Can you identify some in **Figure 1**?

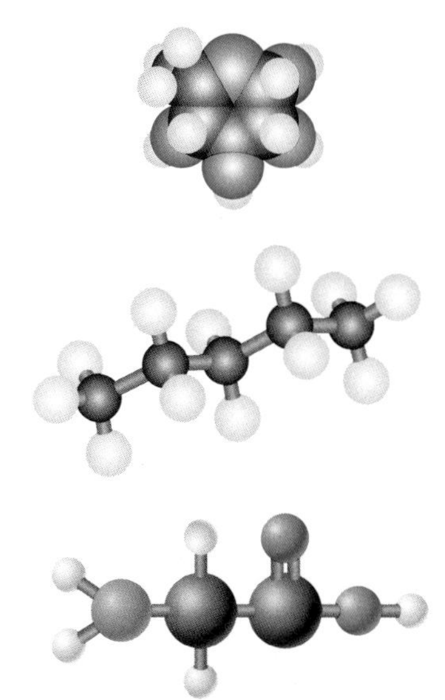

Figure 1
Organic molecules

functional group a cluster of atoms that gives compounds specific chemical properties

Table 1 Five Important Functional Groups

Functional group	Structural formula	Ball-and-stick models
hydroxyl	—OH	O—H
carboxyl	—C(=O)—OH	C, O, O—H
amino	—N(—H)—H	N, H, H
sulfhydryl	—S—H	S—H
phosphate	—O—P(—OH)(=O)—OH	O, P, O—H, O—H, O

Biological Compounds

Many organic molecules in living organisms are large molecules containing dozens of carbon atoms and many functional groups. Because of their large size, these molecules are called **macromolecules**. There are four major groups of biologically important macromolecules: carbohydrates, lipids, proteins, and nucleic acids (**Figure 2**, on the next page).

macromolecule a very large organic molecule

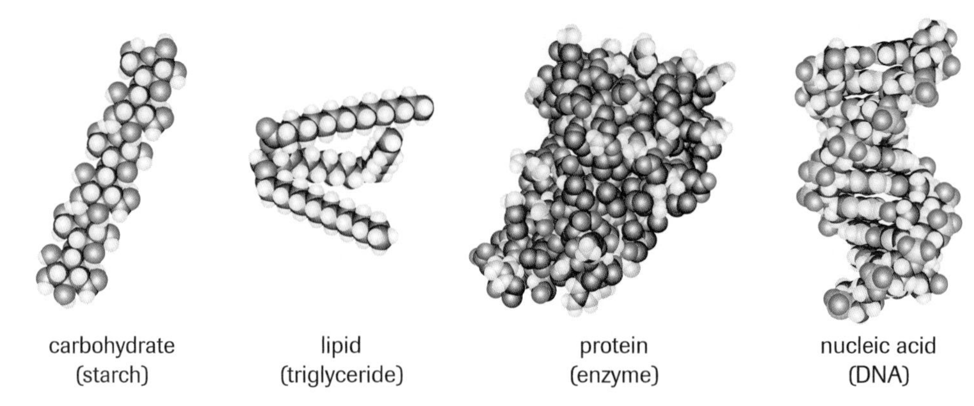

Figure 2
Space-filling models of examples of the four classes of biological molecules: carbohydrates, lipids, proteins, and nucleic acids

carbohydrate a molecule containing carbon, hydrogen, and oxygen, primarily used by living organisms as a source of energy

Carbohydrates

Carbohydrates are molecules that contain carbon, hydrogen, and oxygen, and are primarily used by living organisms as a source of energy. Plants and bacteria called cyanobacteria produce carbohydrates by the process of photosynthesis (see section 1.17).

Figure 3
Glucose ($C_6H_{12}O_6$)

Monosaccharides (Simple Sugars)

The simplest carbohydrates are sweet-tasting sugars called simple sugars, or **monosaccharides**. (In Greek, *monos* means "single" and *saccharon* means "sweet thing.") The most important monosaccharides have six carbon atoms and a number of hydroxyl groups. For example, the primary energy-storage carbohydrate used by living things is glucose ($C_6H_{12}O_6$) (**Figure 3**). Other six-carbon monosaccharides are fructose (fruit sugar) and galactose (found in milk). Fructose is much sweeter than glucose.

monosaccharide the simplest type of sugar molecule

disaccharide a sugar molecule composed of two monosaccharides linked by a covalent bond

Disaccharides (Double Sugars)

Two monosaccharides may link together to form a **disaccharide**, or double sugar. Sucrose (table sugar) is a disaccharide formed by bonding a molecule of glucose to a molecule of fructose. This occurs when a hydroxyl group of glucose and a hydroxyl group of fructose react with each other (**Figure 4**).

glucose + fructose → sucrose + H_2O

Figure 4
The reaction of glucose (a monosaccharide) and fructose (a monosaccharide) to form sucrose (a disaccharide)

Polysaccharides (Complex Carbohydrates)

Monosaccharides such as glucose and disaccharides such as sucrose are soluble in water because their hydroxyl groups form hydrogen bonds with the hydrogen and oxygen atoms of water (**Figure 5**).

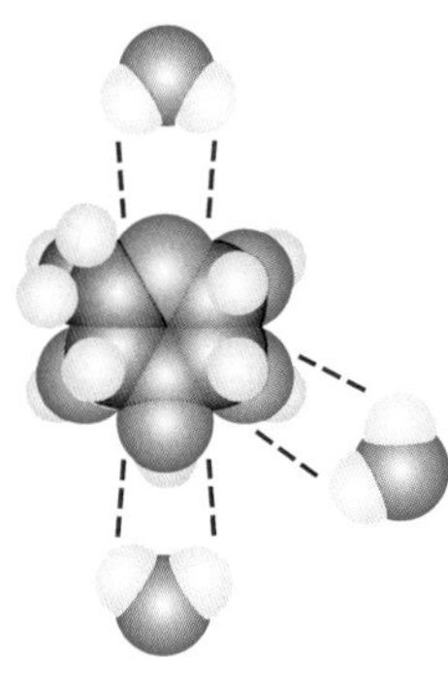

Figure 5
Glucose is water soluble because hydrogen bonds form between glucose and water.

However, sugars are insoluble in water when attached to each other in long chains called **polysaccharides**, or complex carbohydrates. Plants store chemical energy in the form of a polysaccharide called starch. Starch is actually a mixture of two polysaccharides, amylose and amylopectin. Both amylose (**Figure 6**) and amylopectin are composed of repeating subunits of glucose. Animals store some chemical energy in the form of highly branched chains of glucose called glycogen. Humans store glycogen in muscle cells, liver cells, and certain types of white blood cells.

polysaccharide a complex carbohydrate composed of long chains of monosaccharides

Figure 6
Amylose

Molecules composed of many linked subunits are called **polymers**. Polysaccharides such as amylose, amylopectin, and glycogen are polymers of glucose. The individual subunits of a polymer are called **monomers**. Thus, glucose is the monomer in amylose, amylopectin, and glycogen.

Plant starch (amylose and amylopectin) is a common component of food. It is found in large quantities in rice, wheat flour, cornstarch, and potatoes. Polysaccharides are too large to be absorbed by the cells in your digestive system. When complex carbohydrates are eaten, reactions in your digestive system break the bonds between glucose subunits. Your cells may then absorb the individual glucose molecules. This process is called digestion.

polymer a large molecule composed of long chains of repeating subunits

monomer the individual component of a polymer

Cellulose

Plant cell walls are primarily composed of a polysaccharide called **cellulose**. Cellulose is chemically similar to amylose, but the glucose subunits are bonded in a way that humans cannot digest. However, animals such as sheep and cows may digest cellulose. That is why these animals may graze on pastures and eat hay, but humans cannot. Since cellulose is the major constituent of plant cell walls, it is the major component of wood and paper. Cotton is also composed of cellulose. Thus, cotton clothing is actually composed of long chains of glucose molecules.

cellulose glucose polymer found in plant cell walls

Cellulose fibres pass through the human digestive system undigested. This means that cellulose has no nutrient value for humans. However, studies have shown that a diet containing cellulose (also called fibre or roughage) in fresh fruit, vegetables, and grains is healthy. Cellulose attracts water and mucus in the digestive system, and aids in the elimination of solid wastes, helping to prevent constipation (difficulty in eliminating wastes).

Chitin

The hard exterior skeletons of insects and crustaceans, such as lobsters and shrimp, are composed of a modified form of cellulose called **chitin** (**Figure 7(a)**). Chitin is also found in mushrooms (**Figure 7(b)**). Chitin is a tough material that humans find difficult to digest. However, it is used in contact lenses and stitches used in medical operations that decompose as the wound heals.

chitin a modified form of cellulose that forms the tough exoskeleton of insects, fungi, and crustaceans

(a) 

(b) 

Figure 7
(a) The mantis shrimp *Odontodactylus scyllarus* contains a chitin exoskeleton.
(b) The mushroom *Amanita muscaria* contains chitin in its cell walls.

Lipids

lipid a large molecule composed of carbon, oxygen, and hydrogen, with a higher proportion of hydrogen atoms than carbohydrates

Like carbohydrates, **lipids** are molecules made up of carbon, hydrogen, and oxygen, but lipids have a higher proportion of hydrogen atoms. Lipids store more chemical energy than carbohydrates, and are used by animals as major energy-storage molecules. Cells in fat (adipose) tissue are full of lipid molecules (**Figure 8**). Lipids are soluble in oils and other nonpolar solvents, but are insoluble in water and aqueous solutions. Lipids include oils, fats, waxes, phospholipids, and steroids.

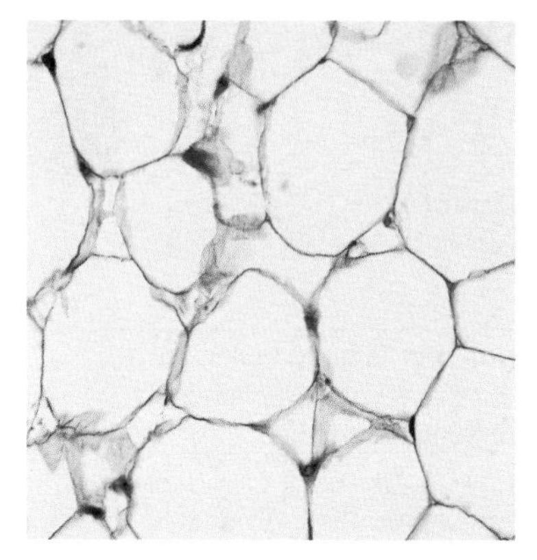

Figure 8
Fat cells

Oils and Fats

triglyceride a lipid molecule composed of glycerol and three fatty acids

Oils and fats are composed of lipid molecules called triglycerides. A **triglyceride** contains four subunits: glycerol and three fatty acids. Glycerol is a three-carbon molecule with a hydroxyl group attached to each carbon atom (**Figure 9(a)**). Fatty acids are long chains of carbon and hydrogen atoms, with a carboxyl group at one end (**Figure 9(b)**).

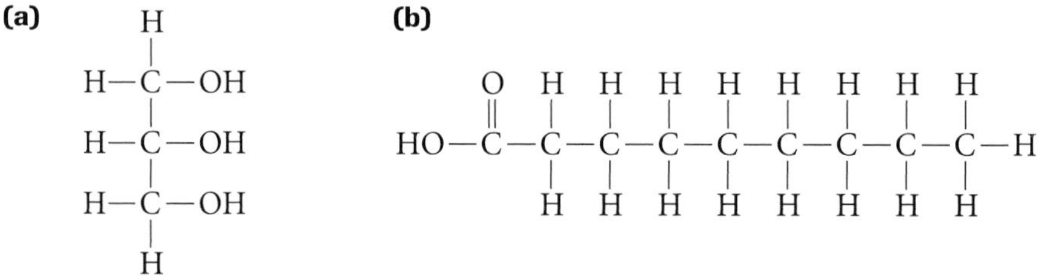

(a)

```
      H
      |
  H—C—OH
      |
  H—C—OH
      |
  H—C—OH
      |
      H
```

(b)

```
     O   H   H   H   H   H   H   H   H
     ‖   |   |   |   |   |   |   |   |
HO—C—C—C—C—C—C—C—C—C—H
         |   |   |   |   |   |   |   |
         H   H   H   H   H   H   H   H
```

Figure 9
(a) Glycerol **(b)** Fatty acid

To form a triglyceride, a fatty acid is attached to each of the three hydroxyl groups of glycerol (**Figure 10**).

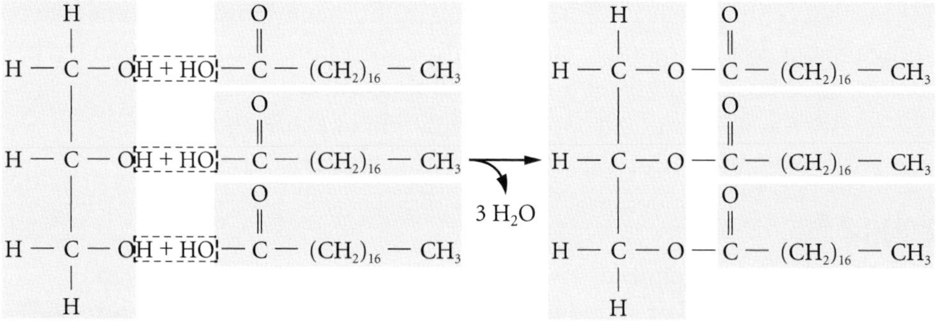

```
    H              O                                     H        O
    |              ‖                                     |        ‖
H — C — OH + HO — C — (CH2)16 — CH3                H — C — O — C — (CH2)16 — CH3
    |              O                                     |        O
    |              ‖                                     |        ‖
H — C — OH + HO — C — (CH2)16 — CH3     ——→        H — C — O — C — (CH2)16 — CH3
    |              O                   3 H2O             |        O
    |              ‖                                     |        ‖
H — C — OH + HO — C — (CH2)16 — CH3                H — C — O — C — (CH2)16 — CH3
    |                                                    |
    H                                                    H
  glycerol   +   3 fatty acids         ——→                 triglyceride
```

Figure 10
The formation of a triglyceride

Saturated and Unsaturated Fats

Fats and oils are mixtures of triglycerides, each differing in the types of fatty-acid molecules attached to the glycerol subunits. Fatty-acid molecules differ in length and in the presence of double bonds between the carbons in their hydrocarbon chains. **Figure 11** shows that a fatty acid with double bonds between its carbon atoms holds fewer hydrogen atoms than a fatty acid with only single bonds.

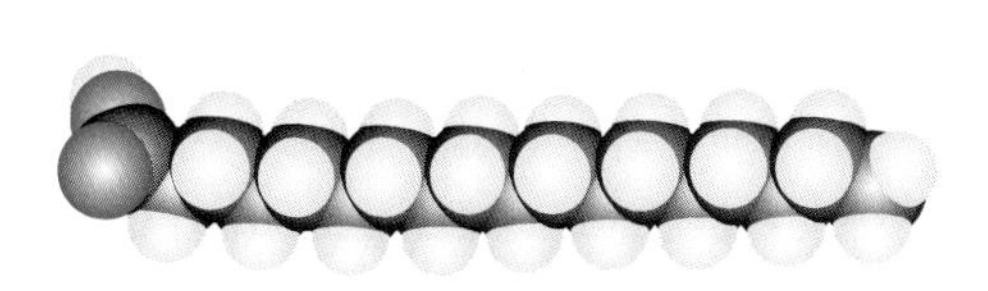

(a) Single bonds between carbon atoms result in a straight fatty-acid molecule.

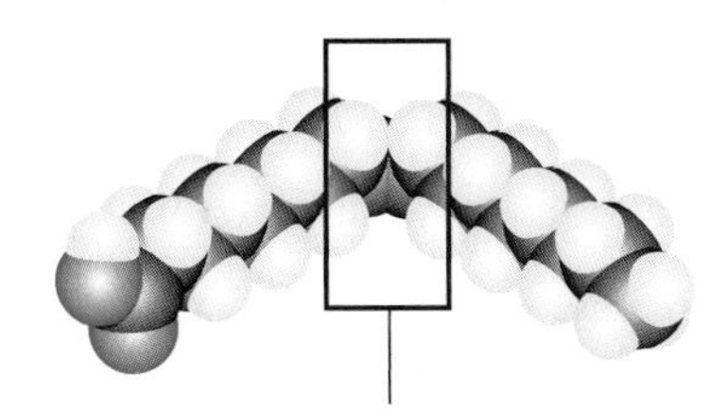

(b) A double bond in the carbon chain results in a bent fatty-acid molecule.

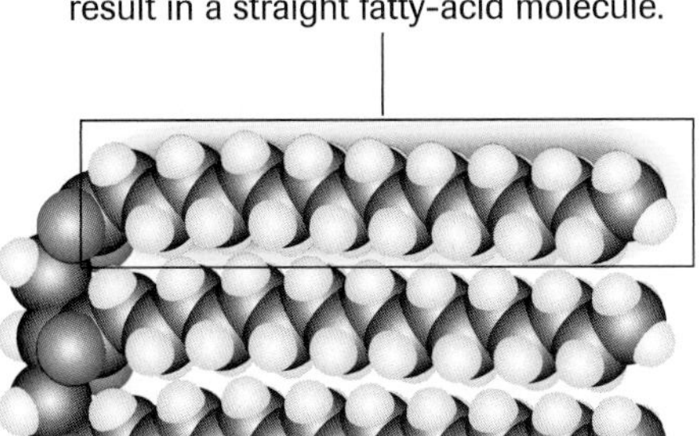

(c) Straight fatty acids fit close together.

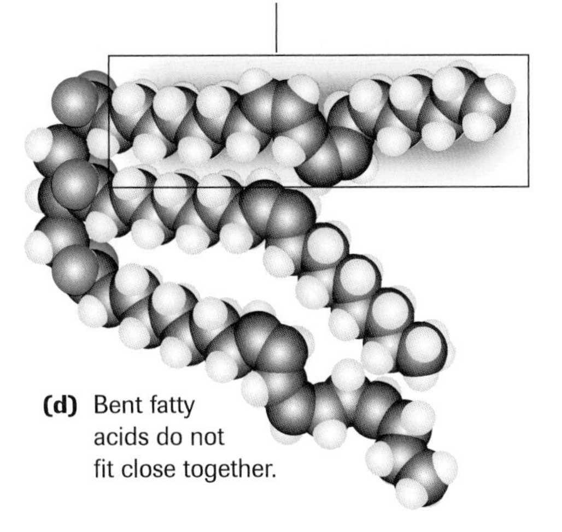

(d) Bent fatty acids do not fit close together.

Figure 11

(a) Stearic acid is a saturated fatty acid containing 18 carbon atoms and only single bonds between carbon atoms.

(b) Oleic acid is an unsaturated fatty acid also containing 18 carbon atoms but with a carbon–carbon double bond between carbon 9 and 10.

(c) A triglyceride containing only saturated fatty acids

(d) A triglyceride containing unsaturated and polyunsaturated fatty acids

Triglycerides containing fatty acids with only single bonds are called **saturated triglycerides** (saturated fats) (**Figure 11(a)**). Triglycerides containing fatty acids with double bonds between carbon atoms are called **unsaturated triglycerides** (unsaturated fats) (**Figure 11(b)**). Triglycerides containing fatty acids with more than one double bond are called polyunsaturated triglycerides. Polyunsaturated triglycerides have low melting points, and are liquid oils at room temperature. Many plants, such as sunflower, canola, olive, and corn, produce large amounts of polyunsaturated triglycerides. These give rise to the many types of oils used in cooking.

saturated triglyceride a triglyceride containing fatty acids with only single covalent bonds between carbon atoms

unsaturated triglyceride a triglyceride containing fatty acids with double bonds between carbon atoms

Margarine is a solid fat produced from vegetable oils. Margarine is made by reacting the polyunsaturated triglycerides in oils with hydrogen gas. The hydrogen atoms of the gas add to the double bonds between carbon atoms, causing the fatty acids to become more saturated. This changes the oil from a liquid to a solid at room temperature. Margarine is usually white when it is produced. Manufacturers add yellow food colouring to make it look like butter.

Animals produce large amounts of saturated triglycerides that occur as hard fats such as lard and fats in butter. Studies have shown that diets containing large amounts of saturated fats may play a part in the development of clogged arteries that increase the risk of heart attack and stroke.

Waxes, Phospholipids, and Steroids

Waxes are lipid molecules commonly used by plants and some animals as waterproof coatings (**Figure 12**).

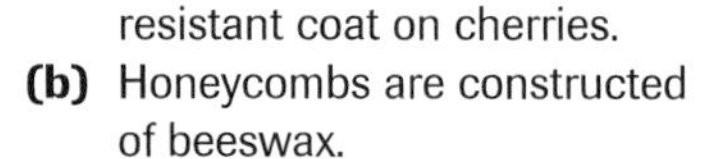

Figure 12
(a) Cutin (a wax) forms a water-resistant coat on cherries.
(b) Honeycombs are constructed of beeswax.

Phospholipids are lipid molecules similar to triglycerides. However, in phospholipids, an additional functional group containing a phosphate group replaces one of the fatty acids (**Figure 13**). Phospholipids play a key role in the structure of cell membranes.

Steroids are lipid molecules composed of four carbon rings. Many hormones, such as the male sex hormone testosterone (**Figure 14 (a)**) and the female sex hormone estradiol, are steroids. Cell membranes contain the steroid cholesterol (**Figure 14 (b)**). Although cholesterol is an essential component of cell membranes, a high concentration of this steroid in the bloodstream has been linked to the development of clogged arteries and an increased risk of heart attack and stroke.

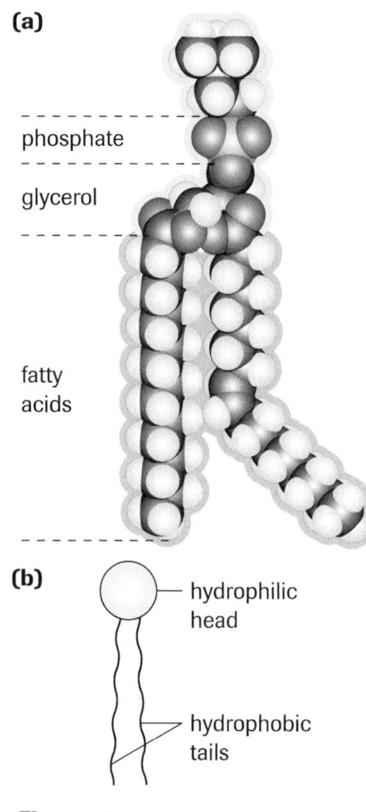

Figure 13
(a) Space-filling model of a phospholipid
(b) Symbol for a phospholipid

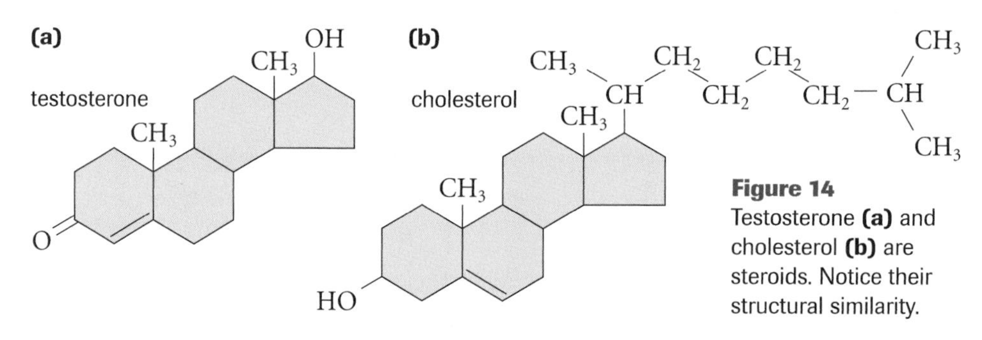

Figure 14
Testosterone **(a)** and cholesterol **(b)** are steroids. Notice their structural similarity.

Proteins

What do Jell-O, hair, antibodies, feathers, blood clots, egg whites, and fingernails all have in common? They are all made of protein. Proteins are the most diverse and among the most important molecules in living organisms (**Figure 15**).

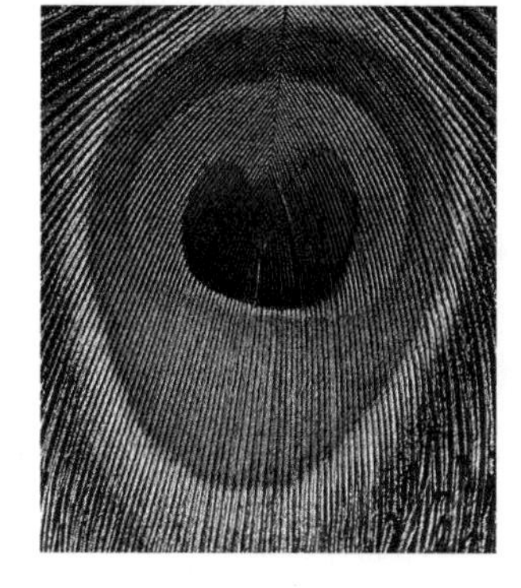

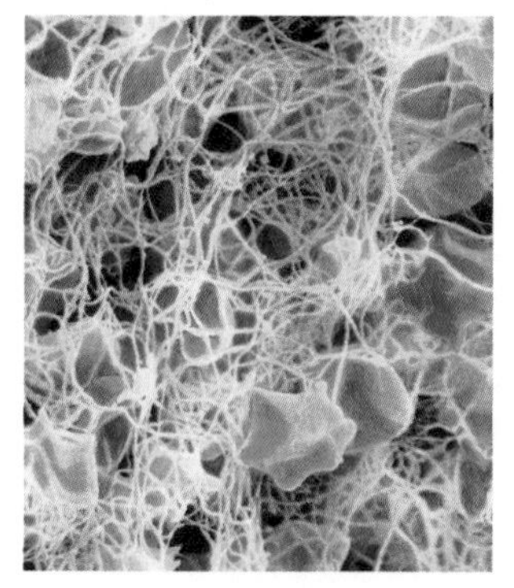

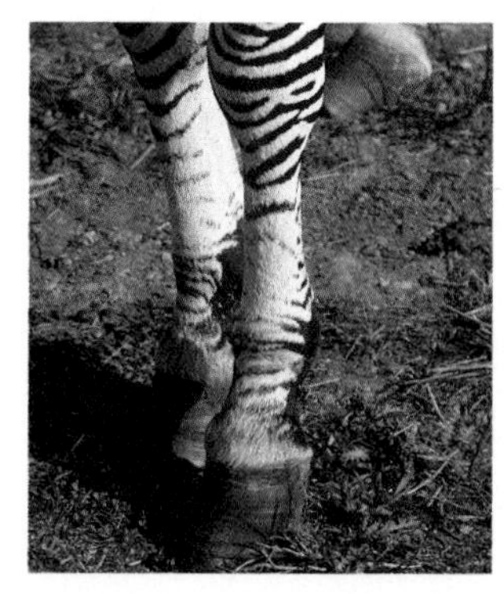

Figure 15
Peacock feathers, blood clots, and zebra hoofs are all made of protein.

Although they have various functions, all proteins have the same basic structure. **Proteins** are unbranched polymers of amino acids. **Amino acids** are small molecules containing a central carbon atom to which is attached an amino group, a carboxyl group, a hydrogen atom, and a side chain (**Figure 16**) (note that the side chain is labelled "R" in the generalized formula in **Figure 16**). The structure of the side chain (R group) distinguishes one amino acid from another. There are 20 different R groups; thus, there are 20 different amino acids. **Figure 17** illustrates the molecular formulas of some common amino acids.

amino group
side chain
carboxyl group

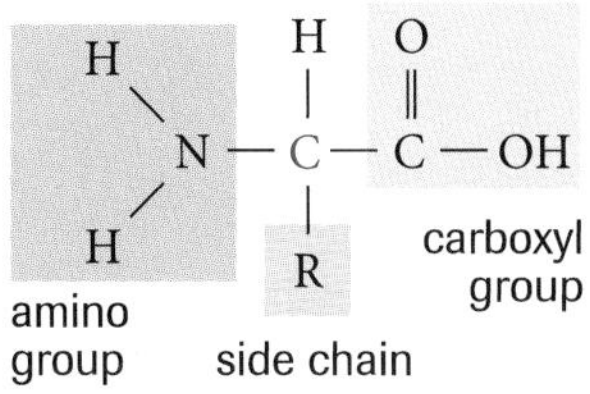

Figure 16
The generalized structure of an amino acid

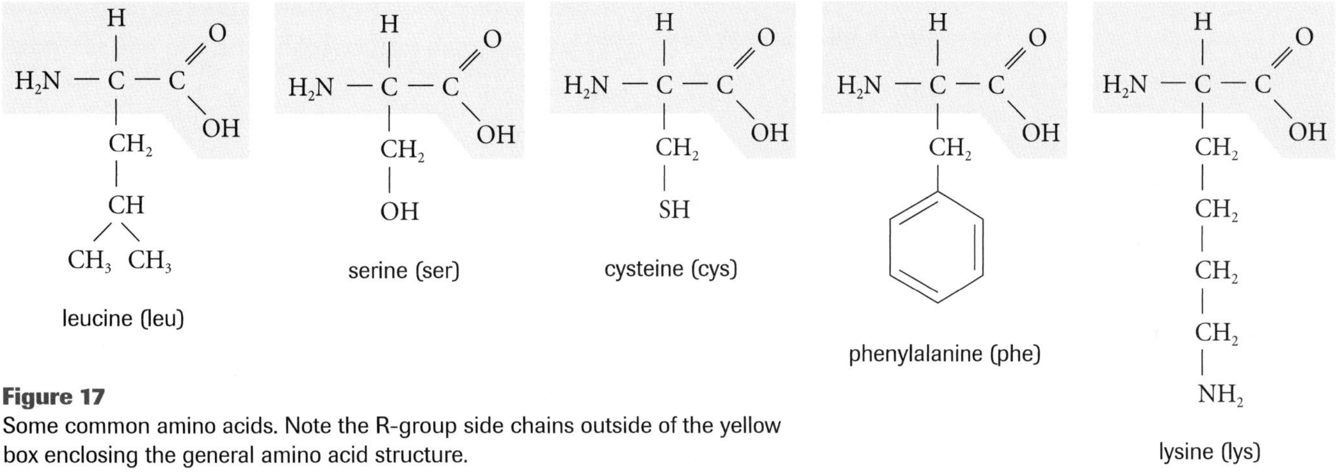

leucine (leu)
serine (ser)
cysteine (cys)
phenylalanine (phe)
lysine (lys)

Figure 17
Some common amino acids. Note the R-group side chains outside of the yellow box enclosing the general amino acid structure.

Of the 20 amino acids found in the food humans normally eat, 8 are considered to be essential amino acids. This means that the body cannot produce them from simpler compounds; they must be obtained from food.

Proteins differ from one another by the number and sequence of amino acids they contain. Proteins are formed by reactions between the amino groups and the carboxyl groups of adjacent amino acids (**Figure 18(a)**). The linkage that forms between amino acid subunits is called a **peptide bond**. A small chain of amino acids is called a **polypeptide**. However, most functional proteins contain hundreds or thousands of amino acids. Proteins may be broken down into individual amino acids (**Figure 18(b)**). This occurs when proteins in your food are digested.

protein an unbranched polymer of amino acids

amino acid a protein subunit

peptide bond a bond between amino acids in a protein molecule

polypeptide a short chain of amino acids

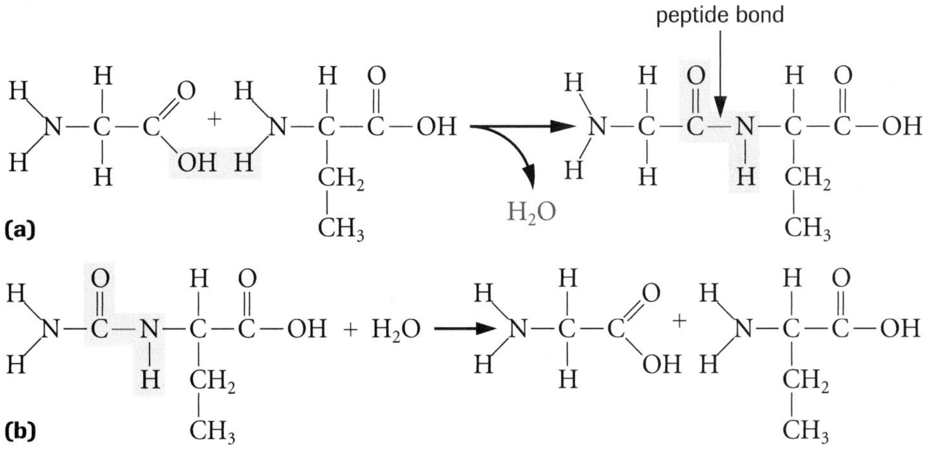

peptide bond

(a)

(b)

Figure 18
(a) Proteins are formed when amino acids are joined by reactions between amino groups and carboxyl groups. This results in the formation of peptide bonds.
(b) A breakdown reaction cleaves the peptide bonds between amino acids.

Proteins perform an enormous number of functions within a living cell and a unique protein is needed for each function. Just as a huge number of English words are created from the 26 letters of the alphabet, a very large number of proteins may be produced from the 20 different amino acids found in living cells.

protein synthesis the production of proteins

Cells make proteins by joining amino acids together in a process called **protein synthesis**. As the amino acid chain lengthens, forces of attraction and repulsion between functional groups cause the growing polypeptide to fold into sheets and wrap into coils (**Figure 19**). Proteins with the same sequence of amino acids will fold into the same shape. If the protein does not assume its correct shape, it will not work properly. A protein's shape may be altered by a single misplaced amino acid.

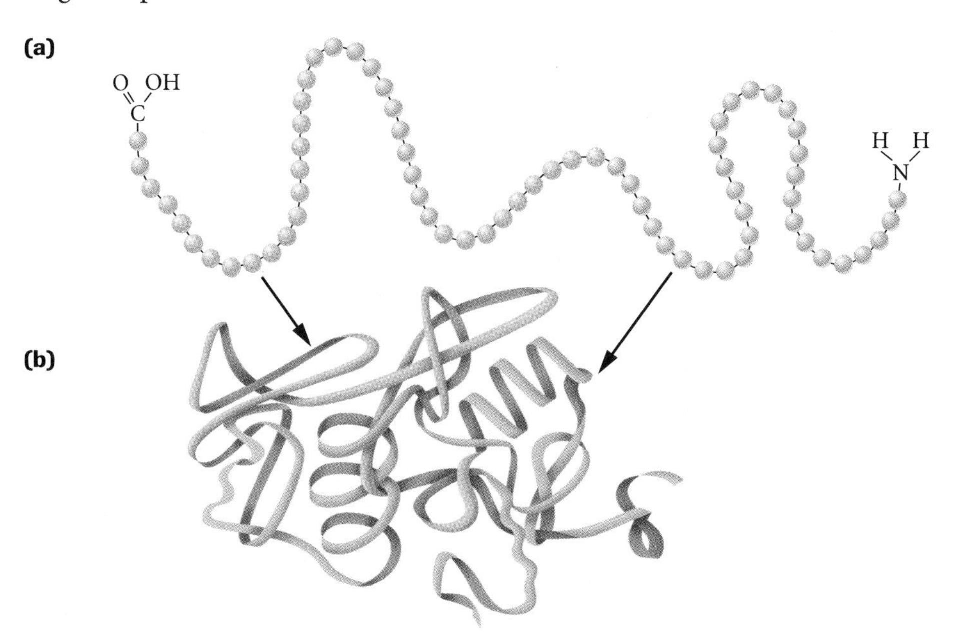

Figure 19
Different parts of a polypeptide chain **(a)** fold into sheets and coils that fold into a functional protein molecule **(b)**.

Proteins with lots of sheet structure form fibres such as keratin in hair and silk in spider webs. Proteins with lots of coils usually fold into round, or globular, proteins such as the subunits of hemoglobin in red blood cells.

enzyme a protein that speeds up chemical reactions

catalyst a chemical that speeds up chemical reactions

Enzymes are one of the most important groups of proteins. They act to speed up chemical reactions. Chemicals, such as enzymes, that speed up reactions are called **catalysts**. Enzymes are needed to catalyze almost every reaction that occurs in living organisms. Other proteins, such as insulin, act as chemical messengers (hormones), while still others, such as collagen, give structural support to bones, cartilage, and tendons.

Denaturation

denaturation a change in the three-dimensional shape of a protein caused by high temperatures or harsh acidic, basic, or salty environments

Proteins may lose their shape if they are subjected to high temperatures (generally above 40°C) or are exposed to acidic, basic, or salty environments. A change in the three-dimensional shape of a protein caused by changes in environmental factors is called **denaturation**. A denatured protein cannot carry out its biological functions but will usually return to its normal shape when the harsh environmental factor is removed, so long as the peptide bonds between amino acids are not broken. If the environmental factor breaks the peptide bonds that hold the amino acids together, the protein will be destroyed.

Protein denaturation can be both dangerous and useful. Fever above 39°C could denature critical enzymes in the brain, leading to seizures and possibly death. Curing meats with high concentrations of salt or sugar and pickling meats or vegetables in vinegar preserves the food by denaturing the enzymes in bacteria that would otherwise cause the food to spoil. Fruits and vegetables turn brown when exposed to air. This brown colour is caused by an enzyme reaction. The practice of blanching fruits and vegetables by quickly dipping them in boiling water denatures the enzymes that would otherwise cause them to turn brown. Heat can be used to denature the proteins in hair, allowing people to temporarily straighten curly hair or curl straight hair.

Humans may consume animal flesh (meat) as a source of protein. However, the fibrous proteins contained in muscle cells make raw meat difficult to chew. Heating denatures the fibrous protein molecules and partially decomposes them. In many cases, the primary purpose of cooking is to denature the proteins in food. Foods are usually a rich mixture of nutrients including carbohydrates, lipids, proteins, nucleic acids, vitamins, and minerals.

Nucleic Acids

Organisms store information about the structures of their proteins in macromolecules called nucleic acids. Nucleic acids are polymers containing subunits called nucleotides (**Figure 20**). Each nucleotide is made up of three smaller components:

1. a five-carbon sugar (ribose in RNA; deoxyribose in DNA)
2. a phosphate group
3. an organic, nitrogen-containing component called a nitrogenous base

There are four types of nitrogenous bases in DNA: adenine (A), guanine (G), thymine (T), and cytosine (C). In a nucleic acid chain, the sugar molecule of one nucleotide is linked to the phosphate group of the next closest nucleotide, and a nitrogenous base protrudes from each sugar.

There are two types of nucleic acids in organisms: deoxyribonucleic acid (DNA) and ribonucleic acid (RNA). DNA is composed of two long nucleotide strands wound around each other like the railings of a circular staircase (**Figure 20**). This "double helix" structure of DNA causes the nitrogenous bases of the two nucleotide strands to face each other. Hydrogen bonds between the nucleotide pairs (dotted lines) hold the double helix together. In DNA, the bases always pair as follows: A–T and C–G.

Unlike DNA, RNA is usually composed of a single chain of nucleotides in the form of a helix. The nitrogenous bases are not paired, and, instead of thymine (T), RNA contains the base uracil (U).

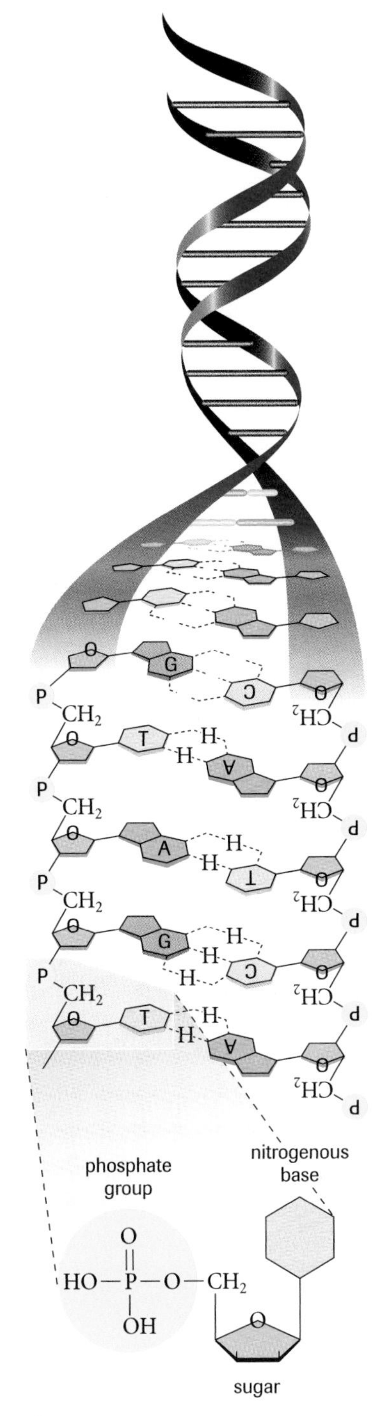

Figure 20
DNA molecule with a nucleotide component highlighted

DNA contains genetic (hereditary) information that is passed from one generation to the next. DNA stores the information for making proteins. The order of nucleotides in DNA forms a code that ultimately specifies the order of amino acids in a particular protein. The set of instructions in DNA that codes a complete protein is called a gene. Results from the Human Genome Project, a research study that has determined the number and sequence of nucleotides in all 46 chromosomes of a human, have shown that there are more than 3 billion base pairs, and between 30 000 and 35 000 genes in the human genome.

RNA molecules perform many important functions in a cell. RNA molecules called messenger RNA (mRNA) take the genetic information in DNA out of the nucleus to ribosomes in the cytoplasm where proteins are produced. DNA molecules remain in the nucleus, where they are protected from enzymes in the cytoplasm that might break them down. RNA molecules also form part of the structure of ribosomes.

Table 2 summarizes the characteristics of biological macromolecules.

Table 2 Characteristics of Biological Macromolecules

Macromolecule	Structure of typical subunit	Examples	Functions
Carbohydrates	CH_2OH, H, O, H, H, OH, H, HO, OH, H, OH		
monosaccharides		fruit sugar (fructose)	store energy
disaccharides		wheat flour (starch)	tough component of plant cell walls
polysaccharides		paper (cellulose)	tough exterior covering
		crab shells (chitin)	
Lipids			
triglycerides	$OH-\overset{\overset{\displaystyle O}{\parallel}}{C}-CH_2-CH_2-CH_2-CH_2-CH_2-CH_3$	butter	store energy
phospholipids		membranes	cell membrane components
steroids		testosterone	act as chemical messengers
Proteins	side chain; R; H, OH; N—C—C; H, H, O; amino group; carboxyl group		
globular		enzymes	help chemical reactions take place (speed them up)
		hemoglobin	transport materials around body
fibrous		hair	support cell and body structures
Nucleic acids	phosphate group; nitrogenous base; O; HO—P—O—CH_2; OH; O; sugar	DNA	store hereditary information (DNA)
		RNA	decode hereditary information (RNA)

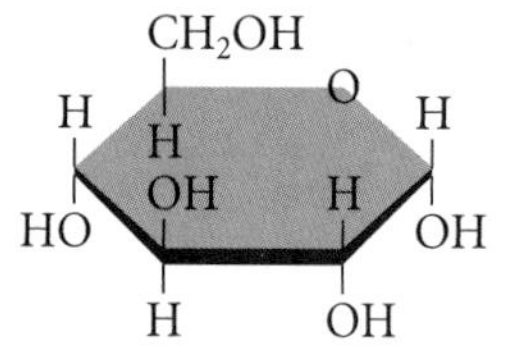

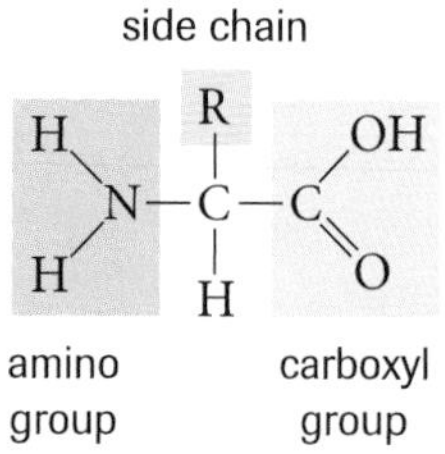

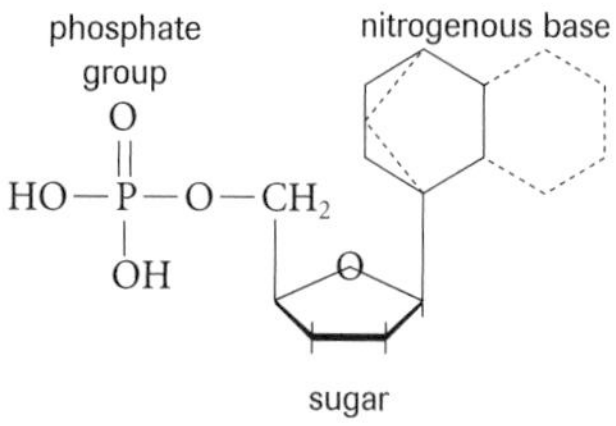

Section 1.5 Questions

Understanding Concepts

1. Why is carbon so common in the molecules found in living organisms?
2. Identify the functional groups in the molecule in **Figure 21**.

```
         O    H    H
         ||   |    |
  OH  —  C  — C  — C  —  S — H
              |    |
            H-C-H  H
              |
        HO—  P=O
              |
              OH
```

Figure 21

3. What is the main difference between a monosaccharide and a disaccharide? Provide one example of each.
4. Why is it not a good idea to consume a diet rich in saturated fats and cholesterol?
5. What is an essential amino acid?
6. Why are prolonged high fevers dangerous?
7. Distinguish between a polypeptide and a protein.
8. Define protein denaturation.
9. (a) Which of the two nucleic acids studied in this section stores genetic information?
   (b) Where in the cell is the nucleic acid referred to in (a) located?
   (c) What is the name and primary function of the other nucleic acid found in cells?
10. Why does the amount of A always equal the amount of T in DNA?
11. What is the relationship between the nucleotide sequence in DNA and the amino acid sequence in a protein molecule?
12. Describe two similarities and two differences between the structure of DNA and the structure of RNA.
13. Why do DNA molecules stay in the nucleus?
14. In your notebook, complete the following chart:

Compound	Macromolecule	Uses
		table sugar
cellulose		
	carbohydrate	medical stitches
		female sex characteristics

Applying Inquiry Skills

15. Egg whites are liquid and are composed almost entirely of a protein called albumin.
    (a) What happens to the albumin molecules in egg whites when an egg is fried?
    (b) Predict whether the process you described in (a) is reversible.
    (c) Design an experiment to verify your prediction.

Making Connections

16. Laxatives are medications that relieve constipation. Metamucil is a laxative that is composed entirely of shredded seed husks of the Plantago plant.
    (a) Why does Metamucil relieve constipation?
    (b) Suggest two nonmedicinal substitutes for Metamucil.
17. (a) How is vegetable-oil margarine produced?
    (b) Why do producers add yellow food colouring to margarine?
18. Hair is composed almost entirely of the protein keratin. When keratin is heated, its molecules are denatured and change their shape. How could this knowledge be put to practical use?
19. (a) How are fruits and vegetables blanched?
    (b) Why would a food-processing company blanch potatoes before packaging them in transparent jars?
20. (a) Describe two household uses of waxes.
    (b) Conduct library and/or Internet research to find two ways in which the food industry uses waxes on fruits and vegetables.

www.science.nelson.com

1.6 Activity

Biological Compounds in 3-D

In this activity, you will build models of biologically important molecules using ball-and-spring model-building kits. You will then view these and other molecules using computer-generated graphics.

Materials

ball-and-spring model-building kit

Procedure

Part 1: Building Biological Molecule Models with Ball-and-Spring Model-Building Kits

1. Open the molecular model-building kit and identify the colours representing carbon (C), hydrogen (H), oxygen (O), nitrogen (N), and sulfur (S). Separate the short springs (single bonds) from the long springs (double bonds).
2. Build a molecule of methane, CH_4.

(a) Describe the molecule's overall shape.

(b) The C–H bonds in CH_4 are essentially nonpolar. Is the CH_4 molecule polar or nonpolar?

3. Replace one H atom with a C atom and fill any open bonding positions (holes) with H atoms.

(c) Sketch the new molecule. This molecule's name is ethane. It is a hydrocarbon, a molecule composed of carbon and hydrogen only. Write its chemical formula and name beside the sketch.

(d) Describe the molecule's overall shape. Is this molecule polar or nonpolar?

4. Repeat step 3 until you have created a straight-chained hydrocarbon containing 6 carbon atoms and 14 hydrogen atoms; this is hexane, C_6H_{14}.
5. Build another hexane molecule and place it alongside the first one you built.

(e) Are the hexane molecules polar or nonpolar?

(f) Will hexane molecules attract water? Explain.

6. Convert one of the terminal methyl groups ($-CH_3$) *on each molecule* into a carboxyl group (**Figure 1**).

Figure 1

(g) What type of biological compound have you created?

7. Orient the two molecules so that portions of the molecules that would normally form intermolecular bonds lie next to each other.
8. Remove hydrogen atoms from the third and fourth carbon atoms of one of the molecules (assume that the carbon atom of the carboxyl group is the first carbon). Replace these atoms with a long spring, forming a double bond between the carbon atoms.

(h) Describe the change in the overall shape of the molecule caused by the double bond.

9. Keeping the carboxyl groups as close to one another as possible, orient the molecules parallel to each other.

(i) How does the presence of the carbon–carbon double bond influence the effectiveness of the intermolecular attractions?

(j) What term describes the presence of a double bond in the hydrocarbon portion of this molecule?

10. Construct eight water molecules and place them around the two molecules you've built, orienting them according to the type of intermolecular bonds they would form.

(k) Around which parts of the organic molecules do the water molecules congregate?

11. Disassemble all molecules and replace the balls and springs in their respective compartments.
12. Using **Figure 2**, build two glucose molecules.

Figure 2

13. Use **Figure 3** to form a bond between the two glucose molecules.

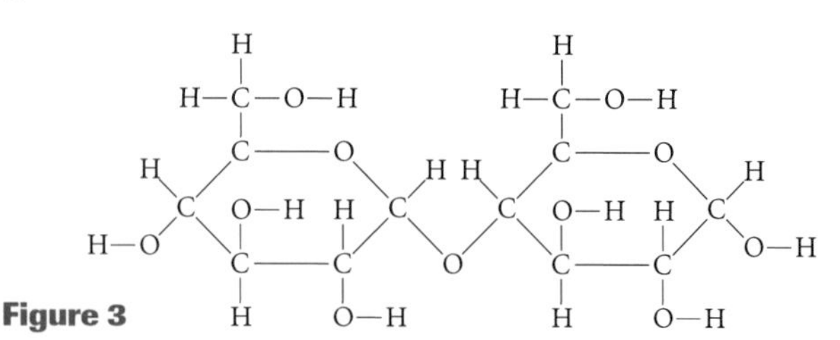

Figure 3

(l) What molecule can be formed with the atoms removed from the two glucose molecules?

14. Disassemble all molecules and return the balls and springs to their respective compartments.
15. Using **Figure 4**, construct the amino acid glycine.

```
       H
 O     |     H
  \\   |    /
    C--C--N
   /   |    \
H-O    H     H
```

Figure 4

(m) Name the functional groups in the molecule.

(n) Identify the R-group side chain.

16. Using **Figure 5**, build the amino acid cysteine. Do not disassemble the glycine molecule.

```
       H
 O     |     H
  \\   |    /
    C--C--N
   /   |    \
H-O    |     H
     H-C-H
       |
       S
       |
       H
```

Figure 5

(o) What functional group is contained in the cysteine R-group?

17. Form a peptide bond between the amino acids.

(p) What functional groups were used to create the peptide bond?

(q) What molecule can you create with the atoms you removed from the two amino acids when you formed the peptide bond?

18. Disassemble all molecules and return the balls and springs to their respective compartments.

Part 2: Viewing Biological Compounds Using Computer-Generated Graphics

www.science.nelson.com

19. Log onto your computer and follow instructions to access the *Biological Compounds* program.
20. Read the introduction, which describes how to view and manipulate the computer-generated images. After practising molecule manipulations, go to the *Exploring Molecules* section.
21. Complete each of the molecule sections as listed and answer the questions on the computer as well as the Analysis questions.

Analysis

Part 2

22. *Carbohydrates*

(r) Observe glucose. What features account for its high solubility in water?

(s) Observe maltose and sucrose. What element forms the bridge between the monosaccharides that make up maltose and sucrose?

(t) Compare the overall shapes of cellulose, amylose, and amylopectin. Which of these molecules forms cell walls in plant cells?

(u) Paper is composed of cellulose. What feature of the cellulose molecule accounts for the water-absorbing qualities of paper towels?

(v) What type of intermolecular bond forms between carbohydrates and water?

23. *Lipids*

(w) Describe the characteristics of all fatty acids.

(x) Compare the shape and molecular structure of saturated and unsaturated fatty acids.

(y) What functional groups are involved in the formation of a triglyceride?

(z) What characteristics of a phospholipid make it attract polar water molecules and nonpolar molecules as well?

(aa) What characteristics of wax make it well-suited to its role as nature's main waterproofing compound?

24. *Amino Acids and Proteins*

(bb) Identify the amino group, the carboxyl group, and the R-group side chain of amino acids.

(cc) Identify the peptide bonds between amino acids in a polypeptide.

(dd) Which functional groups are there at either end of all protein molecules?

25. *Nucleotides*

(ee) Identify the bases that are building blocks of nucleotides in RNA and DNA.

(ff) List some of the structural similarities and differences between RNA and DNA.

(gg) How are the bases paired within DNA?

1.7 Activity

Identifying Nutrients

In this activity, you will use laboratory tests to identify single sugars (monosaccharides), starches, proteins, and lipids. You will then use these tests to identify the nutrients present in an unknown sample.

Question

What nutrients are present in the unknown sample?

Materials

safety goggles, laboratory apron, and gloves
400-mL beaker
8–16 test tubes
test-tube brushes
thermometer
distilled water
test-tube holder
depression spot plate
test-tube brush
10-mL graduated cylinder
wax pencil
medicine droppers
rubber stoppers
detergent solution
utility stand
ring clamp
test-tube racks
hot plate
egg albumin
Benedict's reagent
5% glucose solution
5% fructose solution
5% sucrose solution
5% starch suspension
vegetable oil
whipping cream
gelatin
Lugol's solution
skim milk
Biuret reagent
Sudan IV indicator
unglazed brown paper (2 letter-sized sheets)
1 unknown solution

Wear eye protection, a laboratory apron, and gloves at all times.

Benedict's reagent, Lugol's solution, Sudan IV indicator, and Biuret reagent are toxic and can cause an itchy rash. Avoid skin and eye contact. Wash all splashes off your skin and clothing thoroughly. If you get any chemical in your eyes, rinse for at least 15 min at the eyewash station and inform your teacher. Sudan IV indicator is flammable. Keep away from the hot plate.

Never touch the ceramic surface of a hot plate, as the unit may be hot. Turn off and unplug the unit when finished.

Procedure

1. Construct a table to record your results for all parts of the activity.

Part 1: Monosaccharide Test

Benedict's reagent is an indicator for monosaccharides and some disaccharides, such as maltose. **Table 1** summarizes the quantitative results obtained when a simple sugar reacts with Benedict's reagent.

Table 1

Colour of Benedict's reagent	Approximate sugar concentration
blue	nil
light green	0.5%–1%
green to yellow	1%–1.5%
orange	1.5%–2%
red to red-brown	>2%

2. Prepare a water bath by heating 300 mL of tap water in a 400-mL beaker on a hot plate. Heat the water until it reaches approximately 80°C.
3. Using a 10-mL graduated cylinder, measure 3 mL each of distilled water, fructose, glucose, sucrose, starch, and the unknown solution. Pour each solution into a separate, appropriately labelled test tube. Rinse the graduated cylinder after pouring each solution. Add 1 mL of Benedict's reagent to each of the test tubes.
4. Using a test-tube holder, place each of the test tubes in the hot-water bath.

(a) When you first notice a colour change in any one of the test tubes, record the colour in all test tubes.

Part 2: Starch Test

Lugol's solution contains iodine and is an indicator for starch. Iodine turns blue-black in the presence of starch.

5. Using a medicine dropper, place a drop of water on a depression spot plate and add a drop of Lugol's solution.

(b) Record the colour of the mixture.

6. Repeat the procedure, this time using drops of starch, glucose, sucrose, and your unknown solution.

(c) Record the colour of each mixture in a chart. Which mixtures indicate the presence of starch?

Part 3: Sudan IV Lipid Test

Sudan IV solution is an indicator of lipids, which are soluble in certain solvents. Lipids turn Sudan IV solution from a pink to a red colour. Polar compounds will not cause the Sudan IV indicator to change colour.

7. Using a 10-mL graduated cylinder, measure 3 mL each of distilled water, vegetable oil, skim milk, whipping cream, and the unknown solution. Pour each solution into a separate labelled test tube. Rinse the graduated cylinder with water after pouring each solution. Add six drops of Sudan IV indicator to each test tube. Place stoppers on the test tubes and shake them vigorously for 2 min.

To mix the contents of a stoppered test tube thoroughly, make sure the stopper is on tight. Place your index finger on top of the stopper, gripping the test tube firmly with your thumb and other fingers. Then shake away from others.

(d) Record the colour of the mixtures in a chart.

Part 4: Translucence Lipid Test

Lipids can be detected using unglazed brown paper. Because lipids allow the transmission of light through the brown paper, the test is often called the translucence test.

8. Draw one circle (5-cm diameter) on a piece of unglazed brown paper. Place one drop of water in the circle and label the circle accordingly. Using more sheets, draw a total of seven more circles (5-cm diameter). Place one drop of vegetable oil, skim milk, whipping cream, and the unknown solution, each inside its own circle, labelling the circles as you go. When the water has evaporated, hold all papers to the light and observe.

(e) Record whether or not the paper appears translucent.

Part 5: Protein Test

Proteins can be detected by means of the Biuret reagent test. Biuret reagent reacts with the peptide bonds that join amino acids together, producing colour changes from blue (indicating no protein) to pink (+), violet (++), and purple (+++). The + sign indicates the relative amounts of the peptide bonds present.

9. Measure and pour 2 mL of water, gelatin, albumin, skim milk, and the unknown solution into separate labelled test tubes. Add 2 mL of Biuret reagent to each of the test tubes, then tap the test tubes with your fingers to mix the contents.

(f) Record any colour changes in a chart.

Analysis

(g) What laboratory evidence suggests that not all sugars are monosaccharides?

(h) Answer the Question by summarizing your results for all nutrient tests performed.

Evaluation

(i) Which test tube or depression plate served as a control in each test?

(j) Explain the advantage of using two separate tests for lipids.

(k) Why should the graduated cylinder be rinsed with water after you measure out each solution?

(l) List possible sources of error and indicate how you could improve the procedures.

Synthesis

(m) A drop of iodine accidentally falls on a piece of paper. Predict the colour change, if any, and provide an explanation for your prediction.

(n) A student heats a test tube containing a large amount of protein and Biuret reagent. She notices a colour change from violet to blue. Explain why heating causes this colour change.

(o) Predict the results of a lipid test on samples of butter and margarine.

CAREER CONNECTION

A food technologist looks for better ways to preserve, process, package, and distribute food and beverage products, including the ingredients that go into them. A food technologist also studies the nature, composition, and behaviour of food, including what happens to its flavour, colour, and nutritional properties when cooked or placed in storage. Describe this career: duties, education, job opportunities, and strengths and skills required.

www.science.nelson.com

1.8 Explore an Issue Functional Foods

The delicatessen counter in a grocery store and the prescription counter in a drug store have little in common (**Figure 1**). One provides food, the other medicines. Lately, however, researchers in the food and beverage industry have begun to develop new products that claim to provide health benefits beyond the nutritional value of a food. These products are called functional foods, designer foods, or nutriceuticals (a contraction of the words *nutrient* and *pharmaceutical*). In general, nutriceuticals are foods that claim to prevent or control disease and improve physical or mental performance.

Figure 1
Deli counters provide food, while prescription counters provide medicine.

Types of Functional Foods

You may already have seen or even used products such as SMART(fx) juice with *Ginkgo biloba* or Memory Bars rich in cerocholine in the hope of improving your memory. Other functional foods include Personality Puffs with St. John's wort to "balance your personality," and Kava Kava Corn Chips to help you relax (**Figure 2**). *Echinacea* has become a popular cold remedy, and makers of Straight A Lollipops, which contain sugar, glucose, an unspecified amount of *Ginkgo biloba*, and Canadian ginseng, claim to help children focus their thoughts and strengthen their memory, in hopes of improving their grades at school. Even companies that produce red wine and chocolate now claim that their products contain significant quantities of chemicals called antioxidants that destroy free radicals. Free radicals are chemicals that form naturally in cells and have been found to contribute to the development of cancer, heart disease, and stroke.

Figure 2
Functional food products

Table 1 describes a number of common functional foods.

Table 1 Common Functional Foods

Manufacturer	Product name	Nutriceutical ingredient(s)	Claimed health benefit
Apple & Eve "Tribal Tonics"	Energy Elixir	guarana-seed extract, ginseng	enhances energy, stamina, and endurance
Arizona "Rx Elixirs"	Rx Stress	kava kava	relieves stress
Fresh Samantha "Body Zoomers"	Super Juice (beverage)	echinacea	keeps colds at bay
Golden Temple Cereals	Mango Passion Crisp	St. John's wort, kava kava	supports emotional and mental balance
Hansen's Beverage Co. "Healthy Start"	Immune-ox Juice	echinacea	stimulates the body's production of interferon, a cell-protecting protein
Odwalla	Femme Vitale	vitex, nettles, dong quai	strongly supports women's cyclic nutritional needs
Snapple	Earth	grape-seed extract, ginseng	helps provide your body with the balance it needs

Universal Remedy or Modern Snake Oil?

Producers of functional foods are targeting a number of common conditions for functional food research. These include heart disease, osteoporosis, high blood pressure, diabetes, digestive disorders, menopausal symptoms, and lactose intolerance. In many cases, the health claims made on the packages of functional foods may be exaggerated and confusing to the consumer. For example, there is strong scientific evidence for the claim that soy protein can lower blood-cholesterol levels, but there is little to no evidence for the claim that green tea reduces the risk of cancer. While there is no legal definition for the term *functional food*, or *nutriceutical*, laws are in place to protect consumers. Manufacturers of foods are permitted to make claims about how their products affect the structure and function of the body, but they must obtain government approval before making claims about a food's ability to treat or cure disease. For example, a label may say that a product helps maintain healthy vision, but it can't claim to offer protection against cataracts unless it has gone through Ministry of Health testing and approval.

If functional foods live up to their claims, there is a growing concern that consumers could get too much of a good thing. For instance, makers of the "151 Bar" claim that it contains 151 vitamins and minerals, but it also contains an enormous 11 g of fat and 19 g of sugar. Australia has removed products containing cholesterol-lowering plant steroids from supermarket shelves after obtaining evidence showing that extra-high doses of plant steroids could reduce the absorption of carotene, a nutrient necessary for human health.

Nutriceutical producers would like to adorn their labels with health claims, but dieticians worry that consumption of these products might actually aggravate certain health problems. For example, *Ginkgo biloba* contains ingredients that may act as blood-thinning agents. People who are presently taking blood-thinning drugs, such as aspirin, to control a heart condition could experience additional heart problems if their blood becomes too thin after taking *Ginkgo*. Taking aspirin and *Ginkgo* at the same time could lead to increased bleeding—there have been reports of bleeding behind the eye or bleeding in the brain in patients who take both a blood-thinning drug and *Ginkgo biloba*.

DID YOU KNOW?

Dangers of kava kava

Kava kava is a natural tranquilizer used as an alternative to prescription medications such as Valium. Kava kava has been banned in a number of countries, such as France, Switzerland, and the United Kingdom, because of reports linking it to severe liver damage. Health Canada banned the sale of all products containing kava kava in August 2002. The United States Food and Drug Administration is currently investigating the effects of kava kava on the liver.

Understanding the Issue

1. How does functional food differ from normal food?
2. Describe three common functional foods and the health claims associated with each.
3. How is advertising linked to the problems associated with functional foods?
4. List three medical conditions that developers of functional foods are targeting for functional-food production.
5. How can functional foods provide consumers with "too much of a good thing"?

Take a Stand

Decision-Making Skills

- ○ Define the Issue
- ● Analyze the Issue
- ● Research
- ● Defend a Decision
- ● Identify Alternatives
- ● Evaluate

Should functional foods be regulated?

The functional-food industry and promoters of functional-food products claim that there is so little of any active ingredient in functional foods that the products are not at all dangerous. Worried consumers want package labels (**Figure 3**) to list the names and amounts of all active ingredients in food products, and Health Canada says that new labelling laws will be in place shortly. Nevertheless, protesters claim that better labelling is not enough. They insist that functional foods should be tested in experiments like those that drugs go through before they are sold to the public. Those who disagree say that this would increase the price of these beneficial foods, making them too expensive for people who need or want them.

Statement: Nutriceuticals should be regulated as a drug under the Canada Food and Drug Act.

- In your group, research the issue. Search for information in newspapers, periodicals, and CD-ROMs, and on the Internet.

www.science.nelson.com

(a) Identify individuals, organizations, and government agencies that have addressed the issue.

(b) Identify the perspectives of opposing positions and arrange the different perspectives in a suitable graphic organizer.

(c) Write a position paper summarizing your opinion.

(d) Prepare to defend your group's position in a class discussion.

(e) How did your group reach a decision about which position to defend? Are there things you could have done differently? Explain.

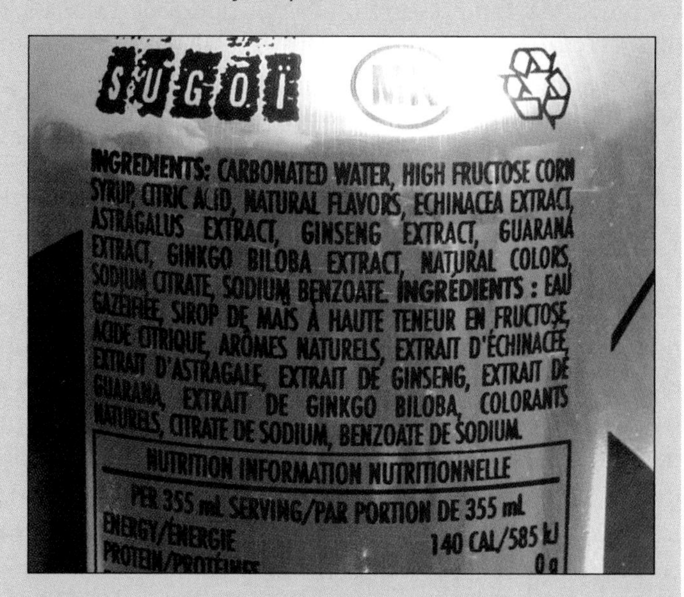

Figure 3
Labels indicate the ingredients contained in the food product.

1.9 Counting Cells

The blood test is a common medical procedure used to diagnose illness. To perform the test, a doctor, nurse, or venipuncturist draws blood from a vein and sends the sample to a diagnostic laboratory for analysis. Counting the number of different blood cells in the sample is one of the most important tests performed. Too many white blood cells could indicate the presence of infection, and an abnormal number of immature white blood cells could be a sign of leukemia (a form of cancer).

Cells are also counted in urine tests and bacteriology (the testing of body fluids and tissues for the presence of disease-causing bacteria). Common laboratory instruments used for counting cells are the hemacytometer and the Coulter Counter.

A hemacytometer (**Figure 1**) is a modified microscope slide that has two counting chambers. The hemacytometer is "loaded" by injecting a known volume of cell-containing liquid (e.g., blood) into the counting chambers and counting the cells under a microscope. A grid is etched into the base of each counting chamber to help the technician keep track of the counted cells (**Figure 2**).

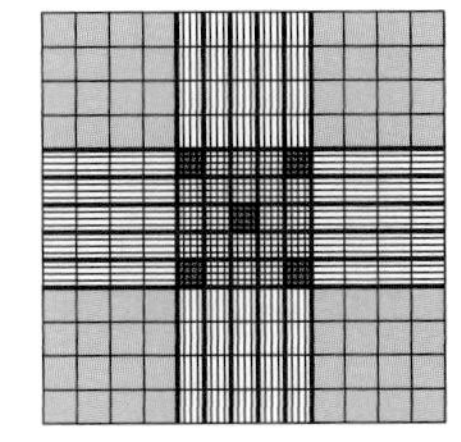

areas of the grid where white blood cells are counted

areas of the grid where red blood cells are counted

Figure 2
A typical hemacytometer grid as seen under low power through a light microscope

Hemacytometers are made of clear, colourless glass, and are relatively expensive and fragile precision instruments. Using a hemacytometer, a technician can distinguish between dead cells and living cells, if the two types of cells can be stained differently.

The Coulter Counter (**Figure 3**) is an electronic cell-counting device that does not require a technician to physically count the cells in a sample. Coulter Counters are commonly used in hematology laboratories (blood-testing laboratories) to count blood cells.

In a Coulter Counter, the cell-containing liquid is drawn with a vacuum pump through a tiny opening into an electrically charged tube. As the cells pass through the hole, each cell within the liquid blocks the electricity for a moment. The force and frequency of the electrical distortions are automatically matched to specific types and numbers of cells.

The Coulter Counter has greatly improved the practice of hematology because it efficiently counts white and red blood cells in minutes when it normally takes technicians at least half an hour to count the cells with a hemacytometer and microscope. The Coulter Counter is not suitable for counting small cells such as bacteria, because the presence of tiny particles of debris in the solution reduces the accuracy of the reading. Coulter Counters provide a total cell count; they cannot distinguish between living cells and dead cells.

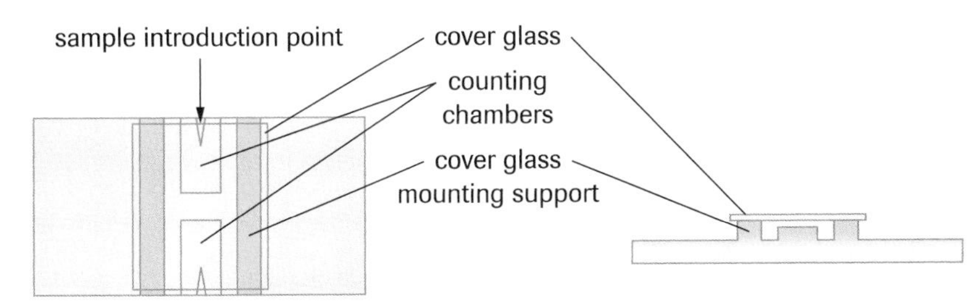

Figure 1
A hemacytometer

Figure 3
A Coulter Counter

Tech Connect 1.9 Questions

Understanding Concepts

1. What types of cells may be counted with a hemacytometer?
2. What is the purpose of the grid etched into the base of a hemacytometer counting chamber?
3. Briefly describe how a Coulter Counter works.
4. Describe one advantage and one disadvantage of using a Coulter Counter rather than a hemacytometer to count cells.

Counting Red Blood Cells and Yeast Cells

A convenient way to count blood cells is through the use of a hemacytometer and microscope. The hemacytometer used in this activity holds 0.1 mm^3 (0.1 μL) of liquid. Precise grid lines are etched into the glass base of the counting chamber (**Figure 1**). The grid lines help technicians keep track of counted and uncounted cells. Note that a blood sample taken from a patient is always diluted before it is placed in a hemacytometer chamber.

Red blood cells (RBCs) are typically counted in only five squares of the grid (**Figure 1**). The number of cells counted is then used to calculate the concentration of red blood cells in the original blood sample. The final result gives the number of red blood cells per cubic millimetre of blood (red blood cells/mm^3). This is called the "red blood cell count," or "RBC count." Diagnostic laboratories usually express RBC counts as red blood cells/mm^3, or red blood cells/μL. Since 1 mm^3 = 1 μL, the two values are the same.

Part 1: Counting Red Blood Cells

Question

What is the concentration of red blood cells in a sample of human blood?

Materials

micrograph of human red blood cells in a hemacytometer chamber (viewed with a compound light microscope, 100× total magnification) (**Figure 2**)
tally sheet

Procedure

(a) Using the micrograph in **Figure 2**, count the red blood cells in each of the five cell counting squares and record your values in a suitable tally chart. (*Note*: Only the centre of the grid shown in **Figure 1** is visible at this magnification.)

Important note: To avoid counting the same cells twice, count cells that are touching the lines at the tops and left sides of the squares, but do not count cells that are touching the bottoms and right sides of the squares.

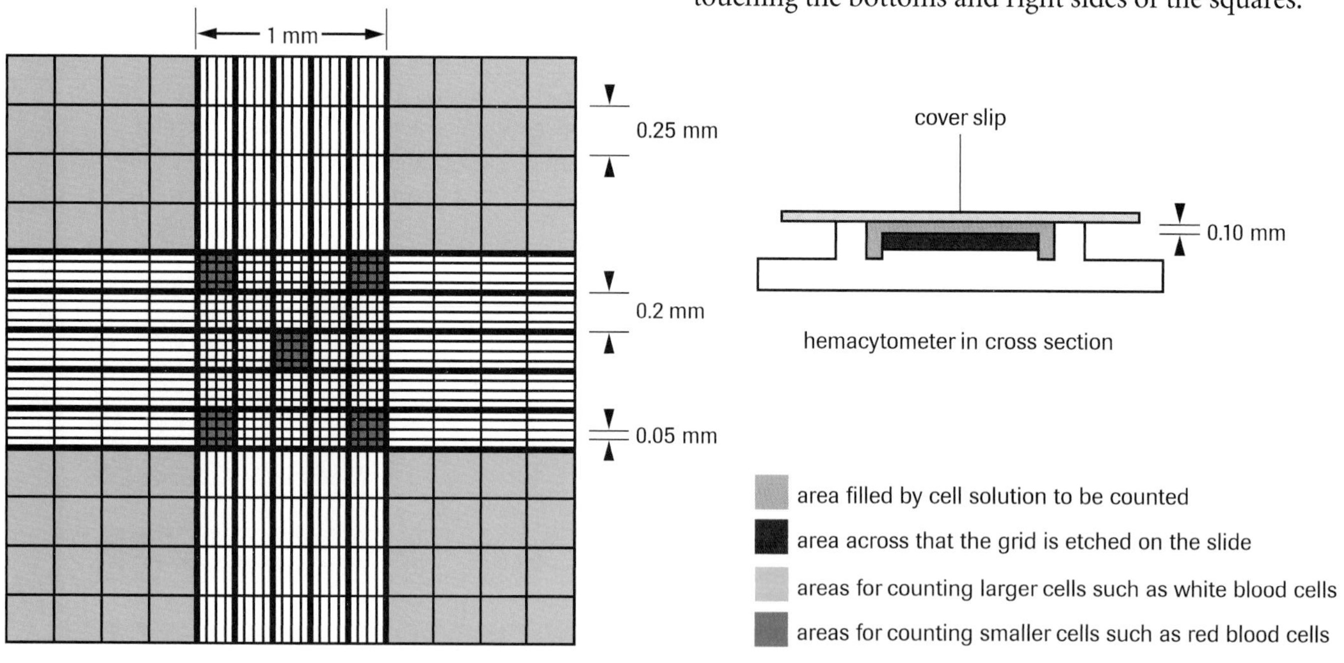

Figure 1
A hemacytometer grid viewed under low power with a compound light microscope. The red and blue colours show the squares normally used to count red blood cells and white blood cells, respectively. The colours are used for illustration purposes only. Normally, a hemacytometer is made from clear, colourless glass, and the grid lines appear black or white when viewed through a microscope.

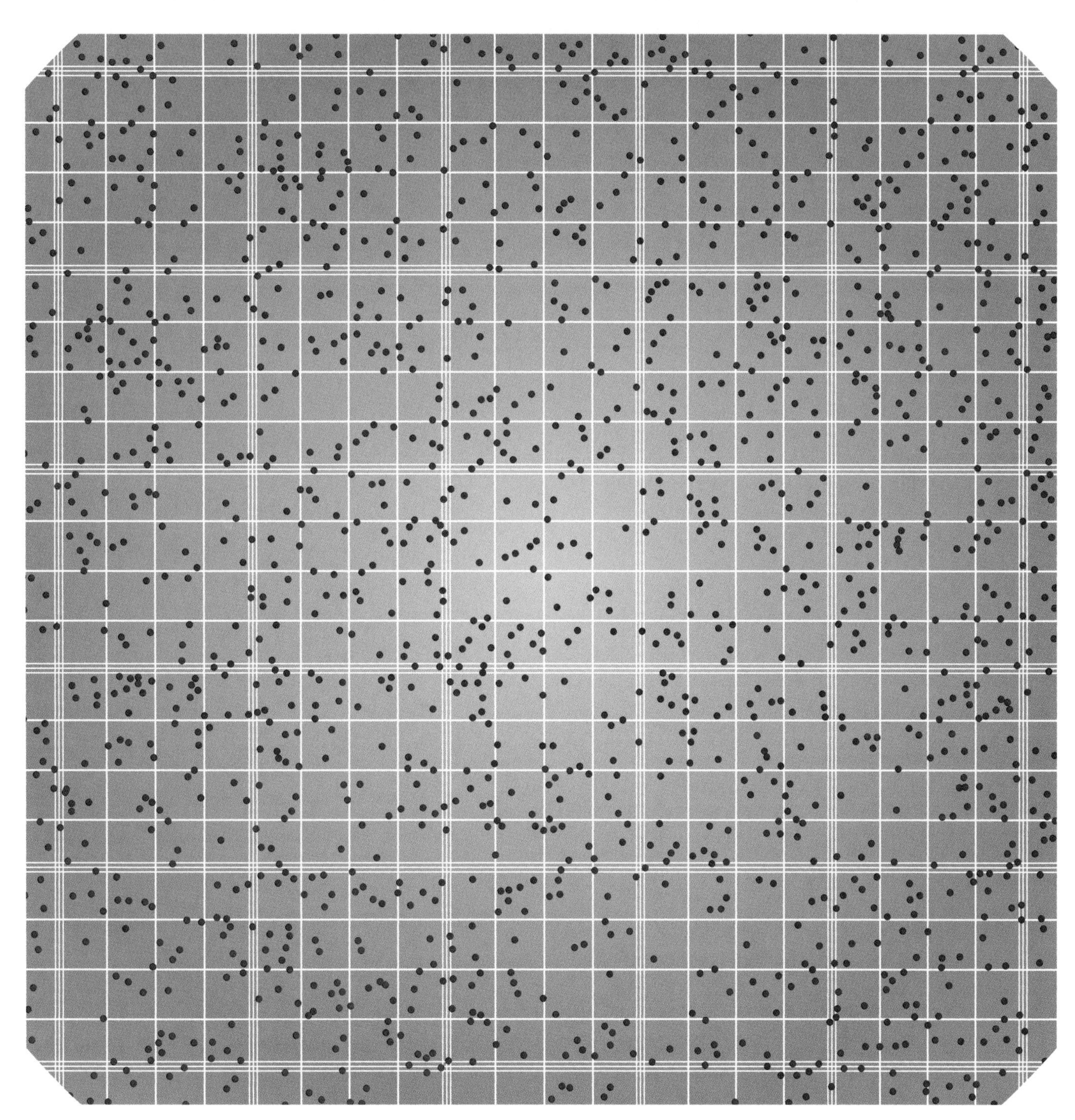

Figure 2
Human red blood cells as seen in a hemacytometer at 100× total magnification

Analysis

(b) Use the following information to calculate the red blood cell count for the blood sample in **Figure 2**.

The central grid consisting of 25 squares is 1.0 mm × 1.0 mm in area and 0.10 mm deep. The total volume may be calculated as follows:

volume = 1.0 mm × 1.0 mm × 0.10 mm

volume = 0.10 mm^3, or 0.10 μL

1. Multiply the number of cells you counted in the 5 red squares of the central grid by 5 to estimate the number of cells in all 25 squares.
2. Since the cells were diluted by a factor of 400 before being placed in the hemacytometer, multiply the value obtained in step 1 by 400 to get the number of cells there would be in the same volume of the original blood sample.
3. Divide the number of cells obtained in step 2 by the volume of the central grid to obtain the RBC count (concentration of RBCs) in the original blood sample.

(c) Record the value in units of RBC/μL.

(d) **Table 1** shows the normal ranges of RBC counts for males and females. Assuming that the blood sample was taken from a male, does the RBC count you calculated fall within the normal range? If not, suggest reasons why.

Table 1 Normal Ranges for Red Blood Cell Counts

Gender	RBC count (RBC/μL)
Female	3.9–5.6 million
Male	4.5–6.5 million

Part 2: Counting Yeast Cells

In this part of the activity, you will use a hemacytometer and microscope to determine the concentration of live yeast cells in a sample. When counting the cells, you will be able to distinguish between dead cells and live cells by using a stain. Living yeast cells contain an enzyme that decolorizes a dye called methylene blue, whereas dead cells do not. When yeast cells are placed in a solution of methylene blue, the dye molecules enter all the cells, but only the living cells decompose the dye molecules. Therefore, dead cells are stained blue and living cells are unstained.

Question

(e) What is the concentration of live yeast cells in a sample of yeast cells?

Materials

gloves and laboratory apron
methylene blue solution
stirring rod
hemacytometer
hand-held mechanical counter (optional)
10-mL transfer pipette (1-mL graduations)
dilute yeast-cell suspension in a stoppered bottle
microscope with 400× total power capability and mechanical stage
kimwipes
40-mL beaker
fine-tip glass pipettes
sheet of paper towel

Wear a laboratory apron and gloves at all times.

Procedure

4. Using a transfer pipette, transfer 9 mL of yeast-cell suspension into a clean, dry, 40-mL beaker. (Swirl the stoppered bottle containing the yeast-cell suspension before pipetting.)
5. Add 20 drops of methylene blue solution. Stir constantly for 2 min, then let sit still for 1 min.
6. Obtain a hemacytometer and cover slip. Position the cover slip on the hemacytometer so that the glass covers both counting areas equally.
7. Place the tip of a glass pipette into the yeast-cell suspension. The liquid will rise up the tip by capillary action. When the liquid stops rising, quickly and gently blot the tip on a sheet of paper towel to remove excess liquid.
8. Fill the hemacytometer chamber by gently setting the pipette tip on the notch at the edge of the chamber (**Figure 3**). The liquid will enter the chamber by capillary action. Keep the pipette tip on the notch until capillary action ends.

Figure 3
Loading the hemacytometer chamber

9. Carefully place the loaded hemacytometer on the microscope stage with the low-power objective lens in place, and focus. Locate the centre grid and change objective lenses until a total magnification of 400× is reached. Refocus until yeast cells and grid lines are clearly visible. Adjust the light intensity using the diaphragm to bring cells and grid lines clearly into view.
10. Count the colourless cells in the grid. (*Remember*: Colourless cells are alive and blue cells are dead.) Count all cells clearly within the squares and those touching the top and left borders, but not those touching the bottom and right borders. Yeast-cell buds emerging from mother cells are counted as a separate cell if the bud is at least one-half the size of the mother cell; otherwise, count as one cell.
    (f) Record your value in a suitable tally chart.
    (g) Count the live cells in the other four grids and record the values.
11. Clean the hemacytometer and cover slip, and replace materials and equipment according to your teacher's instructions.

Analysis

(h) Calculate the concentration of live yeast cells in the sample.

Evaluation

Parts 1 and 2

(i) Describe two benefits and two drawbacks of using a hemacytometer to count cells.

(j) What types of skills do laboratory technicians who count cells using a hemacytometer need to have to be successful?

(k) Describe two sources of error in this method of determining cell concentrations.

(l) Suggest two ways in which the method could be improved.

CAREER CONNECTION

A hematology laboratory technician performs routine blood tests, including blood-cell counts and blood-chemistry tests. Research the training required for this career and programs offered at Canadian colleges.

www.science.nelson.com

1.11 Case Study
Building New Bones with Skin Cells

For the past thirty years, scientists have been developing new technologies and products to help replace damaged organs with parts grown in the laboratory. Biologists have recently been able to change human skin cells into bone and cartilage, the tissues that ears, noses, knees, and hips are made of. Recently, biologists have been studying certain types of cells that can be forced to change their identities. Most of these biologists are focusing on **stem cells**—immature (unspecialized) cells that can, in theory, be transformed into any of the specialized cells of the body. Other researchers are working with certain mature (specialized) cells, such as skin cells, that they have been able to transform into other types of specialized cells, such as bone and cartilage cells.

stem cell an unspecialized cell that may transform into a specialized cell

DID YOU KNOW?

Organ Donation in Canada
Canada's organ-donation rate is among the lowest of all the developed countries. More than 3000 Canadians are waiting for an organ transplant. One organ donor can donate numerous organs and tissues, including lungs, heart, liver, kidneys, pancreas, bowel, eye tissue, skin, heart valves, bone, tendons, veins, and ligaments. You can indicate your wish to become an organ donor on your health card or your driver's licence. Discuss this decision with your family so your wishes are known.

Stem Cells

Every human being begins life as a single cell called a zygote—formed from the combination of a sperm cell and an egg cell. This cell is not a bone cell, a brain cell, or a skin cell. It is not specialized. However, in the cell's nucleus, there are 46 DNA molecules containing all of the instructions (genetic information) necessary to transform it into any of the specialized cells of the mature human body. The information in DNA is organized into sections called genes, each gene coding for a particular characteristic of the cell. Why does a cell like a zygote not use all of the information in its genes? Biologists don't yet have a satisfactory answer to this question, but experiments seem to indicate that some of the chemical reactions controlling the genetic information are blocked. In other words, some of the genes are "turned off." Therefore, in a zygote, the genes for producing the characteristics of a muscle cell or a bone cell are turned off, and only the genes for producing an unspecialized cell are turned on. These are stem cells. As the zygote divides by mitosis, it becomes a collection of unspecialized stem cells called an embryo.

As an embryo continues to develop, various genes turn on in some of its cells, causing them to transform or differentiate into specialized cells. Specialized cells form tissues and organs, such as bone, heart, liver, and skin. However, not all cells specialize. Even in the adult human, stem cells may be found in bone marrow, skin, the spinal cord, and the brain. Stem-cell research has shown that under certain experimental conditions, embryonic and adult stem cells can be encouraged to become almost any kind of specialized cell.

Embryonic Stem Cells and Cloning

The use of tissues from human embryos raises many moral and ethical concerns, and while stem cells are found in adult tissues, they are difficult to isolate and collect because they make up only one out of every 10 000 cells. Using stem cells to produce tissues for transplantation may also cause medical complications. Just as the body rejects transplanted organs from donors, it can reject tissues grown from donated stem cells. Scientists have discovered ways of avoiding this type of tissue and organ rejection through a process called cloning. Cloning produces cells that are genetically identical to the cells of the recipient, and thus are not rejected by the recipient's immune system. Nevertheless, cloning is a highly controversial process. Two forms of cloning are therapeutic cloning and reproductive cloning. In **therapeutic cloning**, the nucleus of a human egg cell is replaced with the nucleus of a body cell (such as a skin cell) from the person who needs the organ transplant. The egg cell is stimulated with a short pulse of electricity, and it then develops into an embryo from which stem cells can be removed. These stem cells are genetically identical to those of the donor of the body-cell nucleus and can be used to produce tissues that will not be rejected by the donor's immune system. Therapeutic cloning raises moral and ethical dilemmas because human embryos are created and then destroyed in the process.

therapeutic cloning the production of cloned embryos for the purpose of obtaining stem cells

reproductive cloning the production of fully formed cloned organisms

Reproductive cloning begins like therapeutic cloning. An egg-cell nucleus is replaced with a body-cell nucleus and stimulated to develop into an embryo. The only difference is that in reproductive cloning the embryo is not used as a source of stem cells; instead, it is implanted in a uterus and allowed to develop into an adult clone—an exact genetic copy of the donor. Dolly, the first cloned sheep, was produced by this method. Reproductive cloning also raises serious moral and ethical dilemmas because it creates embryos artificially, and because it introduces the possibility of creating humans for the sole purpose of providing tissues and organs for other humans.

Skin Stem Cell Research

In 2001, Dr. Freda Miller and colleagues at the University of Montreal Neurological Institute (MNI), discovered that skin contains stem cells that may transform into a number of other cell types, including nerve cells, muscle cells, and fat cells. Further research has shown that some types of stem cells in skin may transform into bone and cartilage cells.

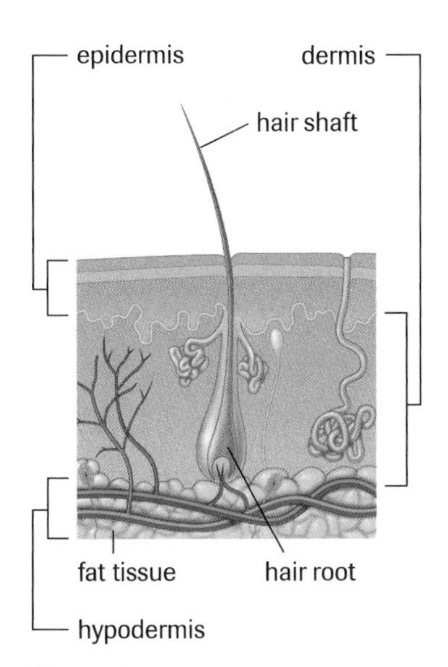

Figure 1
Human skin

Skin is the largest and most accessible organ in the body. Within the dermis, the middle layer of human skin (**Figure 1**), are specialized cells called **fibroblasts**. Recently, a number of researchers have been able to transform fibroblasts into bone and cartilage cells. From a small piece of skin no more than a few cubic millimetres in volume, scientists have been able to produce enough bone or cartilage to fill a cavity 200 to 300 times that size. And because the original fibroblasts come directly from the patient, there is no risk of rejection.

fibroblast an unspecialized skin cell

In order to change fibroblasts into bone cells, scientists attach skin cells to sticky netting made of protein, and surround the cells with phosphate minerals like those normally found in bone. Within weeks, the fibroblasts act just like bone cells and begin to produce the hard material that makes bones rigid. When fibroblasts are packed tightly together and deprived of oxygen (a condition normally found in cartilage tissue), they soon transform into cartilage cells.

Some of the promising applications of this research include the treatment of gum disease (also called periodontal disease) (**Figure 2(a)**) and osteoarthritis, a degenerative disease of the joints (**Figure 2(b)**). In advanced gum disease, the bone in the jaws deteriorates and teeth are lost. In osteoarthritis, a breakdown in the cartilage of hips, knees, and other joints causes thousands of patients every year to undergo painful hip- and knee-replacement surgery. Cartilage and bone replacement using skin fibroblasts may some day help cure these conditions.

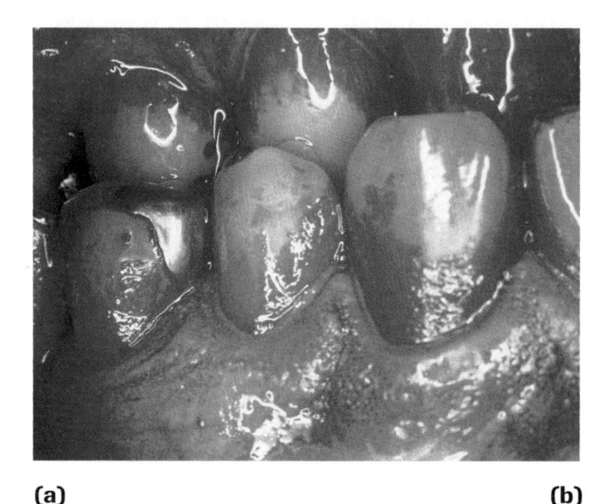

(a)

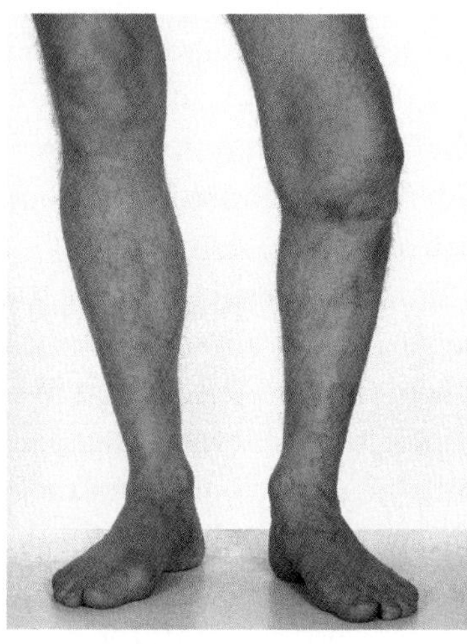

(b)

Figure 2
(a) Advanced gum disease
(b) Osteoarthritis (in the knee on the right)

Case Study 1.11 Questions

Understanding Concepts

1. (a) What is a stem cell?
   (b) How may stem cells be used to help humans?
2. Explain how genes determine whether a cell is a stem cell or a specialized cell.
3. (a) What are fibroblasts?
   (b) Why are fibroblasts of interest to transplantation biologists?
   (c) What medical advantages do fibroblasts have over stem cells in transplantation?
4. Describe the processes biologists use to change fibroblasts into bone and cartilage cells.

Making Connections

5. (a) What are some of the uses of bone and cartilage cells that have been made from fibroblasts?
   (b) Injury to the cartilage in knee joints is common among athletes. A procedure called autologous chondrocyte transplantation (ACT) is sometimes used to replace damaged knee cartilage. Briefly describe the ACT procedure, and explain how the research described in this case study could help improve the process.

www.science.nelson.com

6. Some scientists have been able to change stem cells obtained from fat tissue into cells that resemble nerve cells. Conduct research to determine how newly transformed fat cells can function as nerve cells.

www.science.nelson.com

Enzymes are protein molecules that speed up the chemical reactions of organisms. Without enzymes, most of the reactions in a cell would proceed too slowly to maintain life. When an enzyme is used to speed up a reaction, the reactants are converted into products faster than they would be without the enzyme. At the end of the reaction, the enzyme remains undamaged, ready to catalyze another reaction.

Activation Energy

All chemical reactions require energy for them to start. This energy is called **activation energy** (E_A). Enzymes speed up reactions by lowering the activation energy needed to start a reaction (**Figure 1**). The most common source of activation energy is heat. An increase in temperature will increase the speed of most reactions. However, proteins are denatured at high temperatures and lose their function. This could be devastating for a cell. Living cells cannot rely on high levels of heat as a source of activation energy; instead, cells make and use enzymes that allow reactions to proceed at speeds that keep the cell alive.

activation energy the energy required for a chemical reaction to occur

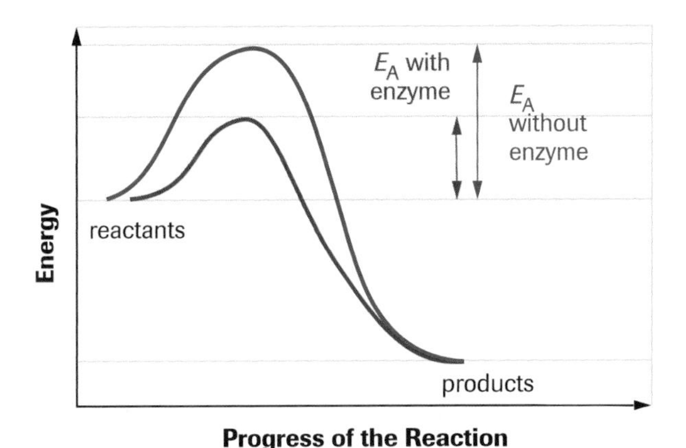

Figure 1
An enzyme speeds up a reaction by lowering the activation energy (E_A).

The names of enzymes usually end in "-ase." Thus, amylase catalyzes the breakdown of amylose into maltose subunits, and maltase catalyzes the breakdown of maltose into individual glucose molecules. An enzyme-catalyzed reaction is usually written with the name of the enzyme over the arrow, as follows:

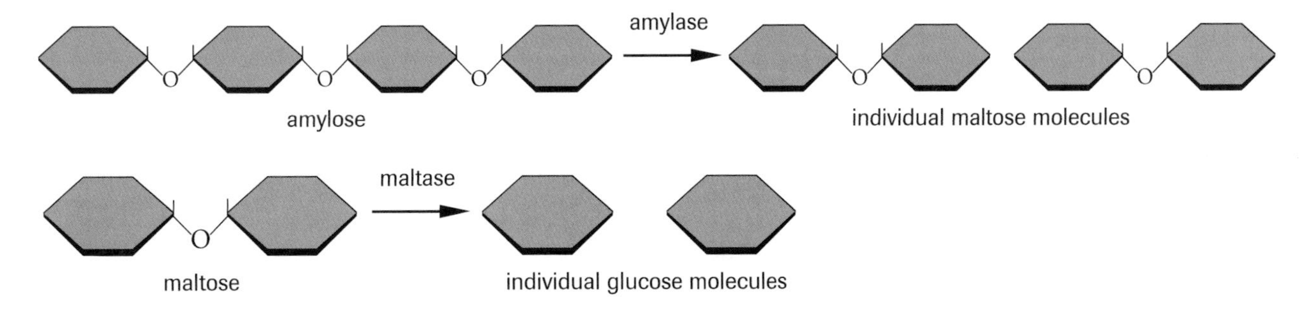

substrate the reactant that an enzyme acts on when it catalyzes a chemical reaction

active site the location where the substrate binds to an enzyme

induced-fit model a model of enzyme activity that describes an enzyme as a protein molecule that changes shape to better accommodate the substrate

enzyme–substrate complex an enzyme with its substrate attached to the active site

Enzyme Activity

In order to catalyze a reaction, an enzyme attaches to the reactants. We call the reactants **substrates.** Substrates bind to a very small portion of an enzyme. The location where a substrate binds to the enzyme is called the **active site,** and is usually a pocket or groove in the three-dimensional structure of the protein (**Figure 2**).

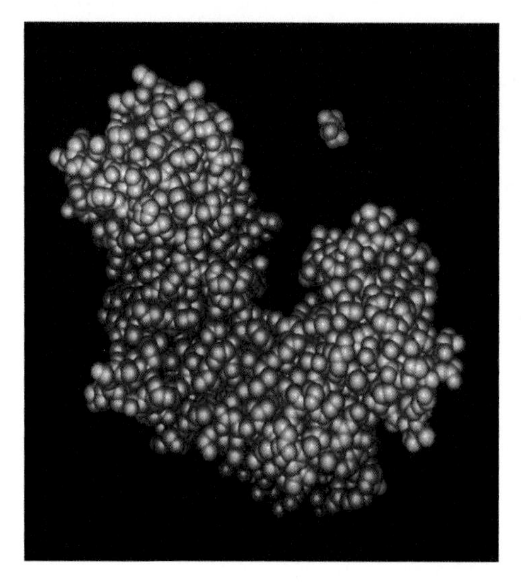

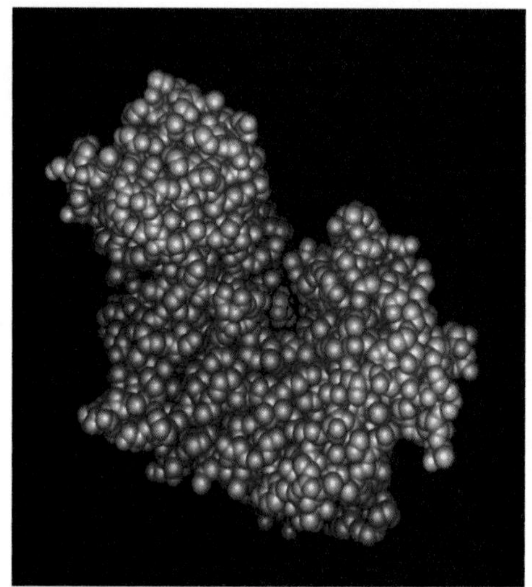

Figure 2
The binding of a substrate to the active site of an enzyme

Enzymes are very specific for the types of substrates they attach. In most cases, they will bind to only one type of molecule. This means that a different enzyme is needed for every reaction. The substrate and the active site must possess well-matched shapes for binding to occur. As the substrate enters the active site, the protein changes shape to better fit the substrate. This is known as the **induced-fit model** of enzyme activity. The attachment of the substrate to the enzyme's active site creates the **enzyme–substrate complex** (**Figure 3**).

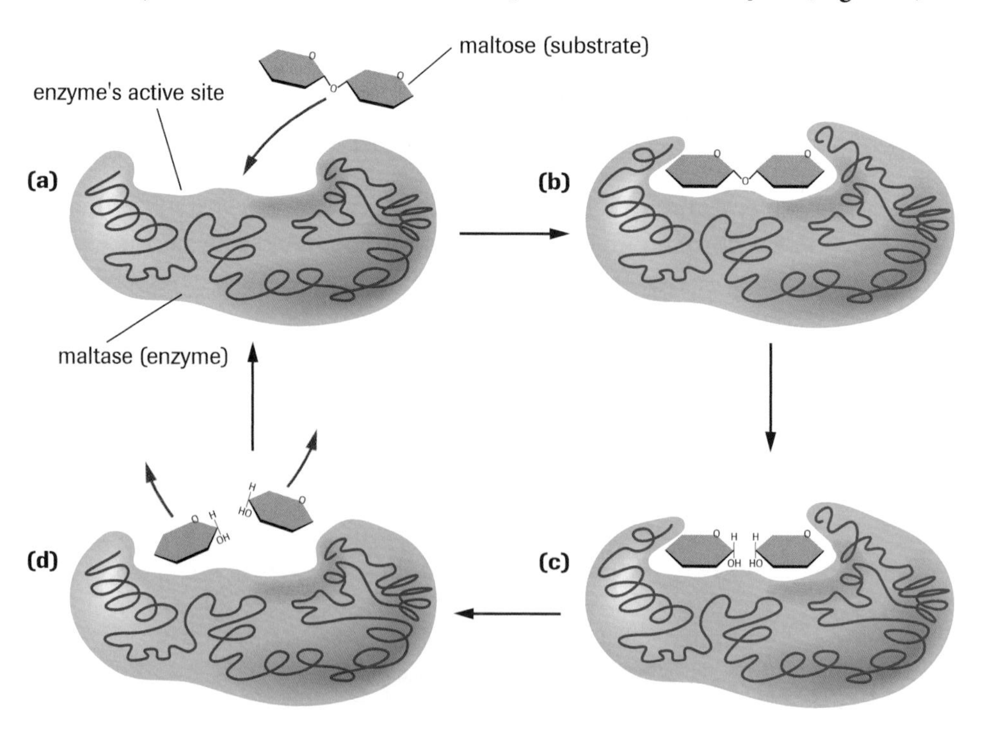

Figure 3
The enzyme maltase catalyzes the breakdown of maltose into two separate glucose molecules by breaking the bond between the individual sugar molecules.
(a) The active site is ready to receive the substrate.
(b) The substrate enters the active site, and the enzyme's shape changes to better fit the substrate (induced fit).
(c) The bond between the two glucose molecules is broken while the substrate is in the active site.
(d) Breaking the bond changes the shape of the protein again; the protein loses its attraction for the product molecules and releases them. The active site now becomes available for another maltose to attach. The recycling of enzyme molecules causes cells to catalyze many reactions with relatively small numbers of enzymes.

▸ TRY THIS activity — Synthesizing a Paper-Clip Polymer with a Paper Enzyme

Materials: coloured paper clips, 5-cm × 21.5-cm strip of paper

1. Prepare the strip of paper as in **Figure 4(a)**.
2. Fold the strip of paper as in **Figure 4(b)**.
3. Place paper-clip substrate 1 on active site 1, spanning the back two layers of the paper enzyme. Place paper-clip substrate 2 on active site 2, spanning the front two layers of the paper enzyme, as in **Figure 4(c)**.
4. Briskly pull the two tabs apart to activate the paper enzyme, as illustrated in **Figure 4(d)**.

(a) Explain how the action of the paper enzyme relates to a real enzyme-catalyzed reaction.

(b) Try to produce a "triclipide" or a "tetraclipide" with one pull of the tabs.

(c) Create a different enzyme simulation.

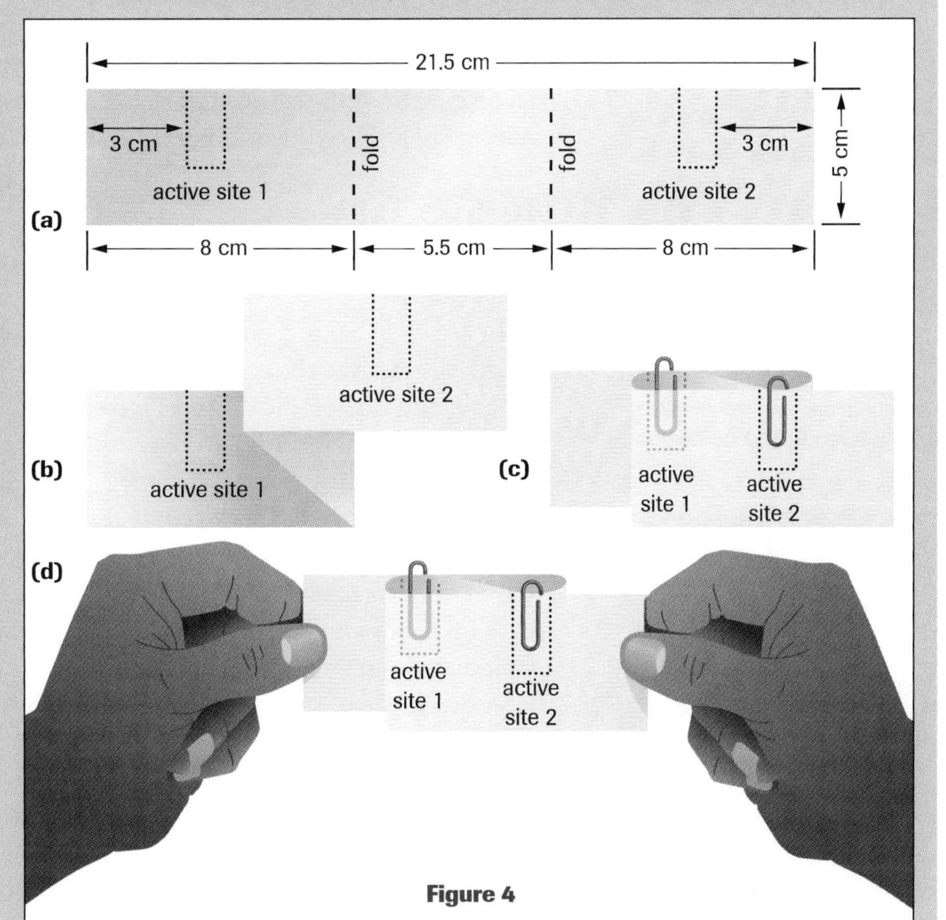

Figure 4

Enzyme Activity and the Environment

Temperature and pH have an effect on an enzyme's activity. As with all other reactions, enzyme-catalyzed reactions increase in speed with an increase in temperature. However, as the temperature increases beyond a particular point, proteins begin to denature and the enzyme becomes less effective. Every enzyme has an optimal temperature at which it works best (**Figure 5(a)**).

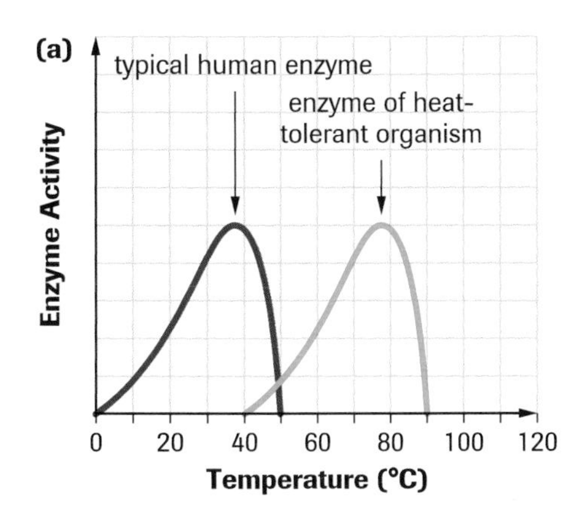

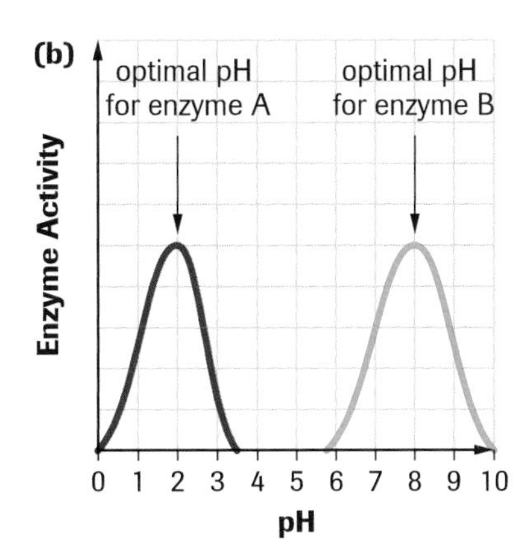

Figure 5

(a) Enzymes have an optimal temperature at which they work best. A typical human enzyme (red line) works best at approximately 37°C; the heat-tolerant organism enzyme works best at approximately 76°C.

(b) Enzymes have optimal pH levels. Enzyme A (pepsin) works best at pH = 2; enzyme B (trypsin) works best at pH = 8.

Figure 6
Enzymes from yeast cells are used in baking, brewing, winemaking, and cheese manufacturing.

Figure 7
Glucose is used as a sweetener in many food products.

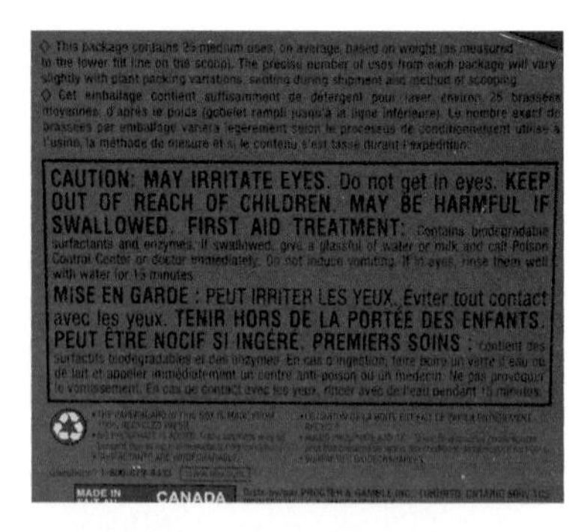

Figure 8
Enzymes are commonly added to detergents to help remove food stains from fabrics. This warning label shows that an active ingredient in this detergent is enzymes.

When temperatures are lower than the optimal temperature, the motion of the enzyme and the substrates is sluggish and reaction speed is reduced. Most human enzymes work best at around 37°C, normal body temperature. Enzymes also have an optimal pH in which they work best (**Figure 5(b)**, on the previous page). The digestive enzyme pepsin works best in the acidic environment of the stomach, pH 2. The digestive enzyme trypsin has an optimal pH of 8 and works best in the basic environment of the small intestine.

Industrial Uses of Enzymes

In addition to the vital role they play in living cells, enzymes have many practical uses.

In brewing, baking, and winemaking (**Figure 6**), enzymes produced by yeast cells catalyze the conversion of glucose (in flour starch or fruit juices) to ethanol and carbon dioxide gas. Ethanol and other compounds give beer and wine their characteristic flavours. In baked goods, the ethanol evaporates away during the heating process, and the carbon dioxide gas creates bubbles that give breads and cakes their spongy texture.

Some of the largest industrial users of enzymes are companies that convert starch from corn, wheat, and barley into glucose. Glucose is used as a sweetener in many foods and beverages, such as candy, biscuits, jams and jellies, sweetened fruit juices, and vitamin preparations (**Figure 7**). In the production of glucose, large quantities of starch are broken down by the enzymes amylase and maltase, which are produced by certain bacteria and moulds (fungi). The glucose may be used directly as a sweetener or may be converted to fructose (a much sweeter sugar) by the action of the enzyme glucose isomerase that is produced by certain species of bacteria.

Enzymes are also used in the cleaning industry. Stains commonly found on clothing, including blood, grass, milk, and perspiration, are usually mixtures of proteins, carbohydrates, and lipids. Although it is possible to remove these stains with soaps and detergents alone, detergent companies sometimes add a number of different amylases, proteases, and lipases to their products to help remove the "tough" protein, carbohydrate, and lipid stains from fabrics (**Figure 8**). Enzymes also allow stains to be removed at lower temperatures, and with less mechanical agitation in a washing machine. **Table 1** lists some additional industrial uses of enzymes.

Table 1 Some Industrial Uses of Enzymes

Product or process	Enzyme function
ethanol fuel	Enzymes convert starch into glucose and glucose into ethanol.
dairy	Enzymes break down the lactose in milk into glucose and galactose to produce "lactose-free" milk for people allergic to lactose.
cloth garments	Cellulases are used to soften cotton garments such as blue jeans.
leather	Enzymes are used to remove fat and hair from hides.
paper	Enzymes are used to de-ink paper in recycling programs.

Section 1.12 Questions

Understanding Concepts

1. (a) How does an enzyme affect the activation energy of the chemical reaction it catalyzes?
   (b) Why can a living cell not use large amounts of heat to speed up its chemical reactions?
2. Use diagrams to illustrate the induced-fit model of enzyme activity.

Applying Inquiry Skills

3. An enzyme called polyphenol oxidase causes the sliced surface of some fruits to become brown when exposed to air. This reaction is called the browning reaction, and it occurs when you cut an apple or potato and leave it exposed for a while.
   Describe an experiment you could conduct to determine how temperature affects the browning reaction.
4. The enzyme amylase catalyzes the breakdown of amylose to maltose. The graph in **Figure 9** was obtained in an experiment in which amylose was exposed to amylase in environments of various pH.

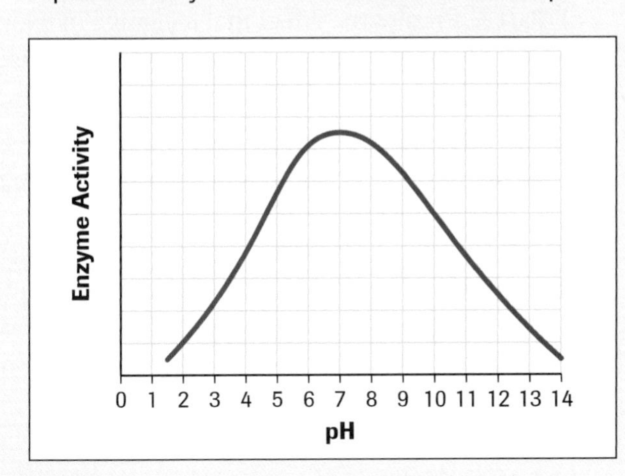

Figure 9

   (a) Provide a hypothesis for the low enzyme activity at pH 10.
   (b) What is the optimal pH for amylase activity? Explain.

Making Connections

5. (a) Describe two industrial uses of enzymes, one in the food-making sector, the other in the cleaning sector.
   (b) Why would a large diaper-cleaning company be interested in a new cleaning powder containing enzymes even though the new powder is slightly more expensive than regular detergent powder?

www.science.nelson.com

6. Papain and bromelain are the two most commonly used enzymes in commercial meat tenderizers. Conduct research to answer the following questions about these enzymes:
   (a) What are meat tenderizers? What are they used for?
   (b) What type of enzymes are papain and bromelain? (What is their substrate?)
   (c) What is the source of papain and bromelain found in commercial meat tenderizer preparations?
   (d) The antemortem (before death) method of tenderizing meat involves the physical injection of a solution of papain or bromelain into the living animal. The enzyme tenderizes muscle tissue while the animal is alive. Discuss this method with fellow classmates and write a brief position paper on the ethics of this procedure.

www.science.nelson.com

1.13 Investigation

Inquiry Skills

- ● Questioning
- ○ Hypothesizing
- ● Predicting
- ● Planning
- ● Conducting
- ● Recording
- ● Analyzing
- ● Evaluating
- ● Communicating

Enzyme Activity

Hydrogen peroxide, H_2O_2, is a poisonous chemical produced by most cells as a result of normal cell activity. If it accumulates, hydrogen peroxide damages cell components and leads to cell death. Hydrogen peroxide slowly decomposes into harmless water and oxygen gas according to the following equation:

$$2H_2O_{2(aq)} \longrightarrow 2H_2O_{(l)} + O_{2(g)} \quad \text{(SLOW)}$$

Cells produce an enzyme called catalase to speed up this reaction and prevent the accumulation of hydrogen peroxide.

$$2H_2O_{2(aq)} \xrightarrow{\text{catalase}} 2H_2O_{(l)} + O_{2(g)} \quad \text{(FAST)}$$

In this investigation, you will conduct experiments to determine how different environmental conditions affect the speed (rate) at which catalase converts hydrogen peroxide into water and oxygen gas.

In this investigation, you will use catalase from liver. Remember that catalase is the enzyme and hydrogen peroxide is the substrate. The speed of the reaction will be determined by measuring the rate at which oxygen gas is produced.

A filter-paper disk coated with liver extract containing the enzyme (catalase) is placed into a test tube containing the substrate (hydrogen peroxide solution). As catalase decomposes hydrogen peroxide into water and oxygen gas, bubbles of oxygen collect on the underside of the disk, making it rise to the surface of the hydrogen peroxide solution. The time it takes for the filter-paper disk to rise to the top of the solution will be used as a measure of the rate of enzyme activity.

There are four parts to this investigation:

- Part 1: Temperature and the Rate of Catalase Activity
- Part 2: pH and the Rate of Catalase Activity
- Part 3: Enzyme Concentration and the Rate of Catalase Activity
- Part 4: Substrate Concentration and the Rate of Catalase Activity

The procedure for Parts 1, 2, and 3 will be provided. You will design the procedure for Part 4.

Question

How do changes in temperature, pH, enzyme concentration, and substrate concentration affect the rate of catalase activity?

Prediction

(a) Part 1: Predict the effect that a change in temperature will have on the rate of catalase activity.

(b) Part 2: Predict the effect that a change in pH will have on the rate of catalase activity.

(c) Part 3: Predict the effect that a change in enzyme concentration will have on the rate of catalase activity.

(d) Part 4: Predict the effect that a change in substrate concentration will have on the rate of catalase activity.

Materials

catalase (liver extract)
1% hydrogen peroxide solution
pH 3 buffer solution
pH 5 buffer solution
pH 7 buffer solution
pH 9 buffer solution
pH 11 buffer solution
ice
graduated cylinder
identical filter-paper disks (cut with a hole punch)
distilled water
hot plate
metric ruler
beakers
marking pencils
safety goggles
10 test tubes
test-tube rack
stopwatch or timer
gloves
forceps

Wear eye protection and gloves at all times.

Hydrogen peroxide is corrosive and may cause burns to respiratory tract, skin, and eyes. Do not inhale vapours.

Perform this activity in a well-ventilated area.

Procedure

The following technique will be used to measure the rate of enzyme activity in this investigation:

1. Using forceps, dip a filter-paper disk into a test tube of liver extract (catalase enzyme).

Never touch the filter-paper disks with your fingers.

2. Gently touch the edge of the filter-paper disk onto a clean paper towel to remove excess liver extract.
3. Using forceps, place the filter-paper disk into the test tube containing hydrogen peroxide solution. The disk must be placed just below the surface of the solution so that it sinks to the bottom of the test tube when released. Do not use disks that do not sink to the bottom of the test tube.
4. Start timing when the disk is released. The disk will fall, and then begin to rise. Stop timing when the disk reaches the surface of the solution.

(e) Record the time in a suitable chart. Note that the rate of enzyme activity is higher for disks that reach the surface of the solution faster. The rate of enzyme activity is proportional to 1/time.

Part 1: Temperature and the Rate of Catalase Activity

5. Prepare the following five water baths: an ice bath (0°C), a room-temperature water bath, a 37°C water bath, a 45°C water bath, and a boiling water bath (100°C).
6. Label five clean test tubes: 0°C, room temperature, 37°C, 45°C, and 100°C.
7. Place 5 mL of stock liver extract, provided by your teacher, into each of the five labelled test tubes, and immerse one test tube into each water bath. Let the liver extract (catalase) incubate at each temperature for approximately 5 min.
8. Using a marking pencil, place a mark on each of five clean test tubes approximately 6 cm from the bottom and fill to the mark with 1% $H_2O_{2(aq)}$. Place one test tube in the ice bath, and leave the other four in a test-tube rack at room temperature.

Do not heat hydrogen peroxide.

9. Using the rate technique (steps 1 to 4), measure the rate of enzyme activity at each temperature. Use the hydrogen peroxide in the ice bath with the liver extract in the ice bath, and the hydrogen peroxide at room temperature with the other samples of liver extract.
10. Wash and dry your test tubes for use in Part 2.

Part 2: pH and the Rate of Catalase Activity

11. Label five clean test tubes: pH 3, pH 5, pH 7, pH 9, and pH 11. Place 5 mL of stock liver extract into each of the test tubes.
12. Place 5 mL of each of the following buffers into the appropriately labelled test tube prepared in step 10: pH 3 buffer solution, pH 5 buffer solution, pH 7 buffer solution, pH 9 buffer solution, and pH 11 buffer solution.
13. Place a mark on each of five clean test tubes approximately 6 cm from the bottom and fill to the mark with 1% $H_2O_{2(aq)}$.
14. Using the rate technique (steps 1 to 4), measure the rate of enzyme activity at each pH.
15. Wash and dry your test tubes for use in Part 3.

Part 3: Enzyme Concentration and the Rate of Catalase Activity

16. Label five clean test tubes: 100% catalase, 80% catalase, 50% catalase, 20% catalase, and 0% catalase.
17. Place the following mixtures into the test tubes prepared in step 14:

    100% catalase: 10 mL stock liver extract

    80% catalase: 8 mL stock liver extract + 2 mL room-temperature distilled water

    50% catalase: 5 mL stock liver extract + 5 mL room-temperature distilled water

    20% catalase: 2 mL stock liver extract + 8 mL room-temperature distilled water

    0% catalase: 10 mL room-temperature distilled water only.
18. Place a mark on each of five clean test tubes approximately 6 cm from the bottom and fill to the mark with 1% $H_2O_{2(aq)}$.
19. Using the rate technique (steps 1 to 4), measure the rate of enzyme activity at each enzyme concentration.
20. Wash and dry your test tubes for use in Part 4.

Part 4: Substrate Concentration and the Rate of Catalase Activity

Experimental Design

(f) Design a procedure to determine the effect that changes in substrate concentration have on the rate of catalase activity. Write a list of the materials you need, indicating any precautions that must be taken. With your teacher's permission, perform the experiment and record your results.

Analysis

(g) Summarize your results in tables and sketch suitable graphs for Parts 1, 2, 3, and 4.

(h) Analyze all of your results for trends and patterns.

(i) Answer the Question at the beginning of the investigation.

Evaluation

(j) Evaluate your predictions, taking into account possible sources of error. Draw reasonable conclusions.

(k) Describe how you could improve your experimental methods and the rate technique.

(l) Suggest other experiments you could perform to extend your knowledge of enzyme activity.

Passive Transport 1.14

An intravenous (IV) drip is a common sight in a hospital's intensive care unit (ICU) (**Figure 1**). This piece of equipment allows liquids to flow directly into a patient's veins. In most cases, the liquids are aqueous solutions containing a variety of different solutes. Common solutions used in IV drips include aqueous sodium chloride, $NaCl_{(aq)}$, and aqueous glucose, $C_6H_{12}O_{6(aq)}$. An aqueous solution of sodium chloride is administered when a patient's blood and other body fluids (extracellular fluids) have low concentrations of water and minerals. Minerals include the positive and negative ions of common salts such as sodium ions, $Na^+_{(aq)}$, potassium ions, $K^+_{(aq)}$, calcium ions, $Ca^{2+}_{(aq)}$, chloride ions, $Cl^-_{(aq)}$, and phosphate ions, $PO_4^{3-}{}_{(aq)}$. Aqueous solutions of ions are also called **electrolytes.**

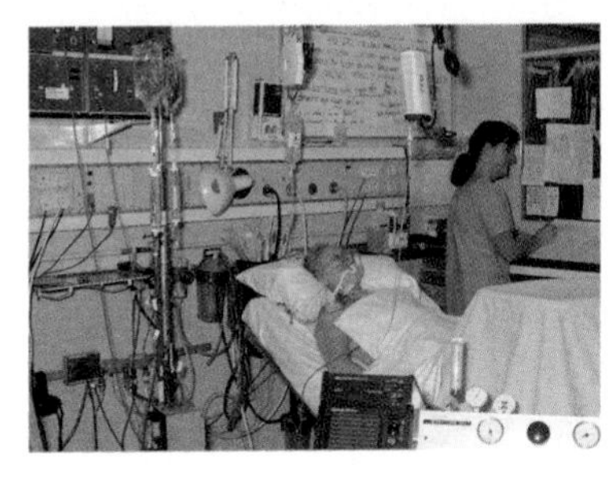

Figure 1
A hospital intensive care unit (ICU) with a patient receiving an IV drip.

electrolyte a solution composed of positive and/or negative ions dissolved in water

Careful inspection of labels on typical IV solutions reveals that the concentration of solute in the solution is always stated (**Figure 2**). Doctors and nurses are careful to administer a solution with a suitable solute concentration to their patients. Giving an IV solution with the wrong concentration could be fatal. Why is it important for patients to receive a solution of a particular concentration in an IV drip? The reasons have much to do with the structure of cell membranes and how substances move across these membranes.

CAREER CONNECTION

A venipuncturist specializes in obtaining blood from patients' veins using a syringe and evacuated tube system.

Simple Diffusion

Cell membranes are **selectively permeable**—only certain substances are able to pass through them. As mentioned in section 1.2, cell membranes are largely composed of a phospholipid bilayer and proteins. Many small, uncharged molecules, such as water, oxygen, and carbon dioxide, pass through the cell membrane freely (**Figure 3**). These substances either go through the phospholipid bilayer directly or through channels formed by proteins in the membrane. Since the cell membrane has a hydrophobic middle section, small lipid molecules such as fatty acids are also able to pass through.

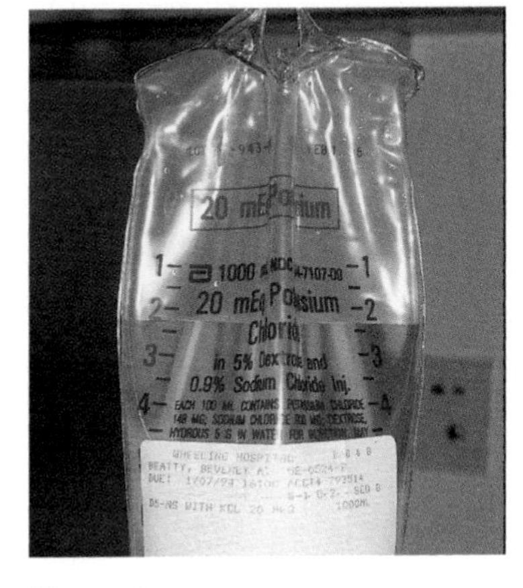

Figure 2
A 0.9% sodium chloride IV solution

selectively permeable allows only certain substances to pass through it

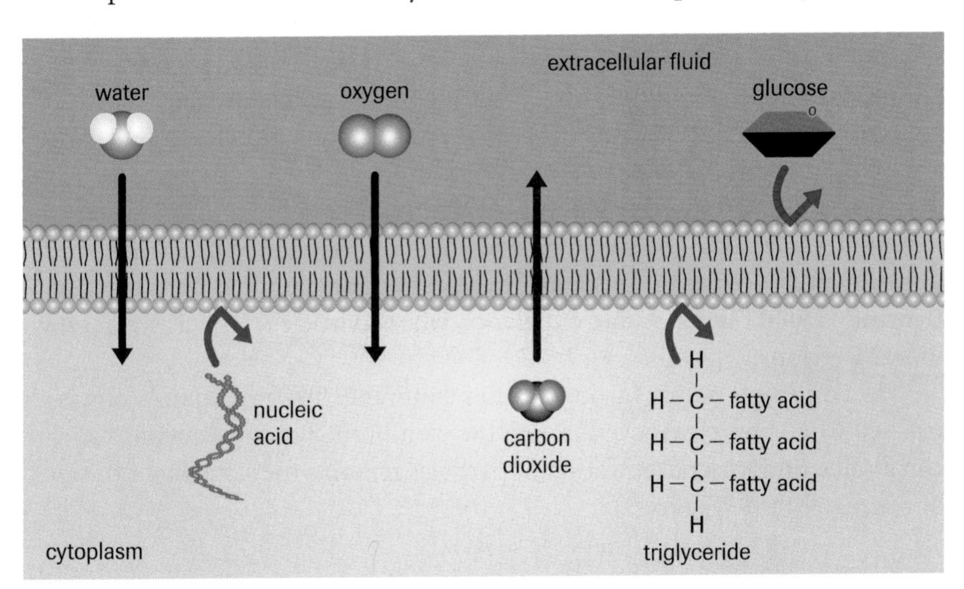

Figure 3
Cell membranes are selectively permeable—they let only certain substances through.

simple diffusion the movement of particles from an area of higher concentration to an area of lower concentration until particle concentration is equal throughout

concentration gradient a difference in concentration between two areas

However, ions, small charged molecules, and large molecules such as amino acids, carbohydrates, nucleic acids, and large lipids (triglycerides) cannot pass through easily (**Figure 3**, on the previous page).

Substances that can pass through the membrane do so by a process called **simple diffusion**. Simple diffusion is the movement of particles from an area where they are more highly concentrated to an area where they are less highly concentrated (**Figure 4**). A difference in concentration between two areas is called a **concentration gradient**, and diffusion always occurs down a concentration gradient (from high concentration to low concentration).

(a)

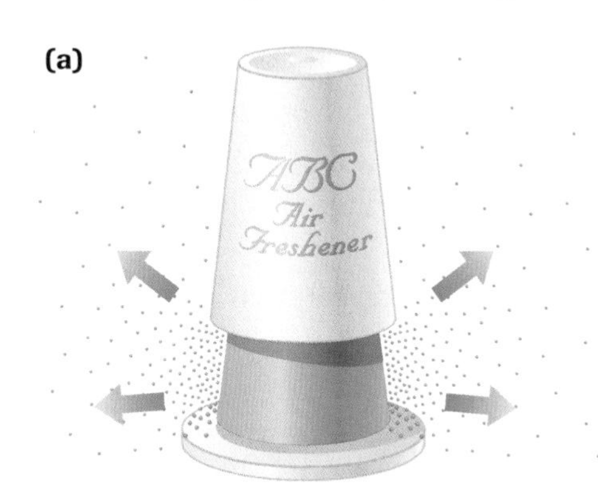

(b)

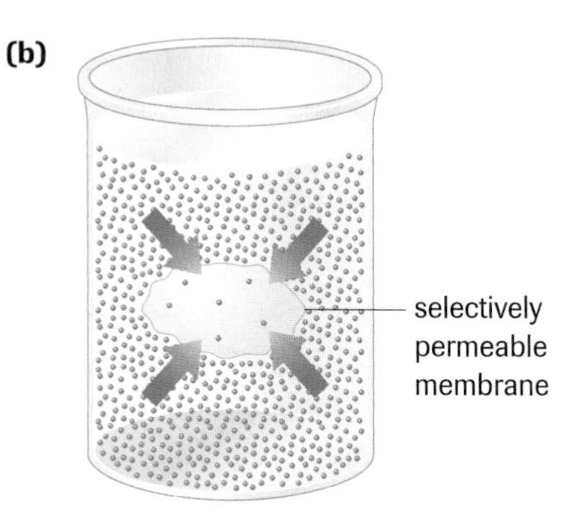

Figure 4
(a) Simple diffusion of air-freshener molecules from an area of higher concentration (in the paste) to an area of lower concentration (in the air surrounding the paste).
(b) Carbon dioxide molecules diffuse through the cell membrane from an area of higher concentration in the extracellular fluid to an area of lower concentration in the cytoplasm.

▶ *TRY THIS* activity — *Tea Infusion Diffusion*

Brewing tea and making coffee involve the diffusion of tea or coffee solutes into water.

Materials: 3 tea bags, 3 250-mL beakers, cold water, room-temperature water, hot water

1. Place 200 mL of hot water, room-temperature water, and cold water into three separate beakers.
2. Gently place a tea bag on the surface of the water in each beaker at the same time. Do not disturb.
3. Observe the diffusion process in each beaker and describe what happens.

(a) Does the filter paper of the teabag act as a selectively permeable membrane? Explain.

(b) How did the temperature of the water affect the diffusion of solutes in each of the three beakers?

(c) Describe two other examples of diffusion that occurs while preparing food or cleaning clothes.

Diffusion is a natural process that occurs because particles are in constant random motion and have a tendency to spread throughout a given volume. Simple diffusion does not use any of a cell's energy, and ends when the concentration of particles becomes equal everywhere within the given volume. However, this does not mean that the particles stop moving. In fact, particles continue moving randomly in all directions all the time. During diffusion, molecules move more in one direction than any other. When diffusion ends, we say that a state of dynamic equilibrium has been reached. **Dynamic equilibrium** is a state of balance, where particles move at equal rates in all directions.

dynamic equilibrium a state of balance where particles move in all directions at equal rates

When diffusion occurs through a cell membrane, dynamic equilibrium is reached when particles move through the membrane in both directions at equal rates, and the concentration of particles remains the same on both sides of the membrane (**Figure 5**).

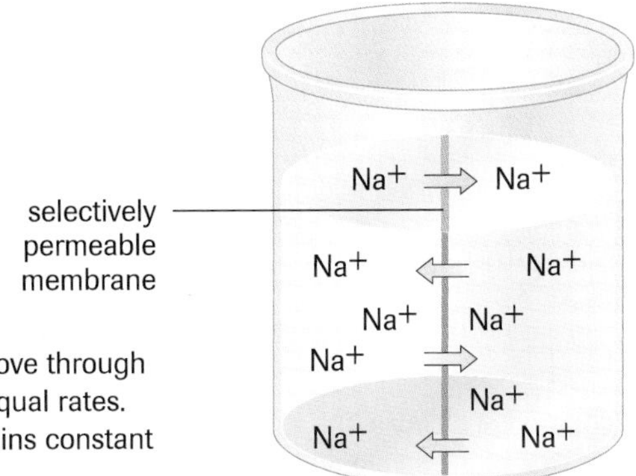

Figure 5
In dynamic equilibrium, particles move through a membrane in both directions at equal rates. The concentration of particles remains constant on both sides of the membrane.

The rate (speed) of diffusion depends on temperature and the concentration of solute molecules in solution. Diffusion occurs faster at higher temperatures because molecules move faster. Faster-moving molecules spread out faster and reach dynamic equilibrium more quickly. Dynamic equilibrium will also be reached more quickly if there are a greater number of solute molecules in solution.

Facilitated Diffusion

Glucose, sodium ions, and chloride ions are three chemicals most cells need to survive. These substances must be able to get across the cell membrane in an efficient manner. However, large polar molecules such as glucose and large ions such as sodium and chloride cannot go through a membrane by simple diffusion because they cannot easily pass through the hydrophobic middle section of the phospholipid bilayer. To compensate, membranes have protein molecules that help some substances through. In general, there are three types of membrane proteins (**Figure 6**). Some membrane proteins go part way through the phospholipid bilayer, some are attached to the outside surface, and others go all the way through. Those that span the bilayer are called **transmembrane proteins.** Some transmembrane proteins act as **carrier proteins** that assist certain substances through a membrane. This method of transporting materials across a membrane is called **facilitated diffusion**, or assisted diffusion. As its name implies, facilitated diffusion is a form of diffusion. This means that particles move down a concentration gradient until dynamic equilibrium is reached. The key difference between simple diffusion and facilitated diffusion is that, in facilitated diffusion, the diffusing particles are assisted through the membrane by transmembrane carrier proteins, whereas in simple diffusion they pass directly through the membrane's phospholipid bilayer or through protein channels. Typically, a given carrier protein transports only one type of substance, or a small group of chemically related substances. An example of carrier protein facilitated diffusion is the movement of glucose into cells of the liver (**Figure 7**, on the next page).

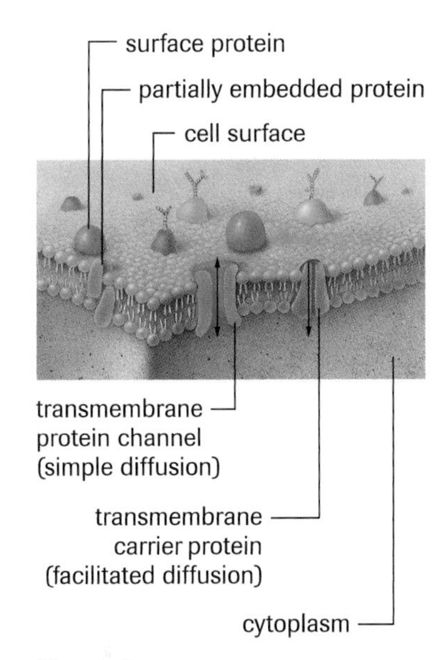

Figure 6
Membrane showing associated proteins

transmembrane protein a protein molecule in a membrane that spans the thickness of the phospholipid bilayer

carrier protein a transmembrane protein that facilitates the diffusion of certain substances through a membrane

facilitated diffusion the diffusion of solutes through a membrane assisted by proteins

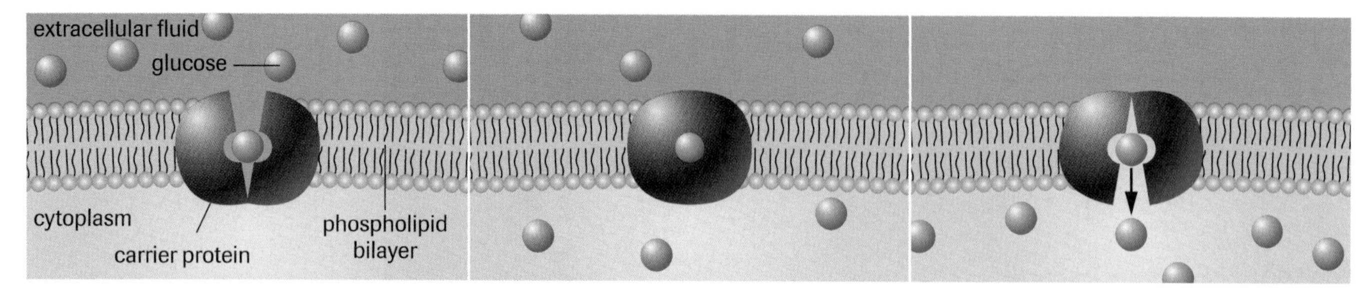

Figure 7
Facilitated diffusion of glucose. The solute (glucose) attaches to a binding site on a carrier protein at one side of the membrane. The attachment causes the carrier to undergo a series of structural changes that have the effect of carrying the solute to the other side of the membrane. The carrier then releases the solute and, through another structural change, transforms itself to its original state, ready to accept another glucose molecule.

Osmosis

Water molecules move freely through cell membranes. Under normal conditions, large quantities of water molecules move into and out of a cell by simple diffusion. The cell remains the same size because equal amounts of water go into and out of the cell. There are, however, many cases in which a net amount of water flows into or out of a cell. This means that more water may enter the cell than leaves the cell, so that the cell gains water, or more water may leave the cell than enters it, so that the cell loses water. In such situations, water still moves through the cell membrane by simple diffusion, but the process is important enough to warrant a special name—osmosis.

Osmosis is the net movement of water across a selectively permeable membrane from the side where water is more concentrated to the side where it is less concentrated. Note that solutions have a high concentration of water when they have a low concentration of solute, and vice versa. Osmosis occurs because more water molecules strike the membrane on the side with a higher concentration of water molecules (i.e., a lower solute concentration) than on the side with a lower concentration of water molecules (i.e., a higher solute concentration). More strikes result in more water molecules passing through the membrane and a net diffusion of water from one side to the other.

The key point to remember about osmosis is that water moves through a membrane from the side with a lower solute concentration to the side with a higher solute concentration. Dynamic equilibrium is reached when sufficient water has moved to equalize the solute concentrations on both sides of the membrane, and at that point, net movement of water (osmosis) ceases (**Figure 8**).

DID YOU KNOW?

Osmosis in Red Blood Cells
It has been estimated that an amount of water equal to 250 times the volume of the cell diffuses across the membrane of a red blood cell every second.

osmosis the net movement of water across a selectively permeable membrane from an area of high concentration to an area of low concentration

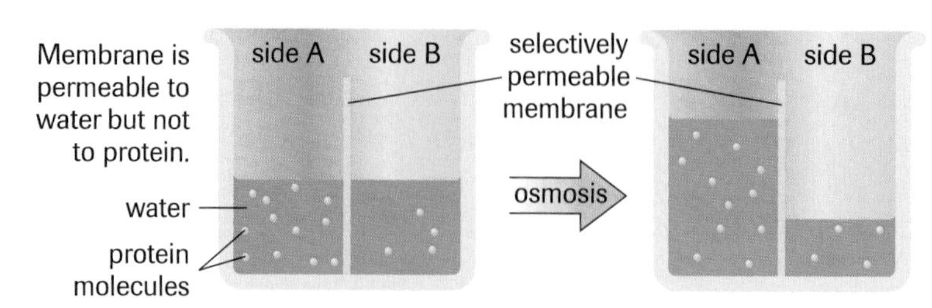

Figure 8
Osmosis. Two containers of equal volume are separated by a selectively permeable membrane that allows free passage of water but totally restricts the passage of large solute molecules such as proteins. Since side B has a lower solute concentration (higher concentration of water) than side A, a net amount of water will move (by osmosis) from side B into side A.

Osmosis occurs whenever there is a difference in solute concentration across a selectively permeable membrane. A number of special terms are commonly used to describe differences in solute concentration (**Figure 9**). **Isotonic solutions** are solutions that have equal solute concentrations. When a selectively permeable membrane separates isotonic solutions, osmosis does not occur. A **hypertonic solution** is one with a higher concentration of solutes than another solution. The solution with the lower concentration of solutes is called a **hypotonic solution.**

isotonic solution a solution of equal solute concentrations

hypertonic solution a solution that has a higher solute concentration than some other solution

hypotonic solution a solution that has a lower solute concentration than some other solution

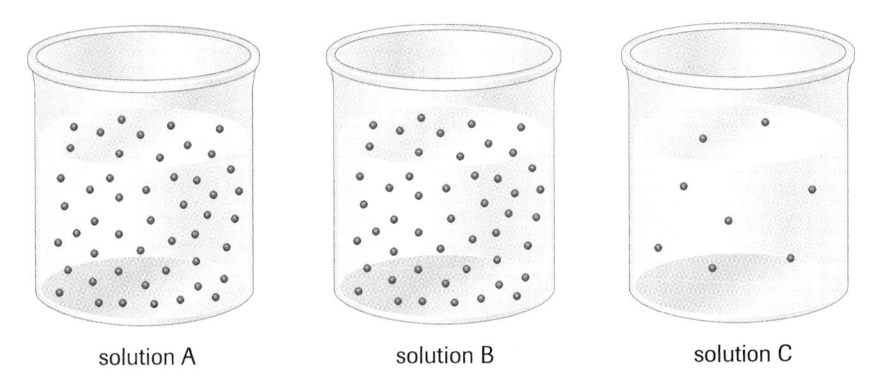

Figure 9
Solutions A and B are isotonic; solution C is hypotonic to solutions A and B; and solutions A and B are hypertonic to solution C.

We may now understand why it is important for patients to receive a solution of a particular concentration in an IV drip. Blood serum (the liquid part of blood) is normally isotonic with respect to red blood cell cytoplasm. Under normal conditions, osmosis does not occur into or out of red blood cells. The cells maintain their normal size and shape (**Figure 10(a)**).

When a patient receives an IV drip, the IV solution mixes directly with blood serum. If the solution contains a lower solute concentration than blood serum (i.e., a hypotonic solution), it may dilute the blood serum until it is hypotonic to blood cell cytoplasm. If so, osmosis will occur into the red blood cells, causing the cells to swell, and maybe burst (**Figure 10(b)**). This condition is called **hemolysis**, and may be fatal because the blood cells will be unable to transport oxygen to body tissues efficiently.

If a patient receives an IV solution that is hypertonic to blood serum, it may concentrate blood serum until it is hypertonic to blood cell cytoplasm. Osmosis will occur out of the blood cells. The cells will lose water and become small and scallop-shaped (**Figure 10(c)**). Scallop-shaped cells have a tendency to stick to one another and clog small veins and arteries, preventing oxygen from reaching body tissues. This condition, called **crenation**, may also be fatal.

hemolysis swelling and bursting of red blood cells placed in a hypotonic solution

crenation clumping of cells (usually red blood cells) that have become scallop-shaped when placed in a hypertonic solution

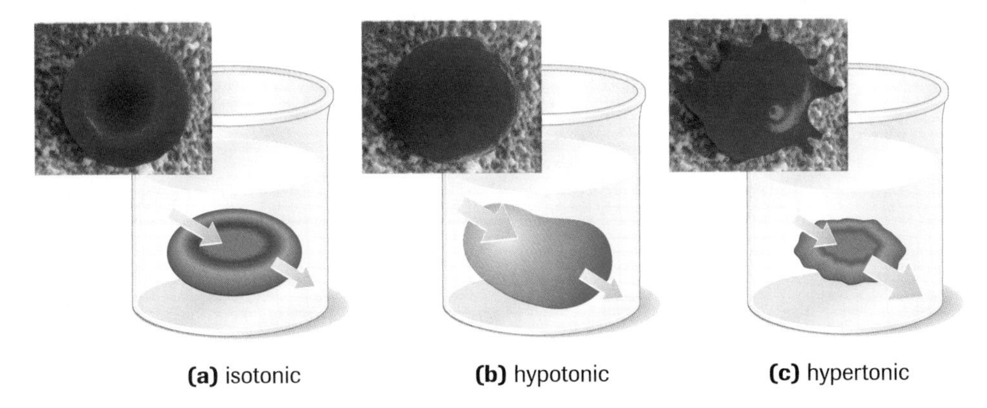

Figure 10
(a) Red blood cells in an isotonic solution. The cells are normal in shape and size.
(b) Red blood cells in a hypotonic solution take on water by osmosis and burst. The empty cells that remain are called red-blood-cell ghosts.
(c) Red blood cells in a hypertonic solution. The cells lose water by osmosis and are scallop-shaped and smaller than normal.

Any solution injected directly into veins and arteries must be isotonic with blood serum. Typical isotonic IV solutions include 5% glucose (also called 5% dextrose, or D5W) and 0.9% sodium chloride (also called isotonic saline).

Section 1.14 Questions

Understanding Concepts

1. (a) What are two similarities and two differences between facilitated diffusion and osmosis?
   (b) Name one substance that moves through cell membranes by facilitated diffusion.
2. Why do some people add salt to a steak only after it has been cooked?
3. Chemical fertilizers are composed of salts that dissolve into soil moisture and then are absorbed by the root cells of plants. Why should you never overfertilize a plant?
4. A particular fish egg may be released in a saltwater environment where its size remains constant because the concentration of salt inside the egg is equal to the concentration outside the egg. If you put the egg in fresh water, what would happen and why?
5. How is the rate of diffusion affected by
   (a) an increase in temperature?
   (b) a decrease in the concentration gradient?
6. In **Figure 11**, assume that water molecules pass easily through the selectively permeable membrane, but starch molecules cannot pass through the membrane.
   (a) Will oxygen molecules diffuse through the membrane? Explain.
   (b) Will osmosis occur? Explain.

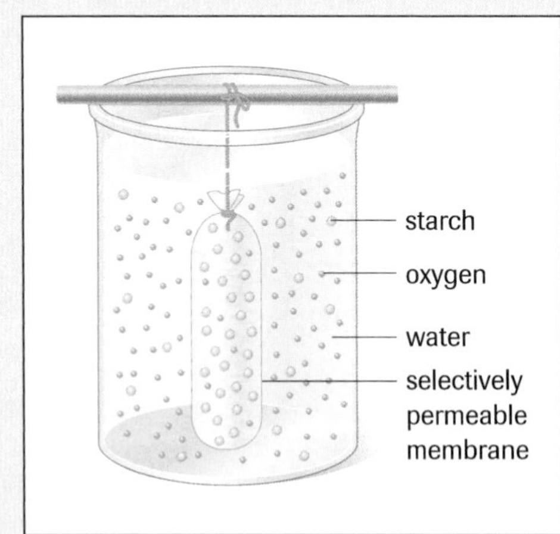

Figure 11

7. Why is it safe to inject a solution of 5% glucose directly into a vein, but not a solution of 20% glucose?
8. Why does salted popcorn dry your lips?

Applying Inquiry Skills

9. Like normal human kidneys, an artificial kidney machine works by the process of diffusion. During dialysis, blood is pumped from a person's artery through a selectively permeable membrane called dialysis tubing. The dialysis tubing is bathed in a solution similar to actual blood plasma. As the blood circulates through the tubing, waste materials diffuse from the tubing into the surrounding solution. The cleaned blood continues moving through the tubing and back into the person's vein.
   (a) What must be done to the surrounding solution in order for dialysis to continue?
   (b) Predict what will happen to the person's blood if your suggestion in (a) is not carried out.
10. (a) How does the filter paper of a tea bag act as a selectively permeable membrane when the bag is placed in a cup of hot water?
    (b) What visual evidence is there that diffusion comes to an end during the tea-brewing process?

Making Connections

11. How do the produce departments of grocery stores help keep vegetables looking fresh and feeling crispy?
12. Conduct library and/or Internet research to determine how "freezer burn" is caused, and how it can be prevented.

www.science.nelson.com

13. Why do many birds possess salt glands? Using the Internet and other sources, find out more about salt glands.

www.science.nelson.com

Inquiry Skills

○ Questioning ● Hypothesizing ● Predicting ● Planning ● Conducting ● Recording ● Analyzing ● Evaluating ● Communicating

Factors Affecting the Rate of Osmosis

In this investigation, you will determine how environmental factors such as solute concentration and temperature affect the rate of osmosis through dialysis tubing. The investigation is divided into two parts, Part 1 and Part 2. In Part 1, you will investigate the effect changes in solute concentration have on the rate of osmosis. In Part 2, you will investigate the effect of changes in temperature. A step-by-step procedure is provided for Part 1. You will design and carry out the procedure for Part 2.

Question

How do changes in solute concentration and temperature affect the rate of osmosis?

Prediction

(a) Predict how solute concentration and temperature affect the rate of osmosis.

Materials

safety goggles
distilled water
40% sucrose stock solution
5 250-mL beakers
100-mL graduated cylinder
Erlenmeyer flasks
10 pieces of nylon string
dialysis tubing
paper towel
transfer pipettes
indelible ink marker

Wear eye protection throughout this activity.

Procedure

Part 1: The Effect of Solute Concentration on the Rate of Osmosis

1. Cut five strips of dialysis tubing (about 15 cm long) and soak them in a beaker of tap water for approximately 2 min.
2. Use an indelible ink marker to label each piece of tubing as follows (one label per tube):
   0% sucrose (distilled water)
   5% sucrose
   10% sucrose
   20% sucrose
   40% sucrose
3. Using transfer pipettes, Erlenmeyer flasks, and distilled water, dilute the sucrose stock solution provided to produce at least 15 mL of the following three sucrose solutions:
   5% sucrose
   10% sucrose
   20% sucrose
   Keep 15 mL of the 40% sucrose solution.
4. Rub the dialysis tubing between your fingers to find an opening. Tie a tight knot near one end of the dialysis tubing with a piece of nylon string. Repeat with the other pieces of tubing.
5. Using a graduated cylinder, pour 10 mL of the solutions you prepared in step 3 into correspondingly labelled dialysis tubes (rinse the graduated cylinder with tap water and shake dry between solutions) (**Figure 1**).

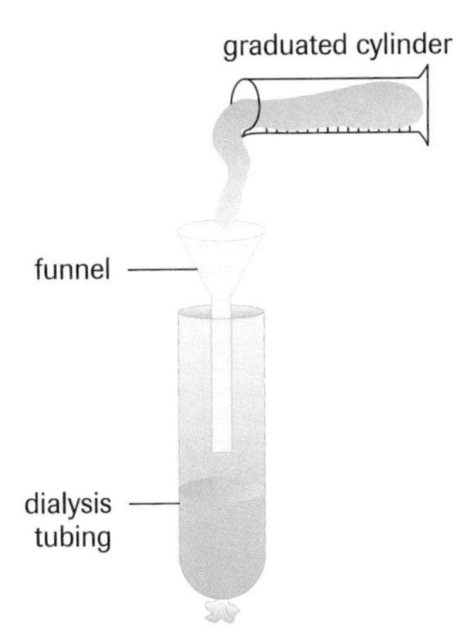

Figure 1

6. Fold over and twist the open end of each dialysis tube and tie with nylon string. Remove excess air from the tube, making sure that it remains limp. Dry the outside of the tubes with paper towel.
7. Measure the mass of each dialysis tube.

(b) Record these masses in your notebook.

8. Label five beakers: 0% sucrose, 5% sucrose, 10% sucrose, 20% sucrose, and 40% sucrose. Fill each beaker with 200 mL of distilled water at room temperature.
9. Place the corresponding dialysis tube into each beaker and wait 20 min.
10. After 20 min, remove the dialysis tubes, blot them dry, and measure their masses.

(c) Record your values in a suitable chart. Use the recorded values to determine the percent mass change for each dialysis tube using the following equation:

$$\% \text{ mass change} = \frac{m_{after} - m_{before}}{m_{before}} \times 100\%$$

11. Discard the solutions and dialysis tubes according to your teacher's instructions, and wash your hands with soap and water.

(d) Sketch suitable graphs of your results.

Part 2: The Effect of Temperature on the Rate of Osmosis

Experimental Design

(e) Design a procedure for determining the effect that a change in temperature has on the rate of osmosis. Include as many controls as possible, and list safety precautions. Have your teacher approve your list of materials and procedure, then carry out your experiment. Record your results in a suitable chart, and sketch suitable graphs of your results.

Analysis

(f) Analyze your graphs for patterns and answer the Question.

(g) Did all dialysis tubes increase in mass? If not, explain.

(h) Which dialysis tube had the greatest increase in mass in Part 1? Why?

(i) Which dialysis tube had the greatest increase in mass in Part 2? Why?

(j) Describe the five liquids in the dialysis tubes in Part 1 as hypotonic, hypertonic, or isotonic with respect to the distilled water in the beakers.

Evaluation

(k) Evaluate your predictions for each part of the investigation.

(l) Describe any sources of error in each of the parts of the investigation.

(m) Suggest possible improvements to both of the procedures.

Synthesis

(n) Many parts of the world such as California, U.S., lack fresh water but have an abundance of salty seawater. Governments pay a lot of money to import fresh water for household use and crop irrigation. Why would it not be wise to save money by irrigating crops with seawater?

Active Transport and Bulk Transport 1.16

Simple diffusion (including osmosis) and facilitated diffusion are methods used by cells to move substances through membranes from areas of high concentration to areas of low concentration until concentrations are equal. These methods for moving materials are useful in many situations, but may be wasteful in others. The primary purpose of eating is to absorb nutrient molecules into the cells of your body. Nutrients include amino acids from proteins, fatty acids from fats, nucleotides from nucleic acids, and glucose from complex carbohydrates such as starch. Glucose is a particularly important nutrient because it is used as a source of energy by all the cells of your body. To ensure that the body absorbs the maximum amount of nutrients from food, energy may be used to "pump" nutrients across cell membranes. A compound called **adenosine triphosphate (ATP)** provides the energy needed in this process, and in many of the other energy-requiring processes of living cells.

adenosine triphosphate (ATP) a compound used as a source of chemical energy in cells

Energy, Cells, and ATP

Living organisms need a continuous supply of energy to power the energy-requiring processes of life. Movement, reproduction, protein synthesis, and certain forms of transport across cell membranes all require energy. This energy is provided by the breakdown of ATP. The molecular structure of ATP is similar to that of the DNA nucleotide containing adenine, except that ATP contains three linked phosphate groups instead of just one (**Figure 1**). When ATP reacts with certain compounds in a cell, the reactions release energy that the cell can use to power energy-requiring activities (**Figure 2**).

(a)

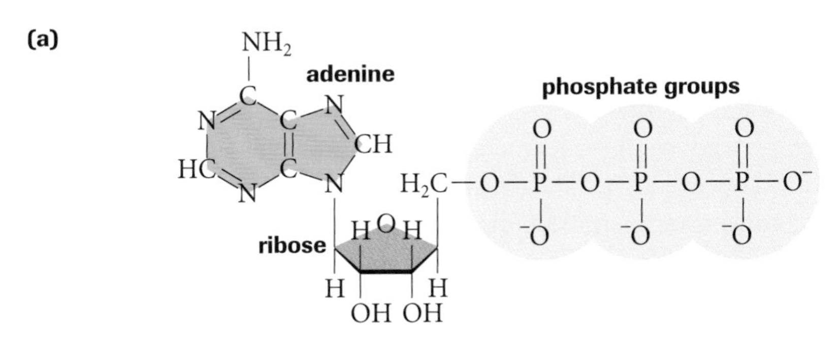

(b)

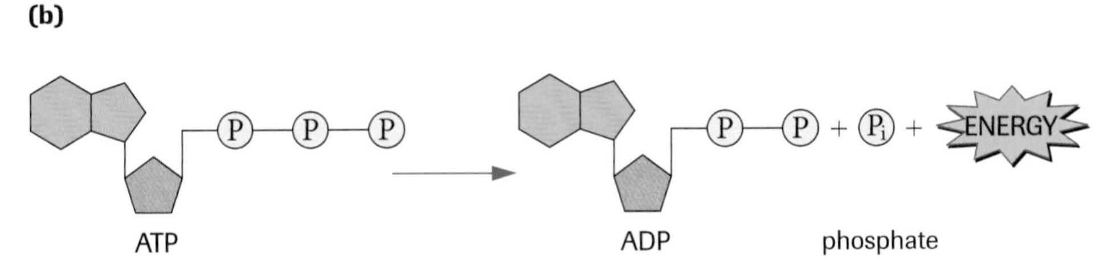

Figure 1
(a) Adenosine triphosphate (ATP)
(b) Release of energy when ATP is broken down into adenoside diphosphate (ADP) and phosphate

(a)

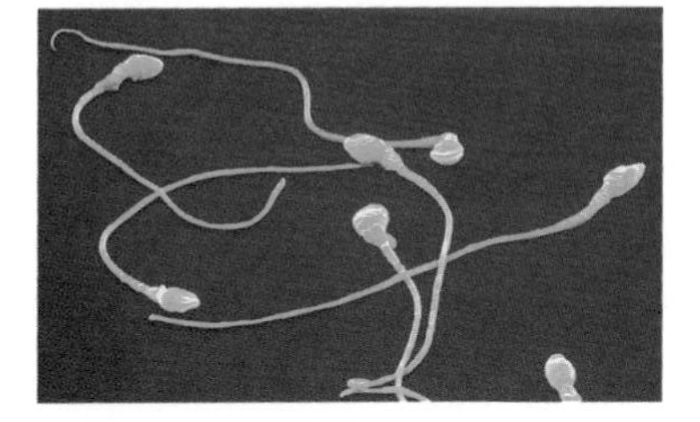

(b)

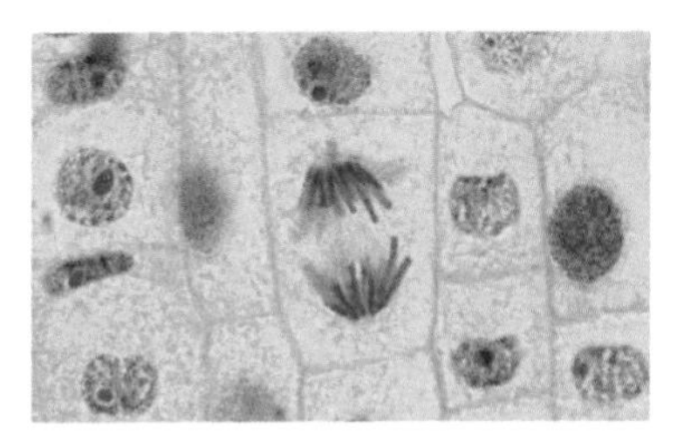

(c)

Figure 2
Muscle contraction **(a)**, locomotion **(b)**, and cell division **(c)** are examples of energy-requiring activities.

Active Transport

After a meal, partially digested food travels from the stomach into the small intestine, where most of the nutrient molecules are absorbed into cells (**Figure 3**).

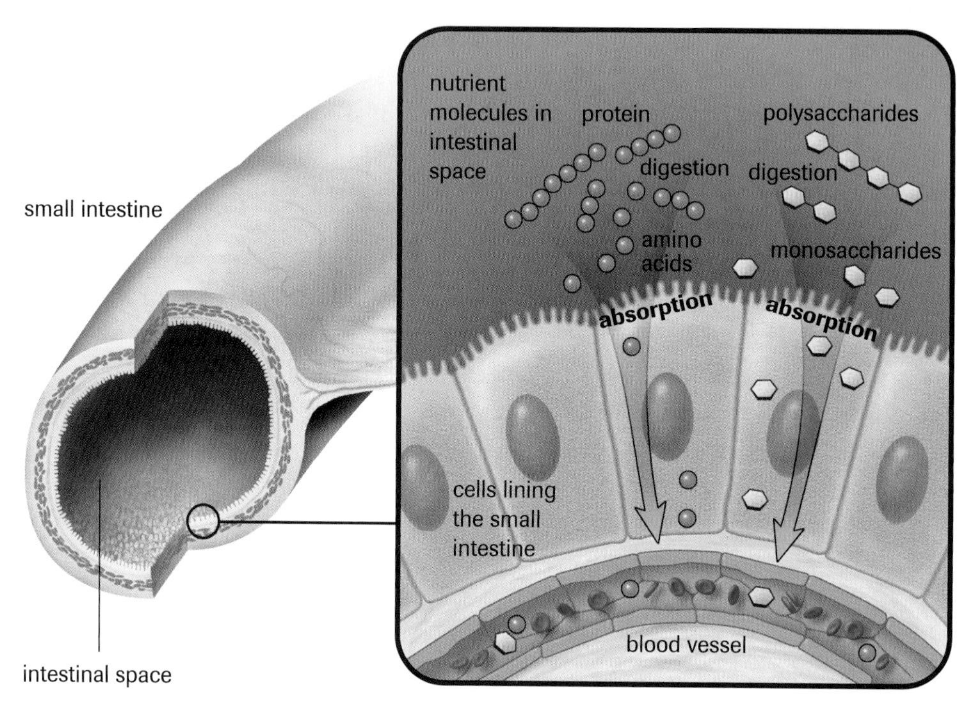

Figure 3
Absorption of nutrient molecules by cells of the small intestine

As the nutrients move through the small intestine, they must be absorbed efficiently to avoid being excreted with waste products that move through the digestive system. If glucose were absorbed into intestinal cells by simple diffusion or facilitated diffusion, only about half of the molecules would be absorbed, since diffusion ends when solute concentrations are equal on both sides of the cell membrane. To maximize the absorption of such an important nutrient, cells of the small intestine may "pump" glucose through their cell membranes against a concentration gradient. This means that glucose continues to enter intestinal cells even when the concentration inside the cytoplasm is greater than the concentration outside the cytoplasm (in the intestinal space). In order to accomplish this, the cells have special protein carriers that use chemical energy from ATP to transport substances such as glucose through their cell membranes. This process is called **active transport** (**Figure 4**). It is important

active transport the movement of substances through a membrane against a concentration gradient using membrane-bound carrier proteins and energy from ATP

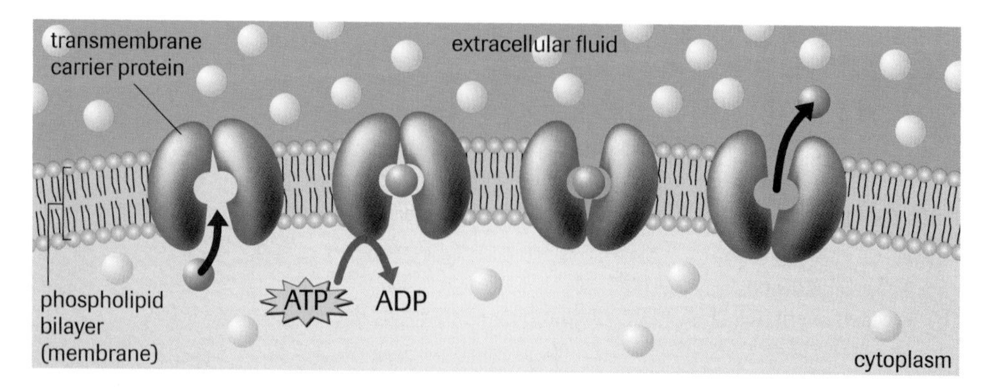

Figure 4
Active transport. The molecule to be transported attaches to an open binding site on one side of the carrier protein. ATP is converted to ADP on the carrier protein and releases energy. The energy causes a change in the shape of the protein that carries the solute to the other side of the membrane.

to note that simple diffusion and facilitated diffusion (unlike active transport) are processes that do not require a source of cellular energy (ATP).

Various types of active-transport "pumps" are found in the membranes of different cells. Potassium ions and sodium ions are moved into and out of cells by a pump known as the **sodium-potassium pump** (**Figure 5**). Without this pump, your nerve cells and muscle cells could not function properly. Other substances, such as vitamins, amino acids, and hydrogen ions, are also pumped across membranes. All of these pumps require cellular energy (ATP) to operate.

sodium-potassium pump an active-transport mechanism that pumps sodium and potassium ions into and out of a cell

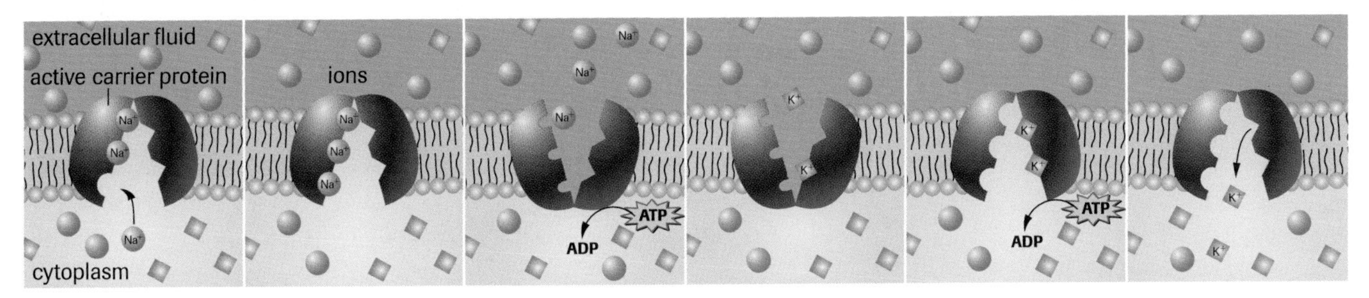

Figure 5
The energy of ATP is used to actively transport three sodium ions out of a cell for every two potassium ions that are transported into the cell.

Bulk Transport

Diffusion and active transport are responsible for moving large amounts of material through cell membranes. However, in both cases, the substances move through the membranes as dissolved particles (atoms, ions, or molecules). Sometimes cells need to move large quantities of materials (bulk) into or out of their cytoplasm all at once. A process called **bulk transport** may accomplish this. As in active transport, all bulk-transport mechanisms use energy in the form of ATP. There are two forms of bulk transport, endocytosis and exocytosis.

bulk transport the movement of large quantities of materials into or out of a cell

endocytosis a form of bulk transport used to bring large amounts of material into the cell from the extracellular fluid

Endocytosis

Endocytosis is the form of bulk transport used to bring large amounts of material into the cell from the extracellular fluid. There are two forms of endocytosis, phagocytosis and pinocytosis. **Phagocytosis** (cell eating) is the bulk transport of solids into the cell, and **pinocytosis** (cell drinking) is the bulk transport of (liquid) extracellular fluid into the cell.

phagocytosis the bulk transport of solids into the cell

pinocytosis the bulk transport of (liquid) extracellular fluid into the cell

Phagocytosis

Phagocytosis begins when a solid particle comes in contact with the plasma membrane of a cell (**Figure 6**, on the next page).

The cell membrane sends out fingerlike projections called **pseudopods** that surround and eventually enclose the particle in a vesicle that is within the cell's cytoplasm. Such a vesicle is called a **phagocytotic vesicle**. Lysosomes containing digestive enzymes may fuse with the phagocytotic vesicle to digest the particles it contains. Nutrients formed by this digestion process move through the vesicle's membrane into the cell's cytoplasm. White blood cells called macrophages frequently engulf invading harmful bacteria by phagocytosis, removing them from the bloodstream and other body tissues (**Figure 7**, on the next page).

pseudopod a fingerlike projection of the cell membrane that surrounds and encloses solid particles in the extracellular fluid that are brought into a cell by phagocytosis

phagocytotic vesicle a vesicle that is formed within a cell's cytoplasm when cells engulf solid particles by phagocytosis

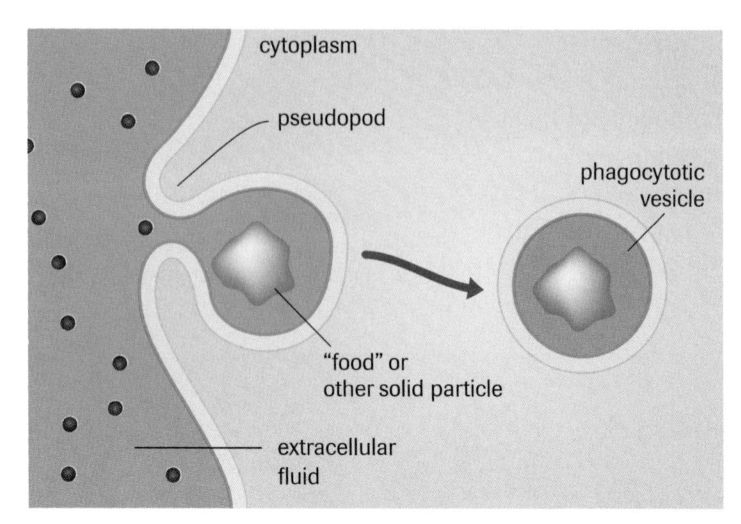

Figure 6
Phagocytosis

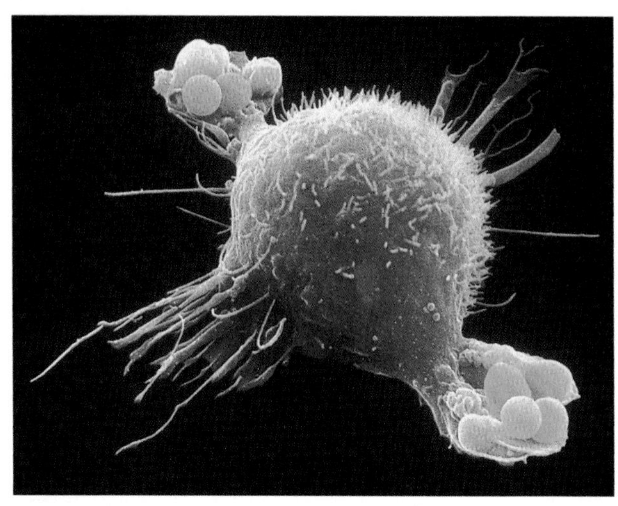

Figure 7
A macrophage phagocytosing harmful cells

pinocytotic vesicle a vesicle that is formed within a cell's cytoplasm when cells engulf extracellular fluid by pinocytosis

Pinocytosis

Pinocytosis occurs when a cell's plasma membrane engulfs a drop of extracellular fluid in a process similar to phagocytosis (**Figure 8**). This results in the formation of a **pinocytotic vesicle**. Cells bring cholesterol molecules into their cytoplasm by a special form of pinocytosis called receptor-mediated pinocytosis. In this process, cholesterol molecules in the extracellular fluid attach to receptor molecules on the external surface of the cell membrane. A pinocytotic vesicle then forms that brings the cholesterol molecules into the cell.

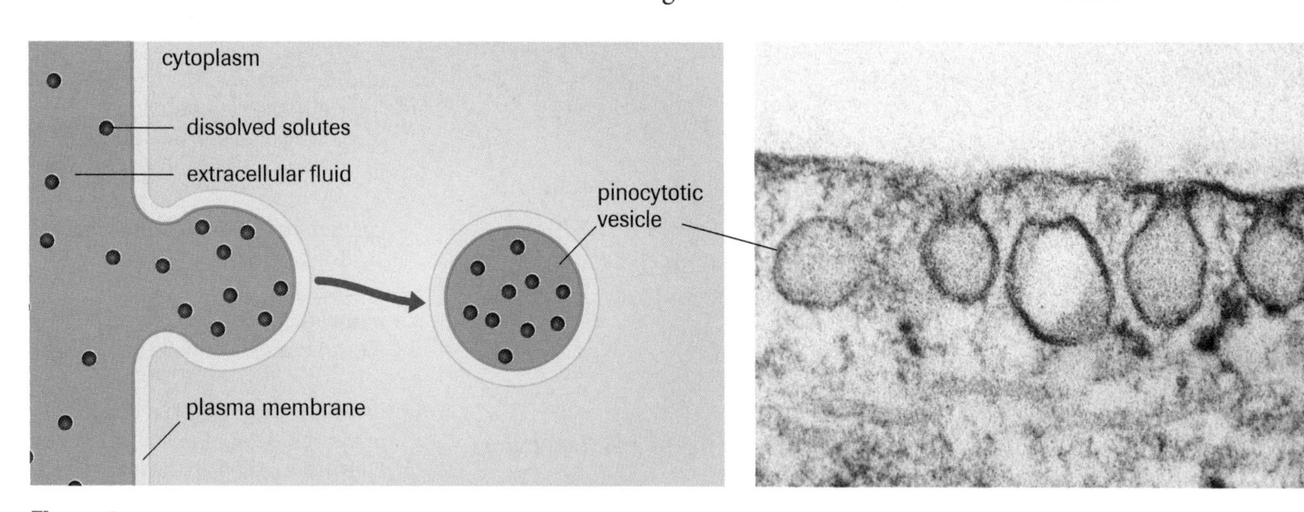

Figure 8
Pinocytosis

exocytosis the movement of large amounts of material out of a cell by means of secretory vesicles

secretory vesicle a membrane sac containing substances that are transported out of a cell

Exocytosis

In **exocytosis**, cells move large amounts of material out of their cytoplasm by a process that is essentially the reverse of endocytosis. In some cases, cells produce substances, such as hormones or enzymes, that must be exported out of the cytoplasm. In these cases, the material is enclosed in a membrane sac called a **secretory vesicle** (**Figure 9**). The secretory vesicle fuses with the cell membrane and spills its contents into the extracellular fluid. In some cases, proteins produced in the rough endoplasmic reticulum are packaged into secretory vesicles by the Golgi apparatus and are subsequently transported out of the cell by exocytosis.

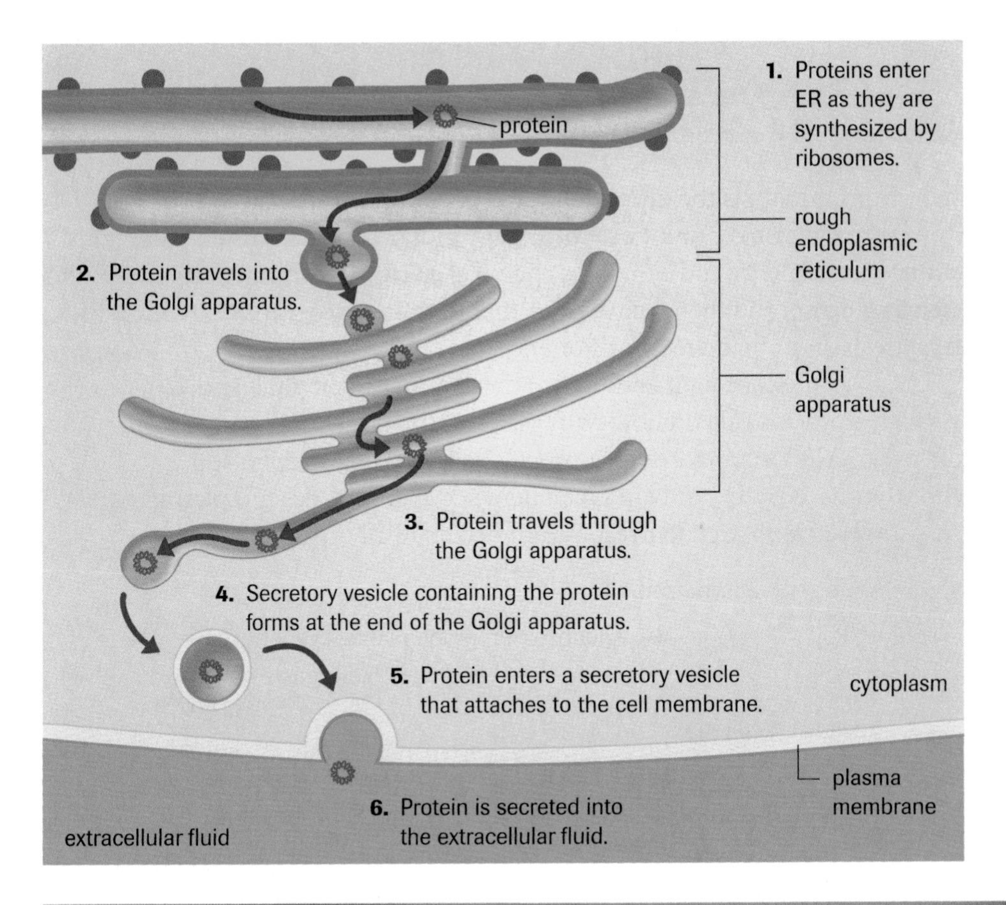

Figure 9
Exocytosis

Section 1.16 Questions

Understanding Concepts

1. (a) Describe one similarity and two differences between facilitated diffusion and active transport.
   (b) Name one substance that is actively transported through the cell membrane of an animal cell.
2. Describe the two types of endocytosis.
3. Contrast endocytosis with exocytosis.
4. (a) What cellular process is illustrated in **Figure 10**?
   (b) Identify the names of the labelled components.
   (c) What role may lysosomes play in this process?

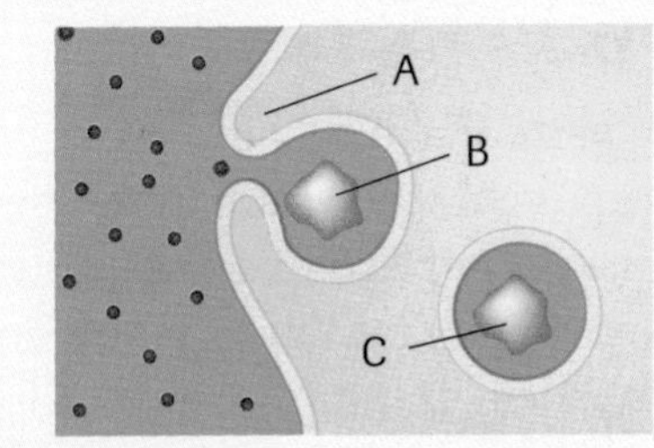

Figure 10

5. What would happen to the size of the cell membrane if a cell were to undergo a lot of endocytosis and no exocytosis?
6. List two substances that cells export to the extracellular fluid by exocytosis.

Applying Inquiry Skills

7. A student places a living cell into a beaker containing a 5% sugar solution. She measures the concentration of sugar in the solution after 30 min, and notices that the concentration has dropped substantially. She draws before and after diagrams (**Figure 11**) to illustrate her conception of the process that caused the change in sugar concentration.

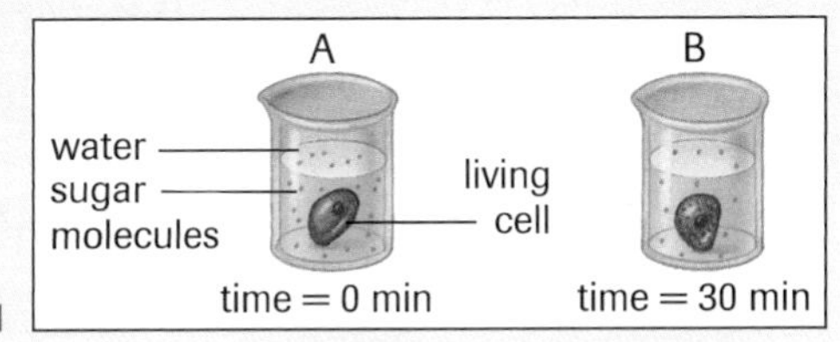

Figure 11

   (a) Based on the diagrams in **Figure 11**, describe the student's hypothesis regarding the cellular process that may have caused the change in sugar concentration inside and outside of the cell.
   (b) Inspecting the cells under a microscope, the student notices that the cells have far more mitochondria than most other animal cells have. How does this finding affect the confidence you have in your answer to (a)?

1.17 Energy Flow and Photosynthesis

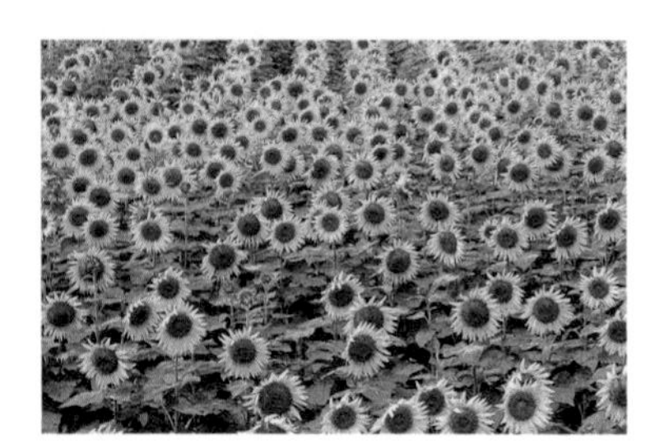

(a)

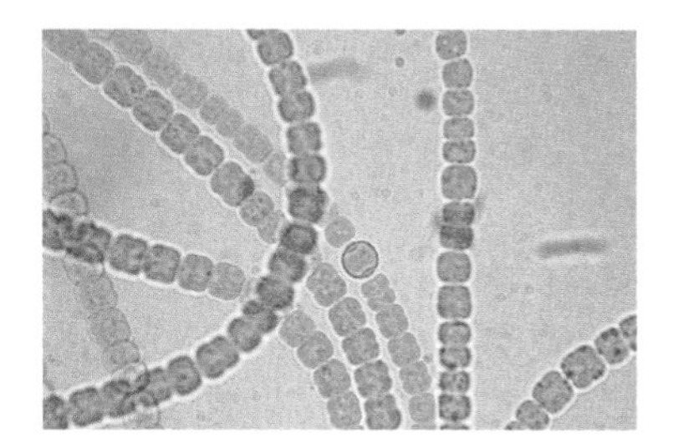

(b)

Figure 1
(a) Sunflowers
(b) *Anabaena,* a cyanobacterium

autotroph an organism (such as a plant) that obtains energy directly from light

heterotroph an organism (such as an animal) that obtains energy by eating other organisms

Light from the sun is the ultimate source of energy for most living things. However, organisms cannot use the energy of light directly. Instead, they first capture solar energy and store it as chemical energy in carbohydrate molecules such as glucose, and then transfer the energy from glucose to ATP, which all cells use as an immediate source of energy.

The absorption of light energy and the production of glucose occur through a process called photosynthesis, and the transfer of energy from glucose to ATP occurs through a process called cellular respiration. Photosynthesis occurs only in green plants, cyanobacteria, and plantlike protists (**Figure 1**). Cellular respiration occurs in all organisms.

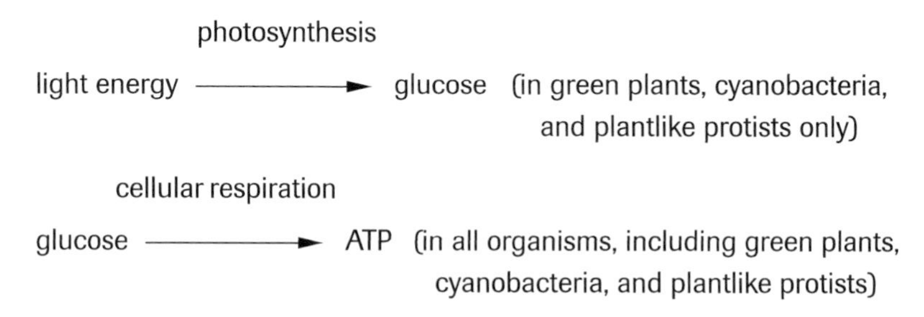

Green plants, cyanobacteria, and plantlike protists are the only organisms on Earth able to carry out both photosynthesis and cellular respiration. This means that they are self-sufficient for their energy needs. Because of this, they are called **autotrophs** (from Greek *auto*, meaning "self," and *troph*, meaning "to feed"). All other organisms, including humans, are **heterotrophs**—organisms that rely on other organisms for energy. In other words, heterotrophs obtain all of their energy by eating other organisms. Some heterotrophs, such as sheep, horses, and squirrels, obtain energy by eating only plants, while others, such as humans, bears, and raccoons, eat plants and animals for energy. Regardless of the type of food eaten, heterotrophs convert the nutrients they absorb into glucose and, through cellular respiration, transfer the energy in glucose to ATP. Cellular respiration is discussed in section 1.18.

Photosynthesis: The Process

Photosynthesis occurs in all green parts of a plant, but the leaves are particularly specialized for this purpose. You will learn more about leaves in Unit 4. Photosynthesis occurs in the chloroplasts of green plants (**Figure 2**). Chloroplasts contain a double outer membrane (an envelope) and inner membranes that form an intricate system of compartments called thylakoids. Thylakoids stack on top of one another to form structures called grana (singular: granum). A protein-rich fluid called stroma fills the space between the envelope and the thylakoid membranes. Molecules of the green, light-absorbing pigment called chlorophyll are embedded within thylakoid membranes (**Figure 2(c)** and **(d)**). Chlorophyll molecules absorb light energy and begin the process of photosynthesis.

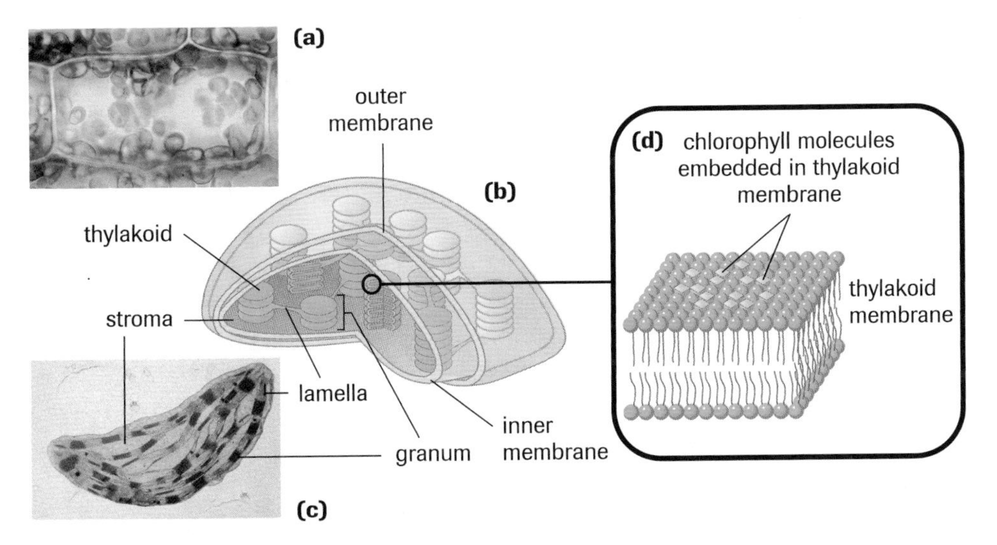

Figure 2
(a) Chloroplasts within plant cells (*Elodea*)
(b) An artist's representation of a chloroplast, showing key components
(c) An electron micrograph of a chloroplast
(d) Chlorophyll molecules in the thylakoid membrane

The photosynthetic process occurs in a series of reactions that uses light energy, carbon dioxide, and water to produce glucose and oxygen (**Figure 3**).

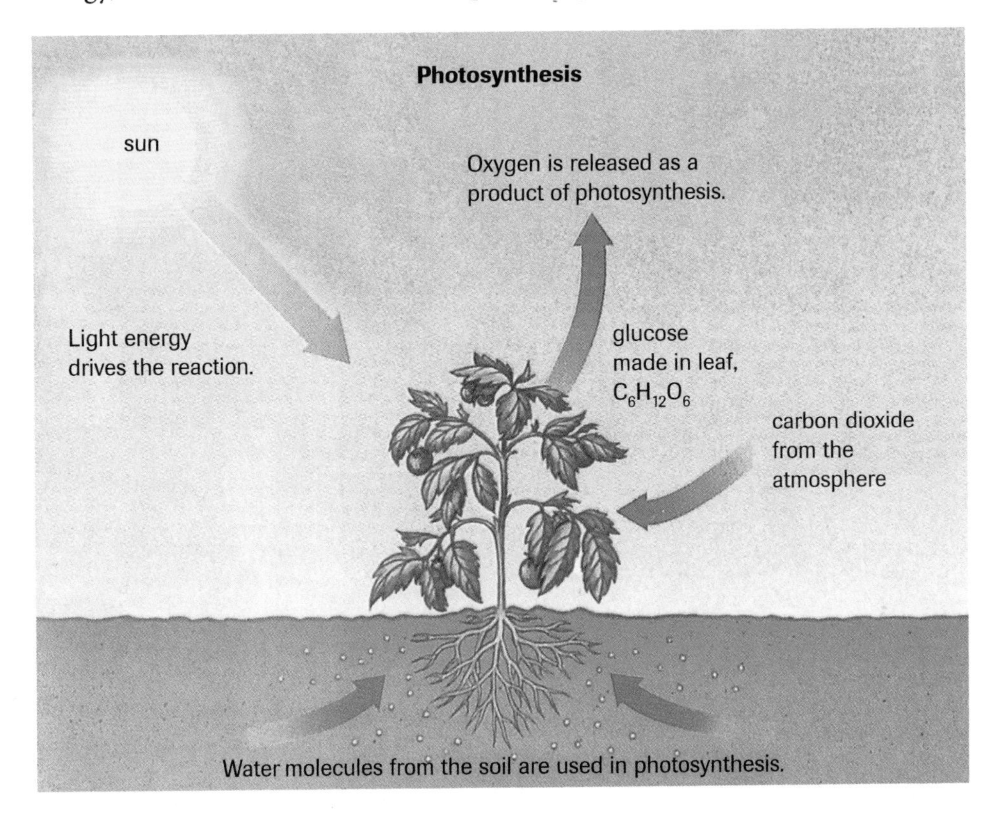

Figure 3
Plants use light energy to form glucose from water and carbon dioxide.

The following equation summarizes the photosynthetic process:

$$\underset{\text{carbon dioxide}}{6CO_{2(g)}} + \underset{\text{water}}{6H_2O_{(l)}} + \text{light energy} \xrightarrow{\text{chlorophyll}} \underset{\text{glucose}}{C_6H_{12}O_{6(aq)}} + \underset{\text{oxygen}}{6O_{2(g)}}$$

Photosynthesis does not occur in one big step as shown in the overall equation. Instead, it takes place through many reactions that may be summarized in two steps: the light reactions and the Calvin cycle.

Step 1: The Light Reactions of Photosynthesis

light reactions of photosynthesis reactions of photosynthesis in which light energy is absorbed by chlorophyll and transferred to ATP

Photosynthesis begins when chlorophyll molecules in thylakoid membranes trap light and transfer its energy to ATP molecules (**Figure 4**). Because light is required for these reactions to occur, they are called the **light reactions of photosynthesis**.

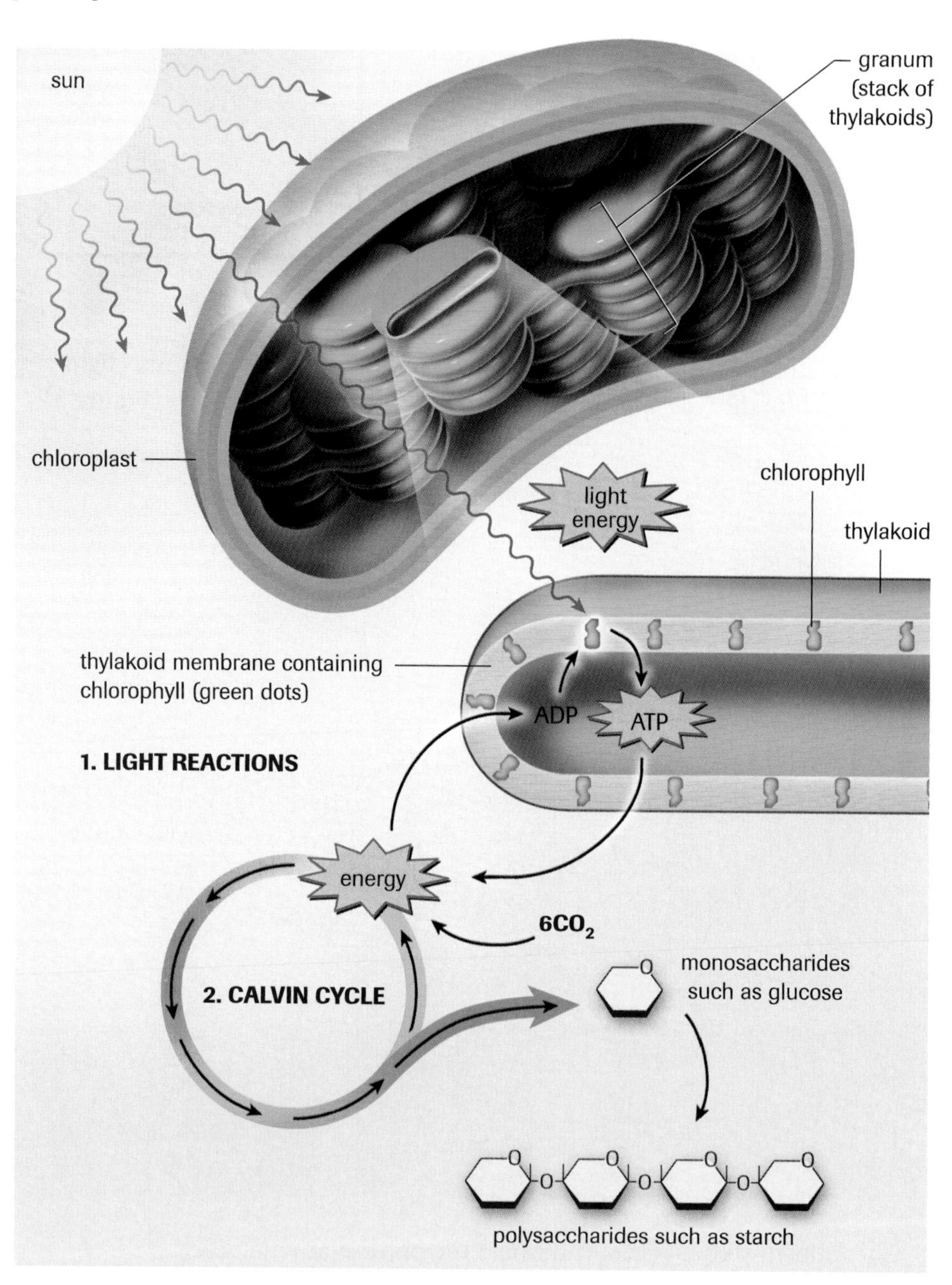

Figure 4
The light reactions of photosynthesis and the Calvin cycle

▶ TRY THIS activity Photosynthesis and Light

Green plants capture sunlight and transfer the energy to carbohydrates through the process of photosynthesis. When plants photosynthesize, they absorb carbon dioxide and produce oxygen. The oxygen produced is released into the environment. In this activity, you will observe the production of oxygen in photosynthesizing plant cells.

Materials: living green plants with leaves, baking soda, liquid soap or detergent, medicine dropper, water, drinking straw, 35-mm film canister with lid, 5-mL syringe

1. Add enough baking soda to barely cover the bottom of a film canister. Fill the canister with water (almost to the top), replace the lid, and shake to dissolve the baking soda.
2. Remove the lid, add one small drop of liquid soap, replace the lid, and gently swirl the contents to dissolve the soap. Do not shake. The soap will help prevent static electricity.
3. Use a new straw like a cookie-cutter to cut four leaf discs from a plant leaf. The leaf discs will accumulate inside the straw.

Never share straws with others. Always use a new straw. When finished using the straw, discard it according to your teacher's instructions.

4. If the syringe you are using has a cap on the tip, remove the cap. Pull the plunger out of the syringe. Blow the leaf discs out of the straw and into the syringe. Replace the plunger.
5. Draw 4 mL of baking-soda solution (prepared in steps 1 and 2) into the syringe. Invert the syringe so that the tip end is pointing up. Gently push the plunger to remove the air near the tip.
6. Put your finger over the syringe tip and pull the plunger. This will create a vacuum, which will pull air and oxygen from the leaf discs.
7. Tip the end of the syringe down so that the leaf discs are in the solution. Release the plunger and remove your finger. Turn the syringe back up and tap the side repeatedly until all (or most) of the discs sink.
8. Place the syringe, open end up, in bright sunlight.
9. As the leaf discs photosynthesize, they will float to the top.

(a) (i) What causes the leaf discs to float to the top while they're in sunlight?

(ii) Would the discs float to the top if the syringes were kept in the dark? Explain.

(b) Why is baking soda added to the solution in the syringe?

(c) Did the leaf discs all float to the top at the same time? Explain why or why not.

(d) How could this procedure be used to investigate whether or not different colours of light cause plants to photosynthesize equally well? Design a procedure for such an experiment (include a list of materials).

Step 2: The Calvin Cycle

You can see from the overall equation of photosynthesis that the carbon atoms in glucose come directly from the carbon atoms in carbon dioxide. The chemical reactions that produce carbohydrates from carbon dioxide are called the **Calvin cycle** (**Figure 4**, on the previous page). This process is named after Melvin Calvin, the biochemist who discovered the reactions. The energy of ATP molecules produced in the light reactions is required for the Calvin cycle to occur. This energy (originally from the sun) is transferred to the carbohydrate molecules that are produced.

Calvin cycle reactions of photosynthesis in which carbon dioxide molecules are used to produce carbohydrates such as glucose

Accessory Pigments

Chlorophyll is not the only light-absorbing molecule in chloroplasts. A number of **accessory pigments**, including orange carotenoids and yellow xanthophylls, help chlorophyll absorb a broad spectrum of light energy. In spring and summer, leaves appear green because of the high concentration of chlorophyll in chloroplast membranes. When temperatures cool off in the

accessory pigment a multicoloured pigment in chloroplast membranes that assists chlorophyll in absorbing light energy autumn, leaves stop producing chlorophyll and break apart the molecules of chlorophyll already there. This causes the bright yellow and orange colours of the accessory pigments to show, giving rise to some of the characteristic colours of leaves in the fall (**Figure 5**). The bright red colour in autumn leaves is caused by the production of a red pigment called anthocyanin.

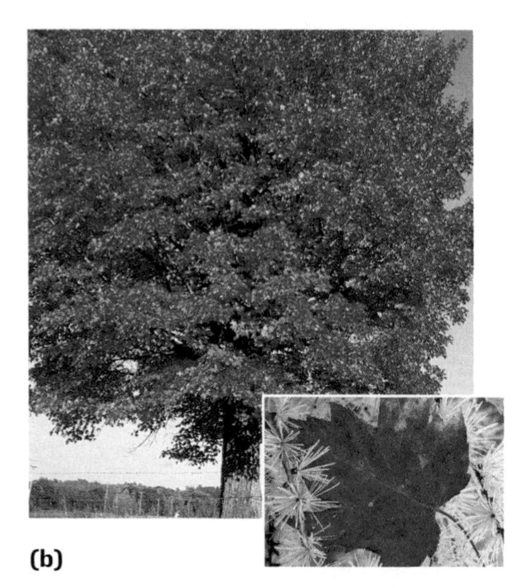

Figure 5
(a) Summer leaves contain chlorophyll and accessory pigments, but the green colour of chlorophyll overwhelms the colour of the accessory pigments.
(b) Autumn leaves contain less chlorophyll, causing the colours of the accessory and other pigments to become visible.

Section 1.17 Questions

Understanding Concepts

1. Name three large groups of organisms that carry out photosynthesis.
2. Distinguish between autotrophs and heterotrophs. Provide one example of each.
3. (a) What is the name of the pigment molecule in green plants that begins the reactions of photosynthesis?
   (b) What are accessory pigments? What is their role in photosynthesis?
   (c) Name two accessory pigments and their associated colours.
4. (a) What is the name of the organelle illustrated in **Figure 6**?
   (b) Name the labelled parts of the organelle in **Figure 6**.
   (c) In which part of the organelle do the reactions of photosynthesis begin?

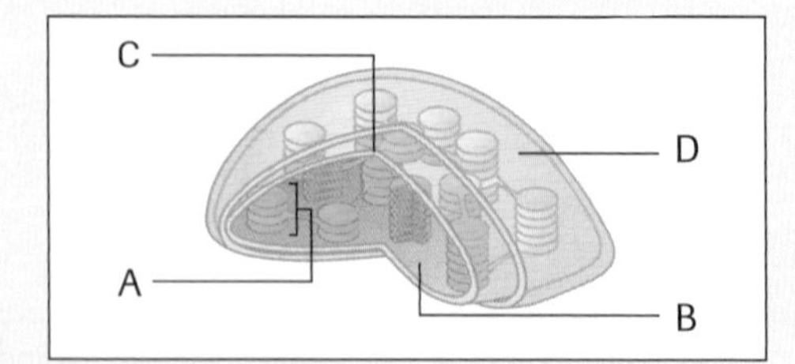

Figure 6

5. (a) Write the overall equation for photosynthesis.
   (b) What is the source of carbon dioxide for land plants? water plants?
   (c) What happens to the oxygen molecules formed during photosynthesis?
6. Why are the leaves of deciduous trees, such as maple trees, green in the summer and yellow and red in the fall?

Making Connections

7. On a sheet of blank paper, draw a labelled diagram with a single caption that you would use to teach the process of photosynthesis to a grade 4 student who has never heard of the process.
8. Several biotechnology companies are experimenting with the possibility of producing a "green" (environmentally friendly) plastic from plants. One procedure converts sugar from corn to polylactide (PLA), a plastic similar to the plastic polyethylene terephthalate (PET), which is used to make pop bottles and a variety of synthetic fabrics. Conduct research to complete the following tasks:
   (a) Describe one or two other "green" plastics and their potential uses.
   (b) Describe some of the costs and benefits of producing green plastics on a large scale.

www.science.nelson.com

On a warm summer day in 1974, 8-year-old Sarah suddenly felt pins and needles in the muscles of her legs as she walked. Within a year's time, she could no longer walk without experiencing muscle pain, shortness of breath, and a racing heart. By the age of 16, Sarah was attending school in a wheelchair. Tests showed that she had high levels of acidity in her blood. Her doctors were puzzled. After extensive study, a team of researchers traced Sarah's problem to a defect in her mitochondria. Her muscles could not make sufficient amounts of ATP and instead produced large amounts of a compound called lactic acid. In time, doctors discovered that a simple mixture of vitamin C and vitamin K reversed the condition. After taking the medication, Sarah began feeling better almost instantly. By the age of 20, she no longer used a wheelchair and was completely free of symptoms. The vitamin mixture did not cure Sarah's mitochondrial defect; it simply filled a gap in the chemical reactions that produce ATP. Without a team of investigators and laboratory technicians who understood the reactions that produce ATP (cellular respiration), Sarah would have been denied the opportunity to lead a normal life.

Cellular Respiration

Virtually all organisms (autotrophs and heterotrophs) transfer chemical energy from glucose to ATP by a process called cellular respiration. As you know, autotrophs, such as green plants, produce glucose by photosynthesis, and heterotrophs, such as humans, obtain glucose by eating other organisms.

There are two forms of cellular respiration, aerobic cellular respiration and anaerobic cellular respiration. **Aerobic cellular respiration** uses oxygen, and produces large amounts of ATP. **Anaerobic cellular respiration** does not use oxygen, and produces small amounts of ATP. Large organisms like humans rely on aerobic cellular respiration to provide the large amounts of ATP they need to maintain their daily activities.

aerobic cellular respiration a form of cellular respiration that uses oxygen and produces large amounts of ATP

anaerobic cellular respiration a form of cellular respiration that does not use oxygen and produces small amounts of ATP

Aerobic Cellular Respiration

Aerobic cellular respiration occurs in two stages, glycolysis and oxidative respiration. **Glycolysis** is a series of 10 enzyme-catalyzed reactions, occurring in the cytoplasm, that essentially breaks one molecule of glucose ($C_6H_{12}O_6$) into two molecules of pyruvate ($C_3H_6O_3$). The reactions of glycolysis are exergonic reactions, meaning that they release energy. Some of the released energy is captured by producing two molecules of ATP. This is accomplished by attaching a phosphate group to each of two molecules of ADP. The formation of ATP absorbs the chemical energy released by glucose. A reaction that absorbs energy, like the formation of ATP, is called an endergonic reaction. The equation on the next page summarizes the process of glycolysis.

glycolysis a series of reactions occurring in the cytoplasm that breaks a glucose molecule into two pyruvate molecules and forms two molecules of ATP

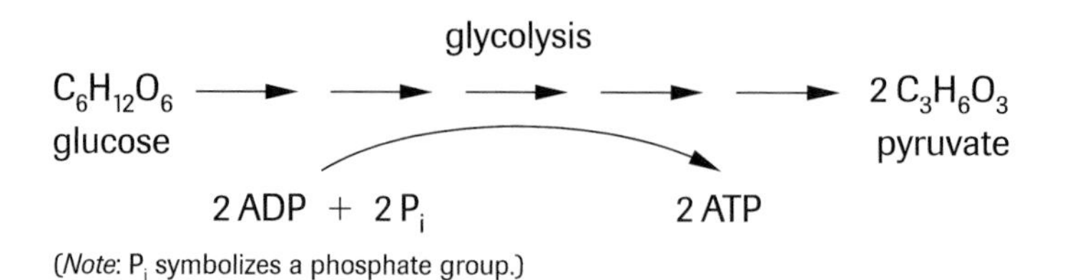

(*Note*: P_i symbolizes a phosphate group.)

The two molecules of pyruvate formed in glycolysis are transported into the matrix of a mitochondrion where, through a series of exergonic reactions involving oxygen, they are broken down into six molecules of carbon dioxide (CO_2) and six molecules of water. Because oxygen is used, this series of reactions is known as **oxidative respiration**. This process is also called the Krebs' cycle in honour of Hans Krebs, the biochemist who discovered these reactions. Much of the large amount of energy released in this process is captured by the production of 34 molecules of ATP. The following equation summarizes the process of oxidative respiration.

oxidative respiration a series of reactions occurring in the mitochondrion that uses oxygen to convert pyruvate into carbon dioxide, water, and ATP

$$\underset{\text{pyruvate}}{2C_3H_6O_3} + \underset{\text{oxygen}}{6O_2} \xrightarrow[34\ ADP\ +\ 34\ P_i\ \rightarrow\ 34\ ATP]{\text{oxidative respiration}} \underset{\text{carbon dioxide}}{6CO_2} + \underset{\text{water}}{6H_2O}$$

DID YOU KNOW?

Cyanide

Cyanide blocks the reactions of oxidative respiration. This disruption severely reduces the amount of ATP that is produced in an animal's body, resulting in coma and death. This is why cyanide is a poison.

The oxygen used in oxidative respiration is obtained from air or water; that is why land animals have lungs and fish have gills. Carbon dioxide is released to the environment as a waste product.

The overall process of aerobic respiration (glycolysis + oxidative respiration) uses oxygen to convert glucose into carbon dioxide, water, and ATP. The following reaction summarizes the overall process of aerobic respiration:

$$\underset{\text{glucose}}{C_6H_{12}O_6} + \underset{\text{oxygen}}{6O_2} \xrightarrow[36\ ADP\ +\ 36\ P_i\ \rightarrow\ 36\ ATP]{\text{aerobic respiration}} \underset{\text{carbon dioxide}}{6CO_2} + \underset{\text{water}}{6H_2O}$$

The Mitochondrion and Aerobic Cellular Respiration

The mitochondrion (**Figure 1**) is the site of oxidative respiration, and is therefore the location in a cell where most ATP is produced. That is why the mitochondrion is sometimes called the powerhouse of the cell. Some of the enzymes that catalyze the reactions of aerobic respiration are free floating in the matrix of the mitochondrion, while others are embedded in the inner mitochondrial membrane. Mitochondria are so important to the cells of larger organisms that they possess their own DNA molecules, called **mitochondrial DNA**, or mDNA. Mitochondrial DNA allows mitochondria to reproduce.

mitochondrial DNA the DNA in mitochondria

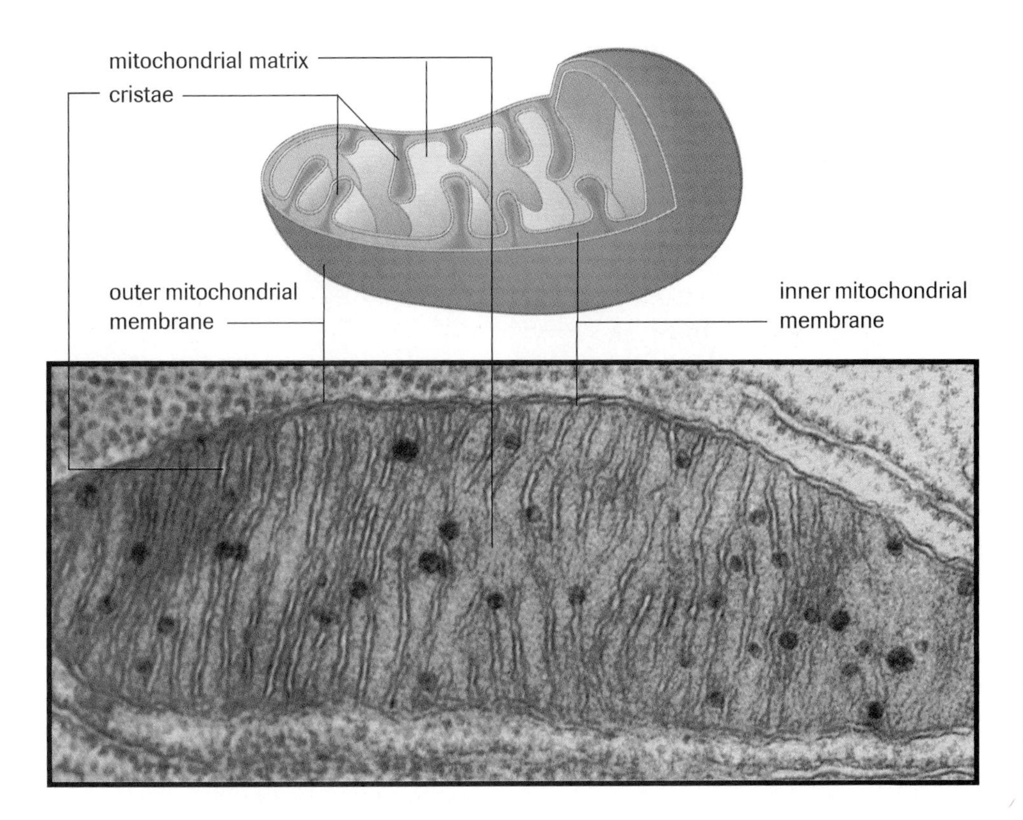

Figure 1
Oxidative ATP synthesis occurs in the matrix and the cristae of a mitochondrion.

Anaerobic Respiration

When oxygen is not available, organisms may still produce ATP from glucose through the process of anaerobic cellular respiration. Some cells accomplish this through a series of reactions called fermentation. There are two forms of fermentation, ethanol fermentation and lactate (lactic acid) fermentation. Ethanol fermentation occurs in yeast cells (single-celled fungi), and lactate fermentation occurs in human muscle cells during periods of strenuous exercise.

Ethanol Fermentation

In **ethanol fermentation**, glucose is first broken down by the reactions of glycolysis (**Figure 2**). This produces two molecules of pyruvate and two molecules of ATP. An enzyme in the cytoplasm called pyruvate decarboxylase removes a carbon dioxide molecule from pyruvate, converting it into ethanol, the alcohol found in alcoholic beverages. The reactions of ethanol fermentation all occur in the cytoplasm of the cells. The two ATP molecules produced satisfy the organism's energy needs, and the ethanol and carbon dioxide are released as waste products. Humans have learned ways of making use of these cellular wastes. Ethanol fermentation carried out by yeast cells is of great historical, economic, and cultural importance. Breads and pastries, wine, beer, liquor, and soy sauce are all products of fermentation (**Figure 3**, on the next page).

ethanol fermentation a form of anaerobic respiration in which glucose is converted into ethanol, ATP, and carbon dioxide gas

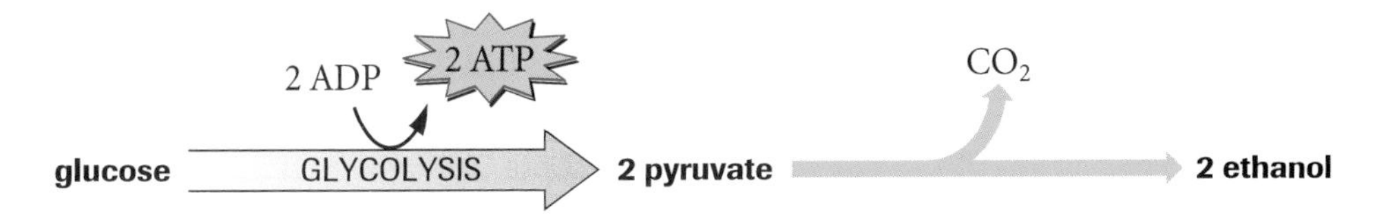

Figure 2
Ethanol fermentation creates ethanol, carbon dioxide, and ATP from glucose.

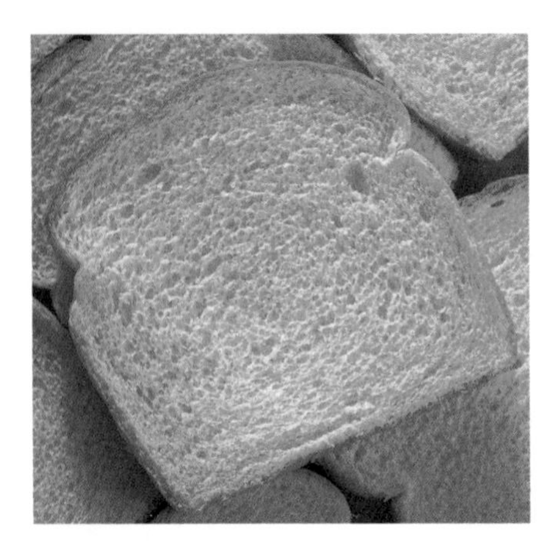

Figure 3
Ethanol fermentation is used in the production of baked goods, wine, and beer.

Bread may be leavened (assisted to rise) by mixing live yeast cells with starches (in flour) and water. The yeast cells ferment the glucose (from the starch in flour) and release carbon dioxide and ethanol. Small bubbles of carbon dioxide gas cause the bread to rise, and the ethanol evaporates away when the bread is baked. In beer making and winemaking, yeast cells ferment the sugars found in sugar-rich fruit juices such as grape juice. The mixture bubbles as the yeast cells release carbon dioxide gas and ethanol during fermentation. In winemaking, fermentation ends when the concentration of ethanol reaches approximately 12%. At this point, the yeast cells die as a result of ethanol accumulation and the product is used as a beverage. Plants with flooded roots also undergo ethanol fermentation (in the roots) and may die if the roots are not exposed to oxygen.

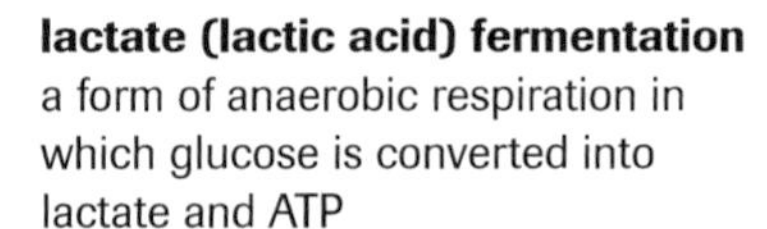

lactate (lactic acid) fermentation a form of anaerobic respiration in which glucose is converted into lactate and ATP

oxygen debt the extra oxygen required to clear the body of lactate that accumulates in muscle cells during vigorous exercise

Lactate (Lactic Acid) Fermentation

When oxygen is plentiful, animals such as humans respire glucose by aerobic respiration. However, during strenuous exercise, muscle cells break down glucose faster than oxygen can be supplied. Under such low-oxygen conditions, oxidative respiration slows down and lactate fermentation begins. Like ethanol fermentation in yeast cells, **lactate fermentation** begins with glucose being broken down in the reactions of glycolysis (**Figure 4**). As usual, this produces two molecules of pyruvate and two molecules of ATP. An enzyme in the cytoplasm called lactate dehydrogenase then converts pyruvate into lactate.

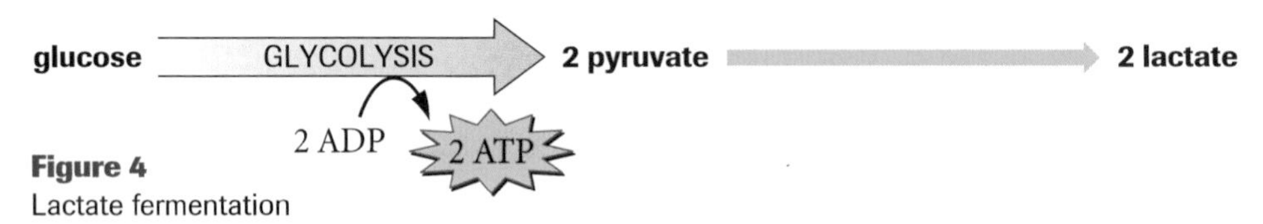

Figure 4
Lactate fermentation

The accumulation of lactate molecules in muscle tissue causes stiffness, soreness, and fatigue. When vigorous exercise ends, lactate is converted back to pyruvate, which then goes into mitochondria and proceeds through the reactions of oxidative respiration. As usual, this produces lots of ATP, but requires oxygen. The extra oxygen needed to clear the body of lactate that accumulates during vigorous exercise is referred to as the **oxygen debt.** Panting after bouts of strenuous exercise is the body's way of "paying" the oxygen debt (**Figure 5**).

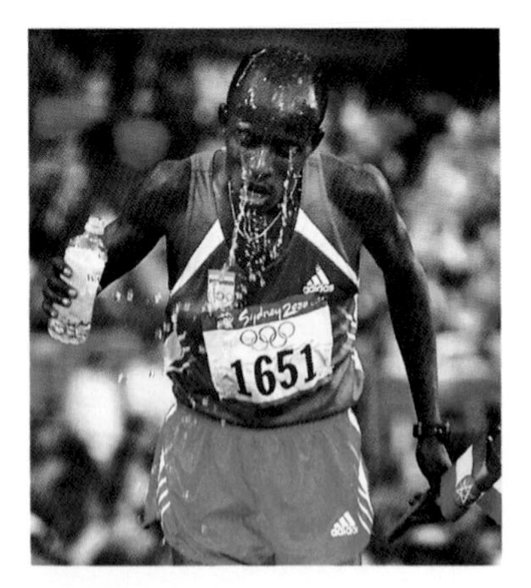

Figure 5
Marathon runners are fatigued after a race because of the accumulation of lactate in their muscles. Panting provides the oxygen needed to respire the excess lactate.

▸ TRY THIS activity Clothespins and Muscle Fatigue

In order to contract, your muscles require energy in the form of ATP. Muscles can produce ATP by using oxygen (aerobic respiration) or not (anaerobic respiration). Anaerobic respiration in muscle cells produces lactic acid. When muscles do a lot of work quickly, the buildup of lactic acid reduces their ability to contract, until eventually exhaustion sets in and contraction stops altogether. This is called muscle fatigue.

Materials: clothespin, timer

1. Hold a clothespin between the thumb and index finger of your dominant hand (the one you use the most).
2. Count the number of times you can open and close the clothespin in a 20-s period while holding the other fingers of the hand straight out. Make sure to squeeze quickly and completely to get the maximum number of squeezes for each trial.
3. Repeat the process for nine more 20-s periods, recording the result for each trial in a suitable table. Do not rest between trials.
4. Repeat steps 1, 2, and 3 for the nondominant hand.

(a) What happened to your strength as you progressed through each trial?

(b) Describe how your hand and fingers felt near the end of your trials.

(c) Were your results different for the dominant and the nondominant hand? If yes, explain why.

(d) Your muscles will likely recover after about 10 min of rest. Explain why.

Photosynthesis and Cellular Respiration

Photosynthesis and cellular respiration are closely related. Plants carry out photosynthesis and cellular respiration because they contain chloroplasts and mitochondria. Animals can carry out cellular respiration but not photosynthesis because they possess mitochondria but not chloroplasts. Nevertheless, both types of organisms rely on each other for the raw materials of both processes. Plants use carbon dioxide for photosynthesis, and animals, plants, fungi, and bacteria give off carbon dioxide as a waste product of cellular respiration. Plants give off oxygen as a byproduct of photosynthesis, and all aerobic organisms use oxygen in cellular respiration. Thus, the products of one process are the raw materials of the other (**Figure 6**).

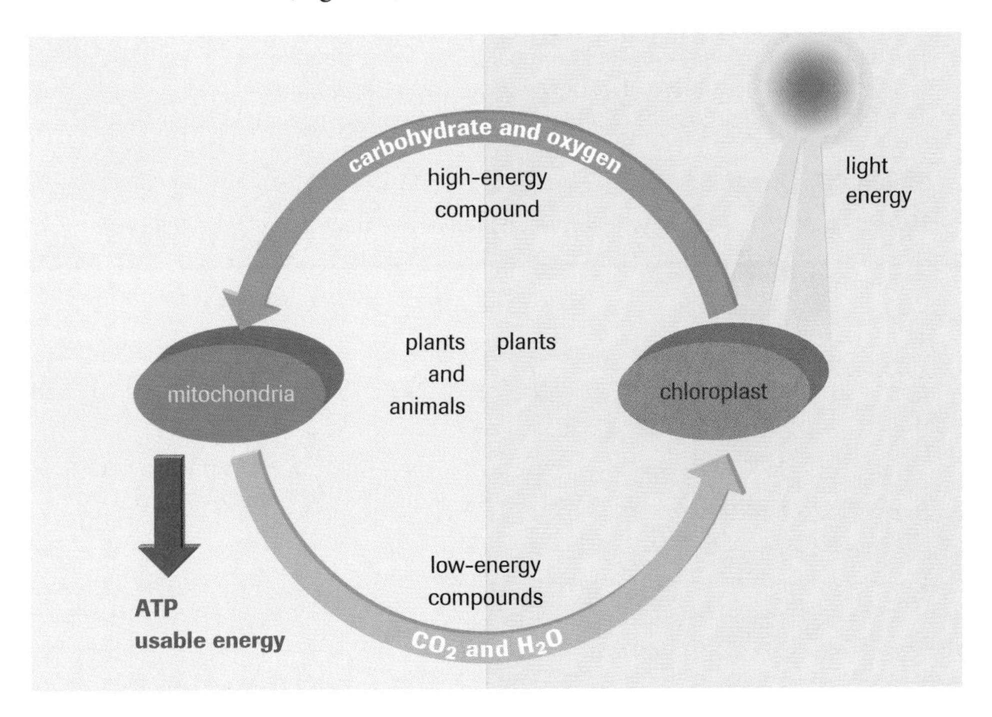

Figure 6
Photosynthesis uses the products of cellular respiration, and cellular respiration uses the products of photosynthesis.

DID YOU KNOW?

Rigor Mortis

When an animal dies, all cells of the body do not die at the same time. Immediately after death, the temperature of the body begins to drop and muscles become stiff. This is called rigor mortis. However, rigor mortis is not caused by the drop in body temperature but by the fermentation of glucose in muscle cells, leading to high levels of lactic acid. The lactic acid causes muscle tissue to become rigid. Rigor mortis sets in much sooner if death occurs immediately following strenuous activity such as running.

The relationship that exists between photosynthesis and cellular respiration reveals a dependency between autotrophs and heterotrophs. Autotrophs produce almost all of the oxygen in the environment, and heterotrophs produce almost all of the carbon dioxide in the environment.

Section 1.18 Questions

Understanding Concepts

1. Why is it possible for organisms to carry out photosynthesis and cellular respiration but not photosynthesis only?
2. (a) What is the key difference between aerobic respiration and anaerobic respiration?
   (b) Which form of respiration is more suited to large organisms such as humans, cows, and fish? Which form is adequate for single-celled organisms such as yeast cells and bacteria? Explain.
3. Describe three cell processes that require the use of ATP.
4. (a) What is meant by the term *exergonic reaction*?
   (b) Is the formation of ATP during the reactions of glycolysis an endergonic reaction or an exergonic reaction? Explain.
5. (a) Write an overall chemical equation for glycolysis.
   (b) Where in a cell do the reactions of glycolysis occur?
   (c) In which organelle do the reactions of oxidative respiration occur?
6. (a) How many ATP molecules are created in glycolysis from one glucose molecule?
   (b) How many ATP molecules are created in the overall process of aerobic respiration from one glucose molecule?
7. (a) What is the cause of the pain and stiffness felt after strenuous exercise?
   (b) Why does the pain eventually subside?
8. How many ATP molecules does a yeast cell obtain from every glucose molecule that undergoes ethanol fermentation?

Applying Inquiry Skills

9. A sealed, flexible container of apple juice is left on a counter at room temperature. After one week, the container looks inflated.
   (a) State a hypothesis that may explain this observation.
   (b) Describe an experiment you could perform to test your hypothesis.

Making Connections

10. Describe two ways in which humans make use of ethanol fermentation.

Unit 1 SUMMARY

Key Understandings

1.1 Cell Theory and the Microscope

- living organisms are made of cells, obtain and use energy, maintain a constant internal environment, and reproduce
- the cell is the basic structural and functional unit of life
- the compound light microscope uses glass lenses and light to magnify an image; transmission electron microscopes (TEMs) and scanning electron microscopes (SEMs) use beams of electrons to produce a magnified image

1.2 Cells: The Basic Units of Life

- cells are composed of a cell membrane and cytoplasm that contains a number of organelles such as mitochondria, endoplasmic reticulum, ribosomes, a nucleus, a nucleolus, a Golgi apparatus, lysosomes, vacuoles, and chloroplasts (in plant cells only); cell membranes are composed of a double layer of phospholipids with embedded proteins
- ribosomes are the site of protein synthesis; the Golgi apparatus packages proteins; mitochondria produce energy; lysosomes digest materials that enter the cell; vacuoles store water and minerals; chloroplasts are the sites of photosynthesis; and the nucleus stores genetic information in the form of DNA and RNA

1.3 Activity: Observing Cells

- cell size can be estimated using a transparent millimetre ruler or a slide micrometer
- cells from different organisms are different shapes and sizes

1.4 Chemistry Basics

- charged atoms called ions are attached to one another and form ionic bonds; noncharged atoms form covalent bonds by sharing electrons
- the uneven sharing of bonding electrons forms polar covalent bonds
- hydrophilic (polar) molecules are attracted to other hydrophilic molecules, and hydrophobic (nonpolar) molecules are attracted to other hydrophobic molecules; like dissolves like
- concentration is a ratio of the quantity of solute to the quantity of solution
- the pH of acidic solutions is less than 7; the pH of basic solutions is greater than 7; and the pH of neutral solutions is equal to 7

1.5 Biological Chemistry

- carbohydrates are an important source of energy: monosaccharides are single sugar molecules; disaccharides are double sugars; polysaccharides are composed of many sugar molecules linked together
- triglycerides are lipids composed of one glycerol and three fatty acids
- plant fats or oils are unsaturated because they do not have all of the hydrogen atoms that they could have; animal fats are saturated because they have the maximum number of hydrogen atoms possible
- cholesterol is an important part of cell membranes; it is also associated with heart disease and circulatory problems
- phospholipids have a hydrophilic end and a hydrophobic end
- proteins are made of amino acids; DNA determines the number and sequence of amino acids in proteins
- there are two nucleic acids, DNA and RNA
- DNA is double stranded, is composed of sugar, phosphate, and paired nitrogenous bases, and provides the instructions for building proteins; RNA is single-stranded and is composed of sugar, phosphate, and single nitrogenous bases
- nitrogenous bases pair up: adenine-thymine (adenine-uracil in RNA) and guanine-cytosine

1.6 Activity: Biological Compounds in 3-D

- ball-and-spring molecular model-building kits and interactive computer software may be used to visualize molecular structures (shapes) and interactions

1.7 Activity: Identifying Nutrients

- various laboratory tests may be used to identify nutrient molecules in food: Benedict's solution identifies simple sugars, iodine identifies starch, Biuret reagent identifies proteins, and the translucent paper test identifies lipids

1.8 Explore an Issue: Functional Foods

- functional foods (nutriceuticals) are food products that claim to produce health benefits beyond the nutritional values of the nutrients in the food

1.9 Tech Connect: Counting Cells

- a hemacytometer is a modified microscope slide used to count many different kinds of cells, including blood cells and microorganisms
- the Coulter Counter is an electronic cell-counting device; Coulter Counters are commonly used in hematology laboratories (blood testing laboratories) to count blood cells

1.10 Activity: Counting Red Blood Cells and Yeast Cells

- the concentration of red blood cells in blood and yeast cells in a yeast-cell suspension may be determined using a hemacytometer and microscope

1.11 Case Study: Building New Bones with Skin Cells

- cells become specialized when certain genes are turned on and others are turned off
- stem cells are unspecialized cells
- biologists have been able to transform skin cells called fibroblasts into bone and cartilage cells

1.12 Enzymes

- enzymes are protein catalysts (i.e., proteins that speed up chemical reactions)
- enzymes have a three-dimensional shape that affects their catalytic activity; they possess active sites where substrates attach and change their shapes to accommodate their substrates
- enzymes possess optimal temperatures and pH, above and below which their activities are reduced
- high temperatures, extremes in pH, and high salt concentrations change enzyme structures, causing a decrease in enzyme activity (denaturation)

1.13 Investigation: Enzyme Activity

- factors such as enzyme concentration, substrate concentration, temperature, and pH affect the rate of enzyme activity

1.14 Passive Transport

- passive transport is the movement of materials across a cell membrane without the use of cellular energy
- diffusion is the movement of molecules from a region of higher concentration to a region of lower concentration until concentrations are the same throughout (dynamic equilibrium)
- osmosis is the diffusion of water molecules across a selectively permeable membrane
- isotonic solutions have the same solute concentrations: a hypotonic solution has a lower solute concentration than another solution; a hypertonic solution has a higher solute concentration than another solution
- in facilitated diffusion, protein carriers move molecules across the cell membrane; cell energy is not used

1.15 Investigation: Factors Affecting the Rate of Osmosis

- factors such as solute concentration, temperature, and pH affect the rate of osmosis

1.16 Active Transport and Bulk Transport

- energy-requiring processes such as movement, reproduction, protein synthesis, active transport, and bulk transport may be powered by a compound called adenosine triphosphate, or ATP
- in phagocytosis, cells engulf solid particles
- in pinocytosis, cells engulf liquid droplets
- in exocytosis, large molecules within the cell are transported to the external environment

1.17 Energy Flow and Photosynthesis

- photosynthesis occurs in chloroplasts; light energy is converted to chemical energy in carbohydrate molecules
- photosynthesis occurs in two steps: the light reactions of photosynthesis and the Calvin cycle

1.18 Cellular Respiration

- cellular respiration occurs in mitochondria; energy in carbohydrates is transfered to ATP
- glycolysis begins cellular respiration; glycolysis breaks glucose into two molecules of pyruvate, with the production of two ATP molecules
- if oxygen is available, pyruvate molecules formed in the cytoplasm by glycolysis enter mitochondria, where oxidative respiration reactions convert the molecules to low-energy carbon dioxide molecules and form 34 ATP molecules

- if oxygen is not available, one of two possible types of anaerobic respiration (fermentation) may occur:
  - in lactic acid fermentation (in animal muscle cells), pyruvate is converted to lactic acid
  - in ethanol fermentation (in yeast), pyruvate is converted to carbon dioxide and ethanol (an alcohol)
- fermentation is used to leaven baked goods, brew beer, and make wine and soy sauce

Key Terms

1.1
magnification
resolution

1.2
extracellular fluid
cytosol
phospholipid
nuclear envelope
nucleoplasm
rough endoplasmic reticulum (RER)
smooth endoplasmic reticulum (SER)
vesicle
apoptosis
cellular respiration
crista
matrix
plastid
chloroplast
photosynthesis
thylakoid
granum
chlorophyll
stroma
turgor pressure

1.4
ionic bond
covalent bond
polar covalent bond
nonpolar covalent bond
hydrogen bond
hydrophobic molecule
hydrophilic molecule
buffer

1.5
functional group
macromolecule
carbohydrate
monosaccharide
disaccharide
polysaccharide
polymer
monomer
cellulose
chitin
lipid
triglyceride
saturated triglyceride
unsaturated triglyceride
protein
amino acid
peptide bond
polypeptide
protein synthesis
enzyme
catalyst
denaturation

1.11
stem cell
therapeutic cloning
reproductive cloning
fibroblast

1.12
activation energy
substrate
active site
induced-fit model
enzyme–substrate complex

1.14
electrolyte
selectively permeable
simple diffusion
concentration gradient
dynamic equilibrium
transmembrane protein
carrier protein
facilitated diffusion
osmosis
isotonic solution
hypertonic solution
hypotonic solution
hemolysis
crenation

1.16
adenosine triphosphate (ATP)
active transport
sodium-potassium pump
bulk transport
endocytosis
phagocytosis
pinocytosis
pseudopod
phagocytotic vesicle
pinocytotic vesicle
exocytosis
secretory vesicle

1.17
autotroph
heterotroph
light reactions of photosynthesis
Calvin cycle
accessory pigment

1.18
aerobic cellular respiration
anaerobic cellular respiration
glycolysis
oxidative respiration
mitochondrial DNA
ethanol fermentation
lactate (lactic acid) fermentation
oxygen debt

▸ *MAKE* a summary

Construct a large pictorial concept map on a blank sheet of paper with a diagram of a typical plant cell as the central concept (make sure that the cell contains all of the cell organelles discussed in the unit). Branch out with individual organelles on the next level (include cytoplasm, cell wall, and cell membrane), and then continue branching with organelle components, up to and including the macromolecules that make up the components. At each level, describe the structures and functions associated with each organelle and/or its parts.

PERFORMANCE TASK

Criteria

Assessment

Your completed task will be assessed according to the following criteria:

Process

- Ask questions.
- Make and evaluate hypotheses and predictions.
- Design appropriate experimental procedures.
- Choose and safely use laboratory materials.
- Analyze results using qualitative and quantitative methods.
- Evaluate experimental procedures and suggest improvements.

Product

- Demonstrate an understanding of the concepts presented in this unit.
- Use terms, measurements, and SI symbols and units correctly.
- Prepare a formal laboratory report and a marketing report to communicate the questions, predictions, materials lists, experimental designs, procedures, results, analyses, and evaluations of your experiments.
- Carry out procedures, measurements, and calculations with accuracy and precision.

Food Technician for a Day

You are a senior laboratory technician in the Biologicals Division of a company that assesses new products for the food industry. A microbiologist in the mycology (fungus study) department has discovered a new type of yeast (called K-19) that she believes will revolutionize the bread and pastry industry. While viewing the newly discovered yeast cells under a microscope, she noticed that the cells were budding (reproducing) faster than *Saccharomyces cerevisiae* (baker's yeast), the yeast most commonly used around the world to make breads and pastries. She has brought a sample of K-19 to your laboratory for testing.

Your task is to conduct controlled experiments to compare the activity of K-19 with the activity of common baker's yeast. Your tests will consist of a number of controlled experiments comparing K-19 activity to *S. cerevisiae* activity. You may, for example, test for cell-reproduction rate, cell-fermentation rate, cell-fermentation quality, or any other indicator of yeast-cell activity that may help determine whether K-19 is a better yeast than *S. cerevisiae* for making breads and pastries.

Experiments for Reproduction Rate

Design and conduct experiments that compare the reproduction rate of baker's yeast with the reproduction rate of K-19 in controlled environmental conditions. Reproduction rate may be measured by counting cells that have been incubated in various environments using a hemacytometer and microscope, or by other methods approved by your teacher. Although you may test for reproduction rates in various environmental conditions, a comparison must be made to conditions normally used for bread making.

Experiments for Fermentation Rate

You may study the rate or quality of carbon dioxide gas production by comparing the speed of gas production or the average size of individual gas bubbles in a fermenting yeast-cell mixture. Although you may test for gas production in various environmental conditions, a comparison must be made to conditions normally used for bread making.

Other Experiments

You may conduct other experiments that may produce useful results. Propose suitable questions, experimental designs, and procedures to your teacher for approval, then conduct the experiments and report your results and analyses.

Possible Technique for Measuring Fermentation Rate

The following procedure provides you with a possible technique for measuring fermentation rate. However, you do not have to use this method. You may research a different method, or develop your own.

Materials

flour
sugar
measuring spoon
colourless, transparent drinking straws
bowls
clothespins
measuring cup
yeast
ruler
marking pen
water
timer

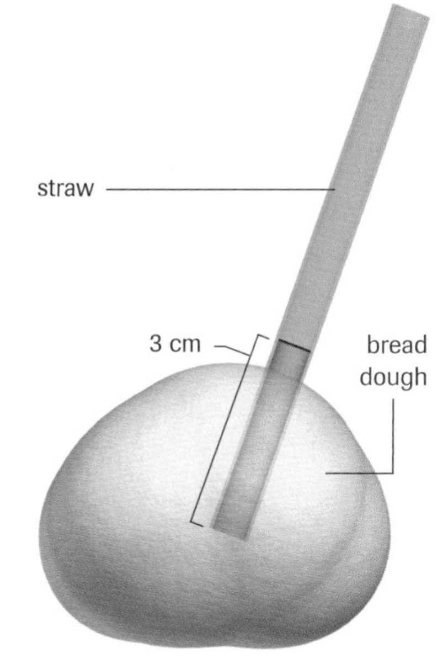

Figure 1

Suggested Procedure

1. Using a ruler, place a mark 3 cm from one end of a colourless, transparent drinking straw.
2. Mix flour, sugar, yeast, and water in a bowl, and knead the mixture to make dough. You may use a bread-making recipe of your choice.
3. Push a straw into the dough until the dough inside the straw reaches the 3-cm mark (**Figure 1**).
4. Pinch the bottom of the straw, pushing the dough up from the bottom enough to clip a clothespin to the end of the straw. Mark the new height of the dough on the straw.
5. Stand the straw upright using the clothespin as a base, as in **Figure 2**, and place the straw in experimental environmental conditions.
6. Mark the height of the dough on the straw every 10 min for your experimental period of time (e.g., 1 h). Record measurements in a suitable chart. Graph your results and analyze.

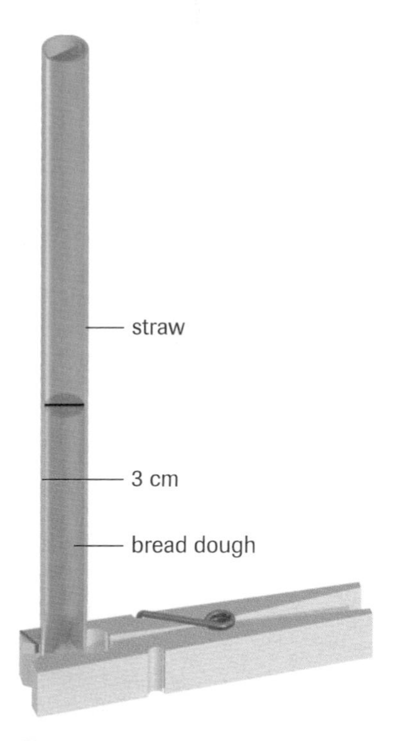

Figure 2

Reporting Guidelines

When you have completed your experiments, you must prepare two reports: a formal report to be submitted to the international journal *Mycelium*, and a marketing report for your company's marketing department discussing whether or not K-19 should be mass produced.

Follow the laboratory report guidelines in Appendix A4 to complete the formal laboratory report. The following list describes some of the components of a suitable marketing report. Your teacher will provide some of the information.

- Title
- Introductory paragraph: explain the purpose of the study, including the names of the organisms involved.
- Describe the organisms (diagrams may be useful).
- Briefly explain the experiments performed and their outcomes, without getting into details regarding materials and procedure.
- Outline the advantages and disadvantages of marketing one type of yeast over the other. Identify the type of yeast you think the company should market, supporting your decision with scientific evidence.
- Conclude with a final paragraph summarizing the scientific findings and stating your recommendation to the company.

Understanding Concepts

1. Relate each of the following statements to one of the three statements of cell theory.
   (a) Viruses are not considered to be alive because they lack structures that allow them to produce energy or reproduce on their own.
   (b) Humans have never seen living organisms created from nonliving matter.
   (c) Multicellular organisms are always composed of a number of cells working together to keep the organism alive.
2. (a) In your notebook, complete **Table 1** on microscopes.

Table 1

Type of microscope	Magnifying energy source	Maximum magnification	Maximum resolution
	electrons	300 000×	
compound light microscope		2000×	
			0.2 nm

   (b) Calculate the total magnification of a compound microscope when a 15× ocular lens and a 100× objective lens are in place.
3. Complete **Table 2** in your notebook.

Table 2

Cell component	Diagram of cell component	Plant/ animal/ both	Main function
			cell rigidity

4. Using a well-labelled diagram, describe how an animal cell such as a macrophage takes in a bacterial cell and digests it into nutrients. Name all organelles and processes involved.
5. Pretend you are a yeast cell looking for a job making bread. Write a resume describing yourself, your background, your skills, and why you feel you're qualified for the job.
6. Describe two similarities and two differences between the structures of a chloroplast and a mitochondrion.
7. (a) What are the functions of amyloplasts and chromoplasts in plant cells?
   (b) Look at **Figures 1–3**. Which ones illustrate plant parts that contain amyloplasts? chromoplasts?

Figure 1

Figure 2

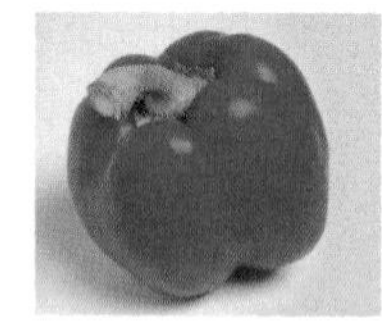

Figure 3

8. (a) Name one type of cell that contains lots of mitochondria, and one type that contains few mitochondria.
   (b) Explain why the cells described in (a) have different numbers of mitochondria.
9. Describe the location of chlorophyll molecules in a photosynthesizing plant cell. Be specific.
10. (a) If skin cells contain all of the genetic information to be nerve cells, why don't skin cells act like nerve cells?
    (b) How are stem cells unlike skin cells and nerve cells?
11. What are the differences between an oxygen atom and an oxygen molecule?
12. List the following chemical bonds in order of increasing strength, from left to right:
    ionic bond, covalent bond, hydrogen bond

13. (a) The molecules in **Figure 4** contain only polar covalent bonds. Which molecules are polar?
    (b) Describe each molecule in **Figure 4** as hydrophobic or hydrophilic.

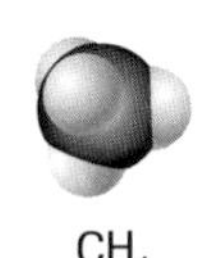
CH_4

H_2S

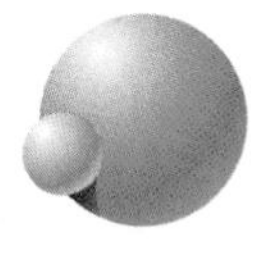
HBr

Figure 4

14. (a) Why is water called the universal solvent?
    (b) What does the saying "like dissolves like" mean? Provide one example that illustrates the saying's meaning.
15. (a) What is an acid-base buffer?
    (b) What are the components of an acid-base buffer?
    (c) Describe a buffer that exists in the human body.
16. How do plant cells and humans use cellulose?
17. (a) What are the differences between the fatty acids in the triglycerides of oils and the fatty acids in the triglycerides of butter?
    (b) What is done to the fatty acids in vegetable oil to convert them into margarine?
    (c) Describe one similarity and one difference between a steroid and a fatty acid.
18. How do plants and animals use waxes?
19. Molecule X in **Figure 5** is a small portion of a much longer polymer chain. Answer the following questions about the molecule in **Figure 5**:
    (a) What type of macromolecule is molecule X?
    (b) Name the functional groups at either end of molecule X.

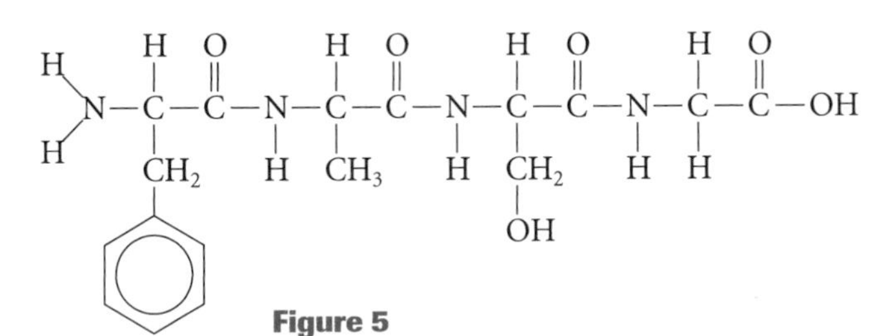

Figure 5

20. What do the objects in **Figure 6** have in common?

Figure 6

21. (a) Explain how so many different proteins can be made with only 20 different amino acids.
    (b) How is the information in DNA used in protein synthesis?
22. (a) Why is the three-dimensional shape of a protein important?
    (b) What happens to proteins in the human body when they are heated above 40°C? Why is this dangerous?
    (c) How does cooking affect the proteins in a steak?
23. (a) Briefly describe the role of DNA and RNA in the processing of genetic information.
    (b) What does mRNA stand for? What does it do?
24. (a) What role does ATP play in living cells?
    (b) What eventually happens to a cell if it does not produce adequate amounts of ATP?
25. Describe the induced-fit model of enzyme activity. You may use a labelled diagram.
26. (a) Identify the three methods used by cells to move substances across the cell membrane illustrated in **Figure 7**.
    (b) What does the purple blob through the membrane in method B represent?
    (c) What two characteristics distinguish method C from methods A and B?

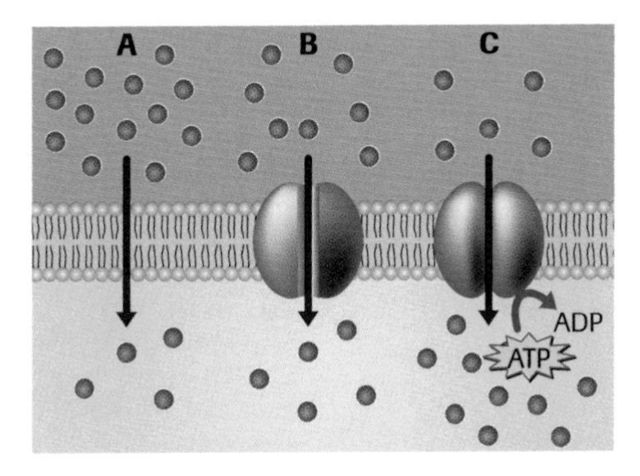

Figure 7

27. (a) When does simple diffusion end?
    (b) Why does simple diffusion end?
    (c) Does active transport reach a state of dynamic equilibrium? Explain.
28. What is the key difference between the transmembrane protein carriers of facilitated diffusion and the transmembrane protein carriers of active transport?
29. The cells in **Figure 8** are in three different solutions in three different test tubes.

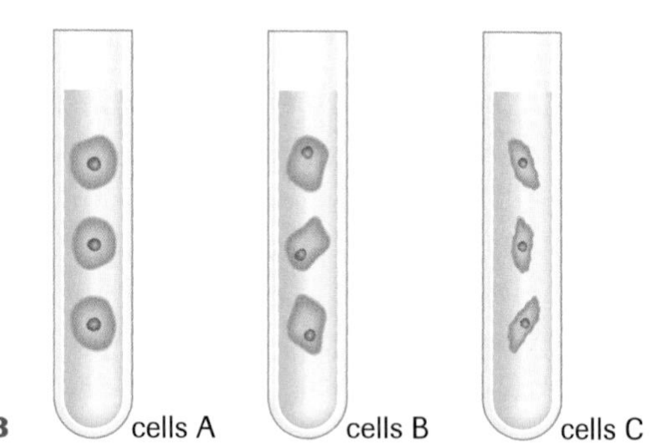

Figure 8

    (a) Identify the cells in solutions that are hypotonic, hypertonic, and isotonic to the cell's cytoplasm.
    (b) What will eventually happen to cells A if they are left in the test tube? Why?
    (c) What will eventually happen to cells C if they are left in the test tube? Why?
30. (a) Using a labelled diagram, explain the process of pinocytosis.
    (b) Describe two differences between pinocytosis and simple diffusion.
31. If oxygen is available, what happens to pyruvate after it is formed by glycolysis in the cytoplasm of a cell?
32. (a) What series of reactions is common to both aerobic and anaerobic respiration?
    (b) How many ATP molecules are produced from one molecule of glucose in anaerobic respiration?
33. Where and under what conditions does lactate fermentation occur in humans?
34. (a) What is the oxygen debt?
    (b) How does the body "pay" the oxygen debt?

Applying Inquiry Skills

35. In your notebook, place into a reasonable sequence the following steps for preparing and viewing a wet mount with a compound light microscope.
    (a) Place a drop of stain on one edge of the cover slip.
    (b) Gently lower a clean cover slip over the specimen.
    (c) Gently touch the opposite edge of the cover slip with the edge of a piece of paper towel or filter paper.
    (d) Place a drop of liquid containing the specimen on a clean microscope slide.
    (e) Focus on the specimen using the fine-adjustment knob.
    (f) Focus on the specimen using the coarse-adjustment knob.
    (g) With the low-power objective lens in place, clip the wet mount onto the microscope stage.
    (h) Bring the medium-power objective lens into position.
    (i) Turn the microscope light switch on.
36. Use the information in **Figure 9** to estimate the size of the cells in **Figure 10**.

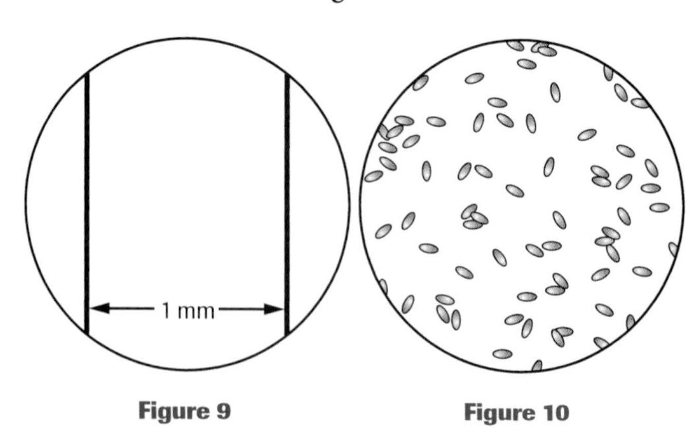

Figure 9 **Figure 10**

37. Describe two advantages and two disadvantages of using a hemacytometer instead of a Coulter Counter to count cells.

38. A hemacytometer is loaded with a yeast-cell suspension that was diluted by a factor of 200. **Figure 11** shows the hemacytometer grid, as seen with a compound light microscope at 450× magnification. Determine the concentration of yeast cells in the suspension in units of cells/μL.

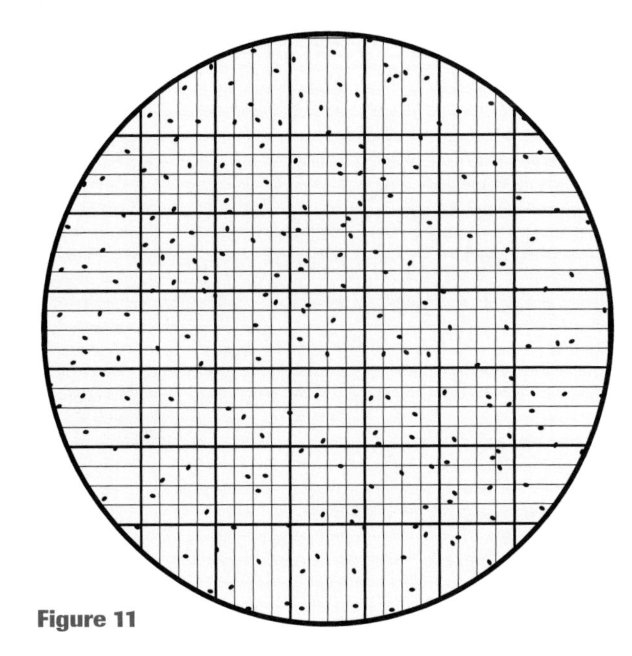

Figure 11

39. Amylase is an enzyme in human saliva that breaks down amylose into maltose. The values in **Table 3** were obtained in an experiment to determine how the activity of amylase varies with changes in temperature.

Table 3

Temperature (°C)	Amylase activity (mg maltose produced per min)
0	0
15	0.42
25	0.54
35	0.61
45	0.48
55	0.12

(a) Sketch a graph of the experimental values.
(b) Determine the optimal temperature of amylase.
(c) Why does amylase activity drop sharply above 35°C?

Making Connections

40. Describe the job of a cytologist.

 www.science.nelson.com

41. (a) Describe one medical and one ethical dilemma that are avoided by using fibroblasts instead of stem cells to make bone and cartilage cells.
(b) Which method of producing bone and cartilage cells do you support? Why?

42. Describe two practical uses of chitin.

43. Visit your local grocery store, pharmacy, or supermarket, or conduct research to discover a functional food not described in section 1.8.
(a) What health claims are listed for this functional food?
(b) Are the claims credible? Explain.
(c) Would you use this food on a regular basis? If so, why? If not, why not?
(d) Write a brief paragraph describing the food to a friend, indicating your assessment of the food's health claims.

 www.science.nelson.com

44. (a) Describe two industrial uses of enzymes.
(b) How have the two industrial uses of enzymes you described in (a) benefited society?

45. In some cases, humans use ethanol fermentation to produce useful carbon dioxide gas (not ethanol), and in other cases they use ethanol fermentation to produce ethanol (not carbon dioxide gas).
(a) Describe an industrial process in which ethanol fermentation is used specifically for the production of carbon dioxide gas.
(b) Conduct library and/or Internet research to investigate two nonfood/nonbeverage uses of ethanol. Explain how industrial ethanol is produced, and describe how society benefits from its uses.

 www.science.nelson.com

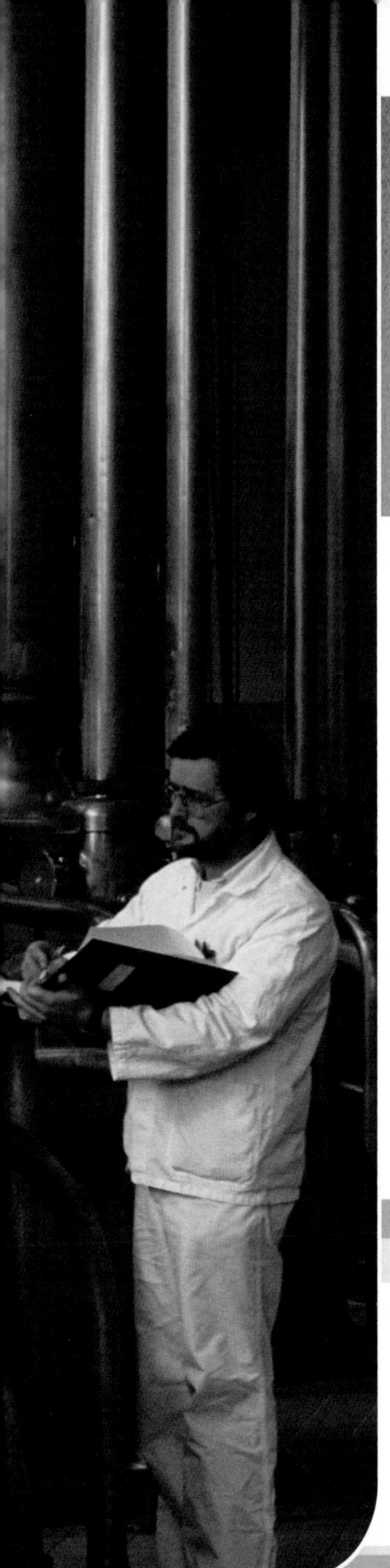

Microbiology

What do beer, wine, bread, and some cheeses all have in common? These popular products all have their characteristic properties determined by the growth of a common fungus (plural: fungi) called yeast. There are more than 700 known yeast species, and almost as many ways of using yeast. Yeast is used nutritionally as a dietary supplement; agriculturally in a process called effective microorganisms (EM) technology; industrially in the production of ethanol; and medically in gene therapy.

Even though many different types of fungi, like yeast, are useful, we are more likely to associate the word *fungus* with athlete's foot, ringworm, thrush, and sick building syndrome (SBS)—only some examples of illnesses caused by fungi. However, the illness and the cure can be closely related; consider that a fungus similar to one living on old food at the back of a refrigerator might be used to produce the antibiotic penicillin.

Consider also that microorganisms—bacteria, protists, and fungi—play an essential role in matter cycles. In a world without microorganisms there would be no plants, little oxygen, no food for animals, and no space unoccupied by waste and dead organisms.

Scientists say we've discovered only 1% of the existing microorganisms, and even their functions are not completely understood. This unit provides a sample of a small portion of the world unseen by the human eye.

Overall Expectations

In this unit, you will be able to

- demonstrate an understanding of the characteristics of various microorganisms, of their role in the environment, and of their influences on other organisms, including humans;
- analyze the development and physical characteristics of microorganisms, using appropriate laboratory equipment and techniques;
- explain the role of microorganisms with respect to human health and in technological applications in medicine, industry, and the environment.

ARE YOU READY?

Prerequisites

Concepts

- cell structure
- DNA, genes
- mitosis and meiosis
- trophic levels

Skills

- design and conduct controlled experiments
- select and use laboratory materials accurately and safely
- prepare and view a wet mount slide
- organize and display observations in appropriate tables or graphs
- present informed opinions about personal lifestyle choices related to disorders and treatments

Knowledge and Understanding

1. Microorganisms contain cell structures similar to those studied in Unit 1. You learned about specialized structures of microorganisms in Grade 8 science. In your notebook, match each cell structure in **Table 1** with its major function.

Table 1

Cell structure	Function
1. nucleus	(a) directs cell activities
2. mitochondrion	(b) whiplike tail, used for motion
3. Golgi apparatus	(c) regulates movement of materials into and out of cell
4. chloroplast	(d) packages proteins
5. chromosome	(e) directs cells toward light
6. contractile vacuole	(f) provides rigid support for plant cells
7. eyespot	(g) expels excess water
8. cell membrane	(h) a sticky coating surrounding some bacteria
9. cell wall	(i) helps bacteria attach to each other and to surfaces
10. capsule	(j) contains chlorophyll and is used in photosynthesis
11. pellicle	(k) contains genes
12. flagellum	(l) provides energy for cell processes
13. pilus	(m) firm covering that surrounds the membrane

2. **Figure 1** presents a typical bacterium and a common protist, the *Euglena*. Each cell structure from **Table 1** is found on the diagrams below. Identify each structure by name.

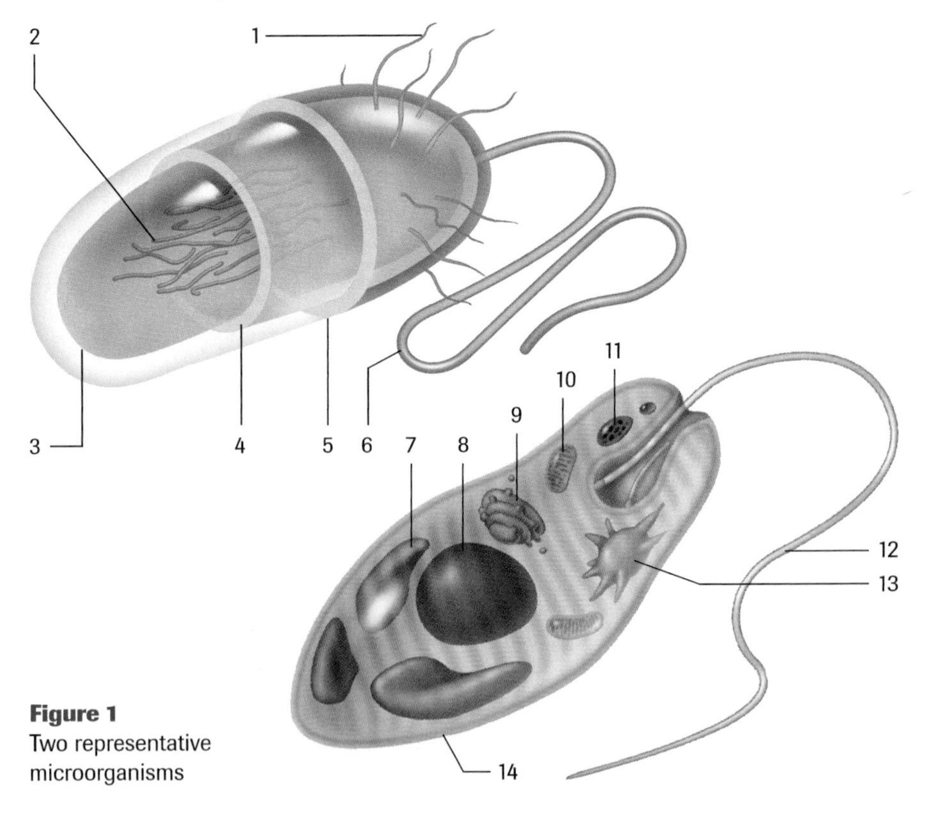

Figure 1
Two representative microorganisms

Inquiry and Communication

3. An experiment was performed to determine the effect of temperature on yeast cell growth in a sugar solution. As yeast grows, carbon dioxide (CO_2) is released. **Table 2** records the volume of CO_2 collected at timed intervals for three different temperatures. The experimenter predicted that the greatest amount of yeast cell growth would occur at 45°C.
   (a) Identify the independent and dependent variables.
   (b) Analyze the data and evaluate the prediction.
   (c) List other variables that could be tested.

Table 2 Yeast Growth

Time (min)	Volume of CO_2 (mL)		
	10°C	37°C	45°C
5	0.0	4.0	2.0
10	1.0	5.5	3.5
15	2.5	7.0	4.5
20	3.5	9.0	7.0
Total	7.0	25.5	17.0

Making Connections

4. Some mosquitoes carry microorganisms that cause diseases such as malaria. Study the sequence of events in **Figure 2** and answer the following questions:
   (a) List possible measures that can stop this cycle.
   (b) What other diseases can humans get from other animals?
   (c) Malaria killed approximately one million people worldwide in 2001; drugs no longer work against some strains of the disease; and mosquitoes in some parts of Central America now carry the disease. Should Canadians be worried about malaria? Explain your answer.

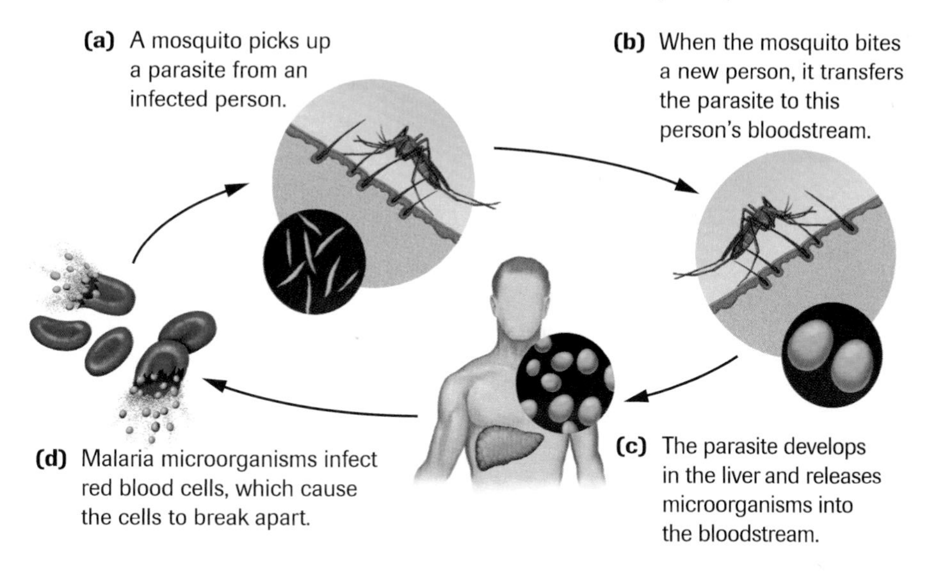

Figure 2
The microorganism that causes malaria cycles between humans and mosquitoes.

Technical Skills and Safety

5. Viewing microorganisms requires good microscope skills. Test your knowledge on the True/False questions below. If False, rewrite the statement to make it true.
   (a) The ocular lens usually magnifies 100 times.
   (b) Objects under the microscope appear upside down and backward.
   (c) Both coarse- and fine-adjustment knobs are used every time a new lens is slid into place.
   (d) The higher the magnification, the smaller the field of view.
   (e) The microscope is stored with its longest objective lens in place.

GETTING STARTED

From the time of the ancient Romans to the late 19th century, it was believed that some life forms arose spontaneously from nonliving matter. Decaying matter was the primary source of such "spontaneous generation." For example, a 17th-century recipe for the spontaneous production of mice involved placing sweaty underwear and husks of wheat in an open-mouthed jar for 21 days. Of course, what was really being observed was the attraction of mice to the contents of the jar. Although we may laugh at this concept today, it is consistent with the beliefs of the time.

It wasn't until people realized that microorganisms caused infections that they could hope to find a cure for the terrible diseases that brought death and destruction around the globe. Initial steps toward the development of modern germ theory—the idea that infectious diseases are caused by microorganisms that enter the body—began in the mid-1850s. A physician named Ignaz Phillip Semmelweis puzzled over the deaths of women, from what was called childbed fever, after they gave birth. He observed that his medical students often moved directly from dissecting a dead body to delivering a baby without washing their hands. When a colleague cut himself during an autopsy and developed childbed fever, it confirmed that childbed fever was caused by an infectious agent. Semmelweis immediately insisted that medical students wash their hands in chlorinated lime before entering a maternity ward. The physician was ridiculed for his ideas, but fewer of his patients died of childbed fever once the students began washing their hands.

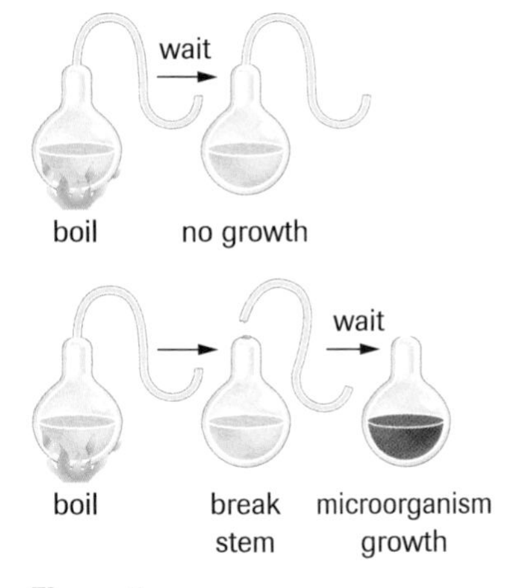

Figure 1
Pasteur used swan-necked flasks to determine that microorganisms were contained in the air and could be killed by heat.

Around the same time, Louis Pasteur used a series of experiments to prove that air contained microorganisms that would grow under the right conditions (**Figure 1**). When asked to investigate a wine-spoilage problem that threatened the French wine industry, Pasteur detected microorganisms in the wine that caused it to spoil. He then found that using heat to kill the organisms before they multiplied—a process still called pasteurization—could prevent spoilage. This experience led him to hypothesize that microorganisms might also cause plants and animals, including humans, to become sick.

Upon hearing Pasteur's hypothesis that microorganisms cause illness, English surgeon Joseph Lister began sterilizing instruments and spraying disinfectants into the air of his operating rooms. Improved survival of Lister's surgical patients provided support for Pasteur's idea.

REFLECT on your learning

1. Are antibiotics effective in treating all infectious diseases? Explain.
2. (a) How are fungi used by humans?
   (b) Describe a human disease caused by a fungus.
3. Four groups of microorganisms are studied in this unit–viruses, bacteria, protists, and fungi. In four columns, list what you already know about these microorganisms.
4. List two methods used to prevent the spread of infectious microorganisms.

During the late 1800s, German physician Robert Koch isolated and cultured the bacterial species responsible for causing anthrax in sheep, and, later, the species responsible for causing tuberculosis in humans. Koch's research eventually supported Pasteur's hypothesis so well that it became known as the germ theory of disease and is still followed today.

We now know that not all diseases are caused by bacteria. For example, protists cause some dysenteries and malaria, which are major public health problems in parts of the world. Mould cells cause many fungal infections, such as athlete's foot.

Viruses can infect plants, animals, and even bacteria. Viral-borne diseases such as influenza or human immunodeficiency virus/acquired immunodeficiency syndrome (HIV/AIDS) are generally the most difficult to treat or control. Viruses are not considered to be alive, but their activities have an enormous impact on living things. Electron micrographs show why. The bacterium in **Figure 2** was originally invaded by a single virus, which then converted the bacterial cell into a factory for manufacturing new viruses. These viruses destroyed the cell when they burst through its wall, ready to invade neighbouring bacteria. An invasion like this is fatal to unicellular organisms and can make multicellular organisms extremely sick as the viruses burst out of infected cells to infect many others.

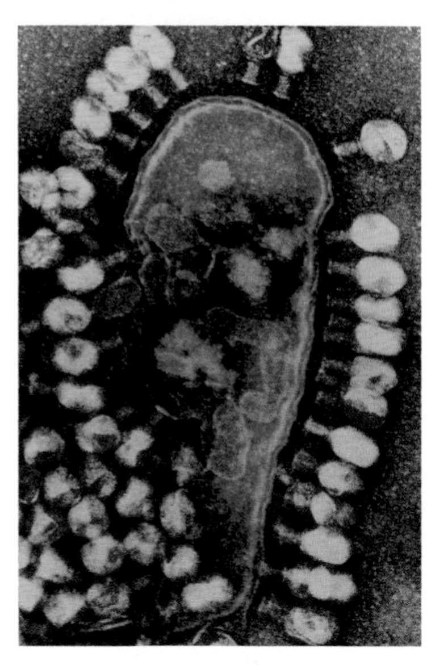

Figure 2
Viruses replicate inside host cells such as this bacterium (*E. coli*).

Electron microscopes are far too large and costly for use in high schools, but you can culture bacteria, protists, and fungi, and use a compound light microscope to view them in much the same way as Koch and Pasteur did.

▶ *TRY THIS* activity — *How "clean" is your mouth?*

In this activity, you will capture and culture microorganisms normally present in your mouth and on your lips. A sterile agar growth medium provides nourishment for microorganisms, which will appear in colonies (distinctive growths) over time.

Materials: agar, petri dishes, sterile swabs, tape, wax pencil

Use only disposable petri dishes.

Do not place tape around the circumference of the lid. Once dishes are taped, do not remove the tape.

When storing cultures, do not stack petri dishes more than three high on incubator shelves.

Wash your hands thoroughly with soap before and after the activity.

If you have a cut or abrasion on your hands, wear gloves.

1. Lick your lips and then swab them with a sterile swab. Brush the swab lightly on the surface of sterile agar in a petri dish.
2. Put the lid on the petri dish and tape it shut across the top and the bottom in the form of an "X." It will remain sealed during this activity.
3. Label your dish with your name and date of the experiment. Set it aside and leave it to incubate upside down at about 30°C.

(a) Create a data table and make observations on your sealed dish daily, over a 5-day period. Consider the shape, colour, texture, and size of colonies. Estimate how much of the surface has microorganism growth.

4. Dispose of your petri dishes as instructed.

(b) Based on your observations, what conditions are necessary for the growth of microorganisms?

(c) What steps could you take to identify the microorganisms on your culture plate?

(d) Hypothesize on what you might observe if the dishes were cultured for another 5 days.

(e) Why is it important to keep the petri dishes sealed?

(f) Compare the growth on your dish with others in the class. What factors could account for differences? Why would there be uniformity?

2.1 Taxonomy

binomial nomenclature a method of naming organisms by using two names—the genus name and the species name. Scientific names are italicized.

species a group of organisms that resemble one another physically, behaviourally, or genetically and that can interbreed under natural conditions to produce fertile offspring.

taxon a category used to classify organisms

Protista a kingdom originally proposed for all unicellular organisms. More recently, multicellular algae have been added to the kingdom.

What is the difference between a bacterium and a virus? A fungus and a lichen? Living organisms are impossible to classify without a system. The classification system used today was originally developed 250 years ago by a Swedish botanist named Carl Linnaeus. He sorted organisms by their physical and structural features on the principle that the more features organisms have in common, the closer their relationship.

Linnaeus created rules for assigning names to plants and animals. He was the first to use **binomial nomenclature**, the system of naming organisms using a two-part scientific name. Latin (and sometimes Greek) is still used today for naming organisms, and provides a common worldwide language for all scientists. Many scientific names are based on some characteristic such as colour or habitat. An example is *Castor canadensis*—*castor* meaning "beaver," and *canadensis* meaning "from Canada." The first part of any scientific name is called the genus (plural: genera). Its first letter is always capitalized, and it can be written alone. For example, the *Acer* genus refers to maple trees. The second part is the **species** name, the category in the classification of similar individuals that can produce fertile offspring. Breeding between different species is rare, and does not produce fertile offspring. The species name is written in lowercase letters and is never used alone. Together, both the genus and species names are italicized. For example, *Acer rubrum* refers to the red maple.

The two-name system provides an added advantage of showing similarities in the anatomy, early development, and ancestry of organisms. For example, binomial nomenclature suggests that the North American black bear (*Ursus americanus*) and the grizzly bear (*Ursus horribilis*) are closely related because of the genus names. Similar organisms are grouped into the same genus. The giant Alaskan brown bear (*Ursus arctos*) and the polar bear (*Ursus maritimus*) are other relatives belonging to the same genus. By contrast, the koala and panda do not belong to the genus *Ursus* and are not considered true bears.

DID YOU KNOW?

Order of Taxa Tip
You can use this mnemonic to remember the order of taxa: King Philip Came Over From Germany Swimming.

Levels of Classification

Linnaeus's use of the terms *genus* and *species* provided a framework for seven main levels of categories or **taxa** (singular: taxon) in use today. **Figure 1** shows the relationship between levels. All living organisms can be classified using this system. Common examples are given in **Table 1**.

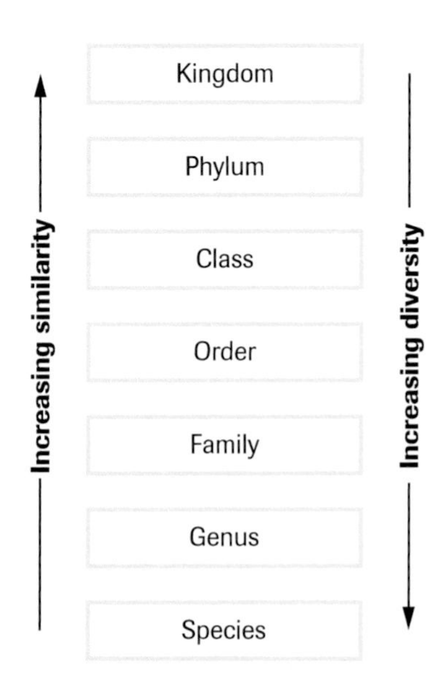

Figure 1
The sequence of classification levels

Table 1 Levels of Classification

Level of classification	Dandelion	Human	A common bacterium
Kingdom	Plantae	Animalia	Eubacteria
Phylum	Tracheophyta	Chordata	Proteobacteria
Class	Angiospermae	Mammalia	Gamma Proteobacteria
Order	Asterates	Primates	Enterobacteriales
Family	Compositae	Hominidae	Enterobacteriaceae
Genus	*Taraxacum*	*Homo*	*Escherichia*
Species	*officinale*	*sapiens*	*coli*

More than 2000 years ago, Aristotle placed all living things into two large groups: plants and animals. Linnaeus was the first to call these groups kingdoms. This two-kingdom system was used until the invention of the microscope. Disagreements occurred as microorganisms began to be placed into either the plant or the animal kingdom. All microorganisms were placed in a new kingdom called **Protista**, resulting in a three-kingdom system. Soon after, it was noted that certain organisms classified as protists lacked a true nucleus. This distinction resulted in the establishment of a fourth kingdom: **Monera**. Later, taxonomists acknowledged that mushrooms and moulds are sufficiently different from plants to be placed in a separate kingdom, called **Fungi**. This resulted in the five-kingdom system that served taxonomists well for more than 30 years.

Eventually, the bacteria (Monera) were divided into two kingdoms: **Archaebacteria** and **Eubacteria**. Archaebacteria can thrive in harsh habitats. Archaebacteria possess cell walls and ribosomes that make them very different from eubacteria. All eubacteria with cell walls possess an unusual polymer, peptidoglycan, that provides a rigid matrix for their surfaces.

The six-kingdom system of classification is summarized in **Table 2**.

Monera in a five-kingdom system, organisms that lack a true nucleus

Fungi a group of organisms formerly regarded as simple plants lacking chlorophyll, but now classified in a separate kingdom

Archaebacteria in a six-kingdom system, a group of prokaryotic microorganisms distinct from Eubacteria that possess an unusual cell-wall structure and that thrive in harsh environments such as salt lakes and thermal vents

Eubacteria in a six-kingdom system, a group of prokaryotic microorganisms that possess a peptidoglycan cell wall and reproduce by binary fission

Table 2 A Six-Kingdom System of Classification

	Kingdom	General characteristics	Cell wall	Representative organisms
	1. Eubacteria	• simple organisms lacking nuclei (prokaryotic) • either autotrophs or heterotrophs • all can reproduce asexually • live nearly everywhere • can cause disease	often present (contains peptidoglycan)	bacteria, cyanobacteria
	2. Archaebacteria	• prokaryotic • heterotrophs or autotrophs • live nearly anywhere including salt lakes, hot springs, animal guts • do not cause disease	present (does not contain peptidoglycan)	methanogens, extreme thermophiles, extreme halophiles
	3. Protista	• most are single celled; some are multicellular organisms; some have nuclei (eukaryotic) • some are autotrophs; some are heterotrophs; some are both • reproduce sexually and asexually • live in aquatic or moist habitats	absent	algae, protozoa
	4. Fungi	• most are multicellular • all are heterotrophs • reproduce sexually and asexually • most are terrestrial	present	mushrooms, yeasts, bread moulds
	5. Plantae	• all are multicellular • most are autotrophs • reproduce sexually and asexually • most are terrestrial	present	mosses, ferns, conifers, flowering plants
	6. Animalia	• all are multicellular • all are heterotrophs • most reproduce sexually • live in terrestrial and aquatic habitats	absent	sponges, worms, lobsters, starfish, humans

phylogeny the history of the evolution of a species or a group of species

Most scientists believe that organisms have changed over time. The history of the evolution of organisms is called **phylogeny**. Relationships are often shown in a diagram called a phylogenetic tree. The ancestral form starts at the trunk, and branches lead to all the descendants. **Figure 2** shows an overall picture of the relationships between the six kingdoms.

Recently, the six-kingdom system of classification has been modified to a three-domain system (**Figure 3**). The three domains in this new system are Bacteria, Archaea, and Eukaryota. In this book, we will use the six-kingdom classification system.

Classification systems continue to evolve as new DNA evidence emerges.

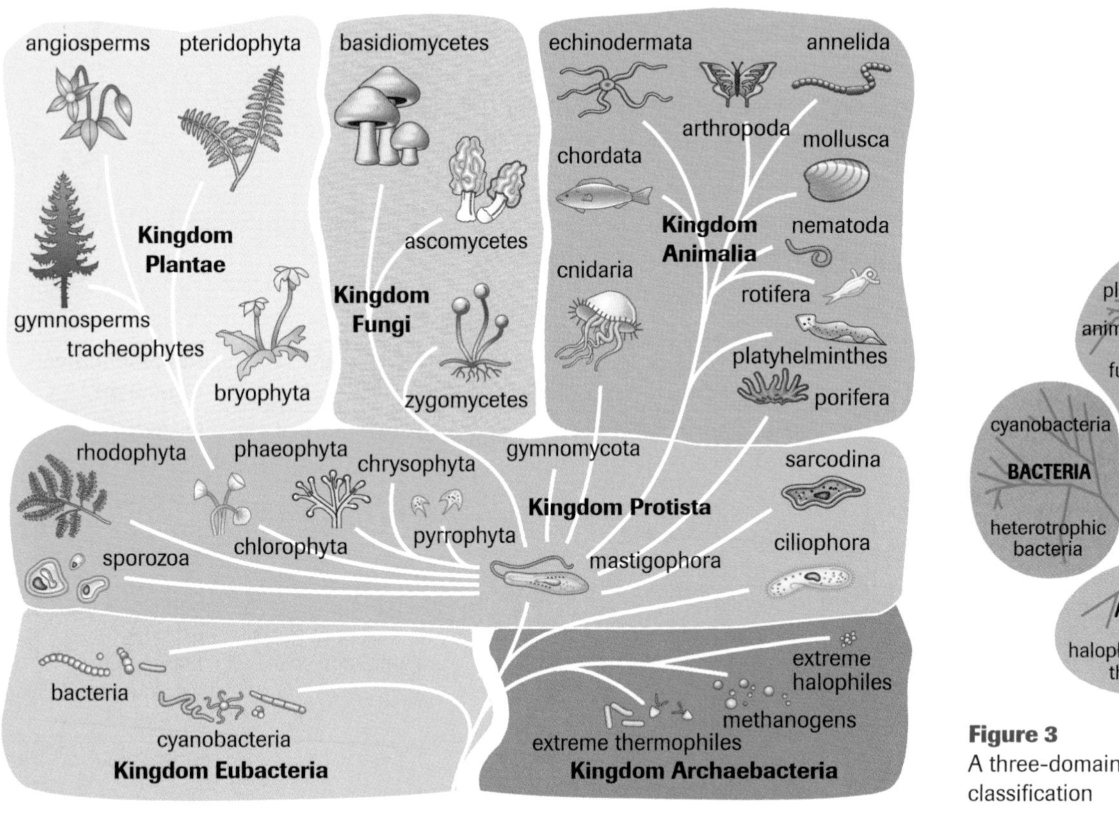

Figure 2
This phylogenetic tree shows relationships within the six kingdoms.

chromists
plantae
animalia
alveolates
fungi
rhodophytes
EUKARYOTA
cyanobacteria
flagellates
BACTERIA
heterotrophic bacteria
basal protista
ARCHAEA
halophiles
thermophiles

Figure 3
A three-domain system of classification

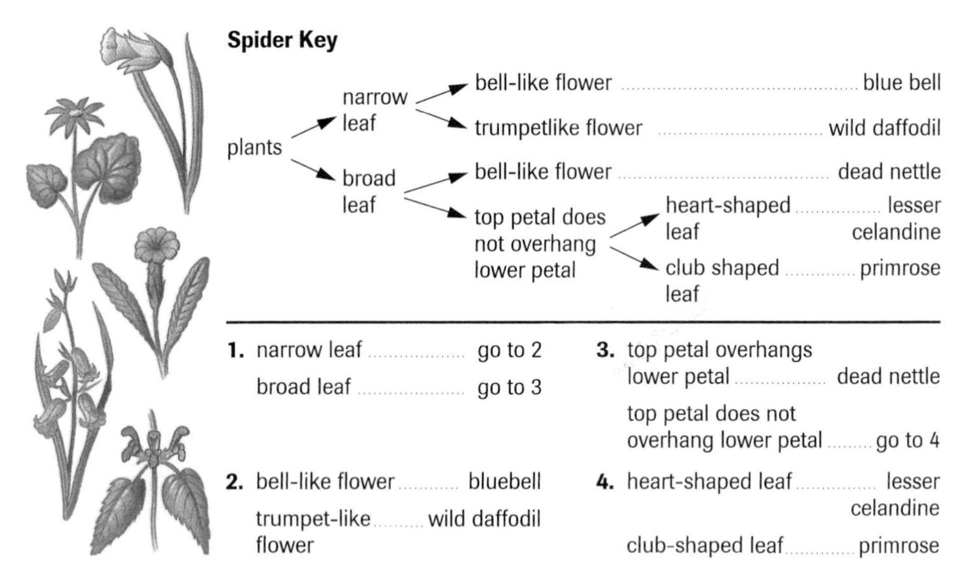

Figure 4
Sample dichotomous keys. The top key is sometimes called a spider key because of its shape.

Many scientists use classification manuals to identify organisms. Usually it involves completing a two-part or **dichotomous key** (**Figure 4**). The key is constructed so that a series of choices must be made, and each choice leads to a new branch of the key. If choices are made accurately, the end result is the correct name of the organism being identified.

dichotomous key a two-part key used to identify living things. *Di-* means "two."

CAREER CONNECTION

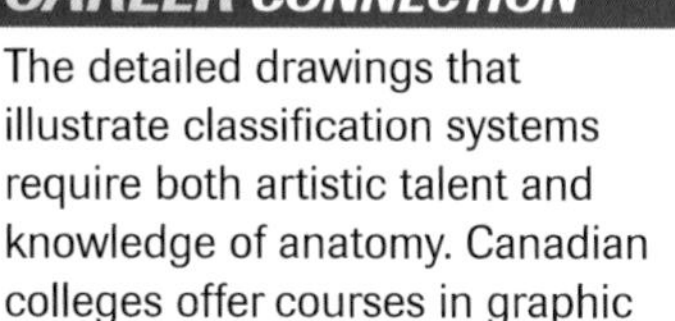

The detailed drawings that illustrate classification systems require both artistic talent and knowledge of anatomy. Canadian colleges offer courses in graphic art for biology disciplines.

▸ TRY THIS activity — Designing a Dichotomous Key

Any collection of objects can be classified using a dichotomous key. Practise this skill by designing a key to classify Canadian money.

1. Consider a sample of the following denominations: $10 bill, $5 bill, $2 and $1 coins, a quarter, a dime, a nickel, and a penny.
2. Study the similarities and differences in the denominations. Here are some to consider: material, images, edges, size, colour.

(a) Sort the eight denominations into a spider key (see **Figure 4**).
(b) Exchange your key with a partner.
(c) Complete the exercise. How well does the key work? What recommendations would you make to improve your partner's key?
(d) What four features should an effective dichotomous key have?

▸ Section 2.1 Questions

Understanding Concepts

1. Why is the classification of organisms important?
2. What is meant by the term *binomial nomenclature*?
3. Indicate the advantage of using a Latin name over a common (e.g., English) name. Provide examples.
4. What change resulted in the six-kingdom system replacing the five-kingdom system?
5. List, in order from most to least general, the seven levels of classification.
6. What do the kingdoms Animalia and Protista have in common? Fungi and Plantae? Eubacteria and Archaebacteria?
7. Use the information in **Table 3** to answer the following questions:
   (a) Which of the species are the most closely related? Explain.
   (b) Is the river otter more closely related to the muskrat or to the weasel? Why?
   (c) Is the groundhog more closely related to the chipmunk or to the ferret? Why?
   (d) Which of the species is (are) the closest relative(s) of the squirrel? Explain.

Table 3

Common name	Family	Scientific name
red squirrel	Sciuridae	*Tamiasciurus hudsonicus*
short-tail weasel	Mustelidae	*Mustela erminea*
groundhog	Sciuridae	*Marmota monax*
mink	Mustelidae	*Mustela vison*
eastern chipmunk	Sciuridae	*Tamias striatus*
river otter	Mustelidae	*Lutra canadensis*
fisher	Mustelidae	*Martes pennanti*
muskrat	Cricetidae	*Ondatra zibethica*
black-footed ferret	Mustelidae	*Mustela nigripes*

Making Connections

8. Suppose that, by conducting a massive coordinated effort, biologists manage to discover, describe, and classify every living animal on Earth by 2020. Do you think the work of taxonomists would then be complete? Explain.

2.2 Activity

Using a Classification Key

You are a biologist who has just collected a water sample from the Mer Bleue Bog near Ottawa. After carefully examining the sample with a compound light microscope, you have drawn the organisms below and listed their identifying features.

In this activity, you will identify various species of pond protists using a dichotomous key.

Procedure

1. Use the vocabulary terms in **Table 1** to help you recognize protist structures.
2. To identify each protist in **Figures 1–8**, start by reading **Table 2**, row 1 (a) and (b). Then follow the "go to" direction at the end of the appropriate sentence.

(a) Identify the protist in each figure by name.

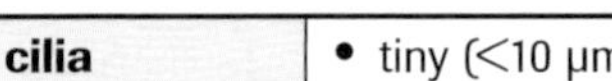

Table 1 Vocabulary Terms

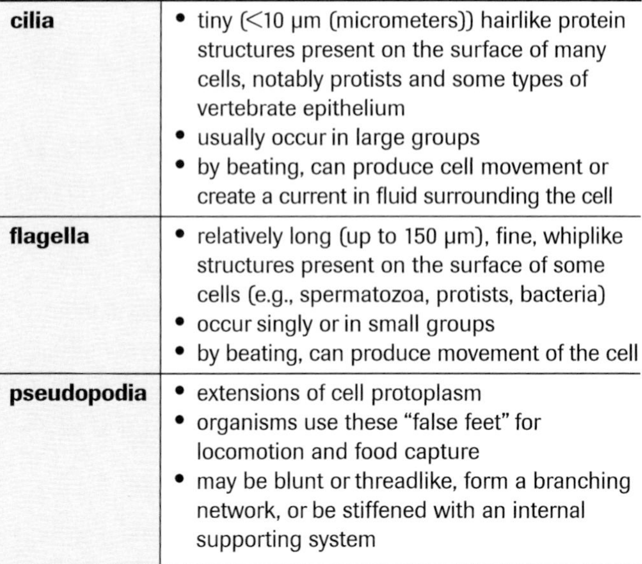

cilia	• tiny (<10 µm (micrometers)) hairlike protein structures present on the surface of many cells, notably protists and some types of vertebrate epithelium • usually occur in large groups • by beating, can produce cell movement or create a current in fluid surrounding the cell
flagella	• relatively long (up to 150 µm), fine, whiplike structures present on the surface of some cells (e.g., spermatozoa, protists, bacteria) • occur singly or in small groups • by beating, can produce movement of the cell
pseudopodia	• extensions of cell protoplasm • organisms use these "false feet" for locomotion and food capture • may be blunt or threadlike, form a branching network, or be stiffened with an internal supporting system

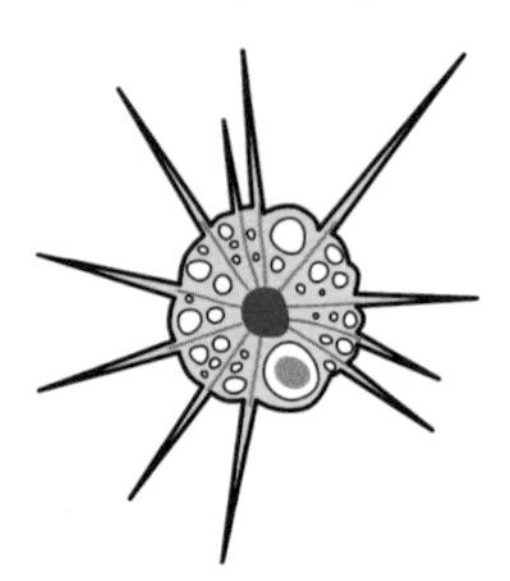

Figure 1
- length 40–50 µm
- spherical
- pseudopodia extend in ray pattern

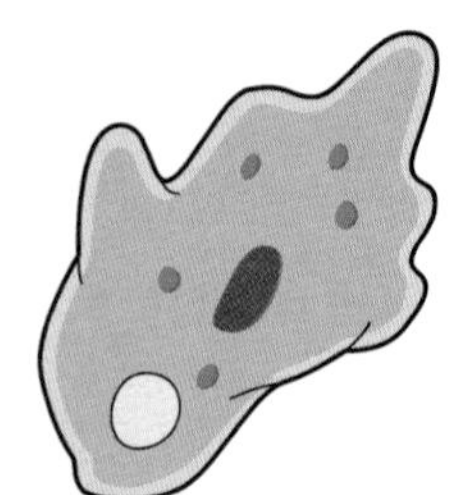

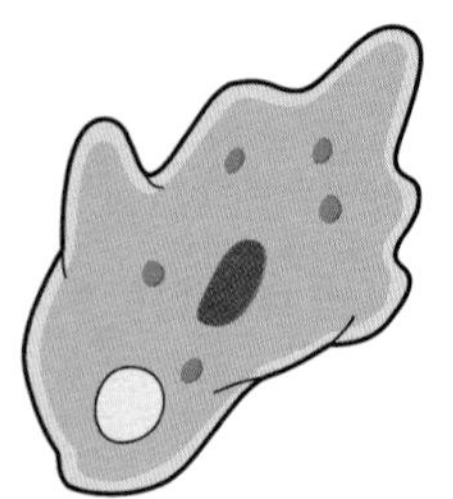

Figure 2
- length ~ 600 µm
- no fixed shape
- uses pseudopodia for movement and to engulf food

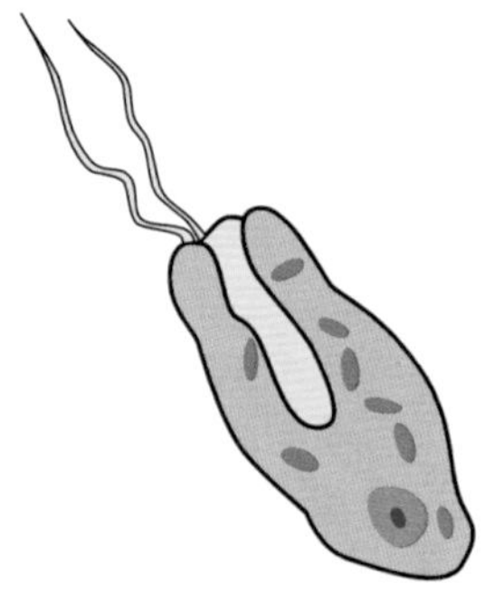

Figure 3
- length 20–40 µm
- oval shape
- has two visible flagella

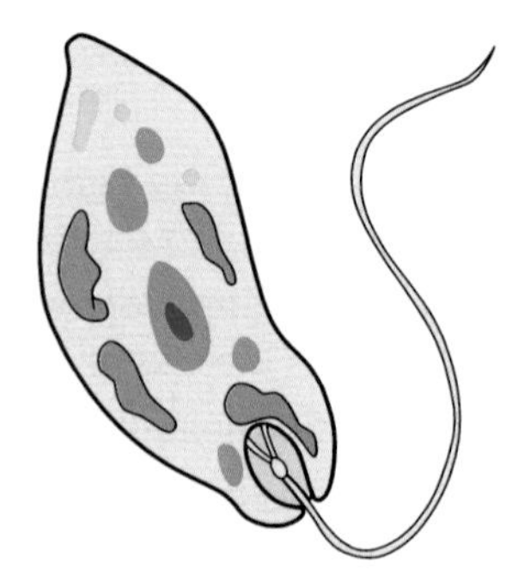

Figure 4
- length 40–50 µm
- slipper shape
- whiplike flagellum
- uses chloroplasts to make food through photosynthesis

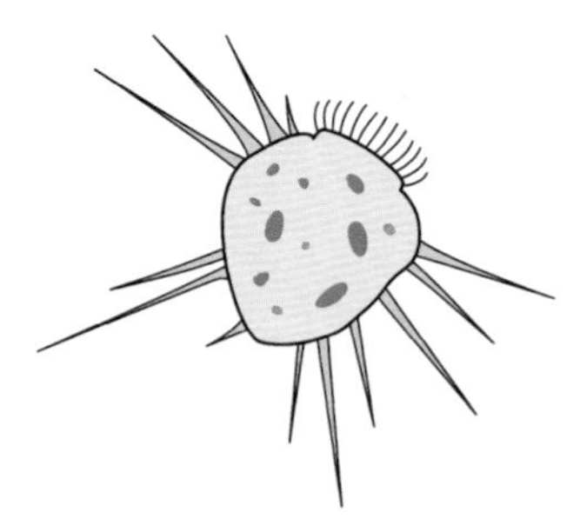

Figure 5
- length 20–40 µm
- acorn shape
- has cilia around anterior opening
- has projections around rest of body

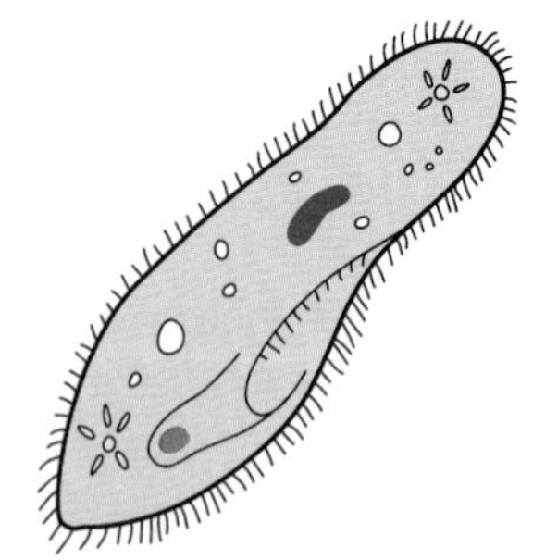

Figure 6
- length 200–350 µm
- slipper shape
- body is surrounded by cilia

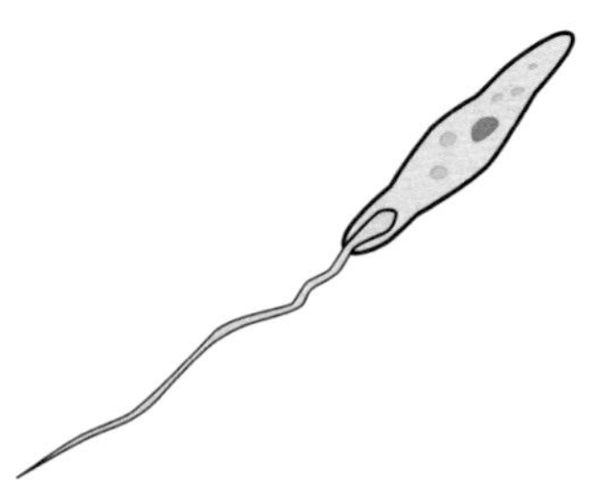

Figure 7
- length 20–70 µm
- slim oval shape
- uses visible flagellum to pull rather than push itself through water
- second, nonvisible flagellum is a short rod used to capture food

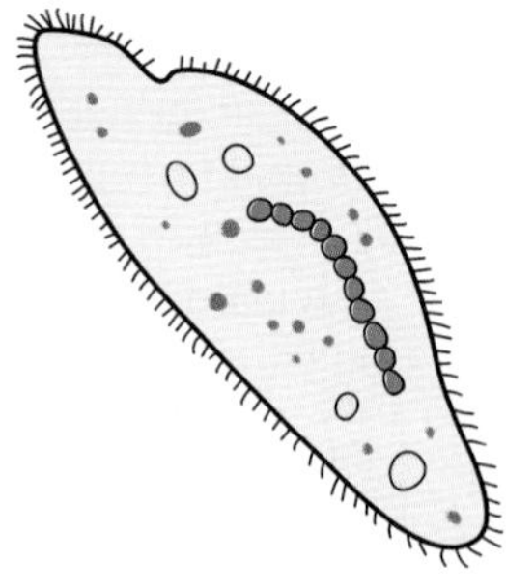

Figure 8
- length 500–800 µm
- wormlike appearance
- cell is covered with cilia in spiral pattern

Table 2 Dichotomous Key for Protist Identification

1	(a) Cilia are present. (b) Cilia are absent.	Go to 6 Go to 2
2	(a) Flagella are present. (b) Flagella are absent.	Go to 4 Go to 3
3	(a) Organism is asymmetrical and has pseudopodia. (b) Organism is spherical and has needlelike pseudopodia.	*Amoeba* *Actinophrys*
4	(a) Organism has a single flagellum. (b) Organism has two or more flagella.	Go to 5 *Chilomonas*
5	(a) Organism has chloroplasts. (b) Organism lacks chloroplasts.	*Euglena* *Peranema*
6	(a) Cilia are present around entire body; no projections around body. (b) Cilia are present only around anterior opening; projections around body.	Go to 7 *Halteria*
7	(a) Organism is 200–350 µm long. (b) Organism is 500–800 µm long.	*Paramecium* *Spirostomum*

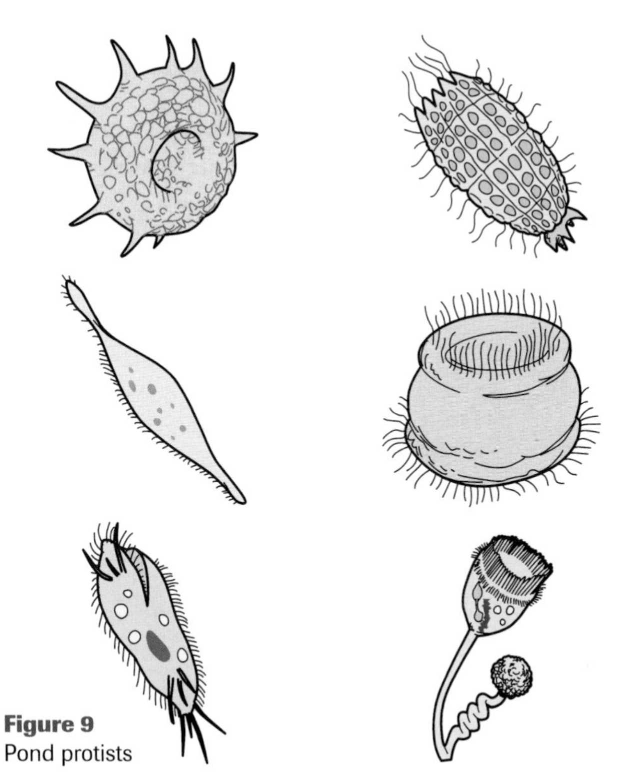

Figure 9
Pond protists

Analysis

(b) What are three characteristics used to classify protists?

(c) Study the features of other pond protists (**Figure 9**) and make a list of additional characteristics that could be used to classify these organisms.

Synthesis

(d) Working in a group, give each organism in **Figure 10** an imaginary genus and species name. Record these names on one piece of paper. On a separate piece of paper, make a dichotomous key that allows others in the class to identify each of the protists.

(e) Exchange your key with another group. How successfully did another group use your key? What changes would you make?

(f) List criteria that could be used to evaluate a dichotomous key.

CAREER CONNECTION

Most pond-water microorganisms are harmless. Other microorganisms (e.g., some *E. coli*) are harmful to human health if present in sufficient quantity. The Walkerton tragedy in Ontario in 2000 emphasized the need for monitoring microorganisms in drinking water. Investigate the opportunities available in this field. What programs are available in Canadian colleges for water-quality technicians?

www.science.nelson.com

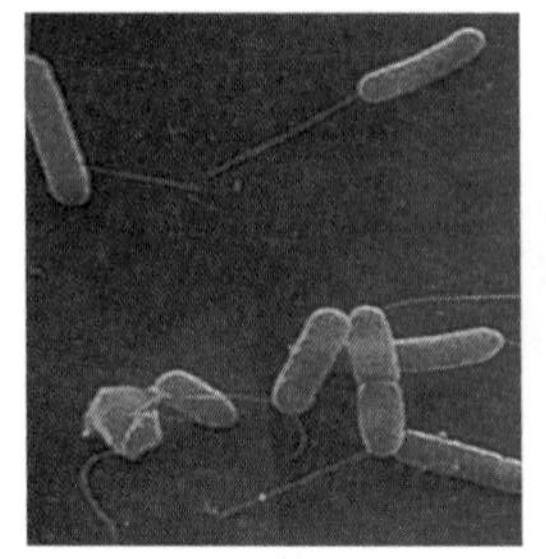
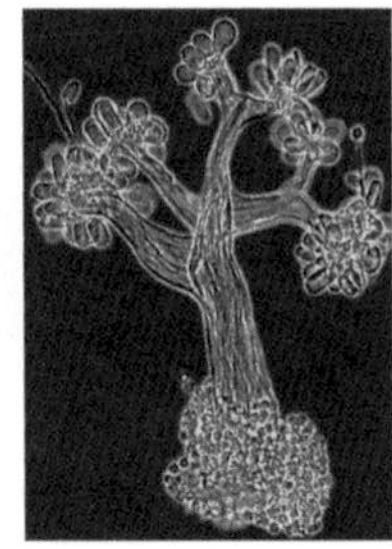
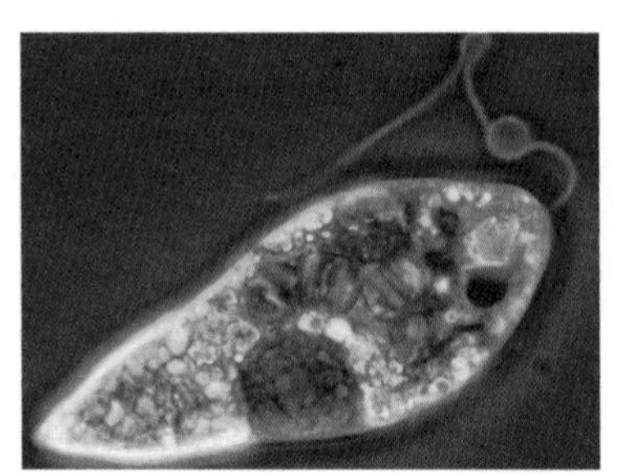
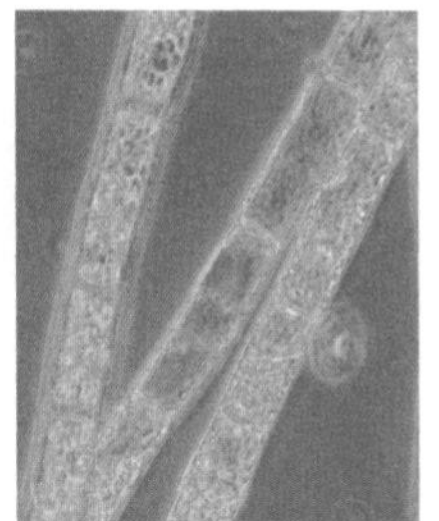
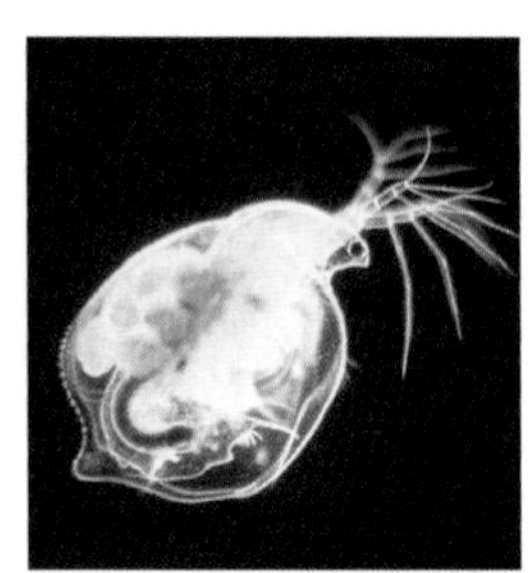

Figure 10
Unidentified pond microorganisms

2.3 Viruses

virus a microscopic particle capable of reproducing only within living cells

Did you notice that viruses do not appear in the six-kingdom classification system presented in section 2.1? Taxonomy provides a framework for examining existing living systems and comparing modern organisms with extinct forms. Viruses (Latin for "poison") do not display the essential characteristics of living cells. A **virus** is a lifeless particle that carries out no metabolic functions on its own and cannot reproduce on its own; however, once it invades a living cell, it is capable of reproduction. On this basis, viruses occupy a position between nonliving and living matter.

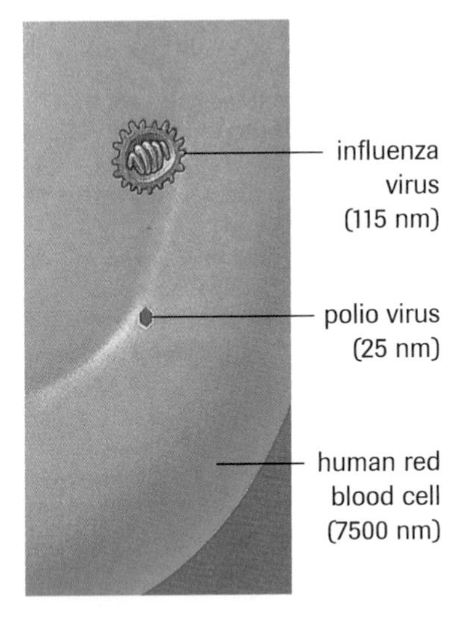

Figure 1
Comparative sizes of viruses and the human blood cell. Each virus is composed of hundreds of thousands of large protein molecules, which surround a core of nucleic acids.

It was not until the 1950s that an early electron microscope enabled scientists to view viruses (**Figure 1**). A virus is so small that it must be measured in units called nanometres (nm). One nanometre equals one-billionth of a metre (10^{-9} m). Viruses range in size from about 20 nm to 400 nm in diameter. Because it is difficult to visualize such small measurements, imagine more than 5000 influenza viruses fitting on the head of a pin.

A virus is far less complex than the simplest living organism. It consists of an inner nucleic acid (DNA or RNA) core or strand, surrounded by an outer protective protein coat called a **capsid**. The capsid accounts for 95% of the total virus and gives the virus its particular shape (**Figure 2**). **Bacteriophages,** also referred to as phages, are a category of viruses known as "eaters of bacteria" that have a unique tadpole shape with a distinct head and tail region.

capsid the protective protein coat of viruses

bacteriophage a category of viruses that infect and destroy bacterial cells

host range the limited number of host species, tissues, or cells that a virus or other parasite can infect

While viruses must enter cells to carry out life processes, not every virus is a killer. The tobacco mosaic virus, which infects the leaves of the tobacco plant, does not appear to destroy plant tissue on any large scale. Generally, viruses are selective; specific viruses enter only specific cells. This is called the **host range**, the number of host species, tissues, or cells that can be infected by a virus or other parasite. Some animal viruses have a broad host range. For example, the swine flu virus can infect hogs and humans; the rabies virus can infect many mammals including rodents, dogs, and humans. Other viruses have a very narrow host range. For example, the human cold virus usually infects only the cells of the upper respiratory tract, and HIV affects the immune system because it attaches only to a specific site on the surface of certain types of white blood cells.

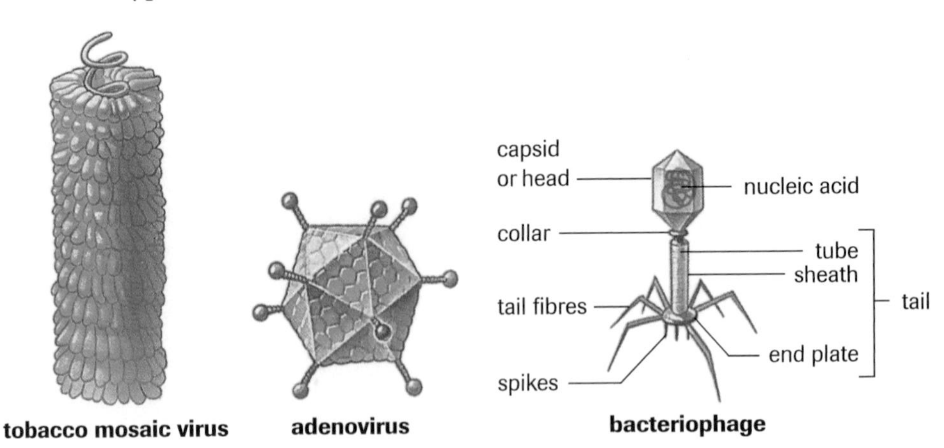

Figure 2
The protein coat, or capsid, of a virus may display various geometric shapes.

Viral Replication

Replication, the process by which the virus and its genetic material are duplicated before a cell divides, occurs in a variety of ways, but there are generally four basic steps:

1. Attachment and entrance: The virus chemically recognizes a host cell, attaches to it, and enters it (**Figure 3**). Either the whole virus or only its genetic material (for certain viruses, this can be DNA or RNA) enters the cell's cytoplasm.
2. Synthesis of protein and nucleic acids: Molecular information in the viral DNA or RNA directs the host cell in replicating viral components (nucleic acids, enzymes, capsid proteins, and other viral proteins).
3. Assembly of the units: The viral nucleic acids, enzymes, and proteins are brought together and assembled into new virus particles.
4. Release of new virus particles: The newly formed virus particles are released from the infected cell, and the host cell dies. This is called **lysis**.

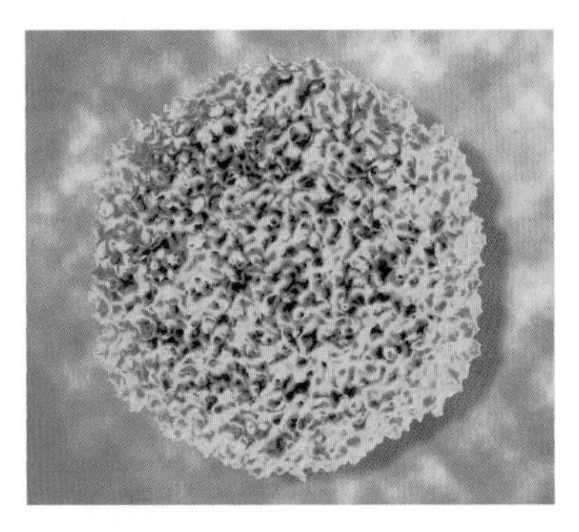

Figure 3
The common cold virus, rhinovirus, contains 60 sites capable of connecting to a receptor on human cells.

lysis the destruction or bursting open of a cell. An example is when an invading virus replicates in a bacterium and many viruses are released.

The entire process, known as the lytic cycle, may be completed in as little as 25 to 45 min and produce hundreds of new virus particles under laboratory conditions.

Certain types of viruses, such as those that cause cancer or those that infect bacteria, have a lysogenic cycle. The virus does not kill the cell outright. It may coexist with the cell and be carried through many generations without apparent harm to the host. As above, the virus enters the host cell. Instead of taking control, its nucleic acid becomes integrated into the bacterium's DNA and acts as an additional set of genes on the host chromosome. Then it is replicated along with the host DNA and passed along to all daughter cells. During this period of normal replication, the virus appears to be in a dormant state, called **lysogeny**. Often the dormant virus may be activated by a stimulus, such as an environmental change to a bacterium (e.g., temperature or pH), or some other event such as changes in available nutrients. This triggers the lytic cycle, and once again new virus particles are formed and released. **Figure 4** shows both the lytic and the lysogenic cycles.

lysogeny the dormant state of a virus

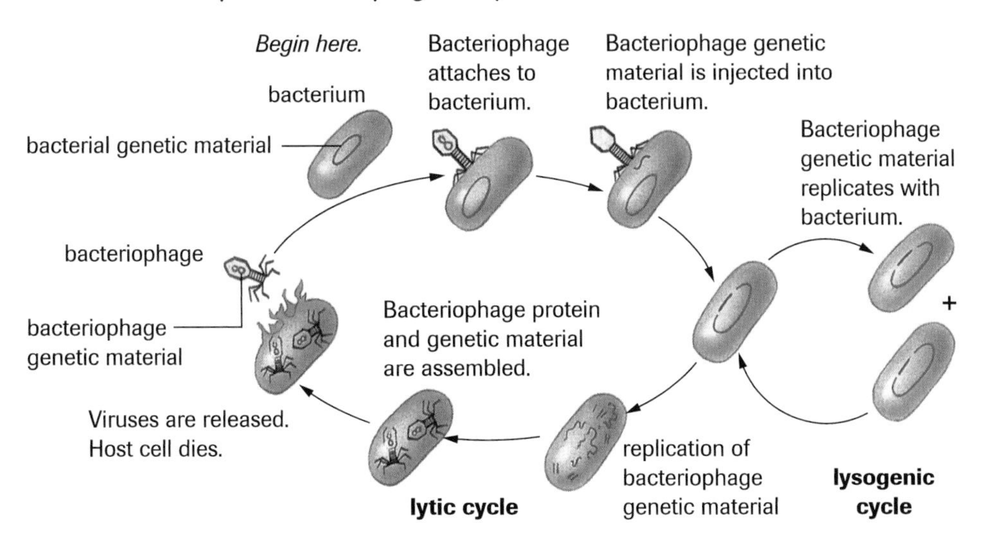

Figure 4
Viral replication

Viruses and Human Health

In many diseases caused by viruses (**Table 1**), the virus attacks cells as it reproduces. The destruction of these cells causes symptoms of the disease. Unfortunately, most viral infections are difficult to treat, as they are not destroyed by **antibiotics**, the substances that inhibit the growth of microorganisms. As well, some viruses remain dormant in the body for years before disease symptoms appear. Certain viruses cause cancer by adding specific genes to an infected cell, thereby turning it into a cancer cell. **Vaccines** are liquid preparations of dead or weakened viral or bacterial cells that stimulate the body's immune system to fight back. Some viral (and bacterial) diseases can be prevented with vaccines (e.g., polio, smallpox, hepatitis B). Vaccines are given orally or by injection (inoculation).

antibiotic a chemical produced synthetically or by microorganisms that inhibits the growth of or destroys certain other microorganisms

vaccine a suspension prepared from dead or weakened viral or bacterial cells

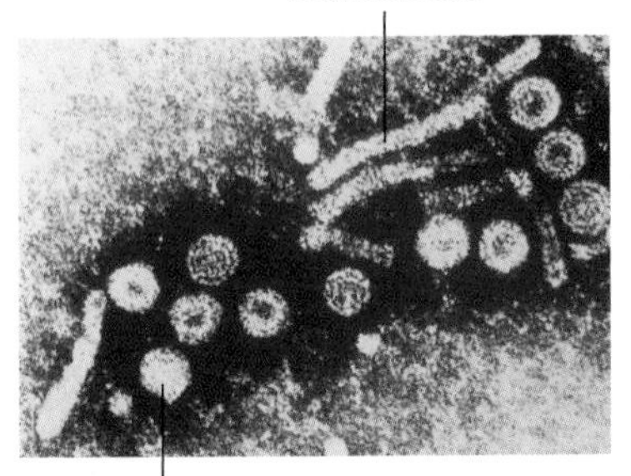

Figure 5
An electron micrograph of two forms of the hepatitis B virus, isolated from the blood of an infected patient. Only the polyhedral forms are infectious.

Table 1 Examples of Human Viral Diseases

Organism	Disease(s)	Transmission
DNA viruses		
Epstein-Barr	infectious mononucleosis	direct contact, airborne droplets
poxviruses	smallpox	direct contact, airborne droplets
varicella-zoster	chicken pox	direct contact, airborne droplets
RNA viruses		
enteroviruses	polio, infectious hepatitis (**Figure 5**)	direct contact, fecal contamination
rhinoviruses	common cold	direct contact, airborne droplets
paramyxoviruses	measles, mumps	direct contact
rhabdoviruses	rabies	bite by infected animal
orthomyxoviruses	influenza	direct contact, airborne droplets
retroviruses (HIV)	AIDS, some cancers	direct contact
flaviviruses	West Nile	mosquito vector
caliciviruses	Norwalk	direct contact, fecal contamination
coronavirus	SARS (severe, acute respiratory syndrome)	direct contact, airborne droplets

Influenza—A Representative Virus

Influenza, or "flu" for short, has affected humans for centuries. Derived from the Italian word for "influence," the flu was originally thought to be caused by a bad influence from the heavens. Although this viral infection is usually considered to be more annoying than dangerous, flu, accompanied by pneumonia, is the sixth leading cause of death in North America.

The influenza virus was identified in the 1930s, and its structure was seen with the electron microscope in 1943. Like all viruses, the influenza virus consists of a nucleic acid core, with single-stranded RNA, inside a capsid (**Figure 6**). Rigid protein spikes radiate from a spherelike body about 20 nm in diameter. Physical and chemical differences result in three flu types: A, B, and C. Strains of influenza are described by the protein coat, the year of isolation, and the geographic location.

Figure 6
Micrograph of the influenza virus, magnified about 150 000 times

The influenza virus is spread by direct contact and can live for hours in dry mucus. Respiratory tissues of the lungs and the throat are the first hosts for the flu virus (**Figure 7**). If ciliated cells in the upper respiratory passages are destroyed, the lining can no longer sweep mucus and foreign particles out of the body. A sore throat and congested lungs may be the first signs of the infection. Accompanying symptoms may include chills, fever (up to 40°C), muscle aches and pains, sweating, fatigue, and nausea and vomiting. Pneumonia, bronchitis, and sinus or ear infections are three examples of complications from the flu.

The flu is highly contagious from one day before symptoms appear to up to seven days later. Symptoms start one to four days after the virus enters the body and may persist for up to two weeks. Treatment for the flu is rest, liquids, and medication to relieve symptoms. Remember: Antibiotics are ineffective against viruses.

The influenza vaccine can be up to 90% effective in preventing illness from a specific influenza virus. The influenza vaccine is highly recommended for the young, the elderly, health and child-care workers, and people of any age with chronic medical conditions. Because the protein in the viral coat evolves as the virus moves from country to country, the vaccine can become less and less effective. New vaccines must be developed and administered each year.

Virus destroys ciliated cells.

Sore throat results from damage to cells.

Without ciliated cells, mucus drains into lungs.

Congested lungs make breathing difficult.

Figure 7
How a flu virus affects the body

Section 2.3 Questions

Understanding Concepts

1. Why aren't viruses placed within the normal classification system?
2. Name two human diseases caused by (a) RNA viruses and (b) DNA viruses. How is each transmitted?
3. The symptoms of many viruses come and go within 24 h. Use viral replication information to explain why this is the case.
4. What is the difference between a nanometre and micrometre? Draw a scale to illustrate this difference.
5. Label the capsid, the core, and the location of DNA in the diagram of the hepatitis B virus (**Figure 8**). Describe the shape and appearance of each.

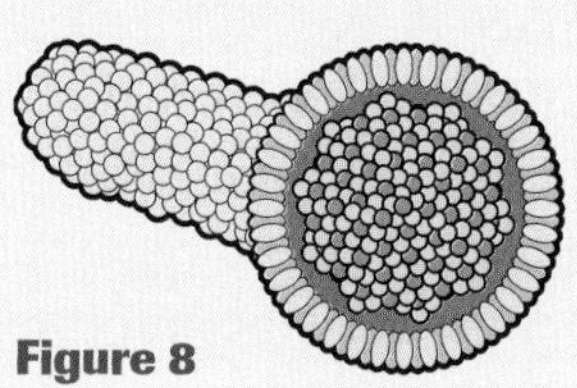

Figure 8

Applying Inquiry Skills

6. During the summer and fall of 2002, the West Nile virus advanced north into Canada. This virus replicates in birds, is transmitted by mosquitoes, and can be spread to humans. Canadian Blood Services impounded all blood supplies collected during this time. Hypothesize on why this action was taken.
7. The numbers of deaths attributed to AIDS in various countries in 1999 are listed in **Table 2**. What conclusions can you draw from these figures? What accounts for the difference in numbers? What additional information would help you analyze these data?

Table 2

Country	Number of deaths attributed to AIDS in 1999
Brazil	18 000
Canada	400
China	17 000
India	310 000
Nigeria	250 000
Russia	850
United Kingdom	450
U.S.A.	20 000

Making Connections

8. "Five million Canadians get sick with the flu every year, resulting in 1.5 million workdays lost. The flu costs the Canadian health care system about $1 billion a year." Interpret these statements by identifying the costs and suggesting solutions to this problem.
9. Current research indicates that multiple sclerosis (MS) may be caused by a slow-acting virus. If this is the case, why might environmental factors (e.g., sunlight, stress) have an effect on the severity of MS symptoms?
10. The hepatitis B virus (HBV) is transmitted through activities that involve contact with blood or body fluids. Research this disease to prepare a fact sheet. Cover the following topics: signs and symptoms; risk and occurrence; prevention; treatment; trends and statistics; connection between HBV and HIV.

www.science.nelson.com

2.4 Bacteria

In section 2.1, you learned that bacteria can be classified into two kingdoms: Archaebacteria and Eubacteria. Before we examine the differences, consider the similarities all bacteria possess:

- All are unicellular; some stick together in colonies.
- Cells are prokaryotic. Bacteria usually lack membrane-bound internal structures, having no organized nucleus, vacuoles, mitochondria, or chloroplasts.
- Cells usually have a single chromosome in the form of a DNA loop.
- Cells reproduce asexually by binary fission.
- Cells thrive only in moist environments and become inactive if the environment dries up.

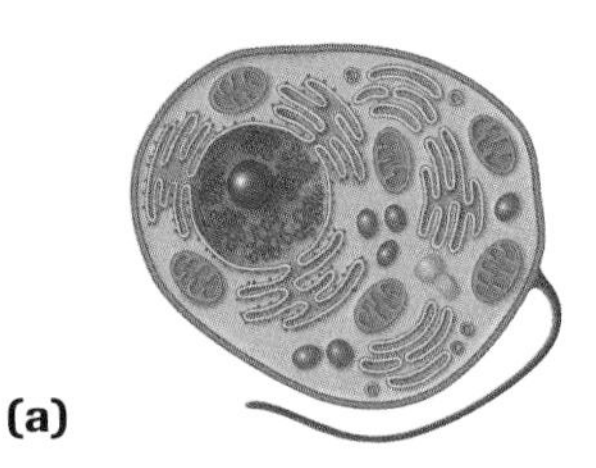
(a)

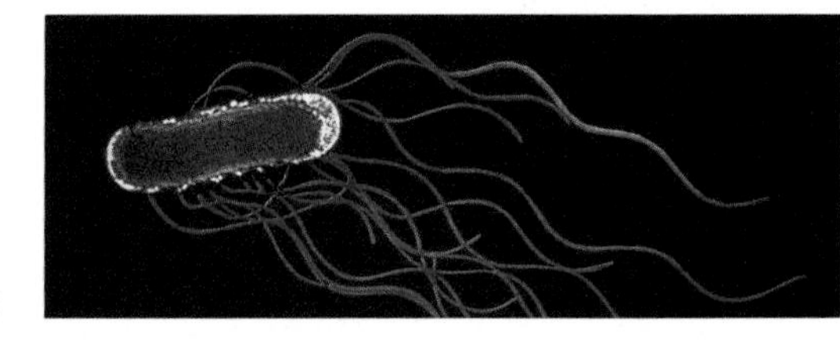
(b)

Figure 1
(a) Eukaryotic animal cell
(b) Prokaryotic bacterial cell

There are two types of cells, eukaryotic and prokaryotic. Cells are placed into one of these two categories based on how complex their structure and cell division process is. Plant and animal cells are **eukaryotic**, meaning "true nucleus"; their structures were outlined in Unit 1 (see **Figure 1** for review). Bacterial cells are **prokaryotic**, meaning "before nucleus." Genetic material floats freely inside the cell in the form of either a loop or a small piece of genetic material called a **plasmid.** Bacteria have a rigid outer wall that gives them shape. Underneath this wall is a more fluid membrane called the cell or plasma membrane. There are few recognizable organelles in the cytoplasm, even under the high magnification of an electron microscope. **Table 1** compares some properties of eukaryotic and prokaryotic cells.

eukaryotic a type of cell that has a true nucleus, surrounded by a nuclear envelope

prokaryotic a type of cell that does not have its chromosomes surrounded by a nuclear envelope

plasmid a small piece of genetic material

Table 1 Properties of Eukaryotic and Prokaryotic Cells

Property	Eukaryotes	Prokaryotes
true nucleus	present	absent
DNA	usually many chromosomes	usually single chromosome
cell division	mitosis	binary fission
ribosomes	larger	smaller
mitochondria	present	absent
chloroplasts	can be present	absent
flagella	complex flagella	simple flagella
size	usually $> 2\ \mu m$ diameter	usually $< 2\ \mu m$ diameter
microorganism examples	*Euglena, Paramecium, Amoeba*	*Escherichia coli, Bacillus anthracis*

DID YOU KNOW?

The Missing Link
Some scientists believe that archaebacteria, isolated in the 1970s, are the link between eubacteria and eukaryotes. Over 50% of the genes identified in this kingdom were previously undiscovered in other organisms.

Archaebacteria

Archaebacteria (*archae* means "early" or "primitive") are derived from one of the oldest groups of living organisms, perhaps the first on Earth. These organisms thrive under extreme conditions that other organisms could not tolerate. Many live without oxygen. Three major groups of archaebacteria include the thermophiles, the methanogens, and the halophiles (**Figure 2**).

The thermophiles live in extremely hot environments. Some obtain energy by oxidizing sulfur, and thrive in and around hot sulfur springs. The methanogens grow on carbon dioxide and hydrogen gas to produce methane. They exist in environments such as volcanic deep-sea vents and the intestines of mammals, including humans. The halophiles live in extremely saline environments, such as salt flats and evaporation ponds, and are responsible for the purplish-red colour in these areas. The bright red pigment protects the cells from intense solar radiation, yet allows them to use sunlight for energy.

Scientists are putting this kingdom to work to benefit society. Methanogens help digest sewage and oil spills, producing methane, which can be used as an alternative fuel source. Halophiles are used in cancer research. Enzymes produced by archaebacteria are used in food processing, perfume manufacture, and pharmaceuticals. The most famous archaebacterial enzyme is *Taq* polymerase, a substance isolated from *Thermus aquaticus* and used in molecular biology.

Figure 2
Examples of archaebacteria

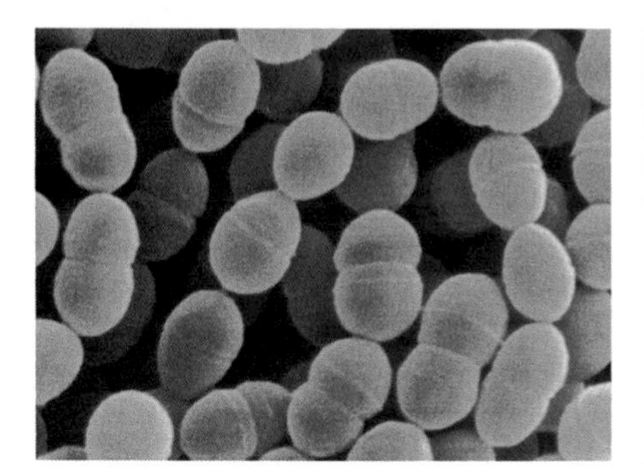

(a) Thermophiles live at temperatures over 45°C.

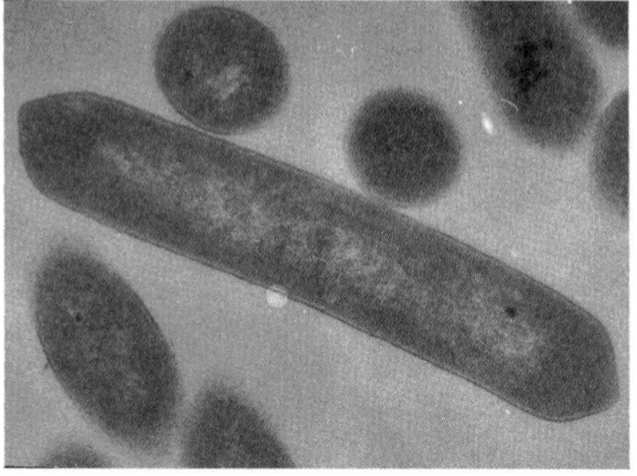

(b) Methanogens are found in environments without oxygen, such as swamps and marshes.

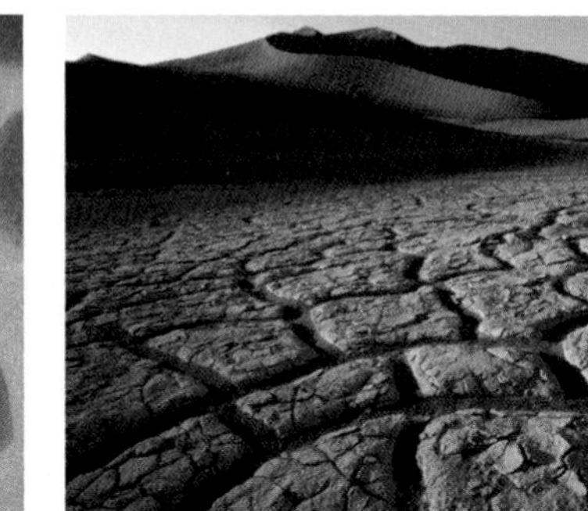

(c) Halophiles colour the landscape around salt lakes.

Eubacteria

The best representative organism of the kingdom Eubacteria, and likely one of the most studied organisms in the world, is *Escherichia coli* (**Figure 3**). As you will discover throughout this unit, this organism is both helpful and harmful to humans. Billions of *E. coli* live in the human intestine, helping with food digestion and the synthesis of vitamin K and B complex vitamins. Human and animal feces also contain billions of *E. coli* bacteria. High counts of such bacteria (coliform bacteria) in water indicate fecal contamination and danger to human health.

Most bacteria range in size from 0.4 μm in diameter to several micrometres in length. As *E. coli* is about 1 μm in length, a line of about 250 organisms could just be seen by the naked eye. The structure of *E. coli* is illustrated in **Figure 4**, on the next page.

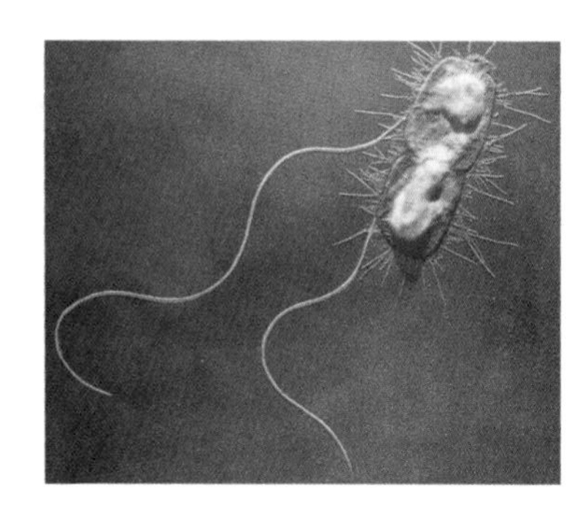

Figure 3
E. coli micrograph

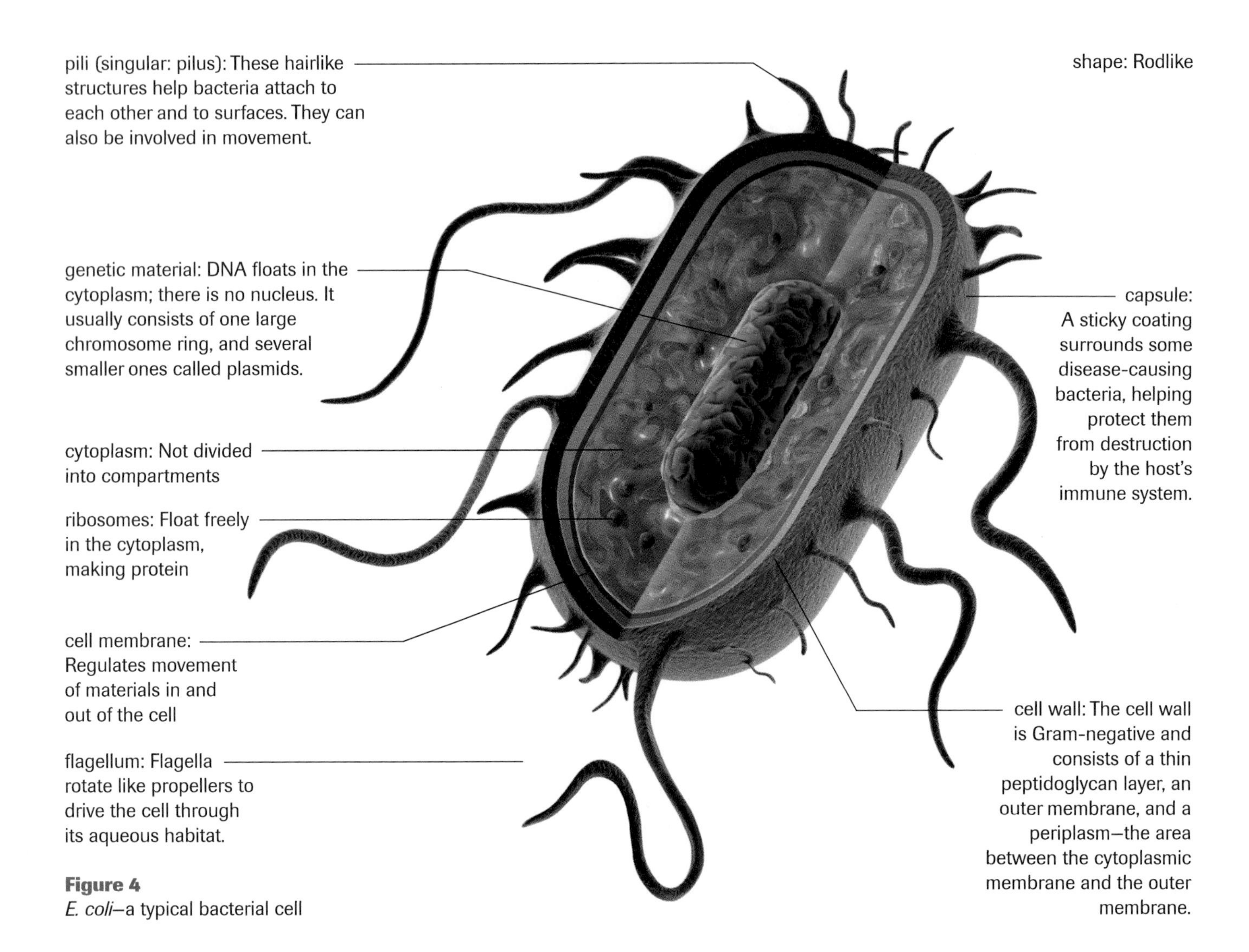

Figure 4
E. coli—a typical bacterial cell

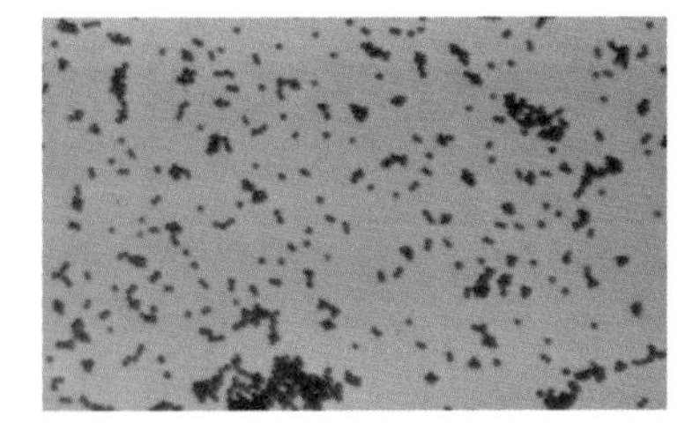

Figure 5
Staphylococcus epidermis is Gram-positive.

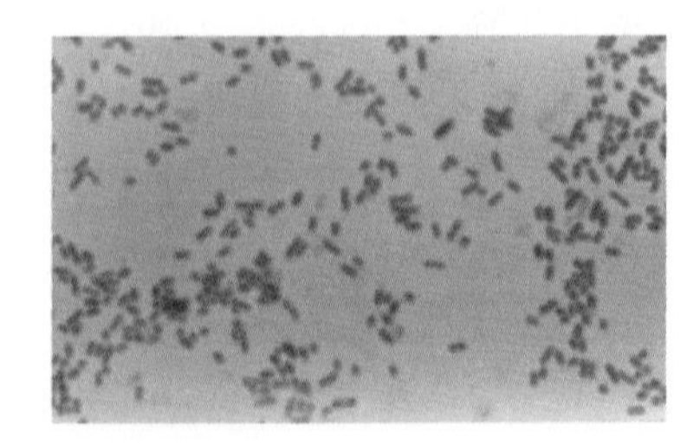

Figure 6
E. coli is Gram-negative.

The composition of the cell wall varies among bacteria species and is an important means of identifying and classifying bacteria. Eubacteria contain a polymer called peptidoglycan in their cell walls. The Gram staining method (developed by Danish bacteriologist Hans Christian Gram in 1884) differentiates between two major cell-wall types. In this method, the bacterial sample is smeared on a microscope slide, and stained with crystal violet dye (purple). The dye is then fixed to cells with Gram's iodine, decolourized with ethanol, and counterstained with safranin (red).

Gram-positive bacteria such as *Staphylococcus* (**Figure 5**) retain the first dye, appearing purple in the light microscope. In Gram-negative bacteria, the decolourizer washes out the violet dye, the counterstain is taken up, and the cells appear pinkish-red. *E. coli* is a Gram-negative bacterium (**Figure 6**). Differences in the amount of peptidoglycan in the wall determine whether or not the crystal violet is washed out. Gram-positive bacterial walls contain more peptidoglycan, so these walls retain the purple dye.

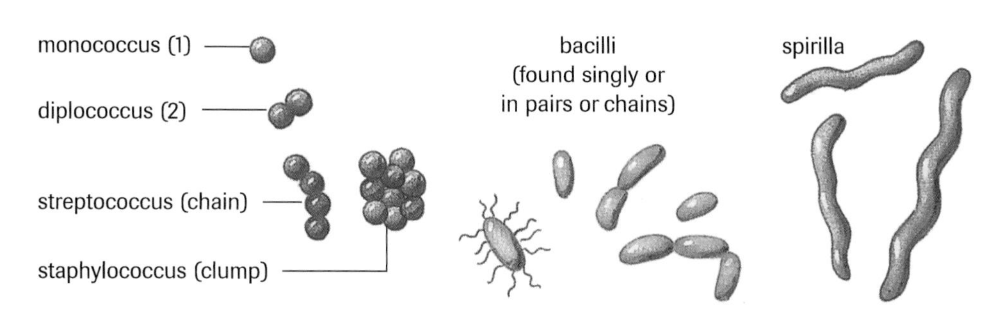

Figure 7
Basic shapes of bacteria

Aside from cell-wall composition, eubacteria can be classified according to shape, configuration, respiration, and type of nutrition.

Most organisms display one of three basic shapes—spherical, rod-shaped, or spiral (**Figure 7**).

After division, many bacteria stay together in groups or clusters rather than remain as individual cells. Cocci (singular: coccus), bacilli (singular: bacillus), and sometimes spirilla (singular: spirillum), form pairs, cluster colonies, or chains (filaments) of cells. For example, *Streptococcus mutans*, the main cause of tooth decay, forms chains. *Staphylococcus aureus*, a common bacterium found on the skin, forms clumps. When large numbers of cells have grown, they become **colonies**. Myxobacteria form specialized colonies in one point of their growth called fruiting bodies (**Figure 8**).

Some bacteria are **aerobic** organisms and must have oxygen to survive. Bacteria that cause tuberculosis are aerobic organisms. Other bacteria are **anaerobic** and can only grow in the absence of oxygen. Organisms causing gangrene, tetanus, and botulism are anaerobic organisms. Many bacteria can survive and grow with or without oxygen. *E. coli*, the bacterium in **Figure 4**, is in this category.

Classifying eubacteria by type of nutrition shows the diversity of this kingdom. Some bacteria are autotrophs, making the food they require from inorganic substances. Photosynthetic bacteria convert carbon dioxide and water into carbohydrates by using energy from sunlight. Chemosynthetic bacteria use chemical reactions rather than sunlight as their energy source. Most bacteria are heterotrophs, obtaining their nutrients from other organisms (e.g., by feeding on dead or decaying matter, or as parasites causing disease by feeding on living tissue).

colony a visible growth of microorganisms containing thousands of cells

aerobic requiring oxygen for respiration

anaerobic conducting respiration processes in the absence of oxygen

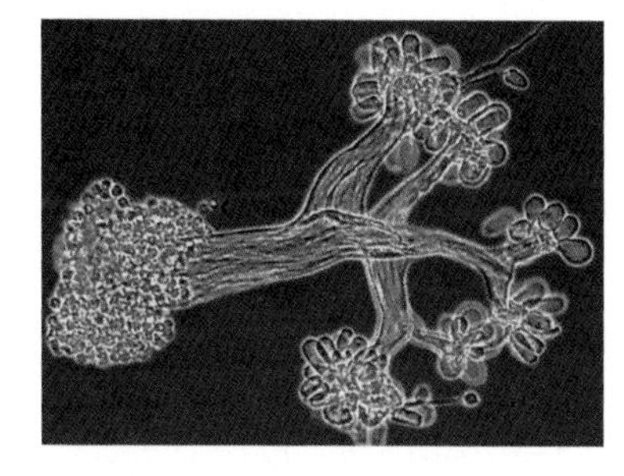

Figure 8
Millions of unicellular myxobacteria form a fruiting body. In this configuration, they can last through harsh conditions for long periods of time.

Reproduction and Growth of Eubacteria and Archaebacteria

Bacteria reproduce asexually by binary fission. Although it bears some resemblance to mitosis, binary fission is much simpler. The single strand of DNA replicates, resulting in identical genetic material being transferred to each new cell. Following replication of the genetic material, the bacterium produces a cross wall, dividing the cell into two identical bacteria, which may separate or remain attached (**Figure 9(a)**, on the next page).

CAREER CONNECTION

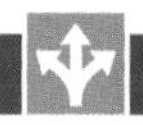

Bacteriology and microbiology technologists work in laboratories culturing bacteria to determine their identity. Many Canadian colleges and institutes of technology offer programs in these fields.

endospore a dormant cell of bacilli bacteria that contains genetic material encapsulated by a thick, resistant cell wall. This form of cell develops when environmental conditions become unfavourable.

Sexual reproduction is not common in bacteria. However, conjugation does occur among some bacteria, such as *E. coli* and *Salmonella.* In conjugation (**Figure 9(b)**), donor and recipient bacteria make cell-to-cell contact by means of a special structure called a sex pilus, where plasmids are transferred, giving the recipient an altered set of characteristics. Following the transfer, the two bacteria separate. Plasmids can also be transferred by other means.

During unfavourable environmental conditions, some bacteria survive by forming dormant or resting cells, called **endospores** (**Figure 10**). Endospores are resistant to heat and other extreme conditions, and cannot be easily destroyed. When suitable growing conditions return, the endospore sprouts or germinates, and an active bacterium emerges.

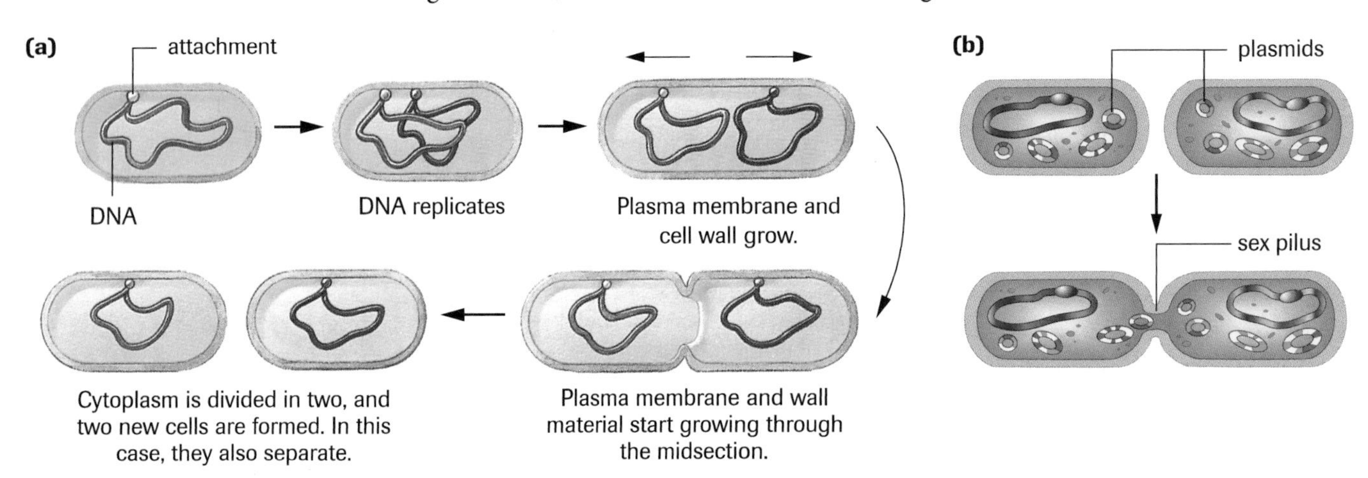

Figure 9
(a) Binary Fission. Features normally associated with mitotic cell division such as centrioles, spindle fibres, and visible chromosomes are not involved in this process.
(b) Conjugation. A sex pilus joins two bacteria cells, and plasmids are exchanged.

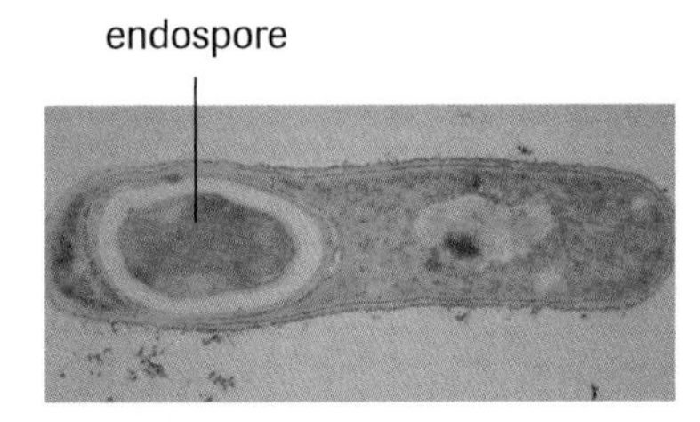

Figure 10
An electron micrograph showing an endospore within a bacterium. A thickened cell wall forms around the genetic material and cytoplasm. The remainder of the original cell eventually disintegrates.

Section 2.4 Questions

Understanding Concepts

1. Explain the statement "archaebacteria thrive on extremes."
2. List six ways in which archaebacteria contribute to human society.
3. Design a graphic organizer that could be used to classify eubacteria according to shape, respiration, and nutrition. Give examples of each.
4. Where would you expect to find anaerobic bacteria in nature?
5. Why is conjugation considered a form of sexual reproduction?
6. How has endospore formation guaranteed the survival of bacteria?
7. Identify the following bacteria types:

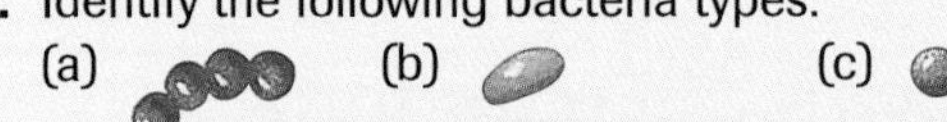

8. In a laboratory experiment, a student grew a colony of bacteria from a fecal sample. She suspected that the organisms were *E. coli.* Provide a description of the anatomy and physiology of *E. coli* to help with the identification.

Making Connections

9. One of the basic techniques in a college microbiology laboratory is Gram staining. Research this practice. Outline the steps in the procedure and list common examples of Gram-positive and -negative bacteria.

www.science.nelson.com

Culturing Bacteria

Growing or "culturing" bacteria in the laboratory is as easy as transferring organisms from tap water to a sterile agar dish. In this activity, you will swab a culture medium to observe and describe the characteristics of bacterial growth.

A bacterial growth curve has a particular pattern over time (**Figure 1**). During the lag phase, bacteria are adjusting to their new surroundings. During the exponential growth phase, the bacteria population grows so rapidly that growth is best expressed by charting it on a logarithmic scale. The stationary phase begins as bacteria run out of nutrients, or when they have produced wastes that limit their growth. A flat curve indicates that the death rate is equal to the birth rate. In the death phase, the number of bacteria dying becomes greater than the number being created. Growth rate is usually measured by the time it takes for a population of bacteria to double in number. This doubling time is a characteristic of each type of bacteria, and it can be used for identification purposes.

Like all organisms, bacteria thrive under optimum conditions such as constant temperature and a standard food source. In this activity, bacteria are grown in a nutrient-rich agar medium. The petri dish containing the medium has been sterilized to remove all contaminants. As you work, be careful to maintain these sterile (aseptic) conditions.

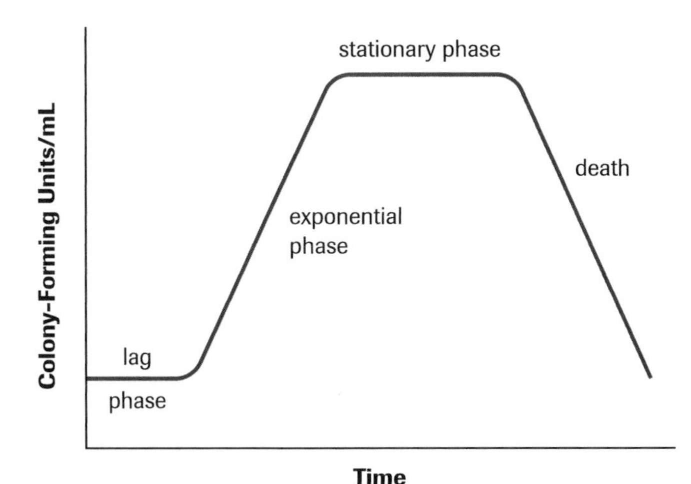

Figure 1
A typical bacterial growth curve

Question

How are bacteria cultured in the laboratory?

Materials

gloves
wax pencil
cotton swabs
tape
disinfectant
hand lens or dissecting microscope
prepared disposable petri dishes
three sources of bacteria: tap water, milk, sour milk, soil solution, aquarium water, bean seed water, OR purchased lab cultures

Some bacteria cause disease.

Wear gloves throughout this investigation.

To prevent sample contamination and student infection, do not remove the tape once the petri dish is taped shut.

Wash your hands thoroughly with soap after the experiment.

Clean your work area with a disinfectant solution.

Procedure

1. Obtain the prepared petri dish containing the nutrient agar for your group. Using a wax pencil, divide the bottom of the dish into quadrants and label them 1, 2, 3, 4 (**Figure 2**).

Figure 2
Marking the petri dish

2. Quadrant 1 is the control, and is not inoculated. Inoculate quadrants 2, 3, and 4 with swabs from the assigned sources. Dip the swab into a liquid, or rub the swab well against dry surfaces. Lightly sweep the swab over the agar to avoid tearing. Work quickly, taking the lid off the petri dish only when required. Place the used swab in the receptacle provided.
3. Tape the petri dish shut across the top and bottom in the form of an "X."

4. After taping your petri dish, label it with your name, the date of the experiment, and the sources of the cultures. Wash your hands and your work area.
5. Incubate each culture upside down at 30°C for 48 h. If an incubator is not available, a warm, dark cupboard can be used. Do not stack petri dishes more than three high on incubator shelves.
6. Remove the petri dish from the incubator.

(a) Sketch the appearance of your dish, showing the relative size and position of the colonies on each quadrant.

7. Sliding the stage clips of the microscope out of the way, place the sealed petri dish under the lens, agar side up. Observe under low-power magnification. Alternatively, a hand lens may be used.

(b) Use the terms in **Table 1** and the shapes in **Figure 3** to describe the colonies on each dish quadrant. Design a table to record these observations.

Table 1 Terms Applied to Bacterial Colonies

Category	Type
Form	punctiform (pinpoint), circular, filamentous, irregular, rhizoid, spindle
Margin (edges)	entire, undulate (wavelike), lobate (lobelike), curled, filamentous
Elevation	flat, raised, convex, craterform
Texture	smooth, rough, granular, shiny, dull, wet, mucoid, dry, wrinkled, powdery
Opacity	transparent, translucent, opaque
Colour	green, grey, orange, etc.

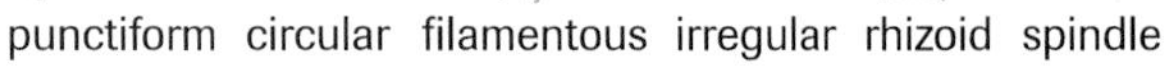

Figure 3
Examples of colony forms

8. Hand in your petri dish to your teacher for disposal. Wash your hands and clean your work area.

Analysis

(c) Answer the Question asked at the beginning of this activity.

(d) The agar plates were incubated at 30°C, not at our body temperature of 37°C. Give a possible reason for this temperature selection.

(e) How similar are the colonies that appear within each quadrant? What would this similarity indicate?

Evaluation

(f) If colonies appeared on your control quadrant, what conclusions could you make about your experiment?

(g) Rank all the bacterial sources used in this activity, from least growth to most growth.

Synthesis

(h) How can you prove that the colonies growing on your culture medium are bacteria, not organisms in other kingdoms?

(i) **Figure 3** translates colony forms into diagrams. Use this same technique to diagram margin types and elevation types.

(j) List the factors that limit the multiplication of bacteria. Consider both the information in **Figure 1** and the conditions in which the petri dishes were incubated.

(k) Under favourable conditions, bacteria may reproduce once every 20 min. At this rate, how many bacteria could be produced from a single bacterium in 6 h, assuming all survive? Show your calculations.

CAREER CONNECTION

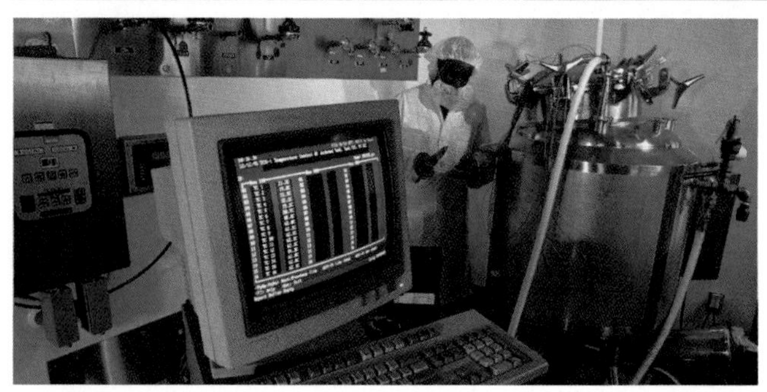

Biotechnology technicians culture bacteria using automated equipment. Research careers in which bacterial culturing is used.

www.science.nelson.com

2.6 UHT Milk

Except for those lucky few who live on dairy farms, the days of drinking milk straight from the cow are over! For food safety reasons, milk has evolved from a raw commodity to a highly processed foodstuff (**Figure 1**).

A few harmless bacteria are present in milk before it leaves the udder; more are introduced through the handling, packaging, and transportation of milk. There are two main groups of milk bacteria. Lactic acid bacteria are responsible for milk souring, which produces the many fermented milk products we enjoy, such as yogurt and cheese. Coliform bacteria are responsible for milk spoilage that can cause sickness. Heat or refrigeration is generally used to control bacterial action in milk.

type of milk	raw, whole	1908 pasteurized, whole	1920 homogenized milk	1960s standardization into various fat contents	1981 UHT milk
type of packaging	1611	1884	1948 1952		1964

Figure 1
Timeline of milk technology in North America

Pasteurization is the usual way that milk is treated. However, on the consumer market today is a premium-quality, Grade-A milk that has been specially processed so that it can be transported and stored without refrigeration. Ultra-high-temperature (UHT) milk has been rapidly heated to 140°C, held for at least 5 s, cooled to 21°C in a pressurized system, then packaged aseptically (free of disease-causing microorganisms). Bacteria are destroyed during this process; no additives or preservatives are added, and the nutritional value of the milk is unchanged. UHT milk has a shelf life of 9 to 12 months. Once opened, it must be refrigerated.

Advances in UHT technology involve electroheating—passing an electric current directly through the milk itself, thus reducing the heating time and eliminating the "cooked flavour" of current UHT products. Look for this new product in your supermarket and consider the advantages it brings.

▶ TRY THIS activity — Observing Milk Souring

Milk changes pH as it sours. In this activity, you will compare changes in the pH of homogenized milk and UHT milk over four days.

1. Measure an equal quantity of each type of milk into separate beakers and allow each to reach room temperature. Leave undisturbed for four days.

(a) Use a table to record daily pH readings and qualitative observations.

(b) Compare pH changes in the two experimental groups.

(c) Relate your qualitative observations to the changes in pH.

(d) Hypothesize on why pH changes as milk sours.

(e) Describe the results you might expect with unpasteurized milk.

▶ Tech Connect 2.6 Questions

Understanding Concepts

1. List the advantages of UHT milk.
2. Use resources to define the following terms as they relate to milk: *pasteurization*, *homogenization*, *sterilization*, and *standardization*.
3. What is the purpose of aseptic packaging in the UHT process?

Making Connections

4. Taking a global view (societal, economic, etc.), list the advantages and disadvantages of UHT technology.
5. What groups of people might buy this product, and why?
6. Using the Internet, research careers related to milk processing and packaging.

www.science.nelson.com

2.7 Helpful Bacteria

Contrary to the popular belief that most microorganisms are harmful, the usefulness of bacteria far outweighs the damage they do. Bacteria recycle dead materials into nutrients for plants and animals; certain bacteria can be used to clean up hazardous wastes; and some types of bacteria are important to human health. In industry, bacteria are used to produce everything from food products (beer, infant formula, soy sauce, cheese, yogurt, gum) to staples (laundry detergent) to clothing (leather goods, stone-washed jeans).

saprophyte an organism that obtains nutrients from dead or nonliving organic matter

bioremediation a process that uses biological mechanisms to destroy, transform, or immobilize environmental contaminants

Bacteria as Recyclers

Many bacteria are saprophytes. **Saprophytes** release digestive enzymes into the organic material around them (e.g., cellulose from trees) and break down the material into nutrient molecules that the bacteria or other organisms can absorb. This process causes the organic material to decay. Decay is a vital link in all nutrient cycles. Without it, the matter from once-living organisms would accumulate and stop the biosphere from functioning normally. Food chains would be broken—producers would not have nutrients to grow; and without producers, consumers would have nothing to eat.

Figure 1
Crews remove an underground tank to clean the contaminated soil left behind by leaking gasoline. Some soil microorganisms break down petroleum products.

Bacteria in Pollution Control

In 1989, the Exxon *Valdez* oil spill brought the realization that humans alone cannot effectively remove poisons from the environment. In one of the strategies used to clean up the oil-soaked coastline, fertilizer was sprayed on the sand to encourage the growth of naturally occurring bacteria that might then consume polluting hydrocarbons. Although the attempt had mixed success, it showed the potential of bacteria to treat polluted sites—a process called **bioremediation.**

A toxic wood preservative called penta-chlorophenol has been known to seep from storage containers, contaminating nearby soil and underground water. The cost of cleaning the soil by the traditional means of excavating and burning contaminants ranges from $200 to $300 per cubic metre. In some cases, hundreds of tonnes of soil must be excavated (**Figure 1**). A bacterium from the genus *Flavobacterium* is a dramatically more economical and efficient alternative to these traditional methods of detoxification. The toxic compound is actually a nutrient for the bacterium. The bacteria require only oxygen and nutrients found in the soil. The treatment is also inexpensive: $30 to $50 per cubic metre of soil. Besides its low cost, another advantage is that the bacterial population grows as long as the toxic molecules remain; once the chemical has been broken down, they die.

DID YOU KNOW?

Bacteria and Headaches
There appears to be a link between headaches and the gut bacterium *Helicobacter pylori*. In tests done with migraine sufferers, treatment with *Lactobacillus* bacteria worked better than the use of antibiotics.

Bacteria as Probiotic Agents

Researchers are discovering that many lesser-known bacteria, grouped as probiotics (literally, "for life"), help maintain health and may have the potential to prevent disease. Although they consist mainly of lactic acid bacteria, bifidobacteria (Y-shaped, found in the human intestine) and yeasts have also been used successfully. In the future, look for these organisms to be included in foods or food additives, or used in the treatment of ear, intestinal, and urinary-tract infections; the reduction of blood cholesterol levels; and the prevention of infections, tooth decay, and perhaps some cancers.

Figure 2
Industrial fermenters are used to produce vinegar from *Acetobacter*.

Bacteria in Industry

During metabolism, bacteria make enzymes that are useful in many industrial processes. Industrial fermenters (**Figure 2**) are used to harvest enzymes, which, in turn, are used to produce consumer products. Examples are summarized in **Table 1**. Even though enzymes can be used, they are costly. Usually, intact bacteria are used because they are less expensive.

Table 1 Examples of Beneficial Effects of Bacteria

Type of bacteria	Beneficial effect
Clostridium	production of butanol and acetone from molasses
Acetobacter	production of vinegar from ethanol
intestinal bacteria	food digestion; synthesis of vitamins in humans (e.g., to regulate blood clotting)
Lactobacillus	production of lactic acid from sugar; firewall function; preventing growth of harmful bacteria production of dairy products (e.g., cheese, yogurt)
Azotobacter, Nitrobacter	nitrogen fixing in soils
Streptococcus	production of dairy products (e.g., cheese, yogurt)
Streptomyces	soil bacteria; source of antibiotics (e.g., streptomycin, terramycin, neomycin, erythromycin)
Pseudomonas	cleanup of wastes in sewage treatment
Bacillus	soil bacteria; natural pest-killer

Section 2.7 Questions

Understanding Concepts

1. *E. coli* bacteria live in human intestines and play a role in digesting food. In which of the helpful bacteria categories would *E. coli* fit, and why?
2. Construct a concept map or spider diagram to summarize key information in this section.

Applying Inquiry Skills

3. A student constructs a composter to decompose garden wastes, and collects the following data.

	Day			
	1	2	3	4
Temperature of compost (°C)	24	25	29	24

Based on these data, on which day were the bacteria most active? Explain why.

Making Connections

4. How are the industrial uses of microorganisms similar to the actions of microorganisms in nature?
5. Evaluate some risks and benefits of the industrial use of microorganisms.

Exploring

6. Research how an industry uses bacteria to make a consumer product. Work with a partner or in small groups to prepare a product summary report. Use these questions to focus your research on your chosen product:
   (a) Describe the consumer product. Include photographs, if possible.
   (b) Describe the bacterium involved in its production: its genus/species name, its shape and mobility, and other relevant characteristics. Use diagrams where possible.
   (c) Using a graphic organizer, outline the steps in producing this consumer product.
   (d) Explain how variations in the process will create variations in the product (e.g., not all cheese is the same).

www.science.nelson.com

2.8 Case Study: Harmful Bacteria

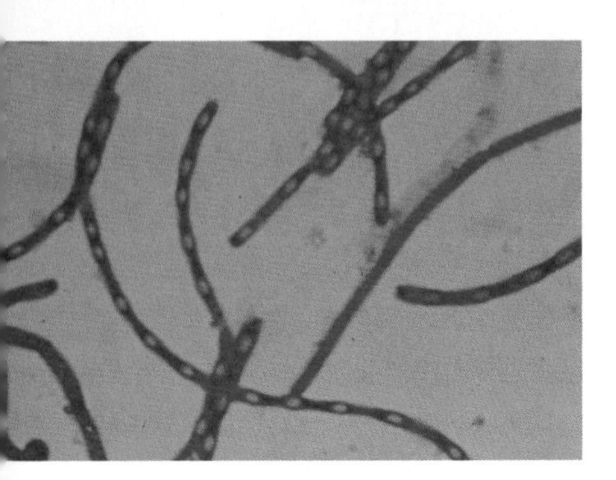

Figure 1
Bacillus anthracis results in the deadly disease anthrax, which affects cattle, sheep, and people. More recently, it has gained the public's attention as a terrorist biological agent after the September 11, 2001, tragedy.

Bacteria are probably best known for causing disease. *Bacillus anthracis* was the first bacterium proven to cause a disease (**Figure 1**). The 1976 Legionnaire's disease outbreak in Philadelphia took scientists many months to identify and was caused by a bacterium named *Legionella pneumophilia.* Some of the best-known bacterial plagues over the years have been diphtheria, typhoid fever, and bubonic plague (Black Death). Tuberculosis still affects many people in Canada each year, particularly in northern communities. It has also become a problem among the homeless in Toronto, Montreal, and Vancouver, as well as among people living with AIDS.

The Walkerton Tragedy

On May 12, 2000, heavy rains washed fecal bacteria from cattle manure into well water near the town of Walkerton, Ontario (**Figure 2**). Within 10 days, hundreds of residents had symptoms of *E. coli* poisoning: vomiting, cramps, bloody diarrhea, and fever. By May 22, the first death directly linked to *E. coli* was reported. Health-unit tests revealed that the drinking water was contaminated with the deadly strain *E. coli* O157:H7.

There are hundreds of strains of *E. coli* bacteria; the combination of numbers and letters in the name of the bacterium refers to the specific markers found on the bacterium's surfac Whereas most live harmlessly in the intestines of healthy animals, *E. coli* O157:H7 produces a dangerous poison or **toxin**. If the toxin is present in small amounts, the immune system of a healthy adult can fight the illness in 5 to 10 days without special treatment or antibiotics. The toxin may have a more dramatic effect, however, in young children and the elderly. It may destroy red blood cells and cause kidney failure, seizures, or strokes. About 2% to 7% of *E. coli* O157:H7 infections lead to these complications.

toxin a poison produced in the body of a living organism. It is not harmful to the organism itself but only to other organisms.

The major source of *E. coli* O157:H7 infections is undercooked ground beef, since fecal matter can contaminate beef as the meat is processed. *E. coli* O157:H7 is found in the intestines of cattle; when the cattle are killed, the bacteria can become mixed into beef when it is ground. Other sources of *E. coli* include processed meats, sprouts and leafy green produce, unpasteurized milk and juice, and contact with cattle. Drinking inadequately chlorinated water or swimming in contaminated lakes or pools may also expose people to this deadly strain. The organism is easily transmitted from person to person. Day-care or long-term-care centres, cattle barns, and petting zoos are prime areas of concern.

The Walkerton tragedy has prompted new government legislation on the testing and treatment of municipal drinking water. Proper chlorination of water would have prevented the death of seven people and the illness of more than 2300 in Walkerton. In addition to chlorination, ultraviolet light can be used for drinking-water purification: large-scale implementation of this technology will come with more research. New biological sensors and imaging techniques are also being developed to identify *E. coli* in minutes rather than days. It is hoped that these new technologies will prevent another tragedy like Walkerton.

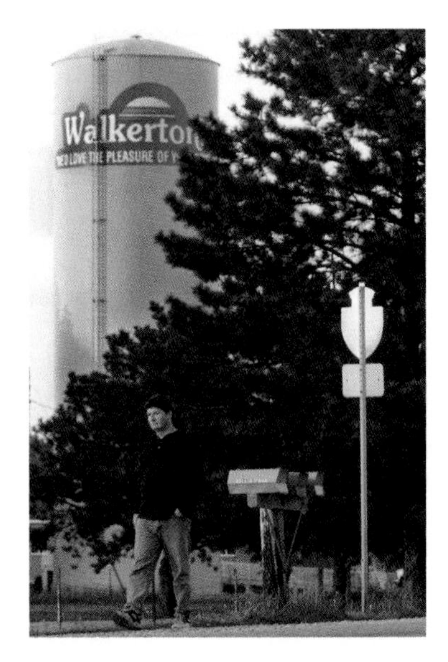

Figure 2
The people of Walkerton, Ontario, suffered from serious contamination of their drinking water by a deadly strain of *E. coli* bacteria.

Bacteria cause illness in a variety of ways. In the Walkerton example, toxins overloaded people's immune systems, reducing their ability to remove poisons from the body. In some cases, the sheer number of bacteria places such a tremendous burden on the host's tissues that normal function breaks down. In other cases, bacteria actually destroy cells and tissues. **Table 1** provides some examples of other diseases and/or destruction caused by bacteria.

Table 1 Examples of Harmful Effects of Bacteria

Type of bacteria	Disease/destruction produced
Clostridium	botulism, tetanus, gangrene
Streptococcus	strep throat, scarlet fever, pneumonia
Staphylococcus	boils, food poisoning, skin infections, pneumonia
Lactobacillus	souring of milk
Pseudomonas	gasoline spoilage, food spoilage, blood infections, eye infections
Bacillus	destruction of silkworms, tuberculosis, anthrax
Staphylococcus	food spoilage, blood infections, eye infections
coliform bacteria	pollution of water sources, soft rot in plants, gastroenteritis, dysentery
Spirillum	syphilis

DID YOU KNOW ?

Bacteria and Ulcers

Most stomach ulcers are caused not by spicy foods, acid, or stress, but by *Helicobacter pylori*. This bacterium is capable of living in the stomach because it secretes an enzyme that neutralizes stomach acids. *H. pylori* then weakens the protective lining of the stomach and intestine, allowing acids to irritate both areas. This causes stomach pain associated with ulcers. Antibiotics, such as amoxicillan and tetracycline can be used to treat ulcers caused by *H. pylori*.

Case Study 2.8 Questions

Understanding Concepts

1. Describe the symptoms of *E. coli* O157:H7 poisoning.
2. List the sources of *E. coli* O157:H7 contamination.
3. Create a graphic organizer for **Table 1**. Draw the three basic bacterial shapes in the centre of a piece of paper. Working out from the middle, name each pathogen and the diseases it produces. (Additional information: clostridium, coliform, and pseudomonads are rod-shaped bacteria.)

Applying Inquiry Skills

4. Pathogenic cases of *E. coli* have been tracked in Canada since 1985 (**Figure 3**). Analyze the trends and hypothesize about the causes.

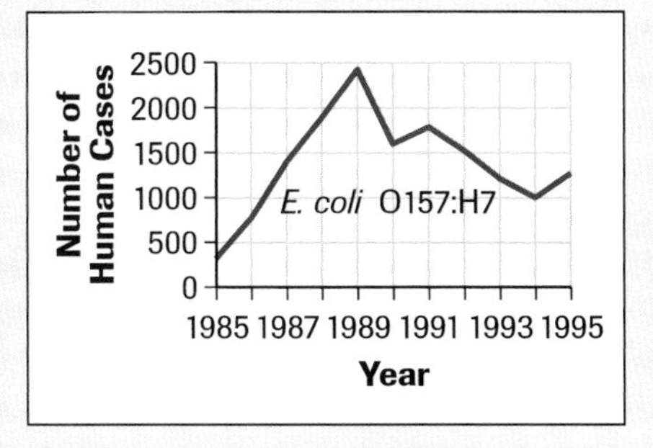

Figure 3 Incidence of *E. coli* O157:H7 in Canada per 100 000 population

Making Connections

5. Explain how *E. coli* contamination can be prevented by the government, and by the population.
6. Could you be a wastewater treatment technician? Describe this career: duties, education, job opportunities, personal strengths and skills required.

 www.science.nelson.com

7. Prepare a fact sheet on any one of the diseases mentioned in this section. Conduct research to answer the following questions:
   (a) Description of disease
   (b) Diagnosis: What parts of the body are affected?
   (c) Organism causing disease: Give the genus and species name, and describe its characteristics.
   (d) Symptoms: List the common initial signs of the disease and the new symptoms that develop as the disease progresses.
   (e) Type of transmission: How is the disease passed from person to person?
   (f) Treatment: Consider all aspects, including those not based on Western medical tradition. Consider their effectiveness, cost, and why they may work.
   (g) Prognosis: What is the cure rate? Can the disease return to the same person?

 www.science.nelson.com

8. If you obtain water from your own well, how can you get it tested, and what tests are performed?

 www.science.nelson.com

2.9 Fighting Disease

pathogen a disease-causing microorganism

DID YOU KNOW ?

Cellular Defences

One out of every 100 cells in your body exists to defend the other 99 cells from foreign materials such as dust and pollen, or from microscopic invaders such as fungi, viruses, and bacteria.

Many species of bacteria are parasitic and cause disease in the host organism. Inside the host, the growth of a **pathogen**—a disease-causing microorganism—causes symptoms that result from tissue destruction and from the body's response to the invasion. All pathogenic bacteria act in much the same way. After gaining entrance to the body, the bacteria begin to produce toxins. As seen in the Walkerton example, these toxins are harmful to the human hosts and are able to travel throughout the body via the circulatory system, causing chills, fever, and other related symptoms.

Sometimes these harmful effects may be observed in tissues distant from the site of infection. If you step on a rusty nail contaminated with *Clostridium tetani* endospores, the dormant endospores that enter deep into the tissue of your foot may find favourable conditions for resuming an active bacterial form. As the bacteria multiply, they produce a neurotoxin that causes body spasms and locking, or tetanus, of the muscles, especially those in the jaw. For this reason, tetanus is often called "lockjaw." The spasms can spread to the breathing muscles, causing convulsions or even asphyxiation and death.

The species that causes diphtheria, *Corynebacterium diphtheriae*, produces a very different toxin, one that prevents cell organelles from manufacturing proteins. The host's cells are disabled because most cell functions involve the production of proteins. Diphtheria is initially contracted in the throat, tonsils, nose, and/or skin, but disease effects are felt far beyond these sites.

The Immune-System Response

First Line of Defence: External

The human body has many defences against bacterial infection. The first line of defence is an external, physical barrier (**Figure 1**). Unless broken, human skin forms an almost impenetrable barrier against bacteria. Although some species may enter hair follicles or sweat glands through skin pores, sweat contains salts and amino acids that are poisonous to most bacteria found on skin.

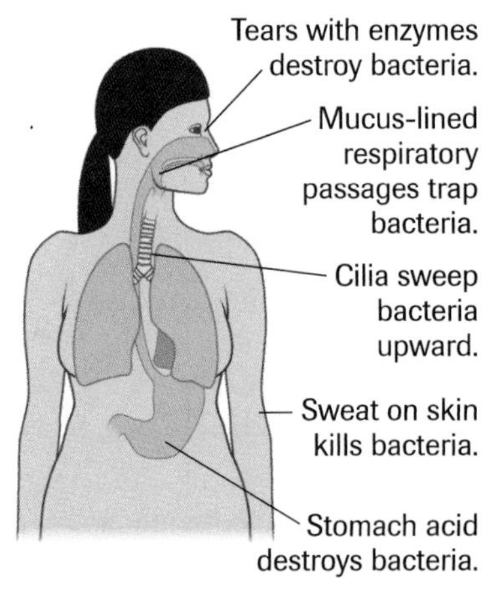

Figure 1
Features of the body fight bacteria before they can enter the internal environment.

Many bacteria enter the body through the nasal passages or the pharynx, both of which are equipped with defensive features. Specialized cells in the lining of these passages secrete mucus, a sticky substance that traps the inhaled bacteria. Mucus also contains digestive enzymes that destroy the cell walls of many bacteria. The lining of the nasal passages and the pharynx also contains specialized cells equipped with cilia. Trapped bacteria and debris are swept upward to the throat by rhythmic beating of the cilia, where they are either swallowed or expelled by coughing.

Bacteria that enter the eyes are attacked by enzymes (e.g., lysozymes, which digest peptidoglycan in bacterial cell walls) contained in tears, which continually bathe eye surfaces, cleaning them of bacteria and dust. Usually, tears drain into the upper nasal passages, where any surviving bacteria are attacked by enzymes in mucus. Bacteria also enter the body through the mouth in food and water, but acids in the stomach kill most of them.

Second Line of Defence: The Lymphatic System

Although the first line of defence keeps many invaders from entering the body, some bacteria penetrate these external barriers. Inside the body, the second line of defence, a complex network of organs (**Figure 2**), works to rid the body of infection. These widely scattered lymphatic organs are connected by a special circulatory system of vessels and nodes (**Figure 3**). The lymphatic vessels circulate lymph, a transparent fluid containing white blood cells, called **lymphocytes**, which are produced in the lymph nodes. Lymph nodes are small round masses of cells throughout the human body that fight infection. As lymph flows through the lymph nodes and the spleen, microorganisms are filtered out, preventing them from entering the bloodstream and causing infection. After travelling throughout the body, lymph is collected and moved from the lymphatic system into blood circulation. Once in the bloodstream, lymphocytes are transported to body tissues where they act as guards against infection.

The core of bones like the pelvis contains bone marrow. This nutrient-filled, spongy tissue produces specialized white blood cells that circulate through the immune system. When a pathogen invades, internal body defences are activated, beginning with white blood cells called **macrophages**. These cells are motile and are able to travel across body membranes to find infection sites. Macrophages engulf foreign bodies such as bacteria and destroy them with digestive enzymes. The white blood cells themselves may also be killed by these enzymes. The mass of dead white blood cells and dead bacteria produced at the site of an infection is called pus. In spite of this second line of defence, some bacteria manage to avoid being digested.

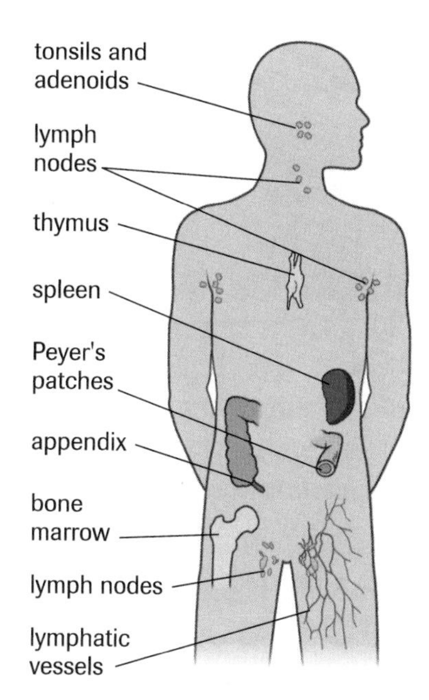

Figure 2
Lymphatic organs and tissues are internal features of the immune system.

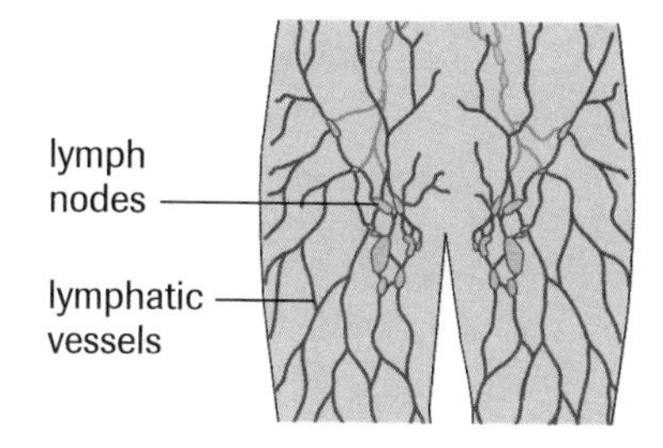

Figure 3
Vessels in the lymphatic system work closely with the circulatory system.

Third Line of Defence: Antibody Formation

Antibodies (**Figure 4**, on the next page) are protein molecules that protect the body from invaders. A foreign material that causes formation of antibodies is called an **antigen**. The surfaces of invaders contain many different antigens, but each antigen stimulates its own specific antibody, meaning that only one specific antibody attaches to an antigen. Antibodies render invaders ineffective, and the antibody-antigen complex provides a signal to macrophages so that these cells can more easily identify and engulf the antigen (e.g., bacterium). It has been estimated that an average human body may contain more than ten million different antibody types, each ready to help repel future antigen invasions.

Active and Passive Immunity to Bacterial Diseases

The production of antibodies to counteract a pathogen is called an immune response. In **active immunity**, the body itself manufactures antibodies to combat a specific disease. This active immunity may be acquired naturally when a person suffers from a bacterial disease and recovers because the body has produced antibodies that identify and help destroy the invader. The same pathogen will trigger an immune-system response in the future, so the immunity is usually lifelong.

lymphocyte a white blood cell (plasma and memory cells) that can make antibodies when stimulated by antigens

macrophage a white blood cell that engulfs and destroys pathogens

antibody a protein that inactivates a foreign substance in the body by binding to its surface

antigen any substance that causes the formation of antibodies

active immunity lasting protection against an invading antigen through the manufacture of antibodies

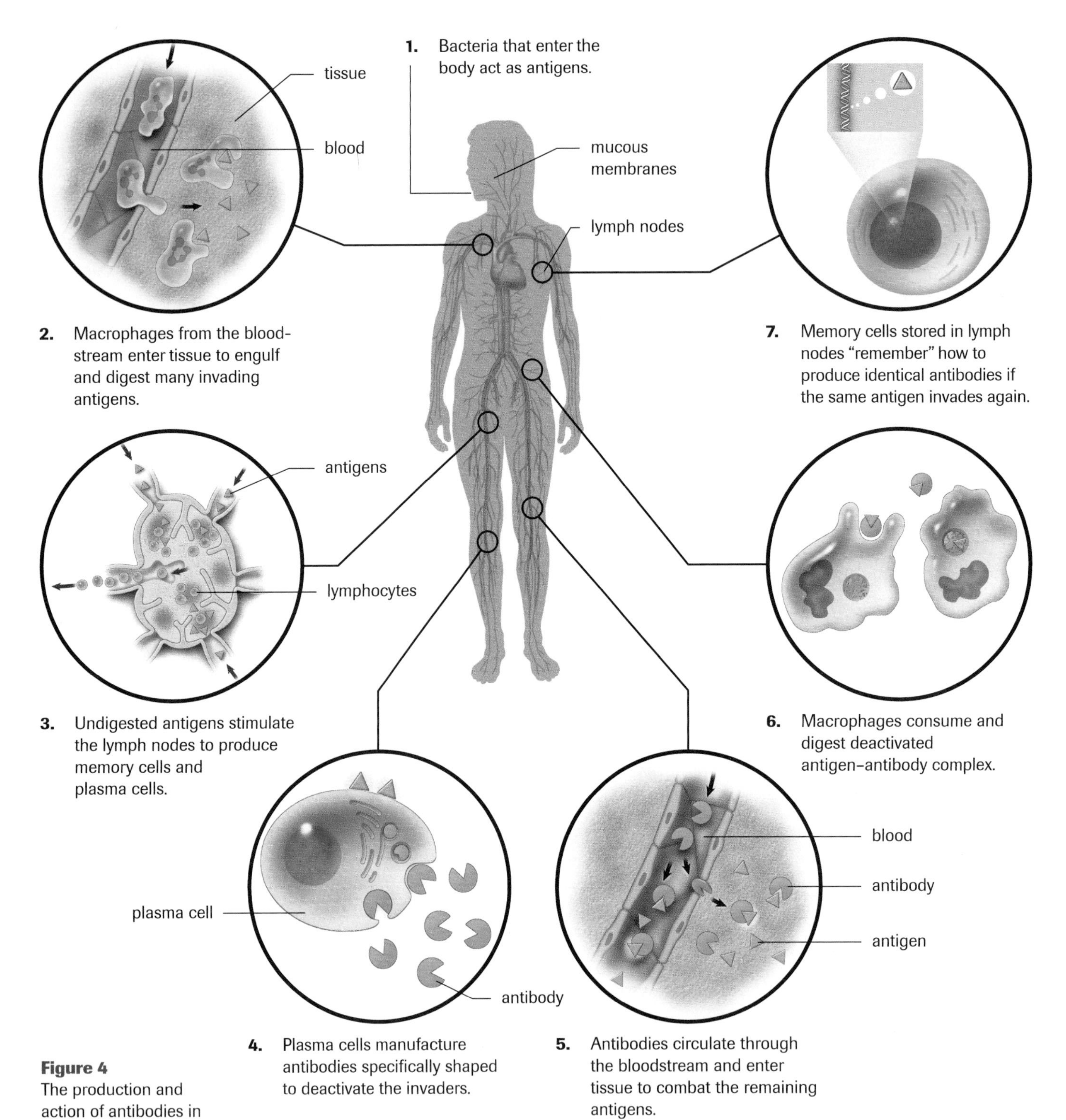

Figure 4
The production and action of antibodies in the circulatory and lymphatic systems

Active immunity may also be induced artificially by the injection of vaccines made of dead or weakened bacterial cells or certain antigens that have been isolated from the bacteria. The body reacts to the vaccine as if it were a real pathogen, and produces antibodies (**Figure 5**). Some vaccines provide lifetime immunity against the disease (e.g., diphtheria), while others must be renewed (e.g., influenza: yearly; smallpox: every three to five years).

In **passive immunity**, the antibodies themselves are introduced directly into the bloodstream of an individual. Passive immunity can be acquired naturally when antibodies pass from a woman to her unborn child during the last month of pregnancy and the initial stages of breast-feeding. The child will then be immune to the same bacteria as its mother for a short period of time. This protection lasts for only a few months, so babies begin receiving immunization shots (vaccinations) in the first year. Passive immunity can also be acquired, on a short-term basis, by the injection of plasma (blood from which the red blood cells have been removed) from an animal that has been exposed to a disease into the blood of another animal. A tetanus shot contains antibodies from a horse against the tetanus toxin. The vaccination protects humans from *C. tetani* for ten years. Booster or refresher shots are needed at this interval to provide continuing immunity.

passive immunity temporary protection against a particular disease by the direct introduction of antibodies

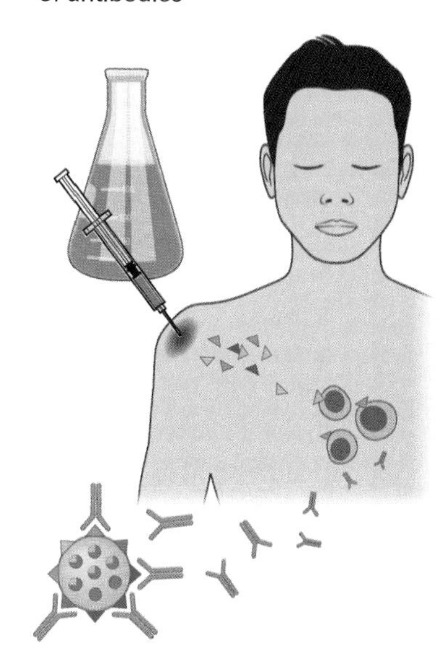

Figure 5
Vaccines stimulate antibody production

Resisting Bacterial Infection

Infectious diseases spread from one individual to another in a variety of ways: moisture droplets in the air, dust, direct contact, fecal contamination, animal bites, and wounds (cuts and scratches). Fortunately for most people, protection from disease is provided by the body's own defence mechanisms. Other methods that provide protection include

- sterilization of instruments (operating rooms and tattoo parlours);
- use of **disinfectants**—chemicals that can kill microorganisms on surfaces (hospitals, restaurant kitchens);
- use of **antiseptics**—chemicals that can be used on the skin (mouthwash, antibacterial soap) to slow the growth of microorganisms;
- extermination of animals that carry disease (foot-and-mouth disease);
- use of antibacterial medication (topical creams for burns and cuts).

disinfectant a strong chemical used to destroy or impede the growth of disease-causing organisms

antiseptic a chemical that destroys or impedes the growth of disease-causing organisms without harming body cells

Section 2.9 Questions

Understanding Concepts

1. Make a list of all body defences that fight bacteria.
2. How are the lymphatic and circulatory systems connected?
3. Describe the location and function of the following immune-system cells:
   (a) macrophage
   (b) lymph
   (c) lymphocyte
   (d) antibody
4. List the main modes of disease transmission.
5. You have fallen off your skateboard and cut your knee. The wound has been cleaned, and, underneath the bandage and inside your cells, healing has begun. Draw a diagram with captions showing how your body is fighting the bacteria that entered your cut.
6. Design a spider diagram to illustrate the types of active and passive immunity.
7. What is the difference between antiseptics and disinfectants?
8. The two photographs in **Figure 6** show a white blood cell and *E. coli* bacteria. Describe what is happening. How does this relate to the immune system?

white blood cell

E. coli bacteria

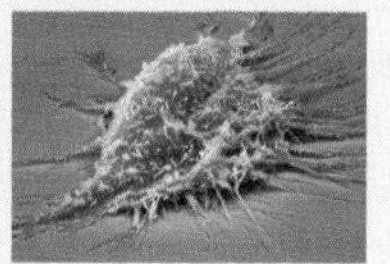

Figure 6

Making Connections

9. Autoimmune diseases occur when the immune system mistakenly attacks the body. Multiple sclerosis and rheumatoid arthritis are examples of autoimmune diseases. Research either example.

www.science.nelson.com

2.10 Future Vaccines?

Tech Connect

Need a vaccination—eat a banana! This sounds far-fetched, but scientists have developed a banana that has been genetically altered to provide a hepatitis vaccine. The hepatitis virus causes liver disease and even death. Similarly, a potato can be used as a vaccine-delivery system by adding genes for the cholera toxin. The potato then carries a cholera antigen that immunizes against cholera. Currently, most vaccines are delivered orally or by injection. Soon people will be able to eat vaccines, inhale them through their noses, or put them on their skin.

There are other areas of research: a vaccine for pregnant women that would protect their babies against group B *Streptococcus* bacteria, which can cause severe illness in infants; and a DNA vaccine that injects genes coding for viral proteins of HIV directly into the body, stimulating the immune system to produce antibodies. There are several classes of vaccines available today (**Figure 1**):

- attenuated whole vaccines (living virus particles or bacteria that have been weakened by mutation or chemicals so that they no longer cause disease);
- killed whole vaccines;
- toxoid vaccines (modified toxins that no longer cause disease);
- subunit vaccines (vaccines that contain only a portion of the pathogen, often only the antigenic protein);
- similar or related pathogen.

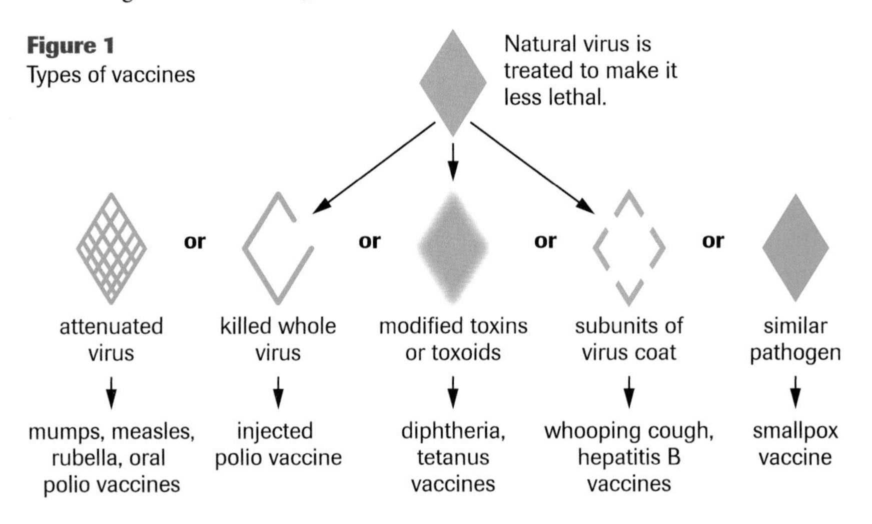

Figure 1
Types of vaccines

Tech Connect 2.10 Questions

Understanding Concepts

1. Describe three ways that vaccines are administered.
2. Review the immune-system information in section 2.9 and draw a diagram showing the response of body cells from the time of vaccination to the production of antibodies.
3. Describe the type of vaccine used to protect against each of the following diseases: smallpox, polio, measles, hepatitis B, and tetanus.

Applying Inquiry Skills

4. Interpret the graph in **Figure 2** that shows an immune-system response to a vaccination and follow-up booster shot.

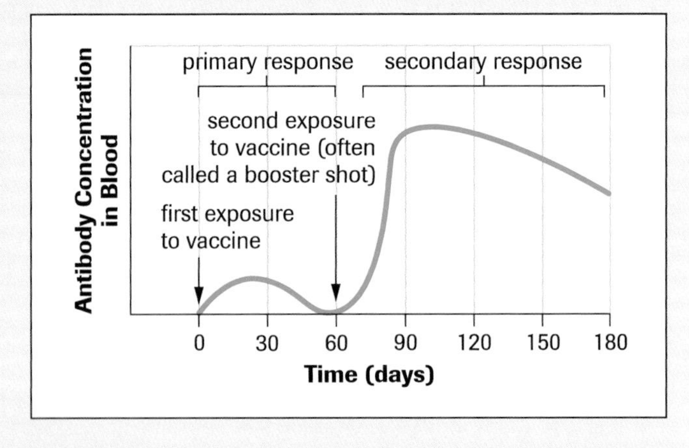

Figure 2
Immune-system response to vaccination

Making Connections

5. Choose a viral disease from the World Health Organization web site. Prepare an information pamphlet on the disease and its control. Describe the vaccine: type, availability, and schedule of doses.

www.science.nelson.com

6. Epidemiological studies track the effectiveness of vaccines. Research the duties of an epidemiological technician and the training requirements for this career.

www.science.nelson.com

2.11 Investigation

Observing the Effects of Antiseptics

Antiseptics act in many ways, all of which inhibit or prevent growth. Most wound-cleansing agents contain an antiseptic to prevent the infection of wounds, scrapes, or burns. Soap, alcohol, and mouthwash are used as antiseptics.

In this investigation, you will determine the effect of antiseptics on a bacterial culture. Your teacher will provide you with a selection of antiseptics (**Figure 1**) and a petri dish containing agar inoculated with a bacterial culture.

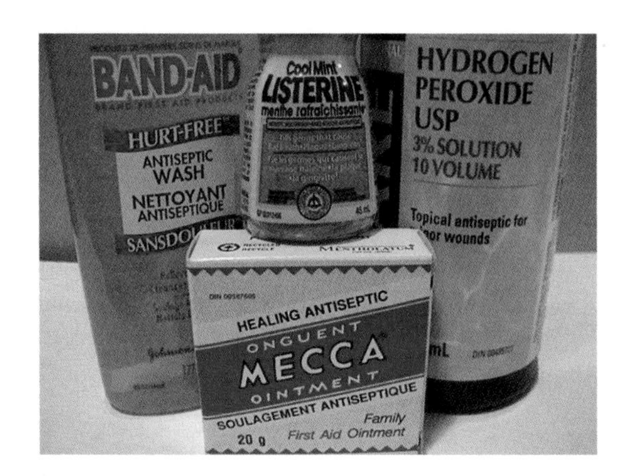

Figure 1

Question

How do different antiseptics affect bacterial growth?

Prediction

(a) Make a prediction about the effect of antiseptics on bacterial growth.

Experimental Design

(b) Design a controlled experiment to test your prediction. Include the following in your design:

- descriptions of the independent, dependent, and controlled variables
- a complete list of materials
- a step-by-step description of the procedure
- a list of safety precautions
- a table to record your observations

Inquiry Skills

○ Questioning	● Planning	● Analyzing
○ Hypothesizing	● Conducting	● Evaluating
● Predicting	● Recording	● Communicating

Procedure

1. Submit the procedure, safety precautions, and observation table to your teacher for approval.
2. Carry out the procedure.
3. Write a report (see Appendix A4 for format). Include in your report ways to improve your design.

Analysis

(c) Answer the Question asked at the beginning of this investigation.

(d) Describe how different antiseptics affected bacterial growth. Which were the most effective? the least effective? Explain.

Evaluation

(e) Was your prediction correct? Explain.

(f) Describe any problems or difficulties in carrying out the procedure.

(g) If you were to repeat this experiment, what new factors would you investigate? Write a brief description of the new procedure.

Synthesis

(h) Predict what will happen if the experiment is repeated using a disinfectant.

(i) Were any of the statements or claims made on the product labels misleading? Explain your answer.

CAREER CONNECTION 

Environmental health and safety technicians perform and/or supervise technical duties in the fields of industrial hygiene, safety engineering, environmental sanitation, fire-protection engineering, and/or radiological health. Research training programs in this field.

www.science.nelson.com

2.12 Explore an Issue
Bacterial Resistance to Antibiotics

CAREER CONNECTION

Quality-control technicians monitor the concentrations of antibiotics in both medications (capsules) and consumer products such as cheese. Certification can be obtained through diploma programs in industrial or manufacturing technology.

Many chemicals, such as cyanide, kill bacteria but also kill human cells. Antibiotics are natural substances that immobilize or destroy microorganisms by attacking metabolic pathways in the bacteria, but not in the host. Fungi or bacteria are used to produce antibiotics. For example, penicillin, an antibiotic derived from the mould *Pencillium notatum*, works by preventing bacteria from constructing a normal cell wall (**Figure 1**). Another antibiotic, streptomycin, works by preventing bacteria from manufacturing proteins, such as the enzymes needed to digest food. Remember that antibiotics are effective only against bacterial infections; they do not destroy viruses.

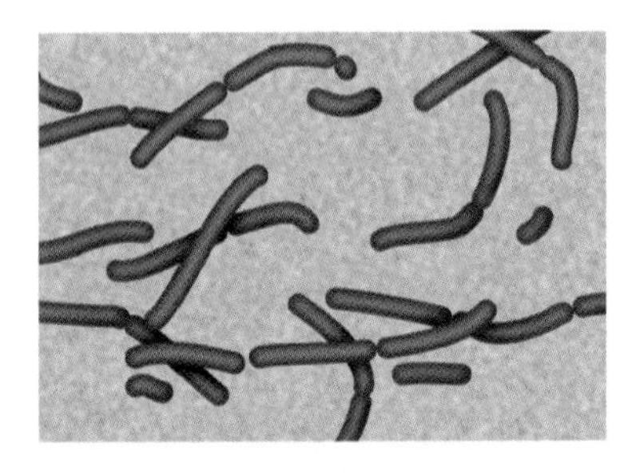

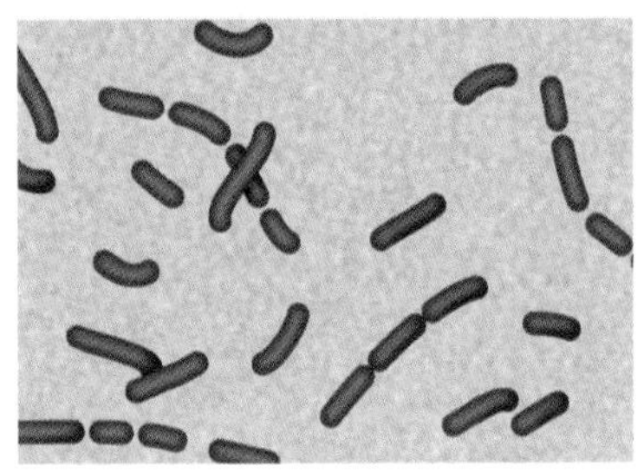

Figure 1
Bacillus cereus, grown with (top) and without (bottom) an antibiotic. Penicillin affects the cross linking within the cell wall, inhibiting growth and division.

Antibiotic Resistance

Since the beginning of the "age of antibiotics" in the mid-1940s, researchers have identified more than 2500 naturally occurring antibiotics that can be used to treat infections caused by some of the most dangerous strains of bacteria. However, over the past 50 years, many disease-causing bacteria have slowly developed resistance to antibiotics.

Antibiotic resistance appears to develop from variations within a bacterial population. When the bacteria are first exposed to an antibiotic, the weaker strains of the bacteria are killed. Other members may have slight variations in their genetic material that allow them to survive the antibiotic. These individuals then reproduce and pass on their resistance.

The number of human pathogens now resistant to antibiotics—even the super-antibiotic called vancomycin—is alarming. Drug-resistant strains are claiming lives in hospitals where antibiotic use is high and immune systems are weakened. Some strains of the agent that causes tuberculosis now resist up to seven separate antibiotics.

Viral infections should be treated with fluids and bed rest instead of antibiotics. With bacterial diseases, follow the directions of treatment: use all doses prescribed, and don't save unused antibiotics for later infections. Keep in mind that antibiotics destroy beneficial bacteria, such as those needed for digestion in the small intestine, as well as harmful bacteria, and that long-term use can suppress the body's natural defence mechanism against infection.

A recent report estimates that more than 70% of all antibiotics manufactured are used in livestock production in one of three ways: therapeutic doses treat specific diseases; prophylactic or preventive doses reduce the risk of disease or death during production; low doses added to feed rations promote growth. Animals receiving daily low doses of antibiotics such as tetracycline, penicillin, and ciprofloxacin gain as much as 3% more weight than they otherwise would. The reason is unclear, but there is evidence that the antibiotics kill the bacteria that normally thrive in the animals' intestines, thereby allowing the animals to use their food more effectively. Poultry growers routinely add the antibiotic fluoroquinolone to drinking water to prevent chickens and turkeys from dying from *E. coli* infection, a disease that is picked up from their own droppings.

Understanding the Issue

1. What is an antibiotic? Define this term and list three examples.
2. How does antibiotic resistance appear to develop within a bacterial population?
3. Divide the following list of diseases into two columns under the appropriate headings below: strep throat, common cold, rabies, smallpox, anthrax, mumps, tuberculosis, AIDS, syphilis, botulism, measles, and influenza.

Effectively treated with antibiotics	Unaffected by antibiotic treatment

4. How could you explain the increase of infections in hospitals by antibiotic-resistant bacteria?
5. What role do agricultural practices play in the development of antibiotic resistance?
6. On a blank sheet of paper, design a poster that could be displayed in a doctor's waiting room to warn of the dangers of antibiotic resistance and to advise on the correct use of antibiotics, based on the information in this section.
7. A mother had some antibiotics left over from fighting a throat infection. She administered them to her young son when he developed a cold. Explain what was wrong with this practice.

Take a Stand

Should non-therapeutic use of antibiotics in farm animals be banned?

Antibiotic digestive enhancers or growth promoters have been approved and used in agriculture for over 30 years. These products increase productivity by controlling intestinal bacteria. Veterinary drugs are an important part of food animal production and contribute to the high level of health in food animals today. They also provide other benefits related to animal welfare, the economy, environment, and society. Discontinuing the use of antibiotic drugs in hog production (**Figure 2**) would initially decrease feed efficiency, reduce production, and raise consumer prices.

There is increasing evidence that the use of antibiotics in livestock production may contribute to antibiotic resistance. Resistant strains of food-borne bacteria such as *Salmonella*, *E. coli*, and *Campylobacter* can be passed from animals to humans; people infected with these organisms may develop illnesses that do not respond to conventional antibiotic therapy. As well, animal bacteria that do not cause disease in humans can transfer drug resistance to human bacteria. The resulting human illnesses do not respond to conventional antibiotic therapy.

Statement: Non-therapeutic use of antibiotics in farm animals should be banned.

- In your group, research the issue. Learn more about the misuse of antibiotics, new strains of drug-resistant bacteria, and the resurgence of diseases such as cholera and tuberculosis.

Decision-Making Skills

○ Define the Issue ● Analyze the Issue ● Research
● Defend a Decision ● Identify Alternatives ● Evaluate

- Search for information in newspapers, periodicals, and CD-ROMs, and on the Internet.

GO www.science.nelson.com

(a) Write a list of points and counterpoints that your group considered.
(b) Decide whether your group agrees or disagrees with the statement.
(c) Prepare to defend your group's position in a class discussion.
(d) How did your group reach a decision? What would you do differently a second time?

Figure 2
Antibiotics used in hog food help prevent illness.

2.13 Protists

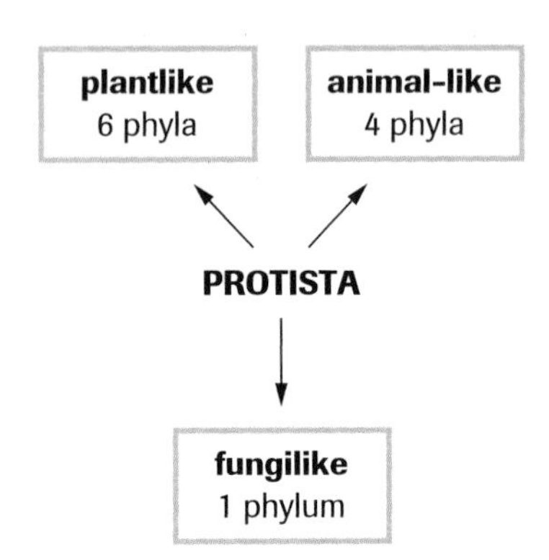

Figure 1
Grouping of protists

In section 2.2, you used a dichotomous key to sort some pond-water protists and, as a result, saw how varied these organisms are. The kingdom Protista is like a junk drawer in that its 11 phyla appear to have little in common. The members of these phyla differ greatly in shape, size, structure, complexity, feeding habits, locomotion, and reproduction. Protists also display great ecological diversity, occupying almost every known niche and habitat.

There are three distinct groups in the Protista kingdom: plantlike protists, animal-like protists, and fungilike protists (**Figure 1**). Even with such diversity, these organisms share the following features:

- Most are unicellular. Some form colonies of identical cells. A few are multicellular but do not form tissues.
- Cells are eukaryotic. There is a membrane-bound nucleus, and most contain vacuoles and mitochondria. Some contain chloroplasts.
- Cells reproduce asexually by binary fission. Some exchange DNA in a form of sexual reproduction.
- Cells thrive in moist surroundings such as fresh water, salt water, animal fluids, or very damp terrestrial environments.

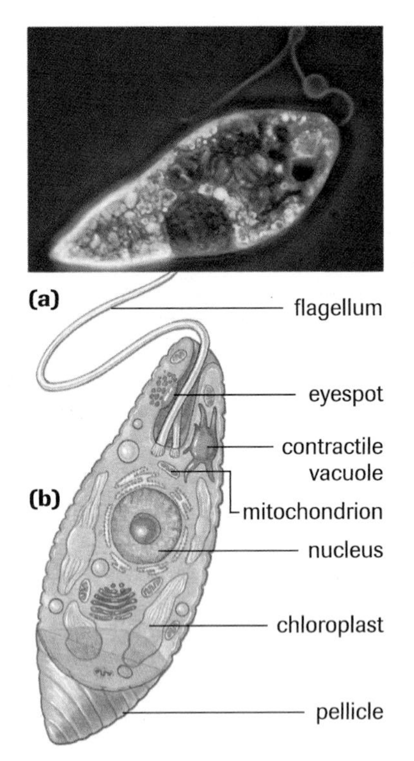

Figure 2
(a) A photomicrograph of a living *Euglena*
(b) This longitudinal section shows the internal organelles of *Euglena*.

Plantlike Protists

This group of protists is plantlike because the organisms contain chlorophyll, the pigment that begins the process of photosynthesis. Energy is obtained through this process, but during periods of darkness, cells become heterotrophic and take in more complex nutrients, often by engulfing solid food, a trait commonly associated with animals.

Euglena is a typical plantlike protist (**Figure 2**). Two striking features of *Euglena* are the eyespot and the flagellum. The eyespot is believed to be part of the organism's sensory-motor system, used to detect light. The flagellum is used to propel the organism through water in a whiplike fashion. Most species of plantlike protists have two flagella.

The entire outer boundary of *Euglena* outside the membrane is surrounded by a firm yet flexible covering called a pellicle. There is no cell wall. Other conspicuous structures are the central nucleus, the large green chloroplasts, and the vacuoles. The chloroplast pigments are identical to those in green algae and land plants. The vacuoles are used to collect and remove excess water. Food is stored in the form of starch granules, a common practice in plants.

Euglena, like many flagellated protists, reproduces asexually. Following nuclear division, the rest of the cell divides lengthwise. This process, called longitudinal fission, involves growth in cell circumference while the organelles are being duplicated. During unfavourable conditions, *Euglena* may form a thickly coated resting cell.

Also in the plantlike-protist category are the green, brown, and red algae (singular: alga). Most algae are multicellular but do not form tissues. This distinguishes them from higher plants that are multicellular and form tissues.

Algae are extremely well adapted to wet or moist environments (**Figure 3**). Normally considered to be aquatic organisms, they can also be found in soils, on the lower trunks of trees, and on rocks. Both brown and red algae, commonly called seaweeds, are generally large multicellular oceanic plants. They contain chlorophyll and other coloured pigments that permit them to carry out photosynthesis with the particular wavelengths of light that occur with changes in water depth. Unicellular green algae, usually referred to as phytoplankton, are found in both marine and freshwater environments, floating on or near the water surface.

DID YOU KNOW?

Algae as Food

Algae are an important food source for many marine mollusks. One mollusk, *Plakibranchus*, incorporates algae into its tissues. The algae continue to function and provide the animal with oxygen.

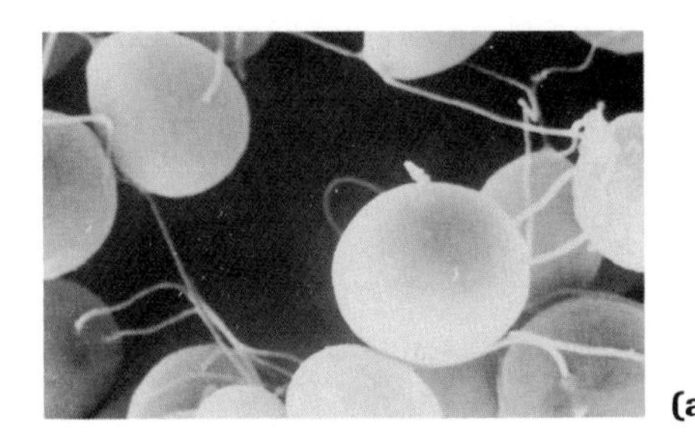

(a)

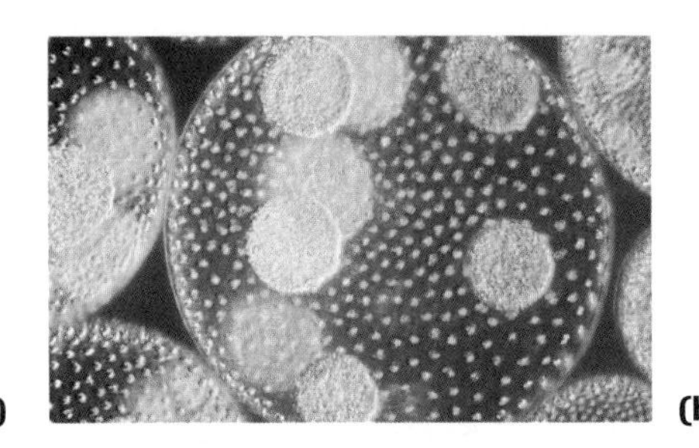

(b)

Figure 3

(a) *Chlamydomonas*, a common single-celled green alga that lives in fresh-water habitats

(b) *Volvox*, a colonial green alga, also lives in fresh water. It is composed of thousands of interdependent flagellated cells that closely resemble the free-living cell *Chlamydomonas*.

There are many variations in reproduction among the algae phyla. Variations range from complex life cycles that include both sexual and asexual reproduction to simple asexual reproduction by fragmentation—the organism simply breaks apart. Most commonly, reproduction is asexual, by binary fission. *Spirogyra* also reproduce by conjugation (**Figure 4**).

Importance of Algae

Algae, especially green algae, play an important role in the overall global environment. They are the primary food producers in aquatic food chains. Through photosynthesis, algae supply about 80% of the global supply of oxygen. It has become a concern that human wastes and industrial contaminants may be reducing algal populations, particularly in oceans. For example, the lakes near Sudbury are crystal clear—no algae can live in them—as a result of the dumping of industrial contaminants. At the other extreme, excessive growth of algae can also be a serious problem: by blocking sunlight, algae can affect photosynthesis and oxygen uptake by marine life.

Humans use algae in a variety of ways. Some algae are consumed directly as food, providing an excellent source of vitamins and trace minerals. Other algae are used as fertilizers. Agar is a mucilaginous material extracted from the cell walls of certain red algae. It is also used in the production of drug capsules, gels, cosmetics, and culture medium. A similar substance, carrageenan, is used in cosmetics, paints, ice cream, and pie fillings.

Algae help create petroleum resources. Brown algae store food as oils. Once filled with oil, the algae die and sink to the bottom of the ocean where they may be buried under mud and sand. Over millions of years and under extreme heat and pressure within the Earth, the oil from the algae can be transformed into crude-oil deposits.

A tube develops, connecting the two cells together.

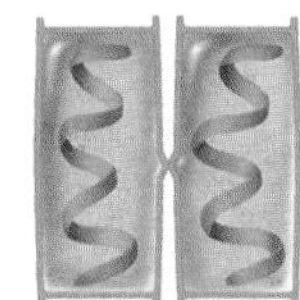

The contents of the left-hand cell move into the right-hand cell.

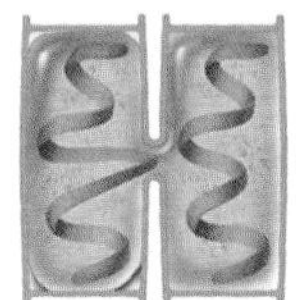

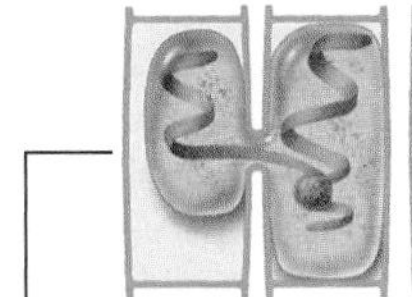

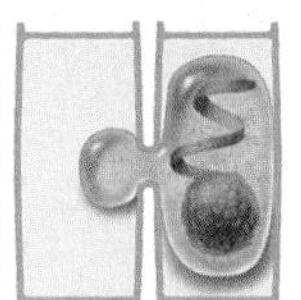

The nucleus of the left-hand cell fuses with the nucleus of the right-hand cell.

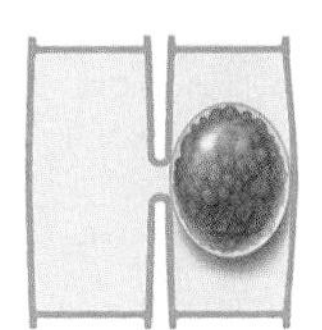

The cytoplasm rounds off and develops a thick wall, becoming a zygospore.

Figure 4

Conjugation of *Spirogyra*, a filamentous green alga that also reproduces asexually by binary fission and by fragmentation

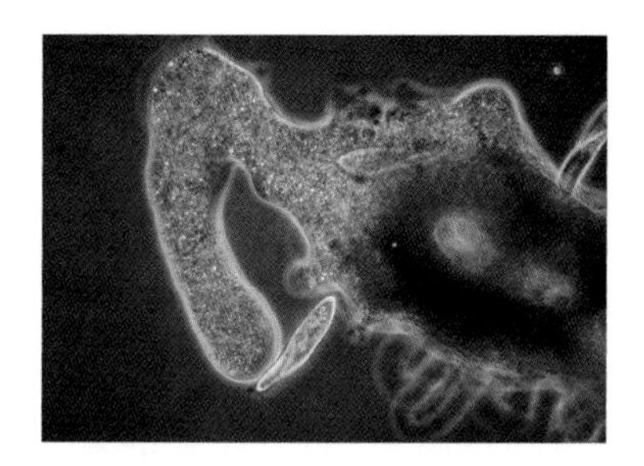

Figure 5
An amoeba engulfing food

ectoplasm the thin, semirigid (gelled) layer of the cytoplasm under the cytoplasmic membrane

endoplasm the fluid part of the cytoplasm that fills the inside of the cell. The endoplasm is responsible for an amoeba's shape as it moves.

Animal-like Protists (Protozoa)

In contrast to plantlike protists, all protozoa are heterotrophs. Generally, they must move about to obtain food. Some engulf their food (bacteria and other microorganisms); others absorb predigested or soluble nutrients directly though their cell membranes.

One of the largest yet least complex protozoans is the amoeba (plural: amoebae). An amoeba moves by repeatedly extending and retracting its pseudopods. Its cytoplasm has two layers: the outer protective layer, the **ectoplasm**, and the inner layer that contains most of the structures, the **endoplasm**. The continuous movement of the endoplasm causes the amoeba to change shape constantly as it moves. No wonder Linnaeus originally named an amoeba *Chaos chaos* when he saw it for the first time!

An amoeba feeds by phagocytosis; its pseudopods simply flow around and engulf food particles (**Figure 5**). The food eventually becomes enclosed in a food vacuole, where the food is digested. Water taken in with the food, along with water taken in via diffusion, collects in a contractile vacuole. As this vacuole fills, it contracts, and the water is discharged though a pore in the cytoplasmic membrane.

Amoebae reproduce by binary fission. Once the amoeba splits, the two organisms then grow to their full size and may split again. Under proper conditions, this could occur once a day.

(a) 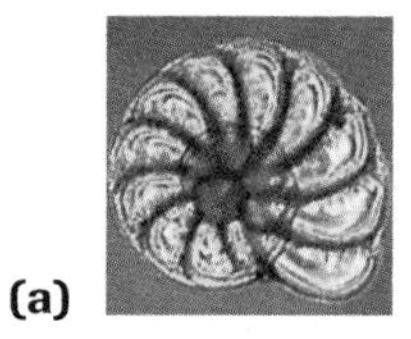 (b)  (c) 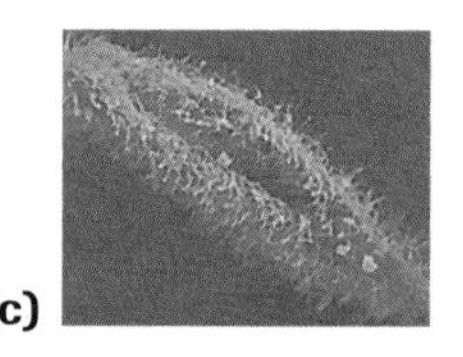 (d) 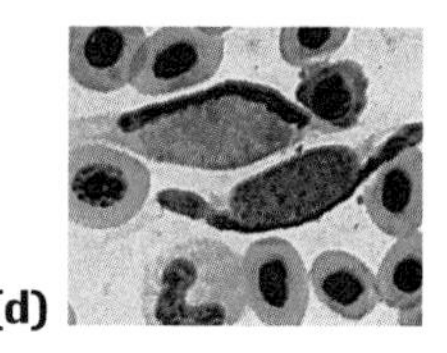

Figure 6
Four phyla of animal-like protists. Protists range in size from 2 μm in length to 5 cm in diameter.
(a) Sarcodina, pseudopod organisms
(b) Mastigophora, flagellated protozoans
(c) Ciliophora, ciliated protozoans
(d) Sporozoa, parasitic protozoans

Figure 6 shows that protozoans are classified according to their type of locomotion. Ciliates (**Figure 6(c)**) use hairlike structures for mobility. In free-moving organisms such as *Paramecium*, cilia are armlike, synchronized for swimming. Other protists use cilia to attach themselves to a surface.

Ciliates are the most advanced of the protozoans, living in both fresh and salt water. The paramecium illustrates many of the group's complex features (**Figure 7**). The oral groove contains a mouth that leads into a cavity called the gullet. Specialized cilia in the gullet sweep bacteria and other food particles into the cavity. From the gullet, food enters a food vacuole where digestion takes place. Wastes are expelled from the food vacuole through an anal pore. Paramecia collect excess water in a contractile vacuole, from which the water is expelled into the environment.

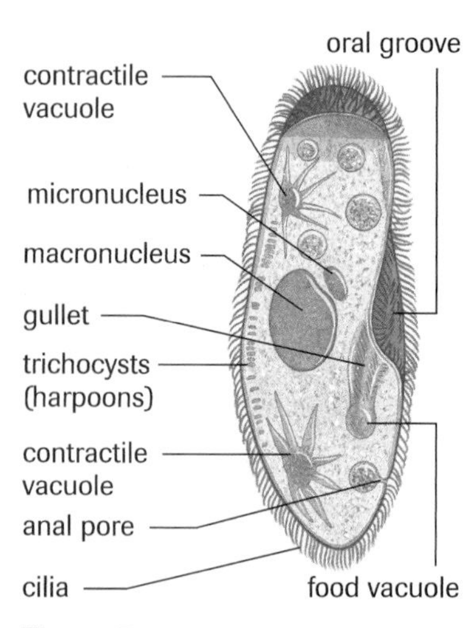

Figure 7
Paramecium is sometimes called the "slipper animal" because of its shape.

The paramecium has an interesting defence mechanism. Hundreds of poison-laden barbs can be discharged either to drive away predators or to capture prey. These structures are called trichocysts.

A paramecium has two nuclei: a large macronucleus that controls the cell's activities and a smaller micronucleus that is involved in reproduction. Reproduction in protozoa is usually asexual, via binary fission. Periodically, sexual reproduction in the form of conjugation may occur. Under adverse conditions, protozoans may form resting cells called **cysts**. If favourable conditions return, the organisms may emerge from their cysts.

cyst a cell that has a hardened protective covering on top of the cytoplasmic membrane

Members of the Sporozoan phylum (**Figure 6(d)**) are unique among the protozoans because they lack any means of independent locomotion; they have no pseudopodia, cilia, or flagella. They are exclusively parasitic and depend entirely on the body fluids of their hosts for movement. Also, they have fewer organelles and specialized structures than other protozoans. A distinguishing feature of sporozoans is the ability of their cells to multiply asexually in animal tissues by forming reproductive cells, called **spores**, that can develop into an individual without fertilization. Spores have a **haploid**, or single, set of unpaired chromosomes.

spore a reproductive cell that can produce a new organism without fertilization. Fungal spores have thick, resistant outer coverings to protect them.

haploid nucleus, cell, or organism with one set of chromosomes. The haploid number is designated as *n*.

Pathogenic Protozoans

The best-known sporozoans are members of the genus *Plasmodium*, which causes malaria in humans. As you can see from the sequence in **Figure 8**, the malaria parasite has a complex life cycle that involves two hosts: humans and the mosquitoes of the genus *Anopheles*. Insects are frequently responsible for transmitting sporozoans from one host to the next and are referred to as insect vectors. Malaria can be treated with drugs, but so far the most effective way to deal with the disease is to eliminate the vector. In many tropical countries where malaria is common, pesticides were widely used during the 1950s and 1960s to eliminate the *Anopheles* mosquitoes. However, pesticide-resistant strains have evolved, and today researchers are attempting to develop a vaccine to control malaria. Meanwhile, the disease continues to be a serious human health problem in many tropical areas.

CAREER CONNECTION

Field directors, working in areas of the world where malaria is prevalent, have backgrounds in health sciences and epidemiology. Malaria inspectors look for areas of stagnant water providing a breeding ground for mosquitoes.

DID YOU KNOW?

Mosquito Food

Only female mosquitoes feed on blood; male mosquitoes feed on plant fluids.

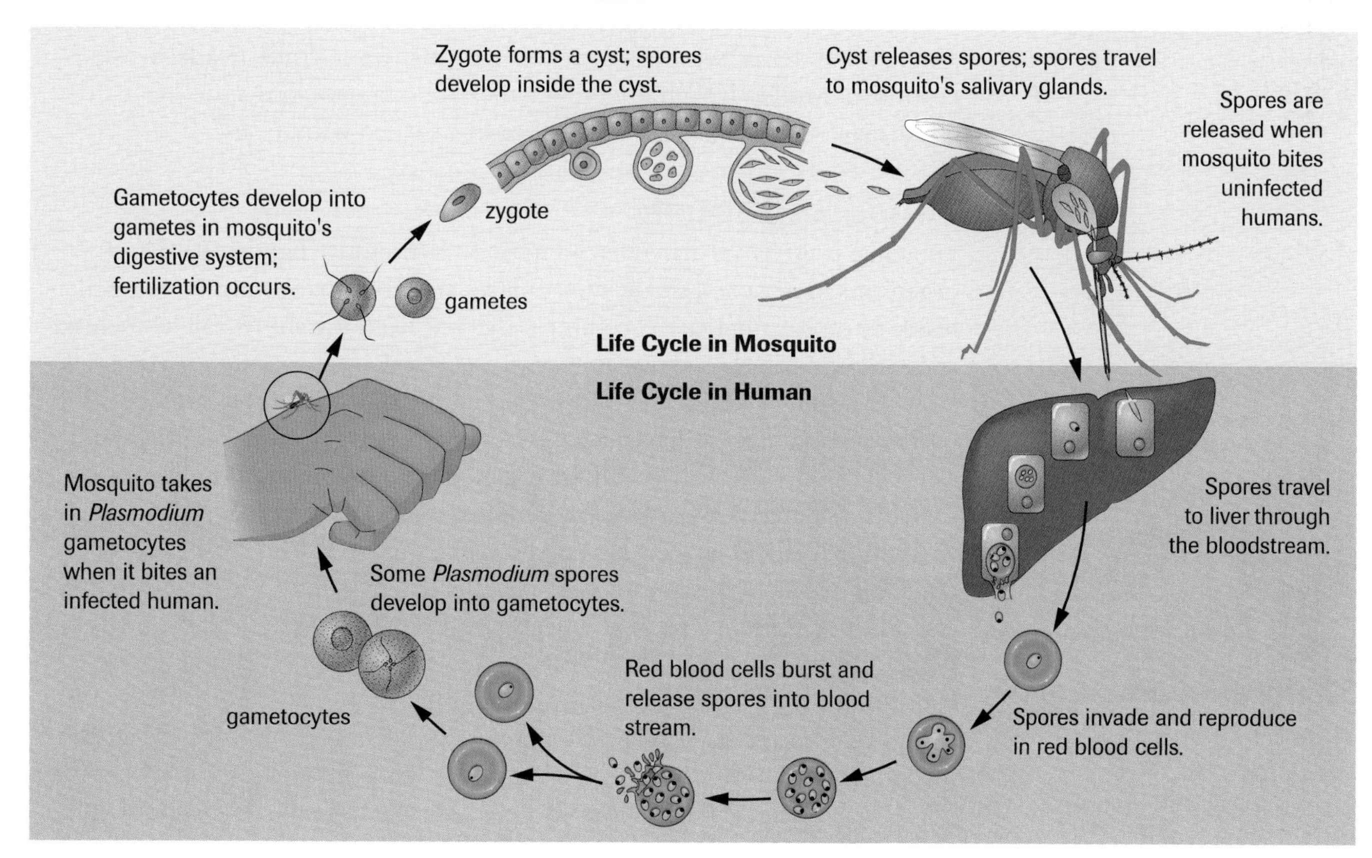

Figure 8
The life cycle of the malaria parasite, *Plasmodium vivax*

carrier an infected individual who carries disease but does not become infected herself/himself

Some other serious human diseases are caused by pathogenic protozoans that live in the human intestine. *Entamoeba histolytica*, unlike its free-living cousin *Amoeba proteus*, is parasitic and causes amoebic dysentery. This amoeba lives in the large intestine of humans and feeds on the intestine walls, causing fever, chills, bleeding ulcers, diarrhea, and abdominal cramps. The disease is spread when some amoebae form cysts that pass out of the infected person in fecal matter. Another person becomes infected by eating food or drinking water contaminated by these cysts. The cysts can survive for long periods of time, making eradication of the parasite very difficult. In some countries where amoebic dysentery is common, up to 50% of the population have the disease-causing organism and can transmit it to others, but are not infected themselves. They are **carriers** of the disease. Sanitary disposal of human waste is the best way to prevent widespread transmission of dysentery. To avoid dysentery in areas where human wastes are not properly handled, any water used for drinking, cooking, or washing food must be boiled. Amoebic dysentery can be treated with drugs, but several follow-up tests are required to ensure that the intestine is cyst-free.

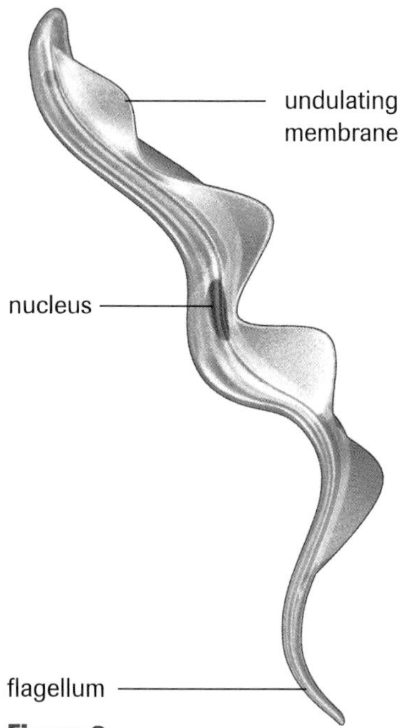

Figure 9
The undulating membrane of *Trypanosoma* allows it to move through the human bloodstream like an eel.

Some flagellated protozoans are also parasitic in humans. *Giardia lamblia* (beaver fever) usually causes stomach upset and diarrhea, but can also have more severe effects on some people. Another common parasite, *Trichomonas vaginalis*, is passed through sexual intercourse and infects the urinary and reproductive tracts. African sleeping sickness is caused by the *Trypanosoma* protozoan (**Figure 9**), and is transmitted by the tsetse fly. This protozoan multiplies in the blood of humans, releasing toxins that affect the nervous and lymphatic systems. Symptoms include irregular fever, chills, headache, skin eruptions, swollen lymph nodes, weakness, and fatigue. If diagnosed at an early stage, African sleeping sickness can be treated with drugs.

Fungilike Protists

Fungilike protists are also referred to as slime moulds (**Figure 10**). These organisms prefer cool, shady, moist places and are usually found under fallen leaves or on rotting logs. The name is derived from the slimy trail left behind as the mould travels over the ground.

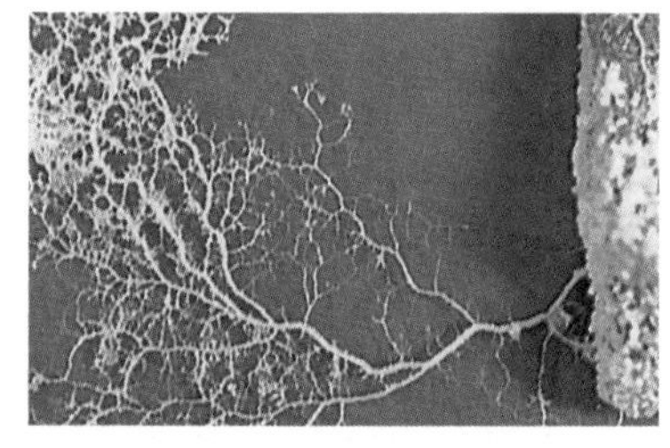

(a)

(b)

Figure 10
(a) A plasmodial slime mould
(b) Fruiting body of the slime mould, where spores are produced

During some stage of their life cycle, slime moulds resemble protozoans and become amoebalike or have flagella. At other times, they produce spores much as fungi do. Unlike other primitive organisms, the slime moulds do not always remain as single-celled organisms. One genus spends some of its life cycle as a single-celled amoebalike organism; however, these cells can converge into a large, slimy plasmodial mass. It then begins to act like a single organism. It extends into a sluglike form and begins to creep, feeding on organic matter as it goes. When it runs into an object, it retracts and slithers around it. The coordinated movements are slow—sometimes only a few millimetres a day.

Biologists recognize that becoming multicellular is a tremendous advancement for an organism. As individual cells begin to work together, the groundwork for cell specialization is laid. In more advanced organisms in higher kingdoms, the specialized cells develop structures or tissues suited for specific tasks.

Overall Importance of Protists

The impact of protists on the biosphere is enormous. Although they cause some of the world's most serious diseases, protists provide aquatic consumers with food and supply both aquatic and terrestrial heterotrophs with oxygen. Large populations of protists that drift in the ocean are called plankton. Photosynthetic plankton, or phytoplankton, produce their own food and help form the base of food pyramids, chains, or webs (**Figure 11**). Heterotrophic protists, or zooplankton, are first-level consumers. If the population of plankton were seriously affected by pollution or other environmental changes, huge filter-feeding whales would starve. Aquatic carnivores such as tuna would also be affected.

Aside from food, protists are involved in a variety of ecological relationships with other organisms, including humans. One species of protozoans (*Trichomonas hominis*) lives in the digestive tract of humans. It feeds on intestinal contents that humans do not use, and it has never been known to be harmful. Another protist lives in the digestive tract of termites, digesting the wood that termites eat. Termites cannot digest cellulose without this protist.

Despite their generally small size and deceptively simple structure, members of the protist kingdom are extremely important to all other life.

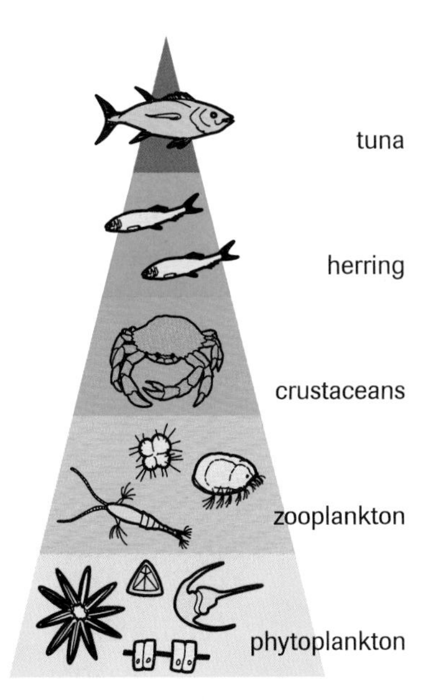

Figure 11
Plankton are an essential part of the food pyramid for many aquatic communities.

Section 2.13 Questions

Understanding Concepts

1. What is the main feature used to classify *Euglena* as a plantlike protist? List five other identifying features of this organism.
2. What is the most important structural difference between algae and *Euglena*? between algae and land plants?
3. List five ways in which algae are useful to humans.
4. "*Amoeba* and *Paramecium* are both animal-like protists, yet their structure is very different." Justify this statement.
5. Use binomial nomenclature to name four pathogenic protists. Explain how each enters the human body, and describe the resulting symptoms.
6. Why do the fungilike protists belong in the kingdom Protista, and how are they different from other protists?
7. Name a protist that feeds by phagocytosis and explain how this happens.
8. Use the details presented in this section to describe how the following organisms reproduce: *Euglena*, *Spirogyra*, *Paramecium*, and *Plasmodium*.
9. Describe the mobility of each of the common protists studied in this section: *Euglena*, alga, *Amoeba*, *Paramecium*, and sporozoan.
10. Explain the differences between a cyst and a spore.
11. How are the sporozoans "parasitic"?

Applying Inquiry Skills

12. An amoeba reproduces by binary fission about once every 24 h. How many amoebae would be produced from an original cell in a month of 30 days, assuming all survived and reproduced?
13. A sample of greenish scum is examined under the microscope. The scum consists of threads made up of identical cells. A single cell is isolated and examined under high power. However, it is difficult to see whether or not the cell has a true nucleus because it contains so many oval-shaped, bright-green organelles. To which kingdom would you assign this organism, and why? Can the organism be classified further without conducting additional tests? Explain.

Making Connections

14. A peregrine falcon in Washington State was diagnosed with malaria. It was thought to be an isolated case until the disease was discovered in other birds of prey. How do you suppose malaria was transmitted to the falcon? What implications could this have for humans?
15. What are harmful algal blooms (HABs)? Draw a flow chart to show their effect on the food chain. Describe a human illness associated with HABs.

www.science.nelson.com

2.14 Activity

Examining Protists

In this activity, you will examine, using a compound light microscope, the cell structure of some common protists, using prepared slides of *Paramecium*, *Spirogyra*, and some other examples. You will also observe live paramecia moving (**Figure 1**), feeding, and responding to their environment.

Paramecium, a ciliated protozoan, is one of the most complex of the single-celled organisms. It reproduces asexually by binary fission (**Figure 2**), and periodically reproduces sexually by conjugation. *Spirogyra*, a filamentous green alga, reproduces asexually by fragmentation from spring to autumn, and sexually by conjugation during the colder months. Zygospores formed during the cold season remain dormant until the following spring.

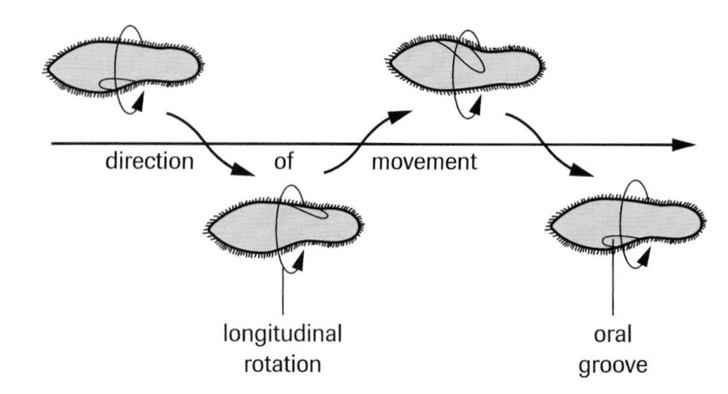

Figure 1
The paramecium swims freely with a spiralling motion. Cilia beat at an angle in coordinated waves, and the organism rotates on its longitudinal axis. Cilia are denser in the oral groove area, and they beat more quickly, causing the front end of the organism to move in a circle.

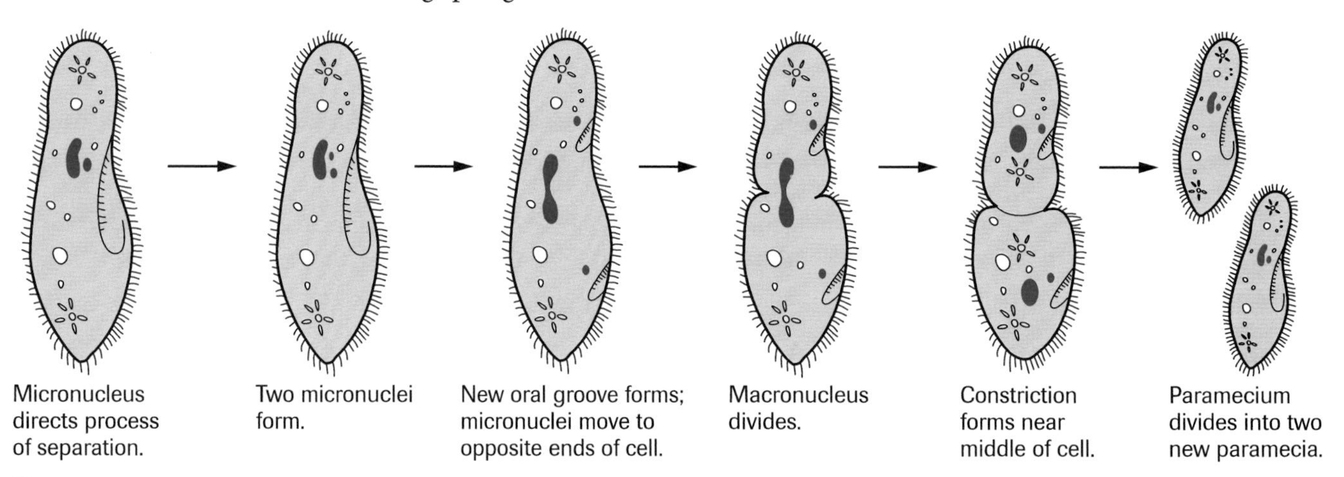

Figure 2
Binary fission in a paramecium

Materials

laboratory apron
compound light microscope
paramecia culture
10% methyl cellulose solution or glycerine
prepared slides of *Paramecium, Spirogyra*, and any of the following protists: *Euglena, Amoeba, Volvox*
medicine dropper
microscope slides and cover slip
stained yeast culture (congo red)

For microscopes without their own light sources, make sure that the sun cannot be directly focused through the instrument. Direct exposure to the sun can damage the retina.

Procedure

Part 1: Examining Protists

1. Place each prepared slide on the stage of the microscope.

(a) Using medium-power or high-power magnification, draw one example of each type of organism you observe. For each drawing, label all the visible structures (e.g., macronucleus, micronucleus, contractile vacuole, food vacuole, cilia, oral groove, and anal pore).

(b) Use information given to you by your teacher to estimate the length and width of each protist.

(c) Describe reproduction in the paramecium if it is observed.

Part 2: Observing Paramecia

2. Using the medicine dropper and microscope slide, prepare a wet mount of the paramecia culture. Add a cover slip.

(d) In your notebook, prepare a table similar to **Table 1**. Record your observations as you observe the movement of paramecia under low- and medium-power magnification.

Table 1 Observation of Paramecia

Feature	Description and/or sketch
anterior/posterior end	
general movement pattern	
reaction to obstacles	
leading end during locomotion	
cilia length and movement	
yeast cells in gullet	
food vacuoles: number, location	
trichocysts	
contractile vacuoles	
other interesting features	

3. Prepare another wet mount by mixing one drop of methyl cellulose solution (or smearing a small amount of glycerine) with an equal-sized drop of culture fluid on a slide. Before applying a cover slip, add a drop of the stained yeast culture to the mixture. Under medium- or high-power magnification, examine the slide for cilia movement, yeast cells, and food vacuoles.

(e) Add your observations to the table.

Analysis

Part 1: Examining Protists

(f) Compare the protists that you examined with those of another group. Were the same protists approximately the same shape and size? If not, what might account for the differences?

(g) What clues to functions are provided in the names of the structures you observed (e.g., oral groove, contractile vacuole, trichocysts, and chloroplasts)?

Part 2: Observing Paramecia

(h) From your examination, summarize how a paramecium ingests and digests food and eliminates waste.

(i) From your examination, summarize how a paramecium moves.

(j) Why was methyl cellulose solution (or glycerine) added to the slide?

(k) From your examination, what might be the function of trichocysts?

(l) Was reproduction evident? If so, describe the stage(s) that you viewed.

Synthesis

(m) *Spirogyra* is similar to plants because the cells contain chloroplasts and have cell walls. Why isn't it classified as part of the plant kingdom?

(n) *Euglena* does not have cell walls. Why is it considered to be plantlike?

(o) Describe the differences between conjugating cells and nonconjugating cells in *Spirogyra* and in *Paramecium.*

(p) What might happen to the contractile vacuoles in a paramecium if the organism were transferred from the culture medium into sea water? Explain.

(q) Research the techniques for preparing fixed and stained slides of protists. With your teacher's permission, prepare slides of various protists.

www.science.nelson.com

CAREER CONNECTION

Biological technicians and technologists collect biological samples, analyze them, and report on what they find. Research the duties, programs, and job possibilities in this field.

www.science.nelson.com

2.15 Fungi

Fungi were once classified as members of the plant kingdom. However, fungi, unlike plants, do not photosynthesize, and enough other differences exist between plants and fungi to place them in separate kingdoms. **Figure 1** summarizes some of the similarities and differences.

The structure of fungi has been adapted for two main functions: absorption of nutrients and reproduction. Like some bacteria, fungi are saprophytes that feed on dead or decaying matter. Digestion is extracellular—nutrients are digested externally before being absorbed. In multicellular forms, nutrient absorption takes place in the **mycelium** (plural: mycelia), a mesh of microscopic branching filaments that are usually on or below the surface where the organism grows or is attached, called the substrate. Each of these filaments is called a hypha (plural: hyphae) (**Figure 2**). Often the only visible parts of a fungus are its reproductive structures, which display a wide variety of sizes, shapes, and colours (**Figure 3**).

mycelium a collective term for the branching filaments that make up the part of a fungus not involved in sexual reproduction

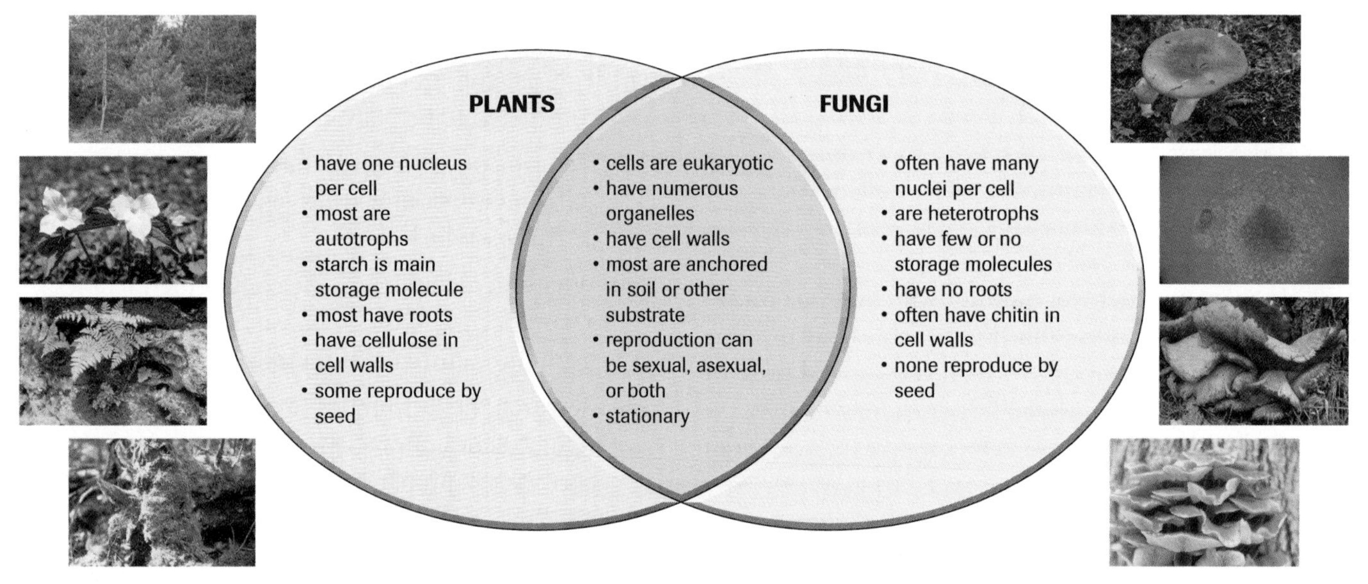

Figure 1
Similarities and differences between plants and fungi

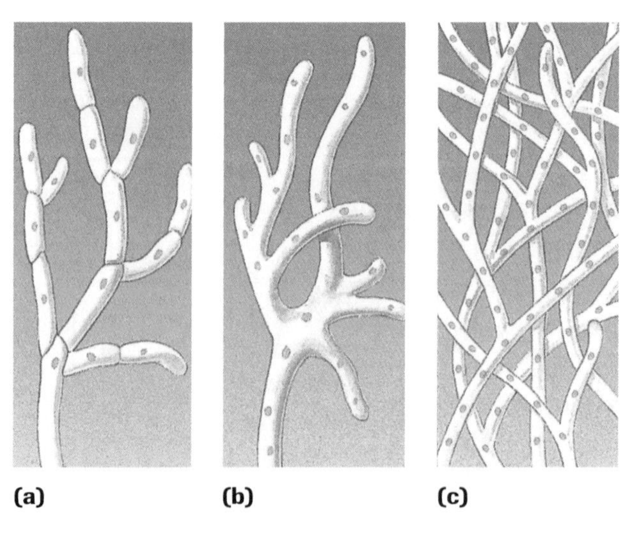

Figure 2
Some fungi have hyphae with cross-walls. Most hyphae have cell walls that are reinforced with chitin. It forms structures of considerable strength.
(a) Hyphae with cross-walls
(b) Hyphae without cross-walls
(c) Mycelium showing many interlocking hyphae

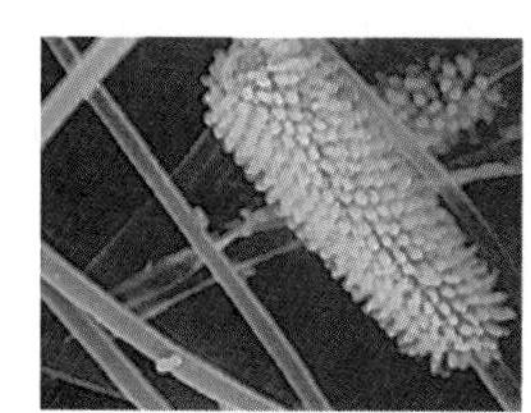

chytrids, water moulds
- mainly aquatic
- cross-walls in hyphae
- some saprophytes, some parasites
- asexual and sexual reproduction

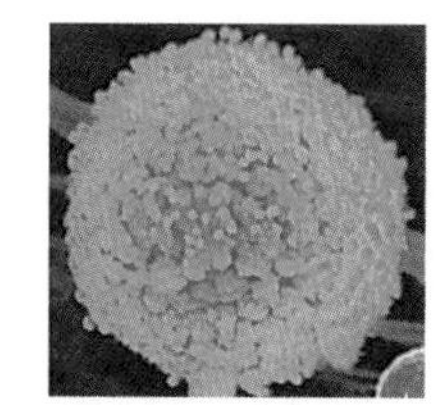

common moulds (bread, dung moulds)
- mainly terrestrial (soil, decaying plant matter)
- mainly multicellular
- no cross-walls in hyphae
- some saprophytes
- asexual and sexual reproduction

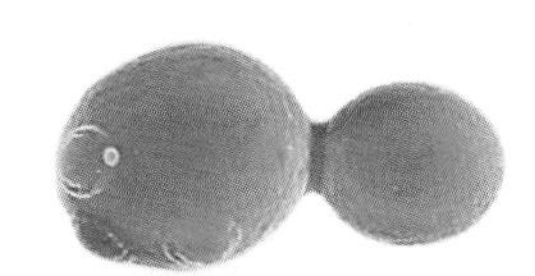

yeast, morels, truffles
- terrestrial and aquatic
- mainly multicellular
- cross-walls in hyphae
- many are pathogens
- asexual and sexual reproduction

mushrooms, shelf fungi
- mainly terrestrial
- mainly multicellular
- cross-walls in hyphae
- many are pathogens
- sexual reproduction

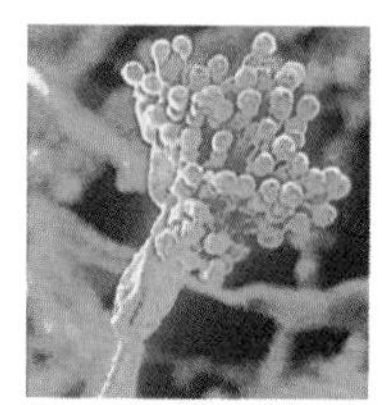

parasitic fungi
- mainly terrestrial
- some produce spores in sacs
- some produce spores in club-shaped structures
- asexual reproduction

Figure 3
Examples of fungi

Life Cycle of Fungi

A life cycle describes the development of an organism from a single cell to its reproductive stage. The cycle is complete when the organism produces the next generation. There is a wide variety of fungal life cycles. Fungi reproduce both asexually and sexually but always produce spores as reproductive cells. Spores have a haploid chromosome number and are produced in specialized reproductive structures called **sporangia** (singular: sporangium). Spores are usually dispersed by air currents, and when in a suitable environment, they sprout or **germinate**. Fungi may also produce asexually by fragmentation, in which portions of the mycelium break apart.

sporangium a reproductive structure in which spores are produced

germinate to grow or sprout; refers specifically to the embryo inside a plant seed

vegetative any growth or development that does not involve sexual reproduction

Field mushrooms grow wild in fields, lawns, and gardens (**Figure 4**). They are the type of mushroom most commonly cultivated for human consumption, and their life cycle is described in **Figure 5**, on the next page.

The mycelium is the nonreproductive or **vegetative** part of the mushroom. It consists of many hyphae and is usually just below the surface of the soil. When compatible haploid parent hyphae fuse, reproduction begins. These new hyphae develop enlargements that increase in size until they break through the soil's surface and can be seen as small white spheres called buttons. As the buttons grow and mature, they form a stalk and a cap that remains spherical. Within this cap, many changes occur. Thin membranous gills form and radiate out from the stalk. The gills consist of many tangled, modified hyphae. At first, the gills are pink, but they darken as they grow and mature. In specialized extensions from the gills, some of the cells that contain two haploid nuclei of different parents, called **dikaryotic**, fuse in a process considered to be a sexual union. The new diploid nucleus then undergoes meiosis to produce four haploid spores.

Figure 4
Field mushrooms showing caps and stalks. Spore-producing cells line the dark, mature gills on the undersurface of the mushroom cap.

dikaryotic describes cells that contain two haploid nuclei, each of which came from a separate parent

The underside of the cap detaches from the stalk and eventually looks like an opening umbrella. Spores are discharged from the mature gills over a period of several days. It is estimated that more than two billion spores may be produced from one average field mushroom. If a spore lands in a favourable environment, its protective coat will eventually split, and the haploid cell will divide by mitosis to produce a new hypha. It will continue to grow and branch into a complex mycelium.

Figure 5
General life cycle of a field mushroom
(a) An early button, consisting of dikaryotic tissue created by the fusion of the haploid hyphae, not the nuclei, of two separate parents
(b) A longitudinal section through an immature field mushroom. When the lower rim of the cap detaches from the stalk, the cap opens like an umbrella, exposing the maturing gills.
(c) A highly magnified view of a small section of one side of one gill consisting of tangled hyphae. Three extensions are shown at sequential stages of development.
(d) An enlarged view of a germinating haploid spore showing the first hypha protruding from the ruptured protective spore coat

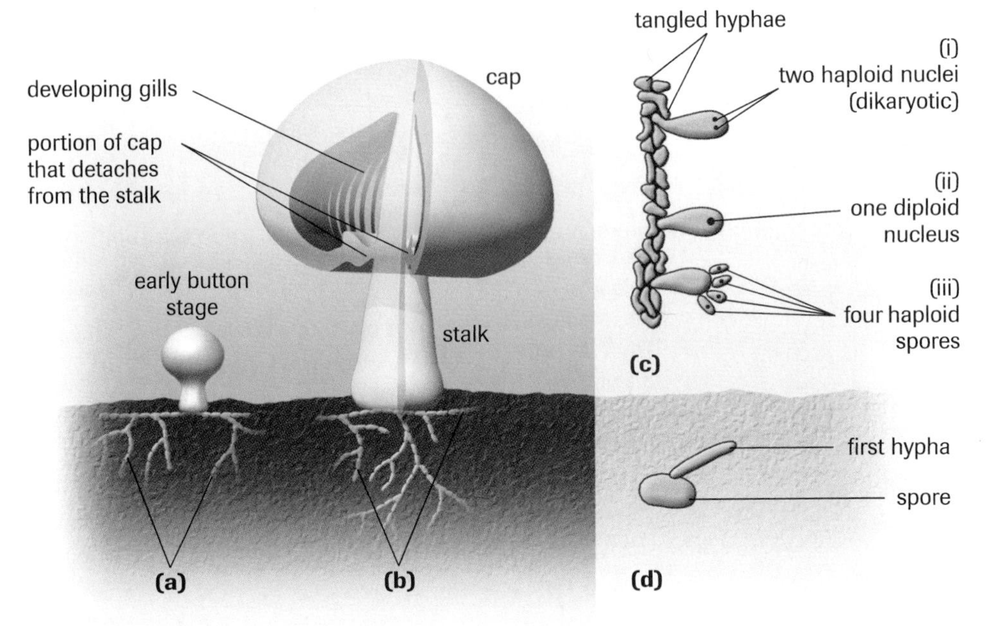

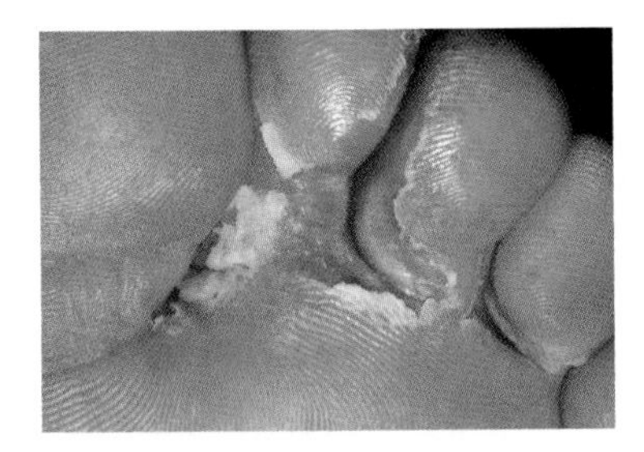

Figure 6
Epidermophyton floccosum is a fungus that causes athlete's foot.

Figure 7
Amanita phalloides, or Death Cap, is the world's most dangerous mushroom.

Connection to Human Diseases

Several human diseases are caused by fungi. Most of these diseases are merely annoying, but some can be deadly.

Ringworm is a skin infection caused by *Microsporum*. This organism can cause severe itching in the groin, scalp, and beard. If left untreated, *Microsporum* can spread rapidly by spore and may be passed to others by direct or indirect contact. Sprays and powders containing fungicides can be used to help control its spread. Athlete's foot (**Figure 6**) is also caused by this genus of fungus, and is contracted from dirty shower floors or running shoes.

Eating mushrooms picked in the wild can be a dangerous undertaking ince many are poisonous. The most harmful and potentially fatal mushrooms belong to the genus *Amanita*. Ingesting only one or two of these so-called destroying angels can lead to abdominal pain and cramps, vomiting, and eventually death. These effects are caused by toxins in the mushroom that enter the bloodstream. Another type of *Amanita* produces poisons that can affect the nervous system and cause hallucinations and behaviour similar to drunkenness. Coma and death may result from these neurotoxins. The first step in effective treatment involves emptying the stomach and intestines before the poisons enter the bloodstream. Further treatment for mushroom poisoning depends on knowing which *Amanita* species was ingested. If *Amanita muscaris* was eaten, atropine is injected to counteract overstimulation of the nervous system. If *Amanita phalloides* (**Figure 7**) was ingested, an intravenous solution of sugar and salt is necessary to combat possible liver damage. If the species is not known, it is difficult for doctors to determine which treatment is needed. Eating only store-bought mushrooms removes this danger.

Importance of Fungi

Humans often overlook the value of the lowly fungus and its role as a decomposer in nature's recycling system. Fungi, along with eubacteria, transform complex organic substances into raw materials that other fungi and plants use for growth and development.

Many fungi benefit humans. Yeast is used to make bread, wine, and beer. *Penicillium* produces an antibiotic; *Aspergillus* is used to flavour soft drinks. Certain mushrooms, morels, and truffles are sought-after food items. Some fungal species are potentially useful in decomposing harmful pollutants. Research is continuing into ways of using fungi to break down some complex hazardous chemicals in toxic dumpsites and wastewater treatment plants.

Many fungi are involved in a beneficial or **symbiotic relationship** with another organism. For example, in **mycorrhizae** (singular: mycorrhiza), an extensive network of fungal hyphae helps the roots of plants absorb nutrients such as phosphorus. In return, the fungi obtain sugars that the plant has made during photosynthesis. Some plant seeds will not germinate in the absence of the mycorrhizal fungi. For this reason, environmental biologists monitor mycorrhizal growth in contaminated soils to predict plant growth. It has even been suggested that the evolution of land plants has been dependent on the presence of mycorrhizae.

symbiotic relationship a relationship between two organisms in which both partners benefit from the interaction

mycorrhiza a symbiotic relationship between the hyphae of certain fungi and the roots of some plants

Lichen (**Figure 8**) is another example of symbiosis, in that two organisms function as one. The photosynthetic partner to the fungus is generally a green alga or cyanobacterium, and the fungus is usually a sac fungus. The fungal mycelium surrounds the photosynthetic cells and provides them with essential minerals, carbon dioxide, and water for photosynthesis. The mycelium also lends structural support to the entire organism. The algae share the carbohydrates they manufacture from carbon dioxide with the fungi. Lichens reproduce asexually by fragmentation. In the arctic tundra and boreal forests of northern Canada, lichens are an important source of food for caribou and other animals.

lichen a combination of a green alga or cyanobacterium and a fungus growing together in a symbiotic relationship

Lichens are important in plant succession. By being able to establish themselves on rocks and in barren areas, lichens help form basic soil material. Researchers have long recognized the value of lichens in detecting air pollutants. Unlike plants that absorb water that has been filtered through the ground, lichens absorb water directly from the air. In this way, they absorb more dissolved toxic substances than do plants. The presence or absence of different lichens in an area can be used to map concentrations of pollutants.

(a)

(b)

Figure 8
Examples of common Canadian lichens are **(a)** reindeer moss and **(b)** British soldiers.

▸ TRY THIS activity — Collecting and Examining Lichens

Materials: ruler, small plastic bags, hand lens, prepared slide of lichen, compound light microscope

1. Use the edge of a ruler to carefully scrape lichens from various trees and rocks. Try to collect the scrapings from separate lichens in separate plastic bags.
2. Examine the lichen pieces under a hand lens.

(a) Classify the lichen according to **Table 1**, on the next page.

3. Examine a prepared slide of a lichen using a compound light microscope.

(b) How are you able to distinguish the photosynthetic partners from the fungi?

DID YOU KNOW?

Absorbent Lichens
Lichens are remarkably resistant to drought. A dry lichen can quickly absorb 3 to 35 times its weight in water!

Table 1 Appearance of Lichens

Lichen type	Description	
crustose	flat or crusty; forms a matlike structure on rocks or tree bark	
foliose	leaflike lobes; spreads outward; has a paperlike appearance	
fruticose	raised structure with stalks and multiple branching threads; may hang from trees	

Section 2.15 Questions

Understanding Concepts

1. What main feature is missing that prevents fungi from being included in the plant kingdom?
2. Study **Figure 1**, on page 136. Summarize the criteria used to classify fungi.
3. For each of the following, provide a definition and describe the function:
   (a) mycelium (c) hypha
   (b) spore (d) sporangia
4. When fungi reproduce asexually, where do the haploid spores form, and what steps lead up to spore formation?
5. List several harmful effects resulting from the ingestion of poisonous mushrooms such as those belonging to the genus *Amanita*. How can such harmful effects be avoided?
6. Describe the role of the fungus partner in each of the following:
   (a) mycorrhizae (b) lichens
7. Classify the lichens in **Figure 8**, on page 139, as crustose, foliose, or fruticose.

Applying Inquiry Skills

8. Hypothesize why a slice of store-bought bread might grow mould more slowly than a slice of homemade bread.
9. Obtain three varieties of edible mushrooms and prepare a visual record: sketch the mushrooms and describe their colour, odour, surface characteristics, and structure. Make spore prints for each type according to your teacher's directions. Use resources to identify these fungi.

Making Connections

10. One of Canada's largest breweries has an in-house brewer training program. Find out more about what it takes to become a brew master. What other programs are available in this field?

www.science.nelson.com

11. If fungi were wiped out, how would this affect our lives? Consider the social, economic, environmental, and medical impacts.

www.science.nelson.com

12. What steps could people who are allergic to fungal spores take to ensure that their living environment is relatively spore free?
13. Research one of the following topics:
    (a) the Irish potato blight of 1845–1847
    (b) the use of fungi in the production of antibiotics
    (c) the physiological effects of poisonous fungi on the human body

www.science.nelson.com

2.16 Investigation

Inquiry Skills

○ Questioning	○ Planning	● Analyzing
○ Hypothesizing	● Conducting	● Evaluating
● Predicting	● Recording	● Communicating

Monitoring Bread-Mould Growth

Moulds usually reproduce by means of spores. Under certain conditions, mould will grow on bread and can be observed after several days. For this investigation, your group will prepare mould cultures to study optimum growth conditions. Eight sealed containers, each containing a cube of bread, are subjected to various conditions. Observations of mould growth are recorded in a table, and the types of mould grown are identified.

Question

Which environmental conditions promote or inhibit the growth of mould on bread?

Prediction

(a) Write a prediction based on which environmental conditions will either promote or inhibit mould growth on bread.

Materials

- safety goggles
- laboratory apron
- eyedropper
- paper towel
- compound light microscope
- water
- small metric ruler
- cardboard box with lid
- cubes of various types of bread (labelled)
- petri dishes or baby-food jars with lids
- masking tape
- hand lens
- prepared microscopic slides of bread mould
- incubator or warm radiator
- refrigerator or cool location
- lamp
- marker or ink pen

Any student allergic to airborne particles should inform the teacher immediately.

Wear eye protection and a laboratory apron during this investigation.

Wash your hands thoroughly at the end of the investigation.

Procedure

1. Line the bottom of eight petri dishes or jars with one layer of barely moist paper towel.
2. Place one cube of bread on the paper towel. Use the same type of bread for all containers.

   (b) Record the type.
3. Label each container with your group number and the type of bread used. Also number your containers from 1 to 8.
4. Use an eyedropper to moisten (but not soak) the bread cube in containers 1, 2, 3, and 4 with water. Pour off any water that may have pooled on the bottom of these containers.
5. Allow all the bread cubes to dry for 10 to 20 min in the open containers, depending upon room humidity.
6. Cover each petri dish and tape the top and bottom of the petri dishes together in the form of an "X" or put the lids on the jars. DO NOT OPEN these containers again. Try to avoid bumping, tipping, or upsetting them as you place them according to step 7.
7. Store all containers for four to seven days as follows:
   (i) Store containers 1 and 5 in a warm incubator with a light source or near a warm radiator with a lamp turned on for at least 10 h per day.
   (ii) Store containers 2 and 6 in a warm incubator with no light source or near a warm radiator in a closed cardboard box so they receive no light at all.
   (iii) Store containers 3 and 7 in a refrigerator with a battery-operated lamp or in a cool place with exposure to light for at least 10 h per day.
   (iv) Store containers 4 and 8 in a refrigerator or other cool place in a closed cardboard box so they receive no light at all.
8. Make daily observations without opening the containers. Carry the samples cautiously, and be sure to return the containers to the same storage site.

(c) Record your observations in a table in your notebook. For each container, include the approximate number of spots or colonies and also the approximate diameters of one or two of the larger colonies on each bread cube.

9. When the mould becomes quite dark or different colours are noticed, examine the bread with a hand lens WITHOUT OPENING the containers. Identify the following types of hyphae: sporangiophores, which develop sporangia at their tips; stolons, which run horizontal to the bread surface; and rhizoids, which develop at the swellings along the stolons and anchor the fungus to the bread (**Figure 1**).

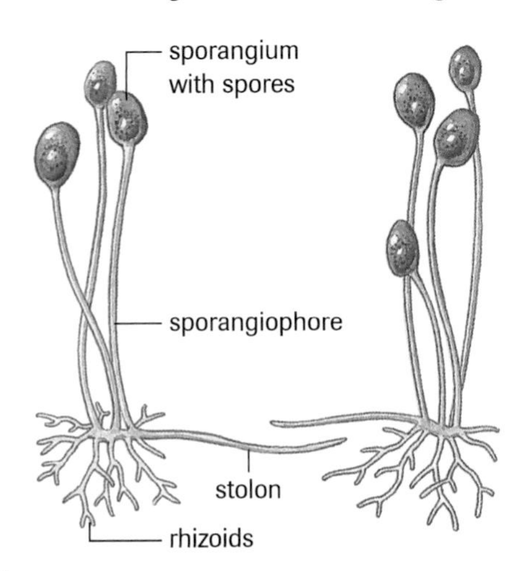

Figure 1
Main structures of bread mould

10. WITHOUT OPENING the containers, make drawings of what you observe.

(d) Draw each bread cube, without its container, and indicate the location of any of the types of mould listed below as identified by colour.

Common black bread mould is *Rhizopus nigricans*; blue-green mould is *Penicillium*; brown mould is *Aspergillus*; and grey mould is *Mucor*.

Also draw enlarged views of any of the parts you identified in step 9.

11. Obtain a prepared microscope slide of bread mould. Examine the slide under the microscope.

(e) Make drawings of the parts listed in step 9, plus the spores.

12. When the experiment is finished, give the closed containers to the teacher for proper disposal.

Analysis

(f) Count the number of colonies (spots) visible to determine the growth rate of each mould over several days. Graph this information.

(g) Using your observations about the mould colours, identify the types of mould that grew on your bread.

(h) List your numbered containers in order, from the one with the most mould at the end of the observation period to the one with the least.

(i) Which conditions favour mould growth?

(j) Which conditions seem to inhibit mould growth?

(k) What is the control in this investigation?

Evaluation

(l) Evaluate your prediction based on your analysis.

(m) Why was it necessary to keep the containers sealed shut?

(n) Compare your results with those of a group that used a different type of bread. Suggest reasons for any differences.

(o) What conclusion(s) could you make if no mould formed in any of the containers?

(p) List possible sources of error. Suggest improvements to the design.

Synthesis

(q) Suggest how you could identify any organisms other than bread mould growing on the cubes.

(r) Based on the results of this investigation, list several precautions you could take to prevent mould from growing on your left-over food.

CAREER CONNECTION

Cheese makers use mould growth factors to produce a standard product—for example, blue, Camembert, and Brie cheeses. Food-preservation companies use techniques to prevent mould growth in their products. Job opportunities exist at both extremes. Research one career that requires a college-based knowledge of microbiological growth.

www.science.nelson.com

Microorganisms in Ecosystems 2.17

No organism lives alone in nature, but interconnections are more direct in some species than in others. Many microorganisms obtain their nutrition through symbiotic relationships with other living things in their environment. There are three types of symbiotic relationships: parasitism, commensalism, and mutualism.

In **parasitism**, two organisms of different species live in a close association that benefits one partner (the parasite) but harms the other (the host). Usually, a parasite consumes only enough food or living tissue to survive without destroying the host, on which the parasite depends for survival. *Streptococcus* organisms in the human mouth digest sugars, producing lactic acid that dissolves tooth enamel, causing cavities. All disease-causing microorganisms, including viruses, are parasites.

In **commensalism**, two organisms of different species live together in close association, but only one partner benefits; the other is neither helped nor harmed. Many intestinal bacteria live in a commensal relationship with humans. Bacteria can cause illness such as diarrhea. *Lactobacillus* can slow diarrhea by crowding out harmful microorganisms, which is why some people take it in capsule form to prevent diarrhea when travelling. *Corynebacterium* organisms inhabit the surface of the human eye. They live on secretions and discarded cells, with no apparent benefit or harm to the host.

In **mutualism**, each organism is metabolically dependent on the other to the extent that part of each one's life cycle cannot be completed without the other. The enormous population of *E. coli* that normally lives in the human intestine uses gut contents for growth, and synthesizes vitamins K and B for humans. Digestive upsets often occur as a side effect of killing *E. coli* during antibiotic treatment. *Rhizobium* bacteria invade the roots of legumes and live in swellings called nodules (**Figure 1**). The plant provides the bacteria with a protected habitat, an energy source for their nitrogen-fixing activities, and the carbon molecules necessary for the production of organic compounds that contain nitrogen. In return, the plant is provided with a form of nitrogen it can use to make proteins. There are many other examples involving protists and fungi. Termites eat cellulose but cannot metabolize it without the action of protozoans that live in the host's gut. As described in section 2.15, the association of fungi and algae in lichens is a mutualistic relationship, as is that in mycorrhizae (plant and fungi).

parasitism a symbiotic relationship in which one organism (the parasite) benefits at the expense of another organism (the host), which is often harmed but usually not killed. It is categorized as a +/− relationship.

commensalism a symbiotic relationship in which one organism benefits and the other organism is unaffected. It is categorized as a +/0 relationship.

mutualism a symbiotic relationship in which both organisms benefit. Since neither is harmed, it is categorized as a +/+ relationship.

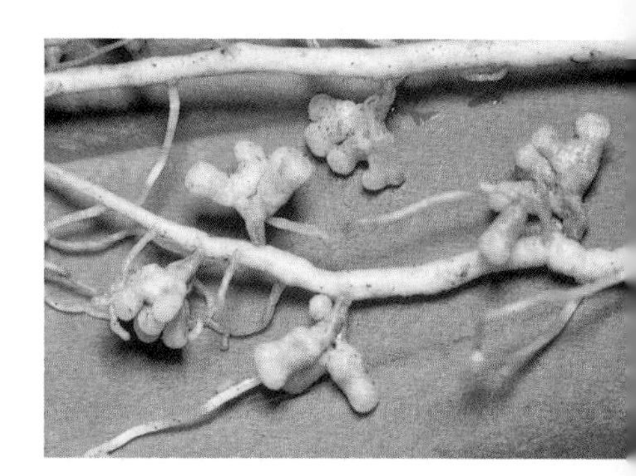

Figure 1
Rhizobium and legumes need each other.

Effective Microorganisms (EM) Applications

In the 1990s, agriculturalists developed a process that applies a cocktail of effective microorganisms (EM) to soil. Three common microorganisms (lactic acid bacteria, yeast, and photosynthetic bacteria) are blended in a sugar medium at low pH. Crop trials have shown that EM application improves soil structure and fertility, reduces disease incidence and the need for pesticides, and increases productivity. Today, in addition to crop applications, EM is used in aquaculture, in livestock production, and in recycling processes.

DID YOU KNOW ?

Bt toxins

Bacillus thuringiensis (Bt) produces a toxin that is a naturally occurring insecticide. It is applied to the foliage of plants infested with leaf- or needle-eating larvae. Within 12 h to 5 days, the Bt toxins paralyze the digestive tract of the larvae, and they die from starvation.

Pesticide Use and Microorganisms

Pesticides are used to control insects, diseases, weeds, and microorganisms on plant and animal material. Fungicides, herbicides, and insecticides are all pesticides.

There are many reasons to use pesticides, both commercially and at home. Pesticides protect crops from diseases (e.g., potato blight), protect humans from diseases spread by other animals (e.g., malaria and West Nile virus), and protect the environment by increasing yield on a shrinking land base. Pesticides are designed to attack particular pests without affecting harmless or beneficial organisms, but widespread use has shown that, when any one member of an ecosystem is affected, so is the diversity of the whole. Any measure that reduces soil microorganisms also reduces nutrient cycling and affects plant growth. Pesticides and soluble fertilizers not only kill organisms within the soil, they also decrease the natural coating of microorganisms on leaf surfaces. Consequently, the plant is more susceptible to disease, and so may require yet more pesticide application.

Many Canadian cities now use Integrated Pest Management (IPM)—a combination of cultural, biological, genetic, and chemical methods—to control pests in their parks and open areas. In this approach, all aspects of the pest problem are considered. Choices are made that present the lowest health risks to humans, have a minimum impact on nontarget organisms, are most specific to target species, and present the least risk to the environment and biodiversity.

CAREER CONNECTION

Agricultural biotechnology technician is an emerging career. Jobs are available in traditional laboratory settings, in greenhouses, or on the farm—working with plants, animals, or microorganisms.

Section 2.17 Questions

Understanding Concepts

1. Using a table with three columns, define each type of symbiosis and classify all the microorganisms described in this section under the appropriate heading.
2. Compare and contrast the practices of EM and IPM.
3. What is the effect of pesticide use on the diversity of microorganisms?

Applying Inquiry Skills

4. Plants were grown on three plots: A = control; B = fertilizer use; C = EM use. Plant growth as indicated by nitrogen uptake is shown in **Figure 2**.
   (a) Summarize the results of this experiment.
   (b) What advantages would there be to choosing soil treatment C over B, even with the difference in growth rates?
   (c) If a fourth test plot were treated with an IPM approach, what might its bar graph look like?

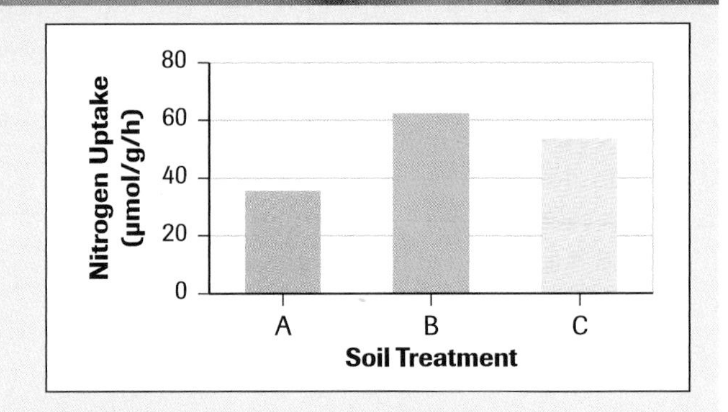

Figure 2
Plant growth, measured by nitrogen uptake

Making Connections

5. Your town council is debating whether or not to pass a resolution banning pesticide use in local parks. Compile a list of pros and cons that could be used in reaching a decision. Be sure to include the effect of pesticides on microorganisms.

Microorganisms in Biotechnology 2.18

Using moulds to produce penicillin is an example of **biotechnology**. The methods for producing yogurt and cheese are examples of biotechnology using bacteria, even though such methods were developed centuries ago. The process of producing silage for animal feed also involves biotechnology. Green plants such as corn, grasses, and alfalfa are harvested, chopped into small pieces, and blown into a silo. If the plant material is packed tightly, anaerobic conditions are created in the silo. Anaerobic bacteria then digest the sugars in the green plant material, converting them to lactic acid. This silage takes on a characteristic smell and taste, and the acid prevents spoilage by other microorganisms. Silage is a major portion of the diet of dairy cattle, especially in the winter.

In the 21st century, biotechnology involves genetic engineering, as described in two processes below.

biotechnology the use of living things in industrial or manufacturing applications

Recombinant DNA Technology

Some of the most exciting developments in biotechnology are based on understanding bacterial reproduction and the structure and function of bacterial DNA. Using this knowledge, microbiologists have perfected a procedure for inserting DNA from other organisms into bacteria. This is called **recombinant DNA technology**.

Recombinant DNA technology is now used to manufacture human insulin for diabetics (**Figure 1**). A section of DNA that controls the production of insulin must be isolated from the chromosomes of a normal human pancreatic cell. A **restriction enzyme** is used as chemical scissors, and the cut-off DNA

recombinant DNA fragments of DNA from two or more different organisms

restriction enzyme an enzyme that attacks a specific sequence of nitrogenous bases on a DNA molecule to create fragments of DNA with unpaired bases

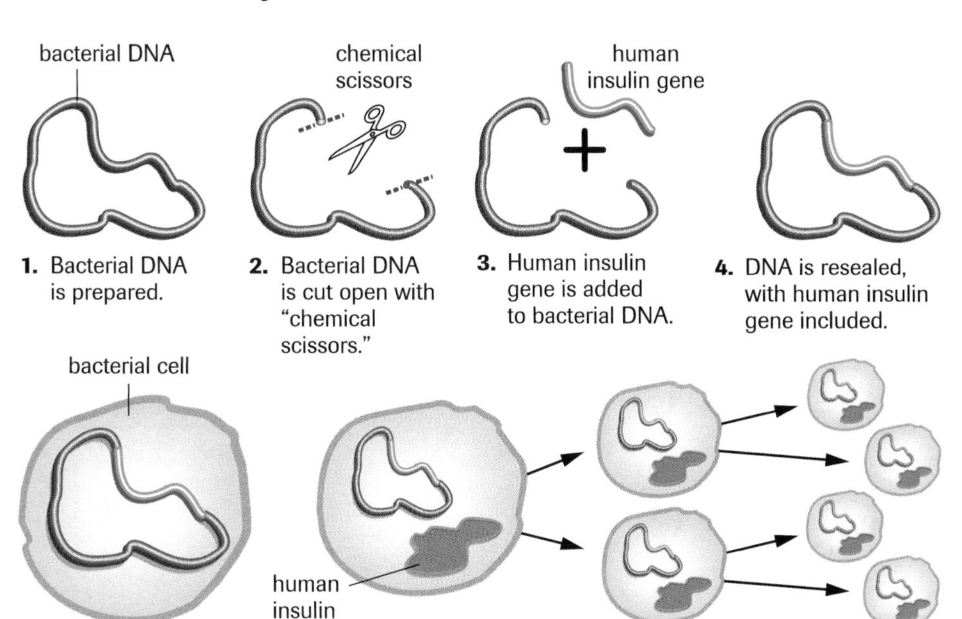

Figure 1
Producing human insulin through recombinant DNA technology

DID YOU KNOW?

Restriction Enzymes
Restriction enzymes are a primitive immune system. These enzymes protect the bacterium from viral DNA.

section is spliced into the plasmid of an *E. coli* bacterium. Under suitable growing conditions, *E. coli* can produce human insulin in substantial amounts. The insulin is then separated from the bacterial cells and purified. Human insulin produced by this method is widely used by diabetics across North America. Use of human insulin has removed the side effects associated with insulin obtained from pork or beef pancreatic tissue.

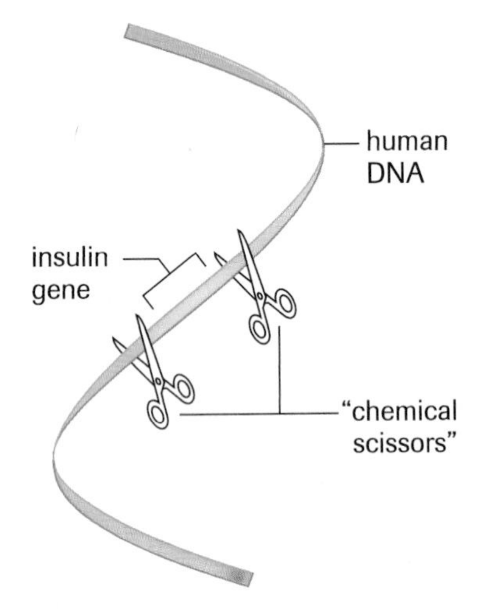

Figure 2
Restriction enzymes at work

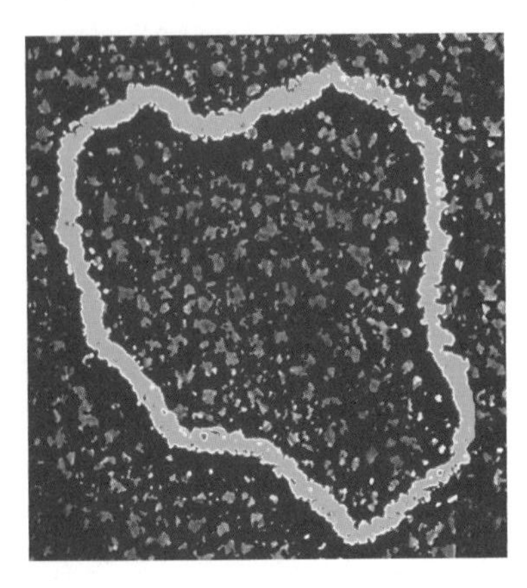

Figure 4
Electron micrograph of the donor plasmid

TRY THIS activity: Cleaving and Recombining DNA

To create recombinant DNA, a restriction enzyme is added to DNA to cut it at specific sites (**Figure 2**). The result is fragments of DNA with unpaired nitrogenous bases.

Materials: a 20-cm strip of red Velcro (both the rough and soft pieces), a 10-cm strip of blue Velcro, scissors, a stapler or masking tape

Cleaving DNA

To simulate the effects of a restriction enzyme on DNA, do the following:

1. Stick the two strips of blue Velcro together to represent two strands in a DNA molecule. Do not match the ends.
2. To demonstrate the effect of the restriction enzyme, use scissors to cut the Velcro (DNA) into smaller strips (about 3 cm long) according to **Figure 3(b)**. This represents an organism's DNA that has been cut.
3. Keep these strips for the next procedure.

Recombining DNA

4. Stick the two strips of red Velcro together, and staple or tape the ends to form a circle, as in **Figure 3(c)**. This circle represents plasmid DNA (**Figure 4**).
5. Cut open the circle using the same procedure you used previously to demonstrate the effect of the restriction enzyme. Remove a strip about 3 cm long. See **Figure 3(d)**.
6. Now take one piece of blue Velcro with sticky ends (the organism's DNA) and attach it to the sticky ends of the red Velcro (the plasmid DNA) to form a circle. See **Figure 3(e)**.
7. You have created a model of recombinant DNA. See **Figure 3(f)**.

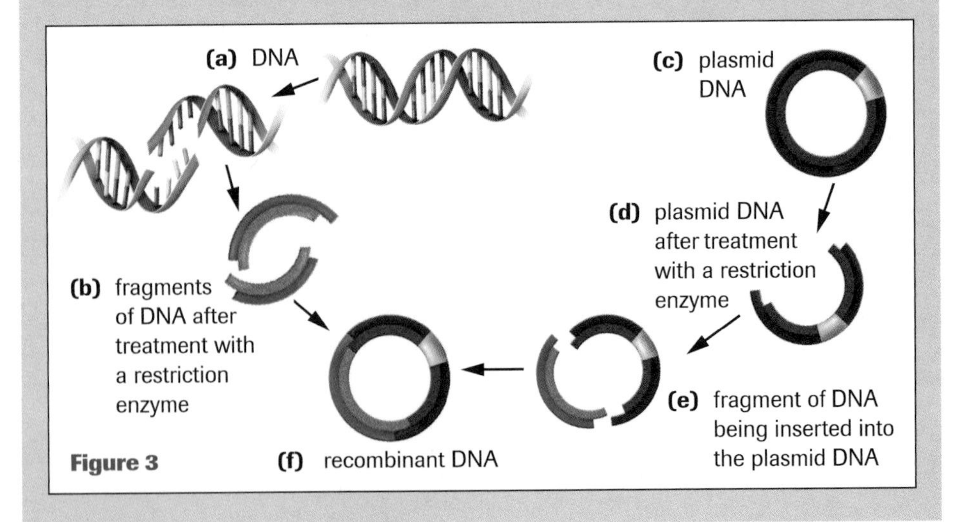

Figure 3

Recombinant DNA technology is one of the most powerful tools ever developed by humans. It is used by plant breeders to develop crops that resist pests, travel to market well, taste better, and contain more nutrients. In the medical field, cultured cells (*E. coli*, yeast, mammalian cells) spliced with a human gene are being used to manufacture products such as the following:

- human growth hormone
- blood factors for treating hemophilia
- vaccines for hepatitis B, diphtheria, and bacterial meningitis
- erythropoietin for treating anemia
- interferons for cancer treatments
- macrophage-stimulating factor for stimulating the bone marrow after a bone-marrow transplant

Inherited genetic diseases such as cystic fibrosis, muscular dystrophy, diabetes, Parkinson's disease, cancer, and Tay-Sachs disease may benefit from recombinant DNA research. As with all new technologies, controversy surrounds gene manipulation in biotechnological processes.

Viral Vectors and Gene Therapy

Viral vectors use viruses to carry altered DNA into cells. The first use of gene therapy in humans occurred in 1990 with an attempt to cure a rare genetic disease called adenosine deaminase (ADA) deficiency in a four-year-old girl. Children develop ADA deficiency when they inherit one defective ADA gene from each parent. These children have little immunity and do not produce antibodies following vaccination against diseases such as whooping cough and diphtheria. If they contract a serious disease, their chances for recovery are slim.

Gene therapy to correct this genetic disease took many years to develop. One step required isolating the normal ADA gene from human T-cell lymphocytes (the white blood cells responsible for fighting infections and cancer). The next step required the development of a method to transfer the normal gene into human immune cells.

To create a viral vector (**Figure 5**), scientists remove some genes in a virus and replace them with the gene to be transferred. Then the vector is mixed with growing cells (tissue cultures) in the laboratory. The vector enters a cell and deposits the new gene in the chromosome of that cell. The gene remains in the cell as long as the cell survives, and is passed on to daughter cells in the tissue culture as the cell divides. Extensive studies with tissue from mice, monkeys, and, more recently, humans, have resulted in the transfer of a new gene into as many as 90% of the cells in the laboratory cultures. These studies have shown this gene technology has a very low risk of causing problems; however, whole-animal studies and clinical trials are necessary.

When the ADA-afflicted child was injected with her own gene-corrected T-cell lymphocytes, her body developed the ability to make appropriate antibodies—the normal immune-system response following infection or immunization. Repeated injections will enable the child to live a relatively normal life. However, this is not an easy or common treatment.

(a) removal of some genes in a virus

(b) insertion of the normal ADA gene (from human T-cell lymphocyte) into the virus

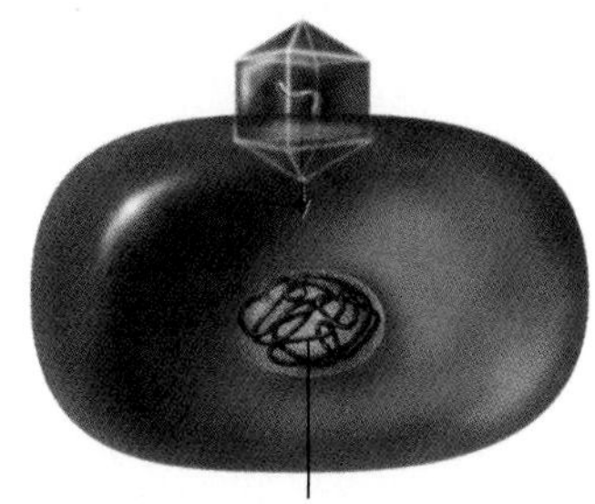

(c) uptake of the altered virus into a human T-cell lymphocyte

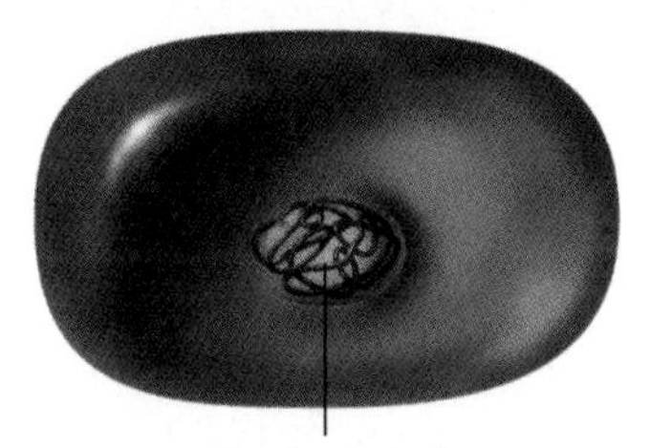

(d) deposit of the new ADA gene into the chromosome in the lymphocyte

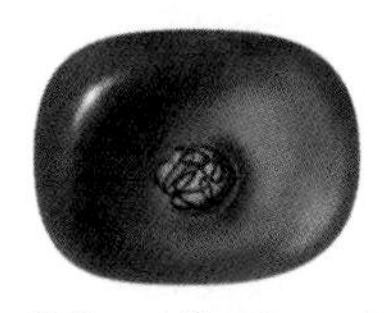
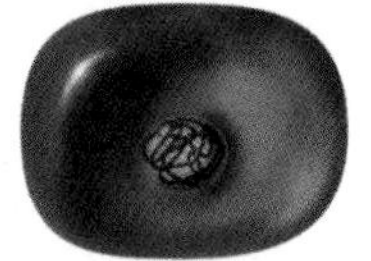

(e) replication of the altered DNA, and division of the lymphocyte

Figure 5
Using viral vectors from genetically altered mouse viruses

Genetic alteration of T-cell lymphocytes has potential in the treatment of many diseases, including AIDS and cancer. Although viral vectors work reasonably well with T-cell lymphocytes, liver cells, and skin, they do not work well with cells that are not multiplying, such as those in the spinal cord and brain. For neurological disorders, such as Parkinson's and Alzheimer's diseases, other types of gene-transfer methods must be developed.

Section 2.18 Questions

Understanding Concepts

1. List 10 products that exist through the use of biotechnology, beginning with text examples.
2. Using a flow chart, describe the silage process.
3. What is recombinant DNA technology? Use the example of *E. coli* bacteria in producing human insulin to describe the role that bacteria play in this technology.
4. List similarities and differences between recombinant DNA technology and the use of viral vectors in gene therapy.
5. What advantages does human insulin have over insulin produced from pigs or cows? What are the disadvantages?

Applying Inquiry Skills

6. Beginning with the following simplified sequence of the steps used to produce human insulin (**Figure 6**), complete the diagram.

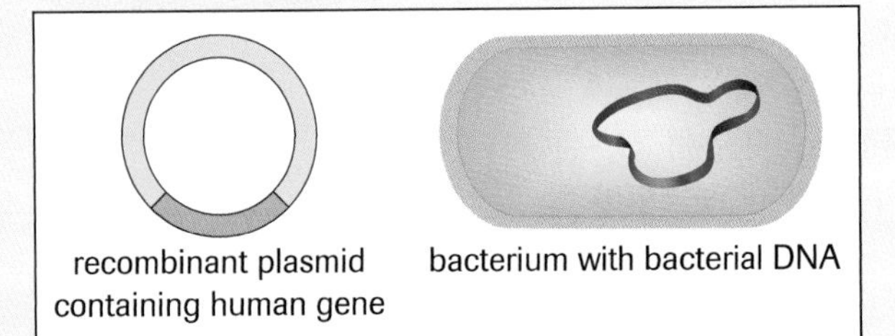

Figure 6

Making Connections

7. Why should you be concerned about the use of genetically altered viruses or bacteria? Identify risks involved with this technology.
8. Write a brief newspaper announcement describing the successful use of viral vectors to treat ADA deficiency in 1990. Answer the questions who, where, when, what, and why. Include a graphic to illustrate your article.
9. What future applications do you see for viruses as vectors of foreign genes? Identify issues of concern: economic, social, political, and so on.
10. Christopher Reeve is raising funds for spinal cord research. Explain why gene therapy is not currently applicable to neurological disorders.
11. Specialists involved in biotechnology are referred to as biotechnologists. Investigate the training and education required for a career in this field.

www.science.nelson.com

12. The Sydney 2000 Olympics were the first Summer Games to test athletes' blood for the recombinant form of erythropoietin (EPO). Summarize the action of EPO and explain why athletes might choose to take this product. List four medical uses of EPO and describe how it is administered.

www.science.nelson.com

13. Cystic fibrosis (CF) is a genetic disorder affecting 1 in every 2500 Canadian children. Research CF: description, current treatment, and the progress of gene therapy in correcting the genetic defect. Present your findings as a table.

www.science.nelson.com

Reflecting

14. "With recombinant DNA technology, the gene pool has become a gene ocean." Explain the meaning of this statement. What are your personal opinions on recombinant DNA technology? What are some of the ethical problems with gene manipulation?

Unit 2 SUMMARY

Key Understandings

2.1 Taxonomy

- taxonomy is used to help biologists identify organisms and recognize natural groupings of living things
- binomial nomenclature assigns each organism a genus name and a species name
- taxa are used to group organisms by their similarities in structure and/or evolutionary history: kingdom, phylum, class, order, family, genus, and species
- this book uses the six-kingdom system: Eubacteria, Archaebacteria, Protista, Fungi, Plantae, and Animalia

2.2 Activity: Using a Classification Key

- like all organisms, protists can be classified using a dichotomous key

2.3 Viruses

- viruses are nonliving particles capable of reproducing only in living cells
- a virus consists of a nucleic acid (DNA or RNA) core and a protein coat
- viruses replicate and produce disease by the lytic cycle; some viruses use a lysogenic cycle as well
- vaccines can prevent some viral diseases; antibiotics are ineffective in treating viral pathogens
- influenza is a representative virus

2.4 Bacteria

- eubacteria and archaebacteria are prokaryotes (DNA is not contained in a membrane-bound nucleus), and are different from all other organisms, which are eukaryotes (possessing a true nucleus)
- bacteria are classified according to cell-wall composition, shape, configuration, respiration, and nutrition
- bacteria reproduce by binary fission
- *E. coli* is a representative Gram-negative bacterium

2.5 Activity: Culturing Bacteria

- bacteria are cultured in a laboratory on nutrient-rich agar, using aseptic techniques
- although individual bacteria are difficult to see, distinctive colony formation aids in classification

2.6 Tech Connect: UHT Milk

- heat processing and aseptic packaging allow milk products to remain bacteria free

2.7 Helpful Bacteria

- bacteria are useful to humans in many ways: as decomposers in the food chain or in pollution control; as probiotic agents within the human body; and as a source of industrial enzymes in the production of consumer goods

2.8 Case Study: Harmful Bacteria

- many bacteria are disease-causing agents; their enzymes or toxins destroy tissues or interfere with normal function
- *E. coli* O157:H7 is a drinking-water contaminant and can cause death

2.9 Fighting Disease

- the human body provides both external and internal defence against pathogens
- the lymphatic system rids the body of infection with the help of special white blood cells called lymphocytes and macrophages
- antibodies are formed in response to antigens; immunity may be either passive or active
- use of antiseptics and disinfectants reduces human exposure to pathogens

2.10 Tech Connect: Future Vaccines?

- different classes of vaccines, administered in a variety of ways, stimulate the immune system to produce antibodies

2.11 Investigation: Observing the Effects of Antiseptics

- antiseptics destroy microorganisms or prevent their development

2.12 Explore an Issue: Bacterial Resistance to Antibiotics

- antibiotics are effective in treating bacterial diseases, but overuse has created bacterial mutations so that resistant bacteria are immune to treatment

2.13 Protists

- the kingdom Protista contains plantlike, animal-like, and fungilike protists; these organisms do not fit into the other five kingdoms
- protists are eukaryotes and may obtain energy through photosynthesis and/or respiration
- plantlike protists vary from simple, one-celled forms to large, multicellular organisms; *Euglena* is a representative example of a one-celled form
- animal-like protists are distinguished by their modes of locomotion; *Paramecium* is a representative example
- fungilike protists have complex life cycles and exist in various forms
- reproduction in protists may be sexual or asexual, with many having complex reproductive cycles
- algae are primary food producers and the source of biological energy for most aquatic food webs; they supply up to 80% of the world's oxygen
- animal-like protists cause a number of diseases in humans

2.14 Activity: Examining Protists

- live specimens provide an opportunity to observe function as well as structure

2.15 Fungi

- fungi have both vegetative and reproductive structures
- fungi can reproduce sexually and asexually but always produce spores
- the field mushroom is a representative example
- some fungi cause disease through skin contact or ingestion
- fungi may have symbiotic relationships with plant roots to form mycorrhizae, or with green algae to form lichens
- lichens are useful in natural succession and as air-pollution indicators

2.16 Investigation: Monitoring Bread-Mould Growth

- microorganisms require certain conditions for optimal growth

2.17 Microorganisms in Ecosystems

- three types of symbiotic relationships give microorganisms a competitive advantage in their environments: parasitism, commensalism, and mutualism
- overuse of pesticides affects all members of the food chain, including microorganisms; new approaches and applications are required to maintain diversity in ecosystems

2.18 Microorganisms in Biotechnology

- recombinant DNA technology inserts DNA from other organisms into bacteria or yeasts; human enzymes and hormones can be produced in this process
- some genetic engineering uses viral vectors to transfer new genes into a host cell; human genetic disorders may be corrected in this process

Key Terms

2.1
binomial nomenclature
species
taxon
Protista
Monera
Fungi
Archaebacteria
Eubacteria
phylogeny
dichotomous key

2.3
virus
capsid
bacteriophage
host range
lysis
lysogeny
antibiotic
vaccine

2.4
eukaryotic
prokaryotic
plasmid
colony
aerobic
anaerobic
endospore

2.7
saprophyte
bioremediation

2.8
toxin

2.9
pathogen
lymphocyte
macrophage
antibody
antigen
active immunity
passive immunity
disinfectant
antiseptic

2.13
ectoplasm
endoplasm
cyst
spore
haploid
carrier

2.15
mycelium
sporangium
germinate
vegetative
dikaryotic
symbiotic relationship
mycorrhiza
lichen

2.17
parasitism
commensalism
mutualism

2.18
biotechnology
recombinant DNA
restriction enzyme

▸ *MAKE* a summary

Design a spider map that compares and contrasts viruses and living organisms (prokaryotic and eukaryotic cells). Place as many key terms in the diagram as possible.

PERFORMANCE TASK

Criteria

Assessment
Your completed task will be assessed according to the following criteria:

Process
- Develop a design.
- Choose and safely use appropriate materials and equipment.
- Carry out the approved investigation.
- Record observations.
- Analyze the results.
- Evaluate the design.

Product
- Compile, organize, and present the research data, using a variety of formats.
- Prepare a formal lab report demonstrating the effects of preservation techniques on cucumber spoilage.
- Demonstrate an understanding of the concepts as they relate to microbiology and food handling.
- Use both scientific terminology and the English language effectively.

Stop the Rot!
Like the microorganisms you have studied in this unit, a cucumber has a life cycle. A seed grows into a plant, which produces a flower, which, when fertilized, forms a fruit. If the fruit is not harvested, it rots. Microorganisms decompose the cucumber into compost, returning its nutrients to the soil.

If you were a farmer growing cucumbers for market, you would be concerned about factors affecting this crop. That list is long: fungi (**Figure 1**), bacteria (**Figure 2**), viruses (**Figure 3**), mycoplasmas (wall-less bacteria—the smallest self-replicating organisms known to science), nematodes (a segmented worm), nutritional deficiencies, and toxic materials all affect cucumber growth and health. Even viroids—infective strands of RNA existing without protein coats—can be responsible for a disease called cucumber pale fruit.

As a consumer, you are interested in maintaining the quality of the fruit once you pick it or bring it home from the market. What are optimum storage conditions for fresh fruit and vegetables? How can a cucumber best be preserved if optimum storage conditions are not available?

Investigation
In this task, you will investigate the cucumber from two angles: decay and preservation.

Part 1: Literature Review
The first part of the Performance Task is a literature review. Use Agriculture and Agri-Food Canada's electronic publications to compile a list of microorganisms that attack the cucumber plant. Prepare a table to include classification (Kingdom), genus/species name, description of the disease, and its treatment or control.

Part 2: Experimental Design
Next, you will design and perform an experiment to determine which method of food preservation best prevents decay while retaining desirable qualities in the fruit (e.g., colour, shape, and texture). As you are dealing with live organisms, flavour is not a consideration, and *no tasting of the products* will be done. Choose some or all of the following standard methods of home food preservation:

Method of preservation and its effect
Freezing: Low temperature inhibits or kills microorganisms.
Canning: Heat kills the organisms.
Dehydration: Removal of moisture kills or deactivates organisms.
Freeze-drying: Water turns from ice to vapour, deactivating organisms and retaining flavour.
Salting: Salt draws out moisture; organisms cannot live in saline environment.
Pickling: Salt and acid are used to inhibit growth of organisms.
Other methods (commercial): pasteurizing, fermenting, carbonating, and the use of chemicals such as benzoates, nitrites, sulfites, and sorbic acids to inhibit or kill microorganisms.

Question

What method of preservation best prevents the rate of decay in a cucumber?

(a) Conduct a literature review to make a list of microorganisms that attack cucumbers. Research how decay caused by certain organisms can be prevented by preservation methods.

www.science.nelson.com

Figure 1
Fungal infection of cucumber

Hypothesis

Choose a method of food preservation.

(b) Formulate a hypothesis for your experiment.

Materials

(c) List the materials you need to carry out the investigation.

Procedure

(d) List the steps you will take to conduct the experiment, and include all necessary safety precautions.

(e) Identify the control and the independent and dependent variables. Have your teacher check the procedure for safety considerations.

(f) Identify the types of qualitative and quantitative observations you will make.

Figure 2
Bacterial infection of cucumber

Observations

(g) Use a data table to record qualitative and quantitative observations.

Analysis

(h) Describe how the variables you have chosen affect the rate of decay.

(i) Provide explanations for what you have observed.

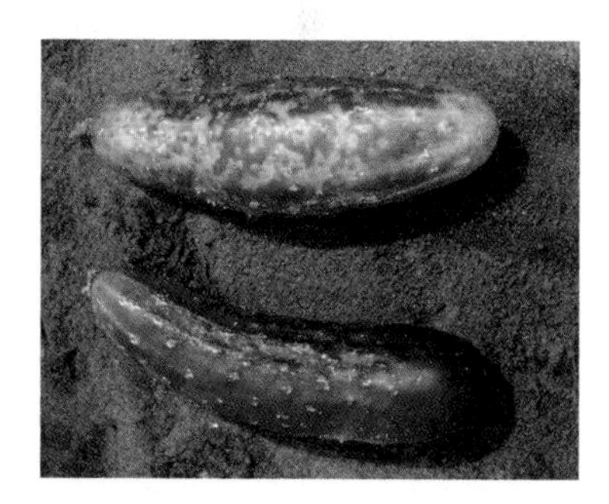

Figure 3
Viral infection of cucumber

Evaluation

(j) Evaluate your hypothesis.

(k) Suggest ways to improve the design of your experiment. Would a taste test allow you to determine the rate of decay better? Why is this not a good idea?

(l) What were some sources of error in this experiment?

Synthesis

(m) Most people prefer to eat cucumber "fresh and raw." From what you have learned, what are the optimum conditions for storing cucumber?

(n) Describe how you could determine the identity of any microorganisms growing on your control cucumber.

(o) If this were research being conducted in a college-level food-technology program, what additional equipment would be available for your experiment?

www.science.nelson.com

Understanding Concepts

1. Approximately 2000 years ago, Aristotle classified organisms into two groups: plants and animals.
   (a) Why is this classification scheme limited?
   (b) Aristotle lived long before the first microscope. If he had been able to view organisms in a microscope, how might his system of classification have changed? What are some examples of organisms that cannot be classified according to his system?
2. Classify the following organisms as Eubacteria, Archaebacteria, Protista, Fungi, or other:
   (a) *Escherichia coli*
   (b) *Plasmodium*
   (c) *Giardia lamblia*
   (d) adenovirus
   (e) methanogens
   (f) *Bacillus anthracis*
   (g) lichen
   (h) yeast
3. Develop an argument to support or refute the following statement: Viruses are nonliving.
4. Why is a host cell unable to make its own proteins while the invading virus replicates?
5. What features distinguish Eubacteria from Archaebacteria?
6. List as many products as you can that rely on a bacterium for part of their production.
7. Prepare a chart to compare viruses, bacteria, protists, and fungi. Consider cell structure, genetic material, locomotion, reproduction, nutrition, habitat, and ecological niche.
8. Describe what is happening in **Figure 1**. If this is the middle photograph in a sequence of events, sketch what comes before and after.

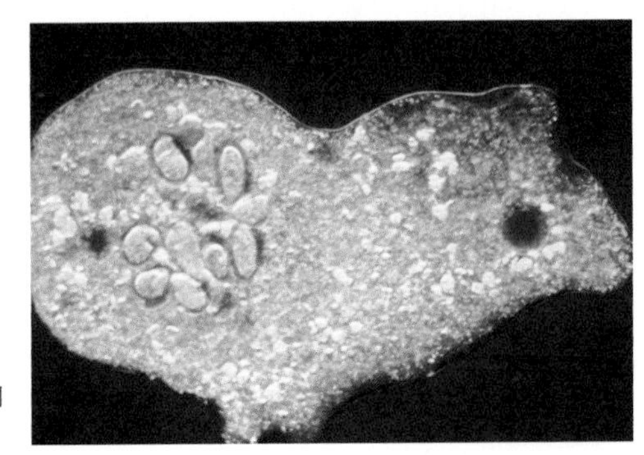

Figure 1
Amoeba

9. Study the food chain in **Figure 2**. Describe how you would redraw it to show the interactions of bacteria.

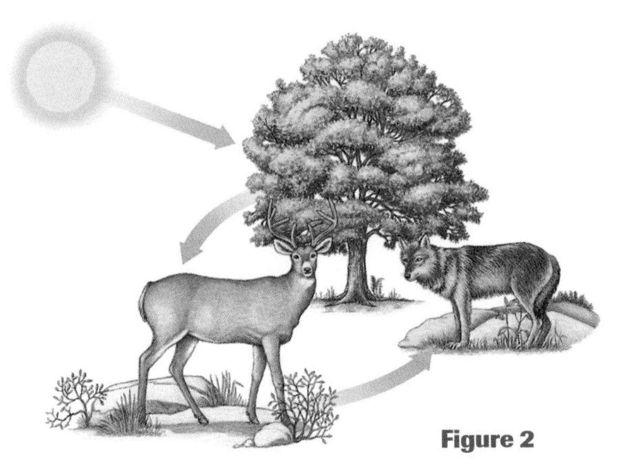

Figure 2

10. List similarities and differences between cilia and flagella.
11. Describe what is happening in **Figure 3**. How does the end product differ from alternative forms of reproduction in protists?

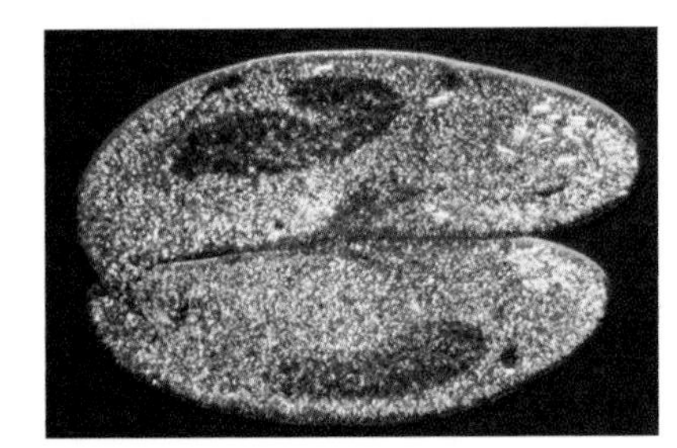

Figure 3
Paramecium

12. What ecological role do algae play?
13. The kingdom Protista has been described as a collection of misfits. Write a short paragraph on the meaning and accuracy of this statement.
14. List six features of fungi that originally placed them in the plant kingdom.
15. In your notebook, label the fungus diagram (**Figure 4**), and explain why the vegetative body or mycelium of a fungus may be present but unnoticed.

Figure 4
Asexual reproduction in fungi

16. Summarize the uses of yeast as described in the Unit Opener on page 93.
17. Explain the differences between using antibiotics and using vaccines to control diseases.
18. In your notebook, classify each of the following vaccines according to the code: 1 = weakened vaccine; 2 = killed vaccine; 3 = toxoid vaccine; 4 = subunit vaccine; 5 = similar pathogen vaccine.

	diphtheria
	whooping cough
	oral polio
	measles
	mumps
	smallpox
	tetanus
	injected polio

19. Identify the green, mosslike structure in **Figure 5**. Explain its symbiotic relationship and how this relationship benefits both humans and the environment.

Figure 5

20. In your notebook, indicate with a Y (yes) or an N (no) whether or not antibiotic treatment is effective in treating the following cases.

	smallpox
	measles
	anthrax
	E. coli food poisoning
	Streptococcus throat infection
	rabies
	malaria
	viral throat infection

21. Define *biotechnology*. Illustrate its use with two traditional methods and two new methods that involve gene manipulation.
22. Humans have applied chemicals to crops for centuries, to stimulate plant growth and to kill pests.
    (a) What effect does this treatment have on microorganisms?
    (b) Describe two other programs that enrich the soil in a more balanced way.
23. A microorganism living in soil produces and releases penicillin (**Figure 6**). Explain how this benefits the fungus.

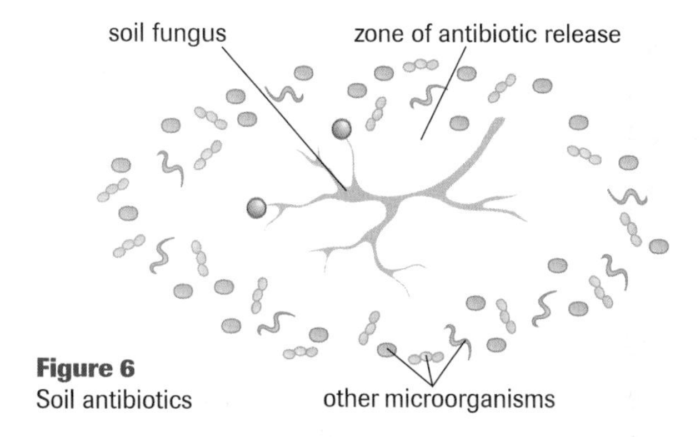

Figure 6
Soil antibiotics

24. Draw arrows and use descriptive terms to link the concepts provided in the concept map (**Figure 7**).

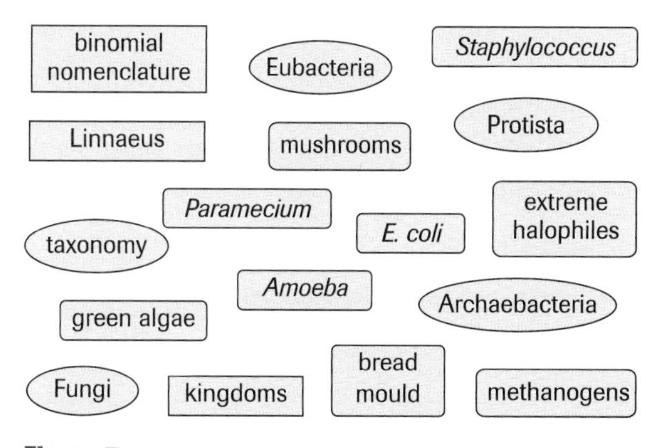

Figure 7

25. In many viral diseases (e.g., smallpox, mumps, and influenza), illness occurs shortly after exposure to the virus. In others (e.g., AIDS), the victim may not show symptoms for many years following the initial infection. How would you explain the difference in these situations?

Applying Inquiry Skills

26. Construct a dichotomous key to classify the viruses in **Figure 8**.

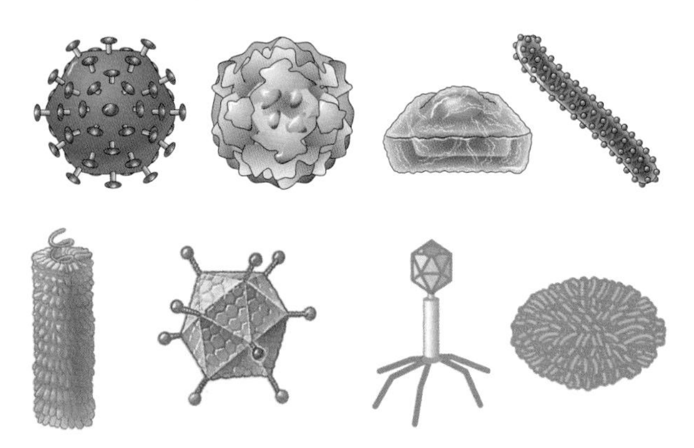

Figure 8

27. While using a microscope to examine pond water, a biology student observed three distinctly different protist species. Although the protists moved too fast for definite identification, the student did observe that protist A had chloroplasts, protist B had flagella, and protist C had a variable shape.
    (a) Classify each protist from the information given.
    (b) What can you conclude about the possible ecological role of each protist?

28. In a mould-growth experiment, bread cubes were moistened with water at three different pH levels. Temperature was kept constant. The number of colonies was counted and recorded daily during a one-week period (**Table 1**). Bread mould is *Rhizopus*.

Table 1

Day	Number of mould colonies		
	pH 4.0	**pH 5.5**	**pH 7.0**
start	0	0	0
1	0	8	1
2	1	14	2
3	2	24	3
4	2	32	3
5	2	40	3
6	3	48	4
7	3	48	4

(a) Arrange the data in a line graph.
(b) According to the data, what pH is optimum for bread-mould growth?
(c) Temperature was controlled in this experiment. Name three other factors that also need to be kept constant in order for the above conclusion to be valid.
(d) Bread mould is a saprophyte. What evidence suggests that bread is a suitable food for this organism?
(e) What environmental conditions would you recommend to prevent mould growth on foods?

29. The following graph (**Figure 9**) shows the growth of microorganisms in raw milk at room temperature as the pH changes over time.

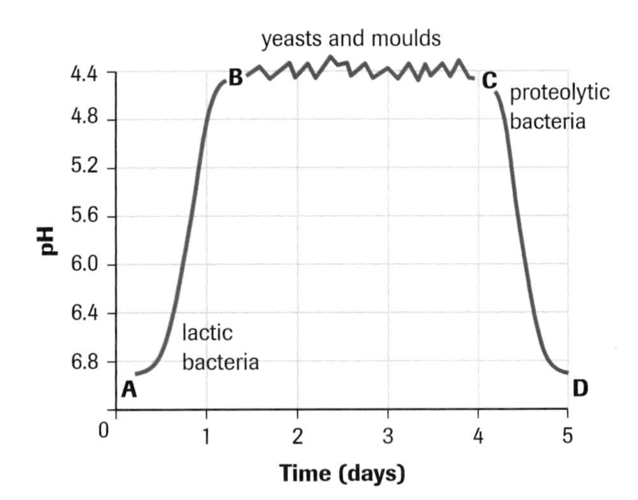

Figure 9
Natural fermentation of raw milk

(a) What four types of microorganisms grow in raw milk?
(b) Use a dictionary to define proteolytic bacteria.
(c) Describe what is happening from A to B, from B to C, and from C to D.
(d) If you were producing a milk product from lactic acid bacteria, at what time period would you halt the growth?
(e) Blue cheese gets its characteristic flavour from moulds. According to the graph, what is the optimum pH for growth of these organisms?
(f) Explain the relationship between pH and time during milk spoilage.

30. Many organic farmers use Integrated Pest Management (IPM) as a strategy to reduce pest problems. Speculate about the effect of this strategy on pest populations and on diversity of species in the soil.

Making Connections

31. *E. coli* is an example of a microorganism that can have beneficial or deadly effects. Use the index of this book to find examples that illustrate this statement. Prepare a table with two columns (helpful and harmful), and describe how *E. coli* fills both roles.
32. You have a sore throat and go to your doctor for treatment. She insists on taking a throat swab and waiting for results of the test before prescribing an antibiotic. Explain why this is the case.
33. Leukemia is commonly known as cancer of the blood. A recent study identified infection as being the major cause of death in leukemia patients. Explain why this is the case, and outline ways that this incidence can be reduced.
34. Many parents do not have their children immunized against polio because "nobody gets polio anymore." What do you think of this reasoning?
35. Dysentery occurs mainly in countries with unsanitary living conditions. However, isolated cases do appear occasionally in countries such as Canada where sanitary conditions are good. What is the most likely explanation for this pattern?
36. Eutrophication is a process where a body of water becomes enriched in dissolved nutrients such as phosphorus. Research this process. Prepare a pamphlet with specific steps that humans can take to reduce eutrophication.

www.science.nelson.com

37. Around the world, many countries are fighting diseases that infect their livestock. Research and report on the infectious agents responsible for mad-cow disease and foot-and-mouth disease.

www.science.nelson.com

38. How might farmers make use of fungi in their fields?
39. Explain how the use of pesticides can have short-term benefits but lead to long-term harm.
40. Gene therapy is used to alter or replace defective genes. For example, ADA deficiency (section 2.18) can be corrected with gene therapy. What is the significance of moving genes from one organism to another? Consider the practical, economic, legal, and ethical aspects of gene therapy.
41. Human growth hormone is a manufactured protein that must be injected frequently into the patient. Research how gene therapy is used to produce human growth hormone and what benefits it could bring to the patient.

www.science.nelson.com

42. Unicellular green-brown algae live symbiotically in coral reefs. These protists supply much of the coral's energy needs by passing on organic carbon (glycerol, sugars, and amino acids) that is produced by photosynthesis. In return, the coral provides the algae with protection, carbon dioxide, and nitrogenous wastes. There is a tight recycling of nutrients in this symbiosis. The waste products of the host go straight to the algae, and not into the water. The presence of the algae seems to increase the rate of calcification (skeletal growth) of the coral. Corals usually grow into a form that maximizes the exposure of algae to sunlight, thus increasing the rate of photosynthesis.
    (a) What form of symbiosis is illustrated in this example? Provide reasons for your choice.
    (b) Use the information given above to explain why coral often grows in water that is nutrient poor.
    (c) Name the other two forms of symbiosis involving microorganisms, and give an equally detailed example of each.

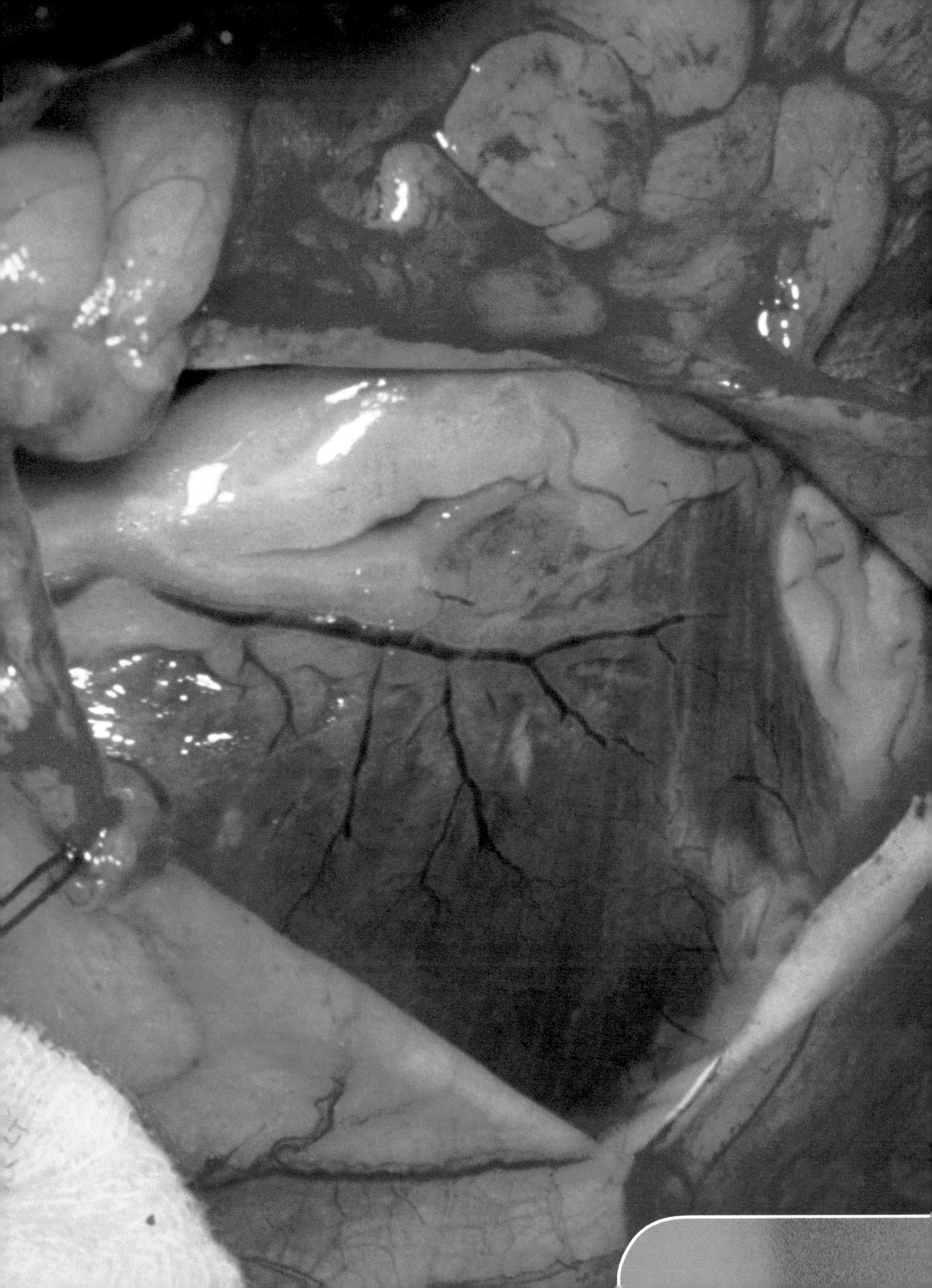

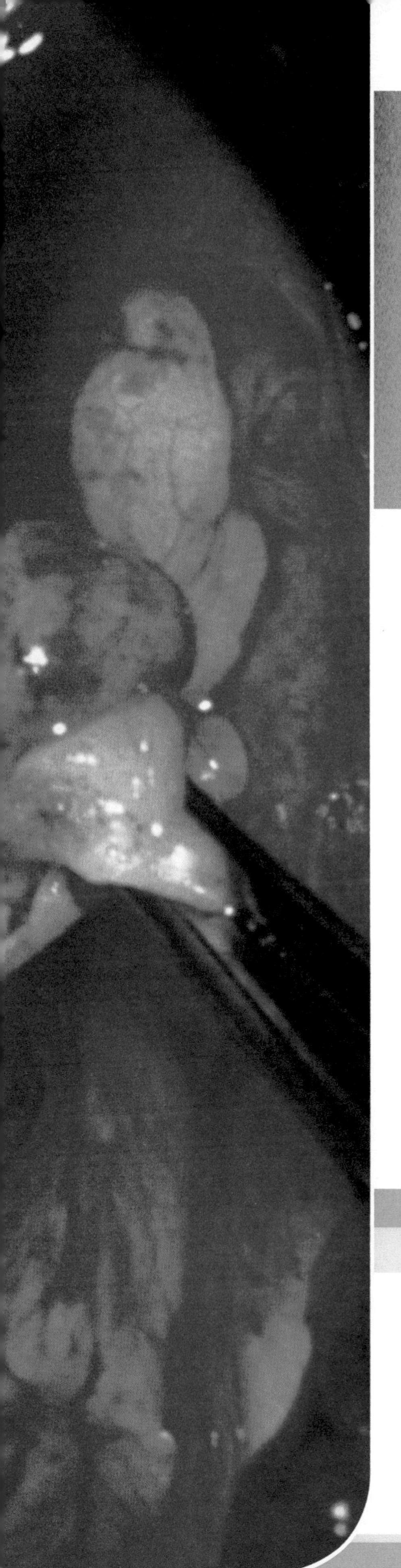

unit

Animal Anatomy and Physiology

In the original *RoboCop* movie, a police officer is killed in the line of duty. Scientists use his remains to construct a cyborg—part human, part machine—that is practically indestructible. This film may have been ahead of its time in showing how science can rebuild human body parts, but it did reflect developments in a rapidly growing field.

Dr. Michael Sefton, director of the Institute of Biomaterials and Biomedical Engineering at the University of Toronto, has proposed a possible solution for heart disease that would provide an almost unlimited number of hearts for transplant. What Sefton calls a "heart in a box" is a transplantable heart that can be grown in the laboratory.

First, researchers create scaffolding—a supporting framework—of biodegradable plastic that the cells will grow around. The next step is to seed the scaffolding with living cells and place it in a sort of incubator that maintains constant temperature and provides the nutrients and oxygen required to support cell division. Although researchers have not yet been able to grow a complete living heart, they have successfully grown components of the heart.

In this unit, you will learn that organs do not work independently; rather, they work in coordinated systems that continuously respond and adjust to changing environments. You will also learn how we use modern technology to diagnose and treat disorders.

Overall Expectations

In this unit, you will be able to

- demonstrate an understanding of the structure, function, and interactions of the main internal systems of humans and other animals;
- investigate, with the aid of laboratory procedures, the physiological mechanisms of animal systems that are responsible for the physical health of the individual;
- demonstrate an understanding of the connections among health, preventive measures, and treatment, and of their social and economic implications.

Unit 3
Animal Anatomy and Physiology

ARE YOU READY?

Prerequisites

Concepts

- levels of organization—cell to organ system
- general structure and function of major human organ systems
- basic cell structure and function

Skills

- design and conduct controlled experiments
- select and use laboratory materials accurately and safely
- organize and display observations in appropriate tables or graphs
- present informed opinions about personal lifestyle choices related to human diseases and disorders

Knowledge and Understanding

1. Use **Figure 1** to complete the following:
   (a) Match the number from the illustration in **Figure 1** with the correct organ name.

Organ name	Organ name
esophagus	brain
testes	intestine
spinal cord	uterus
kidney	heart
lung	

   (b) Match the letter of the illustration in **Figure 1** with the correct body system.
   (i) nervous system
   (ii) reproductive system
   (iii) respiratory system
   (iv) excretory system
   (v) digestive system
   (vi) circulatory system

   (c) Match the body system with its function.

System	Function
1. nervous system	(a) removal of wastes
2. reproductive system	(b) transportation of nutrients, dissolved gases, and wastes to and from body cells
3. respiratory system	(c) response to the environment and control of body activities
4. excretory system	(d) chemical and physical breakdown of food into molecules small enough to pass into cells
5. digestive system	(e) exchange of oxygen and carbon dioxide
6. circulatory system	(f) reproduction of the species

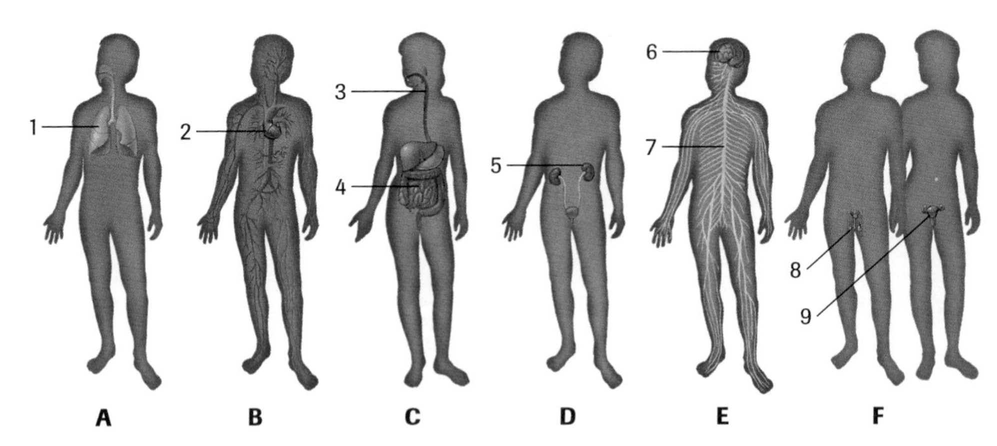

Figure 1

2. Explain the term *cell specialization* as it applies to the development of tissues, organs, and systems. Why is this an important process?
3. What are hormones? Name three hormones and what body systems each affects.
4. Draw **Figure 2** in your notebook and attach one of the following terms to each letter:

   air, carbon dioxide, cell waste, food, food nutrients, food waste, oxygen

   (*Hint:* You will have to use some terms more than once.)
   (a) What do the red arrows represent?
   (b) What do the blue arrows represent?

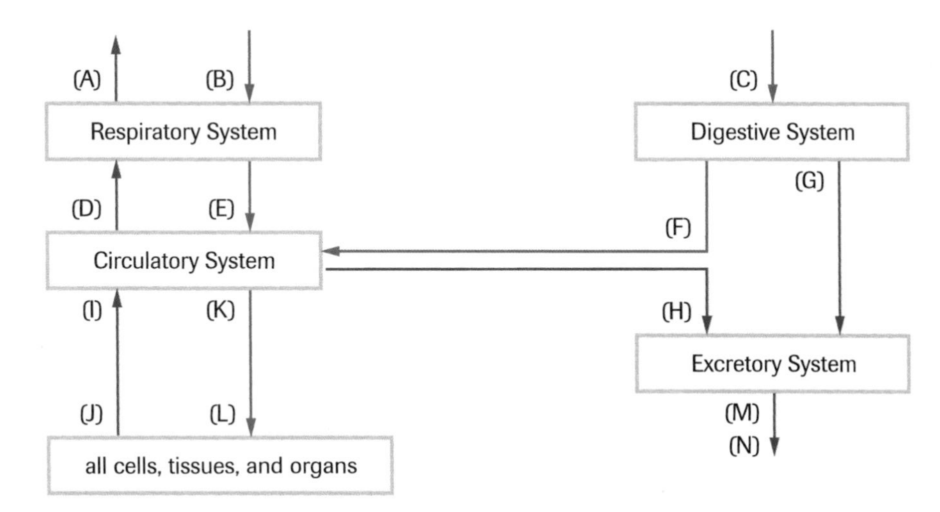

Figure 2
Organ systems working together

Inquiry and Communication

5. **Figure 3** shows a normal red blood cell and one that has been placed in distilled water and examined under a microscope. Explain the changes in the shape of the red blood cell.
6. A group of students conducts an experiment to determine how temperature affects the digestive enzyme amylase. Amylase, found in saliva and gastric juices, helps convert starches into sugars.
   (a) Create a hypothesis for the experiment.
   (b) Identify the dependent and independent variables in the experiment.
   (c) What variables must be controlled in order to obtain reliable data?
   (d) Design an observation table for the experiment.

normal red blood cell

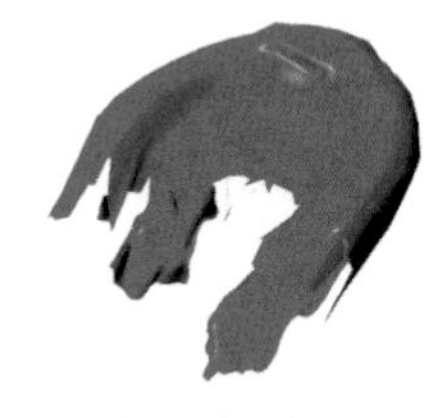

red blood cell after it is placed in distilled water

Figure 3

Technical Skills and Safety

7. The following are materials needed for a worm dissection. Where appropriate, indicate the safety precautions that you should follow when handling each item.

   preserved earthworm
   hand lens (magnifying glass)
   dissecting scissors, scalpel, or single-edged razor blade
   dissecting pan
   dissecting pins
   probes

GETTING STARTED

inactive vs. active youth

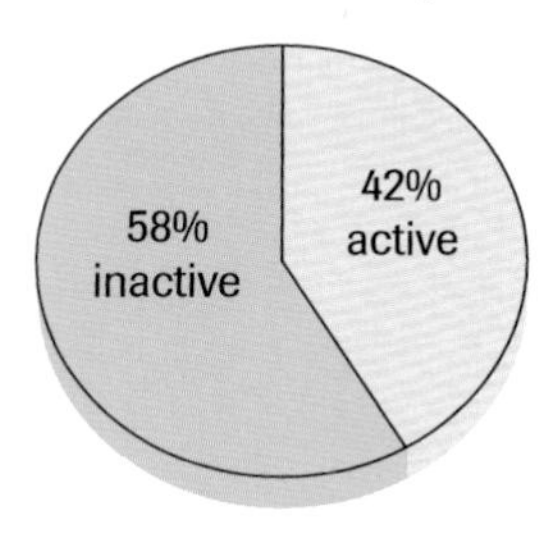

inactive vs. active girls

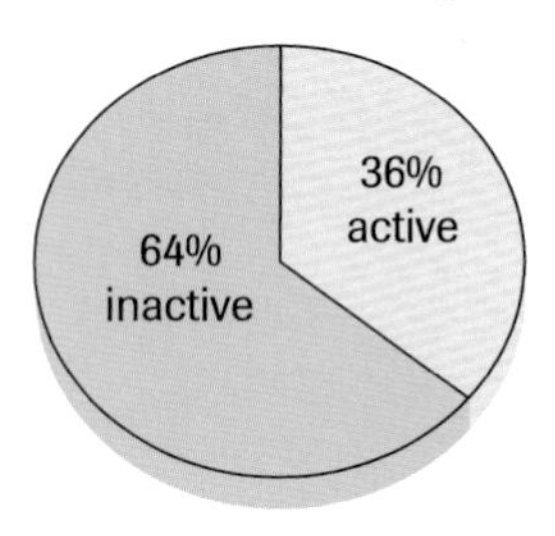

inactive vs. active boys

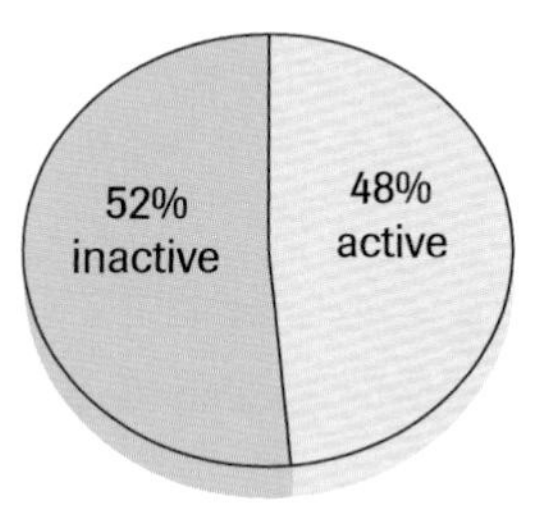

Figure 1
Physical activity in Canadian youth

The functioning of the human body can be compared to a complex machine. Like any machine, it is made up of a number of systems that all work together to enable the machine to function properly. The activities of one system depend on the activities of every other system. If one of the systems malfunctions, other systems are also likely to suffer. And like a machine, the human body has to be fuelled and maintained. We refer to the properly functioning human as being healthy or physically fit. Good physical health means not only being without disease or disability but also being able to participate fully in a variety of physical activities. For most people, lifestyle choices—diet and physical activity—will determine our level of physical fitness.

A recent National Population Health Survey by the Canadian Fitness and Lifestyle Research Institute provided some very interesting statistics.

- Fifty-eight percent (58%) of Canadian youth aged 12–19 were not physically active in the three months before the survey, with significantly more inactive girls (64%) than boys (52%)(**Figure 1**).
- As many as 84% may not have been active enough to meet guidelines for optimal growth and development.
- Youth in higher income families are most likely to be physically active (56% versus between 33% and 43% for other income levels).
- The proportion of those physically active decreases with age. The majority of Canadians (55%) face increased risk of chronic disease and premature death as a result of physically inactive lifestyles.

For the purpose of these statistics, the term *physically active* means using more than 12.56 kJ (kilojoules) of energy per kilogram of body weight per day. A gram of potato chips has approximately 10 kJ of energy. A gram of peanuts has about 20 kJ of energy. A gram of alcohol has 37 kJ. International guidelines for youth recommend a much higher level of activity—25–34 kJ/kg/d. To be considered physically active, a 55-kg person must use 691 kJ of energy per day. That amount of energy can be expended by running at 10 km/h for 18 min; cycling at 20 km/h for 20 min; walking at 7 km/h for 30 min; or swimming at 25 m/min for 37 min.

How much energy would you have to use each day in order to be considered physically active? Are you physically active? **Table 1** provides estimates of the amount of energy required for different activities.

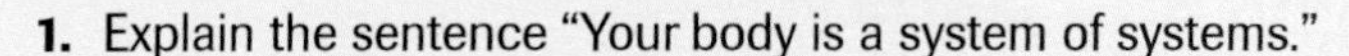

REFLECT on your learning

1. Explain the sentence "Your body is a system of systems."
2. (a) How does your body respond to each of the following situations?
   (i) taking your driving road test
   (ii) passing your driving test

   (b) How are the two responses similar and how are they different?
3. Identify a diet-related disorder and explain the cause(s) of the disorder.
4. Briefly describe how body systems respond to strenuous exercise.

Physical fitness involves more than just physical activity. In order for an individual to be physically active, he or she must eat. Food is the source of energy for all of our physical activities. Our digestive system breaks down food into simpler compounds that our body uses for growth, maintenance, and repair. However, the amount and type of food we eat has an impact on our physical fitness. If we eat more food than we need, the body stores the surplus as fat (**Figure 2**). Fat provides more than twice the energy of an equivalent mass of carbohydrate. So if we eat foods rich in fats, we will likely consume more energy than our body needs, and again, the excess is stored as body fat. Fatty foods aren't the only problem in weight gain; if we consume too many kilojoules of energy from foods high in proteins or carbohydrates, the excess energy is also stored as body fat.

Table 2 shows the amount of energy required daily by humans of various ages and lifestyles. Energy intake beyond these amounts is stored as fat.

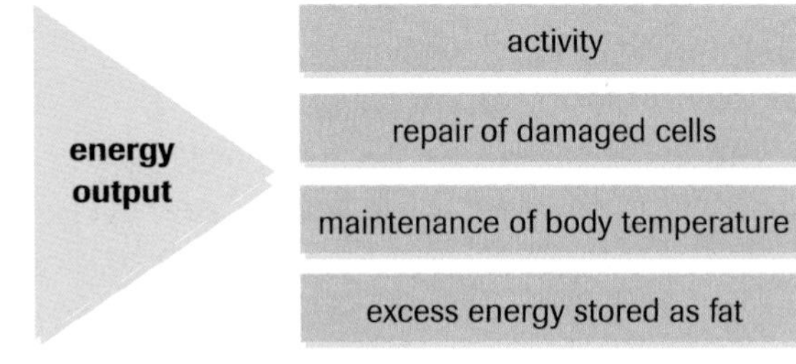

Figure 2

Currently, our national health care system is having trouble keeping up with the demands placed on it. The average age of our population is increasing, causing even more strain on the system. This situation makes the statistics presented on these pages a growing cause for concern. What health problems are associated with physical inactivity? What can you do to reduce or eliminate the risk of developing such health problems? What health problems are associated with excess body fat and with a diet rich in fats? What can you do to improve the nutritional value of your diet? How will these measures affect your overall health?

Table 1 Energy Factors for Various Activities

Type of activity	Energy factor (kJ/kg/h)
sleeping	4.1
sitting	5.2
writing	6.0
standing	6.3
singing	7.3
using a computer keyboard	9.0
cooking	10.5
walking (3.2 km/h)	11.6
bowling	13.6
cycling (13 km/h)	15.8
walking (4.8 km/h)	16.2
walking (6.4 km/h)	20.6
badminton	21.5
cycling (15.3 km/h)	25.8
hiking, fast dancing	27.0
tennis, downhill skiing	36.2
climbing stairs, running (8.8 km/h)	37.5
cycling (20.9 km/h)	40.5
cross-country skiing	42.0
swimming (45.7 m/min)	49.1
running (12.9 km/h)	62.0
competitive cross-country skiing	73.6

TRY THIS activity: Canada's Food Guide to Healthy Eating

Follow the links to Health Canada's Food Guide for Healthy Eating.

www.science.nelson.com

(a) What recommendations does the food guide give? Why do you think it is recommended that you eat large amounts of some foods and smaller amounts of others?

(b) According to the guide, which foods should be eaten in larger quantities? Which foods should be eaten in smaller quantities?

(c) Write down everything you might eat on a typical day. Score yourself using the Healthy Eating Scorecard.

(d) Research the typical daily diet of a person from a country outside of North America. Compare it with your diet.

Table 2 Daily Energy Requirements

Person	Energy requirement (kJ per day)
newborn	2 000
child (2–3 years)	6 000
teenage girl	9 500
teenage boy	12 000
office worker	11 000
manual labourer	15 000

3.1 Organ Systems

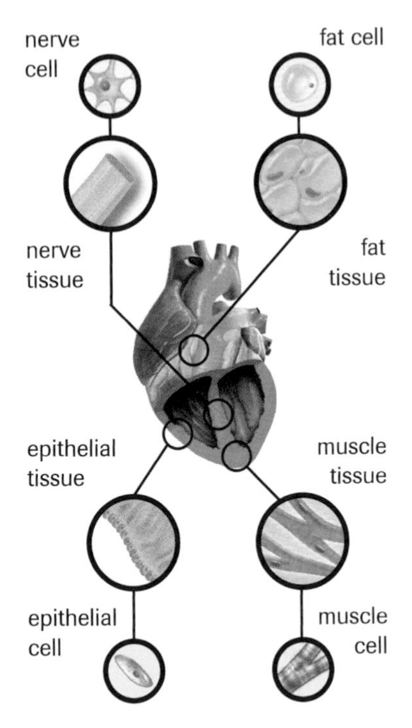

Figure 1
The heart is an organ made up of several different kinds of tissue.

Have you ever used a workshop manual to repair a car? Suppose the headlights were malfunctioning. This problem would be covered in the Electrical Systems section. A knocking noise on acceleration would require an Ignition System diagnosis.

The human body is also structured into systems. Cells are the smallest units of life. Cells similar in shape and function work together as **tissues**. The human body has four primary kinds of tissue: epithelial, connective, muscle, and nervous.

Epithelial tissue is a covering that protects organs, lines body cavities, and covers the surface of the body. Connective tissue provides support and holds various parts of the body together. Cartilage, bone, fat, and blood are examples of connective tissue. Muscle tissue contains sheets or bundles of muscle cells that contract to produce movement. Nervous tissue provides communication between all body structures.

Different types of tissues form **organs** to carry out particular functions. Your hands, stomach, kidneys, and heart (**Figure 1**) are examples of complex organs. Although each organ is composed of a variety of different tissues, the tissues act together to accomplish a common goal.

The heart alone cannot nourish each body cell. It must work together with other organs with related functions (physiology) or structures (anatomy) to form a smoothly running circulatory system. An **organ system** is a group of

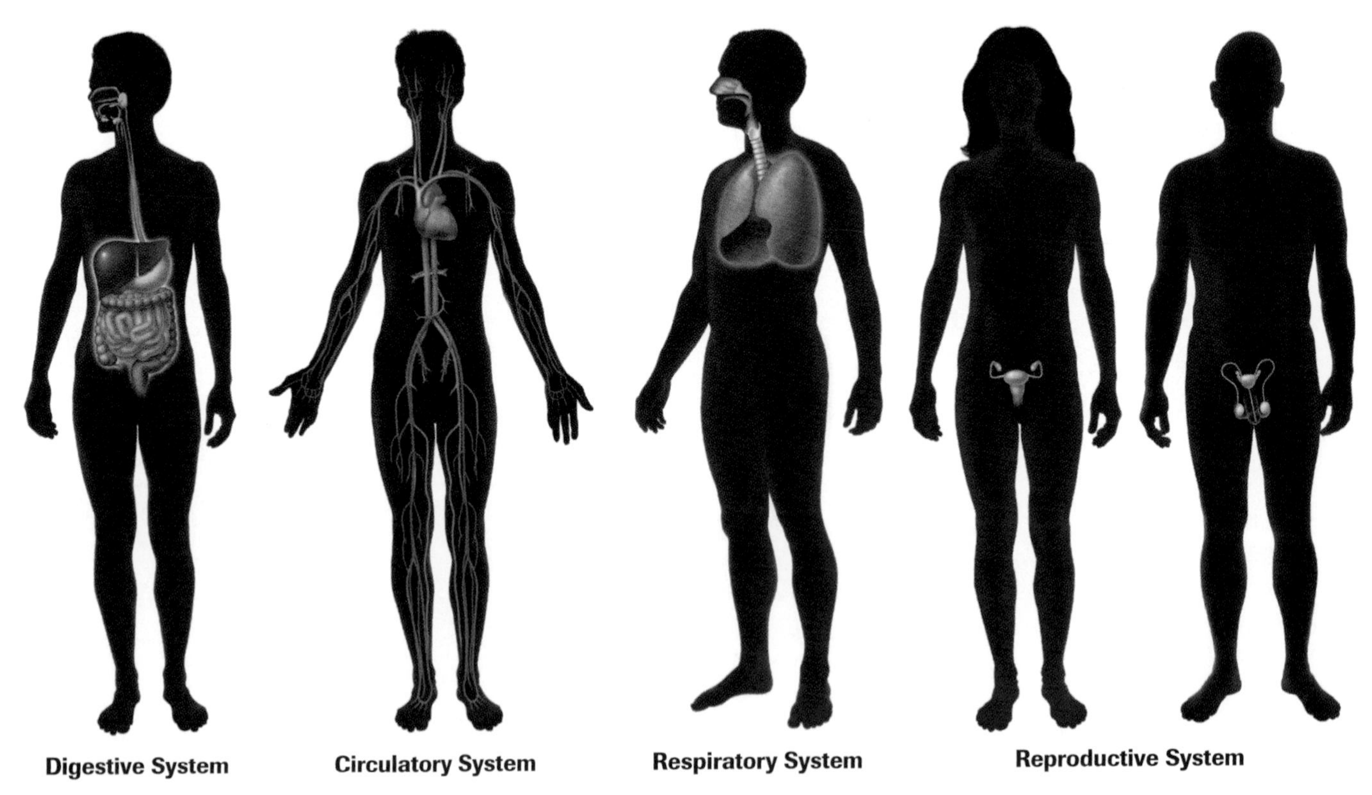

Figure 2
Human organ systems

organs that has related structures or functions. **Figure 2** shows an overview of human organ systems.

Just as a car requires all parts to be synchronized, human organ systems interact closely with each other. For example, the body's digestive system would not be able to function properly if the circulatory system did not allow for the transport of materials and the respiratory system did not provide adequate gas exchange.

Table 1 summarizes the body's important organ systems. 

tissue a group of cells that works together to perform a specialized task

organ a structure composed of different tissues specialized to carry out a specific function

organ system a group of organs that have related functions

Table 1 Human Organ Systems

Organ system	Major organs	Major function
digestive	esophagus, stomach, intestines, liver, pancreas	physical and chemical breakdown of food
circulatory	heart, blood vessels (arteries, veins, capillaries)	transportation of nutrients, gases, and wastes; defence against infection
respiratory	lungs, trachea, blood vessels	gas exchange
reproductive	testes, vas deferens, ovaries, uterus, fallopian tubes, glands	sexual reproduction
excretory	kidneys, bladder, ureter, urethra, liver	removal of wastes
locomotion	bones, muscles	movement of body and body parts
endocrine	pancreas, pituitary gland, adrenal glands	coordination and chemical regulation of body activities
nervous	brain, spinal cord, eyes, ears, nose, tongue, peripheral nerves	response to environment; control of body activities

CAREER CONNECTION 

Medical sonographers operate ultrasound equipment to produce and record images of various parts of the body.

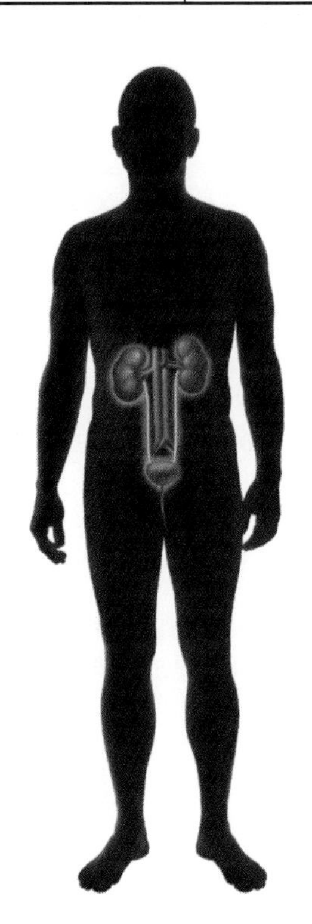

Excretory System

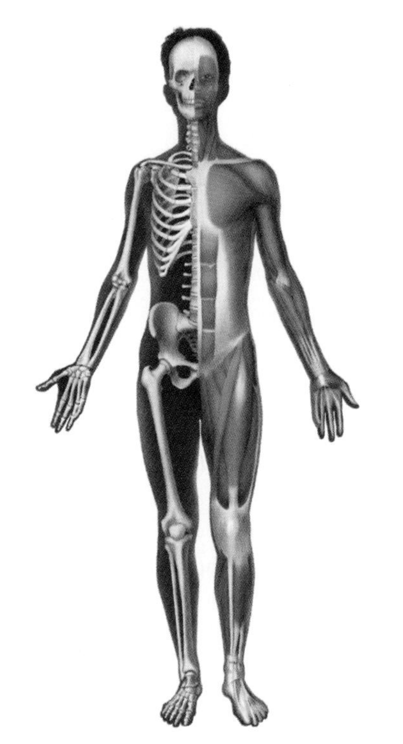

Locomotion: Skeletal and Muscular Systems

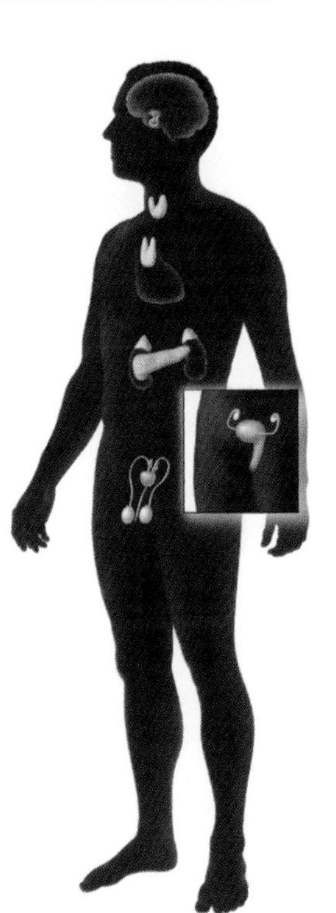

Endocrine System

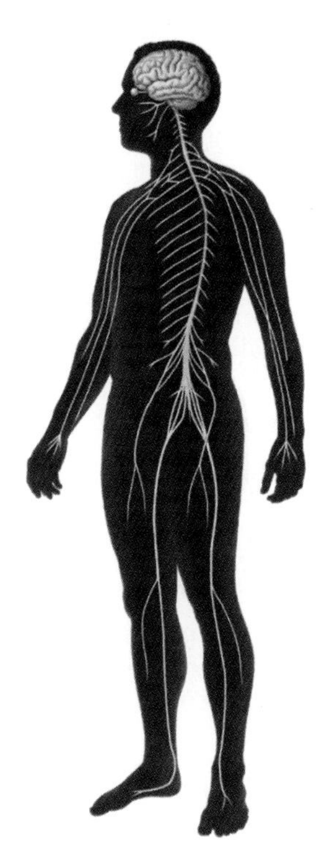

Nervous System

Monitoring Organs

Technology provides many diagnostic tools for examining human organ systems, as shown in **Figure 3**.

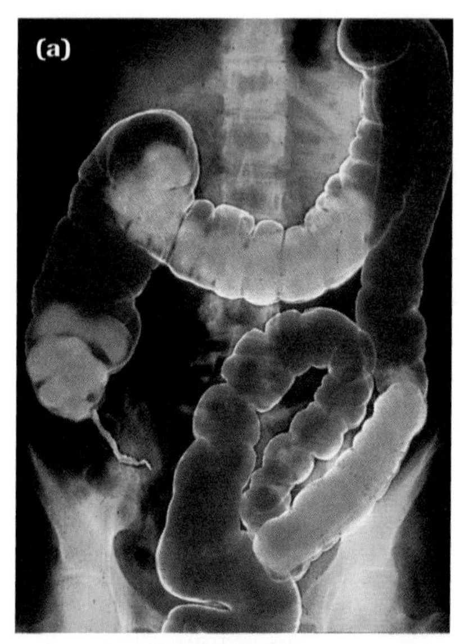

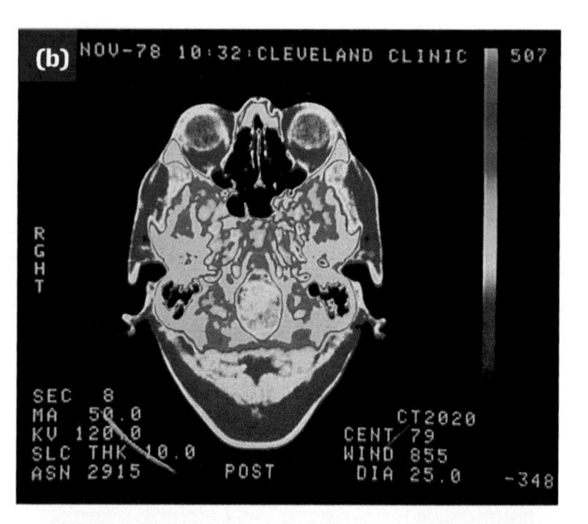

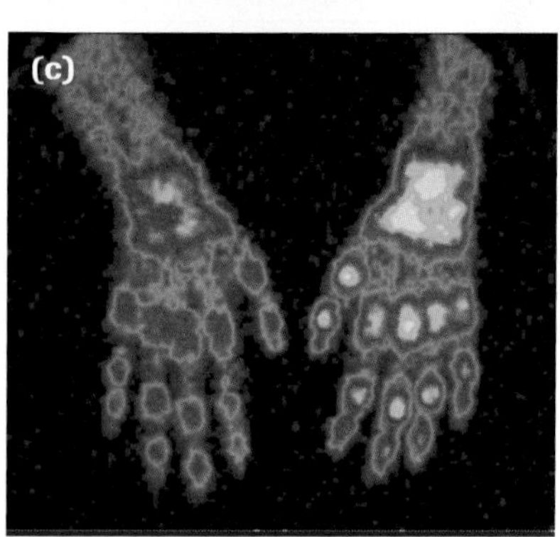

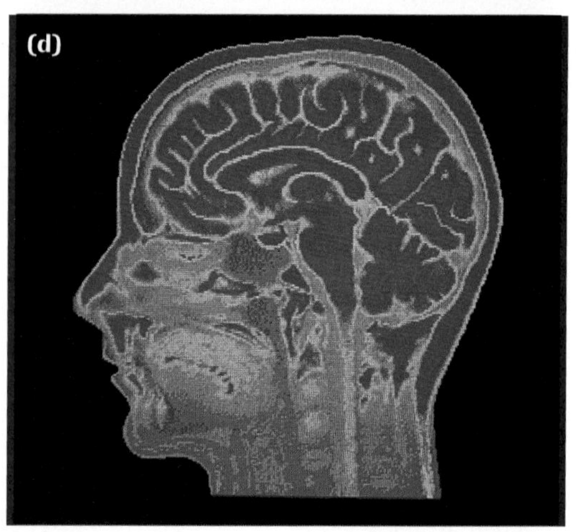

Figure 3

(a) X-rays are high-energy electromagnetic waves that are readily absorbed by bone. This X-ray image also shows the structure of the large intestine, made visible with the ingestion of a metallic compound.

(b) Computed tomography (CT scan) provides information about structure. In a CT scan, many X-ray images are taken from different angles and are reassembled to make 3-dimensional images. This CT scan of a cross section through the skull shows the eyeballs at the top of the image.

(c) Nuclear imaging techniques provide information about function. Radionuclides injected into the body collect in target organs. Arthritic joints appear as brighter areas in the hands of a person with rheumatoid arthritis.

(d) Magnetic resonance imaging (MRI) uses magnetic forces and radio waves to provide information about structure and function. This is a section of a normal female's head.

Section 3.1 Questions

Understanding Concepts

1. What is the difference between a cell, a tissue, an organ, and an organ system? Provide examples of each in your explanation.
2. What two general ways are used to classify organs into organ systems?
3. Give an example of an organ that plays a major role in more than one organ system.
4. Some organ systems are spread throughout the body; others are located together in specific areas. Study the diagrams in **Figure 2**, and classify the eight systems into these two categories.
5. "Organ systems often interact." Use **Table 1** to help you explain what this statement means.
6. List similarities and differences among the four diagnostic tools used to monitor organs.
7. Does a CT scan of the brain indicate anatomy or physiology of the organ? Explain your answer.

Making Connections

8. Research one of the four technologies introduced in this lesson. Describe the process, include a brief history of the technology, and give three specific uses in monitoring organs.

www.science.nelson.com

3.2 Endoscopic Surgery

Tech Connect

An athlete with torn knee ligaments; an elderly man with gallstones; a woman who suffers from chronic heartburn (gastroesophagus reflux disease): What do these patients have in common? Their medical procedures may be performed by endoscopic surgery. Other possible endoscopic surgeries include kidney stone removal; hernia repairs; spleen, appendix, and uterus removal; correction of spinal injuries; and bladder, eye, and colon procedures. Endoscopic surgery can also be used on fetuses in the womb to correct some life-threatening birth defects before birth and can eliminate or postpone the need for traditional heart bypass surgery.

An endoscope (**Figure 1**) is a long, thin, flexible fibre-optic device attached to a viewing tube. On one end it has a tiny lens, light source, and video camera. This instrument may be inserted through any body opening or through a small incision near the body site being repaired. By adjusting the various controls on the endoscope, the surgeon can safely guide the endoscope to a position inside the body. A high-quality picture is projected on a TV monitor, giving a clear, detailed view of the body that is often more precise than an X-ray image. Tiny surgical instruments or biopsy forceps allow the surgeon to perform procedures by remote control.

The compactness and mobility of the device allow it to expose views of crucial areas that would otherwise require extensive cutting to get at the body site needing repair.

Micro-procedures such as endoscopy are classified as minimally invasive surgery. General anesthetic, large incisions, and lengthy hospital stays are rarely necessary with microsurgery, so the patient experiences less trauma than in traditional operations. In North America today, more than half the surgical procedures use endoscopy and are done on an outpatient basis, without an overnight stay in hospital.

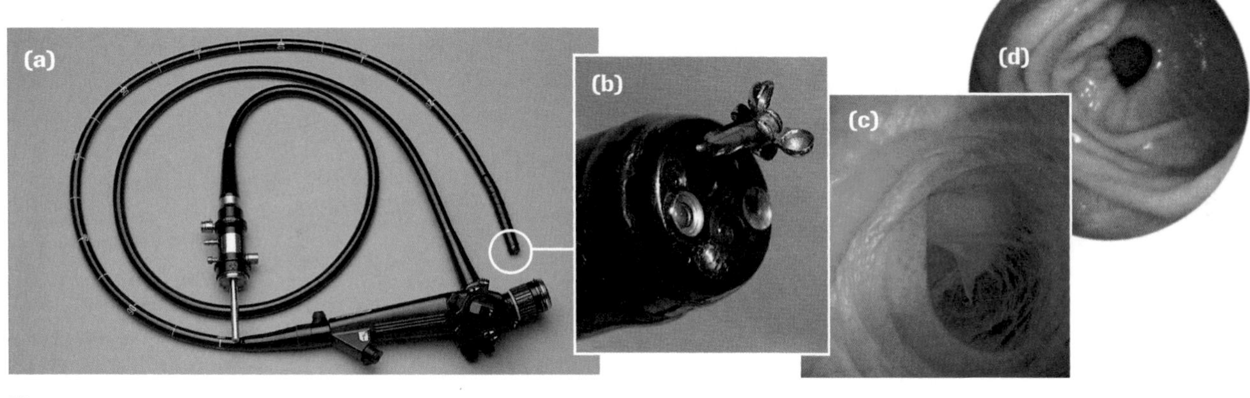

Figure 1
(a) endoscope: control section and flexible tube
(b) internal tip with fittings for surgical tools and pressure-sensitive sensors
(c) endoscopic view of the stomach
(d) endoscopic view of stomach sphincter

Tech Connect 3.2 Questions

Understanding Concepts

1. List 10 medical procedures that can be performed by endoscopic surgery.
2. Describe the features of an endoscope.
3. Why is endoscopy considered microsurgery?

Making Connections

4. What are the advantages and disadvantages of endoscopic surgery? Consider the patient, the surgical team, and Canada's health care system.
5. The surgeon operating the endoscope requires a large support team (e.g., equipment technician, medical record keepers). Describe a career possibility in the medical technology field that might suit your skills and interests. Explain your choice.

www.science.nelson.com

6. Choose any procedure that can be performed by endoscopic surgery. Using the Internet and other resources, describe the medical problem and the solution provided by this technology.

www.science.nelson.com

3.3 Control and Coordination: The Central Nervous and Endocrine Systems

A basic characteristic of all living organisms, not just humans, is their ability to respond to changes in their internal and external environments. The internal environment includes everything inside an organism's body, and the external environment includes everything outside the body.

Homeostasis: A Healthy State of Balance

homeostasis the maintenance of a healthy balance of all chemical reactions in an organism

normal level or **reference level** an average value for measurement of a body structure or a function

You might ask, "How does the body detect changes in its internal or external environment?" The body is able to detect changes because, under normal conditions, it maintains a healthy balance of all chemical reactions—a condition called **homeostasis**. The word *homeostasis* is derived from two Greek words, *homoios,* meaning "same," and *stasis,* meaning "standing still." When a change in the environment upsets this state of balance, the body senses the change and responds by trying to reestablish the balance. This system of active balance requires constant monitoring and feedback about body conditions (**Figure 1**).

Your body is in homeostasis (a healthy balance) when the pH of your blood is 7.35, your body temperature is 37°C, and your heart rate is approximately 72 beats per minute. These values, called **normal levels** (or **reference levels**), are the result of many chemical reactions occurring as they should in a healthy

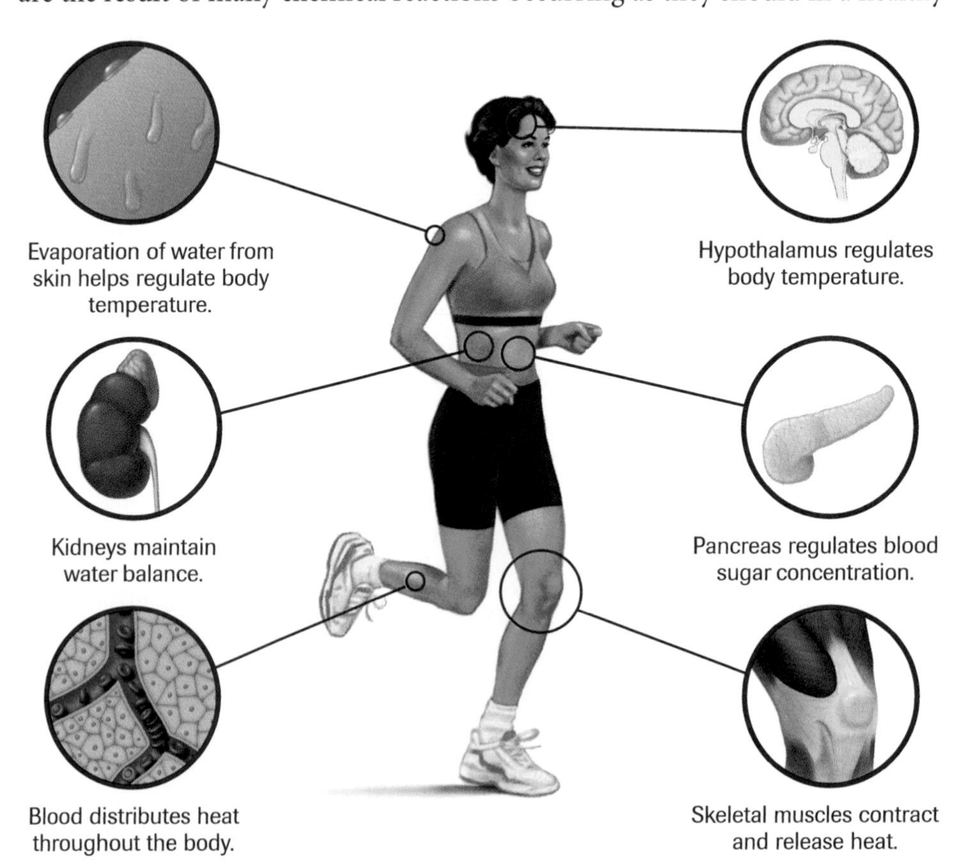

Figure 1
Homeostasis requires the interaction of several regulatory systems. Information about blood sugar, fluid balance, body temperature, oxygen levels, and blood pressure is relayed to a nerve coordinating centre once these factors move outside normal limits.

body. Because people vary in age, gender, genetic makeup, fitness, and lifestyle, normal levels may vary from individual to individual. However, there is a **normal range** (or **reference range**) above or below which a person is suspected of having an illness (**Figure 2**).

normal range or **reference range** the average ranges of values for measurements of body structures or functions above or below which a person is suspected of having an illness

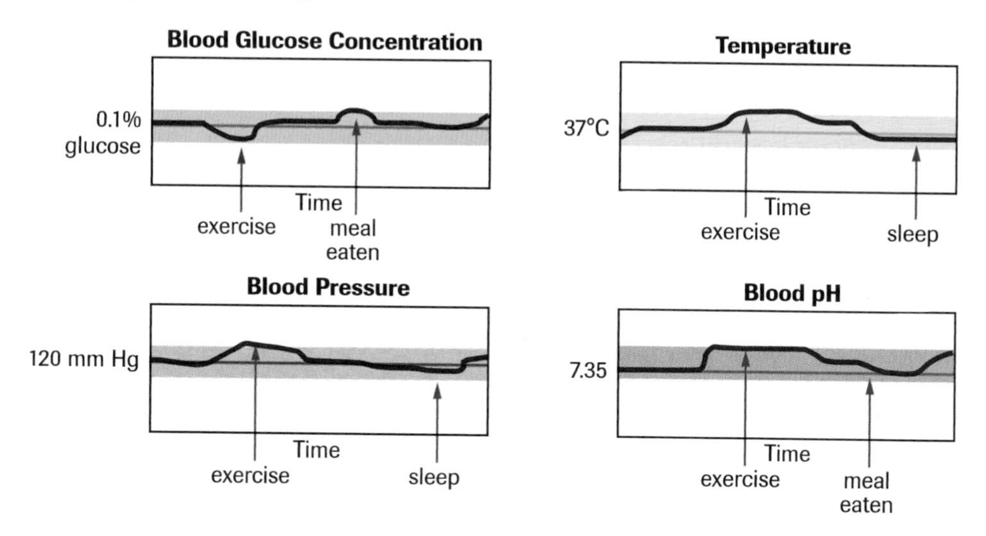

Figure 2
Blood glucose concentration is maintained within a narrow range, and movement outside of the range can signal disease. Body temperature can fluctuate by +2°C with exercise and −2°C with sleep. Blood pressure is usually near 120 mm Hg but can move as high as 240 mm Hg in a very fit athlete for a limited time during strenuous exercise. Blood pH operates within a narrow range, and changes of +/−0.2 can lead to death.

For example, the normal range for body temperature is 36.2°C–37.2°C. When your body is exposed to cold temperatures, causing your body temperature to drop below 36.2°C, it responds by decreasing blood flow to the surface of your body, increasing your heart rate, and twitching your muscles (shivering) in an attempt to warm you up. When you expose your body to hot temperatures, causing your body temperature to rise above 37.2°C, your body increases blood flow to the surface, relaxes your muscles, and increases perspiration in an attempt to cool down.

Physicians (doctors), nurses, and other health care professionals compare a person's measured values with normal ranges to help decide whether the person is healthy or sick. Comparing measured values with normal ranges can also help physicians make a diagnosis. A **diagnosis** is the process of determining the type of illness a person may have, and possibly the cause of the illness. For example, it is well-known that certain microorganism infections cause fever, an increase in body temperature above the normal range. **Table 1** lists the normal range for several components of homeostasis, and the illnesses that may be associated with abnormal values.

diagnosis the process of determining the type of illness a person may have, and possibly the cause of the illness

Table 1 Homeostatic Components and Their Ranges

Homeostatic component	Normal range	Unit	Diagnosis (abnormal levels)
body temperature (oral)	36.2–37.2	°C	fever (high body temperature) hypothermia (low body temperature)
blood pH	7.35–7.45	pH unit	acidosis (low pH) alkylosis (high pH)
resting heart rate (pulse)	50–100	beats/min	tachycardia (fast heart rate) bradycardia (slow heart rate)
resting breathing rate	16–20	breaths/min	hyperventilation (fast breathing rate) hypoventilation (low breathing rate)

Homeostasis and Feedback

negative feedback a process that restores conditions to an original state

Mechanisms that make adjustments to keep body conditions within normal ranges are referred to as **negative feedback** systems. A thermostat-controlled household heating system is an example of a negative feedback system (**Figure 3**). A negative feedback system always contains a monitor, a regulator, and a coordinating centre. The monitor continuously measures the environmental condition being regulated. The regulator can change the environmental condition, and the coordinating centre switches the regulator on or off, depending on the measurements made by the monitor.

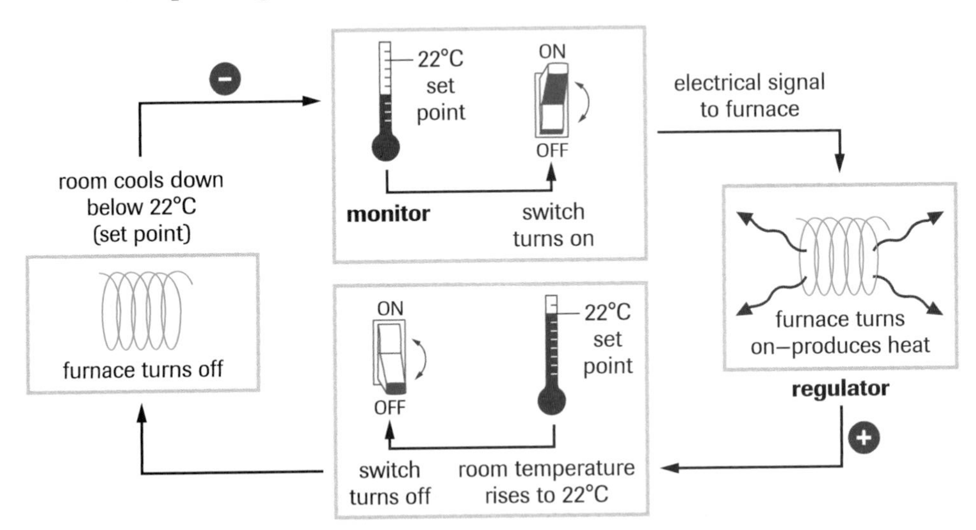

Figure 3
A household thermostat illustrates a negative feedback system. When the temperature of the room (the variable) is lower than the set point (the temperature at which the thermostat is set), an electrical signal is sent to the regulator that turns the heater on. This will increase the temperature of the room. When the room temperature reaches the set point, the electrical signal to the heater is broken and the heater turns off.

In a thermostat-controlled heating system, the monitor is a thermometer, the regulator is a heater (furnace), and the coordinating centre is a thermostat. When the room temperature falls below a particular set point, say 22°C, the thermostat switches on the heater (the regulator). When the thermometer detects a temperature above the set point, the thermostat switches off the heater. This is a case of negative feedback because a change in temperature triggers the control mechanism to make the heater counteract the change ("on" when the temperature is lower than the set point; "off" when the temperature is higher than the set point). Negative feedback mechanisms prevent small changes from becoming large changes. Most homeostatic mechanisms in animals operate on this principle of negative feedback.

There are many examples of negative feedback systems within the human body. In animals, two organ systems are primarily responsible for maintaining homeostasis—the nervous system and the endocrine system.

The Human Nervous System

nervous system the control system that enables animals to detect a stimulus and coordinate a response

A change in the environment that is detected by your body is called a stimulus, and your body's reaction to the change is called a response. When you are hungry and you smell a pizza baking in the oven, saliva is secreted by salivary glands located under your tongue. In this case, the aroma of food is the stimulus, and the secretion of saliva is the response. The control system that enables animals to detect a stimulus and coordinate a response is called the **nervous system**.

Sometimes, changes in your environment are so dangerous that they require the fastest type of response possible. When you accidentally touch a hot iron, you instantly feel pain and quickly pull your hand away. In this case, nerve cells identify the change and cause the body to react very quickly.

Neurons

A **nerve cell** or **neuron** is a specialized cell that uses electrical signals to communicate with other cells of the body. The electrical signal, called an **impulse**, moves through a neuron at a very fast rate, allowing for quick responses. A bundle of neurons is called a **nerve** (**Figure 4**).

nerve cell or **neuron** a specialized cell that uses electrical signals to communicate with other cells

impulse an electrical signal that travels through neurons

nerve a bundle of neurons

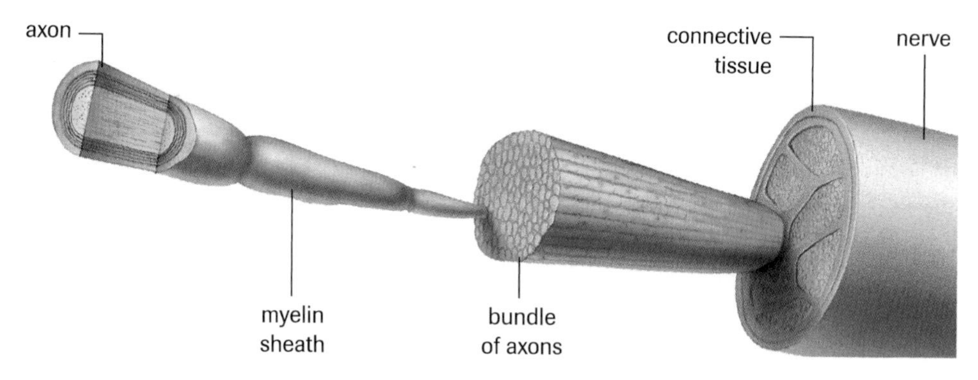

Figure 4
Most nerves are made up of many neurons.

All neurons have the same basic structure, (**Figure 5 (a)**). Each cell has a large cell body and many short **dendrites** that carry signals from outside the neuron toward the cell body. A single, long **axon** carries impulses away from the cell body and toward other cells. In some cases, axons are covered in a fatty white **myelin sheath** that allows impulses to travel very quickly through the axon. **Terminal knobs** at the ends of the axon attach the neuron to cells such as other neurons or muscle cells and allow the impulse to reach its target (**Figure 5 (b)**). The connection between the terminal knob of a neuron's axon and a dendrite of an adjacent neuron or muscle cell is called a **synapse.**

dendrite the part of a neuron that carries impulses toward the cell body

axon the part of a neuron that carries impulses away from the cell body

myelin sheath the fatty white tissue that covers some axons

terminal knob the part of a neuron that attaches it to another cell

synapse the connection between the terminal knob of a neuron's axon and a dendrite of an adjacent neuron

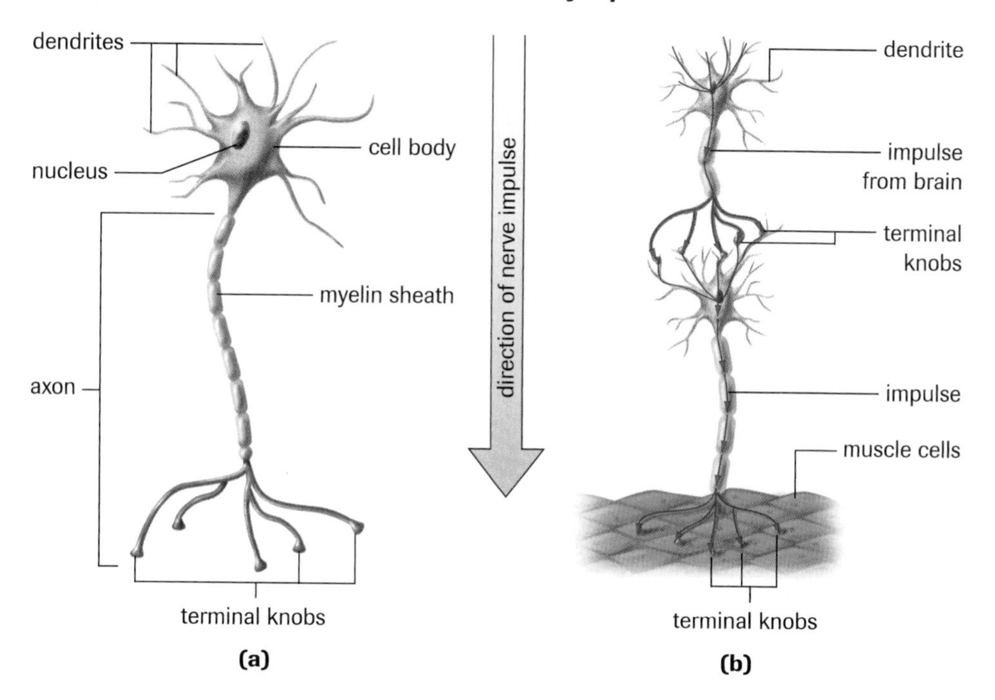

Figure 5
(a) The basic structure of a neuron
(b) A nerve impulse from the brain is relayed to a muscle.

Types of Neurons

sensory neuron a nerve cell that carries an impulse from a receptor to the brain or spinal cord

motor neuron a nerve cell that carries an impulse from the brain or spinal cord to a muscle or a gland

Animals with a brain have three different types of neurons: sensory neurons, motor neurons, and interneurons. **Sensory neurons** carry impulses from receptors in the eye (sight), ear (hearing), tongue (taste), and skin (pressure and heat) to the brain. **Motor neurons** carry impulses from the brain to muscles (for movement) or a gland (for hormone secretion). The muscles and glands are called effectors. Interneurons in the brain or spinal cord link the sensory and motor neurons. When you smell the aroma of pizza baking in the oven, receptors in the nose send an impulse through sensory nerves to the brain. The sensory neurons synapse with interneurons in the brain, which connect with motor neurons that send the impulse directly to salivary glands beneath the tongue. This type of neuron pathway, called a reflex arc, produces very fast responses to stimuli. Reflex actions are predictable, automatic, and unlearned (innate). For example, when you touch the palm of a newborn baby's hand, the hand will automatically grasp in response. The most common reflex arc is shown in **Figure 6**.

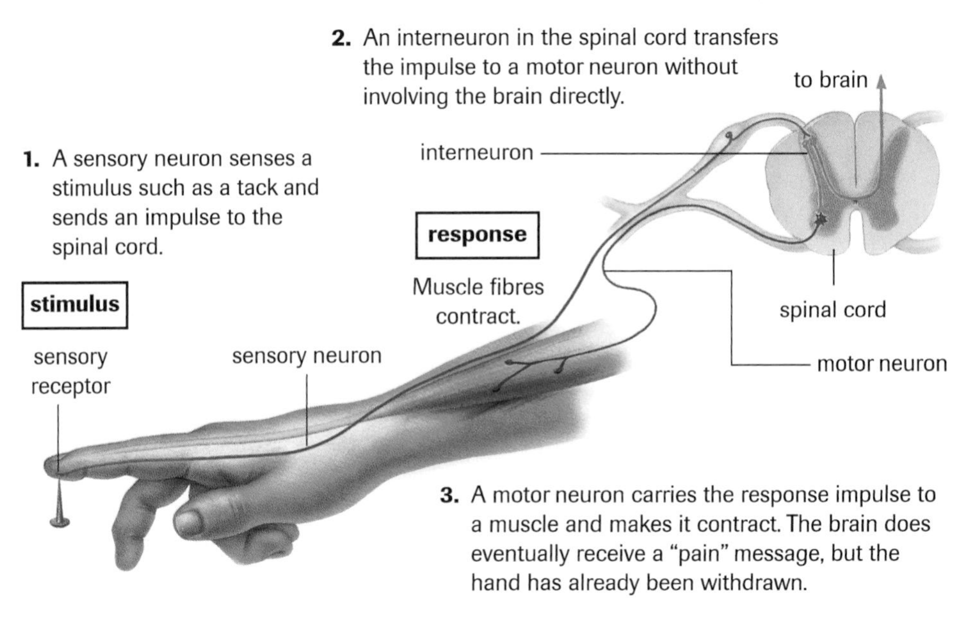

Figure 6
A reflex arc

Nervous Control and Homeostasis

central nervous system the brain and spinal cord

peripheral nervous system all nerves except those in the brain and spinal cord that relay information between the central nervous system and other parts of the body

somatic nervous system motor neurons that are under conscious control

autonomic nervous system motor neurons that function without conscious control

The human nervous system is divided into two parts—the central nervous system and the peripheral nervous system. The **central nervous system** includes the brain and the spinal cord (**Figure 7**), and the **peripheral nervous system** includes all other nerves in the body. It is important, however, to note that all components of the nervous system are coordinated by the central nervous system.

The peripheral nervous system is further divided into two separate parts—the somatic nervous system and the autonomic nervous system (**Figure 8**).

The **somatic nervous system** includes motor neurons that are under conscious control and all of the sensory neurons in your body. The **autonomic nervous system** includes only motor neurons that function without conscious control. It is the sensory neurons that tell your brain that your nose is itchy, and the motor neurons that make your muscles scratch the itch. Both are part of

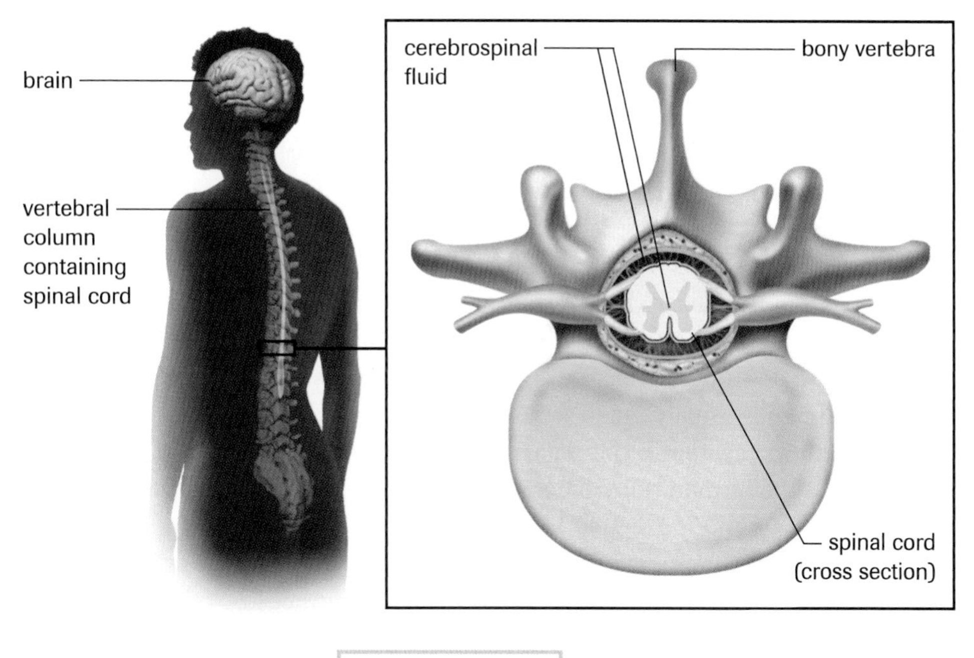

Figure 7
The central nervous system of a vertebrate includes a brain and a spinal cord.

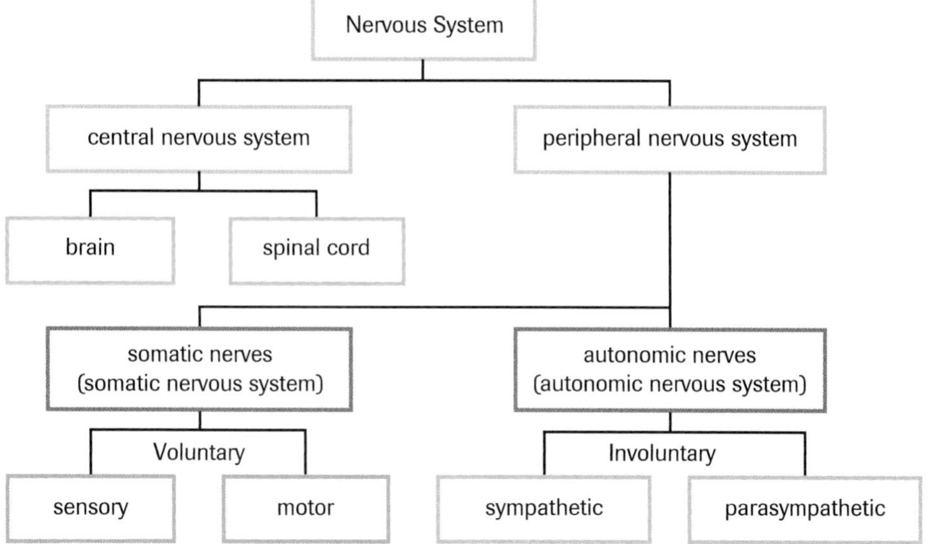

Figure 8
The main divisions of the nervous system

the somatic nervous system. The muscle of your heart, however, is controlled by motor neurons of the autonomic nervous system because, in general, you have no conscious control over heart-muscle contraction. Other functions controlled by the autonomic nervous system are body temperature, blood pressure, pupil dilation, and gland secretions.

The autonomic nervous system includes two completely separate nerves for each body part under its control—a **sympathetic nerve** and a **parasympathetic nerve**. In general, sympathetic nerves counteract the effects of parasympathetic nerves. For example, parasympathetic nerves make the pupils of the eye smaller, slow down heart rate, speed up digestion, and contract bladder muscles. Sympathetic nerves increase the size of the pupils, speed up heart rate, slow digestion, and relax bladder muscles. The sympathetic nerves prepare the body for danger, and the parasympathetic nerves return the body to a normal, more relaxed state.

sympathetic nerve a nerve that prepares the body for danger

parasympathetic nerve a nerve that returns the body to a normal state after a stressful situation

The Human Endocrine System

hormone a compound released by one type of cell that has an effect on other cells of the body

gland an organ that secretes hormones and other useful substances

endocrine system the organ system that regulates internal environmental conditions by secreting hormones into the bloodstream

In humans, special chemicals called hormones play a major role in the maintenance of homeostasis because they help regulate and coordinate the functions of virtually all organ systems. Hormones control many developmental changes. For example, during puberty, hormones help in the development of secondary sex characteristics. A **hormone** is a compound released by one type of cell in the body that is transported throughout the body and has an effect on other cells of the body. Most hormones are manufactured by specialized cells in organs called **glands** and are secreted (released) into the bloodstream. The process of production and release of useful substances such as hormones into the bloodstream is usually called secretion.

The **endocrine system** is the organ system that secretes most of the hormones in the human body. The major glands of the human endocrine system are shown in **Figure 9**.

Although hormones secreted into the bloodstream reach all the cells of the body, they only affect certain target cells. This is possible because cells of the body have different hormone receptors on the outer surface of their cell membranes or within the cell that attach only certain hormone molecules (**Figure 10**). For example, antidiuretic hormone (ADH), secreted by the pituitary gland, acts only on certain cells of the kidney even though it comes in contact with most other cells of the body.

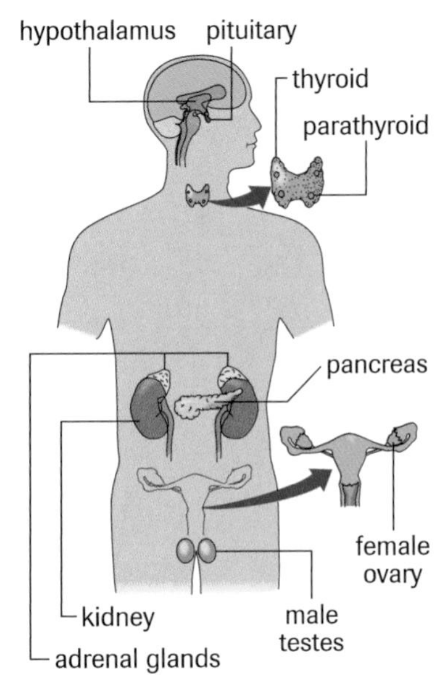

Figure 9
The location and appearance of some important endocrine glands in the human body. Hidden within the thyroid gland are four small glands—the parathyroid glands.

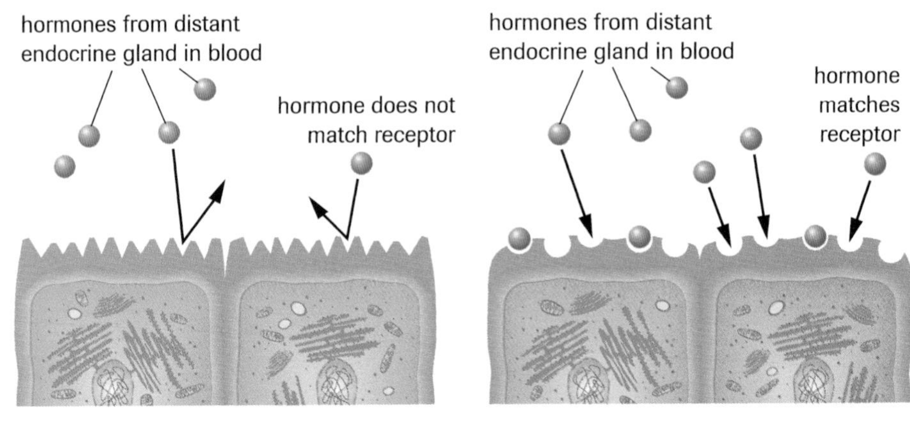

Figure 10
Hormones do not affect nontarget cells because the receptors do not match in shape. However, target cells "recognize" hormones that have the correct shape to be trapped by the receptors.

pituitary gland a gland at the base of the brain that secretes more types of hormones than any other endocrine gland

hormone-releasing factor a hormone released by the hypothalamus of the brain that controls the secretion of various hormones of the anterior lobe of the pituitary gland

Although it is only the size of a pea, the **pituitary gland**, located near the base of the brain, secretes more types of hormones than any other endocrine gland. The hypothalamus and the pituitary gland interact to produce a variety of effects on several distant body parts. The pituitary gland is divided into two lobes—the anterior lobe and the posterior lobe (**Figure 11**).

Neurons in the hypothalamus secrete **hormone-releasing factors** into small blood vessels leading to the anterior lobe of the pituitary gland. The hormone-releasing factors stimulate cells of the anterior pituitary to secrete a number of hormones that control the hormone secretions of various other glands of the body (**Table 2**).

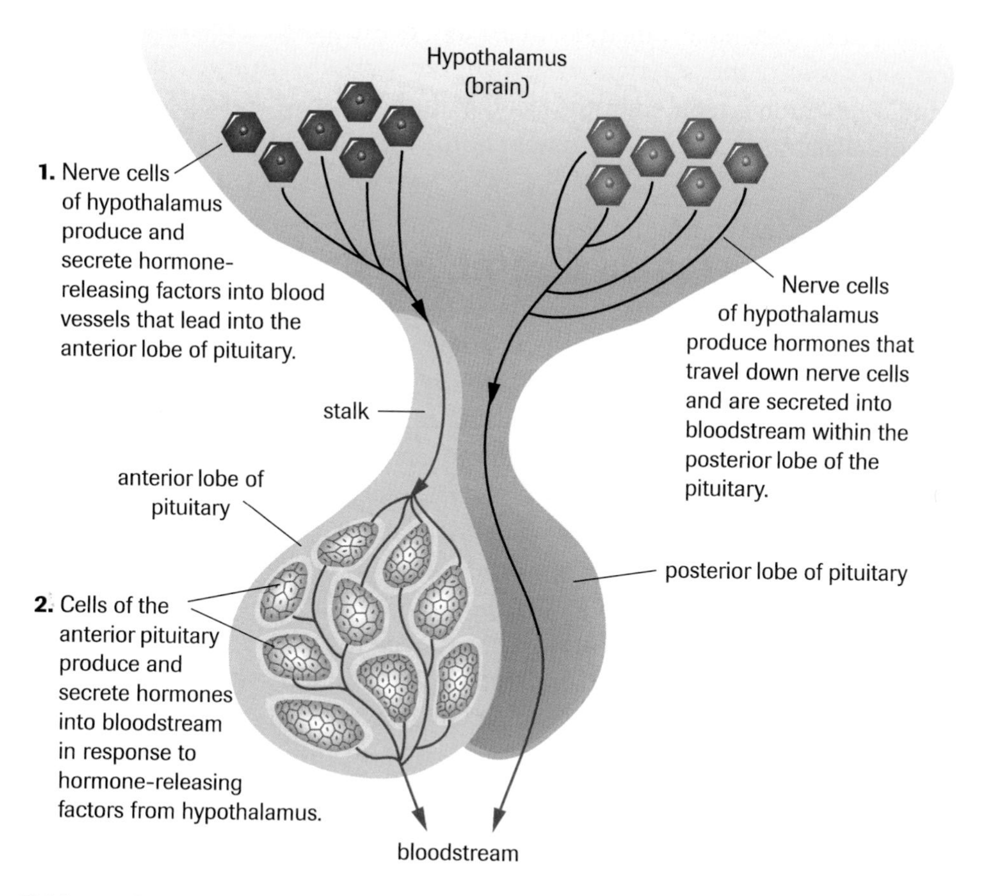

Figure 11
The hypothalamus controls secretion by the pituitary gland. The pituitary gland is composed of two separate lobes: the anterior and the posterior lobe.

Table 2 Pituitary Hormones

Hormone	Target	Primary function
Produced and secreted in anterior lobe of pituitary		
thyroid-stimulating hormone (TSH)	thyroid gland	stimulates release of thyroxine from thyroid; thyroxine regulates cell metabolism
adrenocorticotropic hormone (ACTH)	adrenal cortex	stimulates release of hormones involved in stress responses
growth hormone (GH)	most cells	promotes growth
follicle-stimulating hormone (FSH)	ovaries, testes	in females, stimulates follicle development in ovaries; in males, promotes the development of sperm cells in testes
luteinizing hormone (LH)	ovaries, testes	in females, stimulates ovulation and formation of the corpus luteum; in males, stimulates the production of the sex hormone testosterone
prolactin (PRL)	mammary glands	stimulates and maintains milk production in lactating females
Produced in hypothalamus, secreted in posterior lobe of pituitary		
oxytocin	uterus, mammary glands	initiates strong contractions; triggers milk release in lactating females
antidiuretic hormone (ADH)	kidneys	increases water reabsorption by kidneys

The pituitary gland is sometimes called the "master gland" because its hormones control the secretions of so many other glands. However, its "master gland" reputation is reduced by the fact that the hormone-releasing factors of the hypothalamus control the secretions of the anterior pituitary gland.

The Fight or Flight Response: Nervous-Hormonal Interaction

When you fear for your personal safety or feel stressed in other ways, stimulation by your sympathetic nerves increases your heart rate and redirects blood flow away from your skin (by contracting blood vessels in the skin) and toward your muscles (by dilating blood vessels in muscles), thus improving muscle action. Sympathetic nerve stimulation also improves peripheral vision by dilating your pupils. These are all nervous actions that prepare your body to deal with the potential danger. This is called the fight or flight response or the stress reaction, and it causes your body to either fight the cause of the danger or run away from it. In this situation, sympathetic nerves also stimulate your adrenal glands to secrete the hormone **adrenaline** into the bloodstream. Adrenaline maintains and enhances the changes caused by the sympathetic nerves. Adrenaline increases heart rate, increases blood pressure, causes perspiration, and increases muscle tension. The effects of adrenaline would be harmful to the body if they lasted for an extended period of time. The parasympathetic nerves prevent this from happening by returning the body to normal conditions when the dangerous or stressful situation has ended. It is evident that the endocrine system and the nervous system work together to maintain homeostasis.

adrenaline a hormone secreted by the adrenal glands that increases heart rate, increases blood pressure, causes perspiration, and increases muscle tension

Section 3.3 Questions

Understanding Concepts

1. Explain the concept of homeostasis in your own words.
2. A person wants to remain stationary while walking up the down escalator.
   (a) What will the person have to do if the speed of the escalator suddenly increases? suddenly decreases?
   (b) How is this example similar to the concept of homeostasis?
3. Explain how the body adjusts to cold external temperatures.
4. Explain the difference between sensory neurons and motor neurons.
5. What would happen if neuron I in **Figure 12** was severed?
6. (a) What is a hormone?
   (b) Distinguish between a hormonal response and a nervous response.
7. (a) Draw a single nerve cell and label its parts.
   (b) Describe a function for each of the parts you labelled in (a).

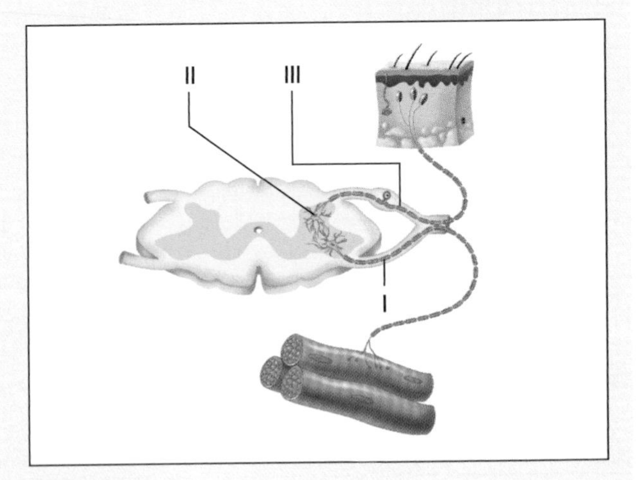

Figure 12
Reflex arc

Section 3.3 Questions *continued*

8. Describe the function of one hormone secreted by the anterior pituitary and one hormone secreted from the posterior pituitary.
9. (a) Use a simple graphic organizer to show how the subsystems of the peripheral nervous system are related.
   (b) Identify each subsystem as either voluntary or involuntary.
   (c) Briefly describe the functions of the sympathetic and parasympathetic nervous systems.
10. Many prescription drugs affect the autonomic nervous system. **Table 3** describes the action of two different drugs. Which drug should not be taken by someone who has high blood pressure? Why?

Table 3 Drug Actions

Drug	Action
pilocarpine	stimulates the parasympathetic nervous system
ephedrine	stimulates the sympathetic nervous system

Applying Inquiry Skills

11. Design an investigation to test the human fight or flight response safely. Ask your teacher to review your plan. What measures have you incorporated in your investigation to ensure safety?

Making Connections

12. Attention Deficit Hyperactivity Disorder (ADHD) is one of the most common brain disorders in children. Conduct library and/or Internet research to answer the following questions:
   (a) What are some of the symptoms of ADHD?
   (b) What are some of the misconceptions about ADHD?
   (c) What do researchers think might be the causes of ADHD?
   (d) What are some of the current treatments for ADHD?
   (e) Describe some societal views about ADHD.

www.science.nelson.com

Exploring

13. Conduct library and/or Internet research to answer the following questions about strokes:
   (a) What is a stroke?
   (b) What are the main symptoms of a stroke?
   (c) What are common treatments for a stroke?
   (d) Describe one example of current research being done on this condition.

www.science.nelson.com

3.4 Psychoactive Drugs and Homeostasis

psychoactive drug a drug that affects the nervous system and often results in changes to behaviour

Psychoactive drugs are a group of legal and illegal drugs that apply their effect on the nervous system, disturbing its ability to receive and process information about the internal or external environment. Because the nervous system is the primary way in which your body receives information about changes in your environment, anything that interferes with the nervous system's operation will create problems and prevent proper homeostatic adjustment. Changes in the nervous system may also affect the functions of the endocrine system, resulting in impaired responses to stimuli.

stimulant a drug that speeds up the action of the central nervous system, often causing an increase in heart and breathing rates

depressant a drug that slows down the action of the central nervous system, often causing a decrease in heart and breathing rates

While a person rests, heart rate and breathing rate are kept within a normal range by homeostatic adjustments, shown in **Figure 1**, graph (**a**). After taking a **stimulant**, such as caffeine, the homeostatic ranges for heart rate and breathing rate increase, as shown in graph (**b**). After taking a **depressant**, such as alcohol, the ranges are lowered, as shown in graph (**c**). Stimulants and depressants cause the homeostatic range to adjust to a different level than is normal.

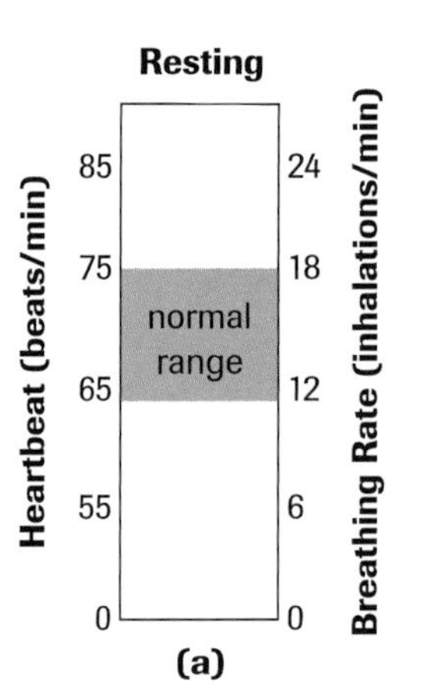

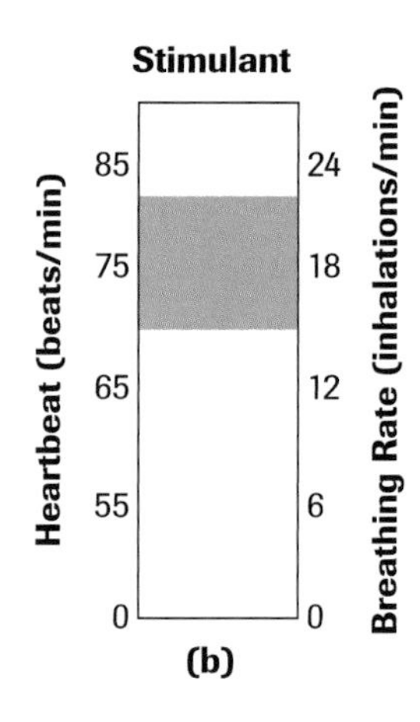

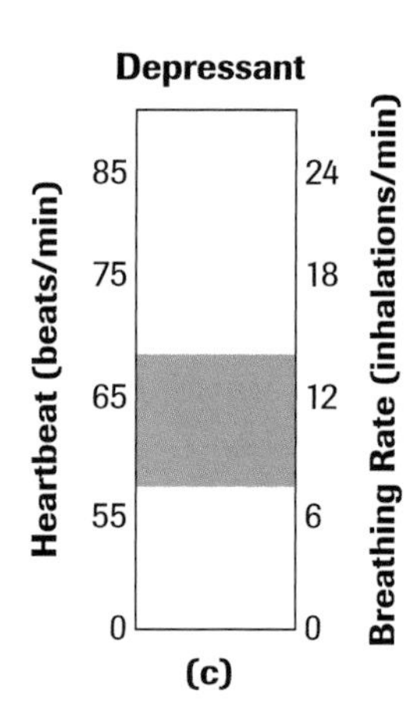

Figure 1
Stimulants and depressants can reset the operating ranges for heart rate and breathing rate.

neurotransmitter a chemical that transmits nerve-cell impulses from one nerve cell to another

Under normal circumstances, impulses are relayed from one nerve cell to another in the brain by chemicals called neurotransmitters. A **neurotransmitter** chemical released from one nerve cell attaches to receptor sites on another nerve cell (**Figure 2**).

When enough receptor sites have been filled by the neurotransmitter chemicals, an impulse is initiated in the nerve cell receiving the neurotransmitters. Psychoactive drugs interfere with either the movement of these neurotransmitter molecules or their attachment to the receptors.

Depressants, such as tranquillizers, barbiturates, and alcohol, are a group of psychoactive drugs that slow down the action of the central nervous system. Some depressants delay the effect of neurotransmitter chemicals by slowing the reaction of connecting nerves. Stimulants, such as cocaine, nicotine, and caffeine, are psychoactive drugs that speed up the action of the central nervous system. Some stimulants prevent the transmitting chemicals from breaking down or recycling. The neurotransmitters remain on their receptors longer than they normally would and keep the receptor sites on the nerve cell full, resulting in more frequent firing of the nerve cell.

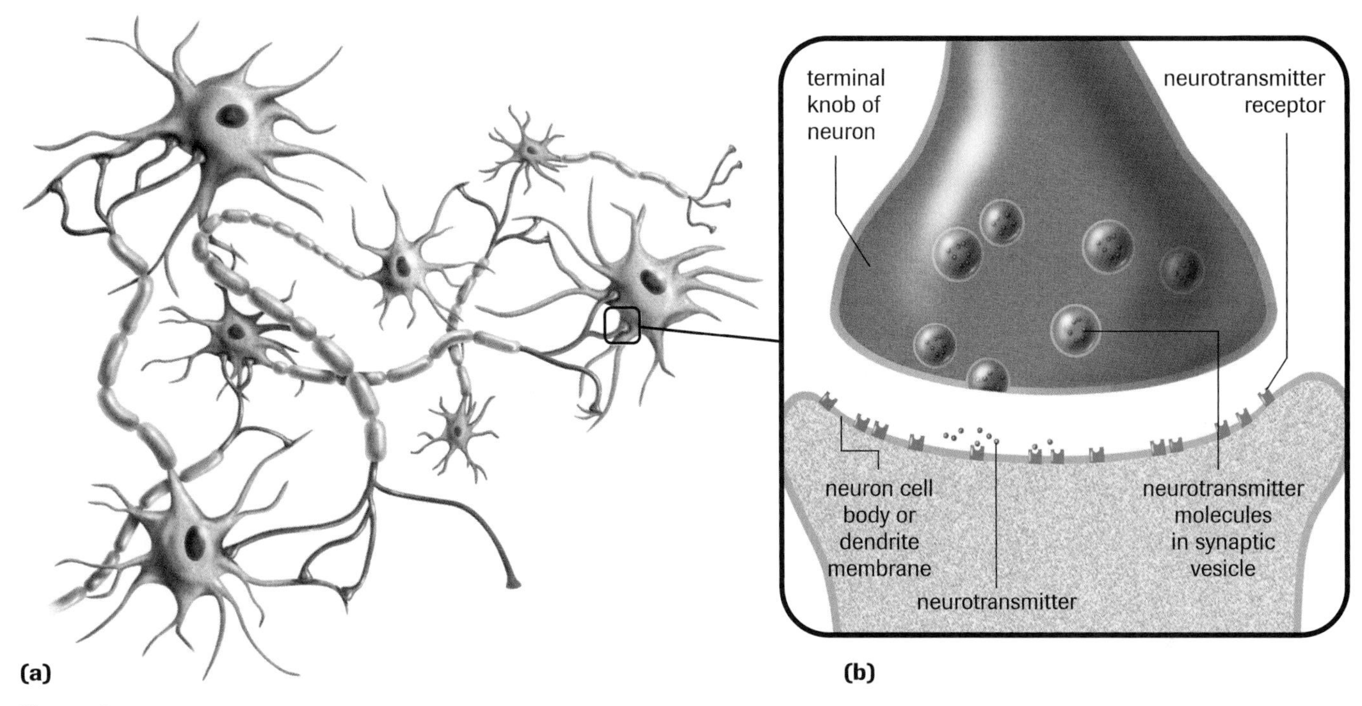

Figure 2
(a) Branching terminal knobs of neuron axons (purple) synapse with cell bodies and dendrites of neighbouring neurons (brown).
(b) Vesicles in the terminal knob of a nerve-cell axon (purple) release neurotransmitter chemicals that transfer an impulse to the neighbouring nerve cell (yellow). The neurotransmitter molecules attach to receptors embedded in the cell membrane of the neuron receiving the impulse.

Alcohol

Alcohol, a depressant, is one of the most widely used and abused of the psychoactive drugs. By interfering with the nervous system, alcohol affects other systems within the body. Nerves that conduct breathing movements from the brainstem are depressed. Alcohol consumption slows the heart and this in turn lowers oxygen delivery to the tissues of the body. Alcohol also affects nerve cells in the brain that control the release of a hormone that regulates water reabsorption by the kidneys. Under the influence of alcohol, the kidneys' ability to reabsorb and store water is impaired, and urine output increases. In turn, the loss of fluids from the body affects blood pressure (**Figure 3**).

One of the most pronounced effects of alcohol on homeostatic mechanisms occurs in the liver. Alcohol is readily broken down and used for energy, preventing the body from using other nutrients such as sugars, amino acids, and fatty acids. These nutrients are converted to fat and stored in the liver. This accumulation of fats in the liver is known as cirrhosis. Normal liver cells are replaced by the fats, and the liver can no longer eliminate chemical wastes from the body, produce needed proteins, or carry out many of its other vital functions. A buildup of toxic wastes or a lack of needed proteins affects every organ system.

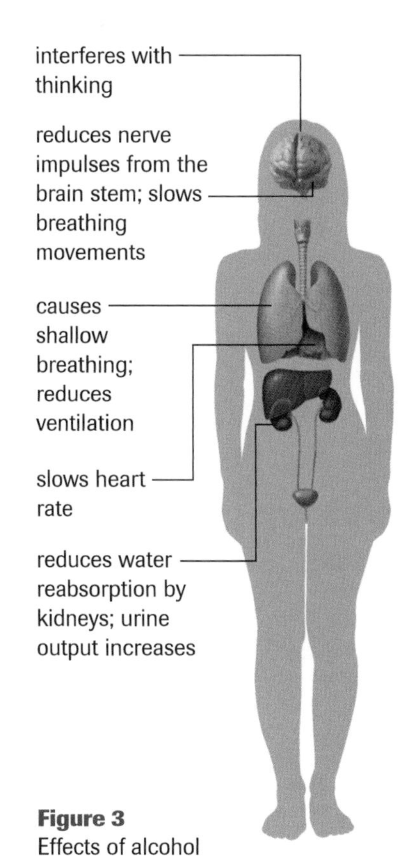

Figure 3
Effects of alcohol

Nicotine

Nicotine is one of the most widely used and most addictive stimulants. A component of the tobacco plant, it is commonly taken in with cigarette smoke. When inhaled with cigarette smoke, nicotine reaches the brain in approximately 10 s. The binding of nicotine to certain receptor sites creates a feeling of pleasure in the brain. This is the same area affected by cocaine, another stimulant. Nicotine increases heart and breathing rates. It also increases the metabolic rate of cells by resetting the rate at which oxygen and sugars are broken down during cellular respiration.

addiction a compulsive need for a harmful substance. Addiction is characterized by tolerance for the substance and withdrawal symptoms.

Addiction is the body's attempt to cope with the chemical disruption caused by a drug. Nerve cells in the brain adjust to prolonged exposure to nicotine by producing fewer neurotransmitter receptor proteins. As a result, the neurons become less sensitive to the drug. As time passes, more nicotine is needed to maintain the same pleasurable feeling.

Marijuana

Marijuana is usually a mixture of dried, shredded leaves, stems, seeds, and flowers of the hemp plant, *Cannabis sativa.* Hashish ("hash" for short) and hash oil are stronger forms of marijuana. All forms of marijuana are psychoactive because they all contain THC (delta-9-tetrahydrocannabinol) as the main active ingredient. Marijuana also contains more than 400 other potentially dangerous chemicals. Its effects on the user depend on the strength or potency of the THC it contains. Some of the short-term effects of marijuana on the brain can include problems with memory and learning; distorted perception (sights, sounds, time, touch); trouble with thinking and problem solving; loss of coordination; increased heart rate; and anxiety.

Ecstasy

Methylenedioxymethamphetamine, or MDMA (commonly called "Adam," "ecstasy," or "X-TC"), is a synthetic, psychoactive drug with hallucinogenic properties. Ecstacy is a stimulant that has been found to produce effects similar to those of amphetamines (speed) and cocaine. These include

- psychological difficulties including confusion, depression, sleep problems, drug craving, severe anxiety, and paranoia (even psychotic episodes have been reported);
- physical symptoms such as muscle tension, involuntary teeth clenching, nausea, blurred vision, rapid eye movements, faintness, and chills or sweating;
- increases in heart rate and blood pressure.

Research shows that ecstasy destroys certain types of brain neurons that play a direct role in regulating aggression, mood, sexual activity, sleep, and sensitivity to pain. It is probably this action that gives ecstasy its reported properties of heightened sexual experience, tranquility, and friendliness.

Narcotics

Narcotics are psychoactive drugs that relieve pain (painkillers) and make you sleepy. These substances act as depressants, reducing heart rate, breathing rate, and blood pressure. Narcotics include natural substances obtained from opium (a chemical found in the opium poppy plant), such as morphine, codeine, and heroin. Codeine is a common prescription painkiller usually taken by mouth in the form of a liquid or a pill. Morphine and heroin are usually injected into a muscle, or directly into the bloodstream. Heroin is the most widely abused narcotic. Narcotics are highly addictive and cause short-term effects such as a feeling of well-being, drowsiness, loss of pain, nausea, and urinary problems. An overdose of a narcotic can result in death, and the sharing of unsterile solutions and needles is responsible for the spread of many infectious diseases, such as hepatitis and AIDS.

narcotic a psychoactive drug that relieves pain and makes you sleepy

Treating Substance Addiction

There are many treatments available for persons who are addicted to drugs. Counselling and medication are two commonly used treatments. In individual and group counselling, patients discuss issues regarding motivation (why they use the drugs) and build behavioural skills that help them replace drug-using activities with rewarding non-drug-using activities. Behavioural therapy helps addicts develop the skills needed to form healthy relationships with friends, family, and the community at large.

In addition to counselling, medication may benefit some people. Prescription and non-prescription medicines are available to treat some forms of addiction, but not all. For example, there is no medicine available to treat habitual use of marijuana. However, alcoholism (addiction to alcohol) may be treated with prescription medicines such as Antabuse and Temposil. Both of these drugs cause patients to experience unpleasant side effects when they drink alcohol. Headaches, nausea, and vomiting are common symptoms experienced by alcoholics who drink alcohol while taking Antabuse or Temposil. These unpleasant feelings are meant to help addicts avoid alcohol as they develop healthier routines in their lives.

Although many people addicted to nicotine break the habit by quitting smoking or tobacco chewing, some benefit from using the nicotine patch or nicotine gum (**Figure 4**). These medical treatments, known as nicotine

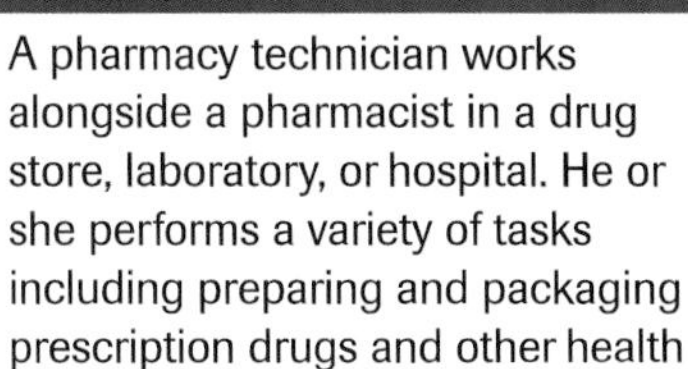

CAREER CONNECTION

A pharmacy technician works alongside a pharmacist in a drug store, laboratory, or hospital. He or she performs a variety of tasks including preparing and packaging prescription drugs and other health products.

DID YOU KNOW?

Antabuse

Antabuse was discovered in the 1930s when workers in a rubber factory became ill after drinking alcoholic beverages. Studies showed that one of the chemicals they were using to produce rubber was causing them to feel sick when they drank alcohol.

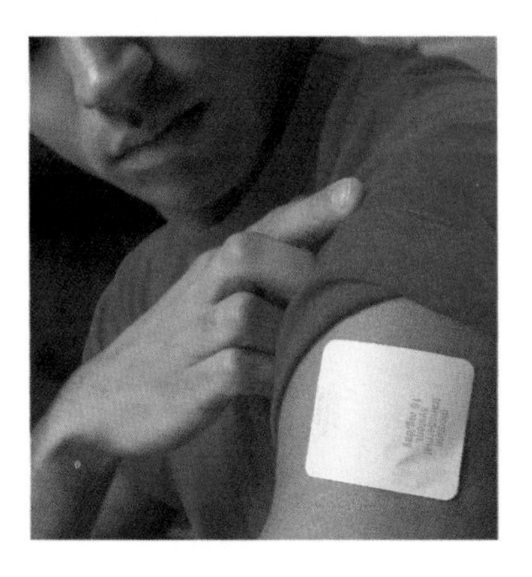

Figure 4
Nicotine patch and gum

replacement therapies, are available in pharmacies without a doctor's prescription. Both the patch and the gum provide a steady, controlled release of nicotine (absorbed through the skin) throughout the day to reduce the cravings associated with quitting smoking or tobacco chewing.

Medical treatment of narcotic addiction includes psychological counselling and, in the case of heroin addiction, the use of a morphine-like synthetic drug called methadone.

In general, no single treatment for drug addiction is appropriate for all individuals. As with other chronic (long-term) conditions, addicts may begin using the drug again during or after treatment. Recovery from severe drug addiction is usually a long-term process requiring many forms of treatment.

Section 3.4 Questions

Understanding Concepts

1. What are the effects of stimulants and of depressants on heart rate and breathing rate?
2. Describe how neurotransmitter chemicals in the brain work, and the effects that depressants and stimulants have on their function.
3. Describe the short-term and potential long-term effects of alcohol on the body.
4. Explain addiction as a response to chemical disruption in the brain.
5. During the mid-1990s, the death of two top basketball players was linked to the use of cocaine. Referring to homeostatic mechanisms, explain why using a stimulant prior to exercise is dangerous.

Making Connections

6. How might an understanding of the effect of depressants and stimulants on homeostasis affect a person's decisions about whether or not to take these kinds of drugs?
7. Methadone is a prescription drug used to treat addiction to heroin. Conduct library and/or Internet research to answer the following questions about methadone treatment:
   (a) How does methadone treatment work?
   (b) How is methadone treatment for heroin addiction similar to nicotine replacement therapy in the treatment of nicotine addiction? How is it different?
   (c) Why can the nicotine patch and nicotine gum be bought in a pharmacy "over the counter" (i.e., without a doctor's prescription), but in methadone therapy, addicts must obtain a doctor's prescription and physically go to the methadone-treatment centre or hospital every day to obtain their dosage of methadone?

www.science.nelson.com

8. (a) Conduct library and/or Internet research, or speak to a pharmacist in a local pharmacy, to identify two common prescription stimulants and two common prescription depressants.
   (b) Describe a condition for which a doctor would prescribe each drug, and list two serious side effects associated with their use.

www.science.nelson.com

9. Marijuana, an illegal drug in most countries, has been getting a lot of attention lately. Several countries around the world are considering changes in their laws that would decriminalize possession and use of marijuana. Write a brief position paper containing your recommendations to a legislature on this issue.

www.science.nelson.com

The Digestive System 3.5

Animals rely on the digestive system to break food down into nutrients, which are then absorbed and transported by the circulatory system. Body cells require nutrients for growth, maintenance, and repair.

There are four stages of food processing:

- ingestion—the taking in of nutrients
- digestion—the breakdown of complex organic molecules into smaller components by physical and chemical means
- absorption—the taking up of digested molecules into the cells of the digestive tract
- egestion—the removal of waste food materials from the body

Most animals have a system of specialized organs to perform digestion (**Figure 1**). In humans, the digestive system is really just one long tube, open at both ends, called the gastrointestinal tract (**Figure 2**).

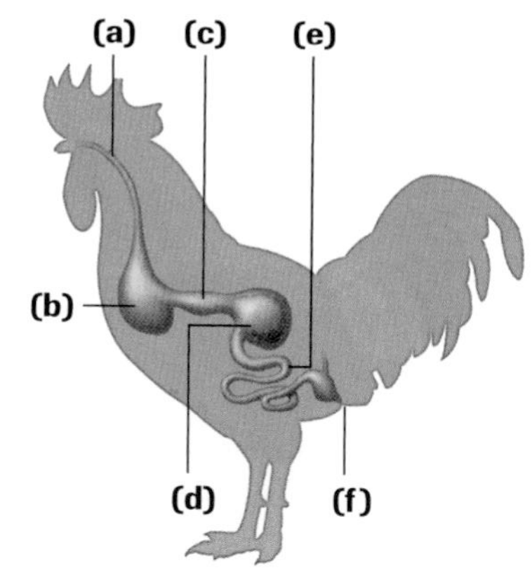

Figure 1
The digestive system of a bird
(a) The esophagus carries food to the crop **(b)**.
(b) The crop stores food.
(c) The proventriculus secretes digestive juices, which break down food.
(d) The gizzard contains gravel to help grind up food. The proventriculus and gizzard form the stomach.
(e) The intestine absorbs digested food.
(f) Waste is expelled from the anus.

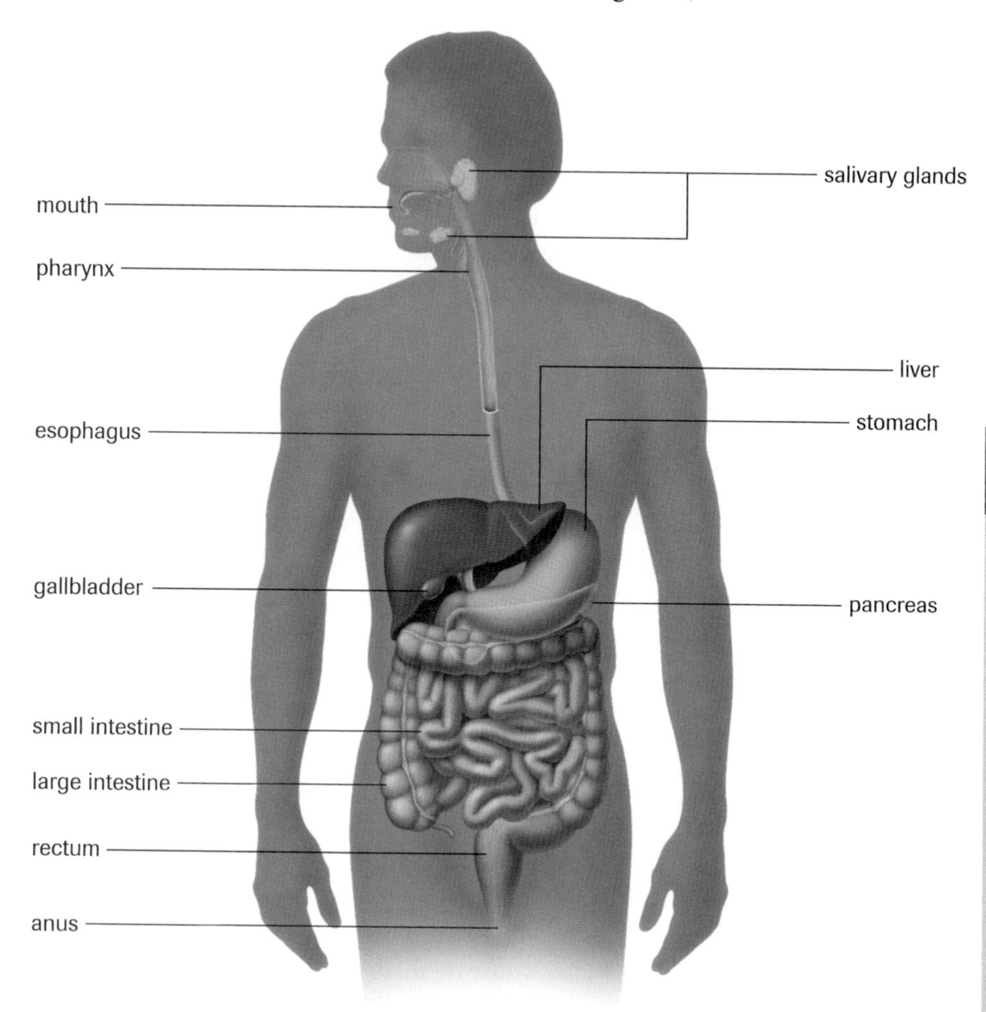

Figure 2
The human digestive system and accessory organs

TRY THIS activity

Beginning Digestion

1. Place an unsalted soda cracker in your mouth and allow it to sit for a few minutes without chewing it.
   (a) Record qualitative observations as soon as the cracker is in place, and one minute later, before swallowing. Describe the cracker's change in texture. Describe the cracker's change in flavour.

amylase an enzyme that breaks down complex carbohydrates

peristalsis rhythmic, wavelike contractions of smooth muscles that move food along the gastrointestinal tract

sphincter a constrictor muscle that surrounds a tubelike structure

mucus a lubricating substance, secreted by glands in the stomach walls, that protects the stomach

pepsin a protein-digesting enzyme produced by the stomach

Ingestion and Digestion—Mouth

In humans, the digestive process begins as soon as food enters the mouth. Teeth begin the physical breakdown of food. The structure of teeth determines their function: incisors cut food into smaller pieces; canine teeth and bicuspids pierce and tear; and molars crush and grind (**Figure 3**).

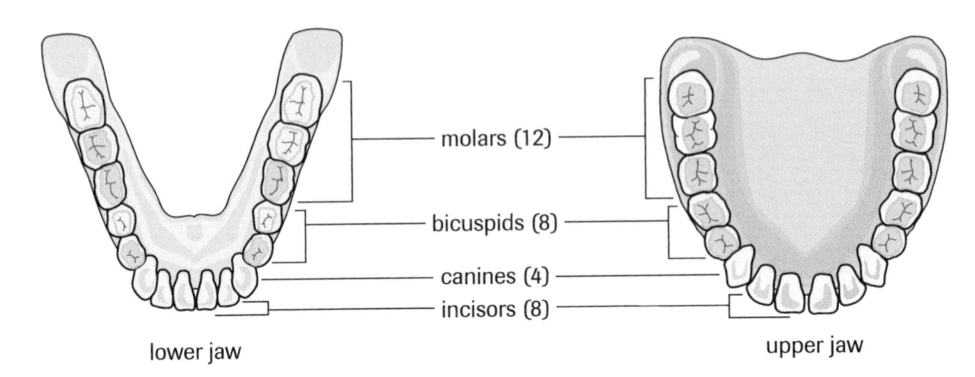

Figure 3
Human teeth

The muscular action of chewing stimulates the salivary glands that surround the mouth. The salivary glands secrete saliva, a mixture of water, mucus, and the enzyme **amylase.** At this stage of chemical digestion, amylase breaks down the carbohydrate amylose into maltose subunits.

The muscular tongue moves food around until it forms a *bolus* (the Greek word for ball) that is pushed back to the pharynx. The pharynx is a dual-purpose passage that receives both food from the mouth and air from the nose. Each time you swallow, a flaplike structure ensures that the food will travel down the digestive tract. The bolus of food stretches the walls of the esophagus, activating smooth muscles that set up waves of rhythmic contractions called **peristalsis** (**Figure 4**).

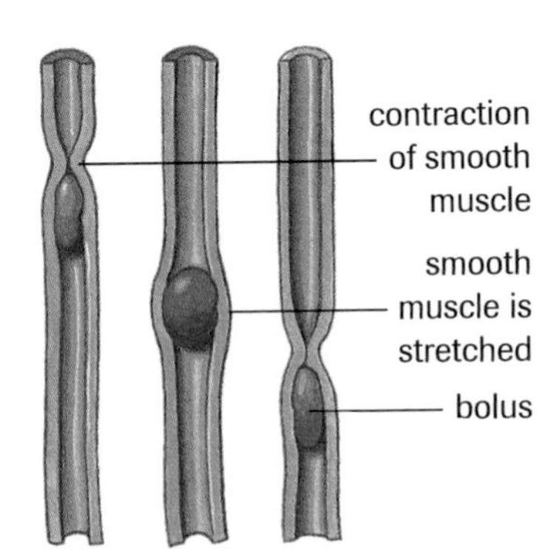

Figure 4
Rhythmic contractions of the smooth muscle move food along the gastrointestinal tract.

Digestion—Stomach

The movement of food into and out of the stomach is regulated by circular muscles called **sphincters.** Sphincters act like the drawstrings on a bag: when the sphincter muscles contract, the opening closes (**Figure 5**).

The J-shaped stomach has three muscle layers that contract and relax, churning the partly digested food. Cells in the stomach lining secrete gastric juice—a mixture of hydrochloric acid and enzymes—and mucus. Hydrochloric acid and enzymes assist in the breakdown of fibrous tissue in the food. Hydrochloric acid also destroys foreign organisms such as bacteria. **Mucus** protects the stomach lining. Without mucus, the gastric juice would digest the stomach lining. Damage to this mucus barrier results in gastric ulcers. Digestive enzymes, such as **pepsin,** need an acid environment to break down large protein molecules into smaller polypeptides.

Although the stomach is primarily involved in digestion, some absorption occurs here: small amounts of water, some medicines (e.g., aspirin), and alcohol.

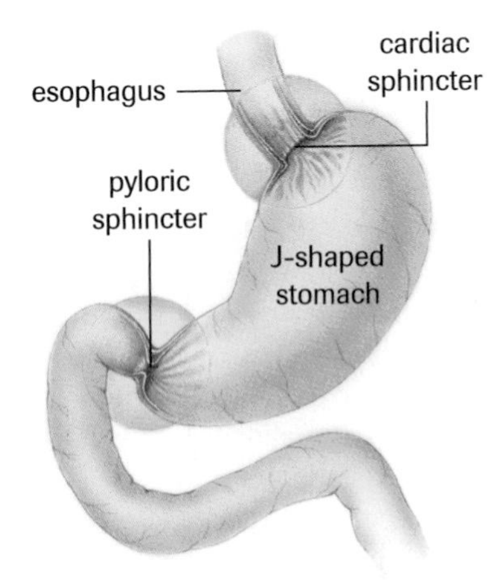

Figure 5
The stomach is the initial area of protein digestion. Sphincters control movement in and out.

Digestion—Small Intestine

The small intestine measures up to 7 m in length, but is only 2.5 cm in diameter (**Figure 6**). Most chemical digestion takes place in the first 25-cm portion, called the **duodenum**. Secretions from the pancreas and the liver enter this portion through the common duct.

An essential organ in the digestive process is the liver, with an average mass of 1.5 kg. The liver produces bile—an emulsifying agent needed for the physical digestion of fats. Bile contains **bile salts** that break large fat globules into small ones. The smaller the size of the globules, the larger the total surface area upon which fat-digesting enzymes work. Bile is secreted continuously by the liver and stored in the saclike gallbladder until it is needed. Aside from producing bile, the liver stores carbohydrates and vitamins and **detoxifies** many harmful substances. To detoxify a substance means to convert it from a harmful to a non harmful substance. If consumed in large quantities, toxins such as alcohol can destroy liver tissue. Damaged liver cells are replaced by connective tissue and fat, in a condition called cirrhosis of the liver.

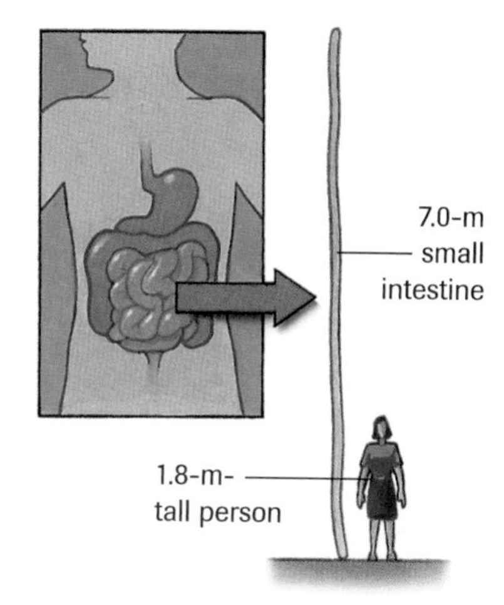

Figure 6
A comparison of the length of the small intestine with the height of a tall person

duodenum the first segment of the small intestine

bile salts the components of bile that break down large fat globules

detoxify to remove the effects of a poison

Simulating Bile Action on Fats

Materials: eyedropper, test tube, test-tube stopper, vegetable oil, liquid soap, hand lens or magnifying glass

1. Fill a test tube one-quarter full of water.
2. Add 10 drops of vegetable oil.

(a) Record the location and appearance of the oil in the test tube.

3. Place stopper in test tube. Shake well, then immediately examine its contents with the hand lens.

(b) Record your observations.

4. Let the test tube stand for 2 to 3 min. Observe any changes.
5. Add about 5 drops of liquid soap to the test tube.
6. Insert the stopper and shake the test tube again. Immediately examine with the hand lens.

(c) Record your observations.

(d) What effect did the liquid soap have on the oil?

(e) How does this activity simulate bile action?

DID YOU KNOW?

The Liver
The liver was once considered to be the centre of emotions. The term *lily-livered*, meaning cowardly, implies inadequate blood flow to the liver.

As the partly digested food passes into the small intestine, it has the same texture as porridge and is called **chyme**. Chyme consists of partially digested food, water, and gastric juice. The acidity of chyme would quickly destroy the cells of the small intestine if it weren't for the action of the pancreas. This organ produces sodium bicarbonate, which raises the pH of chyme from about 2.5 to 9.0. The pancreas also secretes digestive enzymes, such as **lipase**, pancreatic amylase, and **trypsin**.

The small intestine itself secretes **maltase**, which breaks maltose into glucose, thus completing the digestion of carbohydrates that was started in the mouth by amylase in the saliva. **Peptidases**, in the small intestine, complete the digestion of protein.

chyme a mixture of partly digested food, water, and gastric juice

lipase a fat-digesting enzyme

trypsin a protein-digesting enzyme

maltase a carbohydrate-digesting enzyme

peptidase an enzyme that breaks peptides into amino acids

Products of Digestion

As you learned in Unit 1, carbohydrates, proteins, and fats are macromolecules that provide the main nutrients in the human diet. Through the digestive process, these three macromolecules are broken into molecules small enough to be absorbed from the intestine and transported to body cells (**Figure 7**).

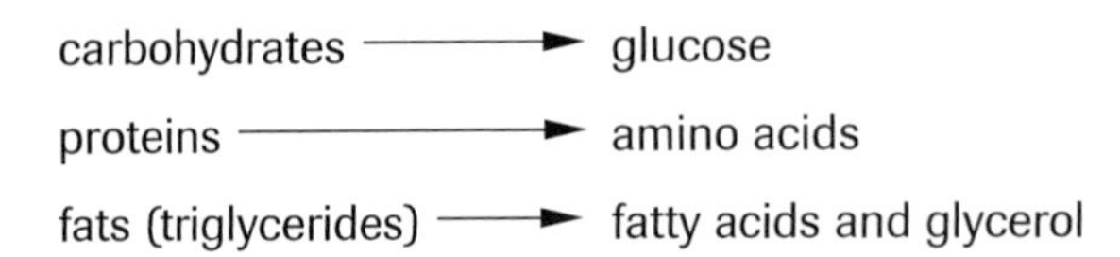

Glucose is the most important energy source for all cells. Fats are also involved in energy supply and storage. As well, they help the body synthesize hormones, insulate the body, protect organs, and provide a protective coating around nerves. Amino acids are used by cells for growth and repair, and to produce enzymes that catalyze chemical reactions. For example, all the digestive enzymes are proteins. Of the 20 amino acids that combine in various ways to form protein, eight cannot be synthesized by the human body and must be obtained from food. These are called essential amino acids.

In addition to the three basic macromolecules, our food supplies micronutrients in the form of **vitamins** and **minerals** to help regulate body processes. These substances are dissolved in the water and fats we eat, and are absorbed along with the digested nutrient molecules. A summary of the digestive process is shown in **Table 1**.

Table 1 Substances Involved in Digestion

Organ	Secretion	Function
mouth, salivary glands	salivary amylase	begins the breakdown of starch to sugars (maltose)
stomach	hydrochloric acid	lowers pH to optimum level; breaks down food; kills microorganisms
	pepsin	begins the digestion of protein
	mucus	protects the stomach from pepsin and HCl
liver	bile	emulsifies fat
gallbladder		stores fat
pancreas	bicarbonate	neutralizes HCl from stomach
	trypsin	converts long-chain peptides into short-chain peptides
	pancreatic amylase	continues starch breakdown to maltose
	lipase	breaks down fats to fatty acids and glycerol
small intestine	peptidases	complete the breakdown of proteins to amino acids
	maltase	breaks down maltose into glucose
large intestine	mucus	helps the movement of food

carbohydrate

amylase
- salivary glands

pancreatic amylase
- pancreas

maltose

maltase
- intestinal cells

glucose

protein

pepsin
- stomach

trypsin
- pancreas

peptides

peptidases
- intestinal cells

amino acids

fat (triglycerides)

lipase
- pancreas

glycerol + fatty acids

Figure 7
Digestion flowchart

vitamin an organic molecule needed in trace amounts for normal growth and metabolic processes

mineral an element (such as copper, iron, calcium, phosphorus, etc.) required by the body, often in trace amounts. Minerals are inorganic materials.

Absorption—Small and Large Intestines

The products of digestion—glucose, amino acids, fatty acids, and glycerol—are now ready to be removed from the digestive tube and transported where needed. Most absorption of nutrients takes place within the small intestine. Small fingerlike projections called **villi** (singular: villus) increase the surface area of the small intestine by as much as tenfold (**Figure 8**). The cells that make up the lining of each villus have **microvilli** (singular: microvillus), which are fine, threadlike extensions of the membrane that further increase the surface for absorption. Each villus is supplied with a capillary network that intertwines with lymph vessels called **lacteals** that transport materials. Some nutrients are absorbed by diffusion; others are actively transported from the digestive tract. Glucose and amino acids are absorbed into the capillary networks; fats are absorbed into the lacteals.

villi small fingerlike projections that extend into the small intestine, increasing surface area for absorption

microvilli microscopic fingerlike outward projections of the cell membrane

lacteal a small vessel that provides the products of fat digestion access to your circulatory system

Villi in the mammalian intestine

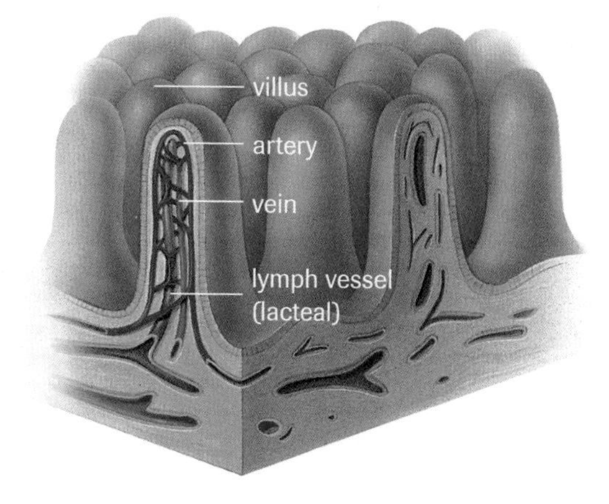

Figure 8
Villi greatly increase the surface area for absorption of nutrients. Each villus has blood capillaries and lymph vessels.

DID YOU KNOW?

Waste Production
An average person eats 1.5 kg of food daily and produces about 0.4 kg of waste in the form of feces.

The large intestine, or **colon**, is 1.5 m long and has about twice the diameter of the small intestine. The main function of the large intestine is to absorb usable materials, primarily water. In its second function, the harmless bacteria that live in the intestinal tract produce vitamin K and some B vitamins. These micronutrients are absorbed by cells lining the walls of the large intestine along with dissolved minerals and water.

colon the largest segment of the large intestine, where water reabsorption occurs

Egestion—Large Intestine

Cellulose, the long-chain carbohydrate found in plant cell walls, reaches the large intestine undigested. Cellulose is the main component of feces, along with living and dead bacteria and water. As wastes build up in the large intestine, receptors in the intestinal walls provide information to the central nervous system that, in turn, prompts the large intestine to void the waste. Through this mechanism, potentially toxic wastes are removed from the body. Individuals who do not eat sufficient amounts of cellulose (roughage or fibre) have fewer bowel movements and may be at risk of developing colon cancers.

CAREER CONNECTION 

Dietary technicians assist nutritionists and dieticians to plan and supervise food service operations.

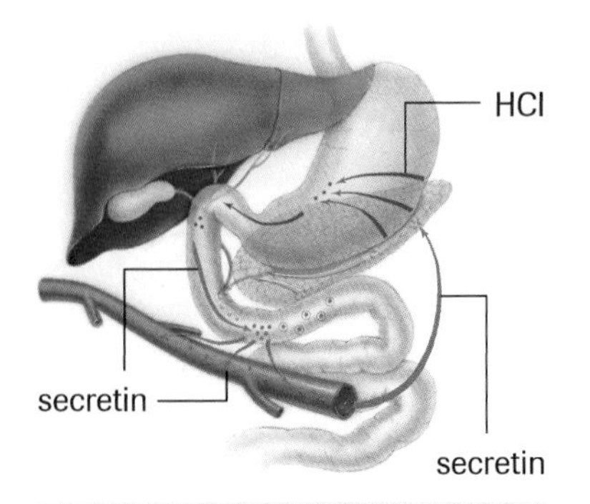

- Secretin is absorbed into the blood and carried to the pancreas where it stimulates the release of pancreatic fluids.
- Bicarbonate ions, released by the pancreas, neutralize the HCl from the stomach.

Figure 9
Secretin acts as a feedback agent.

Digestion and Homeostasis

In section 3.3, you learned that homeostasis is the maintenance of a healthy balance of all chemical reactions in an organism. The nervous, endocrine, circulatory, and digestive systems interact to control digestion. Seeing, smelling, tasting, or thinking about food will stimulate glands to produce hormone secretions. Stimulation of the salivary glands produces saliva. Gastrin, a hormone secreted by the stomach, triggers the release of gastric juices even before there is any food in the stomach.

Hormones play a large role in the maintenance of homeostasis because they help control and coordinate the functions of almost all organ systems. For example, a hormone called secretin is released when acids from the stomach move into the small intestine along with chyme. Secretin is absorbed into the blood and travels to the pancreas, where it initiates the release of neutralizing substances (**Figure 9**).

The speed at which the digestive system processes food is also under homeostatic control. A large meal will activate receptors, causing more forceful stomach contractions and faster emptying. If the meal is fatty, the small intestine secretes a digestive hormone that slows peristalsis, allowing time for fat digestion and absorption. This explains why we feel full longer after a high-fat meal than we do after a low-fat meal.

Section 3.5 Questions

Understanding Concepts

1. What are the main functions of the digestive system? Name the end products.
2. List, in order, the structures food must pass through in a typical vertebrate digestive system.
3. Outline the similarities and differences between digestive organs in humans and birds.
4. State the function of the following digestive enzymes: amylase, pepsin, trypsin, and lipase.
5. Describe bile: formation, storage, and purpose.
6. What are the functions of the macromolecules after they are broken down in the digestive process?
7. List the two ways in which absorbed nutrients leave the intestine and get into body cells.
8. Design a summary table for digestion, listing each organ and its function.
9. Give three examples of homeostasis in the digestion process.

Applying Inquiry Skills

10. The large intestine's job is to absorb water. Over a 4-h period, a person with diarrhea absorbs about 10 mL of water. A constipated person absorbs about 100 mL of water. Interpret these data. What would you expect a normal intestinal absorption rate to be, based on the above data?
11. During a hockey game, Juan (A) consumed a glucose drink, and Jeff (B) consumed a drink containing sugars such as sucrose. Their metabolic rates, as indicated by blood glucose, are graphed in **Figure 10**. Explain the difference.

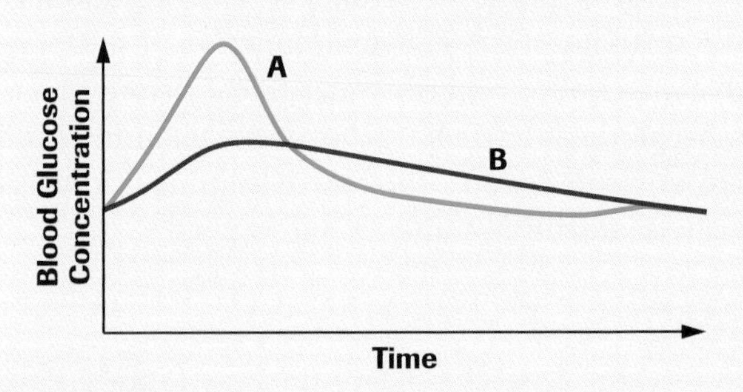

Figure 10
Effect of glucose and sucrose on blood sugar

Making Connections

12. What role do environmental factors play in the development of ulcers? Research the causes, risk factors, symptoms, and treatment of this digestive tract disorder.

www.science.nelson.com

Case Study 3.6 Digestive Disorders

Doctors who specialize in gastroenterology examine and treat patients with the complaints listed in **Figure 1**. All or some of these symptoms indicate a problem in the digestive system.

- indigestion or heartburn (burning sensation)
- discomfort or pain in the abdomen
- nausea and vomiting
- diarrhea or constipation
- bloating after meals
- loss of appetite
- weakness and fatigue
- vomiting blood, or blood in the stool

Figure 1
Common symptoms of digestive disorders

Crohn's Disease

Crohn's disease is a chronic inflammatory disease of the intestines. It primarily causes ulcers in the small and large intestines, but it can affect the digestive system anywhere between the mouth and the anus.

In addition to many of the symptoms in **Figure 1**, Crohn's sufferers often have rectal bleeding, weight loss, and fever. Bleeding may be serious and persistent, leading to anemia. Children with Crohn's disease may suffer delayed development and stunted growth.

Crohn's disease affects men and women equally and seems to run in some families. About 20% of people with Crohn's disease have a blood relative with some form of inflammatory bowel disease (IBD). Certain ethnic groups, especially those of Jewish descent, have a higher incidence of the disease.

Crohn's disease has no known cause. The most popular theory is that the body's immune system reacts to a virus or a bacterium by causing ongoing inflammation in the intestine. People with Crohn's disease tend to have abnormalities of the immune system, but doctors do not know whether these abnormalities are a cause or a result of the disease. Crohn's disease is not caused by emotional distress.

A thorough physical examination and a series of tests may be required to diagnose Crohn's disease. Blood tests may be done to check for anemia, which could indicate bleeding in the intestines. The doctor may do an upper gastrointestinal (GI) examination to look at the small intestine. For this test, the patient drinks barium, a chalky solution that coats the lining of the small intestine. Barium shows up white on X-ray film, revealing inflammation or other abnormalities in the intestine. The doctor may also do a colonoscopy, where an endoscope is inserted into the anus to detect any inflammation or bleeding in the large intestine (**Figure 2**).

The goals of treatment are to control inflammation, correct nutritional deficiencies, and relieve symptoms such as abdominal pain, diarrhea, and rectal bleeding. Treatment may include drugs, nutrition supplements, surgery, or a combination of these options. At this time, treatment can help control the disease, but there is no cure.

DID YOU KNOW?

Crohn's Disease
Crohn's disease can be difficult to diagnose because its symptoms are similar to other inflammatory bowel diseases (IBD), such as irritable bowel syndrome and ulcerative colitis.

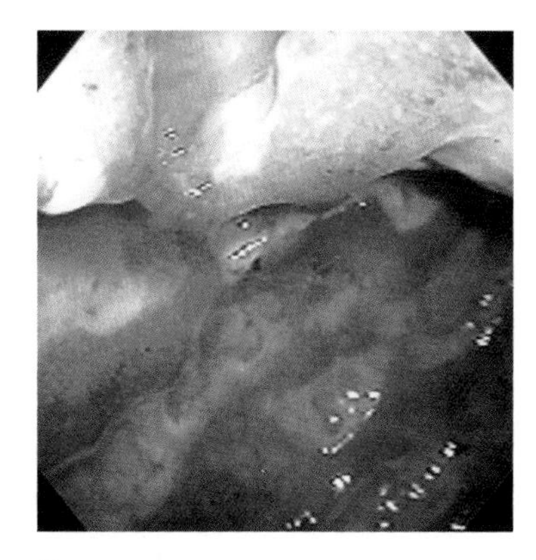

Figure 2
Endoscopy photo of an inflamed bowel

Other Digestive Disorders

Digestive disorders can be classified as structural, malabsorptive, or inflammatory (**Figure 3**, on the next page). **Table 1**, on the next page, lists common examples.

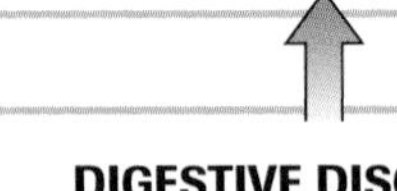

Figure 3
Types of digestive disorders

Table 1 Examples of Digestive Disorders

Disorder	Definition
appendicitis	inflammation and/or infection of the appendix, usually resulting in its removal
celiac disease	an immune (allergic) reaction to gluten, a protein present in cereal grains. The lining of the small intestine is affected, decreasing the amount of surface available for the absorption of nutrients.
diverticulitis	inflammation of the diverticula (outpouchings) along the wall of the colon
dysentery	inflammation of the intestine with ulcers in the colon resulting from infection with an amoeba (a single-celled parasite transmitted to humans via contaminated water and food). The liver and other organs may also be infected.
gallstones	stones that form when substances in the bile harden. Gallstones can block the ducts that carry bile from the liver to the small intestine. Blocked ducts can cause severe damage or infections in the gallbladder, liver, or pancreas.
gastroesophageal reflux disease (GERD)	stomach contents regurgitate and back up (reflux) into the esophagus, occasionally reaching the breathing passages, causing inflammation and damage to the esophagus, as well as to the lungs and the voice box
gingivitis	also known as trench mouth; a progressive, painful infection with ulceration, swelling, and sloughing off of dead tissue from the mouth and throat resulting from the spread of infection from the gums
hiatus hernia	protrusion of the stomach up into the opening in the diaphragm normally occupied by the esophagus. This type of hernia can be present at birth, but is more often acquired through strenuous physical activity or obesity.
pyloric stenosis	narrowing of the outlet of the stomach so that food cannot pass easily from it into the duodenum, resulting in feeding problems and projectile vomiting
ulcerative colitis	a relatively common disease that causes inflammation of the large intestine

Case Study 3.6 Questions

Understanding Concepts

1. Which digestive organs are associated with each symptom listed in **Figure 1**, on the previous page?
2. Prepare a fact sheet on Crohn's disease. Using a full piece of paper, organize all case study information in blocks under the following headings: description, symptoms, population data, cause, diagnosis, and treatment.
3. On an outline of the human body, locate the site of each digestive disorder described in **Table 1**, on the previous page. Label the organs involved in each disorder.
   (a) Colour-code to indicate whether the disorder affects ingestion, digestion, absorption, or egestion.
   (b) Classifiy each disorder as structural, malabsorptive, inflammatory, or other.

Making Connections

4. Obesity in children is a growing concern in Canada today. Research this problem. Consider risk factors such as diet and lifestyle. How effective are surgical procedures such as stomach stapling or removal of an intestinal section? Propose ways to reduce the incidence of obesity in the young.

www.science.nelson.com

5. If a gallbladder is removed, what kind of dietary changes would a person need to make? Why?

Exploring

6. Crohn's disease, irritable bowel syndrome, and ulcerative colitis are all inflammatory bowel diseases. How are they similar? How are they different?

www.science.nelson.com

Perch Dissection

Perch, like most fish, are carnivores. They consume zooplankton, smaller fish, and invertebrates such as insects and crayfish. The long, straight stomach of the perch allows it to swallow a fish almost the same size as itself (**Figure 1**).

Question

How is the structure of a vertebrate digestive system suited to its function?

Materials

safety goggles, gloves, and laboratory apron
dissection tray with pins and probe
scissors
preserved perch
ruler

Wear eye protection and a laboratory apron at all times.

Wear gloves when handling preserved specimens.

Cut away from you rather than toward you.

Wash dissecting instruments in a disinfectant solution.

Cuts and scrapes should receive immediate medical attention.

Partially dissected specimens are susceptible to disease organisms; do not store them for any longer than necessary.

Dissected remains should be disposed of as directed by your teacher.

Wash thoroughly with soap and water after completing the dissection.

Procedure

1. Hold the fish with the ventral (lower) side up and the head pointing away from you. Insert the point of the scissors through the body wall in front of the anus and cut up the midline of the body to the space between the gill covers. Lay the fish on its right side in the dissecting tray.
2. Continue to cut up around the back edge of the gill chamber to the top of the body cavity. Make another incision from the starting point of the ventral incision close to the anus and cut upward to the top of the body cavity. Be careful not to disturb the internal organs.
3. Remove the gills and the lateral (side) body wall by cutting along the top of the body cavity. This

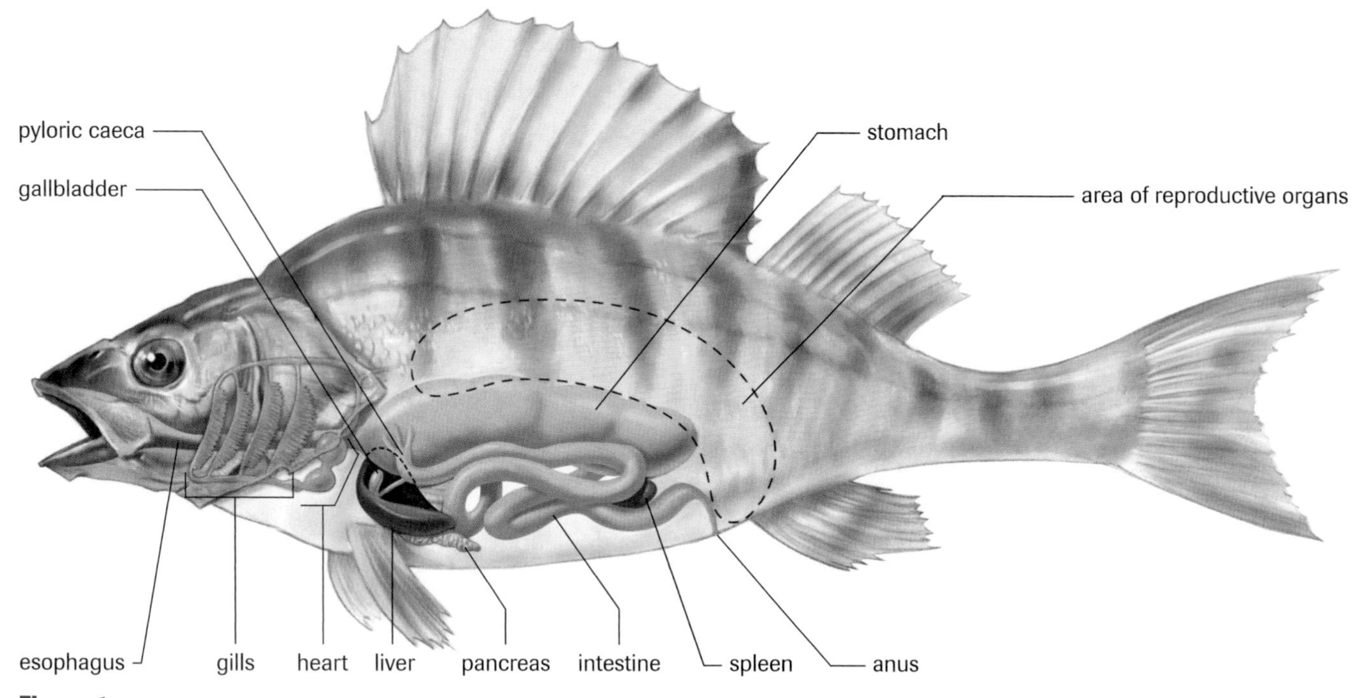

Figure 1
The digestive system of a perch

procedure will expose the body organs in their normal position (**Figure 1**).

(a) Draw a large outline of the perch digestive system as you see it (to scale) on a separate piece of paper. Label each structure identified in the procedure steps, and use bullets to record observations of each structure.

4. Locate and record the esophagus—a short tube running from the mouth to the stomach. This may be hidden by the gills.
5. Locate and record the stomach—a larger, thick-walled, U-shaped tube. Mechanical and chemical breakdown of food begins here. The size of the stomach varies according to how much food it contains.
6. Locate and record the pyloric caeca—three blind-ended tubes at the junction of the stomach and intestine. Secretion and absorption take place here.
7. Locate and record the intestine—an S-shaped tube near the stomach that constricts and straightens toward the anus. Because the perch is a carnivore, its intestine is shorter than the length of its body. Herbivorous fish have an intestine that is 2 to 15 times their body length. A longer intestine is necessary to provide greater digestive and absorptive surfaces for the herbivores.
8. Locate and record the liver—the noticeable brown organ in front of the stomach. Bile is produced here.
9. Locate and record the gallbladder—a small, round structure attached to the right lobe of the liver. The gallbladder drains bile from the liver and opens by a number of ducts into the intestine.
10. Locate and record the pancreas—usually found along the front border of the intestine. In some fish, it may be embedded in the liver. The pancreas secretes digestive enzymes into the intestine and hormones into the blood.
11. Locate and record the spleen—a dark, football-shaped organ found in the connective membranes above the intestine, near the stomach end. The spleen helps produce and maintain blood cells.
12. Cut the intestine about 1 cm forward from the anus and begin gently lifting it out of the body cavity, using the probe to ease it away from the connective membranes. Cut the tract forward from the stomach and spread it out on the tray.

(b) Measure the lengths of the stomach and intestine. Record the results on your diagram.

Figure 2
Digestive organs exposed

Analysis

(c) What type of tissue permits the stomach to change in size? (*Hint:* Review section 3.1.)

(d) Contrast the nutrients present in a carnivorous and herbivorous diet. What major nutrient appears to require greater digestive and absorptive surfaces in fish?

(e) The spleen is investigated here as a digestive organ, yet its function appears to involve blood. What is its connection to digestion?

Synthesis

(f) Describe the relationship between the food-gathering methods of the perch and its digestive structures.

(g) Compare and contrast the perch and human digestive systems. Use the four components of digestion presented in section 3.5. Consider organ structure, relative size, and function.

CAREER CONNECTION

Fish farming provides excellent career opportunities for college biology graduates. Using the Internet, research programs and job possibilities in this field.

www.science.nelson.com

Explore an Issue
Diet and the Media 3.8

The media constantly bombard us with dietary advice (**Figure 1**). Advice on what to eat and what to avoid eating is often given by diet gurus or pharmaceutical companies that have much to gain. Scientific research studies sometimes contradict earlier, equally reputable studies.

Diet is in the media spotlight more than ever before as Canada's population grows heavier. The incidence of obesity in Canada is said to be at epidemic levels and steadily increasing. By the turn of this century, it was estimated that 35% of men and 27% of women in Canada were overweight, using the criteria of a **Body Mass Index (BMI)** greater than or equal to 27. Canadians spend over \$2 billion, or 2.5% of the health care budget, on treating obesity and diseases related to obesity. Indirect costs may be double that figure. Obesity is known to contribute directly and indirectly to conditions such as coronary artery disease, stroke, osteoarthritis, cancers, diabetes, respiratory diseases, gastrointestinal abnormalities, and psychosocial difficulties. A reduction in weight of as little as 5% to 10% can reduce obesity-related risk factors and contribute to overall health.

Less well-known is the fact that being underweight can also cause problems, such as fatigue and increased susceptibility to illness and injury.

Adults of normal weight have a BMI of 18.5 to 24.9 kg/m^2. Be aware that BMI is not a foolproof way to define obesity, as it does not distinguish between weight due to muscle and weight due to fat. With a BMI of 30, is Arnold Schwarzenegger obese?

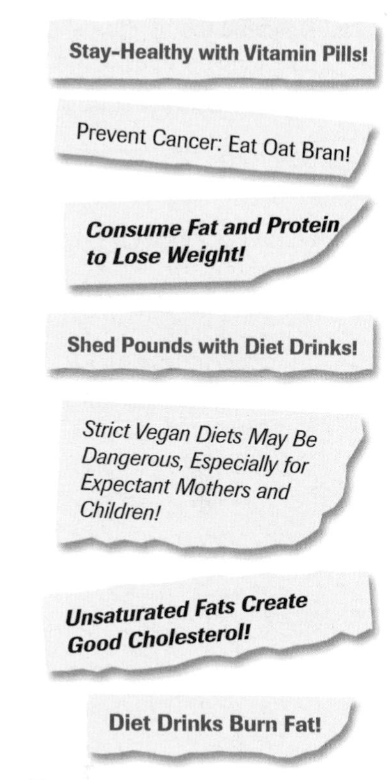

Figure 1
Dietary headlines in the media

Body Mass Index (BMI) a calculation obtained by dividing an individual's weight in kilograms by his or her height in metres squared

Dieting

During the past 30 years, society has promoted thinner ideal body types for women. Even if their weight is in the proper range, women feel pressure from friends, TV—and sometimes parents—to be very thin. Diets, for the most part, do not work. Ninety-five percent of people who diet gain the weight back within two years. Many follow an endless cycle of dieting and gaining, dieting and gaining. Studies also show that weight is regained more quickly after each dieting phase. This constant weight fluctuation may be more dangerous to health than keeping a greater, but stable, body mass, as fluctuation subjects the body to repeated stress. Some weight-loss plans include appetite suppressants such as amphetamines or laxatives, which are both very dangerous if abused.

A slightly different problem exists for many young males. They may wish to emulate the muscular heroes they see in movies, sports, and advertising (**Figure 2**). Most adolescent boys, however, do not have muscular bodies. Muscle development becomes most accelerated only after height growth has slowed. In an effort to look larger, men may use protein diets, which are aimed at increasing muscle development by increasing food intake beyond the body's needs. For some people, anabolic steroids have become a quick but dangerous way to gain muscle mass. Negative side effects include mood swings and prematurely halted growth. Testicular shrinkage, infertility, brain cancer, and premature heart disease have also been associated with anabolic steroid use.

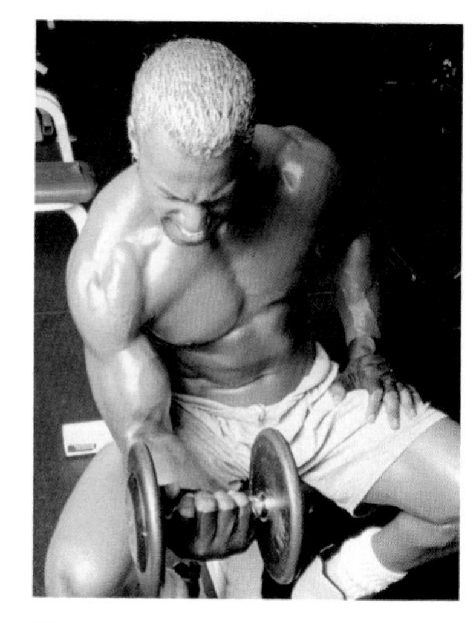

Figure 2
How appealing or healthy is this body type?

Understanding the Issue

1. What calculation is used to define normal body weight? What are its strengths and limitations?
2. Describe the health risks and costs of being substantially heavier than ideal body weight.
3. List dangers associated with weight-control diets.
4. List the dangers associated with bulking-up regimes.
5. Refer to section 3.5 and hypothesize why constant weight fluctuation stresses body systems.

Take a Stand

What Is Healthy?

Decision-Making Skills

- Define the Issue
- Analyze the Issue
- Research
- Defend a Decision
- Identify Alternatives
- Evaluate

Statement: The media should show bodies of different shapes, not just the stereotypical "ideal" figures (**Figure 3**).

- In your group, define the issue. Where is the balance between the medical profession's idea of a healthy body type and the social standards of ideal body types as reflected by the diet food industry and media?
- Research the issue. Search for information in newspapers, periodicals, and CD-ROMs, and on the Internet.

www.science.nelson.com

You might consider these aspects:
- the role of genetics in influencing weight;
- the prevalence of eating disorders such as anorexia nervosa and bulimia;
- factors that you think contribute to society's image of the ideal body;
- other weight-reduction options (aside from the diet-food industry) for people whose obesity is dangerous to their health;
- other muscle-building options (aside from the food-supplement industry) for those who choose this lifestyle.

(a) Identify the perspectives of each of the opposing positions.
(b) Develop and reflect on your opinion.
(c) Brainstorm suggestions for a solution.

Figure 3
Body stereotypes in the media

(d) Write a position paper summarizing your stand.
(e) How did your group reach a decision? What would you do differently if you were to do this again?

The Circulatory System 3.9

The heart works tirelessly, with most hearts beating continuously and rhythmically throughout a lifetime. Some hearts, however, experience problems and require assistance in one form or another. Sometimes a pacemaker must be attached to help the heart maintain its regular rhythm. In extreme cases, the heart may have to be replaced by another human heart or by an artificial heart.

The first totally artificial heart, the Jarvik-7, was implanted in a human patient in 1982. The patient lived for 112 days. Since that time, significant advances have been made in technology. Both Canadian and U.S. research teams are developing artificial hearts, with the first artificial heart transplant in nearly 20 years being performed in July 2001 at Jewish Hospital in Louisville, Kentucky. A group of Ottawa researchers, under the direction of Dr. Tofy Mussivand, is close to production of an artificial heart called the Heartsaver Ventricular Assist Device (HeartSaverVAD™) (**Figure 1**), which is intended for long-term circulatory support. Unlike other artificial hearts, both the Canadian and U.S. hearts are powered and controlled externally, eliminating the need for a permanent opening in the skin and allowing patients freedom of movement, as they would no longer need to be attached to a machine.

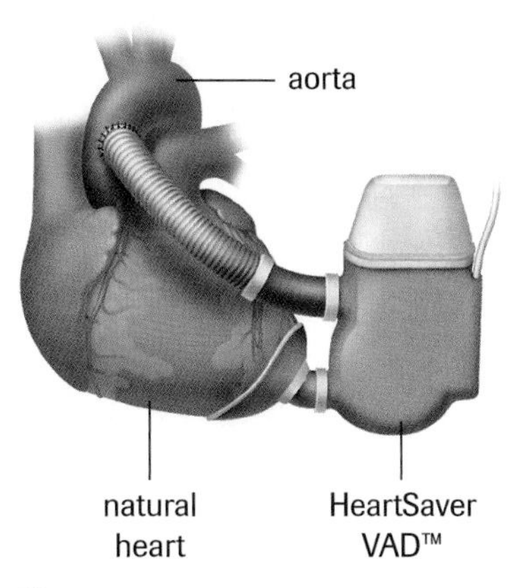

Figure 1
The HeartSaverVAD™

The Human Circulatory System

People once believed emotions came from their hearts, because the heart beats faster when a person is scared or excited. We now know that emotions come from the brain, and that the brain controls the heart.

Is one body system more important than the others? The proper functioning of all body systems is essential for our health and well-being, so no one system is the most important. There are many interactions among the various systems, but the circulatory system appears to be connected to all other systems.

The **circulatory system** is the transport system of the body and has four principal functions:

- transportation of oxygen and carbon dioxide
- distribution of nutrients and transport of wastes
- maintenance of body temperature
- circulation of hormones

circulatory system the body system consisting of the heart, blood vessels, and blood that is responsible for the transportation of materials around the body and plays a major role in the regulation of body temperature

Structure of the Circulatory System

A circulatory system consists of three general components: (1) a fluid in which materials are transported, such as blood; (2) a system of blood vessels or spaces throughout the body in which the fluid moves; and (3) a pump, such as the heart, that pushes the fluid through the blood vessels or spaces. The heart and blood vessels are commonly known as the **cardiovascular system** (from *cardio*, meaning "heart," and *vascular*, meaning "vessels" or "tubes").

cardiovascular system part of the circulatory system comprising the heart and blood vessels

Your circulatory system is a complex network of tissues and organs distributed throughout your body (**Figure 2**). No body cell is farther than two cells away from a blood vessel that can carry nutrients. There are 96 000 km of blood vessels in your body to sustain your 100 trillion cells. No larger than the size of your fist and with a mass of about 300 g, the heart beats about 72 times/min from the beginning of life until death. During an average lifetime, the heart pumps enough blood to fill two large ocean tankers.

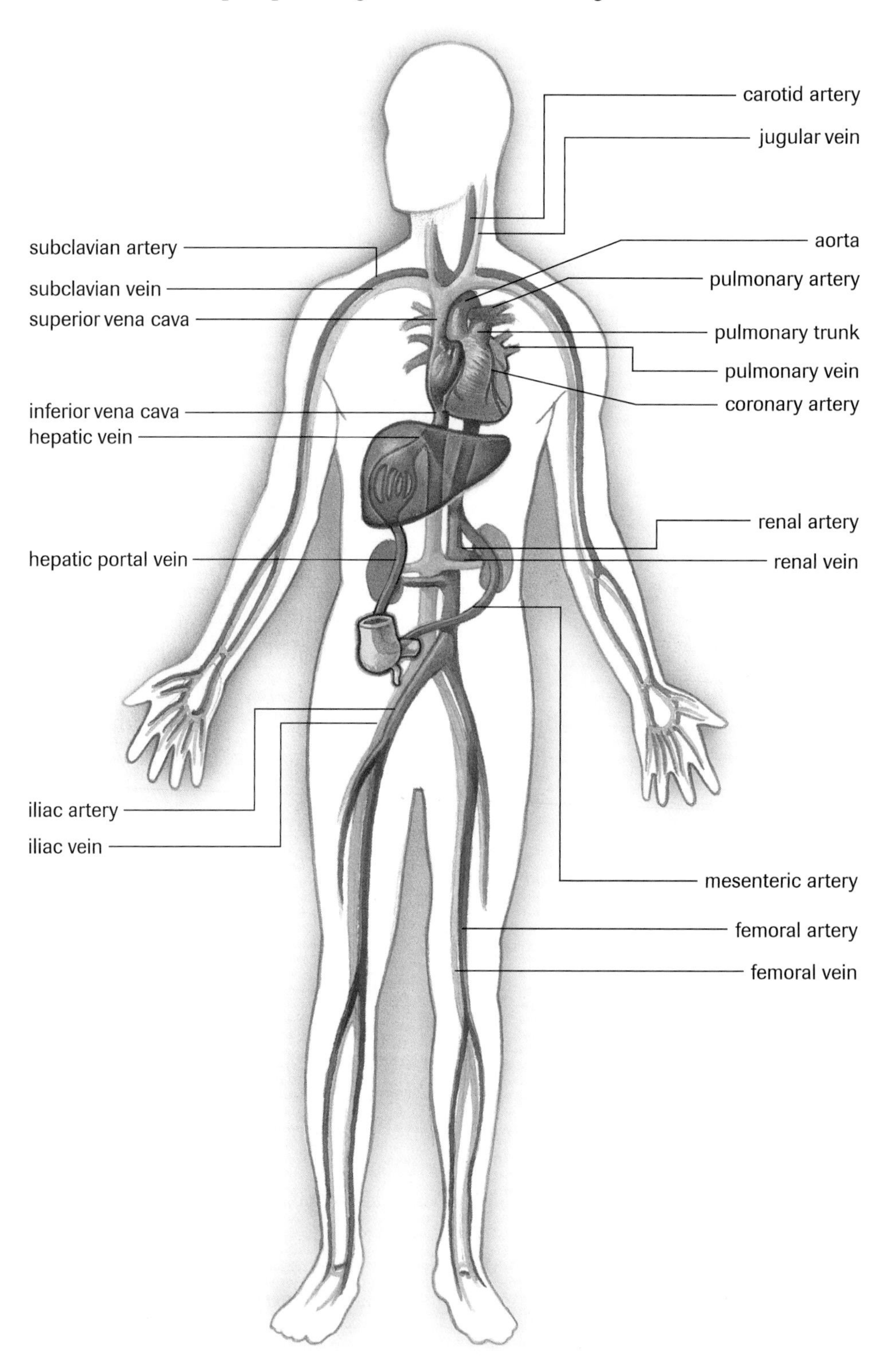

Figure 2
The circulatory system

Every minute, 5 L of blood cycles from the heart to the lungs, picks up oxygen, and returns to the heart. Then, the heart pumps the oxygen-rich blood and nutrients to the tissues of the body. The circulatory system plays a central role in providing the oxygen that the body's cells require. This oxygen is used to produce energy for a variety of processes including building new materials, such as proteins, and repairing tissues.

The circulatory system is actually two systems in one (**Figure 3**). The right side of the heart pumps blood to the lungs, where the blood picks up oxygen and returns to the left side of the heart. This constitutes the **pulmonary circuit**. The left side of the heart then pumps the oxygenated blood out to all parts of the body, where the blood vessels (arteries) deliver oxygen and other materials. This part of the system is known as the **systemic circuit**.

pulmonary circuit the system of blood vessels that carries deoxygenated blood to the lungs and oxygenated blood back to the heart

systemic circuit the system of blood vessels that carries oxygenated blood to the tissues of the body and deoxygenated blood back to the heart

The Pulmonary Circuit

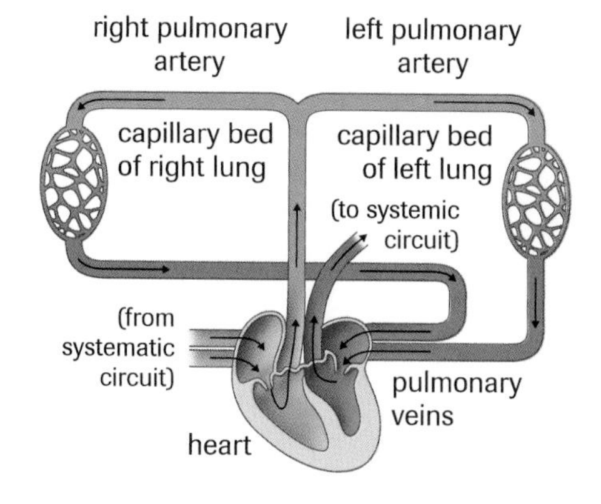

The Systemic Circuit

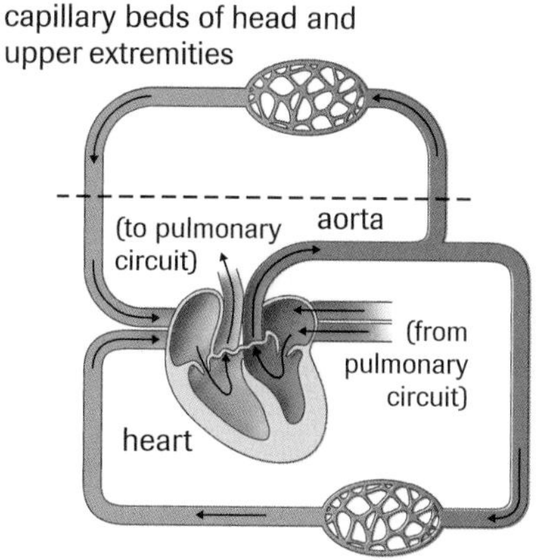

Figure 3
The pulmonary and systemic circuits of the circulatory system. The blood vessels carrying oxygenated blood are in red; the vessels carrying deoxygenated blood are in blue.

Heart Structure

The mammalian heart consists of a double pump separated by a wall of muscle called the **septum**. The pump on the right powers the pulmonary circuit, and the pump on the left powers the systemic circuit. Each pump consists of a thin-walled **atrium** (plural: atria) and a thick-walled **ventricle**. The atria receive blood from veins and pump it into the ventricles; the more muscular ventricles then pump the blood to distant tissues. When the heart muscles contract, the valves of the heart prevent blood from flowing back into the ventricles (**Figure 4**). Follow the path of blood through the heart (**Figure 5**).

septum a wall of muscle that separates the right heart pump from the left

atrium a thin-walled chamber of the heart that receives blood from veins

ventricle a muscular, thick-walled chamber of the heart that delivers blood to the arteries

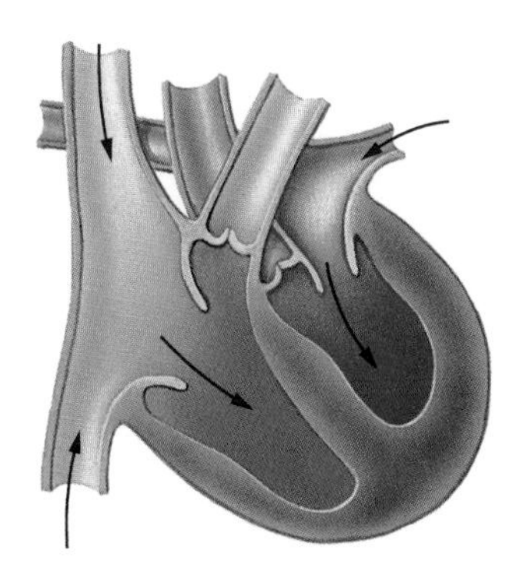

atria contracted, ventricles relaxed

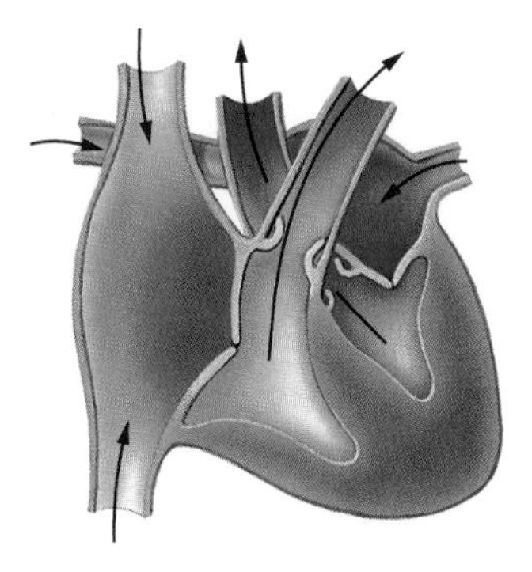

atria relaxed, ventricles contracted

Figure 4
The right and left atria contract in unison, pushing blood into the right and left ventricles. Ventricular contractions close the atrioventricular valves and open the semilunar valves. The relaxation of the ventricles lowers pressure and draws blood back to the chamber. The closing of the semilunar valves prevents blood from reentering the ventricles.

DID YOU KNOW?

Pulmonary Artery

The only artery that does not carry oxygenated blood away from the heart is the pulmonary artery. The pulmonary artery carries deoxygenated blood from the heart to the lungs.

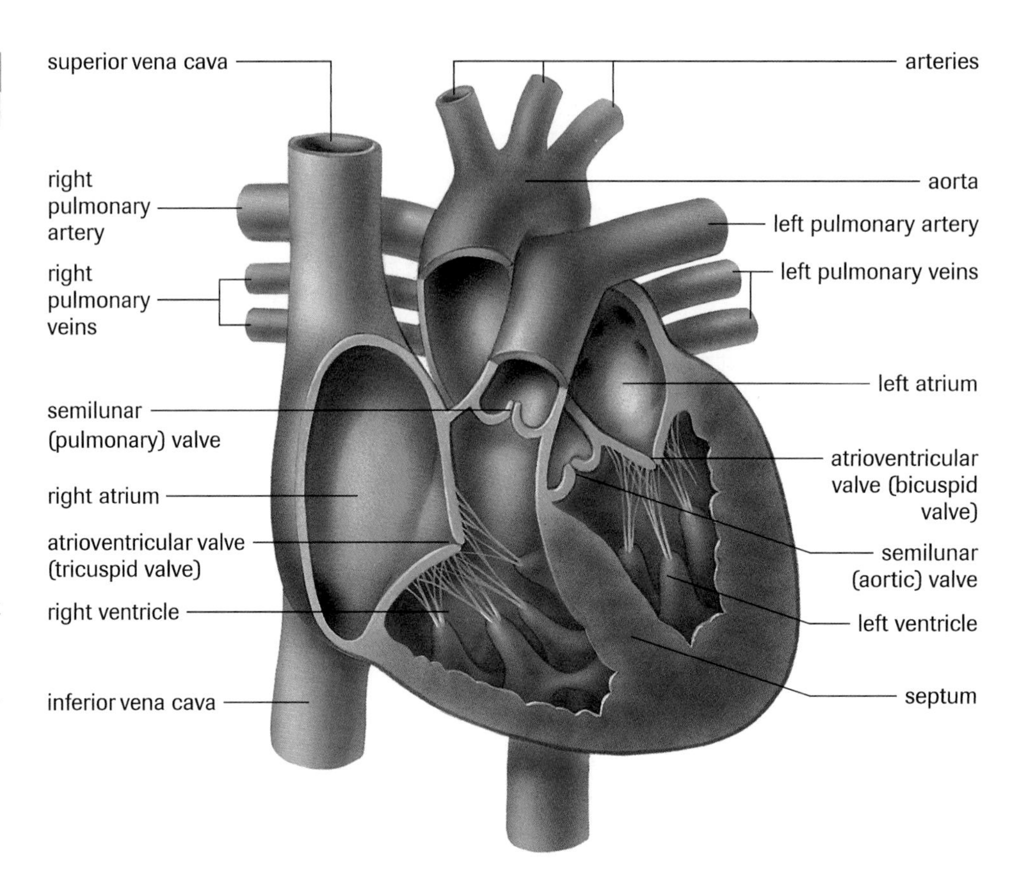

Figure 5

Deoxygenated blood enters the right atrium. Blood from the right atrium is pumped to the right ventricle, which, in turn, pumps it to the lungs. Oxygenated blood moves from the lungs to the left atrium. Blood from the left atrium is pumped to the left ventricle, which, in turn, pumps it to the cells of the body.

sinoatrial (SA) node a small mass of tissue in the right atrium that originates the impulses stimulating the heartbeat

atrioventricular (AV) node a small mass of tissue through which impulses from the sinoatrial node are passed to the ventricles

diastole relaxation (dilation) of the heart, during which the cavities of the heart fill with blood

systole contraction of the ventricles, during which blood is pushed out of the heart

pulse change in the diameter of the arteries that can be felt on the body's surface following heart contractions

blood pressure the pressure exerted on the walls of the arteries when the ventricles of the heart contract. Blood pressure is measured in millimetres of mercury, or mm Hg.

sphygmomanometer a device used to measure blood pressure

Heart Rhythms, Sounds, and Pressure

The heart has two specialized bundles of nerves that control its beating. The first bundle, called the **sinoatrial (SA) node**, acts as a pacemaker and sets a rhythm of about 72 beats per minute. The second bundle, the **atrioventricular (AV) node**, passes the nerve impulses to the ventricles, a message that causes them to contract in unison.

The familiar *lubb-dubb* heart sounds are caused by the closing of the heart valves (**Figure 4**). When the atria are relaxed, they begin to fill with blood. This relaxation is called **diastole.** As the atria push blood into the ventricles, the ventricles contract, pushing blood into the arteries. This contraction is known as **systole.** The increase in pressure during systole forces the AV valves shut, causing the *lubb* sound. As the ventricles relax, the pressure inside decreases and blood tends to flow into them. This decrease in pressure causes the semilunar valves to close, preventing blood from flowing back from the arteries into the ventricles. The closing of the semilunar valves creates the *dubb* sound.

Every time the heart beats, blood surges through the arteries. In response to this increase in pressure, the arteries stretch and increase in diameter. Where the arteries are near the surface of the body, this increase in diameter can be felt as the **pulse.**

Blood pressure is the force of the blood on the walls of the arteries. It can be measured with an instrument called a **sphygmomanometer.** A cuff with an air bladder is wrapped around the arm, and the air bladder is inflated with a pump. This cuts off blood flow to the major arteries of the arm. A stethoscope is used to listen for the sound of blood entering the artery as air is gradually released from the cuff. The pressure on the gauge of the sphygmomanometer at which the sound is first heard is called the systolic blood pressure. Normal systolic blood pressure for young adults is about 120 mm Hg. The cuff is then deflated even more until the sound disappears as a result of blood flowing into the artery as the ventricles fill. At this point, the pressure is called the diastolic blood pressure, which is normally around 80 mm Hg for young adults. This normal blood pressure would be reported as 120/80 (120 over 80): a systolic pressure of 120 mm Hg and a diastolic pressure of 80 mm Hg.

There are many factors that can raise or lower blood pressure from its normal range. Low blood pressure reduces the body's capacity to transport blood around the body. High blood pressure is equally serious because it can weaken an artery, which might result in its rupture.

aorta the largest artery in the body leading directly from the heart and carrying oxygenated blood to the tissues of the body

Blood Vessels

Arteries are blood vessels that carry blood away from the heart. They have thick walls with layers of muscle sandwiched between layers of connective tissue (**Figure 6**). The **aorta** leads directly from the left ventricle and branches into many arteries that carry blood around the body. From the arteries, blood flows into smaller blood vessels called arterioles. Arterioles control blood flow into much smaller vessels, called capillaries. Capillaries are very small blood vessels, normally between 0.4 mm and 1.0 mm long with a diameter of less than 0.005 mm. Their diameter is so small that red blood cells must travel through them in single file (**Figure 7**).

Capillaries merge and become progressively larger vessels, called venules, which merge into veins that have greater diameter. As blood flows from arteries to arterioles and then to capillaries, blood flow is reduced. The pressure exerted by the pumping of the heart is not enough to push the blood back to the heart, especially from the lower limbs. The structure of veins helps overcome this problem. Veins have valves that open in one direction, preventing blood from flowing back in the other direction. Contraction of the muscles surrounding the veins also helps push blood toward the heart (**Figure 8**).

(a) artery

(b) vein

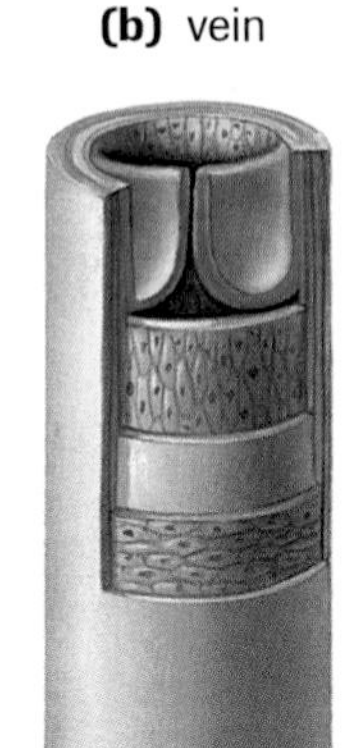

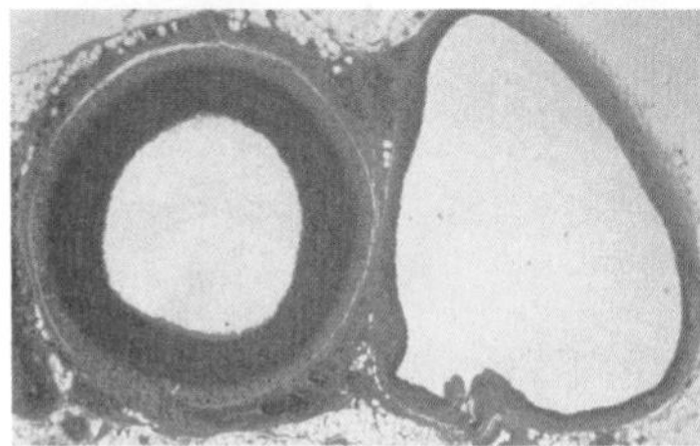

Figure 6
Arteries have strong walls capable of withstanding great pressure. The low-pressure veins have a thinner middle layer. The photo shows a cross section of an artery and a vein.

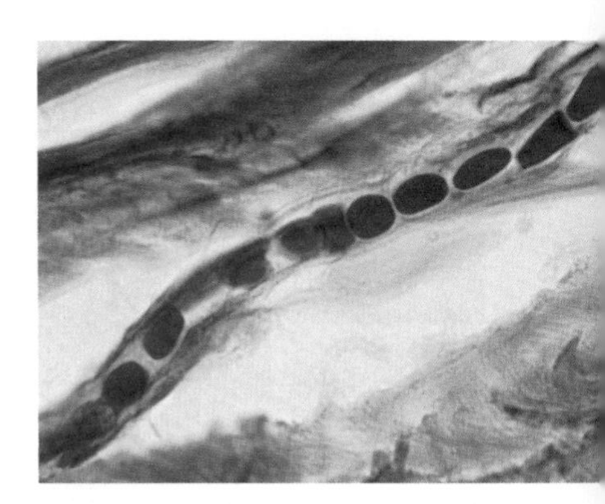

Figure 7
Red blood cells in a capillary. Notice that the capillary is only wide enough for cells to pass through one at a time.

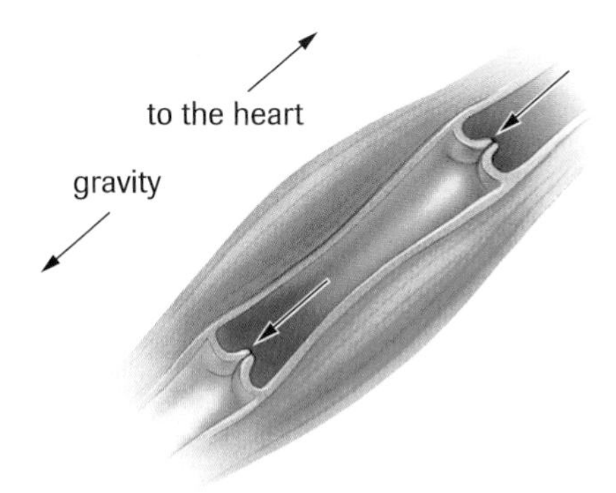

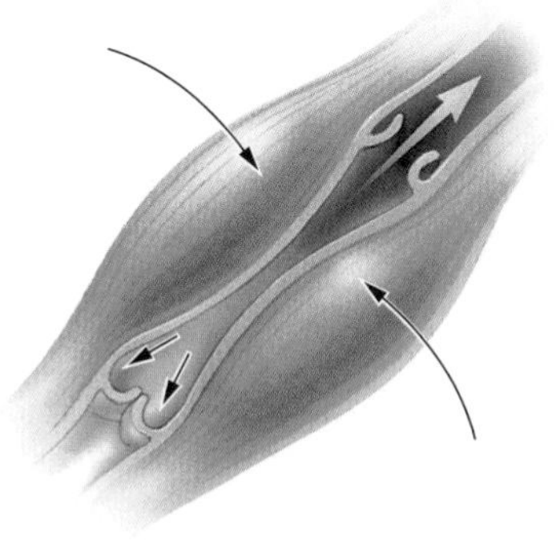

Figure 8
Venous valves and skeletal muscle work together in a low-pressure system to move blood back to the heart.

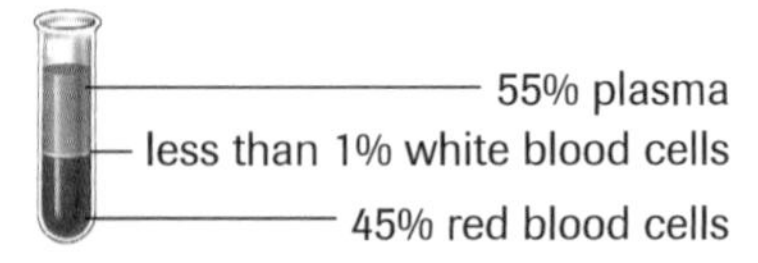

Figure 9
The composition of blood

Blood Components

The average 70-kg individual is nourished and protected by about 5 L of blood. Approximately 55% of the blood is a liquid, called the plasma; the remaining 45% is composed of blood cells (**Figure 9**). The blood cells are considered to be a tissue because they work together for a common purpose. **Table 1** summarizes the components of blood.

Table 1 The Components of Blood

Blood component	Structure and function
plasma	• the fluid medium through which the solid blood components, along with dissolved gases, nutrients, wastes, and hormones, are transported around the body • contains a number of dissolved proteins that have different functions • One protein is essential in blood clotting. Another group contains antibodies that provide immunity against diseases such as measles. Others regulate the balance of fluids in the plasma.
erythrocytes *(red blood cells)*	• contain hemoglobin, a molecule that enables the cell to bind oxygen molecules • without hemoglobin, would supply enough oxygen to support life for about 4.5 seconds. With hemoglobin, life can continue for about five minutes. (This seems like a short time, but blood is continuously transported back to the heart and then on to the lungs, where the oxygen supply is renewed.) • do not have a nucleus in their mature stage, allowing the cell to carry more hemoglobin
leukocytes *(white blood cells–different types)*	• function mainly to protect the body against invading microorganisms and toxins • are much fewer in number than red blood cells • are of several types, and unlike red blood cells, contain a nucleus • Some types destroy microorganisms by engulfing them and then using enzymes to digest the microorganism and the leukocyte itself; others are specialized in producing antibodies that are an important part of the body's immune system. • Fragments of white blood cells and invading microorganisms are called pus.
platelets (a) (b) (c)	• small fragments of larger cells produced from cells in bone marrow • function primarily to initiate blood clotting • are irregularly shaped **(a)** and move easily through smooth blood vessels when resting • become active and rupture **(b)** if they encounter a sharp edge, such as a cut, as they move through a blood vessel • When the platelet membrane breaks, the platelet releases a substance that reacts with proteins in the plasma to create a mesh of fibres **(c)**. • This mesh prevents further blood flow. After a few days, the fibres contract and begin to close the wound to the blood vessel.

Circulatory System of the Fish

The structure of an organism's circulatory system depends on its characteristics and its requirements for life, which are determined by the environment in which the organism lives. While mammals have the most complex circulatory system, that of the fish is similar but less complex. It is a closed system (i.e., the blood is contained within a continuous system of interconnected blood vessels) and is connected directly with the respiratory organs, the gills. The blood vessels near the surface of the gills absorb the dissolved oxygen from the water as it flows over the gills (**Figure 10**).

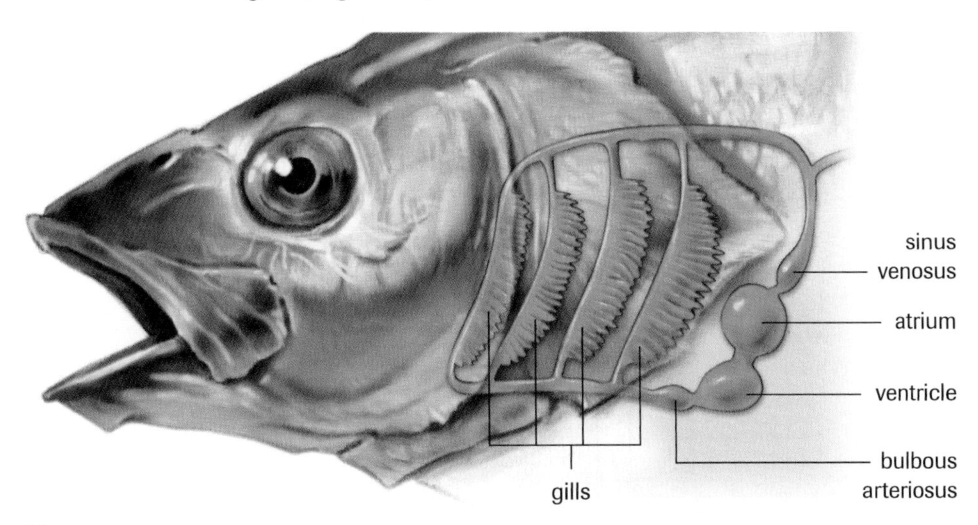

Figure 10
The circulatory system of the fish

Fish have the simplest true heart, with muscular walls and a valve between its chambers. Fish hearts have two chambers, but also have two other structures involved in circulation. For instance, the heart of the perch has four parts—an atrium, a ventricle, the bulbous arteriosus, and the sinus venosus. The four parts are arranged in an S-shaped structure. The sinus venosus is the site of specialized cardiac pacemaker tissue in many fish, and its role is to control the heartbeat. The sinus venosus has a volume that is never more than a fraction of that of the atrium into which it opens. The bulbous arteriosus is not a pumping chamber but may extend the period of the cardiac cycle during which blood is pumped to the gills.

Blood enters the atrium of the fish heart and, as the heart relaxes, moves from the atrium through a valve into the ventricle, much as in a mammal. As the ventricle contracts, the blood is pumped from the heart to the gills, where it receives oxygen and gets rid of carbon dioxide. The blood then moves on to the organs of the body via capillaries, where nutrients, gases, and wastes are exchanged. However, unlike the mammalian circulatory system, there is no separation of the circulation between the respiratory organs and the rest of the body. In other words, the blood travels in a continuous circuit from heart to gills to organs and back to the heart to start its journey again.

Section 3.9 Questions

Understanding Concepts

1. What are the main functions of the circulatory system?
2. Distinguish between the circulatory system and the cardiovascular system.
3. Assuming that the average human lifespan is 76 years, calculate approximately how many times the heart beats in a lifetime.
4. If shown a blood vessel attached to the heart, how could you tell if it was an artery or a vein?
5. Describe the pulmonary and systemic circuits of the circulatory system, and explain how the two circuits are interconnected.
6. Draw and label the major blood vessels and chambers of the human heart. Trace the flow of deoxygenated and oxygenated blood through the heart.
7. Define the terms *systolic pressure* and *diastolic pressure*.
8. Outline the similarities and differences between the human and the fish circulatory systems. Which is more efficient? Why?

Applying Inquiry Skills

9. Cancer of the white blood cells is called leukemia. Like other cancers, leukemia is associated with rapid and uncontrolled cell division. Examine **Figure 11** and predict which subject might be suffering from leukemia. Explain your reasoning.

Making Connections

10. Athletes can take unfair advantage of the benefits of extra red blood cells. In a process called "blood doping," a blood sample is taken two weeks prior to a competition, and the red blood cells are separated and stored. A few days before the event, the red blood cells are injected into the athlete. Why would athletes remove red blood cells only to return them to their body later? Using the Internet and other resources, research the practice of blood doping.

www.science.nelson.com

Exploring

11. An individual injured in a car accident was being examined by emergency medical personnel. Since both forearms were injured, they could not be used to determine the individual's pulse. Use the Internet and other resources to find other points on the body where the pulse can be measured in such circumstances.

www.science.nelson.com

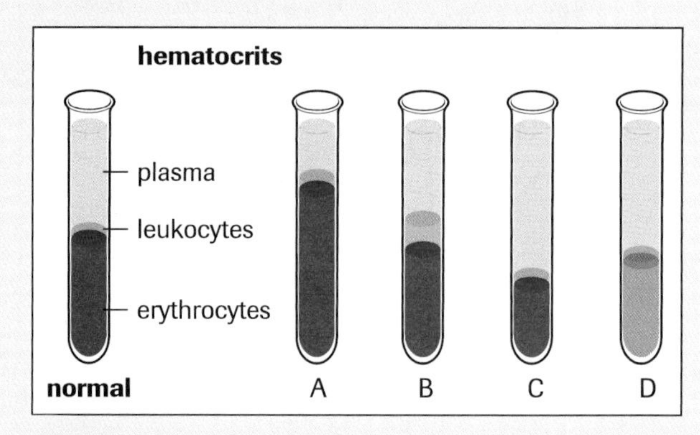

Figure 11
Blood samples from subjects A, B, C, and D

Circulatory Disorders and Technologies 3.10

In the early morning, John Smith, a 47-year-old male, enters the emergency ward of St. Michael's Hospital in Toronto complaining of chest pains. He had been shovelling snow and suddenly experienced sharp pain in his chest, shortness of breath, and nausea.

As the emergency ward staff prepares to diagnose and treat the problem, they note that this is the sixth patient with similar symptoms since the heavy snowfall the previous evening. Based only on Mr. Smith's description of his symptoms, the obvious suspicion is that the patient is having or has had a heart attack, and the priority is to stabilize his condition and prevent further damage.

electrocardiograph an instrument that detects the electrical signals of the heart

electrocardiogram the graphical record produced by an electrocardiograph

Diagnosis and Treatment

Once some basic personal information is received from the patient, the emergency ward medical professionals attempt to determine the exact nature of the health problem that the patient is experiencing, in order to determine what treatment should be given.

Hospital emergency wards have established procedures for such situations. Within 10 minutes, a blood sample is taken and sent for analysis, and the patient is connected to an **electrocardiograph** (**Figure 1(a)**). The blood test will test for the presence of specific proteins that are produced when muscle tissue is damaged. The electrocardiograph will produce a graphical record, called an **electrocardiogram** (ECG), that will show if the heart rate and rhythm are normal or if heart damage has occurred (**Figure 1(b)**).

CAREER CONNECTION

In addition to a cardiologist (a doctor who specializes in the heart), other health care professionals would be involved in the diagnosis and treatment of a heart-attack patient. These would include nurses, medical equipment technicians, lab technologists, and X-ray technologists.

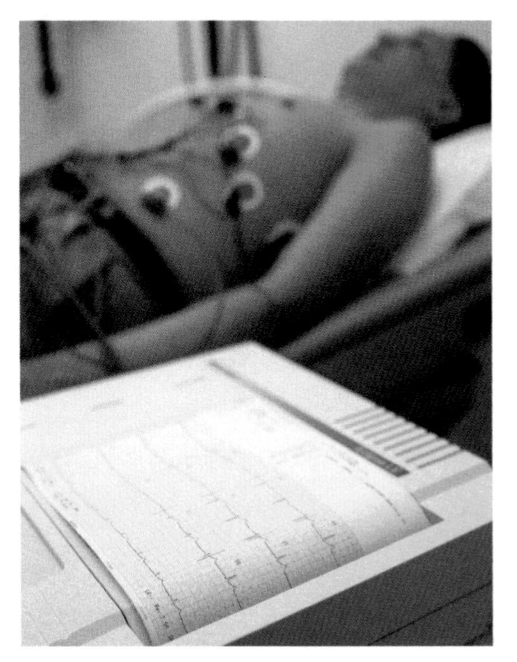

(a)

(b)

Figure 1
(a) Patient connected to an electrocardiograph
(b) An abnormal electrocardiogram

DID YOU KNOW?

Cardiac Catheters
Cardiac catheters can also be equipped with a fibre-optic camera to view the blockage or with a laser that can burn away the material causing the blockage in the coronary arteries.

The tests confirm that Mr. Smith has indeed had a heart attack. The first step in the treatment is to administer an anti-clotting drug to prevent any further blockage of the coronary arteries. For low-risk patients, aspirin is given. For more serious cases, a stronger anti-clotting drug is administered. Oxygen is generally administered through a tube in the nose. This will increase the concentration of oxygen in the lungs and will help compensate for any shortage of oxygen caused by decreased blood flow.

The ECG shows the cardiologist where the damage has occurred, but not the extent of the damage. It is also important to know the condition of the coronary arteries and the status of the blockage or blockages. In order to obtain further information, the cardiologist must perform a procedure called **cardiac catheterization**. In this procedure, a narrow tube (catheter) is threaded into an artery in the groin area. A dye that is visible on X-ray film is then injected into the catheter. Guided by the X-ray images on a monitor, the catheter is threaded through the artery to the blocked area in the coronary artery. This procedure can be used to pinpoint the exact location of the blockage.

cardiac catheterization the insertion of a catheter into the coronary artery through an incision in an artery in the groin

angioplasty a procedure in which a blocked artery is opened by inflating a small balloon at the end of a catheter

stent a small metallic mesh cylinder that is inserted into a blocked artery to allow blood to flow freely and to prevent the recurrence of the blockage

While this catheter is in place, a treatment procedure called **angioplasty** can be conducted (**Figure 2**). A second, thinner catheter with a very small deflated balloon at the tip is threaded through the first catheter. When the balloon is pushed through the blockage, it is inflated, opening the passage to allow blood flow. The balloon is then deflated and removed. Today, a stent is usually used. A **stent** is a fine metallic mesh tube that remains in place in the artery and prevents the artery from narrowing again.

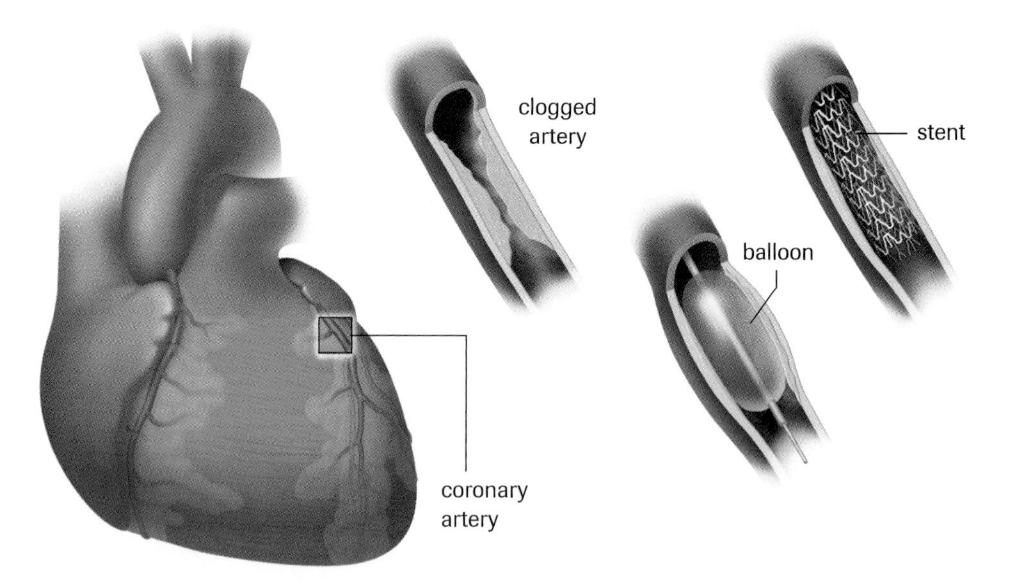

Figure 2
Cardiac catheterization and angioplasty with a stent

Alternative Treatment—Bypass Surgery

coronary bypass surgery a surgical procedure in which a vein graft is used to shunt the blood supply around the blocked area in a coronary artery

If angioplasty is not effective or appropriate, there are other procedures that can be used to treat blockages of the coronary arteries. **Coronary bypass surgery** is the most common of these procedures. The name of this procedure is appropriate because the surgery uses a piece of a vein from another part of the body (a vein graft) to bypass the blocked area of the coronary artery in order to supply blood to the area beyond the blockage (**Figure 3**).

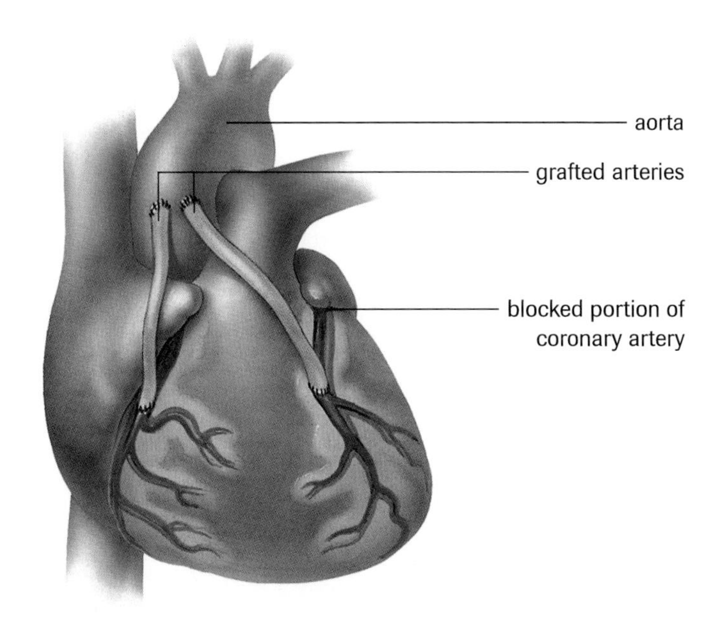

Figure 3
A coronary bypass operation. Blood flow is rerouted around the blockage.

This surgery is much more complicated than angioplasty because it involves opening the chest and operating directly on the heart. This procedure is commonly referred to as open-heart surgery. Both the hospitalization and the recovery periods are much longer than for angioplasty.

Cardiovascular Disease

Cardiovascular disease includes a whole range of conditions that affect various parts of the circulatory system. Together they make up the leading cause of death in Canada and the United States.

Diseases of the arteries include **arteriosclerosis** (from *arterio*, meaning "artery," *scler*, meaning "hard," and *osis*, meaning "condition"), commonly known as hardening of the arteries. *Arteriosclerosis* is a general term that includes a group of conditions in which the walls of the arteries lose their elasticity. Diseases of the veins include such diseases as phlebitis (inflammation of the veins) and varicose veins. Diseases of the heart itself include four categories of conditions, as outlined in **Table 1**.

arteriosclerosis a group of disorders that causes the walls of blood vessels to thicken, harden, and lose their elasticity

Table 1 Categories of Heart Disease

Heart disease category	Description
coronary artery disease	blockage of the arteries that supply blood to the heart muscle
arrhythmias	malfunctions of the electrical system that controls the heartbeat
valve disorders	dysfunction of one or more of the heart valves
heart muscle disease	inflammation or infection of the heart muscle itself

Causes of Cardiovascular Disease

The most well-known of the cardiovascular diseases is coronary artery disease. The blockage of blood flow through arteries is usually caused by **atherosclerosis**, a form of arteriosclerosis. Anyone who has ever washed dishes is aware that fat floats on water. You may have noticed that when one fat droplet meets another, they stick together and form a larger droplet. Unfortunately, the same thing can happen in your arteries. As fat droplets

atherosclerosis a degeneration of the arteries caused by the accumulation of plaque along the inner wall grow into larger and larger blockages, they slowly close off the opening of the blood vessel. Calcium and other minerals deposit on top of the fat, forming a substance known as **plaque**. The buildup of plaque in a coronary artery can narrow the artery, causing an increase in blood pressure, and can eventually block the artery completely, shutting off the supply of oxygen and other nutrients to the muscle tissues of the heart (**Figure 4**).

plaque a sticky yellow substance, composed of fatty deposits, calcium, and cellular debris, that builds up on the inside of arteries

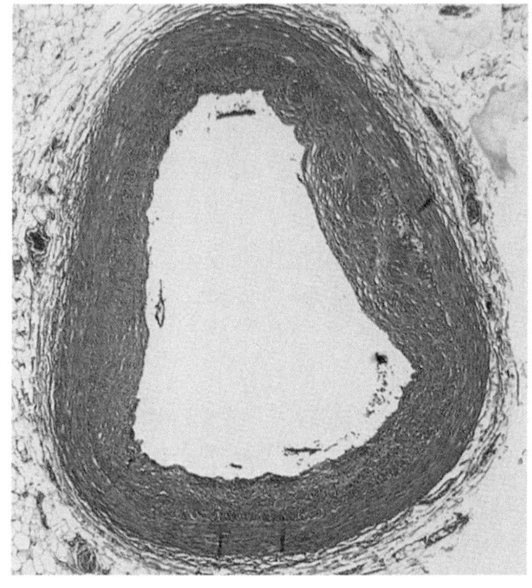

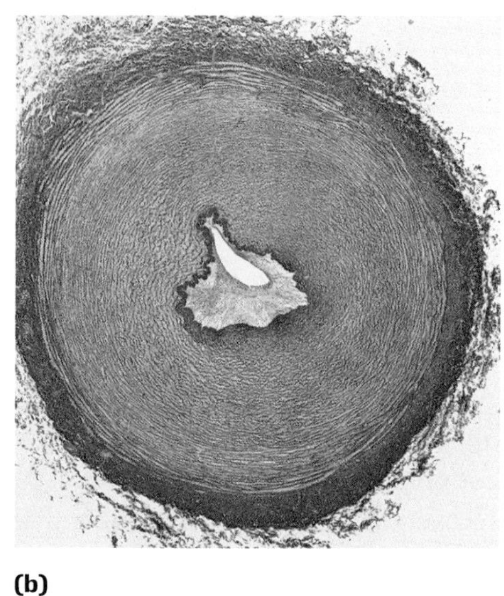

Figure 4
(a) Cross section of a normal artery
(b) Cross section of an artery from a person with atherosclerosis. Note that plaque deposits have narrowed the passageway.

Heart muscles, like other tissues, need oxygen and nutrients to function properly. If the supply is cut off, the tissue dies. When this happens, a person will likely have a heart attack, or **myocardial infarction**. Heart attacks usually occur when existing plaque breaks open and triggers the formation of blood clots on top of it. The common symptoms of myocardial infarction include sharp chest pain, pain in the neck and arms, shortness of breath, and nausea.

myocardial infarction (*myo*, referring to muscle, *cardial*, referring to the heart, and *infarct*, meaning "death") death of heart muscle tissue due to lack of oxygen and other nutrients; commonly known as a heart attack

Even though a heart attack may occur relatively suddenly, the underlying causes may develop over a long period of time. Atherosclerosis is a progressive condition that may begin at an early age.

The exact causes of heart disease are difficult to identify; however, the main risk factors are smoking, physical inactivity, high blood cholesterol, high blood pressure, obesity, poor eating habits, diabetes, and heredity.

Prognosis and Prevention

Many heart attacks can be prevented through a healthy lifestyle, which can reduce the risk of developing coronary artery disease. For patients who have already had a heart attack, maintaining a healthy lifestyle and following their doctor's orders can prevent another heart attack. Most heart-attack survivors see their survival as a second chance and make necessary changes in their lifestyle.

In the continued consultation with Mr. Smith, the heart-attack victim, it is learned that the patient is a business executive whose busy schedule does not allow much time for exercise. He is also a smoker and is approximately 20 kg overweight. The doctor concludes that the patient's coronary artery disease is largely attributable to his lifestyle and offers the logical advice—quit smoking, maintain a low-fat diet and a healthy weight, exercise regularly, and change work habits to manage stress.

TRY THIS activity — Listening to Heart Sounds

1. Place a stethoscope on your own chest and listen for a heart sound (**Figure 5**).

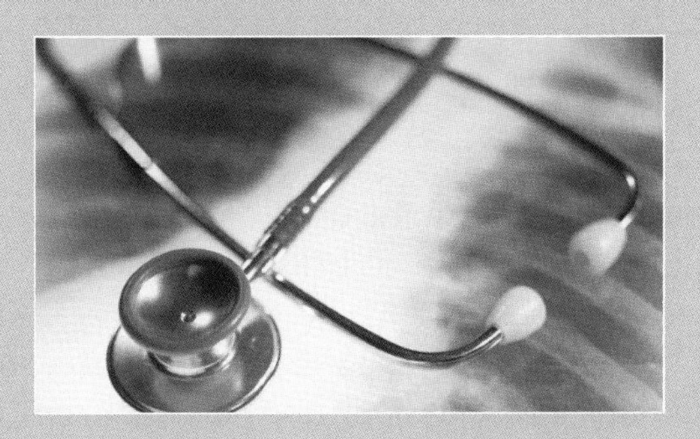

Figure 5

2. After one minute of moderate exercise (e.g., walking on the spot), listen for your heart sounds again.
   (a) Draw a diagram of your chest showing where you located the clearest sound.
   (b) Did the sound of your heartbeat change after exercise? Describe what differences you heard.

Make certain to disinfect the earpieces of the stethoscope with rubbing alcohol before and after use.

Do not perform this activity if you are not allowed to participate in physical education classes.

DID YOU KNOW?

Heart Murmur
A heart murmur is the sound heard when a heart valve is damaged or malformed owing to a birth defect. The hissing sound, detected with a stethoscope, is caused by the leakage through the heart valves.

Section 3.10 Questions

Understanding Concepts

1. Explain what usually causes a myocardial infarction.
2. What is an electrocardiogram? Why is it useful?
3. Why is it important to conduct tests quickly when a patient is complaining of chest pain?
4. Why does atherosclerosis raise blood pressure?
5. Draw a flowchart describing the process of angioplasty and explain why it is an important surgical procedure.
6. What is coronary bypass surgery and why is it performed?
7. Why are plaque deposits in arteries dangerous?
8. Use a graphic organizer to name and briefly describe the four categories of heart disease.
9. Name the risk factors that are associated with cardiovascular disease.

Making Connections

10. Research to find and describe the components of a heart-healthy lifestyle. Evaluate your own cardiovascular health and identify ways in which it can be improved.

www.science.nelson.com

Exploring

11. Choose one of the diseases or disorders from the list below. Gather information about the cause, symptoms, and possible treatment for the disease or disorder you selected. Present your findings in a poster, a short essay, or a presentation to your class.

    congestive heart failure — hypertension
    cerebral thrombosis — phlebitis
    pulmonary embolism — aneurysm

12. Search the Internet and other resources to explain what the different sections of the electrocardiogram in **Figure 6** indicate.

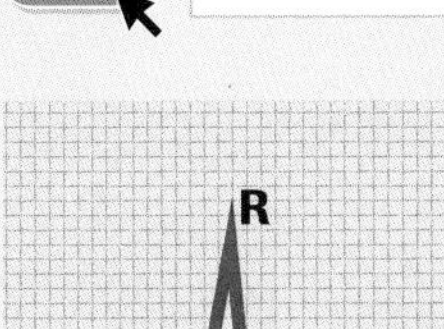

www.science.nelson.com

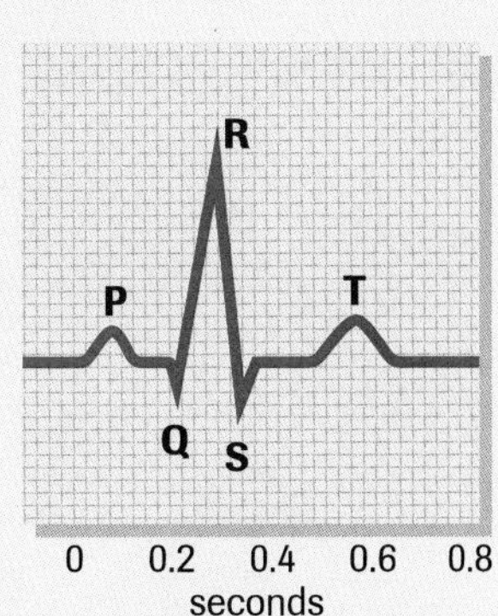

Figure 6

3.11 Investigation

Inquiry Skills

- ○ Questioning
- ○ Hypothesizing
- ● Predicting
- ● Planning
- ● Conducting
- ● Recording
- ● Analyzing
- ● Evaluating
- ● Communicating

The Body's Responses to Exercise

There is overwhelming scientific evidence establishing the relationship between exercise and cardiovascular health. The more physically active you are, the lower your risk of developing a cardiovascular problem such as atherosclerosis, coronary artery disease, or hypertension (high blood pressure). Physical fitness does not guarantee that you will not develop any cardiovascular problem, but it will significantly reduce your risk.

Personal fitness trainers need to understand the structure and function of body systems in order to offer advice to others about an appropriate fitness program. They use blood pressure and heart rate, along with other measures such as endurance, strength, flexibility, body composition (the ratio of fat to bone to muscle), and recovery time, to determine the overall fitness of an individual and to plan an appropriate program for that person (**Figure 1**). Fitness trainers are particularly concerned with aerobic fitness, or the ability of the heart, lungs, and bloodstream to supply oxygen to the cells of the body (especially the muscle cells) during physical activity. They are also interested in how quickly the body returns to normal conditions after the stress of physical activity. This recovery is an indication of the body's ability to maintain homeostasis.

In this investigation, you will use simple exercises to determine how your body responds to physical activity. Changes in blood pressure and pulse rate will be used to indicate this response. You will compare your responses with those of a number of your classmates and to the normal values of a person your age. The results of this investigation may give you a preliminary indication of your level of physical fitness.

Question

How are blood pressure and pulse rate affected during and after exercise?

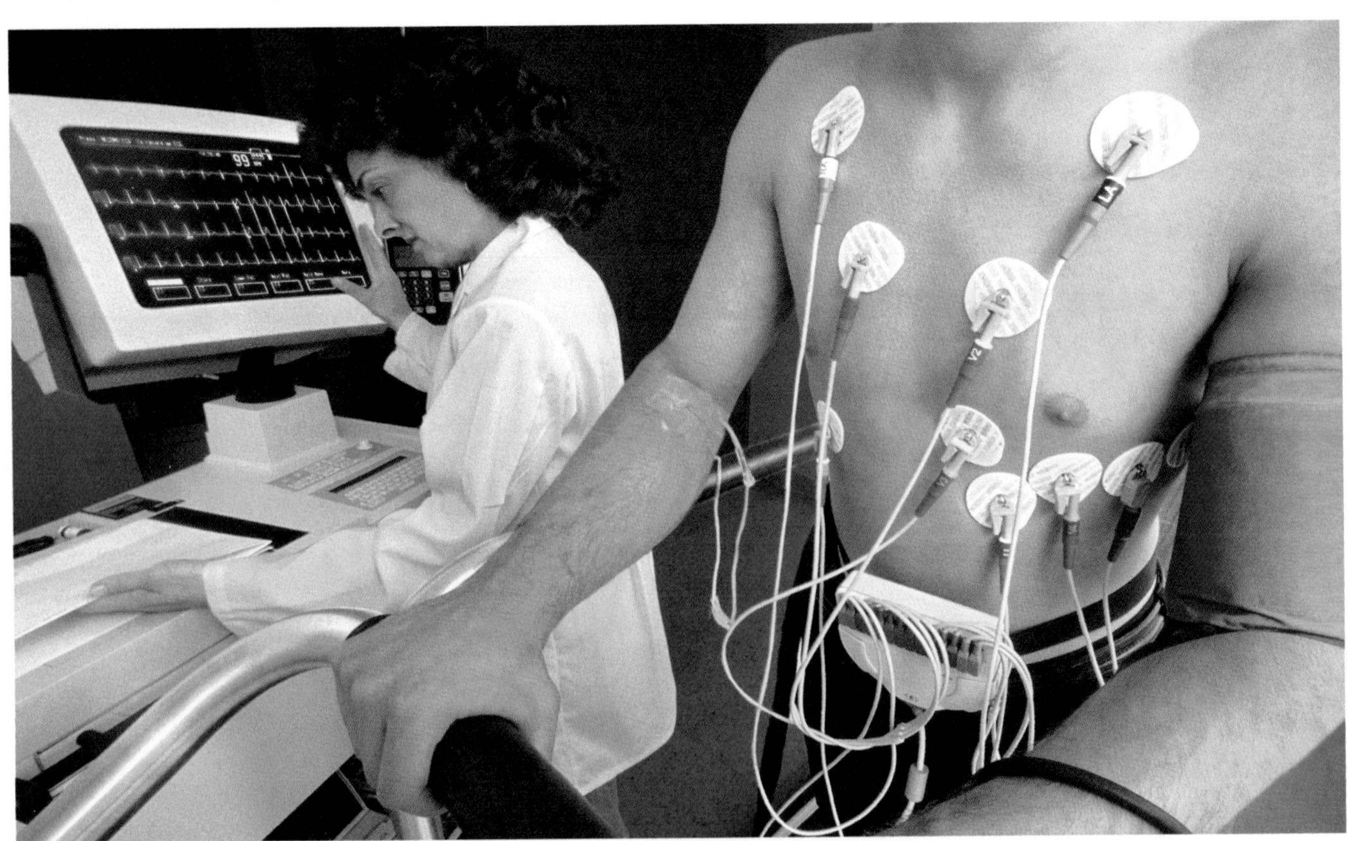

Figure 1
Some personal fitness assessment centres use technology to determine the level of physical fitness.

Prediction

With a partner, discuss how blood pressure and pulse rate are affected during and after exercise.

(a) Predict how blood pressure and pulse rate are affected during and after exercise.

Experimental Design

(b) Design a controlled experiment to test your prediction. Include the following in your design:
- descriptions of the independent, dependent, and controlled variables
- a step-by-step description of the procedure
- a list of safety precautions
- a table to record your observations

Materials

stopwatch, or a watch with a second hand
sphygmomanometer (**Figure 2**)
stethoscope

Do not perform this activity if you are not allowed to participate in physical education classes.

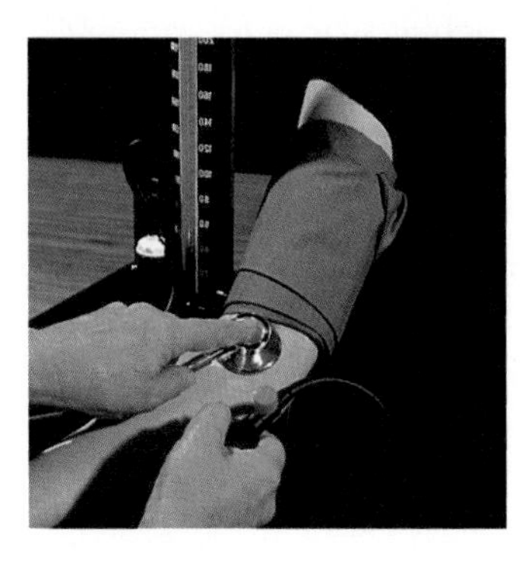

Figure 2
Using a sphygmomanometer and a stethoscope to measure blood pressure

Procedure

1. Submit the procedure, safety precautions, and observation table to your teacher for approval.
2. Carry out the procedure.
3. Write a report (see Appendix A4 for format). In your report, be sure to include ways to improve your design.

Analysis

(c) Graph your own results for both blood pressure and pulse rate. Determine your recovery time for each value. Graph both lines on the same graph. (*Hint:* Use two different vertical axes. Plot blood pressure on the left vertical axis [or *y*-axis] and pulse rate on the right vertical axis.)

(d) How does your recovery time compare with the recovery time of other individual participants and with the average recovery time of the class or group?

(e) Answer the Question on the basis of your analysis in (c) and (d).

Evaluation

(f) Evaluate your prediction.

(g) Describe any problems or difficulties you experienced in carrying out the procedure. What could or should you have done differently?

Synthesis

(h) Explain the reasons for the pulse and blood-pressure changes.

(i) Use the evidence obtained in this investigation to explain how the body maintains homeostasis.

(j) Using the concept of negative feedback, explain how heart rate is regulated during and after exercise.

(k) Explain how recovery time indicates the level of physical fitness.

(l) Research to find the normal resting pulse rate and blood pressure for someone of your age. How does your pulse rate compare? What reasons can you suggest for any differences?

www.science.nelson.com

(m) If you were to repeat this experiment, what new variables would you investigate? Write a brief description of the new procedure.

3.12 The Excretory System

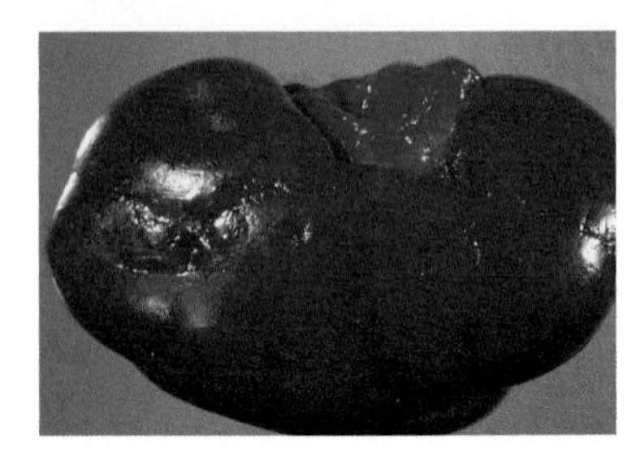

Figure 1
The human kidney is about the size of a fist and weighs approximately 0.5 kg.

Like most machines, the human body is not 100% efficient in converting raw materials into other useful materials or energy. During metabolic processes, the body produces wastes that must be eliminated. Cellular processes produce carbon dioxide that is dissolved in the blood and eliminated through the lungs. The digestive system produces toxic solid wastes that are eliminated through the large intestine. Another significant source of waste is the breakdown of proteins.

Humans have to include protein in their diet because it is the source of amino acids. The body uses amino acids to construct proteins that are used for building and repair. Any amino acids that are not needed for constructing or repairing the body are broken down in the liver. This breakdown process, called **deamination**, supplies some energy for the body, but the amino group ($-NH_2$) is left over. The amino groups are converted into ammonia (NH_3), but this creates a serious problem. Ammonia is extremely toxic—a buildup of as little as 0.005 mg can kill humans. Fish are able to avoid ammonia buildup by continually releasing it through their gills. Humans, however, have a special system that allows them to convert toxic wastes into less toxic substances and temporarily store the less toxic wastes for short periods of time.

deamination the removal of the amino group from an organic compound

urea nitrogen waste formed from two molecules of ammonia and one molecule of carbon dioxide

In the liver, the toxic chemical ammonia is quickly combined with carbon dioxide to produce much less toxic **urea** ($[NH_2]_2CO$). Urea constitutes about 90% of the body's nitrogenous—nitrogen-containing—wastes. Urea is dissolved in the blood and is carried to the kidneys (**Figure 1**), where it is excreted in the urine along with other substances such as uric acid and creatinine.

The Urinary System

The main organs of the urinary system are the kidneys. The kidneys are located in the back of the upper abdomen at either side of the spinal column (**Figure 2**). Since the kidneys are the filters that remove wastes from the blood, they may hold as much as 25% of the body's blood at any given time.

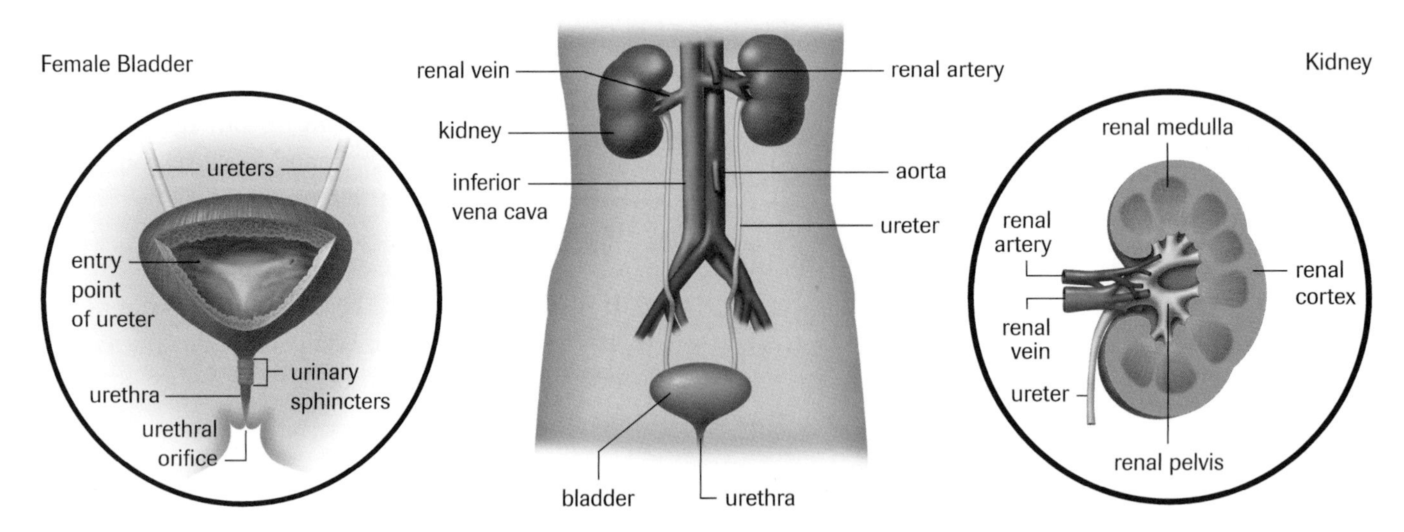

Figure 2
A simplified diagram of the human urinary system

Blood containing wastes is brought to the kidneys by a pair of blood vessels called **renal arteries** (*renal* refers to the kidneys) that branch off the aorta. The kidneys form **urine**, an aqueous waste fluid, which is filtered from the blood without removing blood components or usable nutrients. The purified blood is then returned to the circulatory system by a pair of blood vessels called **renal veins**.

Urine travels from the kidneys to the urinary bladder by tubes called **ureters**. A urinary sphincter muscle located at the base of the bladder acts as a valve, permitting the storage of urine. When approximately 200 mL of urine has been collected, the bladder stretches slightly and nerves send a signal to the brain. This is your signal to go to the bathroom. When the bladder fills to about 400 mL, more nerve endings are activated and the message becomes more urgent. If a person continues to ignore the messages, and the bladder fills to about 600 mL, voluntary control is lost, the sphincter muscle relaxes, and the urine enters the **urethra**, a tube from the bladder to the outside of the body, and is emptied.

renal artery a blood vessel that carries waste-laden blood to the kidneys

urine the solution of dissolved wastes produced by the kidneys and stored in the bladder

renal vein a blood vessel that carries purified blood from the kidneys to the body

ureter one of the tubes that conducts urine from the kidneys to the bladder

urethra the tube that carries urine from the bladder to the exterior of the body

nephron a functional unit of the kidneys

Bowman's capsule the cuplike structure that surrounds the glomerulus

glomerulus a capillary bed that is the site of fluid filtration

The Kidney and Urine Formation

Each kidney is made up of approximately one million slender tubules called **nephrons** (**Figure 3**). Each nephron is surrounded by a network of capillaries, but the closest connection with the circulatory system is at the **Bowman's capsule**, which surrounds a cluster of capillaries called the **glomerulus**, the area of fluid filtration. Fluids to be processed into urine enter the Bowman's capsule from the blood. **Table 1**, on the next page, summarizes the role of each part of the nephron.

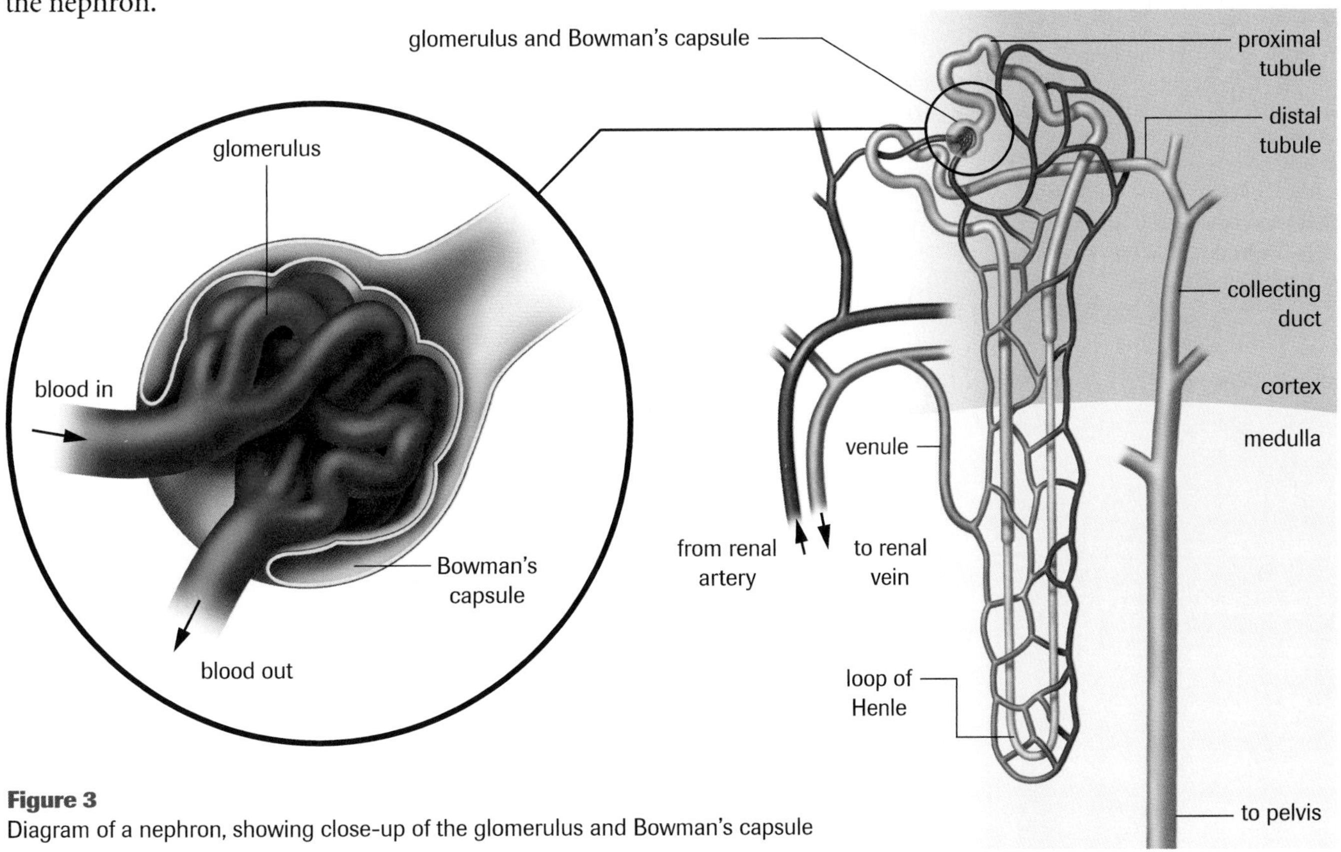

Figure 3
Diagram of a nephron, showing close-up of the glomerulus and Bowman's capsule

Table 1 Structure and Function of the Nephron

Name and structure	Function
glomerulus a cluster of capillaries that accept blood from the renal artery	*filtration.* Pressure of blood forces about one-fifth of its plasma out through the capillary walls, along with everything small enough to pass through: urea, glucose, amino acids, and salt, but no red blood cells, platelets, or large proteins.
Bowman's capsule a hollow, cup-shaped end of the nephron with double walls one cell thick	*filtration.* Plasma with dissolved materials diffuses into the hollow space and is passed on to the tubule.
proximal tubule a hollow, winding, large-diameter tube	*reabsorption.* Glucose, sodium, and amino acids are pumped back into the capillaries. Most urea stays inside the tubule.
loop of Henle (descending) straight part of loop moving away from the proximal tubule	*reabsorption.* Since the concentration of water inside the tubule is greater than that of the surrounding blood, water flows out by osmosis. Urea stays inside.
loop of Henle (ascending) straight part of loop leading to the distal tubule	*reabsorption.* Sodium is pumped out of the tubule. Walls are impermeable to water, so it remains inside.
distal tubule a hollow, winding, large-diameter tube	*reabsorption.* Final adjustment of water and dissolved materials. If the body is dehydrated, permeability is increased to let water return to the capillaries by osmosis. If not, permeability decreases to let water dilute the urine and leave the body.
collecting duct system of urine-collecting ducts that widen as they near the renal pelvis	*secretion.* Other wastes are transferred from the capillary network into the duct. *reabsorption.* Water moves by osmosis from the duct into the surrounding blood.

DID YOU KNOW?

Urine

Despite what you might think, urine has no odour. Once outside the body, the urea in urine undergoes a chemical reaction and is broken down into ammonia and carbon dioxide. The strong smell that you usually associate with a baby's wet diaper is really the smell of ammonia, not urea.

Urine formation depends on three processes—filtration, reabsorption, and secretion. Water containing dissolved materials such as salt, glucose, and urea is transferred through the walls of the glomerulus into the Bowman's capsule. Filtration ensures that plasma protein, blood cells, and platelets that are too large to move through the walls of the glomerulus are retained in the blood.

On average, about 600 mL of fluid flows through each kidney every minute. Approximately 20% of the fluid, or 120 mL, is filtered into the nephron. If all of this fluid were excreted, you would have to drink at least a litre of fluid every 10 minutes just to maintain homeostasis. Fortunately, reabsorption ensures that 119 mL of the fluid, along with some essential materials, are put back into the body. The remaining 1 mL becomes urine. Urine is approximately 95% water, with the remaining 5% consisting of dissolved salts, urea, and a few other substances. A typical adult excretes approximately 25 g of urea per day. Secretion is the transfer of additional nitrogenous wastes, excess minerals, and some other chemicals from the blood into the nephron. Even drugs such as penicillin can be secreted.

The body adjusts for increased water intake by increasing urine output. Conversely, it adjusts for increased exercise or decreased water intake by reducing urine output. The kidneys not only prevent the buildup of wastes but also help maintain homeostasis by controlling the volume, composition, and pressure of body fluids.

Kidney Disease

Proper functioning of the kidneys is essential for the body to maintain homeostasis. Kidney dysfunction affects other body systems, and kidneys are affected when other body systems break down. **Table 2** summarizes some of the more common diseases that are directly or indirectly linked to kidney function.

Table 2 Kidney-Related Diseases

Disease	Description
diabetes mellitus	Blood sugar levels rise because of inadequate secretion of insulin from the pancreas. The excess sugar is transferred to the nephrons, and an excess amount of water is lost through osmosis. People with this condition void large volumes of urine. This explains why they are often thirsty.
diabetes insipidus	The lack of hormones that regulate water reabsorption causes a dramatic increase in urine output. A person with this condition must drink large quantities of water to replace what he or she has not been able to reabsorb.
Bright's disease	This is also called nephritis, which is not a single disease but a group of diseases characterized by inflammation of the nephrons. Proteins and other large molecules are able to pass into the nephron, and this causes more water to remain in the nephron and be excreted as urine.
kidney stones	Dissolved minerals from the blood may precipitate and form stones that can lodge in the renal pelvis or move into the ureter (**Figure 4**). The sharp edges of the stones tear delicate tissues as the stone moves. Stones can move down into the urethra, where they can cause terrible pain as they move.

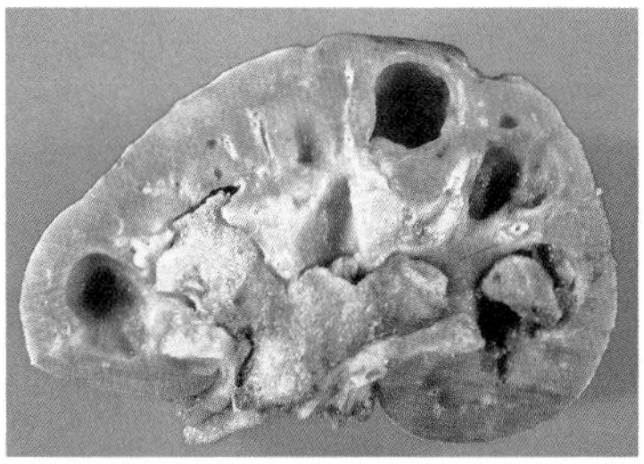

Figure 4
This kidney contains several stones. Stones consist of precipitated minerals.

Excretion in Fish

The excretory system of fish has special adaptations that suit fish to living in an aquatic environment. As in humans, deamination results in the production of ammonia, but unlike in humans, this does not require additional structures and processes to make the ammonia less toxic. Ammonia never builds up to toxic levels in fish because it simply diffuses outward through the gills. Fish have a different problem, however: regulating the concentration of salt.

Because their bodies have a lower concentration of dissolved salts than the surrounding seawater, saltwater fish tend to lose water by osmosis. To compensate for this, they drink seawater. They keep the water and excrete the excess salt through specialized cells in their gills. Marine fish have kidneys with small glomeruli, but very little urine is excreted.

Freshwater fish have the opposite problem. Because their bodies have a higher concentration of dissolved salts than the surrounding water, they tend to gain water by osmosis and absorb salt through their gills. Their kidneys have large glomeruli, and they excrete large volumes of urine to get rid of the excess water absorbed through osmosis.

DID YOU KNOW?

Aquarium Water
Because of the ammonia excreted by fish, the water in an aquarium becomes toxic and has to be changed regularly, depending on the number of fish and the size of the aquarium.

Section 3.12 Questions

Understanding Concepts

1. Excess amino acids cannot be stored. What happens to them in the human body?
2. How does the formation of urea prevent poisoning?
3. Explain the function of the nephrons.
4. Use the diagram in **Figure 5** to identify the structure that
   (a) filters blood
   (b) carries urine from the kidney
   (c) carries blood containing urea into the kidney
   (d) stores urine

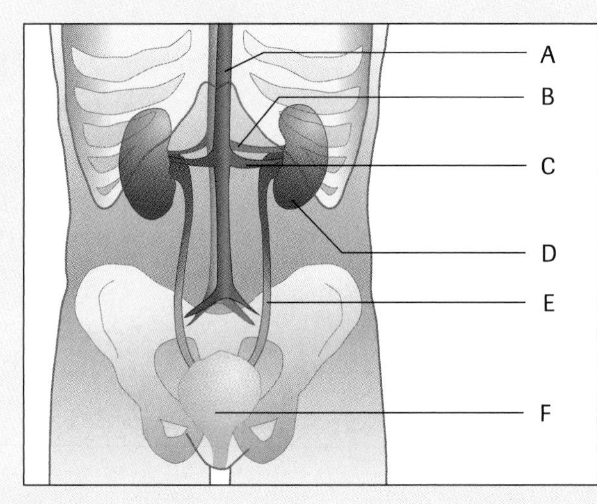

Figure 5

5. Describe the three main processes that are involved in urine formation.
6. What are kidney stones? Why are they a problem?
7. Explain why the chemical analysis of urine can reveal the use of certain drugs.
8. Because of their habitat, fish have different excretory needs than mammals. Compare the functions of the fish and mammalian excretory systems.

Applying Inquiry Skills

9. An adult under normal conditions will eliminate about 1.5 L of urine daily. Describe an experiment that you could conduct to determine how urine output is affected by the consumption of a food containing caffeine (e.g., coffee, tea, chocolate, or cola).

Making Connections

10. Why do you think it is beneficial for humans to have two kidneys rather than one?
11. John Smith has had three separate incidents of kidney stones. In the first, the stones were small enough to be passed through the urinary system. In the second, the lithotripsy procedure was used to break up the stones. And in the third incident, the stones were removed surgically. Because of his tendency to develop kidney stones, his doctor has strongly advised him, among other suggestions, to refrain from drinking alcohol. Explain why the doctor offered this advice.
12. Do you think people's urine should ever be tested for drugs? Do you think it could be more justified in some cases than in others? Use a PMI chart to examine the advantages and disadvantages of urine testing in terms of ethical, health, and technology issues, and make recommendations on its use.

3.13 Activity

Urinalysis

Urinalysis is the chemical and physical analysis of urine. The identification of proteins and sugars or other characteristics (e.g., colour and pH) of urine samples can reveal kidney and related diseases. Urinalysis is a very useful test because urine contains waste products from a number of organs in the body. For example, in addition to providing information about the urinary system, urinalysis can also indicate infections that may not even involve the urinary tract.

Most laboratories use commercially prepared test strips to perform the chemical analysis. These test strips are narrow plastic strips with one or more test pads that have chemicals in them. When a strip is dipped into urine, the test pads absorb the urine, and a chemical reaction changes the colour of the pad.

The medical laboratory technologist compares the colour change for each reaction pad with a colour chart (provided with the test strips) to determine the result for each test. Each reaction pad must be evaluated at the appropriate time. If too little time or too much time has passed since the reaction, the technologist may get incorrect results.

In the laboratory, extreme caution has to be exercised while performing these tests. Urine and other body fluids (e.g., blood) are treated as hazardous material because the technologist does not know what the fluid contains until it is tested.

The most frequently performed chemical tests using test strips are

- specific gravity
- pH
- protein
- glucose
- ketones
- leukocytes
- nitrite
- blood
- bilirubin
- urobilinogen

Table 1 shows normal and abnormal values for urine characteristics and indicates possible causes for the abnormal values.

In this activity, using one or more test strips, you will test samples of synthetic urine for some common diagnostic properties. The results will be compared with normal values and used to diagnose possible disorders.

Table 1 Normal and Abnormal Values for Urine Properties

Property	Normal values	Abnormal values and possible causes
colour	• straw • yellow • golden • amber	• white, foamy—protein • yellow-green—biliverdin • greenish—bacterial infection • brown/green—bilirubin • bright red—blood from ureter, bladder, urethra • brown/dark red—blood, myoglobin from kidney
specific gravity	1.002–1.028	< 1.002 (chronic) fail to make hormones 1.028–1.035 dehydration > 1.035 abnormal—diabetes mellitus, fever, kidney infection
pH	4.6–8.0	low high-protein diet, diabetes mellitus; starvation high high-fibre/vegetarian diet
protein	low-none	trace may be due to normal exercise + hypertension, kidney disease, pregnancy toxemia
glucose	< 0.03 g/100 mL	+ diabetes mellitus
ketones	none	+ starvation or malnutrition, diabetes mellitus
leukocytes	few	+ infection in kidney or urinary tract
nitrite	none	+ bacterial infection
blood	low-none	+ kidney infection, hepatitis, malaria, any disease/injury that causes breakdown of red blood cells
bilirubin	low-none	+ excess red-blood-cell (RBC) breakdown from disease or infection; liver damage or liver infection
urobilinogen	low-none	+ excess red-blood-cell (RBC) breakdown, hepatitis, cirrhosis of liver

Question

How can we use the results of urinalysis to diagnose kidney or other related disorders?

Materials

safety goggles
laboratory apron
6 urine samples (simulated), labelled A to F in dropper bottles
6 small test tubes
wax pencil
paper towel
stopwatch, or a watch with a second hand
urine test strips

Eye protection and a laboratory apron must be worn for the entire laboratory.

Procedure

1. Before use, mix each urine specimen thoroughly by inverting the dropper bottle a few times.
2. Remove one test strip from the vial and replace the cap. Refer to specific instructions for the appropriate reading time.
3. With the dropper, place one or two drops of urine sample A on the test area of the strip. If you are using strips with more than one test pad, complete one test at a time. Hold the strip in a horizontal position to prevent possible mixing of chemicals from neighbouring test areas and/or contaminating the hands with urine (**Figure 1**).
4. Start the timer and touch the edge of the strip on a paper towel to remove the excess urine.
5. Hold the strip close to the colour chart and match carefully. Do not lay the strip directly on the colour chart as this will soil the chart. At the specified time, compare test areas with the corresponding colour on the chart (**Figure 1**).

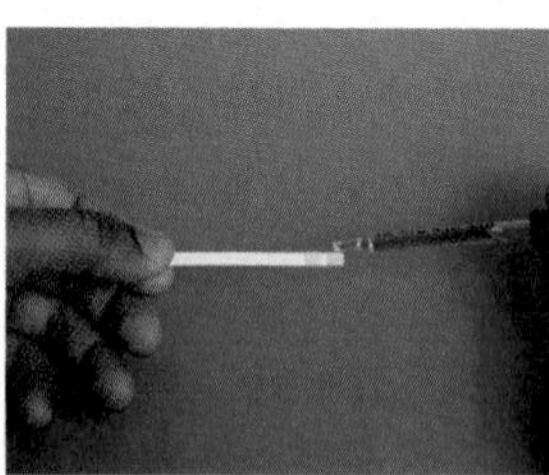

Step 3

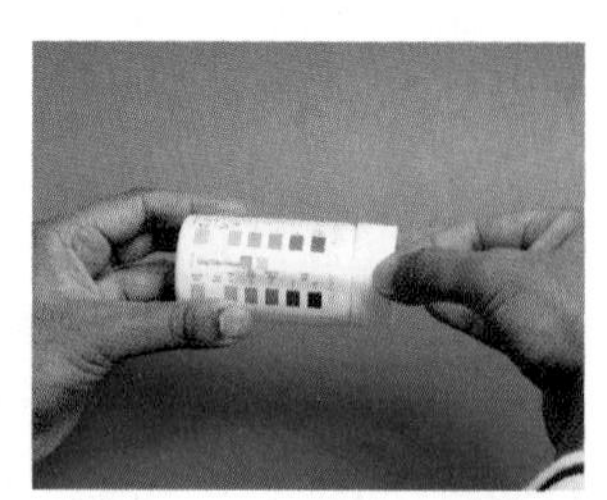

Step 5

Figure 1

(a) Record the results in an appropriate table.

6. Dispose of the used test strip according to your teacher's instructions.
7. Repeat steps 1 to 6 for each of the urine samples.

Analysis

(b) Answer the Question.

(c) Which sample indicates diabetes mellitus? Provide your reasons.

(d) If glucose filters through the glomerulus, why isn't glucose normally found in urine?

(e) Which sample indicates a bacterial infection? Give reasons for your response.

(f) Which sample suggests that the patient may be malnourished? Explain your answer.

(g) Describe any other abnormal values and give possible explanations.

Synthesis

(h) What are recommended treatments for diabetes mellitus?

(i) Select one of the diseases or disorders from **Table 1** and research to prepare a report on the symptoms, causes, and treatments.

GO www.science.nelson.com

CAREER CONNECTION

Anyone whose employment requires him or her to test samples and submit reports to medical professionals must be a registered technologist. The certification and registration of technologists are maintained by the Canadian Society for Medical Laboratory Science (CSMLS). In order to be certified by the CSMLS, candidates must complete a program of training, after which they must successfully pass the certification exam. Research to find out the educational requirements for the medical laboratory technologist program.

GO www.science.nelson.com

3.14 Kidney Dialysis

For people whose kidneys cannot effectively get rid of body wastes, a dialysis machine can restore the proper balance of dissolved materials in the blood. Dialysis is defined as the exchange of substances across a selectively permeable membrane. Like a kidney that is functioning normally, dialysis operates on the principles of diffusion and blood pressure. There are two types of dialysis available: hemodialysis and peritoneal dialysis.

In hemodialysis, the machine is connected to a vein in the patient's circulatory system (**Figure 1**). Blood is then pumped through a series of dialysis tubes that are submerged in fluids.

Because the dialysis fluids have no urea, urea will move from the blood into the dialysis fluid until equal concentrations are established in both. By continually flushing the dialysis solutions and replacing them with fresh solutions, the machine continuously removes urea and other wastes.

In peritoneal dialysis, approximately 2 L of dialysis fluids are pumped into the abdominal cavity (**Figure 2**). The lining of the cavity, the peritoneum, has a large surface area and a rich network of blood vessels. Urea and other wastes diffuse from the blood through the peritoneum and into the fluid in the abdominal cavity. The fluid must be left in the cavity long enough to allow waste materials from the bloodstream to pass slowly into it. Then the fluid is drained out and replaced with fresh fluid. The advantage of this method over hemodialysis is that patients can perform the procedure on their own at home and may continue with less strenuous activities as the dialysis occurs.

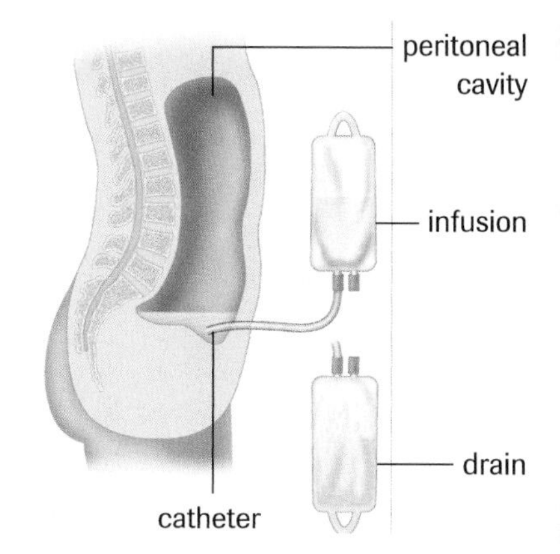

Figure 2
Peritoneal dialysis is done through the peritoneal membrane, which is the lining of the abdominal cavity. In a minor surgical procedure, a catheter is first inserted. A solution, called the dialysate, is then fed into the abdominal cavity. The dialysate remains in the cavity for two to six hours and is then drained from the abdomen via the catheter. Once the fluid is drained, new fluid is fed into the cavity to begin the process again.

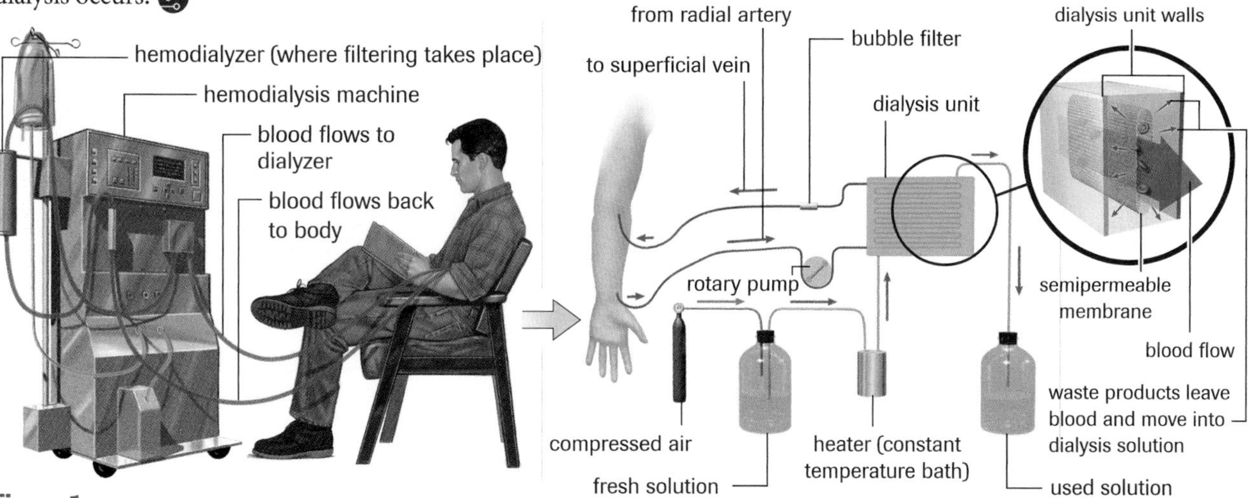

Figure 1
In hemodialysis, a machine mimics the action of the nephrons. A person must first have a minor surgical procedure to create an access, a shunt, for the needles and tubing needed to connect the machine to the circulatory system. Most people need three weekly dialysis sessions of about four hours each.

Tech Connect 3.14 Questions

Understanding Concepts

1. Briefly describe the differences between hemodialysis and peritoneal dialysis.
2. Explain the advantages of peritoneal dialysis over hemodialysis.

Making Connections

3. The operation of a hemodialysis unit requires special expertise. Research to find the qualifications required to assist patients with dialysis.

www.science.nelson.com

Tech Connect

3.15 The Respiratory System

Figure 1
Scuba technology allows our respiratory system to continue to function properly when we are under water for long periods of time.

The intimate link between humans and their environment is clearly demonstrated every minute of every day. Humans, while very adaptable (**Figure 1**), can survive for only a few minutes without oxygen.

All animals exchange gases with their surroundings. Animals take in oxygen gas and release carbon dioxide in the process of **respiration**. The oxygen reacts with nutrients in mitochondria to release energy. Carbon dioxide is a waste product produced by these same reactions. Unlike water and food, gases cannot be stored easily within living tissues, and, therefore, most animals must exchange gases with the atmosphere on a continual basis.

TRY THIS activity — Fitness Test ($VO_{2\,max}$)

One way to measure your level of fitness is to calculate the maximum amount of oxygen that your body can use when you perform a very strenuous activity such as running. In theory, the greater your level of fitness, the greater the quantity of oxygen you can consume in a set time interval. $VO_{2\,max}$ is an estimate of the maximum amount of oxygen (in millilitres) that a person can use in one minute per kilogram of weight while breathing air at sea level.

Do not perform this activity if you are not allowed to participate in physical education classes.

This can be calculated using the following formula:

$$VO_{2\,max} = (\text{speed} \times 0.172) + 10.4$$

$VO_{2\,max}$ is expressed in mL/kg/min. Speed units are metres per minute (m/min).

Example: A student runs a distance of 400 m in a time of 1 min 20 s. Calculate the $VO_{2\,max}$.

$$\text{Speed} = 400\text{ m}/1.33\text{ min} = 301\text{ m/min}$$

$$VO_{2\,max} = (301 \times 0.172) + 10.4 = 62.2\text{ mL/kg/min}$$

A $VO_{2\,max}$ value over 35 mL/kg/min is considered good for females aged 13–19, while a $VO_{2\,max}$ value over 45 mL/kg/min is considered good for males aged 13–19.

(a) Determine your speed running a distance of at least 400 m. Apply the formula above to calculate your $VO_{2\,max}$.

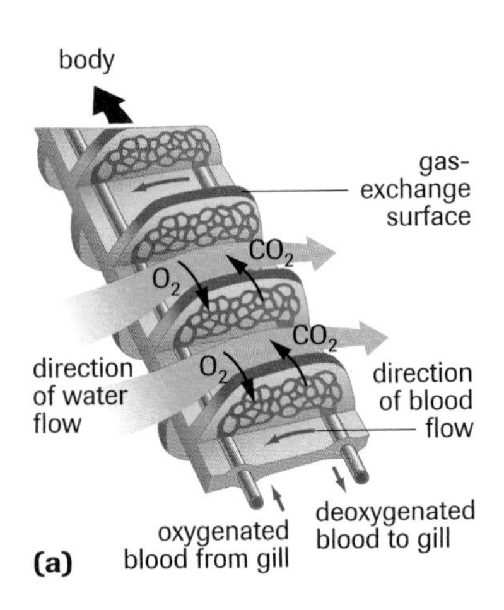

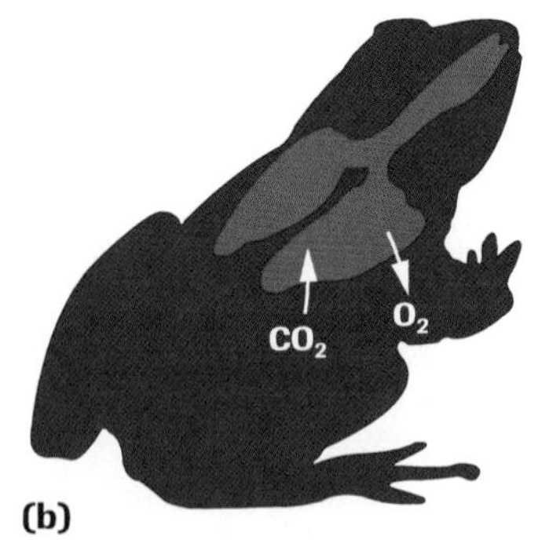

Figure 2
Gas exchange in fish occurs at the gill surface **(a)**, while gas exchange in land animals usually involves a set of lungs **(b)**.

As you learned in section 3.9, a main function of the circulatory system is to transport oxygen and carbon dioxide to and from the cells of the body. Animals use large moist surfaces in order to exchange these gases between the blood and their external environment—the atmosphere or water. At these surfaces, gas exchange occurs by the process of diffusion. In aquatic animals, such as fish, gas exchange occurs as water flows over large gill surfaces. In most land animals, this exchange occurs in lungs (**Figure 2**).

respiration all processes involved in the exchange of oxygen and carbon dioxide between cells and the environment

The Human Respiratory System

Humans are large warm-blooded animals called **endotherms**, with a very high oxygen demand. This demand is met using the structures of the human respiratory system illustrated in **Figure 3**.

endotherm an organism that maintains a near-constant body temperature. All birds and mammals are endotherms.

DID YOU KNOW?

Eat Carefully

Every year people choke to death on hotdogs because wieners have a shape ideally suited for getting stuck in the trachea.

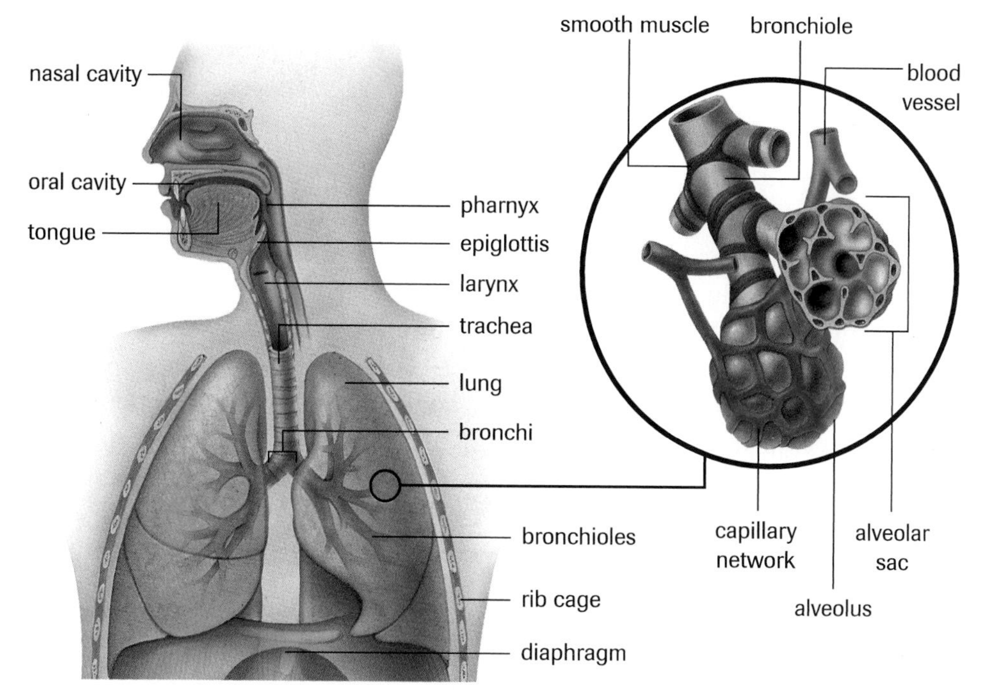

Figure 3
The human respiratory system

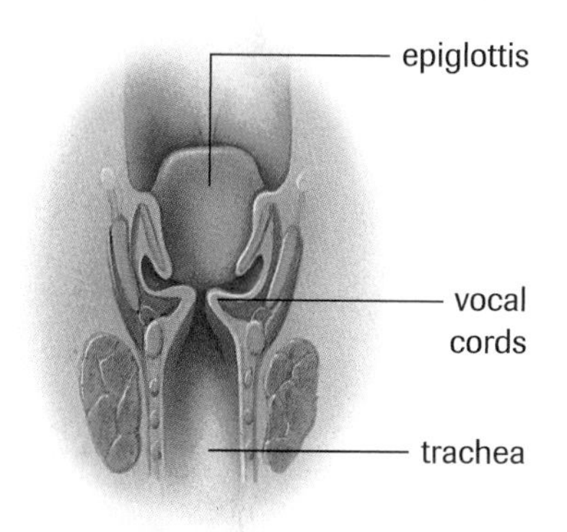

Figure 4
Larynx, showing vocal cords

Air usually enters the respiratory system through your nostrils. Dust and other foreign particles are trapped and filtered out of the air by tiny hairs and a layer of mucus that line these passages. The incoming air is also warmed and moistened by these surfaces.

At the back of the mouth, the nasal and oral cavities join to form a region called the pharynx. From the pharynx, air passes down through the **trachea**, or windpipe, into the lungs, while food enters the esophagus and is passed to the stomach. A very important flaplike structure called the **epiglottis** is located at the top of the trachea. When you swallow, a reflex causes the epiglottis to close over the opening of the trachea to prevent your food from accidentally entering the trachea. If you accidentally swallow too quickly or laugh, food or liquid may get past the epiglottis and enter the trachea. This usually results in another of your body's reflex actions—a cough.

trachea the windpipe through which air passes from the pharynx toward the lungs

epiglottis the structure that covers the opening of the trachea during swallowing

larynx the enlarged portion of the trachea containing a pair of vocal cords

The **larynx**, or voice box, is located in the trachea just below the epiglottis (**Figure 4**). The larynx contains two thin elastic ligaments, called vocal cords, that are under muscular control. When air is forced up from the lungs and through the larynx, the vocal cords vibrate, producing sounds. Different sounds are produced and controlled by adjusting the position and muscular tension of the vocal cords.

DID YOU KNOW?

Laryngitis

When you have laryngitis and "lose your voice," your vocal cords are inflamed (swollen) and unable to vibrate properly.

In males, the larynx increases in size following puberty. The vocal cords vibrate at a lower frequency, producing a lower voice in men.

bronchus one of the air passages from the trachea that goes to the right or left lung

Inhaled air moves down through the trachea and into two **bronchi** (singular: bronchus). The trachea and bronchi are both surrounded and supported by many thin rings of cartilage that hold them open while remaining flexible.

bronchiole one of the small air passages located within and throughout each lung

goblet cell a specialized cell that produces mucus

Air enters the lungs through the bronchi that then divide into an extensive network of smaller tubes called **bronchioles.** Smooth muscles located in the walls of the bronchioles can contract to decrease the diameter of these small air passages. The inner linings of the trachea, bronchi, and bronchioles are covered by cilia and contain mucus-producing **goblet cells** (**Figure 5**). The mucus traps tiny particles such as bacteria, dust, and pollen that are then carried up out of the lungs and trachea by the sweeping action of the cilia. This continuous flow of mucus and trapped material is referred to as the bronchiole escalator.

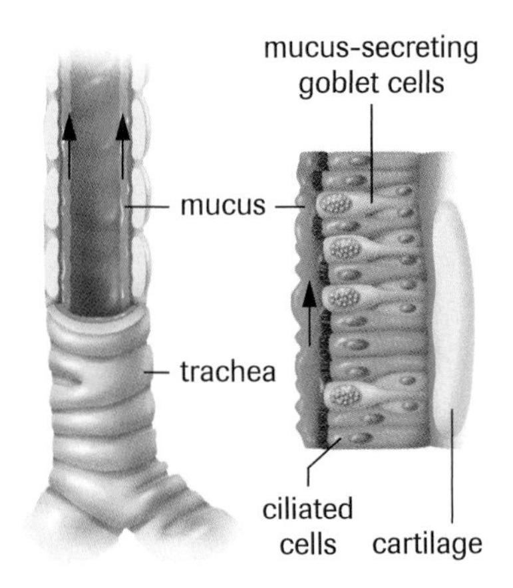

Figure 5
Goblet cells in the lining of air passageways produce mucus, which then traps debris and is carried up and out of the lungs by the action of cilia.

alveolus one of the tiny air sacs within the lungs in which gas exchange occurs between the air and the blood

Leading into the lungs, the branching bronchioles become smaller and smaller until they reach a dead end in the tiny air sacs called **alveoli** (singular: alveolus). It is in the alveoli that the actual exchange of gases between the air and the blood takes place. Each lung is made up of about 150 million alveoli, with each measuring between 0.1 mm and 0.2 mm in diameter. The total surface area of alveoli within a pair of lungs is approximately 80 m^2—about 40 times the surface area of your skin!

A network of capillaries surrounds each cluster of alveoli. Blood entering the network has a low oxygen concentration and a high carbon dioxide concentration. This blood passes through the capillary network, where oxygen gas diffuses from the air space within the alveoli through the single-celled walls of the alveoli and capillaries and into the blood (**Figure 6**). Simultaneously, carbon dioxide diffuses out of the blood and into the air of the lungs following the reverse path. Blood leaving the capillary network is now high in oxygen and low in carbon dioxide.

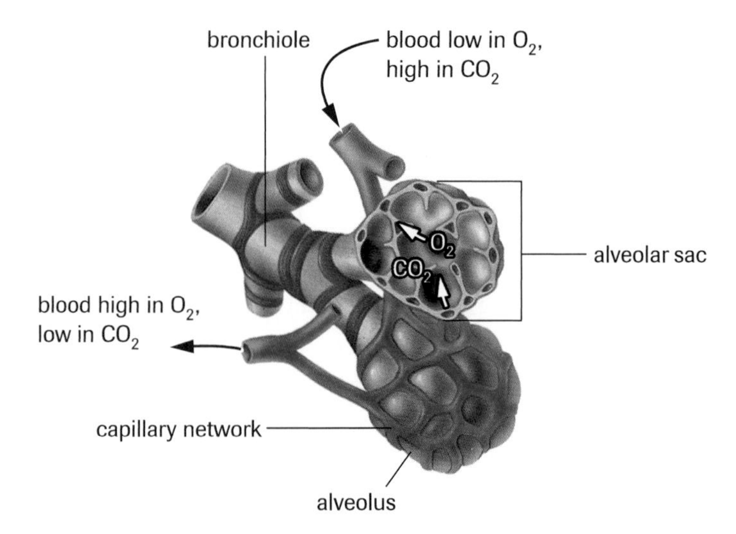

Figure 6
Gas exchange occurs between the air within the alveoli and the blood in the surrounding capillary network.

How We Breathe

Air within the alveoli must be exchanged regularly in order to provide a fresh supply of oxygen and to carry away waste carbon dioxide gas. This involves two main stages of breathing: **inspiration** (inhaling), in which air is moved into the lungs, and **expiration** (exhaling), in which air is forced out of the lungs.

Understanding how such air movement occurs requires an understanding of air pressure. Air will always move from an area of high pressure to an area of low pressure. Therefore, if the air pressure inside the lungs is less than the air pressure outside the body, air will move into the lungs. Similarly, if the air pressure inside the lungs is greater than the air pressure outside the body, air will leave the lungs. How does the body change the pressure within the lungs?

Carefully examine **Figures 7** and **8**. The lungs are located within the chest, or thoracic cavity. This large cavity is surrounded by the rib cage and a thin sheet of muscle called the **diaphragm.**

During inspiration, the diaphragm contracts and moves downward while the rib cage expands upward and outward by the actions of the **external intercostal muscles**. The result is an increase in the volume of the chest cavity. This lowers the air pressure inside the chest. Air then moves down through the trachea into the lungs, filling this extra space.

During expiration, the diaphragm relaxes and moves upward while the rib cage moves inward and downward. This decreases the volume and increases the air pressure within the chest cavity. This results in the movement of air out of the lungs.

inspiration the action of breathing in, or inhaling, air

expiration the action of breathing out, or exhaling, air

diaphragm a sheet of muscle that separates the organs of the chest cavity from those of the abdominal cavity

external intercostal muscles muscles between the ribs that raise the rib cage, increasing volume and reducing air pressure within the chest

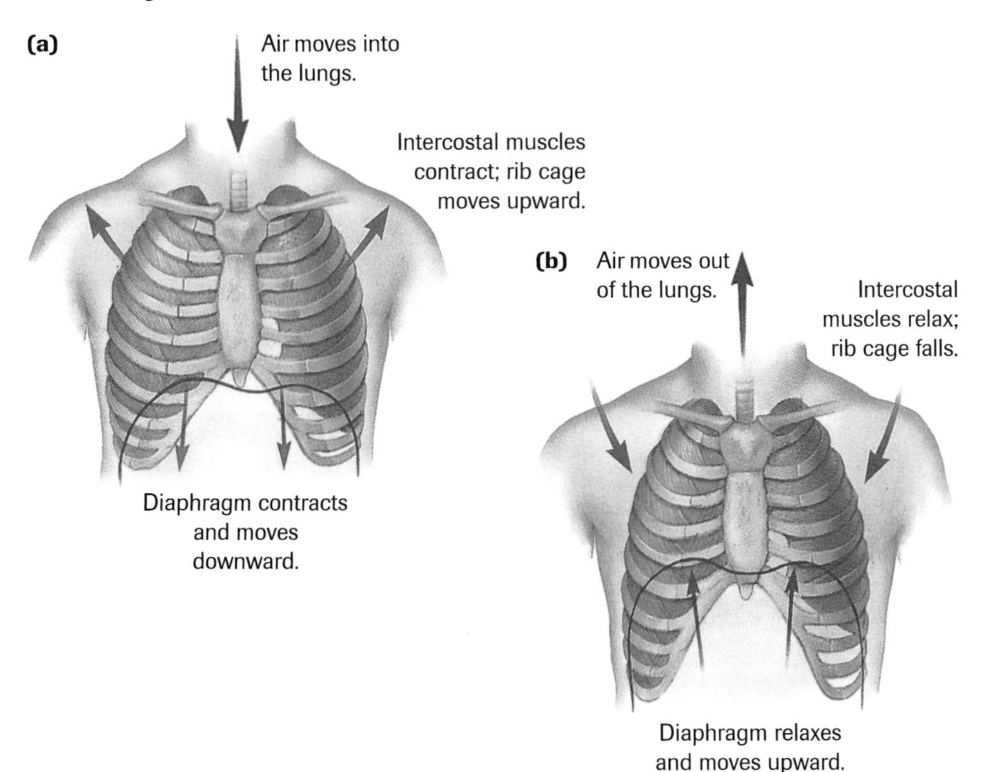

Figure 8
Changes in lung volume during **(a)** inspiration and **(b)** expiration

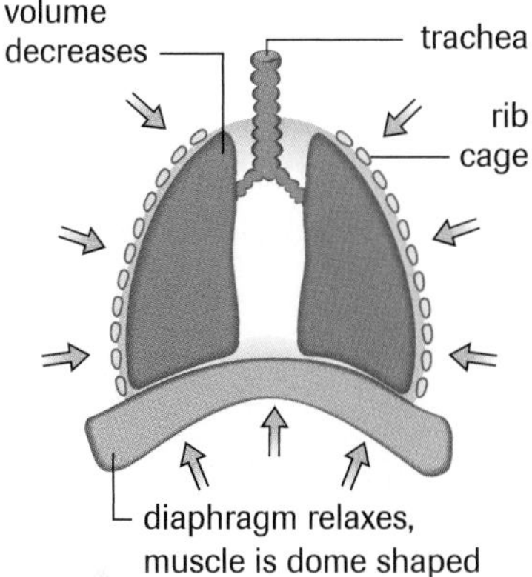

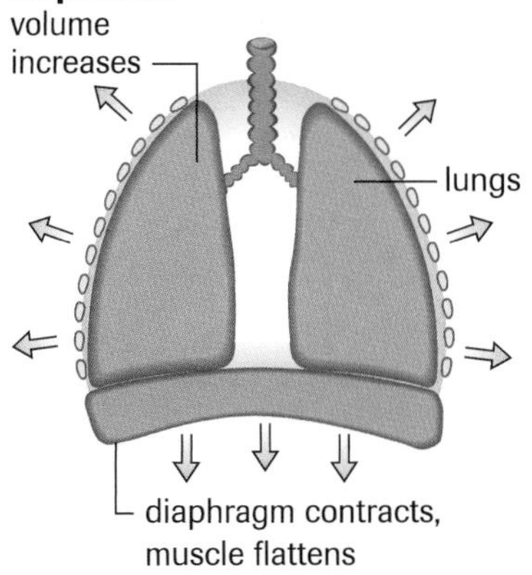

Figure 7
The lungs are contained within the chest, or thoracic cavity.

DID YOU KNOW?

Hiccups
Sometimes the diaphragm becomes irritated, causing it to experience muscle spasms. This forces air rapidly through the larynx, creating the sounds we call the "hiccups."

pleural membrane a thin, fluid-filled membrane that surrounds the outer surface of the lungs and lines the inner wall of the chest cavity

Lungs continuously expand and contract. This constant motion could result in damage to their delicate outer surface. To prevent such damage, special **pleural membranes** cover the lungs and the inner walls of the chest cavity. A thin film of liquid between these membranes allows the lungs to move freely. These same membranes create an airtight seal around each lung. If this seal is punctured, air may be able to get in between a lung and the chest wall, causing the lung to collapse.

CAREER CONNECTION

Paramedics and other medical professionals are trained to seal openings in the chest wall resulting from serious injuries. They also assess and monitor breathing responses and take immediate action to maintain adequate gas exchange within the lungs.

Maintaining fairly constant levels of oxygen in the blood depends on an alternating pattern of inspiration and expiration. This pattern is controlled by the autonomic nervous system, by nerve impulses from a breathing centre in your brain. Interestingly, the brain does not monitor oxygen levels in the blood, but instead monitors carbon dioxide levels. The greater the level of carbon dioxide in your blood, the faster your breathing rate will be. As you exercise vigorously, you consume oxygen quickly and produce a lot of carbon dioxide. Your brain detects this increase in blood carbon dioxide and increases your breathing rate.

Breathing in Extremes

Humans often venture into environments with very unusual conditions (**Figure 9**). Scuba divers explore an underwater environment with much higher pressures than those found at the surface, while mountain climbers and others may live or spend time at very high altitudes where the air pressure is much lower than at sea level. These environments place extreme demands on the respiratory system.

At high altitudes there is less oxygen in the air, and our bodies compensate both by increasing our breathing rate and, over a period of days and weeks, by gradually increasing our number of red blood cells. At extreme altitudes, mountain climbers may have to carry oxygen tanks.

Scuba divers always carry their air supply in tanks and use special devices called regulators to compensate for the changes in pressure at different depths (**Figure 10**). Advanced and professional divers may use specialized mixtures of

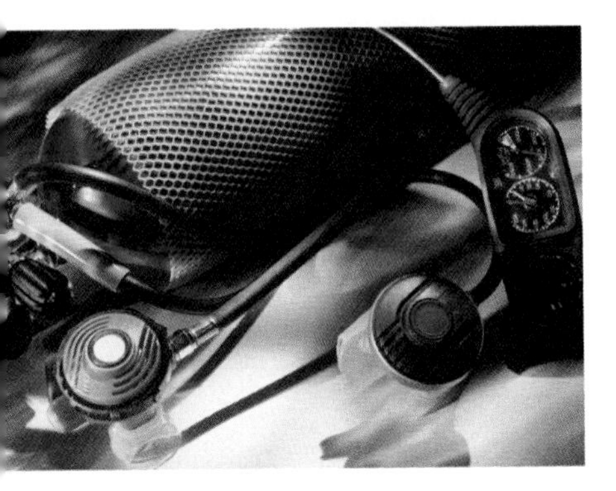

Figure 10
Scuba divers use regulators and high-pressure tanks to ensure a safe air supply.

DID YOU KNOW?

Nitrogen Narcosis
Scuba divers can suffer from a number of effects as a direct consequence of high pressures under water. Nitrogen narcosis is caused when too much nitrogen diffuses into the blood under high pressure. The symptoms may include mental impairment similar to the intoxication from alcohol consumption.

Figure 9
A thorough understanding of the respiratory system and the influence of pressure on gas exchange is essential for safe mountaineering and scuba diving.

gases, such as nitrox (32% oxygen and 68% nitrogen), for added safety and to extend their dive times. Pure oxygen is often deadly when breathed at depths below approximately 7 m and is not used in scuba tanks.

The Respiratory System of Frogs

The respiratory system of the frog changes dramatically during its life. Immature tadpoles live in the water and respire through a set of gills. When these tadpoles become adults, they leave the water and begin using a pair of internal lungs (**Figure 11**). Unlike humans and other mammals, frogs do not inhale by expanding their chest cavity. Instead, they must force air into their lungs by a gulping and swallowing action. Both immature and adult frogs also exchange gases directly through their moist outer skin (**Figure 12**).

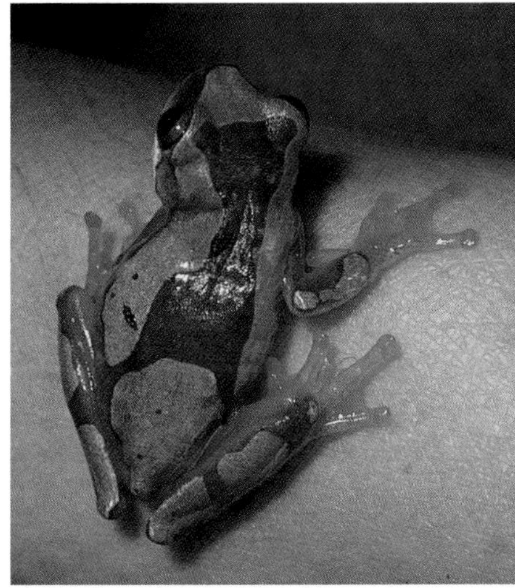

Figure 12
Frogs are able to exchange gases directly through their moist living skin surface—something not possible through the dry skin surface of humans.

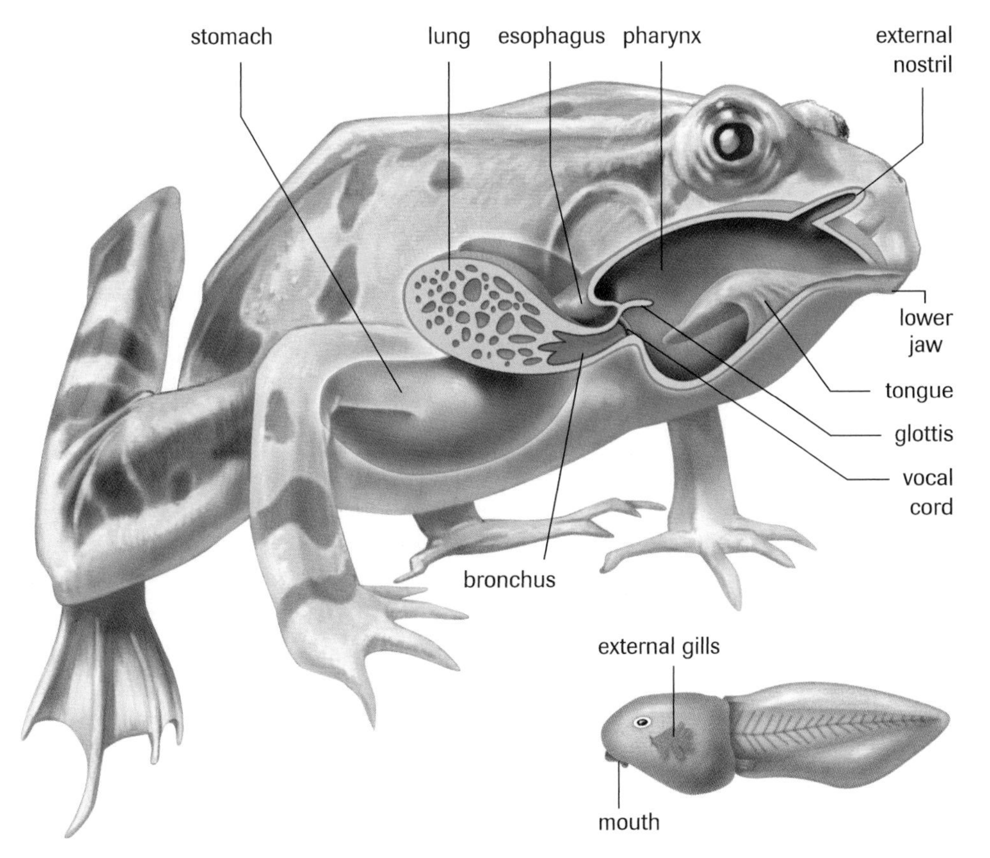

Figure 11
Tadpoles respire through their skin and external gills, while adult frogs breathe through their skin and a pair of simple internal lungs.

Section 3.15 Questions

Understanding Concepts

1. What are the main functions of the respiratory system?
2. How is the respiratory system related to the circulatory system?
3. Humans need to take in food and water only a few times each day. Why, then, is it necessary for us to breathe almost continuously?
4. State the main functions of the following:
   (a) epiglottis
   (b) trachea
   (c) cartilage rings of the trachea and bronchi
   (d) goblet cells
   (e) cilia
   (f) alveoli

5. Trace the path of an oxygen molecule in the air from the time it enters the body until the time it enters the bloodstream. Name each structure it passes or enters in the correct order.
6. Describe the structure and function of the larynx. How does the larynx differ in males and females?
7. Draw a fully labelled diagram of a functioning alveolus, illustrating blood flow and gas exchange.
8. Describe the muscle actions responsible for expiration and inspiration.
9. Briefly compare the respiratory system of the frog with that of the human. Consider the advantages and disadvantages of each.

Applying Inquiry Skills

10. Changes in the partial pressure of gases in arterial blood were monitored over time as a subject began to perform light exercise (**Figure 13**).

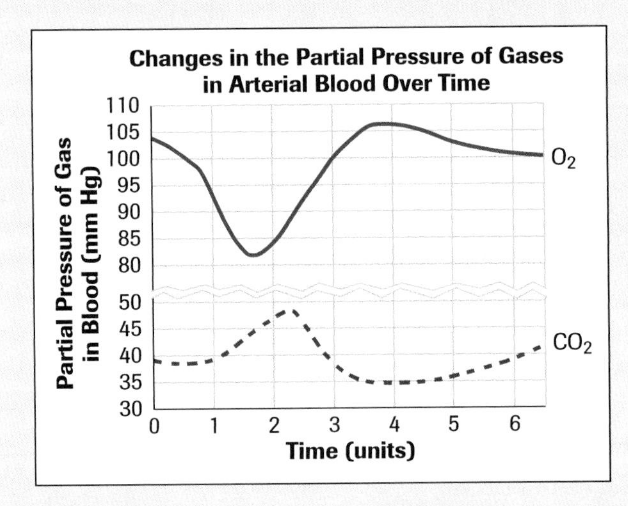

Figure 13

(a) At which time would the breathing rate likely be greatest? Provide reasons for your answer.
(b) Predict when the subject began exercising. Give your reasons.
(c) When would the breathing rate return to normal? Give your reasons.

11. A medical researcher suspects that people with cystic fibrosis (CF) have more respiratory infections than people who don't have CF because they produce thicker mucus in their lungs. In an investigation involving 1000 people over a period of one year, the researcher obtained the following results:

Number	Mucus viscosity + = normal ++ = thick +++ = very thick	Average frequency of respiratory infections (number per year)
750	+	2
235	++	3
15	+++	6

(a) Describe any evidence that supports the researcher's suspicion.
(b) Provide a hypothesis (a possible explanation) to explain how extra-thick mucus in the lungs could affect the frequency of respiratory infection.

Making Connections

12. A number of medical conditions, such as bronchitis, cause air passageways to narrow.
(a) What problems would be caused as the air passages decrease in diameter?
(b) What occupations or working environments are likely to aggravate bronchitis?
13. Research scuba diving and determine what sorts of courses are necessary to become a certified diver. Explore the career opportunities that may be available for professionally trained divers.

www.science.nelson.com

Exploring

14. When someone is choking, an object is blocking his or her trachea. The Heimlich manoeuvre is often used to clear this obstruction. Research this first-aid action and be prepared to describe it to the class.

www.science.nelson.com

Determining Lung Volumes

Many factors can influence lung capacity and breathing rates. In general, the volume of a normal breath and your maximum lung volume can be used as indicators of health. In this activity, you will measure lung volumes that correspond to different depths of breathing.

Questions

(i) What is your normal breathing volume?

(ii) What is the maximum volume of air that you can exchange in your lungs?

Materials

spirometer with disposable mouthpieces
nose plug (optional)

NEVER INHALE through the spirometer mouthpiece. Make sure mouthpieces are not shared.

Procedure

1. Set the spirometer gauge to zero, and then place a new, unused mouthpiece in the spirometer.
2. Before using the mouthpiece, relax and allow yourself to get into a regular, relaxed breathing pattern. Then, AFTER inhaling normally, place the mouthpiece in your mouth, hold your nose closed, and exhale normally through your mouth.

(a) Read the gauge on the spirometer and record the value as your tidal volume in a data table.

3. Reset the spirometer gauge to zero. Inhale and then exhale normally. Now, AFTER exhaling normally, place the mouthpiece in your mouth and forcibly exhale all of your remaining air.

(b) Record this value as your expiratory reserve volume in your data table.

4. Repeat step 3, but this time take a deep breath and then exhale fully, forcing the maximum volume of air from your lungs.

(c) Record this value as your vital capacity in your data table.

(d) Use the relationships displayed in **Figure 1** to calculate your inspiratory reserve volume. Record this value in your data table.

(e) Collect data from at least two other students and place them in your data table.

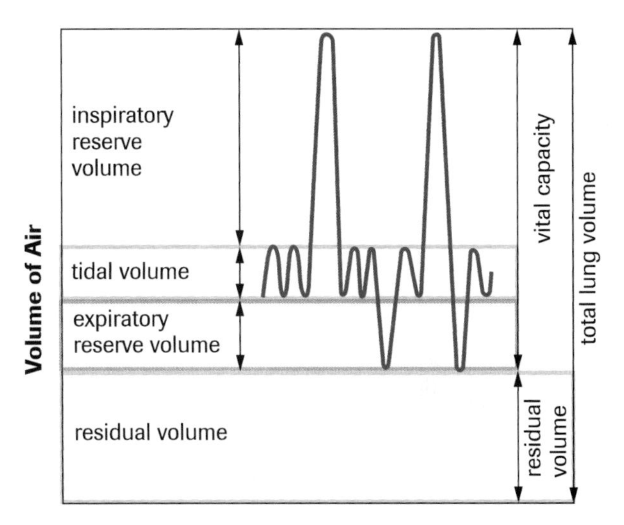

Figure 1
Lung volumes and breathing depths

Analysis

5. Examine **Figure 1** showing the various relationships between lung volumes and breathing depths. Note that the residual lung volume is the volume of air remaining in the lungs even after you have exhaled completely.

(f) What is your "normal" breathing volume?

(g) What is the maximum volume of air you can exchange?

(h) How do your volumes compare with those of your classmates?

(i) How might body size, gender, or health account for these differences?

Synthesis

(j) How might living at high altitudes influence tidal volume and/or vital capacity? Explain.

(k) What types of activities do you think would most contribute to increasing your vital capacity? How might this improve your general health?

3.17 Disorders of the Respiratory System

Figure 1
Our respiratory systems are subjected to natural irritants such as cold viruses, ragweed pollen (top) (responsible for "hay fever"), and human pollutants such as the carbon monoxide in automobile exhaust.

Diseases of the respiratory tract are extremely common and can be very serious. The delicate surfaces of the human respiratory system are in continual contact with air from the external environment. This contact exposes the moist, thin tissues of the respiratory system to possible damage and disease from a wide variety of pathogens and pollutants (**Figure 1**). Pollen and infectious agents such as bacteria and viruses usually occur naturally. Chemical pollutants such as carbon monoxide, tar, and nitrous oxides are generated by humans through such things as auto exhaust, cigarette smoking, and industry. All pollutants and pathogens that damage the respiratory system decrease the oxygen supply to the body.

The Common Cold and Influenza

The two most common infections of the respiratory tract are the common cold and influenza. Both are caused by viruses that can spread easily from person to person. While the common cold usually causes only mild discomfort, influenza, or the "flu," can be much more serious—especially for infants, elderly persons, and those with other health complications. Influenza viruses mutate regularly, making previous immunity ineffective. For this reason, each year health agencies

DID YOU KNOW?

Influenza
The flu is one of the world's leading killers. In 1918–19, more than 20 million people died during an influenza epidemic.

TRY THIS activity — The Spread of Disease

In this activity, you will witness the ease with which a disease-causing organism can spread from person to person. Each student will be given an eyedropper and a single numbered test tube containing a clear liquid that will represent their "bodily fluids." In this case, one student's sample has been secretly "infected" with a clear, colourless, and odourless chemical.

1. At the direction of your teacher, you will randomly seek out another student and each exchange one eyedropper of liquid from each other's test tube.

(a) Create a data table to record the name and the corresponding test-tube number of the person you exchanged fluids with.

2. Perform this exchange at least three or four more times.
3. Make sure to update your data table each time you exchange fluids.
4. After all students have completed their exchanges, your teacher will use a special test solution to see if you have been "infected."

(b) How many students in your class became infected?

(c) How do the results of this activity model infections in real life?

(d) What human activities could the exchanging of liquids with the eyedropper represent?

The chemical used in this activity to model the infectious agent is mildly corrosive. Avoid contact with skin and eyes. Wear eye protection and gloves when carrying the test tube.

Be careful with any spills.

develop vaccines to combat new dangerous strains of the influenza virus. While most healthy people do not require the vaccine, individuals who work with the sick, the young, or the elderly are strongly encouraged to receive the "flu shot" to prevent them from contracting and spreading the disease to those more vulnerable people.

pneumonia a lung infection caused by a virus, bacterium, or fungus that leads to the accumulation of fluid within the lung

Bronchitis

Bronchitis is an inflammation or swelling of the bronchi and bronchioles. The inflammation can be caused by a viral or bacterial infection, or by chemical irritants such as smoke. The infection or irritant also results in the production of additional mucus. The inflammation and excess mucus result in a narrowing of the air passageways, making breathing and gas exchange more difficult. As mucus accumulates in the bronchi, the coughing reflex is triggered to help clear these airways. Bronchitis is associated with an increased risk of **pneumonia**, a potentially serious bacterial, viral, or fungal infection.

Bronchial Asthma

Approximately three million Canadians suffer from asthma. It is the most common serious chronic disease of children. Like bronchitis, asthma also involves the inflammation of the bronchi and bronchioles. Rather than being caused by an infectious agent, however, it is usually triggered by an allergic reaction to a foreign substance. In response to the foreign substance, excess mucus is produced, and the muscles surrounding the bronchi and bronchioles go into spasms, causing the air passages to narrow. This makes breathing very difficult and results in a shortness of breath. Many asthmatics rely on an inhaler (or "puffer") to deliver drugs such as corticosteroids directly to their lung tissues (**Figure 2**). These corticosteroids stimulate the bronchi and bronchioles to increase in diameter. For individuals with serious asthma, it is essential that they always carry their inhaler with them to use in the event of an asthma attack.

Figure 2
An individual experiencing an asthma attack can inhale through a "puffer" to deliver medication directly to the lungs.

Carbon Monoxide Poisoning

Carbon monoxide (CO) is sometimes called the silent killer. You cannot see, smell, or taste carbon monoxide, so you cannot tell if you are breathing it in. This odourless, colourless, tasteless gas is produced during the incomplete combustion of fuel. It can come from many sources including automobiles, barbecues, gas stoves and furnaces, wood-burning stoves, and fireplaces. Carbon monoxide can be deadly because it combines with the hemoglobin in your red blood cells, preventing them from carrying oxygen to the cells of your body. The symptoms of carbon monoxide poisoning are nausea, headaches and dizziness, loss of sensation in the hands or feet, burning eyes, drowsiness, confusion, and loss of consciousness.

In order to protect yourself and your family from possible carbon monoxide poisoning, you should install carbon monoxide detectors in your home and have your furnace and other fuel-burning appliances inspected regularly.

DID YOU KNOW?

Smog
The city of Toronto experienced about 18 smog-alert days in the summer of 2002. These are days when air-pollution levels are considered serious enough to issue a general warning to the public.

CAREER CONNECTION

Certified heating contractors are qualified to inspect heating systems, including the furnace, venting systems, and chimney, to ensure that carbon monoxide will not accumulate in the home.

Smoking and Lung Cancer

Lung cancer is one of our most dreaded diseases. Unlike most cancers, however, it is also one of the most preventable. The single most important cause of lung cancer reaches the body directly through our respiratory system—cigarette smoke. Consider the following statistics regarding smoking and human health:

- Lung cancer is the leading cause of cancer death in Canada for both women and men—killing approximately 18 000 Canadians in 2001.
- Smoking is specifically related to about 87% of lung-cancer cases, and smoking is estimated to be responsible for 30% of all cancer deaths.
- Each year, about 330 nonsmoking Canadians die from lung cancer as a result of exposure to secondhand smoke.
- Tobacco use remains the single most important preventable cause of death in the world—resulting in about 10 000 deaths each day. This figure is expected to rise to 25 000 per day by 2010.

Lung cancer usually begins in the basal cells of the bronchi or bronchioles. Various components in cigarette smoke contribute to the development of the disease. **Figure 3** shows the progression of lung cancer through the following stages:

- Cigarette smoke travels through the bronchi and bronchioles and irritates the cells.
- The cells produce excess mucus to trap the foreign particles.
- Chemicals in the smoke of even a single cigarette stop the cilia from moving for several hours.
- Tar and other toxic materials become trapped in the mucus and can trigger mutations in the basal cells.
- The mutated cells begin uncontrolled growth, resulting in a cancerous tumour.

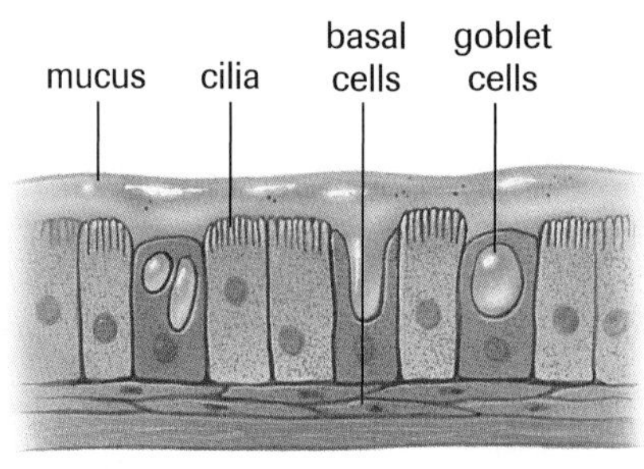

(a) normal

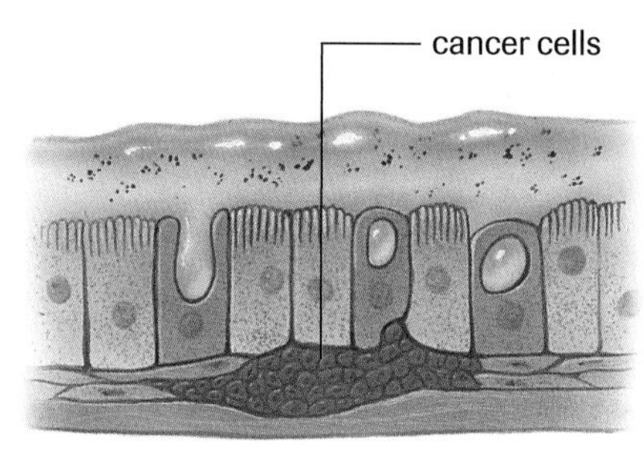

(b) cancer develops

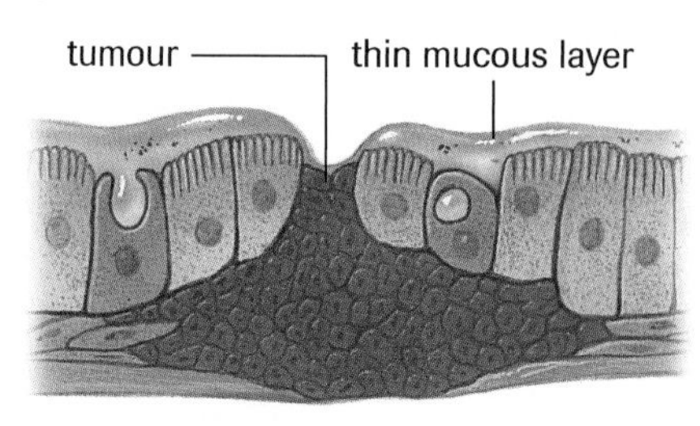

(c) cancer advances

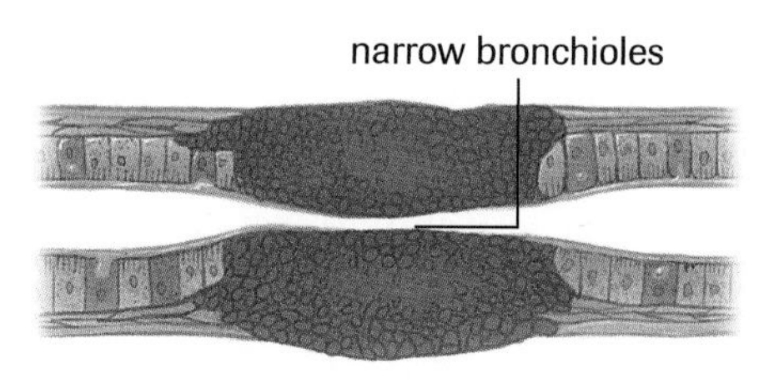

(d) tumour spreads

Figure 3
Development of a lung cancer tumour in the tissues of a bronchiole

When smokers quit their habit, the health of their respiratory system begins to improve immediately, and their risks of developing cancer and other smoking-related diseases decrease over time. **Figure 4** compares the lungs of a nonsmoker with those of a cancer victim.

(a)

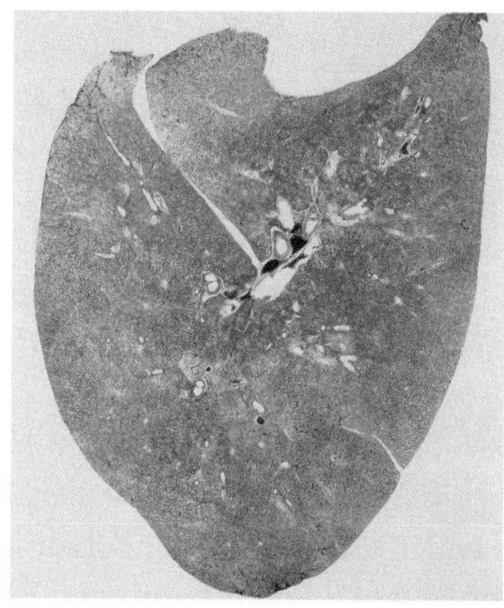

(b)

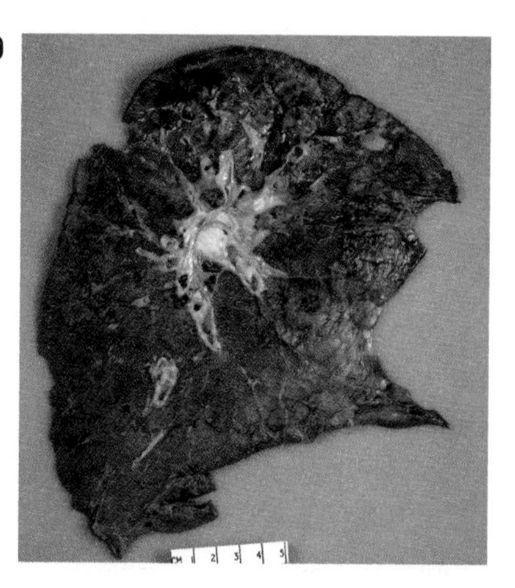

Figure 4
(a) The lung of a nonsmoker is pink and spongy.
(b) The lung of a regular smoker is black. This one also has a cancerous tumour.

Section 3.17 Questions

Understanding Concepts

1. What features of the respiratory system make it particularly vulnerable to pollutants or infection?
2. Outline the sequence of events that occur when the bronchi and bronchioles become irritated or infected.
3. What physical and chemical properties of carbon monoxide make it a dangerous air pollutant?
4. Which respiratory diseases and pollutants result in a decreased oxygen supply to the body?

Applying Inquiry Skills

5. Survey several people who smoke, and calculate the amount of tar taken into their lungs each day. Most cigarettes contain about 15 mg of tar, with about 75% of the tar being absorbed. Show your calculations.
6. A familiar symptom of the common cold is a runny nose. Some scientists hypothesize that this symptom may help remove the virus from your body, while others suggest it may actually help the cold virus spread. Design an experiment that could be used to test these hypotheses.
7. A researcher measures the lung volumes of several adult patients and obtains the data in **Table 1**.

Table 1 Lung Volumes (mL)

Patient	Tidal volume	Expiratory reserve volume	Vital capacity	Inspiratory reserve volume
1	600	120	4700	
2	450		5100	3300
3	280	600	2100	

(a) Calculate the missing values for each patient.
(b) Which patient may have significantly reduced respiratory capacity? What questions would you need to have answered before being convinced of a serious health concern?

Making Connections

8. A health-and-safety inspector determines that there is poor ventilation in a high-school chemistry lab. Would this pose a greater health risk to the teacher or the students? Explain. What other occupations involve exposure to chemicals on a daily basis?
9. What steps can you take to safeguard yourself against carbon monoxide poisoning in the home?
10. Research your local bylaws to determine what restrictions are placed on smoking in public places.

www.science.nelson.com

3.18 The Reproductive System

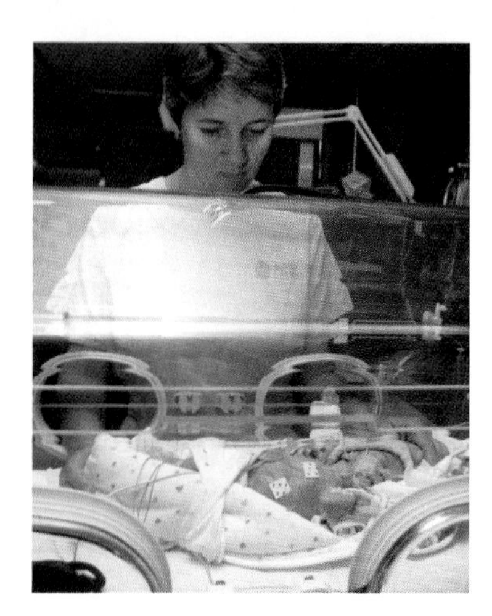

Figure 1
Premature babies are kept warm in an incubator.

CAREER CONNECTION

A neonatal nurse specializes in caring for newborn babies, usually in the neonatal (maternity) ward of a hospital.

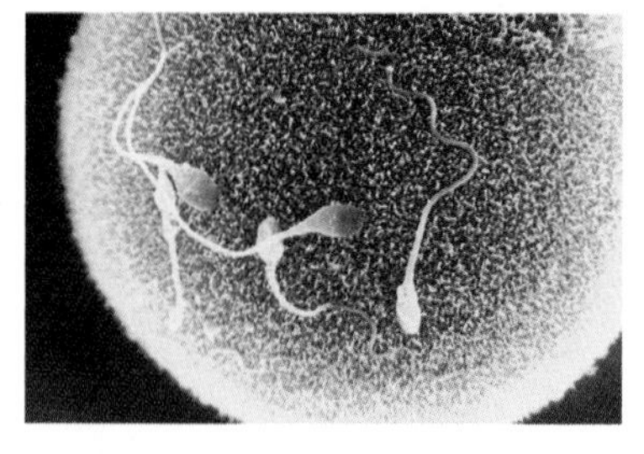

Figure 2
In humans, the egg cell is 100 000 times larger than the sperm cell.

reproductive system the organ system in males and females that performs the functions of sexual reproduction

acrosome a vesicle in a sperm cell containing enzymes that allow the sperm to penetrate an egg cell

Morgan was born 24 weeks after conception. She was 29 cm long and weighed 531 g. Most babies are born 38–40 weeks after conception, are 46–56 cm long, and weigh 2.7–3.9 kg. Morgan was what doctors call a "micro preemie," a very premature baby. About 9% of babies are born prematurely, and a much smaller percentage are micro preemies, like Morgan. Premature babies (**Figure 1**) require special care because their organ systems are not yet fully developed. They lose body heat much faster than typical babies do, so they must be kept warm. If preemies get cold, their bodies use energy to generate warmth and they fail to grow.

Morgan was placed in an incubator in the Neonatal Intensive Care Unit (NICU) of the hospital where she was born. An incubator is a walled plastic box with a heating system to circulate warm air. Babies in incubators are usually dressed in a gown or T-shirt, a diaper, and a hat. A baby can lose large amounts of heat through his or her head. Premature babies are often not well enough or strong enough to breast-feed or take milk from a bottle. Many babies in NICU receive nutrient-rich fluids through a tube in a vein called an intravenous (IV) line. Blood tests are performed regularly to make sure that they are receiving enough nutrition. NICUs are equipped with complex machines and monitoring devices designed for the unique needs of tiny babies. There is technology to monitor nearly every system of a baby's body, including body temperature, heart rate, breathing rate, oxygen and carbon dioxide levels, and blood pressure. Morgan spent a total of four and a half months in the NICU. The day she went home, she weighed 2.0 kg and had grown to 41 cm.

Humans reproduce sexually. This means that a new person is created from the union of two specialized reproductive cells called gametes, one from a sexually mature male and the other from a sexually mature female. The union of gametes and the growth and development of the new individual take place internally, within specialized organs of the female's **reproductive system**.

Gametes

Males and females play specific roles in the process of reproduction.

Males produce sperm cells (male gametes), and females produce egg cells (female gametes). Egg cells are generally round and contain a nucleus with hereditary material (DNA) and a large amount of food material called yolk. Egg cells are much larger than sperm cells (**Figure 2**).

Sperm cells contain a nucleus with hereditary material, a vesicle called an **acrosome** that contains digestive enzymes, a midsection with mitochondria, and a tail called a flagellum. The flagellum allows sperm cells to swim toward an egg cell, and the acrosomal enzymes allow sperm to penetrate an egg cell during fertilization. The mitochondria in the midsection provide energy for the flagellum. Sperm cells and egg cells are fragile and will die quickly if exposed to air.

Gametes are produced in organs called **gonads**. The female gonads are called ovaries, and the male gonads are called testes (singular: testicle or testis). In general, females are born with two ovaries and males are born with two testes. Although all structures needed for reproduction are present at birth, they are nonfunctional until puberty. Hormones produced in the gonads at puberty stimulate specialized cells to produce and release sperm cells, and specialized cells in the ovaries to produce and release egg cells.

gonad the organ of males and females that produces and stores gametes

In the process of fertilization, a sperm cell unites with an egg cell to form a single cell called a zygote that will eventually grow into a new individual.

Male Reproductive Structures

Figure 3 shows the components of the male reproductive system.

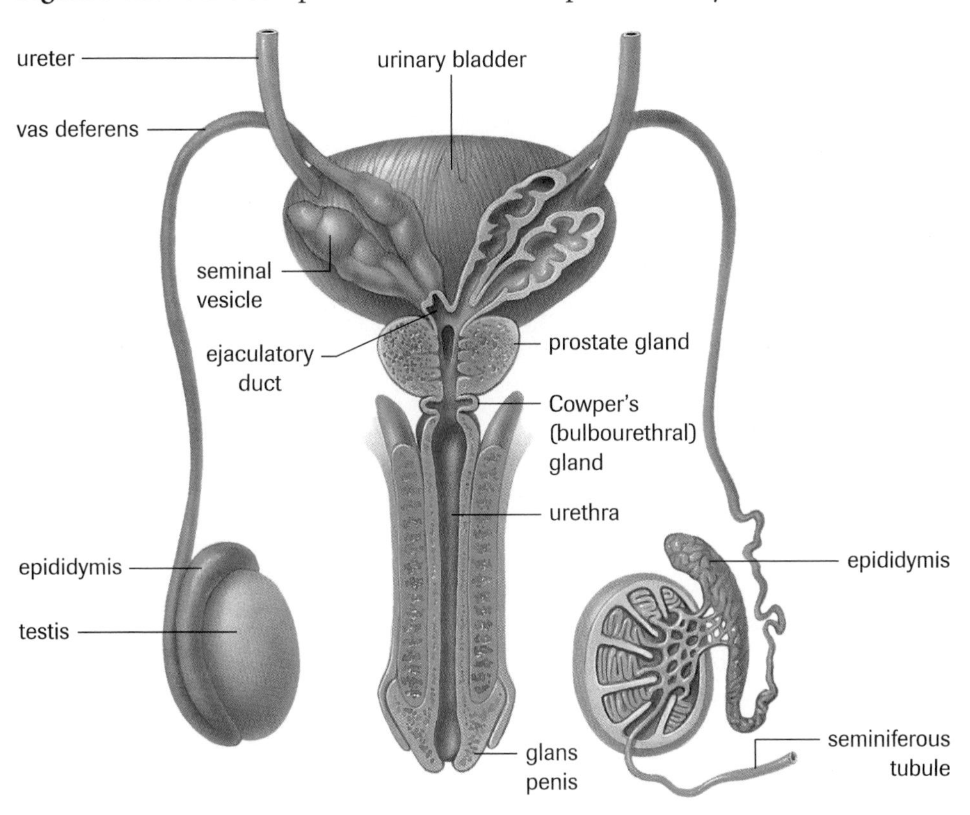

Figure 3
View of the male reproductive system

The testes of a male are outside the body, contained in a sac called the scrotum. The scrotum ensures that the testes are held away from the body and kept at a temperature slightly below 37°C because sperm cells are most efficiently produced at that temperature. Sperm cells are produced in the long, highly coiled seminiferous tubules of the testes and become mature and are stored in the epididymis. Cells in the testes also secrete testosterone, the hormone that stimulates sperm production and the development of secondary sex characteristics such as facial and body hair, muscular build, and a deep voice. As sperm cells pass through the vas deferens, nutrient-rich **seminal fluid** is added. Seminal fluid secreted by the seminal vesicles and the prostate gland contains fructose, a sugar that provides energy for sperm-cell locomotion, and other substances that help sperm cells survive within the female reproductive system. The mixture of sperm cells and seminal fluid is

seminal fluid a nutrient-rich fluid secreted by several glands of the male reproductive system called semen. Semen exits the body through the urethra, the same tube in the penis that urine passes through. The penis is the male organ used for discharging urine as well as depositing semen into the female reproductive tract. Simultaneous contraction of the bladder sphincter prevents urine from mixing with semen. Secretions from Cowper's glands add chemicals that protect sperm cells from acids that may be encountered in the male urethra and in the female system.

Female Reproductive Structures

The main organs of the female reproductive system are the vagina, the uterus, the fallopian tubes, and the ovaries. **Figure 4** illustrates the organs of the female reproductive system.

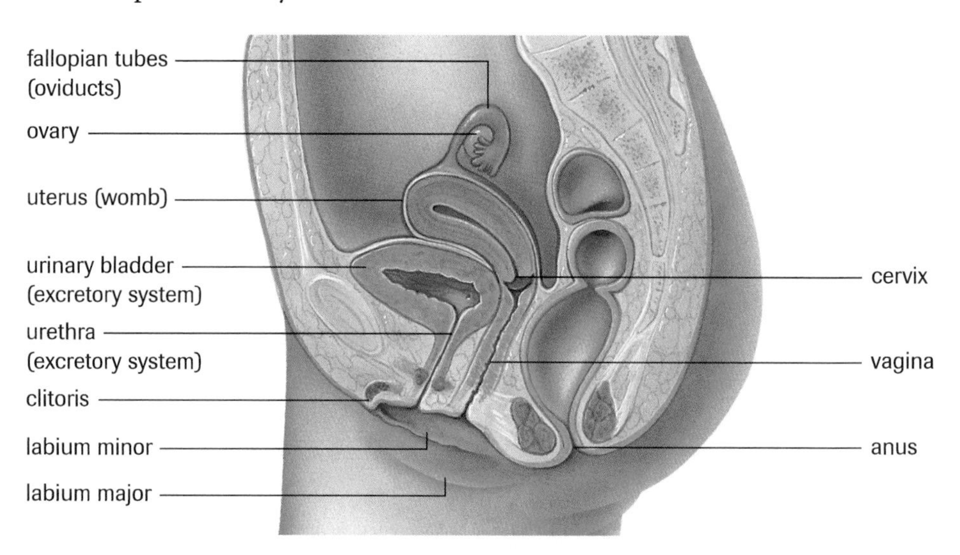

Figure 4
Female reproductive anatomy, side view. Note that there are two ovaries and two oviducts.

The vagina (also called the birth canal) is the smooth muscle-walled tube that connects the uterus to the external environment. The vagina receives the male's semen, and is the canal that the baby passes through during natural childbirth. The fallopian tubes are long, tubelike structures connected to the uterus that receive egg cells released by the ovaries. Normally, fertilization occurs in the fallopian tubes. The uterus is the muscular chamber in which a new baby develops.

At puberty, the ovaries begin to secrete a group of hormones called estrogens. The secretion of estrogens stimulates the development of secondary sex characteristics, including the enlargement of breasts, growth of pubic hair, deposition of extra fat under the skin, and development of broader hips to allow for childbirth. A number of hormones, including estrogens, stimulates egg production in the ovaries.

The Menstrual Cycle

At birth, the ovaries contain thousands of partially developed egg cells, each surrounded by a layer of cells called a follicle. Once a girl reaches puberty, the pituitary gland begins to secrete two hormones, **follicle-stimulating hormone (FSH)** and **luteinizing hormone (LH)**. FSH stimulates one of the many underdeveloped egg cells to mature. When mature, a sudden increase in the

follicle-stimulating hormone (FSH) the hormone secreted by the pituitary gland at puberty that stimulates the development of egg cells in the ovary

luteinizing hormone (LH) the hormone secreted by the anterior pituitary gland that stimulates the development of egg cells in the ovary, the monthly release of mature egg cells, and the development and maintenance of a yellowish mass called the corpus luteum in the ovary after fertilization

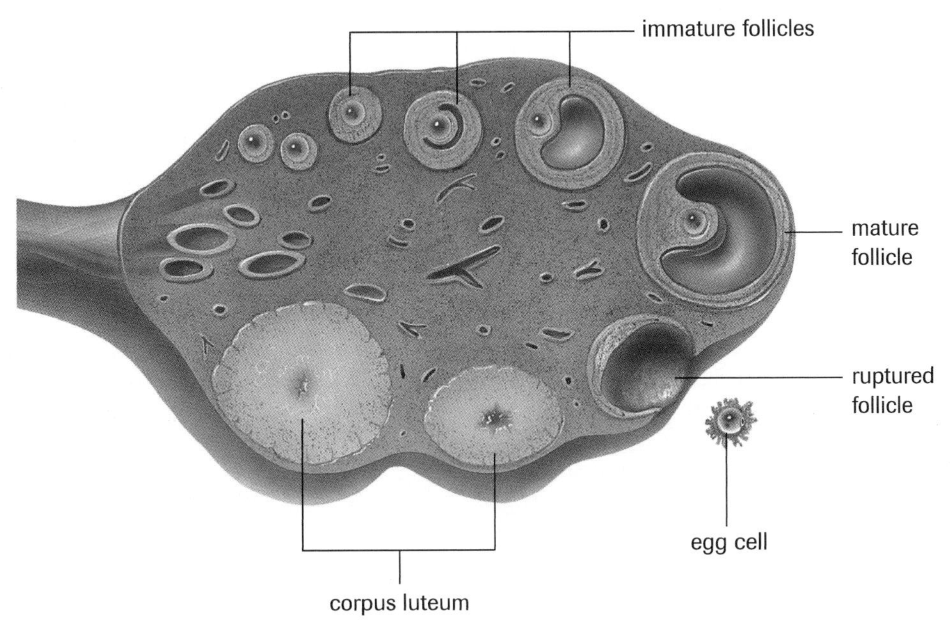

Figure 5
The process of ovulation in an ovary. Pituitary hormones regulate the events of follicle development, ovulation, and the formation of the corpus luteum.

secretion of luteinizing hormone by the anterior pituitary gland causes the release of a mature egg cell from the ovary. This process is called ovulation (**Figure 5**) and occurs approximately once a month.

Right after ovulation, cilia lining the nearby opening of a fallopian tube create a current that carries the egg cell into the tube. The egg continues moving through the fallopian tube toward the uterus. In the meantime, LH stimulates the remaining follicle cells in the ovary to become a yellowish mass called the **corpus luteum**. Cells of the corpus luteum begin to secrete a hormone called **progesterone**. Progesterone acts on the endometrium, the inner lining of the uterus. The zygote will develop into a baby inside the uterus. Progesterone causes the cells of the endometrium to multiply rapidly, glands in the lining to enlarge and become more active, and new blood vessels to form. All of these changes increase the thickness of the endometrium and prepare it for a possible pregnancy. The egg cell travelling toward the uterus must be fertilized by a sperm cell in the fallopian tube in order for the egg cell to attach itself to the endometrium. If fertilization does not occur within about 48 hours of ovulation, the egg cell will disintegrate. The endometrium breaks down and is shed through the vagina.

The resulting flow of blood and tissue through the vagina is called menstruation. As long as fertilization does not occur, the menstruation process repeats itself every month. Because of its monthly occurrence (approximately 28 days), it is often referred to as the menstrual cycle (**Figure 6**, on the next page). The **flow phase** (days 1–5) of the menstrual cycle is the period of time in which menstruation occurs. The **follicular phase** (days 6–13) is characterized by the development of follicles within the ovary, and the **luteal phase** (days 15–28) is the period in which the uterus lining (the endometrium) is prepared to receive a fertilized egg. If the egg cell is not fertilized, a flow phase begins a new cycle.

DID YOU KNOW?

Birth-Control Pills
A common type of birth-control pill contains synthetic estrogen and progesterone. These hormones trick the pituitary gland into thinking that plenty of estrogen and progesterone have been produced (as if a pregnancy exists) and, therefore, ovulation is not needed. When there is no egg to fertilize, conception is prevented.

corpus luteum a yellowish mass of cells in an ovary after ovulation that secretes the hormone progesterone

progesterone the hormone secreted by cells of the corpus luteum that stimulates the cells of the lining of the uterus to prepare for a possible pregnancy

flow phase the phase of the menstrual cycle in which menstruation occurs

follicular phase the phase of the menstrual cycle in which follicles develop within the ovary

luteal phase the phase of the menstrual cycle in which the uterus lining (the endometrium) is prepared to receive a fertilized egg

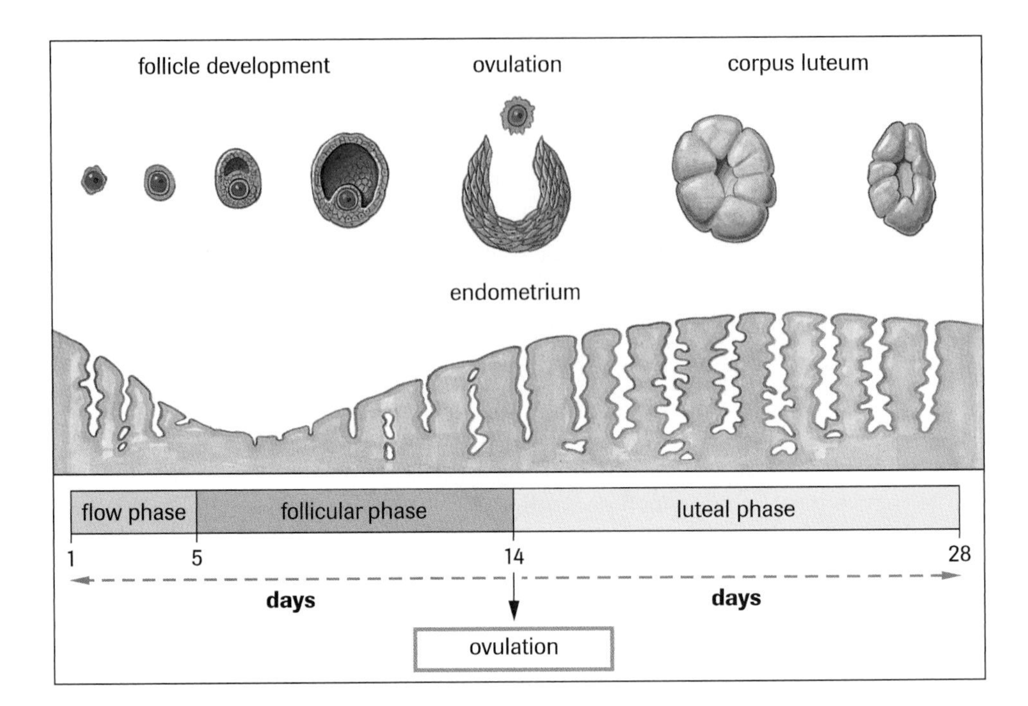

Figure 6
The thickness of the endometrium increases from the beginning of the follicular phase to the end of the luteal phase. The development of blood vessels helps prepare the uterus for a developing embryo. Should no embryo enter the uterus, menstruation occurs, and the menstrual cycle begins again.

Sexual Intercourse, Fertilization, Implantation, and Development

In order for fertilization to occur, sperm cells must be deposited in the vagina. This is usually done through sexual intercourse. Sexual intercourse begins with sexual stimulation that causes blood to flow into the spongy tissue of the male's penis, causing it to become relatively rigid and erect. The erect penis is able to enter the female's vagina and be as close as possible to the opening of the uterus. As sexual stimulation continues, a series of rhythmic contractions of muscles surrounding the male's urethra causes semen to be ejected. This process, called ejaculation, forces semen through the urethra.

During sexual intercourse, millions of sperm cells are deposited into the vagina. The vagina secretes fluids that facilitate entry of the erect penis and allow sperm cells to swim toward and through the **cervix**, the opening of the uterus into the vagina. Fertilization will not occur unless many sperm cells reach the egg cell in a fallopian tube.

cervix the opening of the uterus into the vagina

When a sperm cell comes in contact with an egg cell, the sperm cell's acrosome secretes digestive enzymes that soften the egg cell's membrane. The sperm cell penetrates the egg cell, and the sperm and egg cell nuclei unite to form a zygote. Almost immediately, a membrane forms around the zygote that prevents other sperm cells from entering the egg cell. As the newly formed zygote moves through the fallopian tube toward the uterus, it obtains nutrients from the yolk and divides by mitosis into a solid mass of cells called an embryo.

A small group of embryo cells secretes enzymes that help it attach to the endometrium. This process is called **implantation** and marks the beginning of pregnancy. The embryo undergoes several distinct stages of growth and development during pregnancy.

implantation the attachment of an embryo to the endometrium

Embryonic Development, Pregnancy, and Birth

By the time of implantation, the embryo has consumed virtually all of the original egg's nutritious yolk. From now on, the embryo must rely on the mother's body for nutrition, gas exchange, and excretion. A special organ called a placenta develops between the endometrium and the membrane surrounding the embryo (**Figure 7**).

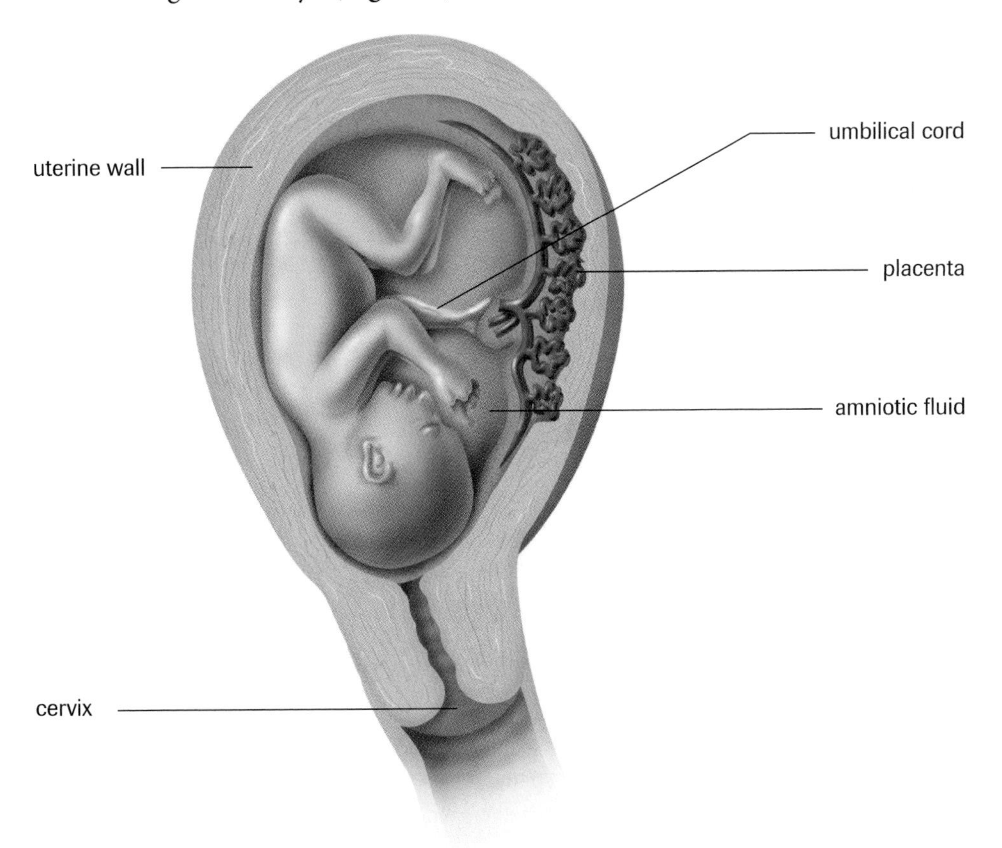

Figure 7
The placenta forms between the uterine wall and the embryo. Gas and nutrient exchanges take place at the placenta.

An umbilical cord carries blood vessels from the embryo to the placenta. The circulatory systems of the mother and the embryo are close together in the placenta, but the blood of the two individuals does not mix. Nutrients, gases, and wastes cross the placenta by diffusion. Although the placenta prevents certain substances from diffusing into the embryo's circulatory system, alcohol, nicotine, and other drugs can pass through and harm the developing baby. The placenta also secretes hormones that prevent menstruation during pregnancy.

At approximately seven to eight weeks of development, many features of an adult become recognizable. From this point on, the embryo is called a fetus. The fetus continues to grow and develop for another 32–33 weeks. In humans, the normal period of time from conception to birth is approximately 266 days (38 weeks). This is called the **gestation period. Table 1** gives the normal gestation period for a number of other animals.

At birth, the baby's circulatory and excretory systems are fully functional, but the nervous system continues to develop and mature throughout childhood, and the reproductive system remains virtually dormant until hormones produced at puberty activate it.

gestation period the normal period of time from conception to birth

Table 1 Gestation Periods in Mammals

Mammal	Average gestation period (days)
opossum	13
mouse	21–28
grey kangaroo	40
cat	60–63
tiger	90
pig	112–115
cow	266
human	266
elephant	600–660

Disorders of the Reproductive System

A number of disorders affect the human reproductive system (**Table 2**).

Table 2 Disorders of the Human Reproductive System

Disorder	Symptoms	Causes	Effects	Treatment
prostate disorders (male)	difficult urination abdominal pain	bacterial infection aging gonorrhea	blocked urethra thickening of bladder walls	antibiotics partial or whole removal of prostate gland
infertility (male)	inability to fertilize an egg	low sperm-cell count abnormal sperm cells exposure to radiation, heat, malnutrition, mumps, STDs	sterility	difficult to treat drugs or surgery in some cases
infertility (female)	inability to produce egg cells	failure to ovulate uterine abnormalities fallopian tubes too narrow hormonal defects tubal disorder (STDs) vaginal or cervical environment hostile to sperm cells	sterility	hormone therapy
impotence (male)	inability to produce and maintain an erection	diseases such as alcoholism or diabetes spinal injury psychological/emotional causes	inability to engage in sexual intercourse	dialogue with family physician psychotherapy drugs
abnormal menstruation (female)	absence of menstruation painful menstruation	infections of reproductive system malfunctioning ovaries or pituitary gland emotional or psychological causes	difficulty in becoming pregnant infections may spread	hormone therapy reduction in emotional and/or psychological pressures
tumours (female)	none in early stages may be detected by Pap test	develop spontaneously individuals may be genetically predisposed	if cancerous, may spread malfunctioning reproductive system	surgery, drugs, or radiation to remove tumour
undescended testes	testes not contained in scrotum poor sperm-cell production	unknown	sterility malignant tumours	surgery to lower the testicle into the scrotum or to remove the testicle
ectopic pregnancy	abdominal pain bleeding from uterus blackouts	implantation of embryo in fallopian tube usually results from previous tubal infection	shock death unless embryo is removed	operation (laparotomy) necessary surgery to remove the tube containing the embryo
premature birth	premature labour	fetal abnormalities, physical stress, hormone imbalance, structural abnormality of the uterus or cervix, infection, age, multiple births, causes unknown	pregnancy lasts fewer than 37 weeks	bed rest, drugs to stop contractions of the uterus, ties to keep the cervix from opening, antibiotics to treat infection

A **Pap test** may detect the presence of abnormal cells in the cervix. The Pap test is a microscopic examination of cells scraped from the surface of the cervix. The test is named after Dr. Papanicolau, who invented the process. If performed on a regular basis, the Pap test can detect cervical cancer early enough for effective treatment.

Pap test the microscopic examination of cells scraped from the surface of the cervix to detect the presence of abnormal cells

Sexually Transmitted Diseases (STDs) and AIDS

Sexually transmitted diseases (STDs) are infectious diseases passed by sexual contact with infected persons. STDs are caused by bacteria or viruses, and can cause serious long-term effects such as sterility or even death if left untreated. **Table 3** describes seven sexually transmitted diseases—their symptoms, causes, and treatments. Syphilis and gonorrhea are also called venereal diseases.

Table 3 Sexually Transmitted Diseases

Disease	Symptoms	Cause	Treatment
herpes genitalis	recurring painful blisters on genitals	herpes virus	not known
gonorrhea	painful urination and yellowish discharge sometimes no symptoms are noticed arthritis or infertility in severe cases	*Neisseria gonorrhoeae* bacteria	antibiotics
syphilis	Three stages of symptoms if left untreated: stage 1: hard sores on genitals, fingers, or lips stage 2: sore throat, headache, ulcers, and/or skin rashes stage 3: neurological damage, including blindness and/or paralysis	*Treponema pallidum* bacteria	antibiotics in early stages
chlamydia	painful urination, abnormal discharge from vagina or penis sometimes no symptoms are noticed	*Chlamydia trachomatis*	antibiotics
genital warts	warts on the outside or inside of the vagina, or tip of the penis sometimes no symptoms are noticed (genital warts are very contagious)	Human Papillomavirus (HPV)	not known; warts removed by drugs, freezing (cryocautery), burning (electrocautery), or surgery
chancroid	open sores on genitals swollen, painful lymph nodes in groin	*Haemophilus ducreyi*	antibiotics
pubic lice	itching in the pubic area	*Pediculosis pubis* (tiny insects that live in pubic hair and feed on human blood)	shampoo containing lindane or other effective pesticide

Acquired Immune Deficiency Syndrome (AIDS) is a disease of the immune system caused by the human immunodeficiency virus, or HIV. The virus prevents the immune system from mounting an effective defence against almost all disease-causing organisms, including HIV itself. Persons who have AIDS may have a number of symptoms, including fever, colds, and weight loss. They are also susceptible to a number of rare cancers. AIDS may be transmitted through sexual intercourse, blood transfusion using blood from an infected person, or the sharing of needles with an infected drug user.

DID YOU KNOW?

DDT

The insecticide DDT accumulates in the bodies of birds, such as eagles, who are high up in the food chain. DDT affects the activity of the shell gland, resulting in very thin, fragile egg shells that break before the chicks hatch.

The Reproductive System of Birds

Like humans, birds reproduce sexually, but unlike humans, birds do not give birth to live young birds. Instead, they lay fertilized eggs that develop outside of the mother's reproductive system. Fertilization occurs internally within the female's oviduct (**Figure 8**).

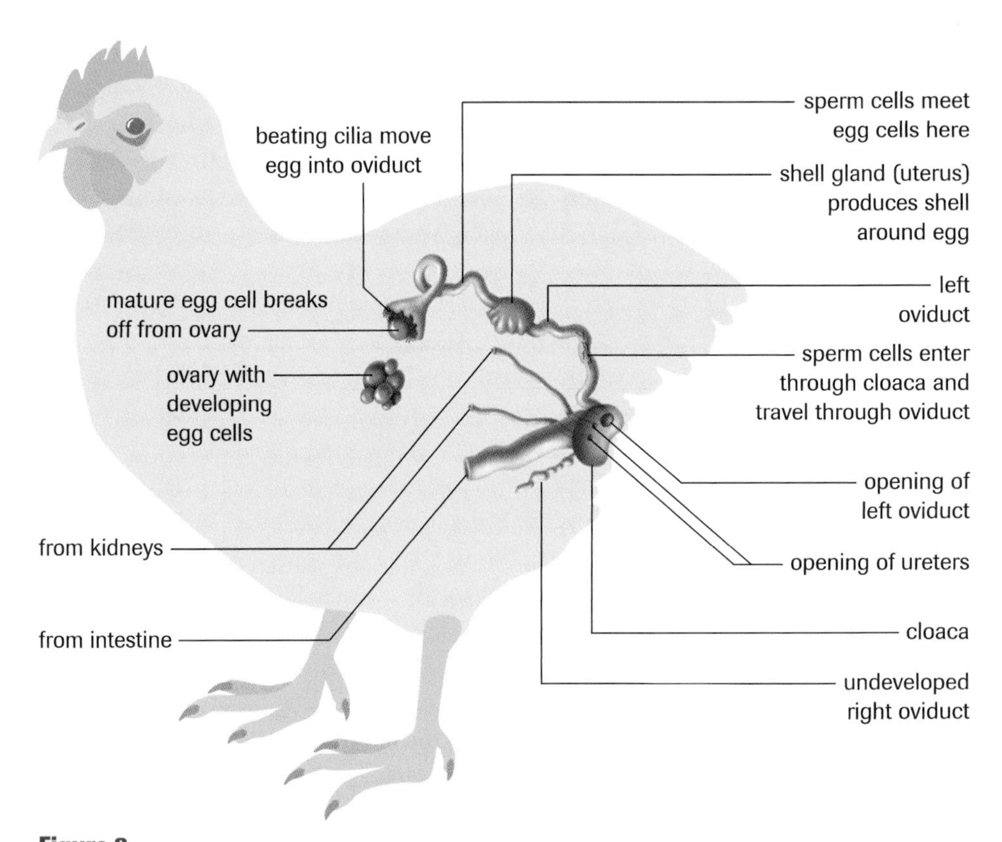

Figure 8
Internal fertilization in birds takes place in the upper part of the left oviduct. The undeveloped right oviduct is a weight-saving adaptation.

cloaca in female birds, the external organ containing the oviduct opening; in male birds, the external organ containing the vas deferens opening

shell gland the gland in the bird oviduct that forms a hard shell around the egg

albumen the protein-rich nutrient surrounding the yolk in a bird egg

Notice that the right oviduct of the bird is nonfunctional. This is an adaptation of birds that helps keep their weight to a minimum, since they have to fly. Much of the male bird's reproductive system is similar to that of mammals; however, most male birds do not have a penis. Sperm cells are produced in testes that are located within the bird's abdomen. Semen travels through the vas deferens and is delivered to an external organ called the **cloaca** that is similar to that of the female system. Male birds mount females from the rear and line up their cloaca with that of the female. The male releases semen into the moist surroundings of the female cloaca, and sperm cells swim up the left oviduct where they may fertilize an egg cell. After fertilization, the egg is surrounded by a hard shell made of calcium salts. This occurs as the fertilized egg passes through the **shell gland** within the left oviduct. Birds' eggs contain two food sources for the developing embryo: yellow-coloured yolk is rich in fatty substances, and **albumen**, or "egg white," contains protein (**Figure 9**).

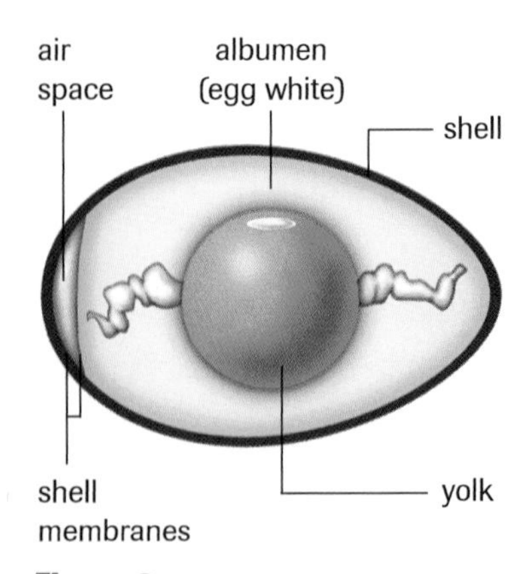

Figure 9
Cross section of a typical bird egg

Wild birds lay between 1 and 20 eggs per year. Once laid, eggs are tended by parents, and hatchlings are cared for and fed until they are able to fly.

Section 3.18 Questions

Understanding Concepts

1. Why are premature infants placed in incubators with hats on their heads?
2. Distinguish between the structures of sperm cells and egg cells.
3. Why are testes contained in a scrotum?
4. Complete the following chart:

Reproductive system part	Male/female/both	Function
		coiled tubes in the testes where sperm cells are produced
urethra	both	
		birth canal
fallopian tubes		

5. What effect do each of the following hormones have on the female reproductive system?
   (a) estrogen
   (b) progesterone
6. Describe the path a released human egg cell takes if it is not fertilized.
7. What is menstruation? Why is it important?
8. (a) Describe how the corpus luteum forms in the ovary.
   (b) What is the purpose of the corpus luteum?
9. What is the function of the placenta during fetal development?
10. What is the Pap test and what is it used for?
11. Distinguish between infertility and impotence.
12. Describe the path a fertilized egg takes in a bird's reproductive system.
13. List two similarities and two differences between the reproductive systems of a bird and a human.

Making Connections

14. A study by Dr. Marcia Herman-Giddens and her associates in 1997 seems to indicate that American girls are reaching puberty at an earlier age than in the past. Conduct library and/or Internet research to determine the possible causes of this trend.

www.science.nelson.com

15. To lessen the symptoms of menopause, many women are prescribed hormone replacement therapy (HRT). HRT involves taking estrogen pills. Additional estrogen reduces the typical "hot flashes" experienced by menopausal women, prevents bone loss, and may even improve memory. However, estrogen therapy raises the risk of blood clots and of ovarian and breast cancer. Conduct Internet and/or library research to investigate lifestyle changes that might help menopausal women.

www.science.nelson.com

Exploring

16. Viagra is a relatively new drug used to treat impotence. Conduct library and/or Internet research to determine how Viagra works and its potential side effects.

www.science.nelson.com

3.19 Movement and Locomotion: The Musculoskeletal System

Figure 1
Christopher Reeve

locomotion the ability to move from place to place under one's own power

musculoskeletal system the organ system composed of the skeleton (skeletal system) and muscles (muscular system)

Actor Christopher Reeve (**Figure 1**), best known for playing Superman, was paralyzed from the neck down when he fell off his horse in a horse-jumping competition in 1995. Unable to move any of his limbs and needing a power-assisted ventilator to help him breathe, he vowed that he would one day walk again. In 2002, after years of intensive therapy, he amazed his doctors by wiggling his fingers and moving many of his joints without assistance. Even though Reeve is not fully recovered, a breakthrough such as this gives hope to the millions of people with spinal cord injuries that they could someday regain the use of their limbs.

Some animals, such as sponges and corals, remain fixed in one location for their entire lifetime. Humans, however, need to move their limbs to obtain food and water, and to deal with dangers in their immediate environment. Under normal conditions, we move our limbs so effortlessly that we take our ability to move for granted. **Locomotion** is the ability to move from place to place under one's own power. Two organ systems, the skeletal system and the muscular system, work together to support the human body and allow it to move. Many consider the two systems together and call the combined system the **musculoskeletal system.**

The Human Skeletal System

The human skeletal system is a framework of more than 200 bones that supports the muscles that allow the body to move. In addition to movement, the system performs three other vital functions: protection, blood-cell production, and mineral storage. **Figure 2** shows the main components of the human skeletal system and indicates its four main functions.

Protection

Bones protect many organs of the body. The skull protects the brain, and the bones of the vertebral column protect the spinal cord. The ribs protect the heart and lungs, and the bones of the pelvis partially protect the female reproductive organs.

Blood-Cell Production

Bone marrow in many large, flat bones, such as the pelvic bones, the ribs, and the sternum, and in the ends of the long bones of your arms and legs produces blood cells and releases them into the bloodstream.

Mineral Storage

Your bones store a large amount of calcium and phosphorus. These minerals are partially responsible for your bones' strength and rigidity. A certain amount of dissolved calcium and phosphorus must circulate in your bloodstream to maintain healthy muscles. If your diet is poor in calcium, your body will dissolve the minerals stored in your bones and will weaken them. A diet rich in calcium and other minerals is important to maintain a healthy musculoskeletal system.

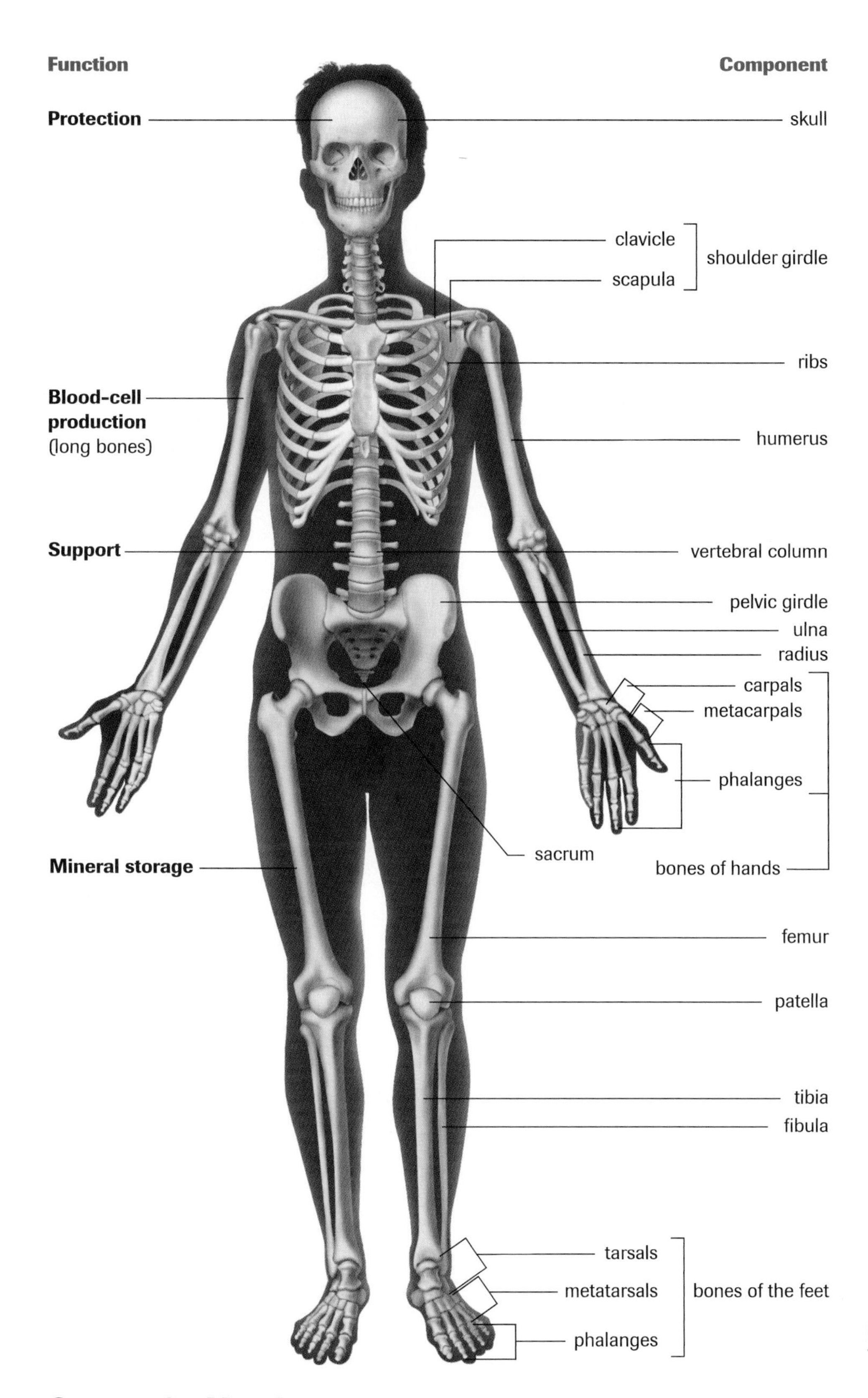

Figure 2
The human skeleton

Support for Muscles

Movement and locomotion depend on the ability of muscles to move the bones of the skeleton. Muscles are attached to bones in ways that allow the bones to move when the muscles contract.

Bones and Cartilage

The human skeleton is like a modern high-rise building in many ways. Like a building, the skeleton contains many structural parts that perform specific functions. In a building, steel posts and beams are arranged and fastened together in a pattern that will support the weight of the building and everything in it (**Figure 3**).

In a skeleton, bones are shaped and attached to one another to perform particular functions. The thin, flat bones of the **skull** (**Figure 4**) are curved and fused to one another to form a strong, hard, protective covering for the brain. The short, cylindrical bones of the **vertebral column** are stacked on top of one another, providing support for the upper body and a hollow, protective casing for the spinal cord (**Figure 5**).

Figure 3
The framework of a building is similar to a skeleton.

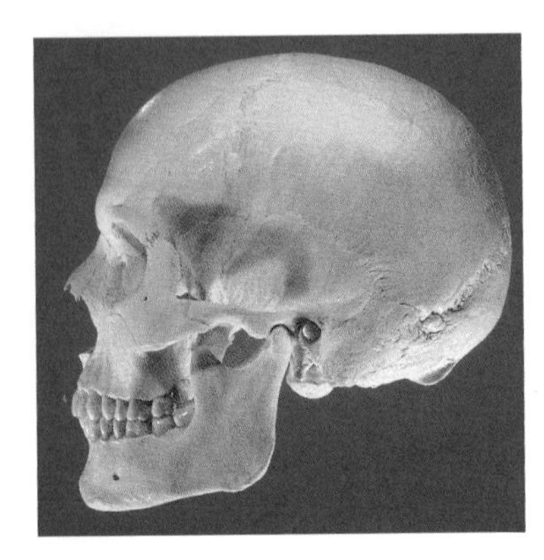

Figure 4
Human skull

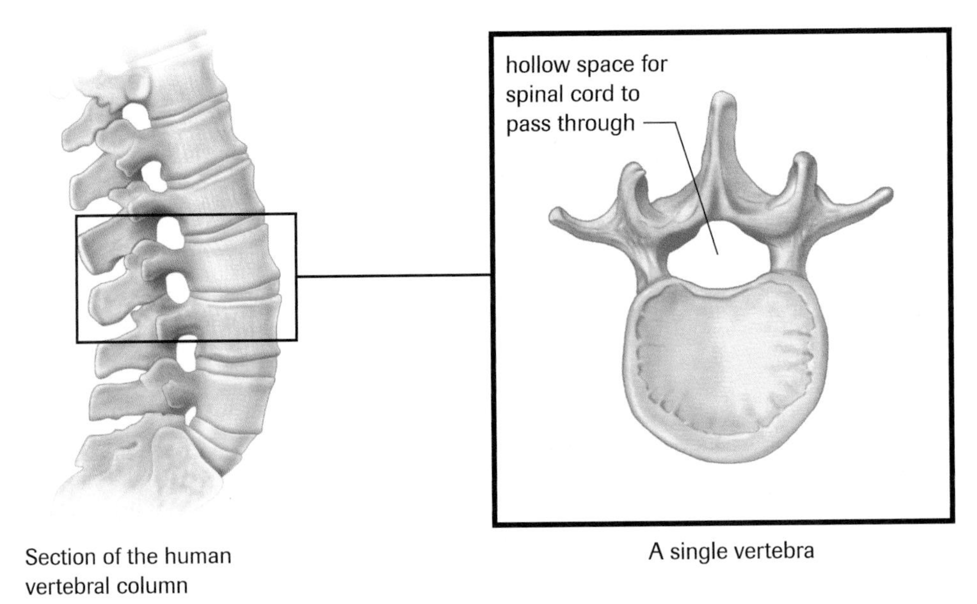

Figure 5
Short, hollow, cylindrical bones called vertebrae are stacked on top of one another to form the vertebral column.

Like a building, the human skeleton is made of many different materials. Bones are covered in a tough membrane called the **periosteum** (**Figure 6**). Below the periosteum is a layer of **compact bone** containing living bone cells called **osteocytes**. The osteocytes are set in a hard, dense mixture, or matrix, made of minerals such as calcium phosphate and a network of protein fibres called **collagen**. In many of the long bones such as the femur, or thighbone, a tissue called **spongy bone** lies below the compact bone layer. Spongy bone is more porous and less dense than compact bone. Bone marrow is a soft spongy material found in the spongy bone and the hollow interior of the long bones of the body. Bone marrow contains immature cells called stem cells that can develop into a variety of mature cells, including red and white blood cells.

Some parts of the skeleton, such as the tip of the nose and the ears, are made of a semi-solid, flexible connective tissue called **cartilage**. Cartilage is

skull the curved, fused bones that form a hard protective covering for the brain

vertebral column a stack of short, hollow, cylindrical bones that provides a protective covering for the spinal cord

periosteum the tough outer covering of bones

compact bone dense bone

osteocyte a bone cell

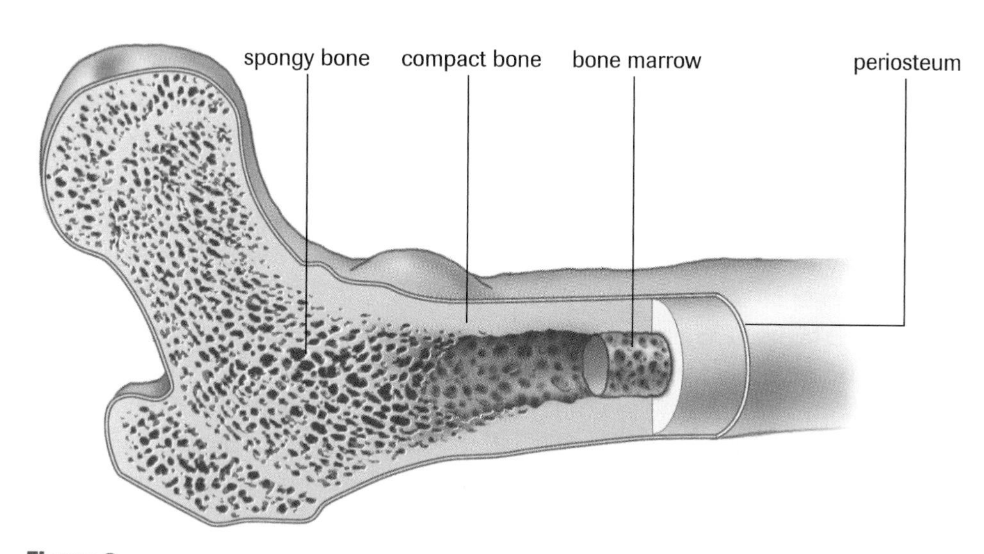

Figure 6
Cross section of a human femur

collagen the connective protein fibres found in bone matrix

spongy bone a porous bone

cartilage the semi-solid, flexible connective tissue

a supporting material containing living cells called **chondrocytes** and a flexible matrix composed of collagen fibres and polysaccharides. The skeleton of a developing embryo begins as a framework of cartilage. As the baby grows, the cartilage is replaced by compact bone and spongy bone in a process called **ossification** (**Figure 7**).

Ossification continues into the late teens and early twenties. During this time, a diet containing adequate amounts of minerals and protein is very important.

chondrocyte a cartilage cell

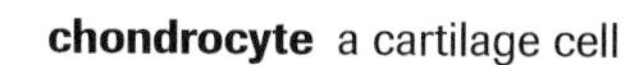

ossification the process in which cartilage is replaced by compact bone

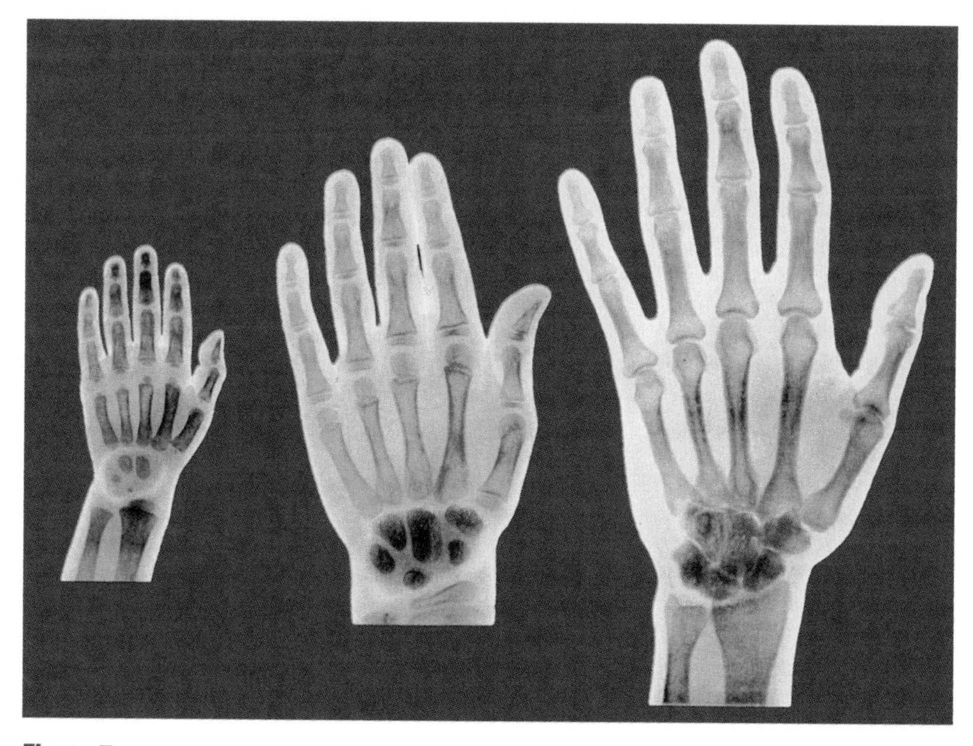

Figure 7
X-ray images of a child's hand, a teenager's hand, and an adult's hand. Notice the decrease in space (cartilage) between the carpals, metacarpals, and phalanges as the hand ages and ossification progresses.

joint any place in a skeleton where two or more bones meet

immovable joint bones that are attached so tightly that little or no movement is possible

suture the immovable joints between bones of the skull

movable joint a joint that allows bones to move

Joints and Ligaments

A building is a structure with many different uses. The parts of the framework holding the walls and the roof together are welded or bolted together to form rigid, immovable structures. Doors, however, are placed on hinges that allow them to move freely. The human skeleton also has movable and immovable components. A **joint** is any place in a skeleton where two or more bones meet. Bones that are attached so tightly that little or no movement is possible form **immovable joints.** The human skull is composed of several immovable joints called **sutures** (**Figure 8**).

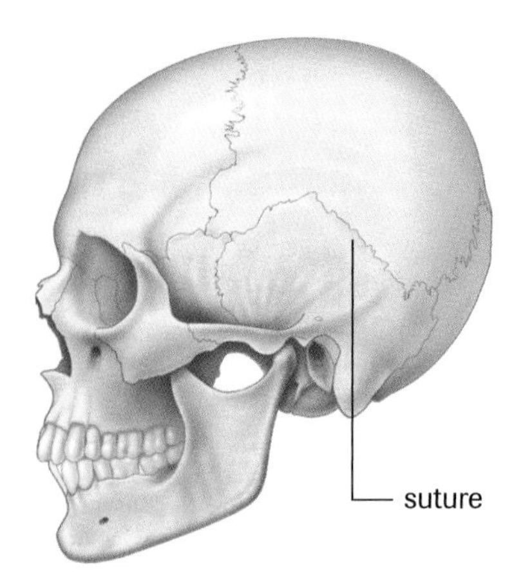

Figure 8
The sutures of an adult human skull are examples of immovable joints.

Sutures do not form in the skull until late childhood. In the fetus, the bones of the skull are loose so that the shape of the baby's head can change during birth. Young children typically have "soft spots" on their heads until the bones of the skull fuse permanently. Other immovable joints are found in the sacrum at the base of the vertebral column.

Most of the joints in the human skeleton allow for movement. These are called **movable joints.** Movable joints may be classified into four types: ball and socket, pivot, gliding, and hinge joints (**Table 1**).

Table 1 Four Types of Movable Joints

Type of joint	ball and socket	pivot	gliding	hinge
Location	shoulder	elbow	vertebrae	knee
Diagram				

ligament a strong, stretchable band of connective tissue that holds the bones of movable joints together

synovial fluid lubricating fluid between the bones of a joint

Ligaments are strong bands of connective tissue that can stretch. They hold the bones of movable joints together. Pads of cartilage at the ends of bones act as shock absorbers that reduce friction and wear and tear. In some cases, the cells of a synovial sac, between the bones of a joint such as the knee, secrete slippery **synovial fluid** that acts as a lubricant (**Figure 9**).

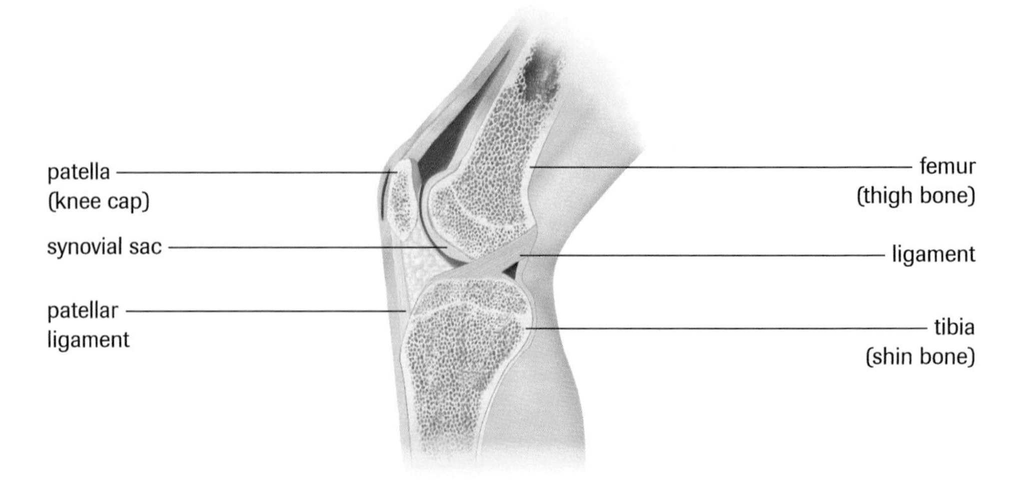

Figure 9
Lubricating synovial fluid is surrounded by a synovial sac in the knee joint.

Although movable joints allow your bones to move, bones cannot move by themselves. Bones can only be moved by the contracting action of muscles.

Muscles

Your body has more than 600 muscles that can be divided into three basic types. **Cardiac muscle** is the muscle that makes the heart beat, and is found only in the heart. Cardiac muscle contracts and relaxes automatically because it is controlled by nerves of the autonomic nervous system. **Smooth muscle** is found in the lining of organs such as the stomach, the esophagus, the uterus, and the walls of blood vessels. Smooth muscle contractions move digested food through the digestive system and help push a baby through the vagina during delivery. Smooth muscle is also controlled automatically. Unlike cardiac muscle and smooth muscle, the muscles that are attached to the bones of the skeleton are under conscious control, and are called **skeletal muscle**. **Figure 10** shows

cardiac muscle the involuntary muscle of the heart

smooth muscle the involuntary muscle found in the lining of many internal organs

skeletal muscle the voluntary muscle that makes the bones of the skeleton move

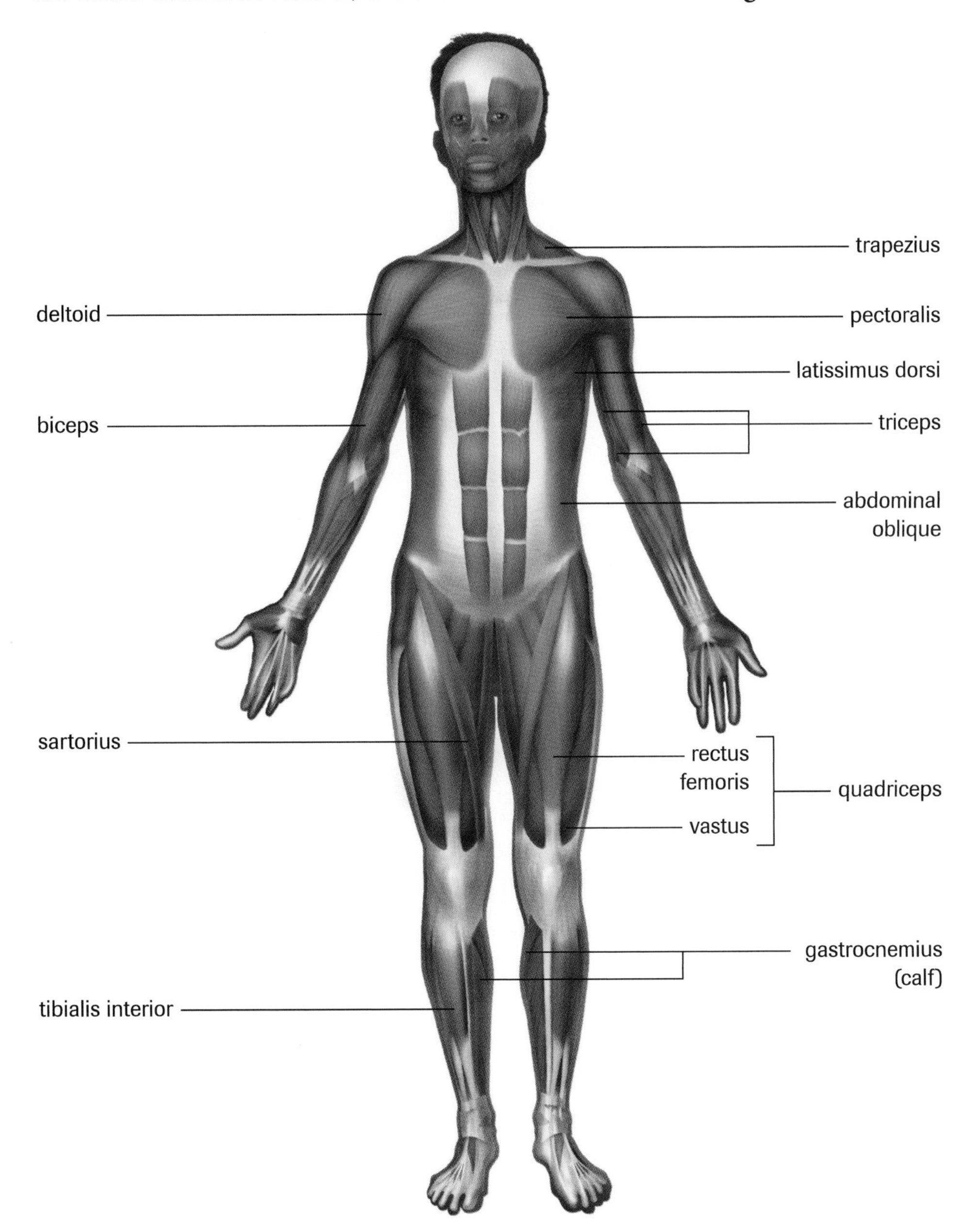

Figure 10
Muscles of the human body

tendon a tough, largely inflexible strip of connective tissue that attaches muscle to bone

origin the place where a muscle is attached to a stationary bone

insertion the place where a muscle is attached to a moving bone

antagonistic muscles the skeletal muscles that are arranged in pairs and that work against each other to make a joint move

some of the main skeletal muscles of the body. These are the muscles that allow you to walk, talk, and hit a baseball with a bat. Skeletal muscles are attached to bones by tough, largely inflexible strips of connective tissue called **tendons.**

Individual muscles are usually attached to two bones: one movable, the other stationary. The place where the muscle is attached to the stationary bone is called the **origin**, and the place where it is attached to the moving bone is called the **insertion** (**Figure 11**). The tendon at the insertion is usually longer than the tendon at the origin. Muscles shorten when they contract and lengthen when they relax. A body part moves only when a contracting muscle pulls it. Many skeletal muscles are arranged in pairs that work against each other to make a joint move. These are called **antagonistic muscles.** The biceps and triceps muscles of the upper arm are an antagonistic pair of muscles (**Figure 11**).

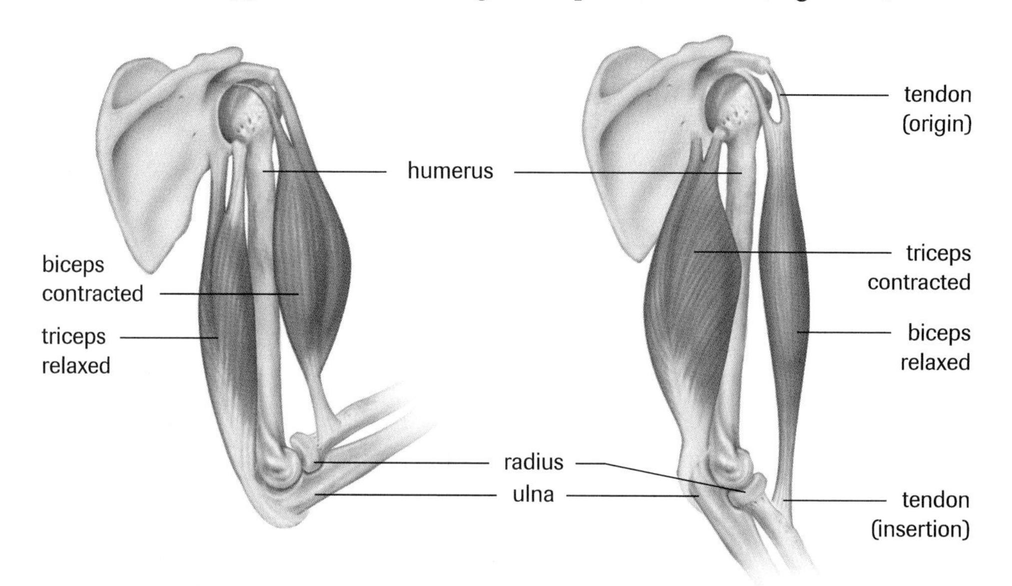

Figure 11
Biceps and triceps are muscles that are attached to bone by tendons. Why are they examples of antagonistic muscles?

▸ TRY THIS activity — Antagonistic Muscles

In this activity, you will study the action of several pairs of antagonistic muscles of your own body. You will need a partner to perform this activity.

Materials: tape measure

1. While standing, push your right arm straight down your side.

(a) Have your partner measure the origin to insertion length of your biceps muscle.

2. Place the index finger of your left hand at the insertion of the muscle. (Remember that the insertion is on the bone that will move.)
3. Without moving your index finger, slowly bend your right arm until it is completely flexed.

(b) Describe what you feel at the insertion point.

(c) Have your partner measure the length of your flexed biceps, and calculate the difference between the extended and contracted muscle length.

4. Repeat the above for your triceps muscle at the back of your arm.
5. Using **Figure 10** (on the previous page), the Internet, and other resources, select another pair of antagonistic muscles in your body and repeat this activity.

www.science.nelson.com

(d) What happened to the biceps muscle when it was contracted? What did you feel at the point of insertion when the biceps muscle was contracted?

(e) What happened to the triceps muscle when the biceps muscle was contracted?

When the biceps contract and the triceps relax, the bones forming the elbow joint are brought closer together. When the biceps relax and the triceps contract, the two bones are moved apart. The muscle that must contract to bend a joint is called a **flexor**, so the biceps muscle is a flexor. The muscle that must contract to straighten a joint is called an **extensor**. The triceps muscle is an extensor muscle.

flexor the muscle that must contract to bend a joint

extensor the muscle that must contract to straighten a joint

homologous structures body parts that are similar in structure but differ in function

The Musculoskeletal System of Birds

The musculoskeletal systems of birds and other vertebrates are adapted to their unique ways of life. Although the wings of birds are adapted for flight, their skeletons show similarities in overall design with the limbs of other vertebrates, including humans. Body parts that are similar in structure but differ in function are called **homologous structures**. A human forearm and a bird's wing are homologous structures (**Figure 12**).

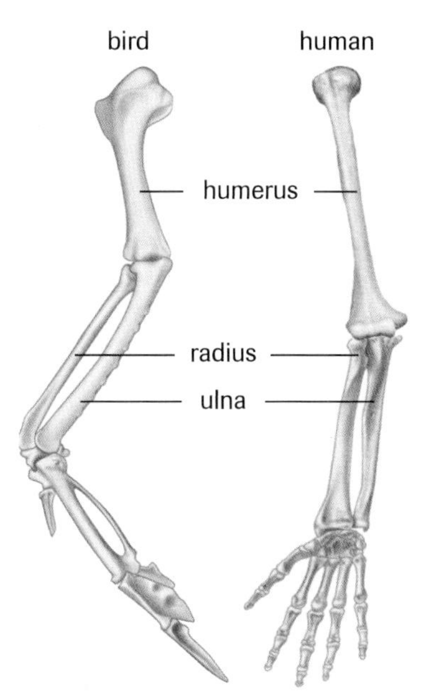

Figure 12
Homologous structures in bird and human

Notice in the diagram of the bird skeleton (**Figure 13**) that individual bones are so much like the bones of the human skeleton that they are given the same names.

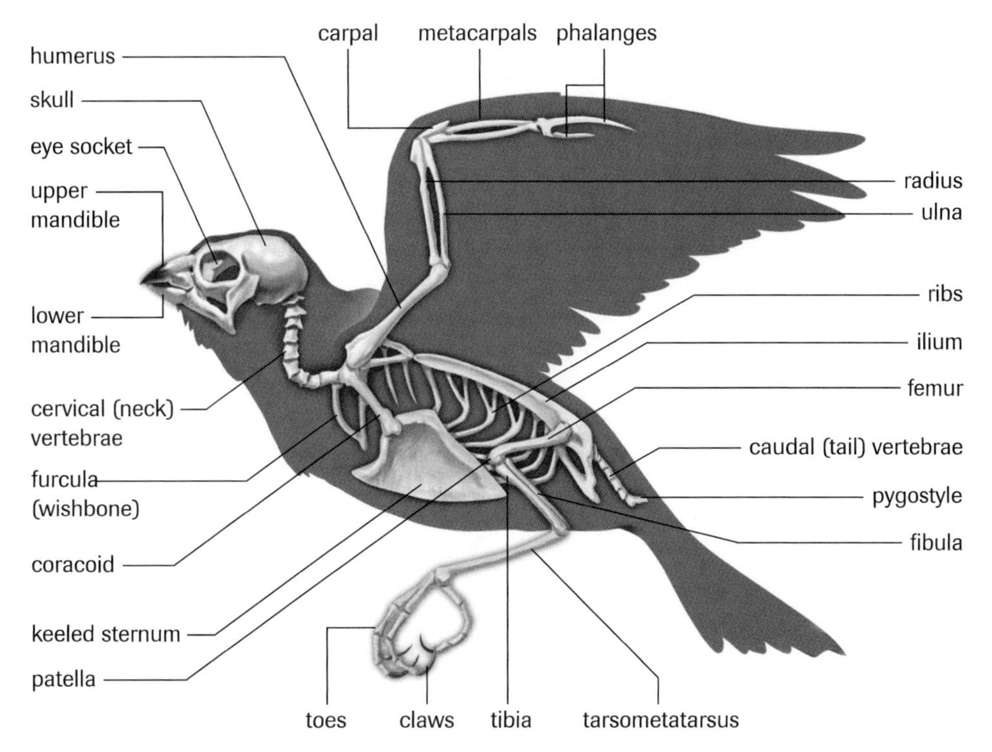

Figure 13
Bird skeleton

Birds have a lightweight skeleton. Most of the bones of flying birds are thin and hollow. Birds have a smaller total number of bones than mammals or reptiles. This is because many of their bones are fused together, making the skeleton more rigid. Birds also have more vertebrae in the neck than many other animals. This makes their necks very flexible and helps them clean their feathers. Birds are the only vertebrate animals to have a fused collarbone (the wishbone) and a very large, keel-shaped sternum (breastbone). The sternum is where the bird's powerful flight muscles are attached.

DID YOU KNOW?

Birds

For many birds, the weight of their feathers can be double the weight of their entire skeleton and as much as 20% of their total body weight. For example, a bald eagle's skeleton may weigh only about 250 g, while its feathers weigh more than twice as much, about 630 g.

CAREER CONNECTION

A paramedic or Emergency Medical Attendant (EMA) is a health care professional specialized in first-aid and emergency treatment. Paramedics are usually the first health care professionals at the scene of a medical emergency and deal with many personal-injury accidents.

Musculoskeletal Injuries

Bones and muscles, like all living tissues, require nourishment from a balanced diet, including an adequate supply of calcium and protein. Regular exercise is necessary for maintaining healthy muscles and bones. Studies done in the near zero-gravity environment of outer space show that astronauts lose bone mass unless they exercise regularly.

Musculoskeletal injuries are common in people who perform heavy work or exercise. Torn muscles, stretched tendons, torn ligaments, joint sprains, joint dislocations, and bone fractures are common sports injuries. A **bone fracture** is a broken bone. In a simple fracture (closed fracture), the broken bone does not break through the skin. A compound fracture (open fracture) occurs when a broken bone sticks out of the skin.

bone fracture a broken bone

greenstick fracture a fracture, commonly found in children, in which the bone does not break all the way through

Children's bones are more flexible than adults' bones. When a force is applied to a child's bone, it will bend and crack, but may not break all the way through. This type of fracture is known as a **greenstick fracture**. The term *greenstick* refers to a young branch of a plant that is flexible and resists breaking when you bend it. Greenstick fractures are difficult to diagnose because they may show up on an X-ray image as a bend in a bone without a clear break. **Figure 14** shows some common types of bone fractures.

A healthy diet, training, and protective equipment can help prevent or reduce the incidence of musculoskeletal injuries.

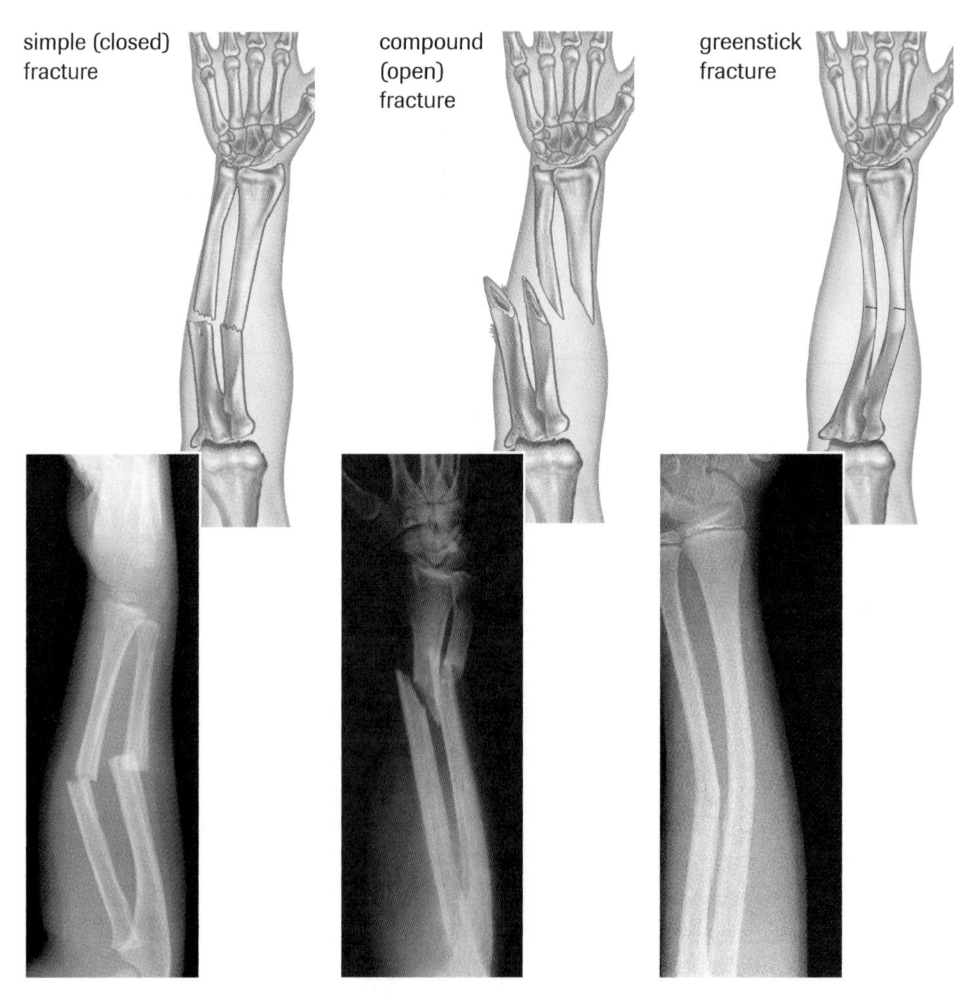

Figure 14
Three common types of bone fractures

Musculoskeletal Disorders

Table 2 lists a number of musculoskeletal disorders, their symptoms and characteristics, causes, and treatments. These conditions may be inherited, or they may develop from a microorganism infection. In some cases, their cause is unknown. The disorders may affect muscles, bones, or joints.

Table 2 Disorders of the Human Musculoskeletal System

Disorder	Characteristics and symptoms	Causes	Treatment
Muscles			
muscular dystrophy (Duchenne)	affects mainly legs affects male children only healthy muscle cells destroyed, replaced by scar tissue and fat	inherited (genetic mutation)	no cure symptoms relieved by surgery, physiotherapy
poliomyelitis (infantile paralysis)	affects children and young adults headache, fever, sore throat, back and limb pain, muscle tenderness muscles atrophy (waste away), leads to paralysis and sometimes death	viral infection of nervous system	can be prevented by vaccination bed rest, nursing care, and isolation muscle exercises with assistance
Bones			
osteoporosis	affects mostly post-menopausal females can also affect adolescent females who over-exercise bone loss (bones become less dense) risk of fractures (especially hip fractures) increases	reduced estrogen levels inadequate calcium intake and lack of exercise	estrogen replacement therapy when appropriate proper dietary intake of calcium increased exercise (for older females); decreased exercise (for younger females)
Joints			
arthritis (rheumatoid)	inflammation of bursa (fluid-filled sac between bones) can be accompanied by fever, anemia, fatigue, and inflamed arteries very painful joints swell, become stiff and deformed	unknown	no known cure physiotherapy anti-inflammatory drugs steroids
gout	affects toes, fingers, knees affects mostly males in first stage, sudden, severe joint pain with pain-free periods if chronic, malformations may result	excess uric acid forms crystals in joints first stage caused by minor trauma, surgery, excessive alcohol consumption often runs in families	anti-inflammatory drugs physiotherapy controlling uric acid levels weight reduction, joint rest, reduced alcohol consumption

Section 3.19 Questions

Understanding Concepts

1. Describe four functions of the human skeletal system.
2. How is the human skeleton similar to a bridge?
3. (a) Distinguish between compact bone and spongy bone.
   (b) Distinguish between a tendon and a ligament.
4. Name and describe the process illustrated by the X-ray images of a hand below.

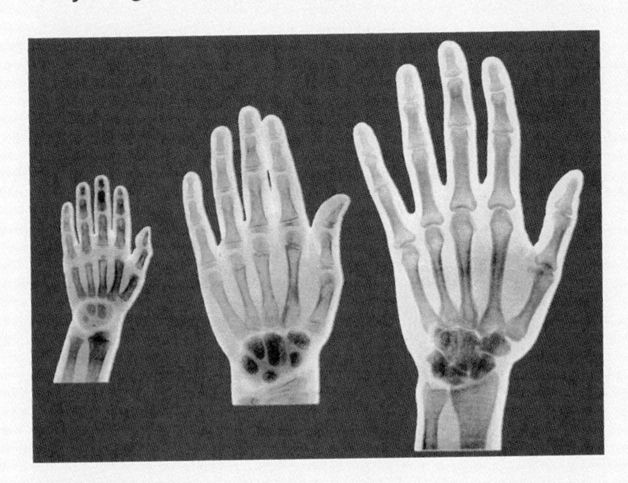

Figure 15

5. Why are the skull bones of a fetus not fused together as they are in an adult?
6. Describe two methods used by the body to cushion the force of movement and prevent wear and tear.
7. Distinguish between smooth muscle and skeletal muscle.
8. What would happen if a pair of antagonistic muscles contracted at the same time?
9. Distinguish between a simple fracture and a compound fracture.
10. (a) What is osteoporosis?
    (b) What causes osteoporosis?
    (c) How is osteoporosis treated?
11. Birds have lightweight bones, many of which are hollow. They also have fewer bones than reptiles or mammals. Why do birds exhibit these skeletal features?

Applying Inquiry Skills

12. A young child goes to the emergency ward of a hospital complaining of severe pain in the shin after falling from her bicycle. The doctor examines the child's shin and notices that there is a lot of pain and a bruise, but no break in the skin. An X-ray image reveals an unusual bend in the tibia (shinbone). What is your diagnosis

Making Connections

13. Bicycle helmets help protect the skull in cycling accidents. British Columbia, Nova Scotia, and New Brunswick have mandatory bicycle-helmet laws for cyclists of all ages. Ontario has bike-helmet legislation for children under the age of 18 years, and Manitoba has bike-helmet legislation for children under the age of 5 years. Many people would like to see all provinces enact laws that require all bicycle riders to wear a helmet. Some prominent groups such as the Ontario Coalition for Better Cycling disagree. Research both sides of this issue and prepare a position paper indicating your recommendations to the Ontario Ministry of Transportation.

www.science.nelson.com

14. Paramedics are health care professionals who specialize in first aid and ambulatory medical care.
    (a) What are the minimum educational requirements for a Primary Care Paramedic in Ontario?
    (b) Describe some of the duties of a Primary Care Paramedic.

www.science.nelson.com

Exploring

15. Casts and splints are commonly applied to broken limbs. What are casts and splints, and why are they used?

Designing and Constructing a Model Hand

The human hand has 27 bones that are jointed in a way that gives it an amazing amount of dexterity (**Figure 1**).

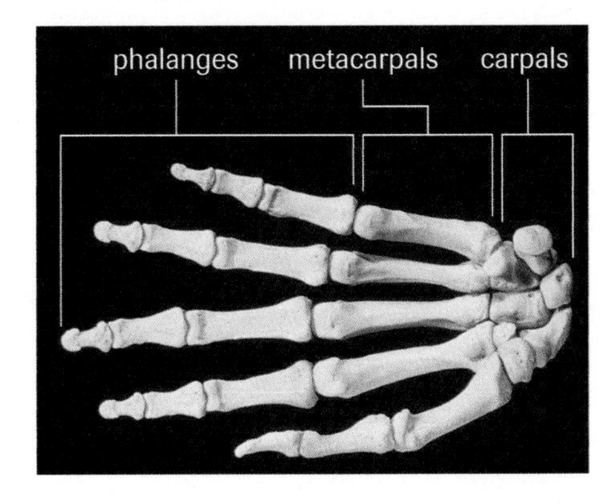

Figure 1
The human hand

Dexterity is the ability to move in a skillful fashion. The hand is connected to the arm by the wrist, which is made up of eight small bones, called carpals. These bones are arranged in two rows of four bones each. The carpals are tied together by ligaments that stop the bones from sliding around. The part of the hand known as the palm is composed of five bones, the metacarpals, that spread out from the wrist or carpus. Each finger has three phalanges, except the thumb, which has only two. The bones of the phalanges are anchored to muscles in the rest of the hand, and in the arms and shoulders, permitting a wide range of movements. The thumb is a unique finger because it is able to touch the other fingers tip to tip. This is an adaptation called the opposable thumb, and it allows you to grip objects.

Experimental Design

In this activity, you will design and build a functioning model of a human hand. Your model hand may contain separate carpals and metacarpals, but must contain 14 phalanges (see **Figure 2**).

You must design and construct a system for moving the bones so that the model hand is able to demonstrate the grasping action of an opposable thumb and the action of antagonistic muscles on each finger.

(a) Design your model hand on paper or computer, and have it approved by your teacher.

(b) Design a procedure for the construction of your model hand.

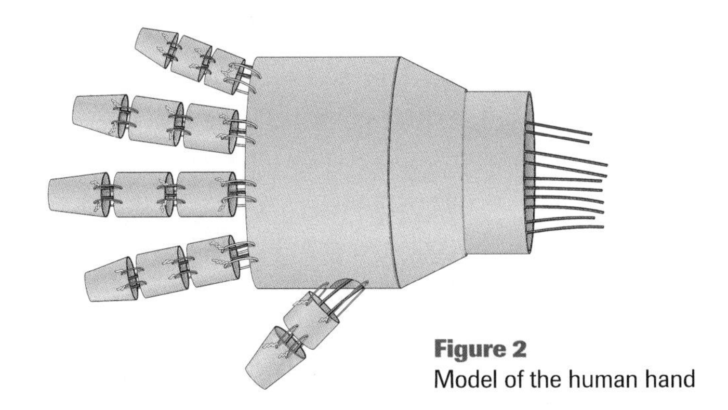

Figure 2
Model of the human hand

Materials

You may use a variety of materials to construct the individual bones and joints of the model hand. Examples include paper or cardboard, plastic sheets, rubber bands, twist ties, staples and stapler, glue, and string.

Procedure

1. Submit your procedure and have it approved by your teacher before beginning construction.
2. Carry out the procedure

Analysis

(c) How is your model hand similar to a real human hand? How is it different?

(d) Where are the flexor muscles in your constructed hand? Where are the extensors?

(e) What differences are there between the joints of the carpals and the joints of the phalanges in real hands?

Evaluation

(f) What functions of a real human hand are demonstrated by your model hand? What functions are not?

(g) How would you improve your model?

CAREER CONNECTION 

A prosthetist designs and constructs prosthetic devices or artificial limbs. Research to find out the education requirements for a prosthetist.

 www.science.nelson.com

Key Understandings

3.1 Organ Systems

- the human body has four primary kinds of tissue: epithelial, connective, muscle, and nervous
- an organ system is a group of organs with related structures (anatomy) or functions (physiology)
- human organ systems all interact with one another
- technology provides diagnostic tools for examining human organ systems

3.2 Tech Connect: Endoscopic Surgery

- an endoscope is a long, thin, flexible fibre-optic device attached to a viewing tube
- endoscopic surgery is classified as minimally invasive surgery

3.3 Control and Coordination: The Central Nervous and Endocrine Systems

- living organisms respond to changes in their internal and external environments
- negative feedback systems make adjustments to keep the body in homeostasis
- in animals, the nervous system is the control system that detects a change in the environment (stimulus) and coordinates a response
- the endocrine system secretes most of the hormones in the human body
- hormones help maintain homeostasis because they regulate and coordinate the functions of almost all organ systems
- the flight or fight response is an example of the nervous and endocrine systems working together to maintain homeostasis

3.4 Psychoactive Drugs and Homeostasis

- the nervous system is the main way in which the body receives information about changes in the environment
- legal and illegal drugs that affect the nervous system interfere with the body's ability to detect changes
- if the nervous system is unable to detect changes, the endocrine system may also be affected
- treating substance addition may involve counselling and/or medication

3.5 The Digestive System

- there are four main stages of food processing: ingestion, digestion, absorption, and egestion
- digestion breaks down the three macromolecules—carbohydrates, proteins, and fats, which are the main nutrients in the human diet—into molecules that are absorbed and transported to the body cells
- body cells require glucose, amino acids, and fatty acids for growth, maintenance, repair, and energy
- the nervous, endocrine, circulatory, and digestive systems interact to control digestion

3.6 Case Study: Digestive Disorders

- digestive disorders can be classified as structural, malabsorptive, or inflammatory
- examples of digestive disorders include Crohn's disease, appendicitis, gallstones, and gingivitis

3.7 Activity: Perch Dissection

- digestive structures of a perch are related to their functions
- similarities and differences exist between the fish and human digestive systems

3.8 Explore an Issue: Diet and the Media

- food intake, body image, and the media are inter-related
- Body Mass Index (BMI) is one measure of determining obesity

3.9 The Circulatory System

- the human circulatory system has four main functions: transportation of oxygen and carbon dioxide; distribution of nutrients and transportation of wastes; maintenance of body temperature; and circulation of hormones
- the circulatory system has three main components: the heart, blood vessels, and blood
- the heart and the blood vessels are also known as the cardiovascular system
- the circulatory system is two systems in one: the pulmonary circuit and the systemic circuit
- the human heart consists of a double pump separated by a septum

- blood has two main components—plasma (55%) and blood cells (45%)—made up of erythrocytes, leukocytes, and platelets
- fish have a closed circulatory system; the blood is contained within a continuous system of interconnected blood vessels connected directly with the respiratory organs

3.10 Circulatory Disorders and Technologies

- circulatory disorders include myocardial infarction, coronary artery disease, and valve disorders
- healthy lifestyles can reduce the risk of developing coronary artery disease
- treating circulatory disorders may include cardiac catheterization, angioplasty, or coronary bypass surgery

3.11 Investigation: The Body's Responses to Exercise

- designing an experiment to test how blood pressure and pulse rate are affected during exercise

3.12 The Excretory System

- in humans, the liver helps eliminate toxic ammonia by combining it with carbon dioxide to produce urea
- urea is dissolved in the blood and is carried to the kidneys where it is excreted in the urine
- the kidneys are the filters that remove wastes from the blood
- the kidneys form urine by filtering these wastes and water from the blood
- urine is stored in the bladder
- urine formation depends on three processes: filtration, reabsorption, and secretion
- kidney-related diseases include diabetes mellitus, diabetes insipidus, and kidney stones
- ammonia does not build up in fish as it does in humans—it diffuses through their gills; however, both saltwater and freshwater fish must regulate their concentration of salt

3.13 Activity: Urinalysis

- analyzing urine to diagnose kidney or other related disorders

3.14 Tech Connect: Kidney Dialysis

- people who cannot effectively rid the body of wastes undergo one of two types of dialysis: hemodialysis or peritoneal dialysis
- comparing hemodialysis with peritoneal dialysis

3.15 The Respiratory System

- all animals exchange gases with their surroundings by taking in oxygen and releasing carbon dioxide (respiration)
- there are two main stages of breathing: inspiration and expiration
- the pattern of inspiration and expiration is controlled by the autonomic nervous system—by nerve impulses from a breathing centre in the brain
- frogs have an external respiratory membrane; they exchange gases through their moist skin

3.16 Activity: Determining Lung Volumes

- measuring lung volumes to determine your normal and maximum breathing volumes

3.17 Disorders of the Respiratory System

- respiratory disorders include the common cold, influenza, and bronchial asthma
- one of the most serious causes of lung cancer is smoking; smoke reaches the body through the respiratory system
- the respiratory system of smokers who quit their habit will improve immediately, and their risks of developing cancer and other smoking-related diseases decrease over time

3.18 The Reproductive System

- the human reproductive systems are those organ systems in males and females that perform the functions of sexual reproduction
- the reproductive system is virtually dormant until hormones produced at puberty activate it
- in humans, fertilization occurs when a male sperm cell penetrates the female egg cell
- the gestation period of human females is 38 weeks
- disorders of the reproductive system include infertility (male and female), abnormal menstruation, and undescended testicles
- a Pap test detects the presence of abnormal cells in the female cervix

- sexually transmitted diseases (STDs) are infectious diseases passed through sexual contact with an infected person
- Acquired Immune Deficiency Syndrome (AIDS) is caused by the human immunodeficiency virus, or HIV; it can be transmitted through sexual intercourse with an infected person, blood transfusions from an infected person, or sharing needles with an infected person
- birds reproduce sexually—fertilization occurs within the female oviduct—but females do not give birth to live young

3.19 Movement and Locomotion: The Musculoskeletal System

- the skeletal system and the muscular system make up the musculoskeletal system
- the human skeleton contains many structural parts that perform specific functions
- joints are places where two bones meet; they may be immovable or movable
- there are four types of movable joints: ball and socket, pivot, gliding, and hinge
- ligaments hold the bones of movable joints together
- bones can be moved only by contracting or relaxing muscles
- the human body has three types of muscles: smooth, skeletal, and cardiac
- the musculoskeletal system in birds is adapted for flight: they have a lightweight skeleton; many of their bones are thin and hollow; and the sternum is where the bird's flight muscles attach to their body
- musculoskeletal injuries are common in people who perform heavy work or exercise
- musculoskeletal disorders include muscular dystrophy, osteoporosis, and rheumatoid arthritis

3.20 Activity: Designing and Constructing a Model Hand

- designing and constructing a model hand and comparing it with a human hand

Key Terms

3.1
tissue
organ
organ system

3.3
homeostasis
normal level (reference level)
normal range (reference range)
diagnosis
negative feedback
nervous system
nerve cell or neuron
impulse
nerve
dendrite
axon
myelin sheath
terminal knob
synapse
sensory neuron
motor neuron
central nervous system
peripheral nervous system
somatic nervous system
autonomic nervous system
sympathetic nerve
parasympathetic nerve
hormone
gland
endocrine system
pituitary gland
hormone-releasing factor
adrenaline

3.4
psychoactive drug
stimulant
depressant
neurotransmitter
addiction
narcotic

3.5
amylase
peristalsis
sphincter
mucus
pepsin
duodenum
bile salts
detoxify
chyme
lipase
trypsin
maltase
peptidase
vitamin
mineral
villi
microvilli
lacteal
colon

3.8
Body Mass Index (BMI)

3.9
circulatory system
cardiovascular system
pulmonary circuit
systemic circuit
septum
atrium
ventricle
sinoatrial (SA) node
atrioventricular (AV) node
diastole
systole
pulse
blood pressure
sphygmomanometer
aorta

3.10
electrocardiograph
electrocardiogram
cardiac catheterization
angioplasty
stent
coronary bypass surgery
arteriosclerosis
atherosclerosis
plaque
myocardial infarction

3.12
deamination
urea
renal artery
urine
renal vein
ureter
urethra
nephron
Bowman's capsule
glomerulus

3.15
respiration
endotherm
trachea
epiglottis
larynx
bronchus
bronchiole
goblet cell
alveolus
inspiration
expiration
diaphragm
external intercostal muscles
pleural membrane

3.17
pneumonia

3.18
reproductive system
acrosome
gonad
seminal fluid
follicle-stimulating hormone (FSH)
luteinizing hormone (LH)
corpus luteum
progesterone
flow phase
follicular phase
luteal phase
cervix
implantation
gestation period
Pap test
cloaca
shell gland
albumen

3.19
locomotion
musculoskeletal system
skull
vertebral column
periosteum
compact bone
osteocyte
collagen
spongy bone
cartilage
chondrocyte
ossification
joint
immovable joint
suture
movable joint
ligament
synovial fluid
cardiac muscle
smooth muscle
skeletal muscle
tendon
origin
insertion
antagonistic muscles
flexor
extensor
homologous structures
bone fracture
greenstick fracture

MAKE a summary

Design a poster summarizing key information about the body systems presented in this unit. In the centre of a piece of paper, draw a large human outline. Without drawing in all the individual structures, show the movement of food, oxygen, and wastes into, through, and out of the body. In bullets around the body, outline the important features of each body system and describe how each functions to maintain homeostasis.

Unit 3 Animal Anatomy and Physiology

PERFORMANCE TASK

Criteria

Process

- Choose and safely use appropriate equipment.
- Record observations in detail and with precision.
- Dissect and expose anatomical structures in a clean and undamaged way.
- Evaluate the usefulness of the dissection as a model for human anatomical features.

Product

- Prepare clearly labelled and accurate diagrams and/or drawings of your dissection.
- Demonstrate a thorough understanding of the key features, functions, and arrangements of mammalian body systems.
- Identify key features of fetal pig anatomy.

Examining the Systems of a Fetal Pig

Like humans, pigs are placental mammals. This means that, before birth, the fetus develops in the uterus and receives nourishment and oxygen from the mother through the umbilical cord. After birth, the young nurse from the mother's mammary glands. All mammals also have a highly developed brain and a four-chambered heart. Pigs are similar to humans in other important ways. Like humans, pigs are omnivores and have little body hair. Because of these many shared features, they have similar internal anatomies. By dissecting a fetal pig, you can directly observe the characteristic features and arrangements of mammalian systems.

Read and follow the procedure carefully. Accompanying diagrams are included for reference only. Use the appropriate dissecting instruments. This activity has been designed to minimize the use of a scalpel.

Materials

safety goggles	string	dissecting pins
laboratory apron	scalpel	scissors
forceps and probe	dissecting tray	ruler
hand lens or dissecting microscope	preserved fetal pig	dissecting gloves

Wear eye protection and a laboratory apron at all times.

Wear gloves when handling preserved specimens.

Always cut away from yourself and others sitting near you, in case the scalpel slips.

Partially dissected specimens are susceptible to disease organisms. Do not store for any longer than necessary.

Wash dissecting instruments in a disinfectant solution.

Make sure cuts and scrapes receive immediate medical attention.

Dispose of dissected remains as directed by your teacher.

Wash thoroughly with soap and water after completing the dissection.

Procedure

Part 1: External Anatomy

1. Place your pig on its side in a dissecting tray. Using **Figure 1** as a guide, observe the major regions of the body: the head, neck, trunk, and tail. Note the anterior (front) and posterior (rear) ends, and the dorsal (upper) and ventral (lower) surfaces of the pig.

(a) Label a diagram of the pig using these terms.

2. The size of the pig can be used to estimate its age. Use a string and ruler to determine the length of the pig from the tip of the snout to the tip of the tail. Compare your value to those in **Figure 2** to estimate the age of your pig.

(b) Record your pig's estimated age.

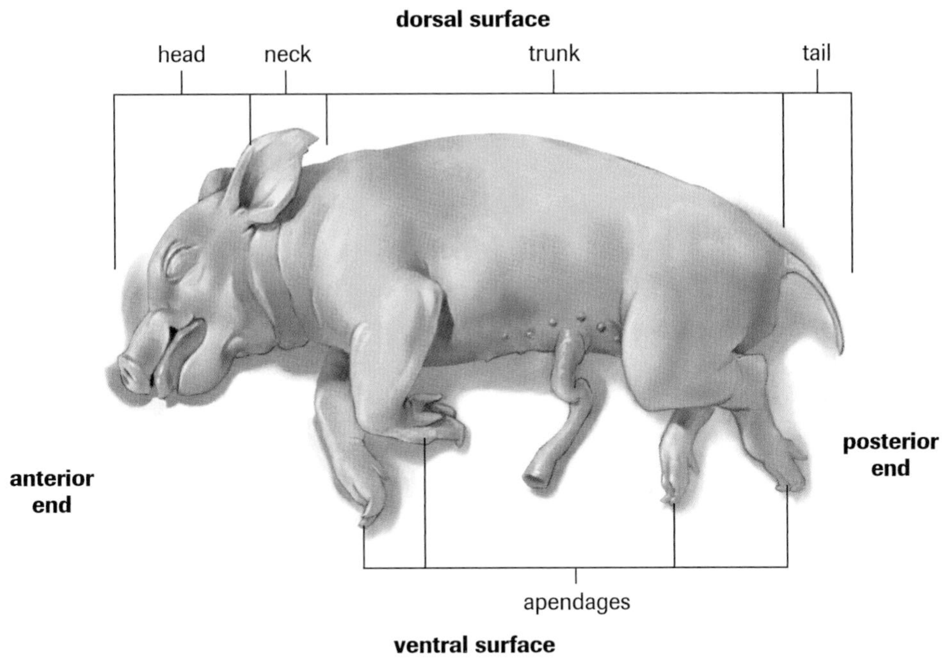

Figure 1
The external features of a fetal pig

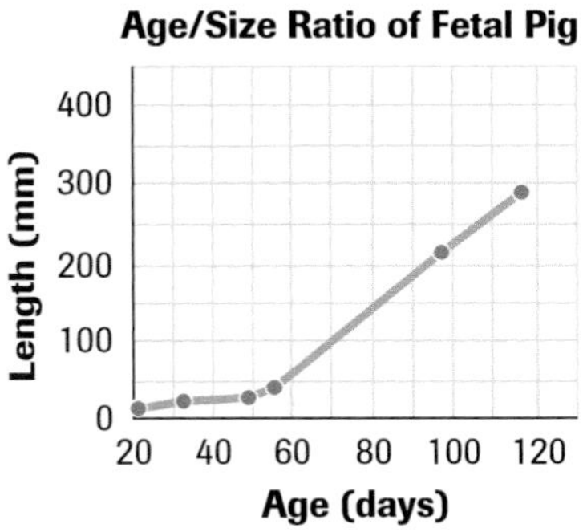

Age/Size Ratio of Fetal Pig

Length (mm): 0, 100, 200, 300, 400

Age (days): 20, 40, 60, 80, 100, 120

Figure 2
Size and age trends in fetal pigs

3. Locate the paired rows of nipples along the ventral surface of the pig. Note the umbilical cord, the closed eyelids, the tongue, the external ears, and the toes. Lift the pig's tail to locate the anus.

(c) Label each of these features on your diagram.

(d) What is the function of the umbilical cord?

(e) The umbilical cord contains two arteries and one large vein. Identify and label these on your diagram.

4. Determine the sex of your fetal pig by locating the urogenital opening (**Figure 3**). In males, this opening for sperm and urine is located just behind (posterior to) the umbilical cord. In females, the opening is located just below (ventral to) the anus. In male fetal pigs, a scrotal sac may be visible. Examine a fetal pig of the other sex to familiarize yourself with its external features.

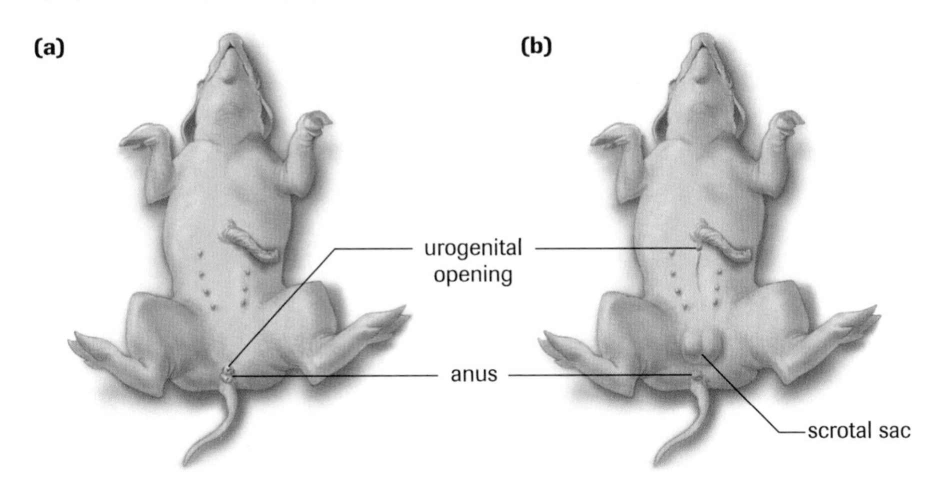

Figure 3
The external features of **(a)** a female and **(b)** a male fetal pig

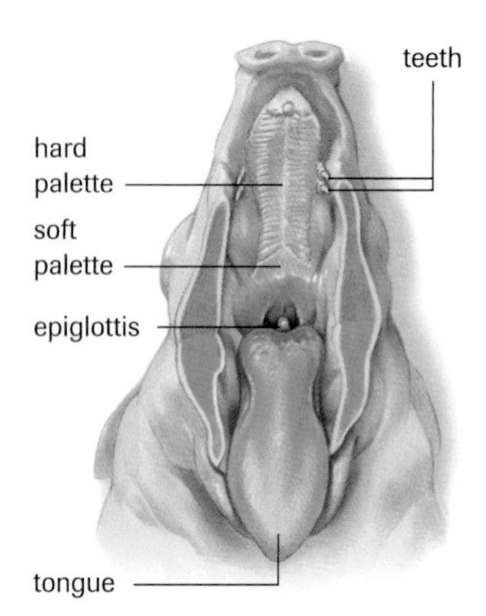

Figure 4
The internal features of the mouth

Part 2: Internal Anatomy

5. *Mouth:* Using a pair of scissors, cut through the corners of the mouth from the opening toward the posterior while following the inner passageway above the tongue but below the roof of the mouth. You will have to pry open the mouth carefully as you cut. The opening will become quite narrow as you reach the very back of the mouth cavity (**Figure 4**).
6. Locate the epiglottis—a stiff flap protruding upward in front of the openings to the trachea and the esophagus.
7. Use your probe to locate the opening to the trachea, the glottis, and the opening of the esophagus.
8. Note the presence of teeth in the upper jaw. The roof of the mouth is divided into a front portion called the hard palate and a rear portion called the soft palate.

(f) On a diagram of the mouth, label the teeth, tongue, hard and soft palates, epiglottis, and opening to the esophagus.

(g) What are the functions of the tongue, esophagus, epiglottis, and trachea?

(h) To what organ is the esophagus attached?

9. *Abdominal Cavity:* Place your fetal pig on its back in the dissecting tray. Tie a string around one of the front legs. Run the string under the tray and tie it to the other front leg. The string should be tight enough to hold the legs well apart. Repeat the procedure with the hind legs.
10. Carefully examine **Figure 5**. With a pair of scissors, make a small cut into the abdomen at position "X," just in front of the umbilical cord. Cut just deep enough to cut through the wall of the abdomen.
11. Insert the tip of the scissors into the small cut and then extend the cut toward the front of the pig along line "1." Just barely insert the tip of the scissors to prevent damaging the internal organs. Stop your incision when you reach the rib cage.

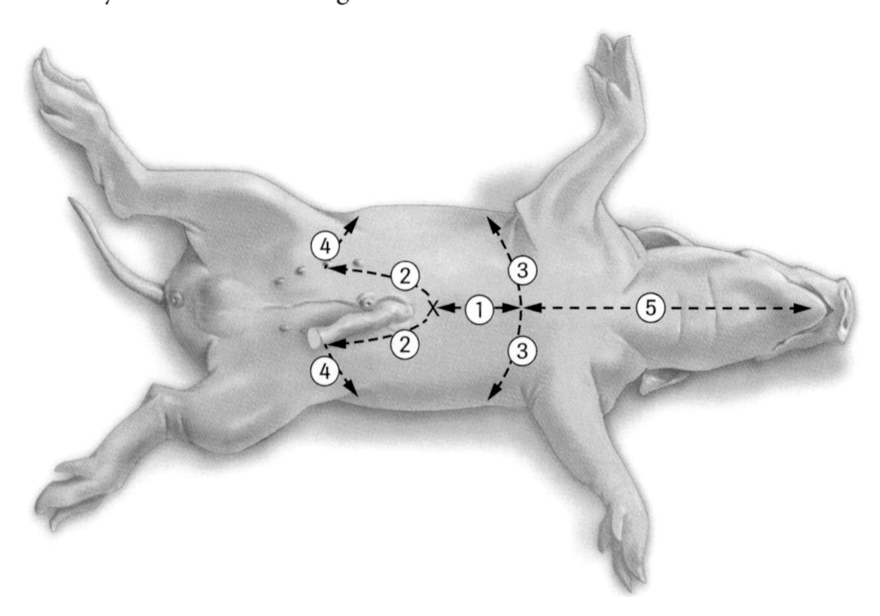

Figure 5
Incisions for opening the abdominal and chest cavities

12. Beginning again at location "X," extend the incision to each side of the umbilical cord and then back along each side indicated by "2."
13. Gently lift up on the umbilical cord. You will see the umbilical vein extending forward inside the abdominal cavity. Cut the umbilical vein using scissors, and fold the umbilical cord and flap of skin back between the hind legs.
14. Next make cut "3," through the wall of the abdomen just below the rib cage around to the back. Make similar cuts at "4," through the wall of the abdomen, keeping just forward of the hind legs.
15. Fold back the flaps of the body wall and pin them to the bottom of the tray. Carefully cut through any membranes that are holding them to the internal organs. Your view should be similar to that in **Figure 6**.
16. Examine the brownish liver, the spleen, and the large and small intestines that are readily visible in the abdominal cavity. Anterior to the liver, locate the sheet of muscle called the diaphragm. Gently lift the liver and locate the soft, saclike stomach.

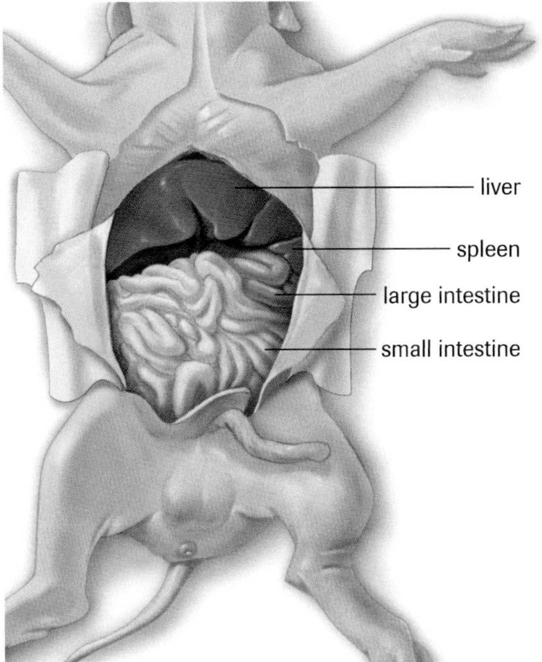

Figure 6
Exposed abdominal cavity of the fetal pig

(i) Identify and label these structures on your diagram.

17. Locate the end of the esophagus as it passes through the diaphragm and enters the stomach. Where the esophagus and stomach meet is a ring of muscle called the cardiac sphincter.
18. Follow the stomach until it joins with the small intestine. A second ring of muscle—the pyloric sphincter—is located at this union.

(j) What are the functions of the cardiac and pyloric sphincter muscles?

19. Try to locate the pancreas. This thin, elongated, cream-coloured organ is found lying between the stomach and the first portion of the small intestine.

(k) What are the functions of the pancreas?

20. Locate the gallbladder. It is a greenish sac found embedded in the under-surface of the liver. It produces bile that passes to the small intestine through the bile duct.
21. Make a cut through the esophagus just below the diaphragm. Gently lift the stomach and intestines up out of the abdomen. Trace the pathway of the large intestine down to the posterior end of the abdominal cavity. Make a second cut through the lowest portion of the large intestine just where it nears the anus. Remove the stomach and intestines and set them down in the dissecting tray.
22. Being very careful not to cut through the intestines, carefully attempt to unravel the entire digestive tract. Find the caecum, a small pouch located at the junction of the small and large intestines.

(l) Measure and record the length of the small and large intestines.

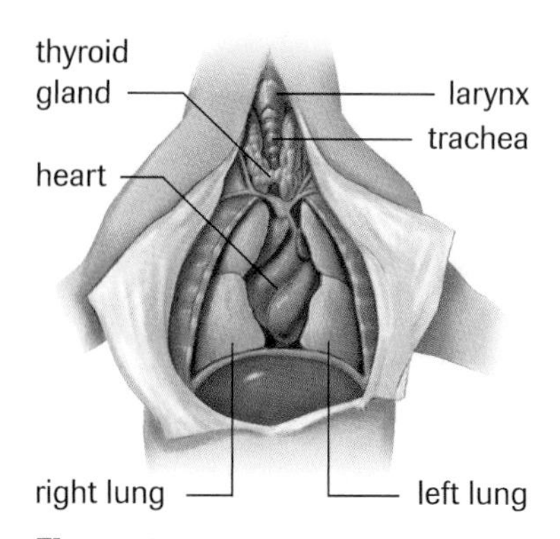

Figure 7
Close-up of the thoracic cavity and throat

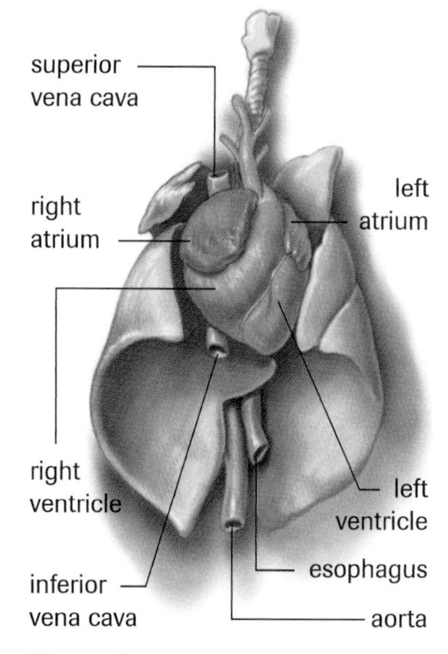

Figure 8
View of the lungs, heart, and main blood vessels of the fetal pig

23. Find the two large, bean-shaped kidneys located along the dorsal surface of the abdominal cavity.

(m) Identify and label your diagrams accordingly.

24. Identify the two ureters that exit from the kidneys and extend posteriorly. These tubes carry urine from each kidney to the urinary bladder. Trace their path until you reach the bladder, located between the umbilical arteries.

25. *Thoracic Cavity:* Refer to **Figure 5** and make incision "5," cutting forward through the middle of the rib cage into the thoracic cavity, finishing just beneath the chin. Carefully cut along the diaphragm separating it from the lower edge of the rib cage.

26. Carefully use scissors to cut away rib and body-wall tissues until the entire thoracic cavity and throat are exposed as in **Figure 7**.

27. Locate the heart. Using forceps and a probe, remove the pericardium membrane covering the heart.

28. Refer to **Figure 8** to locate and identify the right and left atria and ventricles, the aorta, and the inferior and superior vena cavas.

(n) Label these structures on your diagram and note the function of each.

29. Trace the flow of blood through the heart. Blood from the vena cavas enters the right atrium and is pumped into the right ventricle. From there it travels through the pulmonary arteries to the lungs, and back through the pulmonary veins into the left atrium. From there, it is pumped into the left ventricle and out the aorta.

30. Make a diagonal incision across the heart and expose the inner heart chambers. Compare the thickness of the walls of the right and left ventricles.

31. Locate and identify the organs of the respiratory system, including the larynx, trachea, and lungs.

(o) Label these organs on your diagram.

(p) Describe the texture of the lungs.

(q) Describe the texture of the trachea. What is the function of the cartilaginous rings?

Analysis

(r) Outline any special safety procedures that must be followed when performing a dissection.

(s) What is the function of the liver? Why do you think it is so large?

(t) Indicate the location and function(s) of each of the following (summarize your answer in a table):

(i) gallbladder
(ii) spleen
(iii) small intestine
(iv) large intestine
(v) larynx
(vi) kidneys
(vii) stomach

(u) Explain why the small intestine is so long relative to other organs of the body.

(v) For each chamber of the heart, state the type of blood it contains—oxygenated or deoxygenated.

(w) The thoracic cavity is surrounded by a stiff rib cage, while the abdominal cavity has a softer muscular wall. Explain how such structures are beneficial in each case.

(x) In the fetus, only a small amount of blood travels from the heart to the lungs and back. Why is it not necessary to send all the blood to the lungs?

(y) At what point in time will the pig first begin to make full use of the lungs? What will take place in the umbilical cord at that time?

(z) Mammals share many similar features. Use the knowledge you gained from your dissection to identify the organs of the cat in **Figure 9**.

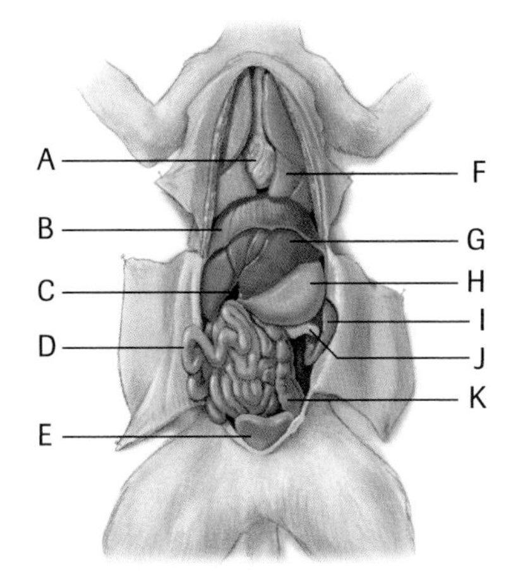

Figure 9

Evaluation

(aa) What advantage(s) is (are) there to dissecting a fetal pig rather than other mammals? Why are pigs similar in anatomical structure to humans?

(bb) What difficulties arose during your dissection? Which system(s) was (were) difficult to view with great detail?

(cc) What other type of equipment or techniques might be used to improve your dissection?

Synthesis

(dd) The anatomy of the fetal pig is similar to that of humans, but there are important differences.

- (i) In what ways do you think we are most similar? Explain your reasoning.
- (ii) In what ways do you think we are most different? Explain your reasoning.

(ee) Some people have a condition known as hiatus hernia. Their cardiac sphincter does not always keep stomach contents from entering the esophagus. What problems might this cause?

(ff) When people "donate a kidney," the incision is usually made through the wall of the back rather than through the front of the abdomen. Based on the fetal pig dissection, explain why this might be preferred.

Extension

(gg) You may wish to dissect the brain and expose the upper portion of the spinal cord of your fetal pig. This is a challenging dissection, as the brain tissue is soft and easily damaged. Seek and obtain teacher approval and additional resources before attempting this dissection.

Understanding Concepts

1. Write a dictionaıy definition of "system," and then explain why this word is used to classify structures and functions within the human body.
2. List eight diagnostic technologies described in this unit. Include a brief explanation of where, how, and why each is used.
3. Describe the functions of the mouth, stomach, small intestine, and gallbladder in the digestion of a hamburger.
4. Why don't the chemicals in the stomach digest the stomach itself?
5. The total surface area of the small intestine is about 300 m^2, approximately the size of a tennis court. Explain this tremendous surface area by describing the structure of the small intestine.
6. Describe how digestive disorders are classified and give two examples of each.
7. Given the information in **Figure 1**, identify the gland and the hormone in each case.

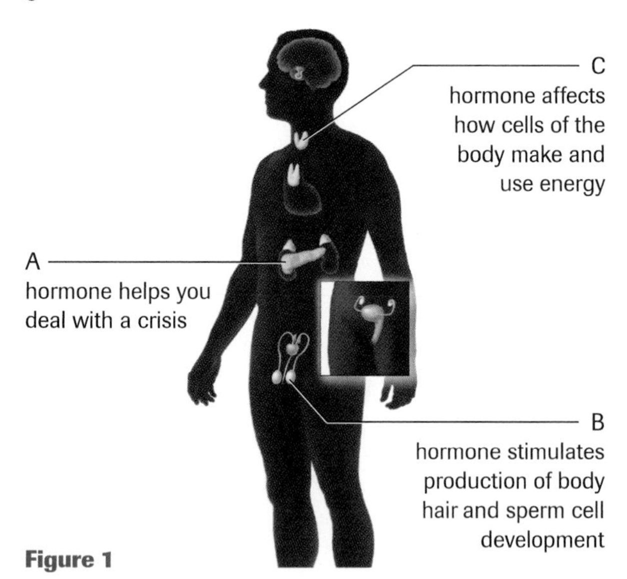

Figure 1

8. (a) What are neurotransmitters and how do they work?
   (b) How do stimulants affect homeostasis?
   (c) How do depressants affect homeostasis?
9. Draw a labelled diagram to illustrate how an impulse travels from one neuron into the next.
10. Determine whether each of the following actions involves the somatic nervous system or the autonomic nervous system.
    (a) A pregnant woman experiences rhythmic uterine contractions once every minute.
    (b) A student walks faster to class upon hearing a bell ring.
    (c) A baby's eyelids close when she falls asleep.
    (d) A teenager holds his eyes shut during a scary scene in a horror film.
11. Describe two similarities and two differences between the endocrine system and the nervous system.
12. Why do we say that seminal fluid is *secreted* by organs of the male reproductive system, but urine is *excreted* by the urinary system?
13. What are the atria and the ventricles? How do they differ in structure and function?
14. "Oxygenated blood is found in all arteries of the body." Is this statement true or false? Give reasons to explain your answer.
15. Name some factors that can result in cardiovascular disorders.
16. Explain what happens in the circulatory system and in the body in general as the heart valves gradually fail.
17. Describe how urea is formed in the liver.
18. Draw and label the following parts of the excretory system: kidney, renal artery, renal vein, ureter, bladder, and urethra. State the function of each organ.
19. What is deamination, and why is it an important process?
20. The following is a random list of processes that occur in the formation and excretion of urine once the blood has entered the kidney. Place these processes in the correct order:
    (a) Urine is stored in the bladder.
    (b) Blood enters the glomerulus.
    (c) Fluids pass from the glomerulus into the Bowman's capsule.
    (d) Urine is excreted through the urethra.

(e) Sodium, glucose, and amino acids are reabsorbed from the nephron.
(f) Urine passes from the kidneys into the ureters.

21. If you drink excess water, your blood will contain more water than usual. What events will occur in the excretory system to compensate for this increased amount of water?
22. Explain how the respiratory and circulatory systems interact within the alveoli.
23. Trace the pathway of an oxygen molecule from the time it enters the body until the time it enters a cell for the purpose of releasing energy from food.
24. What features are common to the gas-exchange surfaces of all large organisms?
25. What are the effects of bronchitis on the respiratory system? How does this affect your ability to obtain oxygen?
26. How does the respiratory system depend on the circulatory system?
27. During mouth-to-mouth resuscitation, exhaled air is forced into the victim's trachea. Exhaled air contains a higher level of CO_2 than atmospheric air. Would the higher level of CO_2 create problems or would it be beneficial? Provide reasons for your answer.
28. The disease emphysema results in a reduction in the number of alveoli within the lungs. How might you expect the person's $VO_{2\,max}$ and their breathing rate to be affected by this disease?
29. Under conditions of heavy exercise, tidal volume is different than during periods of rest. Would it be closer to one's expiratory reserve volume, inspiratory reserve volume, or vital capacity?
30. Describe two endocrine hormones that are secreted only during pregnancy.
31. Briefly describe the following stages in human development: zygote, embryo, and fetus.
32. (a) List three sexually transmitted diseases (STDs), and describe their causes and possible treatments.
    (b) How are the STDs you described in (a) spread from person to person?
33. Describe two components of a joint that help prevent wear and tear.
34. (a) What are tendons used for?
    (b) What are ligaments used for?
    (c) Why are tendons tougher and inflexible?
35. **Figure 2** shows a pair of muscles of the human arm with various components labelled A, B, C, and D.
    (a) Which label points to the flexor muscle?
    (b) Which label points to the extensor muscle?
    (c) Which label points to the origin of a muscle?
    (d) Which label points to the insertion of a muscle?
    (e) What are muscles like the two muscles in **Figure 2** called? What does this name mean?

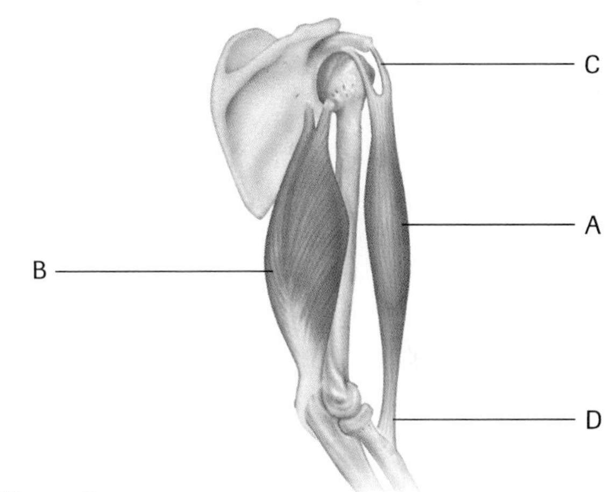

Figure 2

Applying Inquiry Skills

36. People whose pancreas does not make enough insulin have a disease called diabetes mellitus. These people usually have too much glucose in their blood following a meal. The normal range for glucose in blood is 80–90 mg of glucose per 100 mL of blood. A patient with diabetes measures her blood and finds that she has 220 mg of glucose per 100 mL of blood.
    (a) What must the person do in order to get her glucose level in the normal range?
    (b) Insulin is packaged in an amount called a "unit." Each unit will lower the blood glucose level by 3 mg per 100 mL of blood. How many units of insulin does the patient have to take to get her blood glucose level in the normal range?

37. The following experiment was designed to investigate factors that affect fat digestion. A pH indicator is added to the three test tubes, as shown in **Figure 3**, below. (A pH indicator turns pink when the pH is 7 or above and remains clear when the pH is below 7.) The initial colour of each test tube is pink. Predict the colour changes for each of the test tubes shown in the diagram, and provide your reasons.

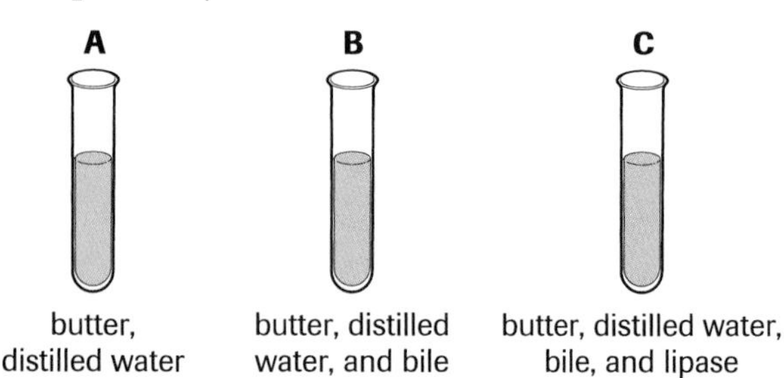

Figure 3

38. The incidence of colorectal cancer in India, where cereal grains form the basic diet, is approximately 4 cases per 100 000. In North America, where people eat the greatest quantities of animal proteins and fat, the incidence rises to about 43 cases per 100 000. What conclusions might you draw from these data? Do you think colon cancer can be eliminated by a change in diet? Justify your answer.

39. Tests were performed on patients A, B, C, and D. Results from the tests are provided in **Table 1**. The results obtained for patient A are considered normal.

Table 1

Patient	Blood pressure (mm Hg)	Cardiac output (L/min)	Glucose in urine (g/100 mL)	Urine output (mL/24 h)
A	120/70	5.0	0.00	1500
B	130/80	5.5	0.00	1700
C	115/70	4.5	0.06	1950
D	90/55	3.0	0.00	500

(a) Which patient could have a circulatory problem?
(b) Explain how a circulatory problem could affect urine output.
(c) Explain why the urine output of patient C is elevated.

40. People who want to have a baby but experience difficulty becoming pregnant sometimes use commercially available ovulation tests. In one type of ovulation test, a chemical is added to a sample of a woman's urine to test for the presence of a particular hormone. **Figure 4** shows changes in the levels of FSH and LH during the menstrual cycle. These changes can be detected in a woman's bloodstream and urine.

Analyze the graph and determine whether the ovulation test should check for the presence of high levels of FSH or LH. Provide reasons for your choice.

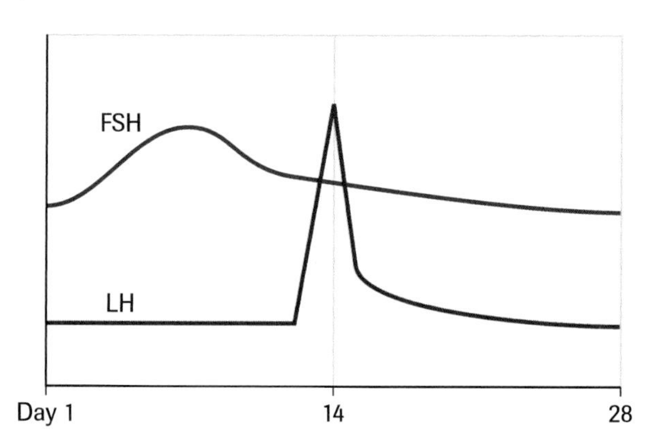

Figure 4

41. The greatest difficulty in developing artificial hands and arms has been a lack of sensory input. Design a safe activity that will allow students to experience the difference between carrying out physical activities with full sensory input and doing so with limited sensory input.

Making Connections

42. Alcohol, marijuana, cocaine, and heroin are sometimes called recreational drugs.
(a) Why do you think these drugs have earned this name?
(b) Is this a good or bad way to describe these drugs? Explain your answer.

43. A popular reducing diet recommends the consumption of meat or fish in increased amounts every day. Individuals who follow this diet do lose several kilograms in only a few weeks. Why would physicians warn that this diet is a dangerous "fad" diet? How can body mass be reduced safely?

44. Bulimia has both similarities to and differences from anorexia nervosa. Research the cause, symptoms, and treatment of this digestive disorder, and compare it with anorexia nervosa.

www.science.nelson.com

45. Heartburn, or acid indigestion, occurs when stomach acids back up into the esophagus, burning its lining. Antacids can be taken to reduce the burning sensation. Some people take an antacid tablet after each meal. What concerns would you have with this practice?

46. Research the progress being made in the area of heart or kidney transplants. Prepare a report that includes how donors and recipients are selected, the kind of preparation that is necessary, how the operation is performed, the postoperative care required, and how organ rejection is controlled.

www.science.nelson.com

47. One of the major functions of skin is excretion. Perspiration or sweat is mainly water, but it also contains inorganic and organic substances. Research the substances that are excreted in sweat.

www.science.nelson.com

48. In some countries, kidneys are sold for transplant. Do you believe that this practice is acceptable? Explain your answer.

49. List three careers that require a thorough understanding of the respiratory system. Explain why this knowledge is needed in each case.

50. Research suggests that air pollution may kill 5000 people a year in Canada. In what occupations and in what living environments might you be most affected by air pollution?

51. You have just moved into an older home. It has a gas furnace and a hot-water heater, as well as a wood-burning fireplace. What precautions should you take to ensure a safe and healthy breathing environment?

52. Registered massage therapists are "drugless practitioners" who treat disorders of the musculoskeletal system.
    (a) Why are these health care workers called drugless practitioners?
    (b) Conduct library and/or Internet research to obtain career information about registered massage therapy. Include information regarding professional duties and educational licensing requirements.

www.science.nelson.com

53. Bone marrow transplants are used to treat leukemia, a form of cancer affecting blood cells. Conduct library and/or Internet research to answer the following questions about bone marrow transplants:
    (a) From which bones of the body is bone marrow usually taken for the transplant?
    (b) How is bone marrow obtained, and what effect, if any, does the removal of bone marrow have on the donor?
    (c) How successful are bone marrow transplants in treating leukemia?

www.science.nelson.com

54. Conduct Internet and/or library research to discover the latest advances in prosthesis research aimed at increasing the sensory input of a replacement limb.

www.science.nelson.com

unit 4

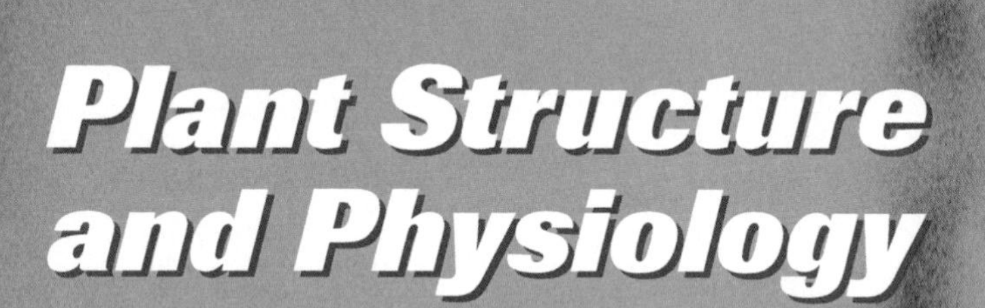

Plant Structure and Physiology

Plants are extremely important to human life. Plants provide us with foods, fibres, medicines, and many other valuable products. To benefit fully from plant products, we need to know about the different kinds of plants in our world. We must understand their life cycles, their structures, and the types of tissues and cells that they contain. We need to know how they grow, and which factors control and affect their growth. We also must be aware of the role plants play in the natural environment.

Most of us are familiar with the use of plants for food, such as potatoes and apples, and for fibres, such as cotton and lumber. Many other valuable products also come from plants. For example, natural rubber (latex) is produced from the milky white sap of rubber trees. Plants produce many medically important substances, such as the cancer-fighting medicines vincristine and vinblastine, which were first isolated from the leaves of the rosy periwinkle in the 1960s. The survival rates of cancers such as Hodgkin's disease and certain types of childhood leukemia increased dramatically following the discovery of these medicines.

In this unit, you will gain a better understanding of and appreciation for the beneficial roles plants play in our lives. You will also learn of the many careers associated with studying, growing, harvesting, and processing plants and plant products.

Overall Expectations

In this unit, you will be able to

- demonstrate an understanding of the diversity of plants, and of their internal transport systems, reproduction, and growth;
- analyze the factors influencing the growth and maintenance of plants, using appropriate laboratory equipment and techniques;
- evaluate the roles of plants in the urban community, in various technologies and industries, and in natural ecosystems.

Unit 4 Plant Structure and Physiology

ARE YOU READY?

Prerequisites

Concepts

- plant cell structure
- plant reproduction
- plant parts
- plant ecology

Skills

- design and conduct controlled experiments
- select and use laboratory materials accurately and safely
- make careful observations and present them in an appropriate format such as images, tables, or graphs
- present informed opinions about the uses and importance of plants

Knowledge and Understanding

1. Plant cells contain many of the same organelles as other living things, which also perform the same important cellular tasks. Like other organisms, plants also have mechanisms for successful reproduction. However, since plants are immobile and photosynthetic, they have a number of unique characteristics. In your notebook, match each plant characteristic listed in **Table 1** with its function.

Table 1 Plant Characteristics and Functions

Plant characteristic	Function
1. nucleus	(a) directs cell activities
2. mitochondrion	(b) structure that produces pollen grains
3. flower	(c) structure in which eggs are formed
4. chloroplast	(d) structure containing a young plant embryo
5. chromosome	(e) structure usually containing seeds
6. cellulose	(f) provides rigid support for cells
7. seed	(g) reproductive organ of many plants
8. ovary	(h) female plant sex cell
9. cell wall	(i) produces the male plant sex cells
10. anther	(j) contains chlorophyll and is the site of photosynthesis
11. egg	(k) contains genes
12. fruit	(l) provides energy for cell processes
13. pollen	(m) main chemical component of plant cell walls

2. **Figure 1** depicts a generalized flowering plant. Draw a similar diagram in your notebook. Add coloured arrows to show the direction of movement of water as it enters, moves within, and exits the plant. Then, label your diagram with as many of the terms given in the word list as you can.

Figure 1
A generalized flowering plant

- light energy
- water
- obtain nutrients and support plant
- location of photosynthesis
- supports above-ground plant
- produces fruit and seeds
- oxygen
- carbon dioxide

3. Many plants are able to reproduce both sexually and asexually. Define and provide an example of each of the following terms:
   (a) vegetative propagation
   (b) cloning
   (c) spore formation
   (d) hermaphrodite
   (e) pollination
   (f) fertilization
4. For each of the following, provide one or more examples of how a plant performs that role in an ecosystem:
   (a) is the main food source for a particular animal species
   (b) provides a home for a particular species of animal
   (c) is competition for other species of organisms
   (d) provides nonfood materials for other species to use
   (e) produces important chemicals to support other life forms
   (f) prevents or reduces soil erosion

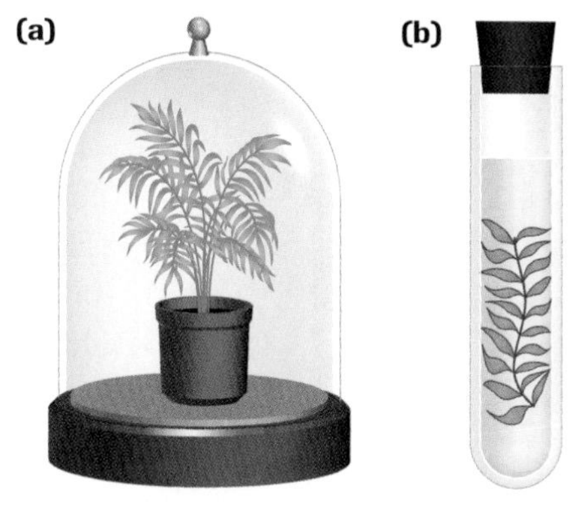

Figure 2
Experimental setup to test for gas production and consumption by **(a)** terrestrial and **(b)** aquatic plants

Inquiry and Communication

5. In an experiment, a potted terrestrial (land-living) plant was sealed in a bell jar (**Figure 2(a)**) and an aquatic plant was sealed in a water-filled test tube (**Figure 2(b)**). Just before the plants were sealed inside, the air in the bell jar and the water in the test tube were tested with limewater. The sealed containers (with plants inside) were then placed in total darkness for 12 h. The air and water were then tested again with limewater; the results of this test are shown in **Table 2**. The sealed containers with plants were then placed under bright lights for an additional 12 h. The air and water were then tested again with limewater; the results of this test are shown in **Table 3**.
   (a) What chemical is tested for using limewater?
   (b) Did the plants in this experiment release this chemical in the dark?
   (c) Did the plants in this experiment release this chemical in the light?
   (d) Relate these observations to the process of photosynthesis.

Table 2 Plants in Dark: Reaction with Limewater

Sample	Before plants added	With plants in darkness for 12 h
air in bell jar	no reaction	milky white
water in test tube	no reaction	milky white

Table 3 Plants in Light: Reaction with Limewater

Sample	With plants before light	With plants under light for 12 h
air in bell jar	milky white	no reaction
water in test tube	milky white	no reaction

Making Connections

6. Recall that Earth's major biomes include boreal forests, tundra, grasslands, temperate deciduous forests, tropical rain forests, and deserts. Each biome is determined by its dominant plant species, which in turn strongly influence the lives of the other organisms, including humans, in that biome.
   (a) How might the plants in each of these biomes directly influence the daily lives of humans?
   (b) How might the plant communities influence the abundance and diversity of the food supply in each biome?
   (c) Provide examples of ways in which humans have altered and/or replaced the main plant species in each of these biomes.

GETTING STARTED

Figure 1
A Maya farmer shares a drink from a freshly cut water vine with a visiting student from Ontario.

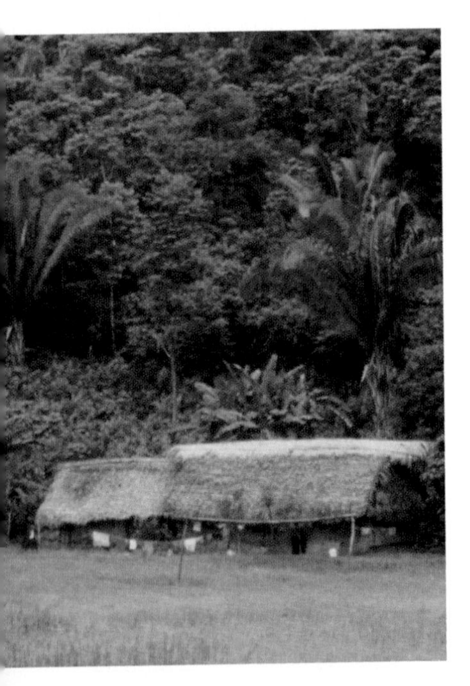

Figure 2
The thatched roofs of Maya homes are covered with leaves of the cohune palm.

In Belize, Central America, several Maya families live in a small village surrounded by dense, tropical rain forest. The daily lives of these people are closely tied to their knowledge of plants. They make their living with traditional farming practices known as *milpa*. *Milpa* involves slashing and burning small patches of forest, then planting food crops in the burned area. At relatively short time intervals, each farm plot is left alone to return to its natural state, which allows regrowth of the forest and renewal of soil nutrients.

The men and older boys in this Maya village spend most of the day working their plots with machetes to clear brush, cut back weeds, and cultivate the soil. They plant a wide variety of crops, including bananas, plantains, pineapples, beans, sugar cane, rice, and corn. They must also watch for animals such as deer that try to take advantage of these rich food supplies.

Conditions in the forest can be difficult. The climate is extremely hot and humid, so the workers must have a steady supply of water. Should they run out, they can use their knowledge of the forest plants to obtain safe drinking water. They know that the tissue of native species of water vine contains relatively large amounts of water, which will flow freely from the end of a stalk that is held upright (**Figure 1**).

The Maya women and girls in the village spend much of their day harvesting, processing, and preparing food. They process corn grain into the flour and starches that they will use to make tortillas. They harvest cacao fruits to obtain cocoa beans, which are used to prepare chocolate drinks. They also gather beans and rice, then process and dry them for storage.

The Maya use plants for many purposes other than food. The outer hard shell of the fruit from calabash trees is hollowed out and used for bowls. Sap from the poisonwood tree may be added to small streams to kill fish, which are then carefully collected, cooked, and eaten. Logwood trees may be harvested for the valuable dyes contained within the innermost wood (heartwood). The walls and floors of Maya homes are constructed from hardwoods cut in the forest, while large leaves of cohune palms are used for thatching roofs (**Figure 2**). Still other palm leaves are used for weaving fine baskets.

Among the most revered members of the Maya village are the bush doctors. Bush doctors have a vast knowledge of the medicinal uses of native plants. For example, they use bark from the gumbolimbo tree to treat the severe chemical burns that result from accidental contact with sap from the poisonwood tree.

REFLECT on your learning

1. Why is it beneficial for ecosystems to have large numbers of different species?
2. List any plants that grow in Ontario that you know how to plant, grow, harvest, and process for food.
3. What might be some of the advantages and disadvantages of growing food crops in the tropics rather than in cooler climates?
4. List some plants grown in Ontario for purposes other than food.

TRY THIS activity: Using Plants in Daily Life

Plants are so much a part of our daily lives that we may overlook them or take them for granted. In this activity, you will examine the rooms of a house owned by a person in Ontario. You will brainstorm items found in the home that are derived in some way from plants.

1. On a full sheet of blank paper, draw an outline sketch that shows the following rooms in a house: bedroom, bathroom, living room, kitchen, dining room, home office, and playroom.

(a) Brainstorm items that are likely to be found in each room and prepare a list. For example, the bathroom likely would contain soap, shampoo, towels, a medicine cabinet, and toilet paper.

2. For each list, identify which items are derived from plants. For each of these items, name the type of plant used to make the item and, if possible, the place you think this plant grows. For example, toilet paper can be produced from conifers grown in northern Ontario. For food items, do not try to list every type of food that comes from plants. Make a short list that includes a few major foods and a few specialty foods, such as flavourings and spices.

3. In the appropriate room outline on your sketch, transfer the names of all the items derived from plants that are used in the room, and any other corresponding information about the items.

(b) Create a table to sort the items on your house outline sketch into the following classes: fibres, wood products, medicines, fragrances, decorative or aesthetic items, nutritional foods, flavourings and specialty foods, and other.

(c) What single plant product is found in the greatest quantity in your sketch?

(d) List the nonfood items on your sketch that you think are extremely valuable. Justify your choices.

(e) List three plants that are used to make many different foods.

(f) List four popular foods that are produced from plants and have low nutritional value.

(g) Describe how each of the products illustrated in **Figure 3** is related to plants.

(h) If they are not already listed, add the following items to the appropriate rooms in your house sketch and to your table: aspirin, cotton, coffee, books, hardwood floors, flour, house plants, and perfume.

(i) In two or three paragraphs, compare the use of plants in daily life by people in Ontario and by the Maya living in Belize.

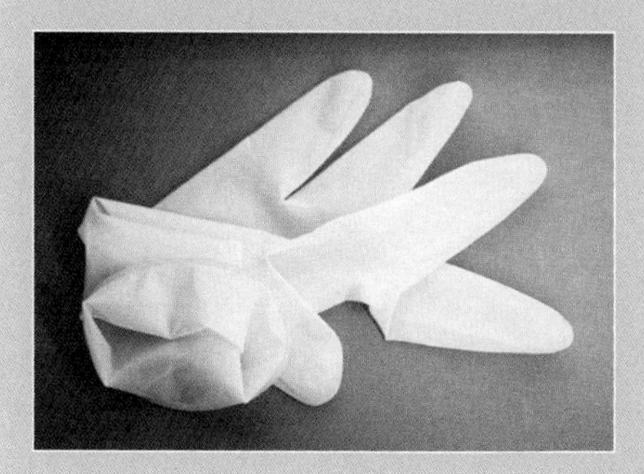

Figure 3
Some common plant products

4.1 Plants and Biodiversity

Plants were the first multicellular organisms adapted to living out of water. More than 500 million years ago, plants first successfully invaded the land. Since then, plants have adapted to a wide range of environmental conditions both in aquatic and terrestrial environments. Today, thousands of different species of green plants carpet much of Earth's land surface. Almost all other life on Earth depends on the ability of plants to use water and carbon dioxide to produce food and oxygen through photosynthesis.

biodiversity a measure of the number and abundance of different species living in a particular ecosystem

Biodiversity refers to the number of different species and the number of individuals of each species that are found in a particular ecosystem. Earth has many different ecosystems, which contain perhaps 15 million species of animals that depend on more than 300 000 species of plants to provide them with food, shelter, and the physical environments in which they live. A single plant species supports an average of 50 different animal species through complex food webs. As a result, loss of even a single plant species can significantly reduce the biodiversity of an ecosystem. The loss of an entire community of plants may threaten other organisms with extinction.

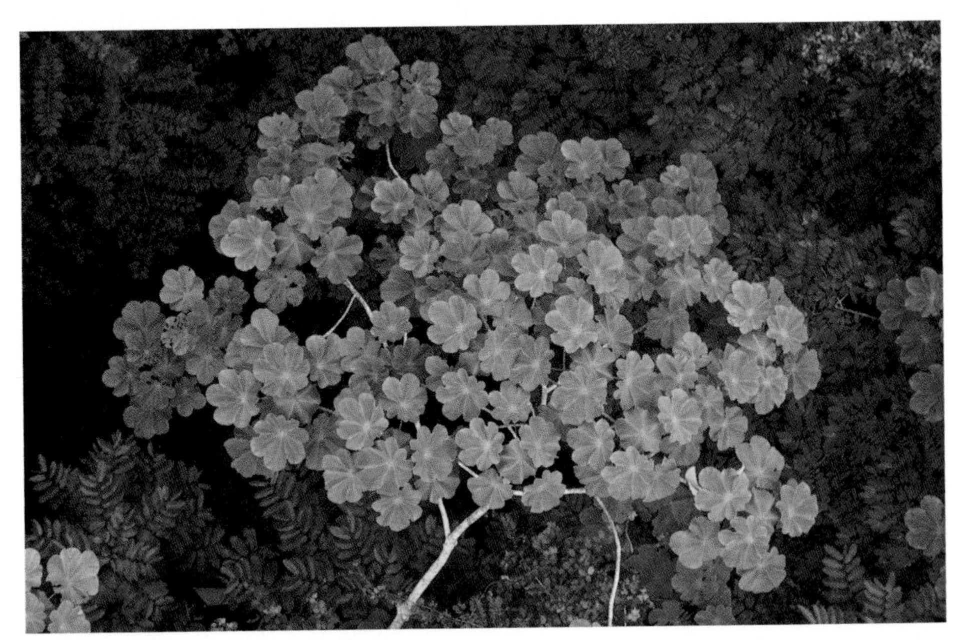

Figure 1
Tropical rain forests, such as this one in Panama, are home to more than half of the world's species.

Figure 2
The green iguana, a common herbivore in tropical regions of the Americas, depends on the biodiversity of plants found in tropical rain forests to survive.

In general, an ecosystem with high plant diversity is able to support a greater number of other species. For example, biological diversity is highest in ecosystems found in tropical rain forests (**Figure 1**). Rain forests are continuously warm and wet, providing ideal environmental conditions for many plant species. As a result, rain forests have a complex physical structure that provides habitat for many different organisms (**Figure 2**). In contrast, biodiversity is lowest in environments with extreme temperatures or low precipitation, such as deserts, mountaintops, or polar regions. Since relatively few plant species can survive such environmental conditions, these natural regions also provide habitat for fewer species overall.

Biodiversity and Human Activity

Biodiversity also benefits human life. We rely on other organisms to provide us with medicines, foods, fibres, and other valuable products. Loss of biodiversity therefore affects us directly, since it reduces the supply of potential new sources of these materials. The primary cause of species loss today is human activity that involves large-scale destruction of natural ecosystems. For example, we grow many of the plants we use in a **monoculture**, a habitat that contains primarily one species of plant. Most monocultures are created through human intervention. A cornfield, a planted lawn, and a plantation of pines (**Figure 3**) are all monocultures. Since overall biodiversity depends on the diversity of plants, monocultures support very few organisms. Other human activities that contribute to loss of biodiversity include clear-cut logging, clearing of land for agricultural use, and flooding of land for hydroelectric power generation. Species loss due to human activity is occurring most rapidly in tropical regions, where biodiversity is greatest.

monoculture a habitat dominated by a single plant species

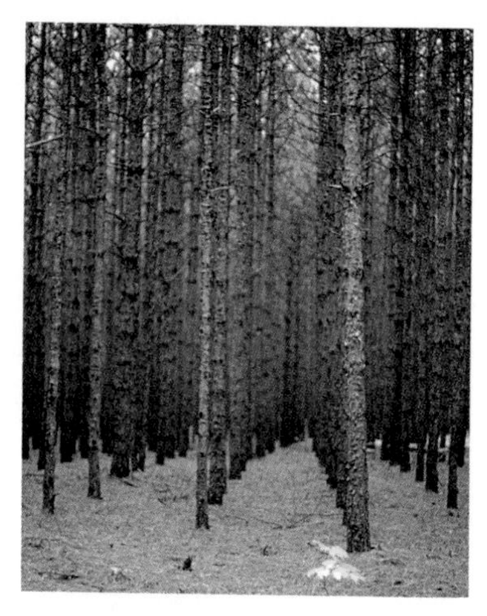

Figure 3
This equal-aged pine plantation supports little biodiversity.

Humans also affect biodiversity by both intentional and accidental introduction of non-native plant or animal species to an ecosystem. In a natural environment, the number of individuals of a species is kept in check by its interactions with other species. When a species is introduced into a new ecosystem, its natural predators, parasites, and diseases may not be present. Local predators and parasites often will not use the introduced species, and it may be resistant to local disease-causing organisms. When these conditions occur, the introduced species out-competes native species, causing them to die out. Introduced species that out-compete native species are called **invasive species**. **Table 1** lists some of the species introduced into North America that have become invasive.

invasive species an introduced species that out-competes native species in an ecosystem

Table 1 Invasive Plants in North America

Purple loosestrife		Eurasian water milfoil	
Japanese knotweed		Kudzu	
Leafy spurge		Water hyacinth	

DID YOU KNOW?

Endangered Plants
In November 2002, scientists reported in the journal *Nature* that as many as one-quarter to one-half of all plant species may be endangered. For example, they estimated that of the 4000 plant species found only in Ecuador, 3500 are seriously threatened.

TRY THIS activity — Exploring Ontario's Biodiversity

Scientists at the Centre for Biodiversity and Conservation Biology conduct research on the interrelationships of species in Ontario's ecosystems and estimate risks of extinction. They also recommend strategies for resource management and conservation programs.

1. Go to the Species at Risk page of the web site for the Centre for Biodiversity and Conservation Biology.

www.science.nelson.com

2. Review the regional checklists of species at risk in Ontario.
   (a) Create a map of the regions shown on the web site. Label the total number of species and the number of plant species that are at risk in each region on your map.
   (b) How many plant species are listed in Ontario's deciduous forests? How many of these are endangered?
   (c) Compare the number of species at risk in southern deciduous forests with the numbers in other regions of Ontario.
   (d) Describe the trend in species at risk as one moves north in the province. Suggest reasons for this trend.
   (e) Prepare a detailed report about two plants of your choice. If possible, include a plant that is endangered in your area.
      (i) Describe each plant and draw a map of its distribution.
      (ii) Outline the main factors threatening the survival of each plant.

Section 4.1 Questions

Understanding Concepts

1. Why are plant species so important to maintaining the biodiversity of an ecosystem?
2. Compare the characteristics of rain forests and monocultures. How do these characteristics influence the biodiversity of these habitats?
3. What is the primary cause of species loss today?
4. What are the potential advantages to the survival of a species that is accidentally introduced into a new ecosystem?

Applying Inquiry Skills

5. Research one of the plants listed in **Table 1**, on the previous page, and find out what criteria scientists used to classify it as an invasive species. Write a brief report outlining your findings.

www.science.nelson.com

Making Connections

6. Compare the advantages and disadvantages of growing plants in a monoculture, and then defend or refute this statement: Monoculture should not be used in agriculture or forestry operations because it is unnatural. Debate your position with a classmate who holds the opposite view.
7. Choose one of the following medicines: atropine, codeine, quinine, digitoxin, aspirin, taxol, *L*-dopa, diosgenin, or colchicine. Conduct research on the medicine you choose and find out its medical use(s) and the name of the plant from which it was first obtained. What are the risks and benefits of using the medicine? Present your findings as an informative pamphlet, such as might be found in a doctor's office.

www.science.nelson.com

Classes of Plants: Nonseed Plants 4.2

In section 2.1, you learned that all the organisms on Earth could be classified according to a system developed by the Swedish botanist Carl Linnaeus. Plants are organisms that are contained in the kingdom Plantae. All of the organisms in this group share the following characteristics: they are multicellular; most are autotrophs; they reproduce both sexually and asexually; and most are terrestrial.

Plants are a very diverse and important group of organisms. However, all the thousands of known plants can be placed into one of two groups: plants that do not produce seeds (nonseed plants) and plants that produce seeds (seed plants). **Seeds** are specialized reproductive structures that contain a plant embryo. This section will focus on the distinguishing characteristics of the two groups of nonseed plants: mosses and their relatives, and ferns and their relatives (**Figure 1**).

seed a specialized plant structure containing an embryonic plant and food supply

(a)

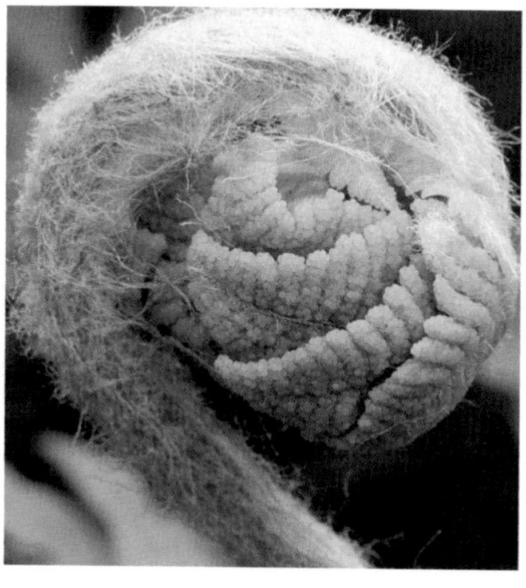

(b)

Figure 1
There are two groups of nonseed plants: **(a)** mosses and their relatives and **(b)** ferns and their relatives.

Mosses and their relatives are nonseed plants that have no **vascular system**, a system of specialized tissues and organs involved in the transport of substances between the cells of the plant body. **Ferns** and their relatives are nonseed plants that do have a vascular system. The plant species within these two groups have adaptations that allow them to survive in very different environments, and so they can look very different from one another.

moss a member of the kingdom Plantae that lacks a vascular system

vascular system a system of specialized cells or tissues that transport water and dissolved materials between cells in the body of an organism

fern a member of the kingdom Plantae that has a vascular system and does not produce seeds

Alternation of Generations

Mosses and ferns also differ in particular aspects of their life cycles. The life cycle of all plants has two phases, one involving sexual reproduction and one involving asexual reproduction. During asexual reproduction, a new individual is produced from one parent plant only, and is genetically identical to that parent. In sexual reproduction, a new individual arises by the union of a male and a female sex cell, and is not genetically identical to the parental generation.

alternation of generations a life cycle that alternates between sporophyte individuals and gametophyte individuals

sporophyte the form of a plant that reproduces asexually

gametophyte the form of a plant that reproduces sexually

Any life cycle that involves sexual and asexual phases is referred to as an **alternation of generations** (**Figure 2**). The **sporophyte** is that form of the plant that is capable of asexual reproduction. The sporophyte produces asexual reproductive cells called spores. The asexual phase in the plant life cycle, when the sporophyte is present, is referred to as the sporophyte generation. A spore develops into a **gametophyte**, which is that form of the plant that is capable of sexual reproduction. Gametophytes can be male or female plants, or a single plant containing male and female parts. The gametophyte generation is the sexual phase in the plant life cycle, when the gametophyte is present. A gametophyte produces either or both sperm and egg cells, which are specialized sexual reproductive cells. When an egg and sperm unite during fertilization, they form a zygote, or fertilized egg. The zygote then undergoes division to become an embryo, which will develop into a new sporophyte.

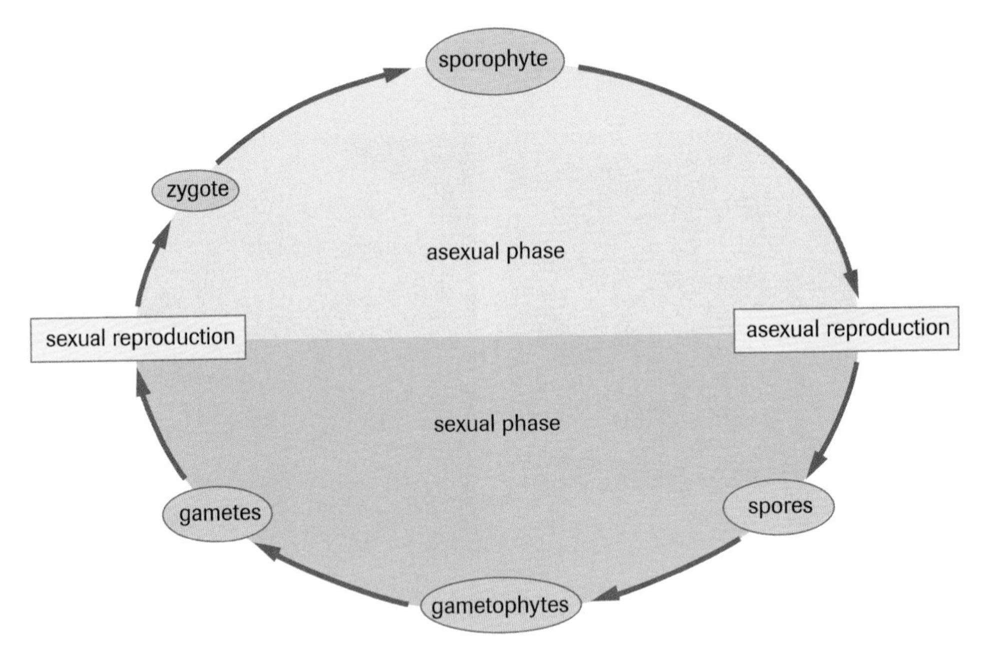

Figure 2
The life cycle of all plants involves an alternation between a sporophyte generation and a gametophyte generation.

Mosses and Their Relatives: Bryophytes

bryophyte a member of a group of nonvascular plants that includes species of mosses, liverworts, and hornworts

rhizoid a simple rootlike filament used to absorb water and anchor mosses to surfaces

The scientific name for plants that lack vascular systems is **bryophytes**. There are more than 20 000 species of bryophytes; among them are the more commonly known mosses and the lesser-known liverworts and hornworts. Bryophytes are relatively simple plants that lack true leaves and roots. Instead, they have **rhizoids**, which are small rootlike structures that absorb water in much the same way as a sponge. Since they have no vascular system, they depend on osmosis, diffusion, and active transport to move water and nutrients between cells. They are therefore unable to lift water to any significant height above the ground, and so bryophytes do not grow very tall. Nevertheless, bryophytes are found in a wide range of ecosystems.

Life Cycle of Mosses

The life cycle of a moss is illustrated in **Figure 3**. The sporophyte produces spores within specialized structures called sporangia. The mature spore begins to grow or germinate, and develops into a gametophyte. Note that the gametophyte plant does not resemble the sporophyte plant. The gametophyte produces both sperm and eggs, which unite during fertilization to form a zygote. The sporophyte then develops within the gametophyte female.

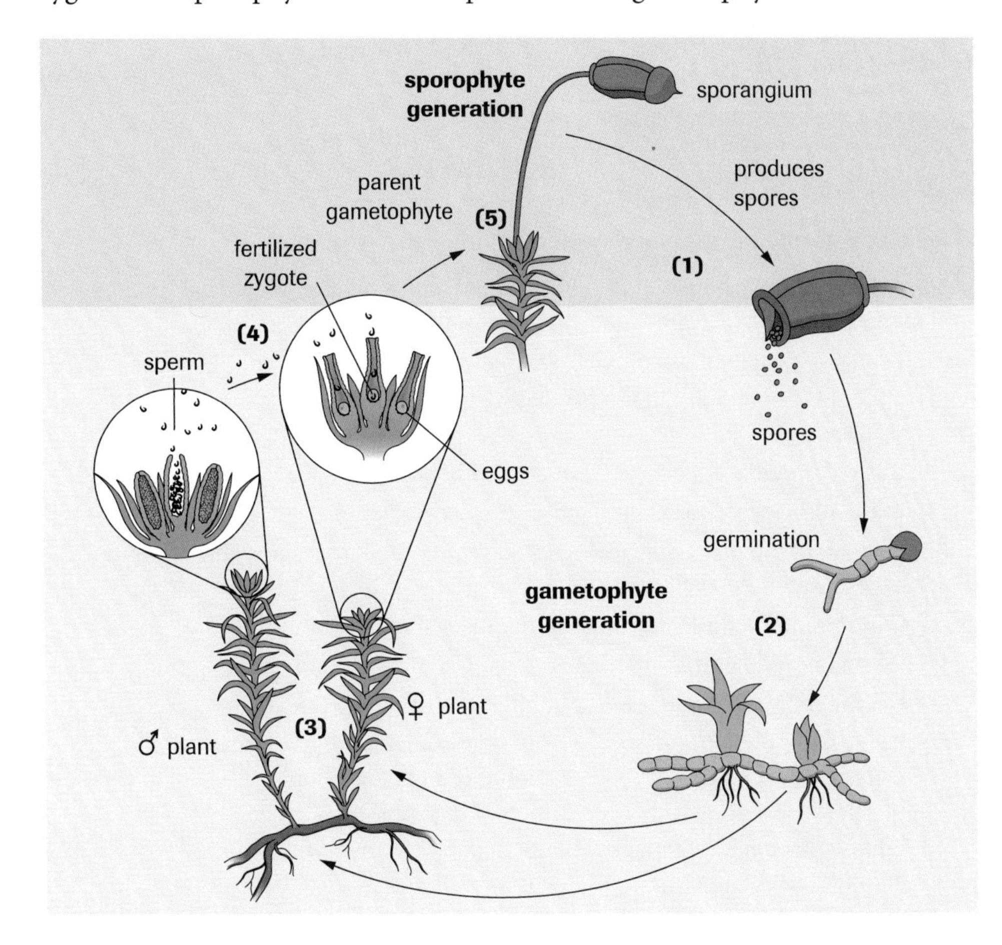

Figure 3
The moss life cycle

(1) Sporophyte produces and releases asexual spores from within a sporangium.
(2) Spores germinate and develop into gametophytes.
(3) The mature gametophyte produces egg and sperm cells.
(4) Swimming sperm reach and fertilize the eggs, producing a zygote.
(5) The zygote develops into a sporophyte.

Note: Although a sporophyte is a separate individual, it remains attached to, and appears to be part of, the female gametophyte.

Before fertilization can occur, the sperm cells must first swim over to the egg cells. They can do this only after a rainfall, when a thin layer of water covers the surface of the plants. Without this moisture, bryophytes are unable to complete their life cycle. However, the spores of mosses and their relatives are easily transported long distances by wind, which helps bryophytes to find new habitats.

Some examples of the bryophyte gametophytes are shown in **Figure 4(a)**. **Figure 4(b)** shows some examples of bryophyte sporophytes. Note that the green individuals we usually identify as moss are gametophytes. Sporophytes are also unique individuals, even though they may appear to be part of the female gametophyte.

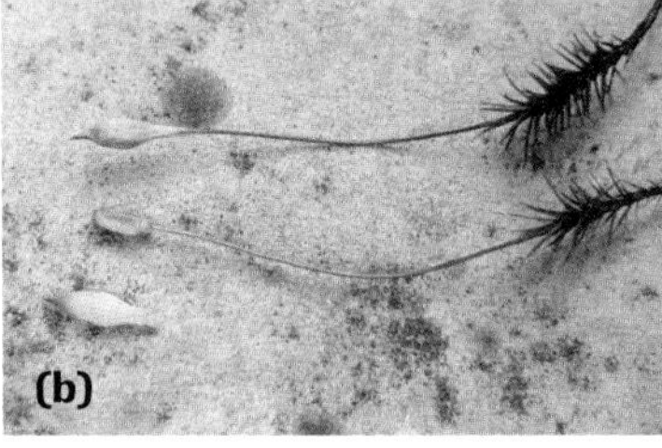

Figure 4
(a) The easily recognized green gametophyte of one species of moss
(b) The asexual sporophyte stage of another species of moss

Peat Moss

The sphagnum mosses that form the vast bulk of material in peat bogs are one of the most economically important bryophyte species. The low pH, cool temperatures, and lack of oxygen in these bogs prevent the moss from decomposing. As a result, large amounts of sphagnum moss accumulate over many years. This material is harvested, dried, and sold as peat moss. Dried peat moss is commonly used as a soil conditioner in gardening and agriculture (**Figure 5**). In parts of northern Europe, Asia, and Canada, peat moss is also used as a fuel.

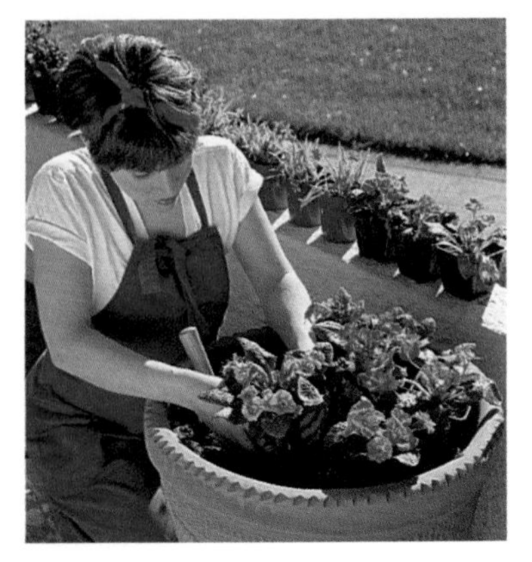

Figure 5
Peat moss is often used to slow down water loss from soil in containers.

CAREER CONNECTION

Garden centre employees (**Figure 5**), landscapers, and farmers may carry out soil tests to determine if they need to improve the physical and chemical characteristics of soil.

TRY THIS activity — Peat Moss

Peat moss is often used in garden and potting-soil mixes. Why?

Materials: 50-mL sample of dried peat moss, small beaker, pH indicator paper, electronic balance, distilled water

1. Obtain a sample of dried peat moss.

(a) Determine and record the mass of the dried peat moss and the pH of distilled water.

2. Soak the peat moss overnight in a small beaker of distilled water.
3. In the morning, pour off and collect the excess water.

(b) Determine the mass of water absorbed by the peat moss and the pH of the excess water. Record your observations.
(c) Calculate the ratio of wet mass to dry mass for the sample.
(d) Is peat moss effective at absorbing and trapping moisture?
(e) Did the peat moss affect the pH of the water?
(f) If you were in charge of trucking large loads of peat moss from northern Ontario to markets in southern Ontario, what actions could you take to save fuel?
(g) What are some problems that might occur if too much peat moss were added to garden soil?

Figure 6
Ferns are common in the temperate forests of coastal British Columbia.

Ferns and Their Relatives

Many millions of years ago, ferns and related plant species were among the first to grow to a large size and dominate Earth's landscape. This was in large part because plants in this group possess a vascular system. This greatly enhances their ability to transport water and nutrients among cells of the plant. Ferns also have large leafy **fronds**, which give them a much larger surface area for gas exchange and photosynthesis than bryophytes. Many ferns also have **rhizomes**, which are stems that run under the surface of the soil. Ferns generally prefer warm and moist environments, and so are most abundant in tropical and temperate rain forests. Some species of ferns grow on another plant, such as on the trunks of larger trees (**Figure 6**). Horsetails are also members of ferns and their relatives (**Figure 7**). Horsetails are commonly found in gardens, in fields, and along roadsides.

frond the leaf portion of a fern

rhizome an underground stem

Life Cycle of Ferns

Figure 8 outlines the general life cycle of a fern, in which there is a sporophyte generation and a gametophyte generation. In ferns, the sporophytes are large, dominant, and photosynthetic, whereas the gametophytes are small (often less than 1 cm in diameter) and may go entirely unnoticed.

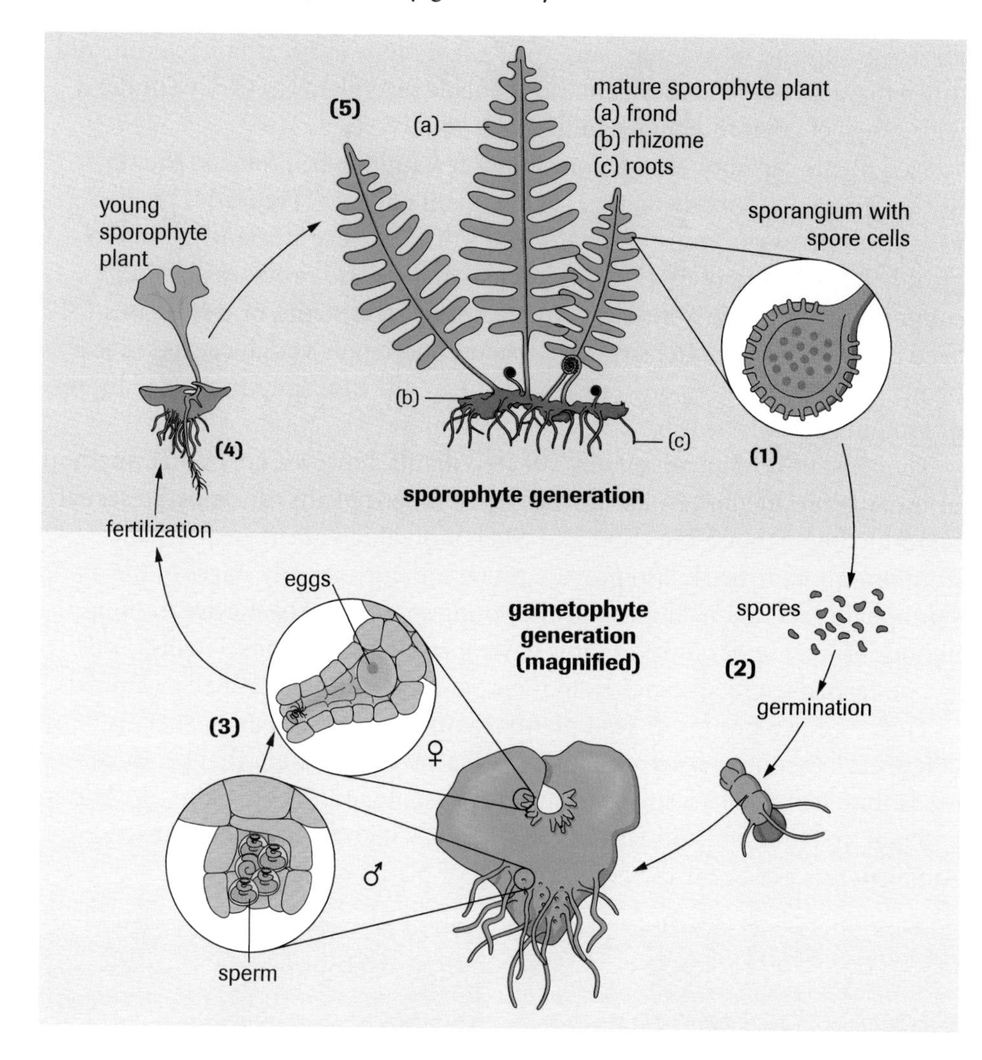

Figure 7
Horsetails have many uses in traditional Native medicine.

Figure 8
The fern life cycle

(1) Spores are produced in sporangia, which are located on the fern fronds of the mature sporophyte.
(2) The spores are released and develop into a small, heart-shaped gametophyte.
(3) The gametophyte produces sperm and egg cells.
(4) Mature sperm cells are released and unite with egg cells, forming a zygote.
(5) The zygote develops into a sporophyte. As the sporophyte matures, the gametophyte is destroyed by the growth of the rhizome and roots.

Section 4.2 Questions

Understanding Concepts

1. Place the following steps in the life cycle of a nonseed plant in the correct order:
   (a) A male gametophyte releases sperm.
   (b) A mature sporophyte produces asexual spores.
   (c) An egg is fertilized by a sperm.
   (d) A young sporophyte grows from a female gametophyte.
   (e) Spores are released and germinate on moist soil.
   (f) A female gametophyte produces eggs.
2. What characteristic limits the size of mosses?
3. Why do mosses require wet conditions in order to reproduce?
4. Explain why farmers or gardeners might add peat moss to their soil.

Applying Inquiry Skills

5. A plant biologist collects a tiny green plant growing in a tropical rain forest. The age and the life-cycle stage of the plant are not clear. Outline the steps you would need to take in order to determine whether the plant is a moss or a fern.
6. Predict whether repeatedly running water through peat moss would change its pH. Give reasons to support your prediction.

4.3 Classes of Plants: Seed Plants

(a)

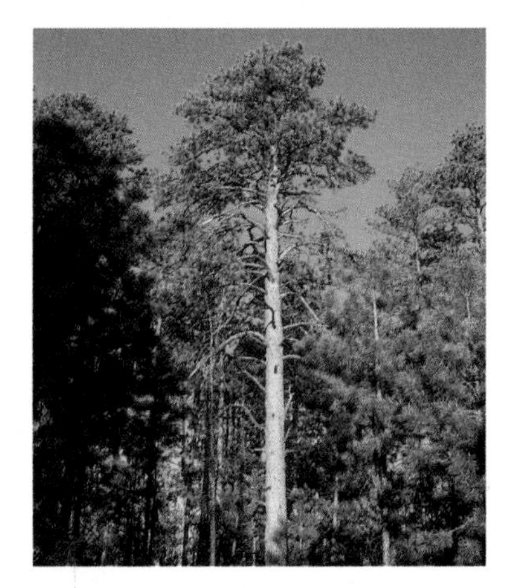

(a)

Figure 1
Seed plants range in size from **(a)** tiny duckweeds, only a few millimetres across, to **(b)** towering conifers such as this white pine growing in Temagami, Ontario.

All seed plants have the structures that people commonly associate with plant life: roots, stems, and leaves. As with nonseed plants, seed plants are also a very diverse group and have a huge variety of adaptations in size, shape, colour, and other characteristics. These adaptations enable seed plants to survive under a wide range of environmental conditions.

Seed plants can vary in size from only a few millimetres, such as the tiny duckweed plant, to many metres, such as giant conifers (**Figure 1**). Their vascular systems can transport water and nutrients very efficiently, often to great heights. In many species, the vascular system also provides physical support and gives rise to woody growth and massive stems, or tree trunks. Seed plants also have extensive root systems and highly variable leaves. Seed plants provide us with many products, and are the most important producers of food in most of the world's terrestrial biomes.

The most important adaptations of seed plants, however, are in their method of sexual reproduction. Unlike nonseed plants, seed plants can undergo sexual reproduction without free water and when individuals are at great distances from one another. Seeds also provide protection for the early stages of the sporophyte generation, and enable the young sporophyte to survive for long periods of time under unfavourable environmental conditions. Finally, seeds can be carried a long distance from their parent by the wind, water, or animals.

There are two groups of seed plants: gymnosperms and angiosperms (**Figure 2**). **Gymnosperms** are plants with a vascular system that produce seeds that are not in an enclosed structure. **Angiosperms** are plants with a vascular system that produce enclosed seeds, which form within flowers. Angiosperms are by far the largest group of plants on Earth.

gymnosperm a member of the kingdom Plantae that has a vascular system and produces naked seeds

angiosperm a member of the kingdom Plantae that has a vascular system and produces enclosed seeds within flowers

(a)

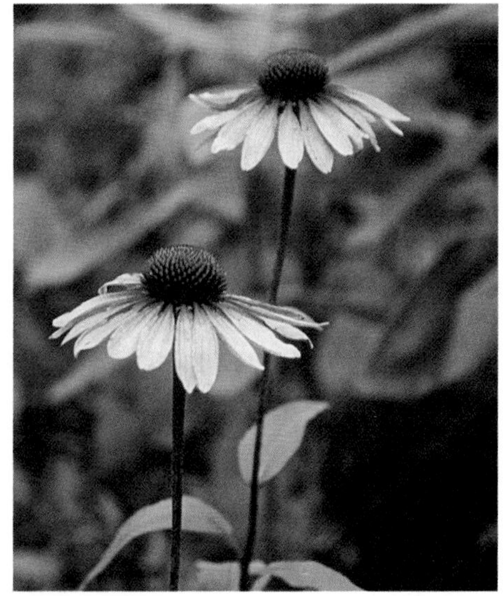

(b)

Figure 2
The two main groups of seed plants are **(a)** gymnosperms, most of which are conifers, and **(b)** angiosperms, which are also called flowering plants.

DID YOU KNOW?

Trees and Flowers
The majority of tree species are flowering plants. Many common trees have small, pale flowers that usually go unnoticed.

Gymnosperms

All living gymnosperms are **woody** (wood-producing) trees or shrubs. Gymnosperms include many species of large **conifers** such as pines, spruces, redwoods, and cedars. Most conifers have needle- or scalelike leaves and are **evergreens**, meaning that they retain their leaves throughout the year. Needles have a small surface area and a thick, waxy coating. These characteristics reduce water loss, helping conifers to survive harsh environmental conditions during long, dry winters. The roots of conifers are relatively shallow and extend over large areas, helping to anchor the trees. This enables them to grow in areas where the soil is thin.

Gymnosperms, such as the species of pine, spruce, fir, and cedar in the boreal forests and temperate rain forests of Canada, are the dominant plant species in many ecosystems. Coniferous trees provide about 85% of all the wood used in building and furniture construction, and the pulp-and-paper industry uses millions of tonnes of conifers each year to make paper products. Conifers also provide us with products such as varnishes, turpentine, disinfectants, and fuels.

woody describes a plant in which some parts are composed of wood

conifer a plant that reproduces sexually using specialized male and female cones

evergreen a plant that retains its leaves throughout the year, such as pines and spruces

DID YOU KNOW?

Taxol

The potent anticancer drug taxol was first obtained from the bark of Pacific yews, a conifer that grows in temperate rain forests.

Life Cycle of Gymnosperms

As with all other plants, the life cycle of gymnosperms involves alternation of generations. The sporophytes are the large trees and shrubs, many of which bear cones. Unlike the gametophytes of mosses and ferns, the gametophytes of gymnosperms are not free-living organisms. Instead, they are tiny, microscopic plants found on specialized structures on the sporophyte. The female gametophyte is found inside the large female cones of the mature sporophyte. The male gametophytes are the **pollen grains**, which are found in the small male cones. Male sex cells are produced within the pollen grains, which are carried by the wind to the female cones. The male sex cells are then released, whereupon they fertilize the egg cells inside the microscopic female gametophytes within the cones. After pollination, each zygote forms a small embryo within a seed. Upon maturity, the seed is released from the cone and develops into a sporophyte.

pollen grain a specialized plant structure from which male sex cells are released

CAREER CONNECTION

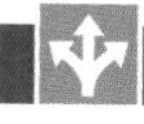

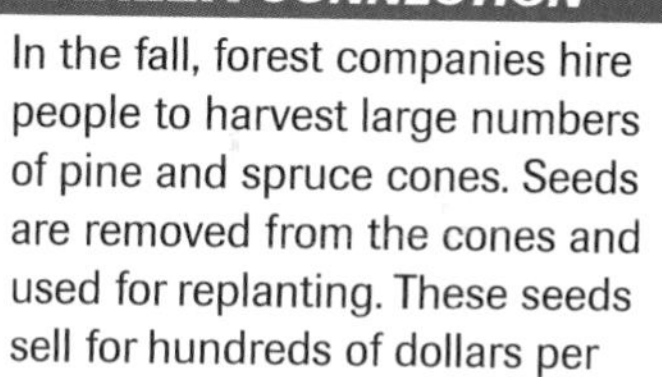

In the fall, forest companies hire people to harvest large numbers of pine and spruce cones. Seeds are removed from the cones and used for replanting. These seeds sell for hundreds of dollars per kilogram.

herbaceous plant a nonwoody plant, such as grasses and tulips

Angiosperms: Flowering Plants

There are more than 250 000 known species of angiosperms—more than all other plant species combined. New species are being discovered almost daily. Angiosperms are commonly known as flowering plants, and are an extremely diverse group of organisms (**Figure 3**). Many angiosperms are **herbaceous plants**, which are plants that lack woody tissue. However, most tree species are also flowering plants.

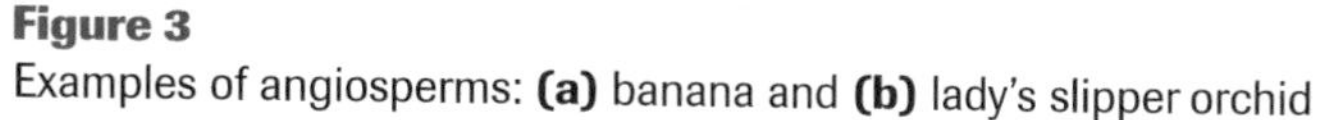

Figure 3
Examples of angiosperms: **(a)** banana and **(b)** lady's slipper orchid

nectar a sweet liquid produced by flowers that attracts insects or other animals

pollinate to transfer pollen from one plant to another

dormant a condition marked by no or very slow growth that is overcome only by specific environmental conditions

Many angiosperms produce large, colourful flowers such as those on tulips, cherry trees, and roses. The large flowers attract insects, birds, and other organisms to feed on the sugary **nectar** produced by the flower. During feeding, these organisms transfer pollen from one flower to another—they **pollinate** the plants. Other angiosperms have small, inconspicuous flowers. These species are usually pollinated by wind. Examples of wind-pollinated angiosperms include tree species such as maple, birch, oak, and poplar. Grasses are also wind-pollinated angiosperms, and include the world's most important food crops—wheat, rice, and corn.

Seeds vary considerably in size (**Figure 4**). Some are barely visible to the unaided eye and can be carried thousands of kilometres by the wind. One of the largest known seeds, the coconut, is so heavy it could kill a human if it were to fall from the tree. Many seeds can stop growing, or remain **dormant**, for long periods of time. Dormancy is broken only under specific environmental conditions. The seed then absorbs water and expands, and the embryo begins to grow rapidly.

Figure 4
Seeds vary considerably in size.

monocot an angiosperm that produces seeds containing embryos with one cotyledon

cotyledon a leaflike structure of the embryo of a seed plant

dicot an angiosperm that produces seeds containing embryos with two cotyledons

endosperm a tissue in angiosperm seeds that stores food

seed coat a protective layer covering a seed

Monocots and Dicots

Angiosperm species can be further subdivided into two important groups, monocots and dicots. **Monocots** produce seeds in which the embryo has only one **cotyledon**, or seed leaf. **Dicots** produce seeds in which the embryo has two cotyledons. **Figure 5** illustrates typical monocot and dicot seeds. Note that the plant embryo makes up only a small portion of the seed. The bulk of most dicot seeds is taken up by the two cotyledons, which store food for the developing embryo. The bulk of most monocot seeds is instead taken up by the **endosperm**, a tissue that stores food material. In both monocots and dicots, a protective layer known as the **seed coat** surrounds the entire seed.

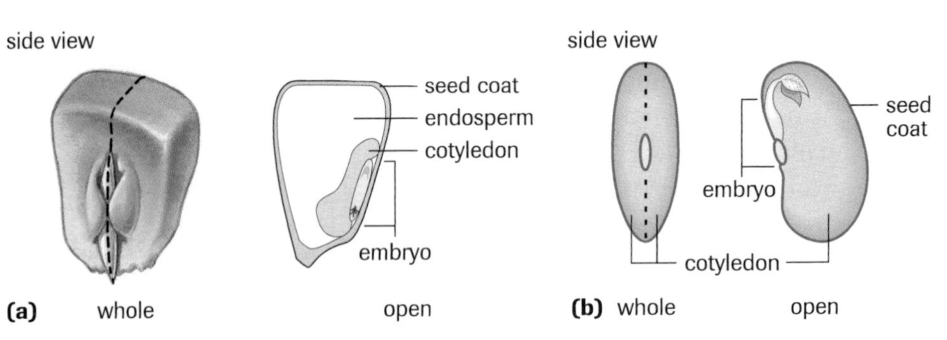

Figure 5
The seeds of **(a)** corn, a monocot, and **(b)** bean, a dicot

DID YOU KNOW?

Grass Seeds
The seeds of grasses are actually a special type of fruit, a caryopsis. Since the seed takes up most of the caryopsis volume, it is often referred to simply as a seed.

Life Cycle of Flowering Plants

The gametophytes of angiosperms are found in the flowers of the plant, and so are hidden from view. The male gametophytes are the microscopic pollen grains, and are produced on the **stamen**, the male part of the flower. Stamens consist of a thin stalk, called a filament, and an **anther**, where the pollen grains are formed. The female gametophyte is found within the **pistil**, the female part of a flower. The pistil consists of three main parts: stigma, style, and ovary. The **style** is a stalk connecting to the ovary, and the **stigma** is the sticky surface of the style. The egg cell is found in the **ovary**.

The pollen germinates and produces a pollen tube, which grows down into the ovary and fertilizes the enclosed egg cell (**Figure 6**). After fertilization, a seed containing an embryo and a food supply forms within a **fruit**, a structure that develops from the ovary wall. Some fruits, such as apples, remain whole and retain the mature seed. Other fruits, such as those of bean plants, open and release the mature seed.

stamen the male reproductive part of a flower, consisting of an anther and filament

anther an organ in a flower that produces pollen grains

pistil the female reproductive part of a flower, consisting of the stigma, style, and ovary

style the stalk on the female reproductive tissue of a flower

stigma the sticky surface at the tip of the style of a flower

ovary an organ that contains egg cells

fruit a structure that develops from the ovary of a pollinated angiosperm, in which the seed develops

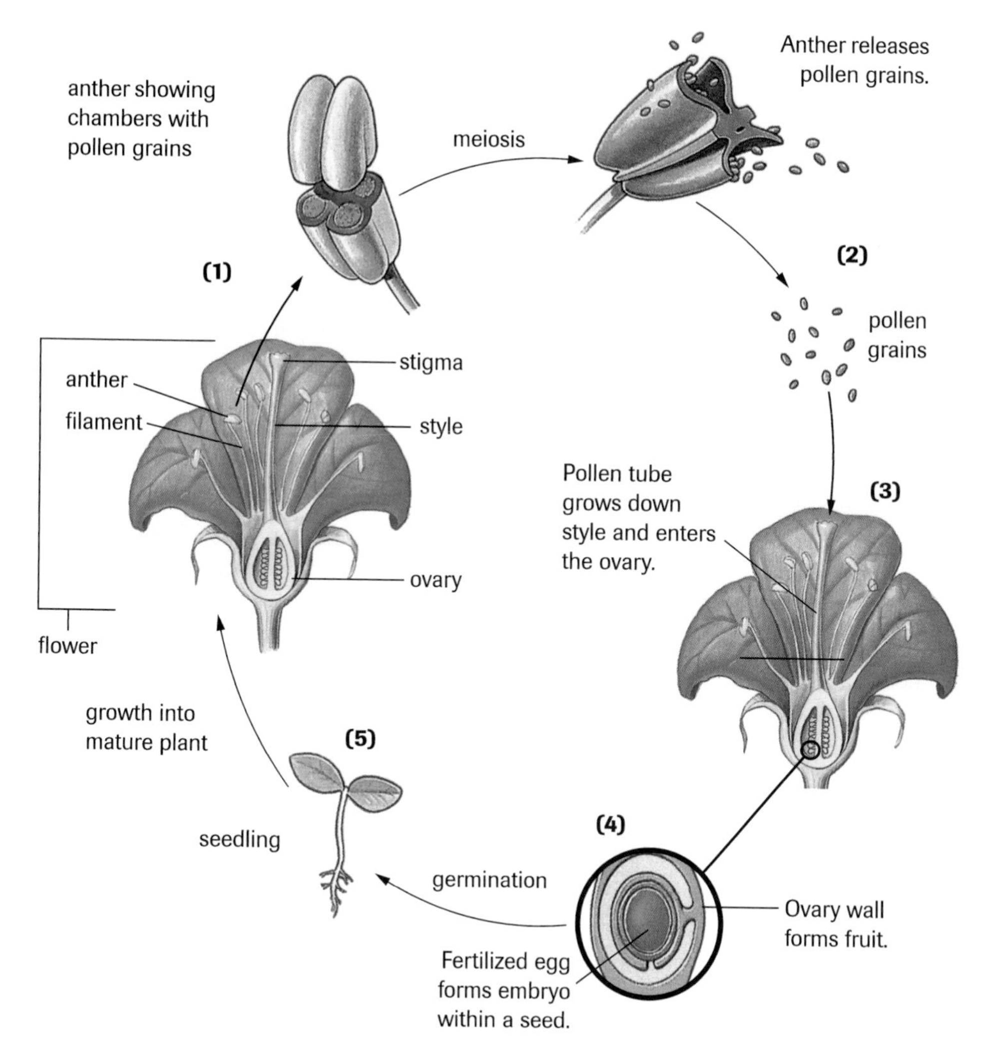

Figure 6
The life cycle of a flowering plant.

(1) Flowers on the sporophyte produce pollen in the anthers.
(2) Pollen is carried by wind or other organisms to other flowers.
(3) The pollen grains grow down through the enclosing tissue to reach the ovary. Sperm within the growing tube then fertilize the egg.
(4) The zygote develops and a seed is formed. The ovary wall thickens to become a fruit.
(5) The seed (with or without the fruit) is released and develops into a new sporophyte.

Biodiversity and Plants

Plants (nonseed and seed) can be found in almost all places on Earth's surface, from dry, hot land to cold lakes and streams. The adaptations that help plants survive in these different conditions also provide a wealth of resources for other organisms, including humans. Despite this amazing diversity, however, scientists have been able to use the similarities and differences in the characteristics of different plants in order to classify them. Classifying plants is a valuable tool for understanding the biology of plants and how they interact with the environment and organisms around them. **Figure 7** is a graphical summary of the similarities and differences of the four main groups of plants discussed in this and the previous section: mosses and their relatives, ferns and their relatives, gymnosperms, and angiosperms.

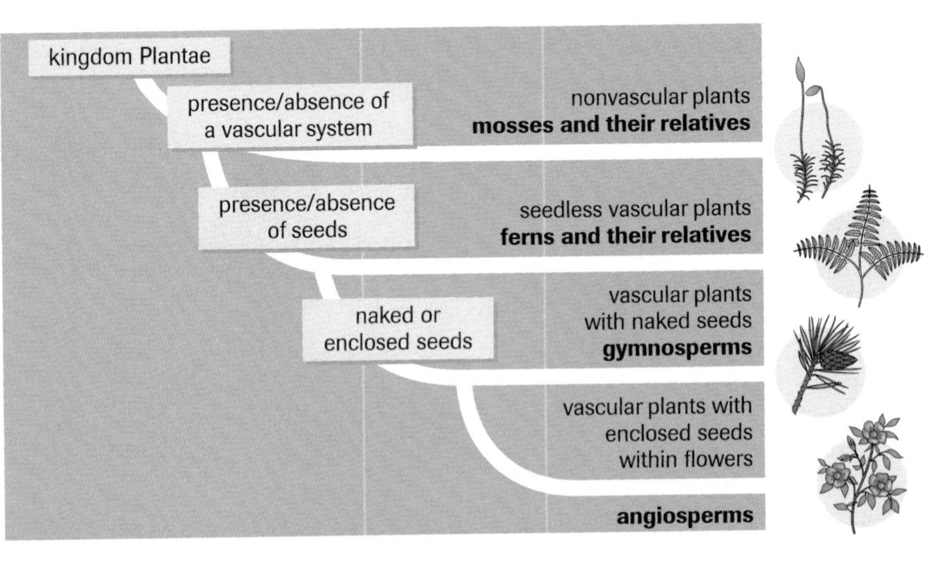

Figure 7
The main distinguishing features of the four major groups of plants

Section 4.3 Questions

Understanding Concepts

1. What is the main way in which the production of pollen and seeds contributes to the survival of plants?
2. Describe the characteristics that make angiosperms the most successful of all plant species.
3. List some products produced from conifers.
4. What are the two main groups of angiosperms? How can you distinguish their seeds from one another?
5. List the parts of an angiosperm seed and their main functions.
6. Many grasses, such as wheat, corn, and barley, are important food crops. Why is it beneficial to plant these crops so that the individuals are close together?

Applying Inquiry Skills

7. Fewer insect species are observed in an area covered with grass lawns than in a similar area covered with different native grass species and wildflowers. Propose a hypothesis to explain this observation.
8. For each of the two endangered species you reported on in the activity "Exploring Ontario's Biodiversity" in section 4.1, identify whether the plant was a moss, a fern, a gymnosperm, or an angiosperm. Carry out any additional research necessary. Outline the methods you used to make your identifications.

www.science.nelson.com

Making Connections

9. During the 1970s, high schools did not have computers. As personal computer and storage disks became more common, some people predicted a future without paper. Survey the number of computers in your high school and the number of photocopies made on a typical day. Based on your survey, is the paper industry at risk? Report your findings in the form of a brief newspaper article, making sure that you justify your stand.

Collecting and Classifying Plants

Plant biologists and ecologists are often interested in the biodiversity of an area. To estimate the number of different plant species, they must collect, classify, and document specimens of every species living in a given area. The scientists also apply appropriate sampling procedures when collecting specimens of plants. They might take samples of the same plant at two different times of the year, or only from plants larger than a particular size. The specimens are then classified by examining their leaves, stems, and any reproductive structures, such as sporangia, flowers, or fruits.

In this activity, you will collect specimens of plants in your area. You will then use your knowledge of the distinguishing characteristics of the four main groups of plants to place each specimen in the correct group.

Materials

- large container such as a box
- gloves
- pruning shears or heavy scissors
- plant identification key or field guide
- hand lens
- plant press (optional)

Inform your teacher if you have allergies to plant material and avoid direct contact with the plants, if necessary.

Use pruning shears with caution.

Procedure

1. Your teacher will instruct you on the type of plant material you are to collect and where you are to work.
2. Collect your specimens as instructed, avoiding damaging any plants or other living organisms unnecessarily. Your samples should be large enough to allow you to see the characteristics of the stems and leaves, and any cones, flowers, or fruits that are present.

(a) Give each specimen an identifying number, and then record the date and location where it was collected.

3. Carefully place each specimen in your container, and then bring your collection back to the classroom.

(b) Classify each specimen as a species of moss, fern, gymnosperm, or angiosperm, using an identification key or field guide to help you.

(c) If possible, identify each plant more precisely. For example, you may also be able to determine the family, genus, or species name.

(d) Prepare a list of the plants you found in your area. If you have access to a plant press, you might preserve your specimens.

Analysis

(e) What specific characteristics were most useful in helping you to classify each of your specimens?

(f) How is the process of plant classification affected by the changing seasons?

Evaluation

(g) Do you think plants are more difficult or less difficult to classify than animals? Explain.

(h) Were you able to collect specimens with all the characteristics that would have been helpful in their classification? Why or why not?

(i) Write a report that describes the biodiversity of plants in the area you studied. Your report should give the number of species you found in each plant group, and state which group was the most plentiful.

Synthesis

(j) Based on your experience in this activity, describe when and how to collect plant specimens in order to document the biodiversity of plant species in an area.

(k) Suggest ways that the following factors might affect the diversity of plant species in your area:
  (i) climate conditions
  (ii) human actions and infrastructure (e.g., roads and sewers)
  (iii) introduction of non-native species

(l) Choose a region in Ontario other than your own local region. Based on the factors you outlined in (k), would you expect your local region to have higher or lower biodiversity than this other region? Justify your choice.

4.5 Vascular Plant Structure

Vascular plants, especially gymnosperms and angiosperms, dominate almost all terrestrial environments. They are the world's most important land plants. Humans are dependent on vascular plants for much of our food, wood, paper, clothing, medicines, and many other products. In order to benefit fully from vascular plants, we need to know about the main parts.

Plants are multicellular organisms, made up of different kinds of cells. These cells are organized into the tissues, which form the organs of the plant body. As in animals, plant parts are specialized to perform important functions that support the life of the organism. These include photosynthesis and respiration, transportation of chemicals, growth, and reproduction. **Figure 1** illustrates and describes the main parts of a growing vascular plant.

shoot typically, the above-ground portions, such as the stem and leaves, of vascular plants

root the portion of the plant that typically grows downward and absorbs water and minerals from the soil

terminal bud the growing tip of a shoot

axillary bud a bud that will form a side branch from a stem

leaf blade the wide portion of a leaf

leaf petiole a stalk that supports a leaf and attaches it to a stem

root tip the end of a growing root

primary root the central, often largest, portion of the root

secondary root a branching root arising from the primary root

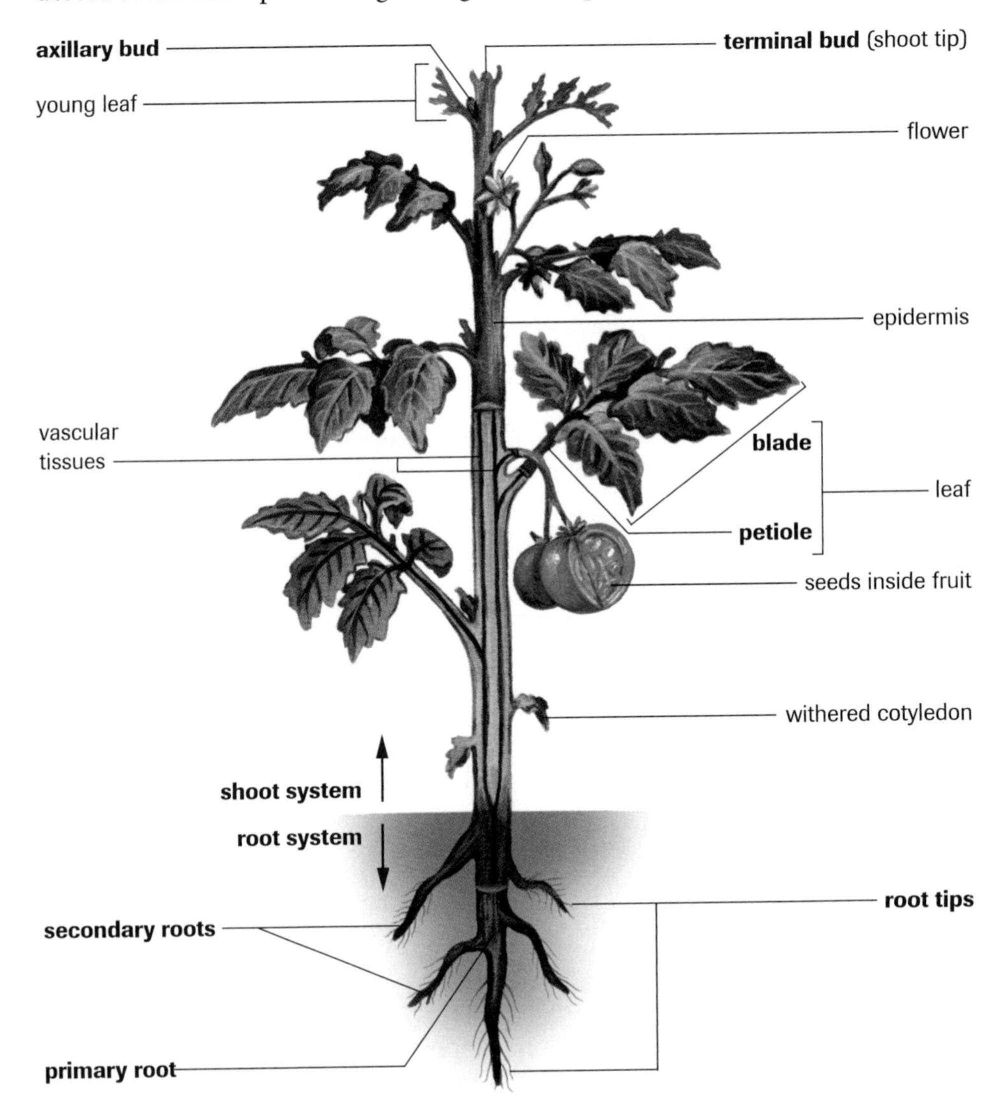

Figure 1
The two main organ systems of a typical flowering plant include the above-ground shoot system and the below-ground root system. Note that plants are highly variable; some plants may lack some parts and plant parts on different species may not look the same.

Plant Tissues

The parts of a vascular plant are made up of four main types of plant tissues. These tissues perform specialized functions such as storage, transportation, photosynthesis, and growth.

Meristematic tissue is composed of cells that undergo cell division by mitosis, the process by which the plant grows. **Growth** is an increase in the size of the organism. Meristems are the only places in a plant where dividing cells are found. Some meristematic cells differentiate after division. **Differentiation** is the process that gives rise to all the specialized cell types that make up the other three primary tissue systems of a plant.

Dermal tissue is composed of the outermost cell layers of a plant. The **epidermis** is the outermost part of the dermal layer. Like our skin, the dermal layer protects the plant body from cuts, invasion by microorganisms, and water loss. On leaves and stems, these cells usually produce a waxy cuticle. This wax is vital to prevent the leaves and other delicate tissues from losing too much water under dry conditions. When you polish an apple, you are polishing this wax. You may be surprised to learn that when you polish a car you may also be polishing plant wax. Many car polishes contain carnauba wax (**Figure 2(a)**), obtained from the cuticle of carnauba palm leaves. Epidermal cells also produce fine hairs on the surfaces of many leaves and stems. In many plants, these hairs discourage herbivores. Some contain painful irritants that are released into the skin when touched. Epidermal root tissue produces large numbers of extremely fine extensions called root hairs. These hairs aid in the absorption of water and minerals. In woody plants, a bark layer forms when the epidermis is replaced by layers of specialized cells that soon die (**Figure 2(b)**).

meristematic tissue plant tissue composed of cells that are able to divide repeatedly by mitosis. This tissue gives rise to all other plant tissue and cell types.

growth the increase in volume or dry weight of an organism

differentiation a change in the form of a cell, tissue, or organ to allow it to carry out a particular function

dermal tissue cells specialized for covering the outer surfaces of leaves, stems, and roots

epidermis the outermost layer of living cells

(a)

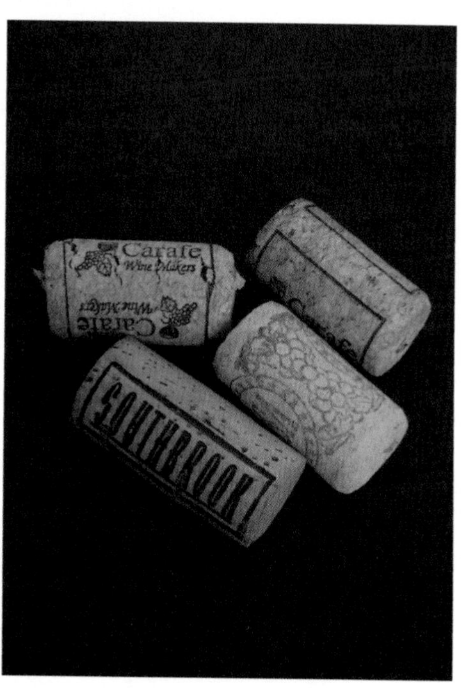

(b)

Figure 2
Dermal tissue can produce a variety of materials such as **(a)** carnauba wax used in car waxes, floor waxes, and shoe polish, and **(b)** cork.

DID YOU KNOW?

Peach Fuzz
Dermal tissue produces the "fuzz" on peaches. This fine, hairlike covering helps the peach shed water and stay dry. This may make it more difficult for yeasts and other fungi to grow on the outer surface of the peach.

ground tissue the internal nonvascular tissues of a plant. They perform many functions, such as photosynthesis and food storage.

vascular tissue tissue composed of cells specialized in transporting water and other substances among cells

xylem vascular tissue that transports water and minerals

Ground tissue is composed of all the internal cells of a plant other than vascular tissue. This tissue is made up of a variety of different cell types that are specialized for storage, support, and photosynthesis. Examples of ground tissue include the fleshy portions of apples, pears, potatoes, and carrots.

Vascular tissue is involved in the transport of substances in the plant. Vascular tissue is composed of two different tissues, xylem and phloem.

Xylem is water-conducting tissue. It is mainly composed of thick-walled cells that conduct water and minerals from the roots to the rest of the plant. When these cells mature, they die, leaving behind hollow cells that are ideally suited for carrying water. In gymnosperms, the walls of these cells have many pits, allowing easy movement of water from one cell to the next (**Figure 3(a)**). In angiosperms, some xylem cells have open ends (**Figure 3(b)**) and are arranged end to end, forming continuous tubes. Wood is composed entirely of xylem tissue and provides the main support for most large plants such as trees.

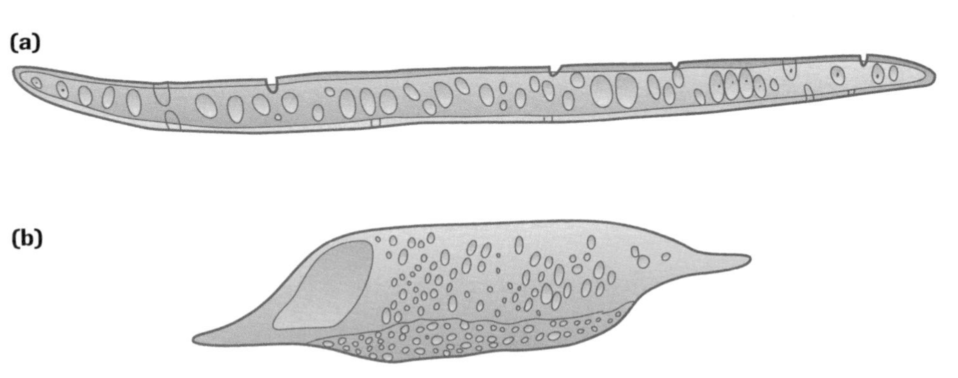

Figure 3
Xylem cells may be **(a)** long, thin cells and **(b)** open-ended cells forming continuous tubes.

phloem vascular tissue that transports sugars and other solutes through the plant

sieve-tube element a cell type in the phloem that lacks nuclei, mitochondria, or vacuoles, and has large pores through the cell walls at either end

sieve tube a long tube formed from many individual sieve-tube elements joined end to end

companion cell a phloem cell that contains a nucleus and other cell organelles, and that controls the cellular activity of sieve-tube elements

Phloem is food-conducting tissue and is composed of thin-walled cells that transport sugars and other plant products from one part of a plant to another. Unlike mature xylem cells, which are dead, phloem cells are living. There are two types of phloem cells. **Sieve-tube elements** are long, thin phloem cells that have large pores through the cell walls at either end. These cells have no nuclei, mitochondria, or vacuoles. The sieve-tube elements are arranged end to end into **sieve tubes** (**Figure 4**). The sieve-tube elements in a sieve tube share cytoplasm. As a result, each sieve tube forms a channel through which sugars and other plant products can flow. **Companion cells** are phloem cells that are found beside the sieve tubes. Companion cells have a cell nucleus and the other cell organelles that are lacking in the sieve-tube elements; they control the activities of the sieve-tube cells.

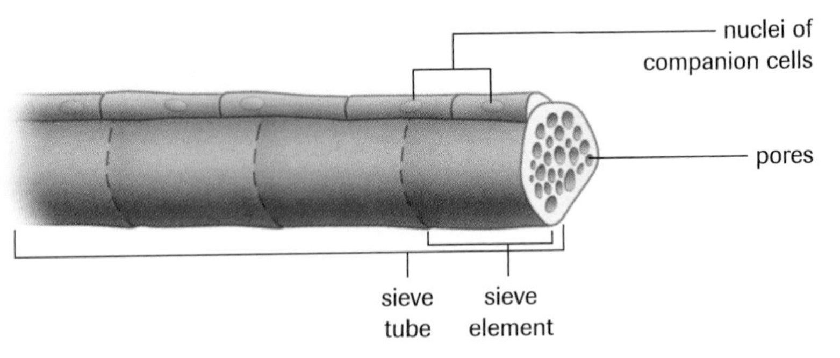

Figure 4
Phloem sieve-tube elements and companion cells

Table 1 summarizes the key functions of nonreproductive plant tissues.

Table 1 Main Functions of Plant Tissue Systems

System	Functions
meristematic tissue	• carries out cell division for plant growth • gives rise to all other tissue and cell types
dermal tissue	• produces a waxy cuticle to limit water loss • has microscopic hairs on root dermal cells to increase water and mineral absorption • protects the plant from elements and herbivores
ground tissue	• carries out photosynthesis • stores nutrients
vascular tissue	• transports water and minerals up from the roots • transports sugars and other nutrients down from the leaves

Section 4.5 Questions

Understanding Concepts

1. In what ways does the epidermis protect plants?
2. Create a table that compares the structure and function of xylem and phloem tissues.
3. Explain how it is possible for dead cells to perform useful functions. Include two examples of such functions.
4. What are the functions of meristematic tissue?
5. Some mammals, such as porcupines and beavers, feed largely on the vascular tissue of trees. Would you expect them to prefer phloem or xylem tissue? Explain.
6. A plant biologist discovers that the shells of nuts contain thick-walled cells of ground tissue. What conclusion might she reach about the function of this tissue?
7. When you "eat an apple a day to keep the doctor away," what type of plant tissue are you consuming? Why might this type of tissue be particularly nutritious?

Applying Inquiry Skills

8. Apples are covered by a thin skin or "peel" of dermal tissue. This layer is thought to protect against infection and water loss. Design and conduct an experiment to test this hypothesis.

Making Connections

9. Many valuable plant products come from roots and stems. Make a short list of four or five nonfood items that we obtain from plant roots and stems.

4.6 Plant Development

At two points in the life cycle of plants, a single cell is produced. This single cell will be either a zygote or a spore. A zygote is the product of sexual reproduction, and it develops into a multicellular sporophyte. A spore is the product of asexual reproduction, and it develops into a multicellular gametophyte. How does this development occur?

Development of any organism involves two processes: growth and differentiation. Growth involves the addition of new cells through cell division and an increase in the size of existing cells. Differentiation is a change in form and/or in function, such as the development of photosynthetic leaf cells. As a plant moves through the stages of its life cycle, it undergoes both growth and differentiation.

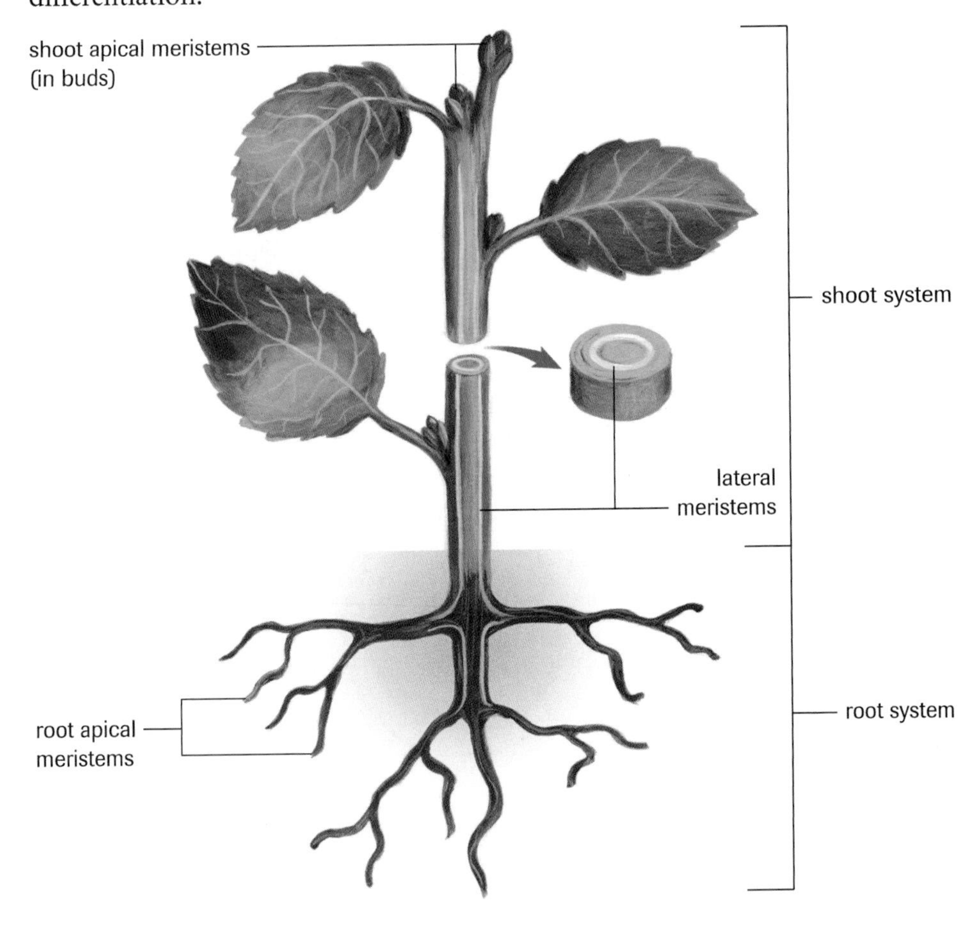

Figure 1
All seed plants have growing regions called meristems. They are located at the tips of roots and shoots and within the shoots and roots.

Growth

apical meristem a meristem located at any growing tip of shoots or roots, which contributes to increases in the length of plant tissues

Apical meristems are found at the tips of roots and shoots. The first stage in the growth and development of a plant occurs when the single cell (zygote or spore) first begins to divide. These divisions establish the shoot apical meristem and the root apical meristem. Further division and differentiation of cells in the shoot apical meristem and root apical meristem form the plant body, composed of the shoot and root (**Figure 1**). All vascular plants continue to grow through cell division in the apical meristems until the plant dies.

TRY THIS activity Observing Growing Plants

Materials: monocot and dicot seedlings, hand lens or dissecting microscope

1. Obtain monocot and dicot seedlings in various stages of early growth. Some seedlings will have just begun to germinate, and others will have leaves.
2. Using a hand lens or dissecting microscope, examine the structure of the seedlings. Note the location, size, and shape of any roots, shoots, and leaves.

(a) Draw each stage of the development of the seedling, using **Figure 2** or **Figure 3** as a guide. Identify and label the emerging roots, cotyledons, root hairs, and first true leaves on your drawing.

(b) What colour are the dicot cotyledons when they have just emerged from the seed?

(c) What happens to the colour of the dicot cotyledons as the shoot grows up above the ground?

(d) What does this suggest about the function of these cotyledons?

(e) Observe and record any differences you find between the monocot and dicot seedlings.

(f) Why do you think plant biologists do not call the green cotyledons "true leaves"?

(g) Find out the approximate ages of your seedlings. Estimate the growth rates of the roots and shoots in centimetres per day. How do you think this compares with your own growth rate when you were one week old?

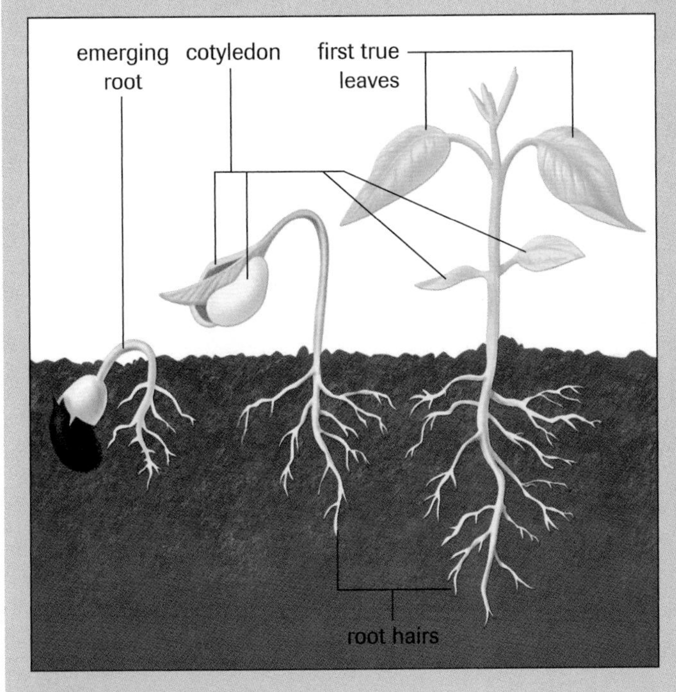

Figure 2
Early stages in the growth of a dicot

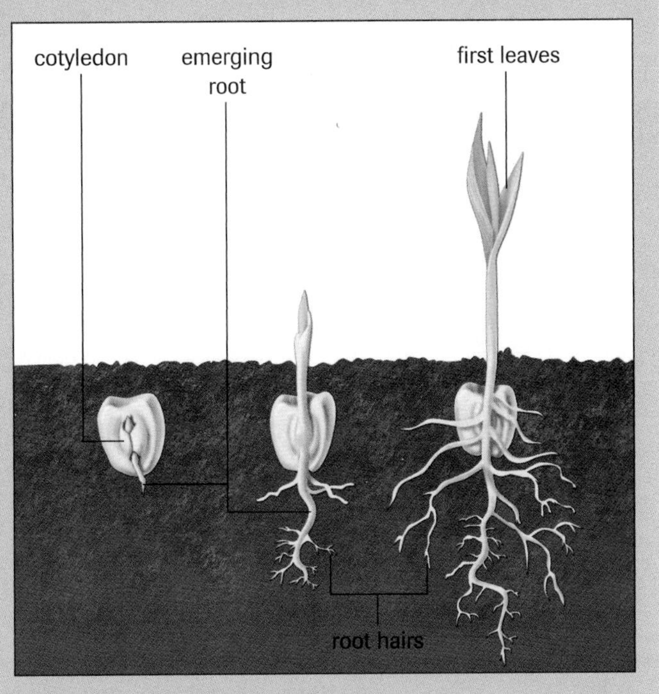

Figure 3
Early stages in the growth of a monocot

Lateral meristems are found within the stems and roots. Recall that meristems are regions that undergo cell division by mitosis. This cell division increases the number of cells in a plant. Plant growth can lead to an increase in the length of a plant or an increase in its diameter, depending on the type of meristems that undergo division. Cell division and elongation from apical meristems increases the length of the roots and stems. Cell division and growth from lateral meristems increases the width of roots and stems.

lateral meristem a meristem located within a stem or root that contributes to increases in the diameter of plant tissues

primary growth all growth of apical meristems and first-year growth of lateral meristems

secondary growth all growth from lateral meristems occurring after the first year

Primary growth includes all growth in the length of roots and stems throughout a plant's life and all growth in the diameters of roots and stems that occurs in the first year of life. **Secondary growth** is any and all growth that occurs by activity in the lateral meristems after the first year of life. The stems of large, long-lived plants such as trees are produced almost entirely by secondary growth. In the majority of plants that live for more than one year, secondary growth produces wood. As a result, even relatively small plants such as raspberry and rose bushes have secondary growth and are called woody, while nonwoody, or herbaceous plants, such as tulips and dandelions, replace their stems and leaves with new shoots each year.

Differentiation

The cells produced by division also undergo differentiation, forming the different cells, tissues, and organs that make up the plant body. Cells from shoot apical meristems give rise to new leaves and stems. Some apical meristems will also produce flowers and fruits under appropriate conditions. The cells of the root apical meristems give rise to new roots and root hairs. Lateral meristems differentiate primarily into vascular and ground tissue.

Section 4.6 Questions

Understanding Concepts

1. What are apical meristems? How do they influence the size of plants?
2. What is the difference between primary and secondary growth?
3. As plants grow taller, they are better able to compete for sunlight. What is the advantage to a plant of growing larger in diameter?
4. Which type of meristem produces the majority of tissue in large tree trunks?
5. One way to make a houseplant grow bushier is to pinch off the upper portion of the growing stems. What tissue would be removed? From what tissue would new growth arise?

Applying Inquiry Skills

6. A student wishes to study the tissues of small house plants. Suggest a way in which he could use food colouring to locate and identify xylem tissue.

Making Connections

7. Unlike animals, plants are not able to run or hide from their enemies.
   (a) Research several ways in which plants defend themselves against attack from humans or other animals.
   (b) How do humans help plants defend themselves?

www.science.nelson.com

Oxygen Consumption by Pea Seeds

Inquiry Skills

- ○ Questioning
- ○ Hypothesizing
- ● Predicting
- ● Planning
- ● Conducting
- ● Recording
- ● Analyzing
- ● Evaluating
- ● Communicating

When seeds are dormant, they are not actively growing. When the right conditions are met, a dormant seed germinates and begins to grow into a seedling. As growth occurs, cells release the energy stored in cellular compounds such as starch and glucose by breaking them down. The cells then use this energy to fuel their growth. When oxygen is present, energy is released through the process of cellular respiration. The following equation summarizes this process:

$$C_6H_{12}O_{6(s)} + 6O_{2(g)} \rightarrow 6CO_{2(g)} + 6H_2O_{(l)} + \text{energy}$$

glucose + oxygen → carbon dioxide + water + energy

The higher its energy demands, the more quickly a cell will consume both glucose and oxygen. In this investigation, you will compare the energy demands of dormant (dry) seeds and germinating (pre-soaked) seeds by measuring the rates of oxygen consumption using an apparatus called a respirometer.

The respirometer you will use consists of a test tube with a two-hole stopper that holds a straight and a bent piece of glass tubing. When assembly is complete, the respirometer is sealed off from the outside air. The sealed respirometer also contains solid potassium hydroxide (KOH) pellets. KOH reacts with carbon dioxide gas according to the following chemical reaction:

$$CO_{2(g)} + 2KOH_{(s)} \rightarrow K_2CO_{3(s)} + H_2O_{(l)}$$

carbon dioxide + potassium hydroxide → potassium carbonate + water

This reaction with carbon dioxide gas produces a solid (K_2CO_3) and a liquid (H_2O). Therefore, any carbon dioxide gas produced during respiration will not contribute to the volume of gas in the respirometer.

Any change in the volume of gas inside the respirometer will cause the food colouring to move within the bent glass tubing. If gases are consumed, the food colouring will move toward the test tube. The rate of oxygen consumption can be determined by measuring the distance the food colouring moves over time.

Question

How does the rate of oxygen consumption by germinating and nongerminating pea seeds vary?

Prediction

(a) Predict whether germinating or nongerminating pea seeds will consume more oxygen in 15 min.

Materials

- pea seeds (dry and pre-soaked)
- water
- paper towels
- nonabsorbent cotton
- laboratory scoop or forceps
- potassium hydroxide pellets (KOH)
- petroleum jelly
- liquid food colouring
- tape
- safety goggles
- laboratory apron
- 2 straight glass tubes
- 2 bent glass tubes
- 2 two-hole test-tube stoppers
- 2 large test tubes
- 2 millimetre rulers
- 2 pinch clamps
- 2 pieces of rubber tubing
- 2 test-tube clamps
- 2 retort stands
- medicine dropper

KOH is highly corrosive.

Avoid any contact with your skin. Wash under cold, running water for 5 min if you get KOH on your skin.

KOH could cause blindness. If KOH comes in contact with your eyes, wash with water for 15 min and seek medical help immediately.

Wear eye protection and a laboratory apron at all times.

Procedure

1. Place 30 dry pea seeds in a large test tube and place a layer of cotton on top of the seeds. Using forceps or a scoop, add approximately 30 KOH pellets on top of the cotton.
2. Assemble the respirometer as shown in **Figure 1**, on the next page. Attach a millimetre ruler to the end of the bent glass tubing, using tape. Seal all stopper openings with petroleum jelly. Do not add the food colouring or the pinch clamp at this time.

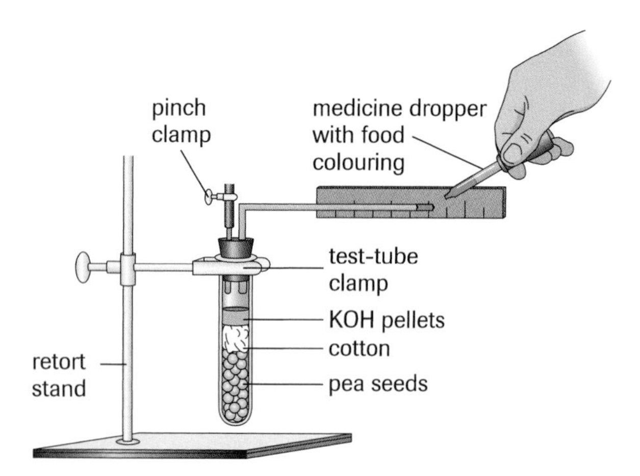

Figure 1
A respirometer

3. Repeat steps 1 and 2 with 30 pre-soaked, germinating pea seeds.
4. Allow both respirometers to stand undisturbed for 5 min.
5. With the medicine dropper, add a few drops of food colouring to the ends of the bent glass tubing.
6. Attach and close a pinch clamp to the rubber tubing on each respirometer.

(b) Record the time at which the pinch clamps were closed.

(c) In which direction does the food colouring first move?

(d) What is the purpose of the ruler?

(e) Why was the cotton used?

7. Monitor the movements of the food colouring in each respirometer for 15 min.

(f) Record the position of the food colouring every minute.

Analysis

(g) Explain how the respirometer works.

(h) Why were the openings in the test-tube stopper sealed with petroleum jelly?

(i) What process in the peas caused the food colouring to move into the glass tubing?

(j) Graph your data by plotting the distance the food colouring moved on the *y*-axis and the time on the *x*-axis. Plot the data set for both the dry and the pre-soaked peas on the same graph.

(k) Determine the oxygen consumption rates for dry and germinating seeds using the following formula:

$$\text{average oxygen consumption rate} = \frac{\text{total distance travelled}}{\text{total time}}$$

(l) Answer the Question.

Evaluation

(m) Evaluate your prediction based on your analysis.

(n) Write a hypothesis that explains your findings, referring to the chemical reaction for respiration.

(o) How would the results of this experiment differ if KOH had not been added to the test tubes?

(p) What plant process produces oxygen gas? Explain why this process does not affect the results of this experiment.

(q) During respiration, glucose ($C_6H_{12}O_6$) is consumed. What was the source of glucose in the germinating seeds?

Synthesis

(r) Seeds will not germinate if they are too wet. Why?

Extension

(s) Design and conduct an experiment that uses a respirometer to investigate one of the following:
   (i) How does temperature affect the rate of respiration in germinating seeds?
   (ii) Do the seeds of all plant species consume oxygen at the same rate during germination?
   (iii) How does exposure to direct light affect oxygen consumption by plants with leaves?

All vascular plants have some form of root and shoot. Whether the plant is a fern, a gymnosperm, or an angiosperm, these structures perform many of the same roles, although they can take many different forms. The most important role of roots and stems is a shared role that depends on the vascular tissues they both contain. Xylem in the root eventually connects to xylem in the shoot, and phloem in the shoot connects to phloem in the root. This connection allows the movement of water up from the roots and the movement of sugars and other dissolved substances away from the leaves.

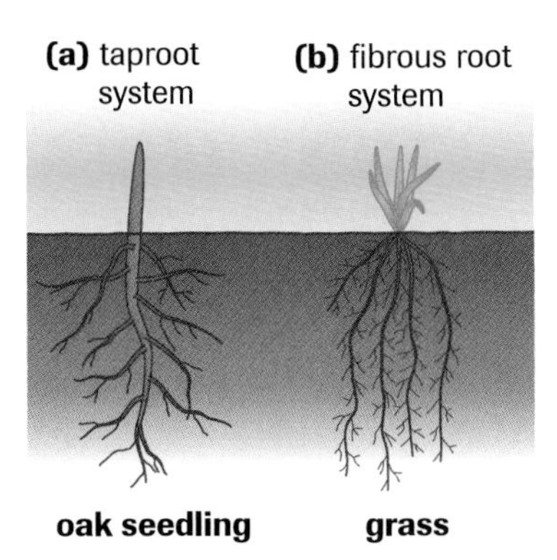

Figure 1
A typical **(a)** taproot and **(b)** fibrous root system

Roots

Although usually hidden below ground, a plant's root system is often larger than the entire shoot system. Roots absorb water and minerals from the soil, and support and anchor the plant. In many plants, roots are also the main storage tissue and provide a good food source for other organisms. Carrots, parsnips, sweet potatoes, and radishes are all examples of edible roots. Roots can take many forms, but the two most common are taproots and fibrous roots (**Figure 1**). **Taproots** have a large, tapering main root that has only slender, short side branches. Taproots are found on carrots, dandelions, and oak plants. Taproots are able to push deep into the soil to reach water. **Fibrous roots** have many smaller roots of about equal size that grow out from the bottom of the plant stem. Although fibrous roots do not grow deeply, they hold the soil in place, preventing erosion. Grasses have fibrous roots.

taproot a root system with a large central root and many smaller side branches

fibrous root a root system in which large numbers of relatively equal-sized roots arise from the base of the plant immediately below the ground

root cap a loose mass of cells forming a protective cap covering the apical meristem of root tips

Root Growth and Differentiation

The tips of most roots have an apical meristem, which is protected by a layer of cells called the **root cap**. Behind the apical meristem is the zone of elongation, in which newly divided cells from the meristem increase in length. Farther back from the root tip is the zone of maturation, in which the root cells differentiate and mature. In this region, xylem and phloem differentiates in the centre of the root, and the outer epidermal cells develop root hairs (**Figure 2**).

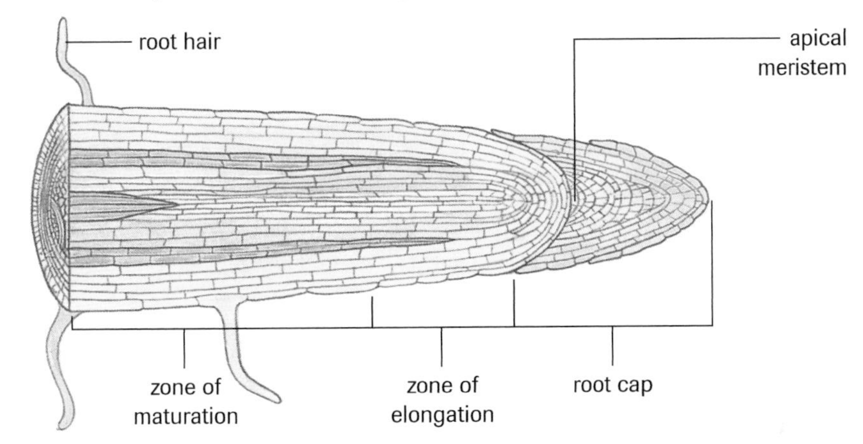

Figure 2
The structure of a growing root tip

Root Adaptations

The roots of many plant species have adaptations that help the plants survive in their particular environments. For example, small aerial roots grow from the above-ground portion of some plants. These roots may serve to absorb nutrients from rainwater on the surface of the plant, or to provide additional support by grasping onto other surfaces. Roots can also help the plant compete for light, water, or nutrients. For example, the aerial roots of the strangler fig slowly kill the large trees on which they grow (**Figure 3**), creating a light-filled hole in the forest canopy.

(a)

(b)

Figure 3
(a) The strangler fig begins life as a small seedling high in a large tree. As the fig grows, it sends roots down and around the larger tree.
(b) Over time, the roots wrap all around the tree trunk, eventually squeezing the vascular system to the point where the tree is killed.

Stems

As you learned in section 4.5, the stems and leaves of a vascular plant make up the shoot. Stems provide support for the above-ground portion of the plant, and link the roots with the leaves and any reproductive structures. Stems may also store water and food. For example, the potatoes we see in grocery stores are part of swollen underground stems of the potato plant.

The stems of herbaceous (nonwoody) plant species are soft and bend very easily. Such stems are relatively weak, so herbaceous plants generally do not grow more than 1 m tall. There are some notable exceptions to this rule, such as palm trees, which grow to great heights even though they lack true woody tissue. Xylem and phloem tissue are usually arranged in bundles in the stems of herbaceous plants. The vascular bundles of herbaceous monocots and dicots are arranged in distinctive patterns (**Figure 4**).

(a) monocot
Vascular bundles are distributed randomly throughout the stem.

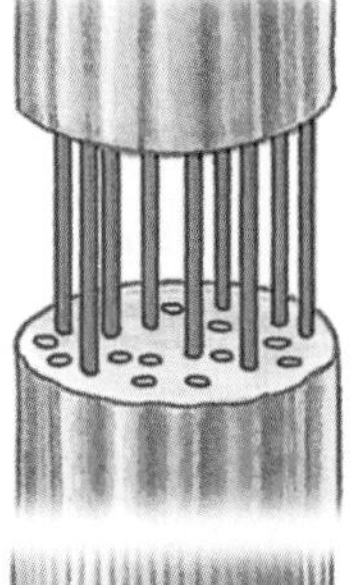

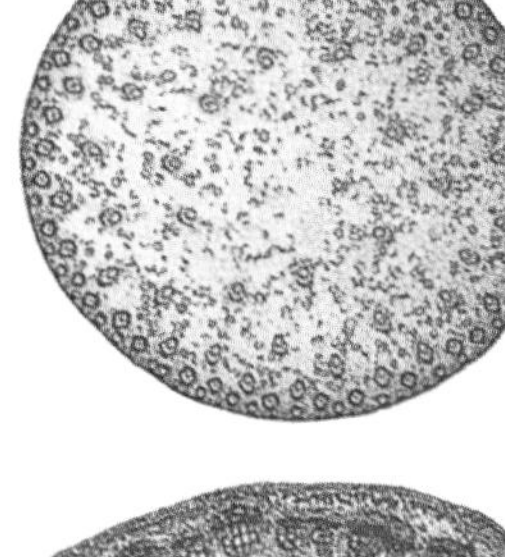

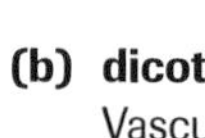

(b) dicot
Vascular bundles in the stem are arranged in a ring.

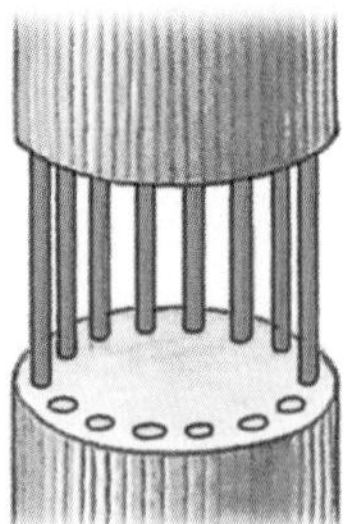

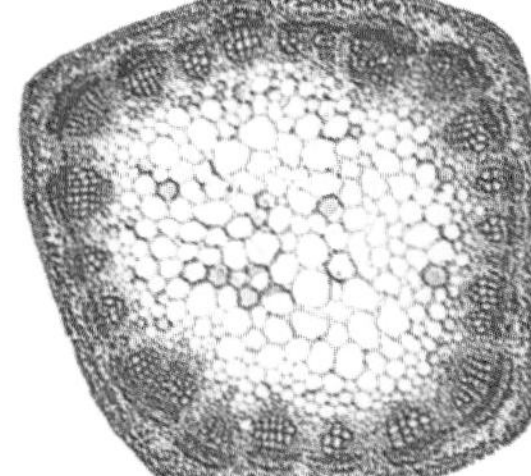

Figure 4
The structure of a young stem of **(a)** a monocot and **(b)** a dicot. Drawings on the left show the distribution of vascular bundles in longitudinal stem sections; photomicrographs on the right show stem cross sections.

The stems of woody plants are hard and do not bend easily. Wood is composed almost entirely of dead xylem tissue, and is extremely strong. It provides enough structural support to allow woody plants to grow into towering trees over 100 m tall. Xylem and phloem tissues form complete rings in woody trees and shrubs.

Stem Growth

Stems increase in length by the division and elongation of cells from apical meristems, and increase in diameter by the division of cells from the lateral meristem or **vascular cambium** (**Figure 5**). Whenever a cell in the vascular cambium divides, it produces a phloem cell toward the outside of the plant stem and a xylem cell toward the inside of the stem. In species that live longer than one year, the new phloem cells tend to crush phloem produced in earlier years. The thick cell walls of the xylem make these cells difficult to crush, so xylem cells accumulate over time, with the oldest xylem tissue being nearest the centre of the stem. The **bark** of a tree is composed of everything from the vascular cambium outward, including the phloem cells.

vascular cambium a lateral meristem that is responsible for creating secondary xylem and phloem tissue

bark the outer layers of older stems, branches, and trunks, consisting of all tissues from the vascular cambium outward

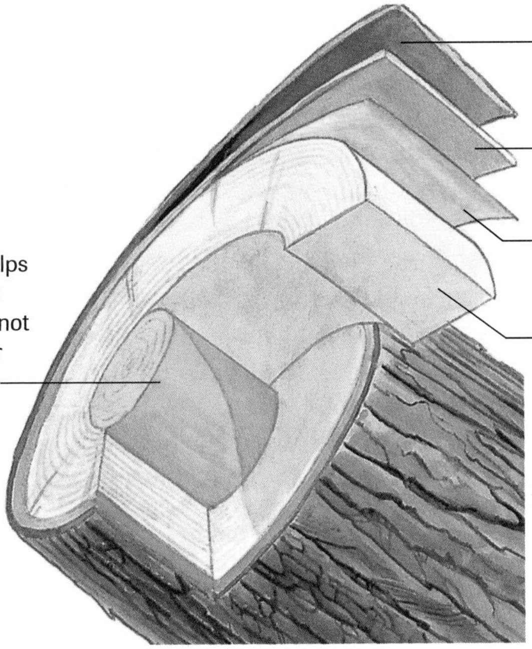

Figure 5
The major tissues within a woody stem. Secondary growth causes an increase in diameter each year.

In woody species, the new xylem matures into wood. When conditions are moist, large-diameter xylem cells are produced. Smaller xylem cells are produced when conditions are drier. Spring tends to be wetter than late summer, so xylem cells tend to be larger each spring. This difference in cell size results in a visible **annual ring** in wood. Each annual ring is composed of one zone of spring wood plus one zone of summer wood. By counting the number of annual rings, we can determine the age of a woody plant. Wood is harvested by people working in the forest industry, which directly or indirectly employs almost 230 000 Ontarians, and supports more than 40 Ontario communities.

annual ring the amount of xylem produced in one year visible in the cross section of a woody stem. The number of annual rings indicates the age of a woody plant.

CAREER CONNECTION

The forestry industry is one of Ontario's largest employers. Many highly trained individuals are needed to harvest and process wood products. Scientists and technicians are also needed to replant and monitor the health of Ontario's valuable forests.

TRY THIS activity Transport Systems

Materials: 1 long-stemmed white carnation with leaves, red and blue food colouring, water, masking tape, two 50-mL beakers, a sharp knife, hand lens

Always cut away from yourself and others sitting near you, in case the knife slips.

1. Place two 50-mL beakers side by side in a well-lit area. Secure them in place with masking tape.
2. Add water until the beakers are three-quarters full. Add 10 to 15 drops of blue food colouring to one beaker and 10 to 15 drops of red food colouring to the other.
3. With the knife, cut 1 cm from the bottom of the carnation stem. This provides a fresh end.
4. Make a cut lengthwise through the lower 10 cm of the stem. This will divide the lower part of the stem into two halves while keeping the stem above the cut intact.
5. Position the carnation so that it is suspended with one stem-half in each beaker, being careful not to pinch, kink, or damage the stem in any way.
6. Let the carnation sit for 24 h. After 24 h, observe and record any changes in the stem or flower.

(a) What happened to the food colouring?

(b) What does the location of the two different colours suggest about the flow of liquids within a stem? Was there any mixing of the colours?

7. Make a cut through the diameter of the stem and observe the cut ends using a hand lens.

(c) Is the dye located in all the cells of the stem? Does this help you explain your answers to (a) and (b)?

(d) How might dyes be used to identify particular types of tissue within plants?

(e) What could you do to test whether fluids travel down stems as rapidly as they travel up?

Extension

(f) Design and conduct a similar experiment to test the effects of salt on the uptake of water through a stem. Report your findings to the class.

Stem Adaptations

Like roots, the stems of some plants perform specialized functions. For example, cacti stems are adapted to store large quantities of water, which helps the plant to survive in its desert habitat. Cacti stems also carry out most of the photosynthesis of the plant, since the leaves of the cactus have been reduced to sharp, protective needles. Other plants, such as potatoes, have modified underground stems for food storage. Stems can also produce new individuals by a form of asexual reproduction. Plants such as strawberries reproduce and spread by sending out above-ground lateral stems. Others, such as poplars and many grass species, produce underground lateral stems that send up shoots that develop into entire new plants.

DID YOU KNOW?

Underground Stems

One of the world's most massive living organisms is an aspen grove in Utah, U.S. The grove started as a single tree, which reproduced asexually from underground shoots. This grove is actually only one individual organism, which now covers more than 80 ha of land.

Using Roots and Stems

The most familiar product from stems is wood, which is the world's most popular construction material. We are all very familiar with the lumber and pulp-and-paper products obtained in Canada from the woody stems of both gymnosperm and dicot trees. Many people, especially those living in rural environments, still depend on wood as a source of fuel for fireplaces and wood stoves.

Another favourite product from stems is maple syrup, which is made by boiling sap obtained from tapping stems of sugar maples (**Figure 6**). Many of our foods come from roots and stems, such as carrots, yams, and sugar from sugar cane. Stems and roots also supply us with some of our favourite flavourings. For example, the flavouring of root beer and ginger ale comes from root extracts, and cinnamon is prepared from bark. The latex sap of the sapodilla tree was the source of *chicle*, the main ingredient of the first chewing gum. Sapodilla trees grow in tropical and subtropical rain forests in Central America. Valuable chemicals such as dyes, resins, gums, and tannins are also produced from stems and roots. One of the world's most important dyes, hematoxylin, comes from the heartwood of the logwood tree. The dye is very valuable in the staining of microscopic specimens, especially animal-cell nuclei. Even the rubber on your pencil eraser comes from the latex sap of a tree.

Figure 6
In early spring, the sap of trees flows up through the xylem and brings stored carbohydrates to developing leaf and flower buds.

Section 4.8 Questions

Understanding Concepts

1. List the main functions of roots and stems. Which functions do both perform? Which function do roots perform almost exclusively?
2. Which function of stems and roots makes some of them valuable food crops? Give examples.
3. Brainstorm and make a list of everyday items you use that come from roots and stems.
4. A mixture of nutrients and seeds is sprayed on the exposed soil along new roadways. What purpose do these plants serve? Would you expect these plants to have fibrous root or taproot systems? Explain.
5. Although many fruits are colourful and sweet, most roots and stems are not. What would be the disadvantage to a plant of having attractive and tasty roots and stems?

Applying Inquiry Skills

6. Before freshly cut wood can be used for construction or burning, excess water must be removed by drying. What characteristic could you easily measure to allow you to determine the amount of water loss from samples of small tree branches over time?
7. Refer to the tree-stem cross section in **Figure 7**.
   (a) Estimate the age of the tree.
   (b) What name applies to the darker area (x)?
   (c) What name applies to the lighter area (y)?

Figure 7
A cross section of a woody stem

Making Connections

8. Most roots and stems store food in the form of starches. Sugar cane and sugar beets are important exceptions. These species store sucrose instead of starch. Research these two important food crops and find out where they are grown and how they are harvested. Is Canada capable of producing its own sugar?

www.science.nelson.com

9. Some plant roots are able to remove pollutants from the soil. Find out how plants are being used to restore mine waste and to treat sewage water.

www.science.nelson.com

4.9 Engineering Wood Products

Although wood is a very useful construction material, even the largest trees have a limited diameter. Most lumber is less than 20 cm wide, but many applications require wider sheets of wood. A number of wood products with larger dimensions than natural wood have therefore been developed.

Plywood was one of the first wood products to be developed (**Figure 1(a)**). To make plywood, large, thin layers of wood are peeled from logs. Only large trees may be used to make plywood. These sheets are then cut and glued together in layers. The grain, or direction of growth, of each successive wood sheet is laid at a 90° angle to the grain of the previous sheet. The finished plywood product, composed of three or more wood sheets, is thin and extremely strong.

A cheaper alternative to plywood is waferboard (**Figure 1(b)**). Waferboard is formed from small, thin flakes of wood that are mixed with glue and pressed into a sheet. These flakes are made from smaller, lower-grade trees than those used to make plywood. Waferboard is not as strong as plywood.

Like waferboard, oriented strand board, or OSB, (**Figure 1(c)**) is made from small-diameter logs. Each OSB sheet has two outer layers and one inner layer. Strands or wafers are first sliced from logs and then bonded with a resin under heat and pressure to create the layers. In the outer layers, the strands (grain) of the wood wafers are aligned (oriented) in one direction. In the inner layer, the strands are randomly aligned. The alignment of the strands in the outer layers greatly increases the strength of the final product, making it comparable to that of plywood. However, since the strands used to make OSB sheets are small, it can be made using the wood from small, fast-growing trees.

CAREER CONNECTION

Wood-processing machine operators require a combination of college or company-run courses and up to one year of on-the-job training.

(a)

(b)

(c)

Figure 1
Common types of wood panels include **(a)** plywood, **(b)** waferboard, and **(c)** oriented strand board.

Tech Connect 4.9 Questions

Understanding Concepts

1. What characteristic of natural wood limits its usefulness in construction?
2. What approaches are used to overcome this problem?
3. List several common uses of large, thin, wood sheets. Which of these are in your home?

Applying Inquiry Skills

4. Obtain two or more small pieces of wood panelling. Design and conduct an experiment to compare one or more of their properties, such as strength, density, or water resistance. Report your findings to the class.
5. Examine the layers in a small piece of wood veneer. Test the strength of the wood layer. In which direction is it easiest to break? How do engineers overcome this difference in wood-panel construction?

Making Connections

6. Since smaller, faster-growing trees can be used to make OSB, there is less pressure on OSB manufacturers to harvest mature forests. However, demand for OSB could create an incentive to replace long-lived tree species with rapidly growing short-lived species. Find out where Ontario's OSB mills are located, and the tree species they are using. How might this affect the environment in the long term? Write a brief report summarizing your findings.

www.science.nelson.com

Leaves 4.10

Green leaves are the major sites of photosynthesis. Their green colour comes from chlorophyll, a green pigment that captures light energy used in photosynthesis. Most leaves have a blade, the flattened main body of the leaf, and a supporting stalk called a petiole that attaches the leaf to the stem. The vascular tissue of the leaf is connected to that of the stem through the petiole. Once it reaches the leaf, the vascular tissue branches out into **veins**, which are comprised of vascular and support tissues. In monocots, the base of the leaf often forms a sheath that wraps around the stem or inner leaves. This sheath is absent in the leaves of dicots (**Figure 1**). Leaves are positioned along the stem at points called **nodes**. The distance between successive nodes is called an **internode**.

vein a vascular bundle that forms part of the conducting and support tissue of a leaf

node the attachment site of a leaf to a stem

internode the space between two successive nodes on the same stem

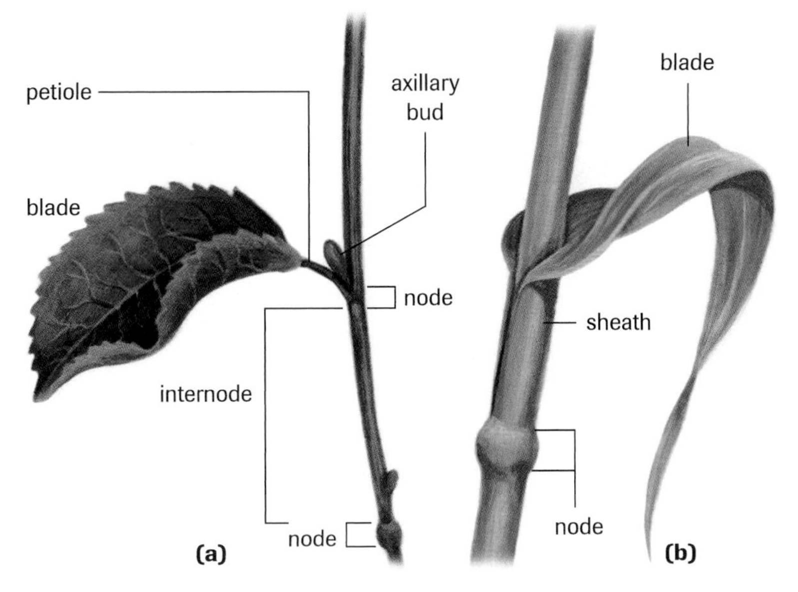

Figure 1
Typical leaf forms of **(a)** dicots and **(b)** monocots. The leaves of dicots usually have smaller veins branching out from a main vein, whereas the veins of monocot leaves usually run parallel to one another.

Although leaves have a wide range of shapes and sizes, there are two main types of leaves. **Simple leaves** have a continuous, undivided blade and margin (edge), whereas in **compound leaves**, the leaf blade is divided into two or more leaflets (**Figure 2**).

simple leaf a leaf that is not divided into leaflets

compound leaf a leaf that is divided into two or more leaflets

Figure 2
Examples of **(a)** simple and **(b)** compound leaves

Leaf Structure

mesophyll cell a photosynthetic ground tissue cell

palisade mesophyll a layer of mesophyll cells at right angles to the leaf surface

spongy mesophyll loosely arranged mesophyll cells within the leaf

The structure of a typical leaf is shown in **Figure 3**. A one-cell layer of epidermal cells covers the upper and lower surfaces of the leaf blade. These cells produce the waxy cuticle that protects the leaf. Most of the cells in a leaf are **mesophyll cells**, which are cells that are rich in chloroplasts, the cellular organelles that contain chlorophyll. Beneath the upper epidermis, mesophyll cells are packed tightly into a layer one or two cells thick called the **palisade mesophyll**. Although the cells in this layer are close together, fairly large air spaces still remain between the cells. Beneath the palisade layer, mesophyll cells are more loosely packed and surrounded by large air spaces. This area of the leaf is referred to as the **spongy mesophyll**. The air spaces between the cells of the palisade and spongy mesophyll allow for efficient gas exchange. Leaf veins run through the mesophyll layers, and conduct water and minerals to the leaves and carry away the products of photosynthesis.

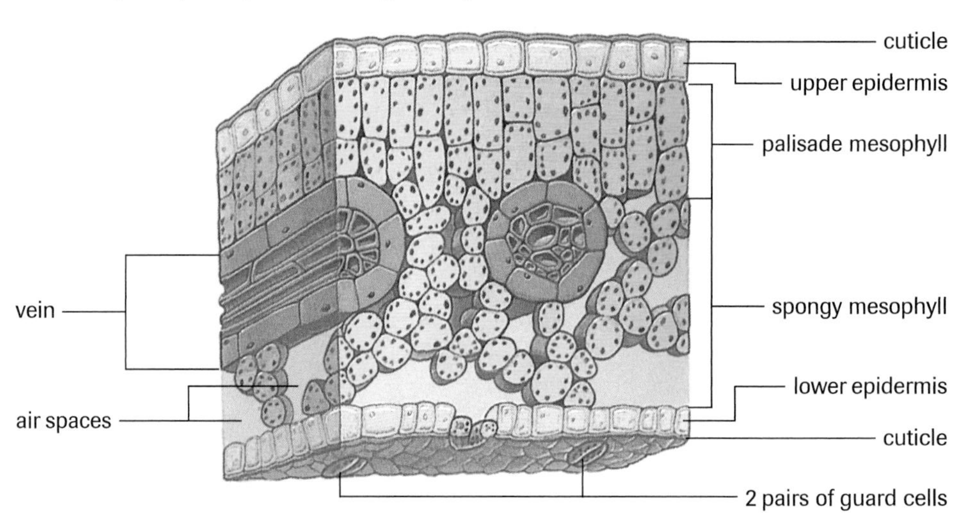

Figure 3
A three-dimensional drawing of a typical leaf showing internal and surface structures

Leaves are the photosynthetic engines of plants. Most of the chlorophyll, the green pigment that absorbs the majority of the light energy used in photosynthesis, is usually found in mesophyll cells in the leaf. Light energy penetrates the epidermis to the mesophyll cells, where it is absorbed by chlorophyll molecules in chloroplasts. The absorbed light energy is then used to convert water and carbon dioxide gas to chemical energy in the form of glucose. Oxygen is also released as a byproduct of photosynthesis.

More sunlight strikes the upper epidermis of a leaf, so the more closely packed palisade mesophyll cells underneath the upper epidermis absorb most of the light energy needed for photosynthesis. Since broad leaves have larger surface areas than thin leaves, they are more efficient at absorbing light.

stoma a small opening in the surface of a leaf controlled by a pair of guard cells

Photosynthesis requires the exchange of gases; photosynthetic cells must take in carbon dioxide gas and release oxygen gas. The gases in the air spaces of the leaf are exchanged with the air outside the leaf through a specialized structure in the epidermis of leaves called a **stoma** (plural: stomata). Stomata are tiny pores that permit the exchange of gases, including water vapour. Stomata function much like windows: when they are open, gases are exchanged with the outside air, and when they are closed, the leaf is effectively sealed off from the outside (**Figure 4**). Stomata are more numerous on the lower leaf surface.

Although the waxy cuticle helps reduce water loss, water vapour escapes readily from the leaves whenever the stomata are open. This loss of water from the leaf is known as **transpiration**. The stomata help regulate transpiration by opening and closing. The opening and closing of a stoma is controlled by **guard cells**, a pair of crescent-shaped cells that surround the pore opening. A stoma opens when guard cells absorb water, and closes when they lose water. As a result, the stomata tend to be open when a plant is experiencing moist conditions, and closed when a plant is experiencing dry conditions. This lost water and that needed for the reactions of photosynthesis must be replaced. Since xylem in the shoots is connected with xylem in the roots, water lost by transpiration from the leaves is replaced by water pulled up through the xylem from the roots. Water is then conducted to each leaf through the xylem of the vascular bundles in the petiole and leaf veins.

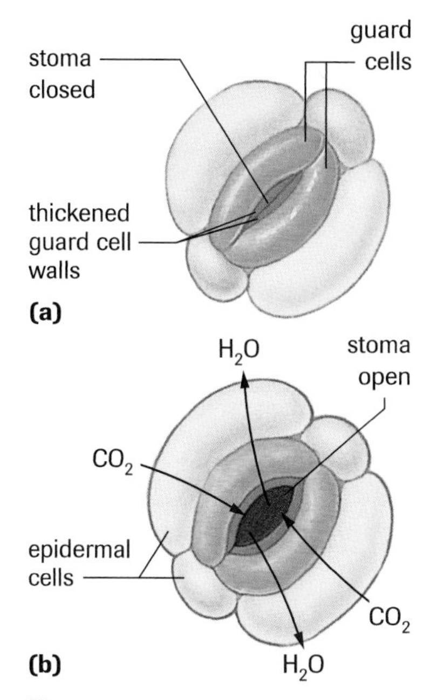

Figure 4
When the stomata are closed **(a)**, the leaf is sealed off from the outside. When the stomata are open **(b)**, gases—including CO_2—are able to enter the leaf while water vapour, H_2O, and oxygen gas escape.

transpiration water loss from a plant surface by the process of evaporation

guard cell one of a pair of cells around each stoma in the epidermis of a leaf or stem that regulates the opening and closing of the stoma

Leaf Adaptations

While most leaves are quite recognizable, some have less-obviously leaflike forms that adapt the plant to a specific environment. For example, the leaves of cacti lack a leaf blade, and the central leaf vein is modified to form a sharp spine. The fleshy layers of an onion bulb are in fact modified leaves, in which food storage has replaced the usual photosynthetic function. Conifer leaves typically take the form of needles or scales to help the plant survive cold, dry conditions.

Some plants are better adapted for growth in direct sunlight, whereas others prefer shade or indirect light. Shade plants typically possess leaves that are thinner, broader, and greener (contain more chlorophyll) than the leaves of sun-loving plants (**Figure 5**). As a result, shade plants are, in general, more efficient at harvesting light at low intensities, whereas sun-loving plants photosynthesize best at high light intensities.

Sometimes, a single plant will have leaves that are modified for both shade and sun, depending on their position. For example, the lower leaves of tall trees are usually in shade conditions, whereas the upper leaves are usually in sunlight. Some species have adapted to this difference by producing lower leaves that are structurally more like those of shade plants, whereas the upper leaves are more like those of sun-loving plants.

(a)

(b)

Figure 5
(a) A shade plant
(b) A sun-loving plant

Figure 6
One common use of plant leaves is as flavourings; leaves of the basil plant are often used in cooking.

Using Leaves

Tea, lettuce, and onions are among the leaves we use as food, while leafy fields of hay supply nourishment to our livestock. The leaves of herbs such as rosemary, basil, and thyme are used for flavourings (**Figure 6**). Many compounds produced within plant leaves are a source of many useful drugs. For example, digoxin, a valuable heart medication, is obtained from the leaves of the deadly foxglove plant (*Digitalis lanata*). Another drug, nicotine, is a natural insecticide produced in tobacco leaves to kill off insect herbivores. Leaves supply a variety of nonmedicinal chemicals, such as the dyes, waxes, and oils used in perfumes and cosmetics.

▸ TRY THIS activity — Counting Stomata

In this activity, you will view impressions of stomata under a microscope.

Materials: plant leaf, clear nail polish, clear cellophane tape, compound light microscope, microscope slides, scissors

1. Paint a thick patch of clear nail polish on the undersurface of a leaf.
2. Allow the nail polish to dry completely.
3. Tape a piece of clear cellophane tape to the dried nail-polish patch.
4. Gently peel the nail-polish patch from the leaf by pulling on a corner of the tape and "peeling" the nail polish off the leaf.
5. Place the peeled impression on a clean microscope slide.
6. Examine the leaf impression with a compound light microscope at a magnification of at least 400×.

(a) Sketch a diagram of your observations of stomata.
(b) Describe the arrangement of stomata.
(c) Are the stomata all the same size, or different sizes?
(d) How many guard cells are there around each stoma?
(e) Repeat the procedure on the upper surface of a leaf. Describe your observations.

▸ Section 4.10 Questions

Understanding Concepts

1. What is the main function of the leaf?
2. Explain the function of each of the following:
   (a) epidermis
   (b) palisade mesophyll
   (c) stomata
   (d) spongy mesophyll
   (e) guard cells
3. What chemicals that enter leaves are reactants for photosynthesis? What chemicals that exit the leaf are products of photosynthesis?
4. List some leaf adaptations that help plants defend themselves against herbivores.

Applying Inquiry Skills

5. Examine a number of house plants. Sketch their leaf outlines and determine if they have simple or compound leaves. Note any other interesting leaf features. Report your findings to the class.
6. A number of plants that live in desert habitats have leaves covered in thick but delicate fuzzy hairs, making the plant appear whitish. How might this feature benefit the plant? Design an experiment that could be used to test your hypothesis.
7. Petroleum jelly can be used to plug the stomata on leaf surfaces. Predict how putting petroleum jelly on a plant's leaves would affect the rate of transpiration. What factors would you need to control to test your prediction?

Making Connections

8. Many plants produce toxins in their leaves to protect themselves from predators. However, humans have found beneficial uses for many plant-leaf toxins. Conduct research and prepare a list of useful toxins obtained from leaves. For each toxin you find, identify the toxin, its source, and how it is used.

www.science.nelson.com

Separating Compounds in Plant Leaves

Plants produce thousands of different chemical compounds, many of which have been found to be useful in medicine, as foods, and in industry. The first step to finding useful plant compounds is to separate the components of a plant tissue. The properties of each separate component may then be tested and the compound identified. Chromatography is a technique that separates different chemicals in a mixture based on their solubility in a particular solvent solution. In this activity, you will use paper chromatography to separate some of the different pigments found in a plant leaf.

Materials

safety goggles	100-mL beaker
laboratory apron	scissors
spinach leaf	pencil
chromatography solvent (90% petroleum ether; 10% acetone, v/v)	chromatography tube with cork stopper
filter paper, 12 cm long	cork stopper with a paper-clip hook
dime	

Petroleum ether and acetone are volatile and flammable.

Use the solvent only under a fume hood.

Do not use petroleum ether in a room with an open flame.

Chemicals should be dispensed only by the teacher.

Wear eye protection and a laboratory apron at all times.

Procedure

1. Obtain a chromatography tube containing 3 mL of chromatography solvent from your teacher. Keep the tube tightly stoppered and standing upright in the beaker.
2. Obtain a 12-cm-long piece of filter paper. Handle the paper by the edges only, to avoid transferring oil from your skin.
3. With the scissors, trim the filter paper to a point at one end. At 3 cm above the point, draw a light line in pencil across the width of the filter paper.
4. Obtain a fresh spinach leaf and place it over the pencil line on the filter paper.
5. Roll the edge of a dime across the leaf, so that the dime edge crushes the leaf tissue onto the filter paper along the pencil line.
6. Repeat step 5 several times until the pencil line has been soaked with a dark, thin band of spinach leaf extract.
7. Obtain the cork stopper with a hook formed from a paper clip. Insert the hook into the upper (straight) edge of the chromatography paper strip.
8. Under the fume hood, remove the cork stopper that is in the chromatography tube. Replace it with the stopper to which you attached the paper strip in step 7, as shown in **Figure 1**. The tip of the paper must just touch the solvent, and the line of leaf extract must stay above the surface of the solvent. Ensure that the chromatography paper is not touching the sides of the chromatography tube. Tightly stopper the chromatography tube and carefully stand the tube upright in the beaker.

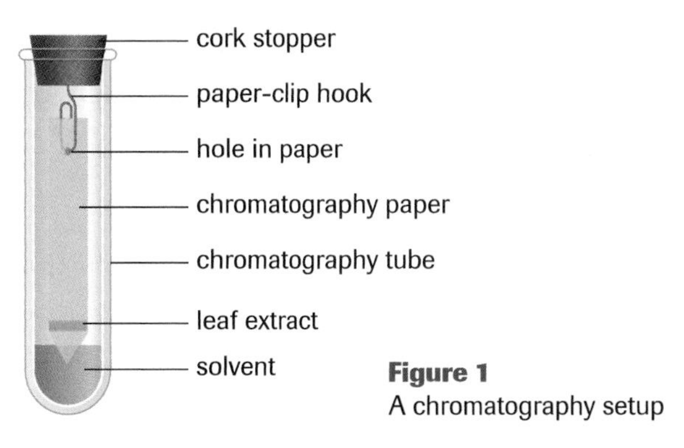

Figure 1
A chromatography setup

9. Observe the movement of solvent and extract for 15 to 30 min. Remove the paper strip before the solvent reaches the top of the paper, and replace the stopper in the tube.

10. Still working under the fume hood, draw a pencil line across the paper strip at the uppermost point reached by the solvent before it dries and becomes invisible. Also mark each pigment band and record its colour before it fades. Keep the paper strip under the fume hood until it is completely dry.

(a) At your desk, measure and record the distance from the original pencil line to the middle of each pigment band. Also measure and record the distance the solvent travelled from the pencil line.

Analysis

(b) After chromatography, compounds may be identified from their R_f values. These values compare the distance travelled by a compound with the distance travelled by the solvent. Calculate the R_f values for each compound on your filter paper according to the following equation:

$$R_f = \frac{\text{distance travelled by compound}}{\text{distance travelled by solvent}} \times 100\%$$

(c) How many compounds were you able to separate using chromatography?

(d) Draw a life-sized diagram of your chromatography strip, showing the precise locations of the pigments and solvent solution.

Evaluation

(e) The more soluble a compound is in a solvent, the farther it will travel during chromatography. **Table 1** gives some properties of pigments commonly found in plant leaves. Which pigments were in your leaf extract? Explain your answer.

Table 1 Common Leaf Pigments

Pigment	Colour	Relative solubility in ether/acetone
chlorophyll *a*	bright green to blue-green	medium
chlorophyll *b*	yellow-green to olive green	medium–low
carotenes	dull yellow-orange	high
xanthophylls	bright yellow to orange	medium–high

(f) List the R_f values for the pigments you identified in the preceding question. Describe the relationship between R_f value and solubility.

(g) Compare your R_f values to those recorded by the other groups in your class. Were the R_f values for each pigment always the same? How might this affect the use of R_f values to identify chemical compounds?

(h) Suggest a step that could be added to the procedure to isolate a specific compound for chemical testing.

Extension

(i) Obtain a black water-soluble marker. Repeat steps 1 to 3, but replace the solvent with water and use the marker to draw a line across a piece of filter paper in step 3. Allow the ink to dry and then draw another line on top of the first. Repeat this several times until you have a very dark ink line across the paper. Perform a chromatography as in this activity. Report your findings to the class.

(j) The colourful Panamanian woven bowls in **Figure 2** were dyed with pigments obtained from tropical plants by the native Embera people. Brainstorm methods these people might use to extract these pigments from leaves, roots, and stems.

Figure 2

Seeds and Fruits 4.12

On a planet with more than 6 billion people to feed each day, food production is clearly one of the most important and challenging of human activities. Surprisingly, although Earth is home to well over 300 000 different plant species, more than half of all food calories consumed by humans are supplied by only a handful of species. Humans consume almost no gymnosperm, moss, or fern plant material. Most of the species we use for food are angiosperms (flowering plants), and the plant structures most commonly used are the seed or fruit of the plant.

For example, the three most important food plants are the angiosperms wheat, rice, and corn (**Figure 1**). The edible portions of these monocot plants are the seeds. The seeds of dicot species such as soybeans, kidney beans, and peas are also important food sources for humans. Seeds are high in protein (up to 30% of their dry weight), carbohydrates or fats (usually up to 50% of dry weight), and fibre (much of the rest), and so are very nutritious. Most seeds can also be stored dry for long periods of time. Humans also consume the fruits of many angiosperm species, including bananas, mangoes, cucumbers, and peppers. Fruits are highly nutritious, but they can be difficult to store for long periods.

DID YOU KNOW?

Success and Grass Seeds

The success of the human species is related to our ability to cultivate, harvest, and process grass seeds! Over 75% of food consumed by humans today is obtained from nine species of grass. Ecologists might fairly describe humans as "monocot seed specialists."

Figure 1
The world's three most important food crops are wheat, rice, and corn; all are examples of monocot seeds.

In everyday life, we use the word *fruit* to refer to sweet fruits such as apples and oranges. However, many of the plant products we call "vegetables" are in fact fruits. Recall that a fruit is a structure that develops within the ovary of a pollinated angiosperm. All fruits therefore develop within a flower and usually contain seeds, but they are not always sweet. *Vegetable* is derived from the biological term vegetative, which refers to any nonreproductive structure of a plant, such as its roots, leaves, and stems. However, the word *vegetable* is commonly used to refer to any edible part of a plant that is not sweet to the taste. The following "vegetables" are in fact fruits: tomato, zucchini, and squash. Examples of vegetables that are not fruits include potato tubers (an underground stem), carrot roots, lettuce leaves, and asparagus shoots.

Fruits help the plant to disperse (spread) its seeds. Plants are stationary, so if they did not have a method of seed dispersal, most seeds would germinate very close to the parent plant. The developing seedlings would then have to compete with the parent and with one another for light, water, and minerals, so the survival rate would be quite low.

DID YOU KNOW?

Pollination in Greenhouses

Some greenhouse crops must be pollinated by hand when natural pollination methods are unavailable. One common technique is to transfer pollen from one flower to another with a small paintbrush.

Figure 2
The fruits of many plants are colourful and sweet, while their leaves are bitter and toxic.

Seed dispersal also helps plants to spread into new environments. Fruits take many different forms to aid in seed dispersal. Maple keys and dandelion "parachutes" are examples of fruits with a form that helps wind carry the seed away from the parent plant. Many popular foods are fleshy fruits. Fleshy fruits are often more colourful and flavourful than the roots, stems, and leaves of a plant (**Figure 2**). The colour and taste attracts animals (including humans) to the fruit. Many plant species produce seeds that pass unharmed through the digestive tract. These seeds are then deposited in the animal's waste in another location. The waste also supplies the young seedling with a rich source of nutrients.

TRY THIS activity — A Slice of Life

Be careful when cutting fruit. Always cut away from yourself and others sitting near you, in case the knife slips.

Do not attempt to cut through harder parts such as the large seeds of peaches.

Do not taste or eat any of the fruit.

1. With a sharp knife, make cross sections of a variety of fruits such as apple, melon, tomato, bell pepper, banana, and avocado. Be sure you cut across the fruit, not along its length.
2. Examine the plant part for any evidence of seeds or ovary structures.

(a) Draw a diagram of the cross section of each fruit, labelling as many structures as possible.

(b) Is there any obvious arrangement or pattern in the structure of fruits?

3. Monocot flower parts occur in multiples of three, while dicot flower parts occur in multiples of four or five. All flowering trees are dicots. Using this information, answer the following questions:

(c) Which of the fruits are from monocots and which are from dicots?

(d) Are monocots or dicots more important in producing fruit crops?

(e) Do bananas grow on trees? Do they have seeds? Use library and Internet resources to find answers to these questions.

www.science.nelson.com

Figure 3
Honeybees are extremely valuable. Placing beehives in orchards dramatically increases both pollination success and fruit production.

Pollination and Fruit Production

Fertilization of the egg cell of a flowering plant requires pollination, the transfer of pollen grains to the stigma of a flower. Most angiosperms will not produce fruit until their flowers have been pollinated. In general, the higher the percentage of pollinated flowers, the greater the fruit production from each plant. Pollination is therefore of vital concern to fruit growers. Many plants, particularly grasses, are pollinated by wind, whereas other plants are pollinated by insects, birds, and other animals. The most important pollinator of our food crops is the honeybee. Honeybees pollinate about 85% of Ontario's apple trees, a crop that is worth $93 million a year. Many farmers place beehives in their fields and orchards to promote pollination (**Figure 3**).

In Canada, many fruits are produced in greenhouses to allow year-round production (**Figure 4**). Greenhouse production of fruit and vegetables is worth more than $500 million to Canada's economy; Ontario is the major producer. Growing fruits in greenhouses poses a challenge to successful pollination. Pollen transfer may require the use of a fan in wind-pollinated plants. Plants that depend on pollinators sometimes must be pollinated by hand, since insects and other pollinators are seldom suited to life in a greenhouse. In recent years, there has been growing use of bumblebees as greenhouse pollinators. Although they have proven to be excellent pollinators, they often escape to the outside. Each bee lost can cost the greenhouse operator $5 to $10!

Sometimes fruit growers act to intentionally prevent pollination. For example, seedless fruits (such as cucumbers) can be produced only from unpollinated flowers. Since wild pollinators would have access to the flowers in open fields, seedless cucumbers can be grown only in greenhouses, where pollination can be prevented.

CAREER CONNECTION

Greenhouse operators must be knowledgeable and highly skilled. They monitor and control the levels of light, humidity, moisture, nutrients, and even gases in the air. They also monitor the health and productivity of their plants.

Figure 4
Greenhouses are used for the commercial production of fruits and vegetables such as peppers and lettuce, as well as of decorative plants such as petunias and geraniums.

Section 4.12 Questions

Understanding Concepts

1. Which group of flowering plants provides most of the food consumed by humans?
2. Which group of flowering plants provides the greatest diversity of foods?
3. Many people use the word *vegetable* when referring to tomatoes, potatoes, squash, cucumbers, onions, and carrots. Which of these are actually fruits? Explain your choices.
4. Why are seeds an important food source for humans?
5. Honeybees are extremely valuable pollinators of food crops. Beekeepers often place their hives in fruit orchards and fields of clover, but never in fields of grain. Explain.
6. How does fruit production benefit the plant?
7. Why are unpollinated plants less likely than pollinated plants to produce fruit?
8. A homeowner with a small greenhouse wishes to grow the following plants: geraniums, lettuce, tomatoes, and cacti. Does the homeowner need to be concerned about pollination? Explain.

Applying Inquiry Skills

9. (a) Make a sketch of **Figure 5** in your notebook. Label the ovary walls and seeds. Identify on your sketch whether each is a monocot or dicot. How did you decide?
   (b) Make a cross section of a fruit of your choice. Sketch and label it as in (a).

Figure 5

Making Connections

10. Honeybees were introduced to North America in the early 16th century. Honeybees are now threatened by a number of diseases, so growers are investigating the use of a different bee, the blue orchard bee. Use the Internet and/or library resources to find out more about how this species is successfully replacing honeybees as a pollinator of fruit trees in parts of Canada. Present your findings in a written report.

www.science.nelson.com

11. You have probably eaten seedless fruits such as seedless grapes and oranges. Use the library and Internet resources to find out how these varieties originated and the methods by which they are propagated. Report your findings as a poster or web page.

www.science.nelson.com

4.13 Activity

Monocots and Dicots

In the previous sections of this unit, you learned that angiosperms are divided into two groups: monocots, which produce seeds with one cotyledon, and dicots, which produce seeds with two cotyledons. Other structures in the plant body also vary between monocots and dicots (**Figure 1**). The stems of monocots are nonwoody, whereas dicots may be woody or herbaceous. The vascular tissue of monocots is arranged in bundles scattered randomly throughout the stem. In dicots, the vascular bundles are arranged in a ring. In woody dicots, the vascular tissue forms complete rings in the stem. The leaves of monocots commonly have unbranched, parallel veins, whereas the leaves of dicots usually branch out from a main vein. In monocots, the parts of the flower commonly occur in groups of three, while in dicots the parts of the flower are usually in groups of four or five.

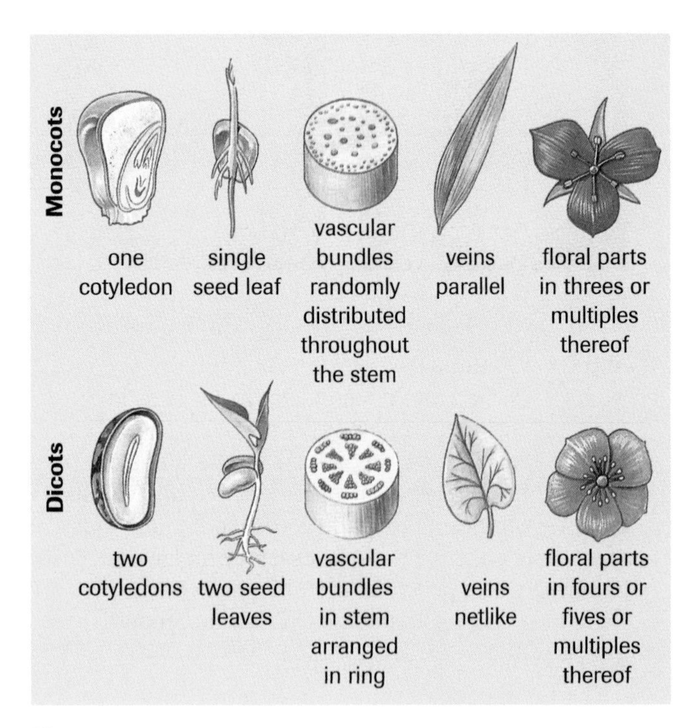

Figure 1
Some of the distinguishing characteristics of monocot plants and dicot plants

In this activity, you will use a combination of techniques to observe and compare aspects of the structure, chemical composition, and metabolic activity of monocots and dicots.

Materials

- leaves and flowers of various monocot and dicot plants
- celery stalks
- corn kernels and bean seeds (dry and presoaked)
- hand lens
- compound light microscope
- paintbrush
- petri dishes
- single-edge razor blade
- microscope slides
- cover slips
- toluidine blue O (TBO) stain
- triphenyltetrazolium chloride (TTC) stain
- Lugol's iodine stain
- gloves
- safety goggles
- laboratory apron

Wear eye protection, a laboratory apron, and gloves at all times.

Stains should be handled cautiously, since they will stain skin and clothing, and are harmful if ingested.

Cut away from yourself and others near you, in case the razor blade slips.

Procedure

Part 1: External Structure

1. Examine the vein patterns of each leaf, using a hand lens, if necessary. Identify whether each leaf is from a monocot or a dicot plant.
2. Examine how the parts of the different flowers are grouped. Determine which flowers are from a monocot and which are from a dicot.

(a) Record your observations in a data table. Include labelled sketches of each plant part you examine.

Part 2: Internal Structure of Seeds

3. Obtain some dry and presoaked bean seeds and corn kernels. Place them in four labelled petri dishes, and then carefully slice each one open lengthwise with a razor blade.
4. Identify the cotyledon(s) in each of the two species of seed, referring to **Figure 2(a)** and **Figure 2(b)**.

(b) Record the number of cotyledons in the embryo of each seed.

5. Cover the cut seeds and kernels with a small amount of TTC stain. Wrap the petri dishes with foil or place them in the dark. You will use these seeds later in step 16.

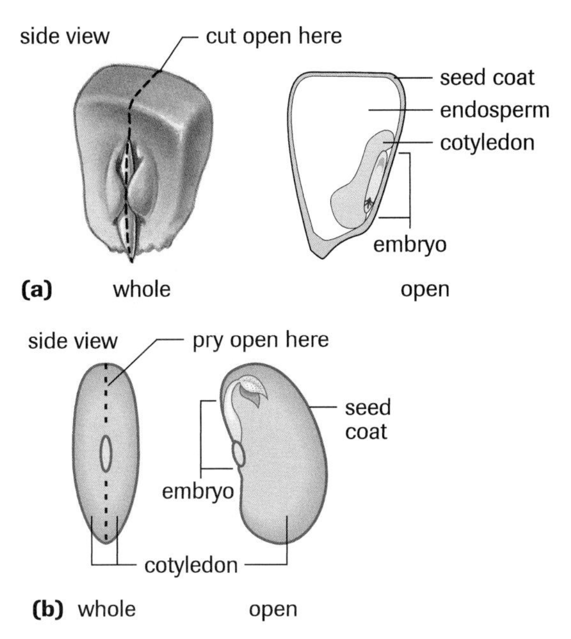

Figure 2
Structure of **(a)** a corn kernel and **(b)** a bean seed

Part 3: Internal Structure of Petioles

6. Obtain a fresh celery stalk. Cut a 3-cm piece from the middle of the stalk and place it in a petri dish containing water.
7. While holding the piece of stalk, cut 10 to 15 very thin cross sections, using the razor blade. Keep the blade at 90° to the stalk at all times.
8. With a paintbrush, transfer a cut section from the water to a microscope slide.
9. Place a few drops of TBO stain directly on the cut section. After 2 min, gently tilt the slide to drain off any excess stain.
10. Place a few drops of distilled water on the stained section to rinse it. Drain or blot the excess water.
11. Repeat steps 8 and 10 to prepare an unstained specimen.
12. Make a wet mount of both the stained and unstained sections by placing a drop of water on each section, then gently lowering a cover slip over the sample.
13. View both the stained and unstained sections with a compound light microscope. Refer to **Figure 3** to help you identify the different tissue types.

(c) Prepare a labelled diagram of the cross section in which you can clearly identify the most tissue types.

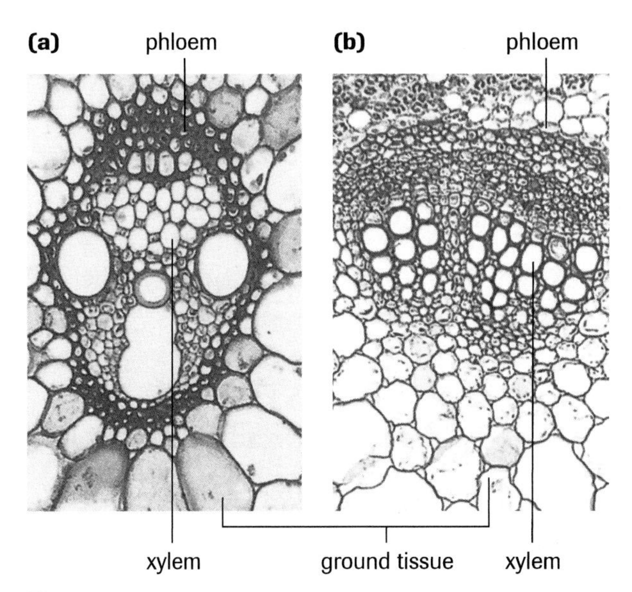

Figure 3
Vascular bundles in **(a)** a monocot and **(b)** a dicot

Part 4: Chemical Composition

14. Obtain a presoaked corn kernel and a presoaked bean seed. Place them both in a petri dish and carefully slice them open longitudinally, using a razor blade, as in step 3 (see **Figure 2**).
15. Add a few drops of Lugol's iodine to the exposed inner surfaces of each section. Lugol's iodine turns blue-black in the presence of starch.

(d) Make a labelled diagram of each section and indicate the location of any starch.

Part 5: Metabolic Activity

16. Observe the sections that you prepared and stained with TTC previously in step 5. This stain turns red when it is in contact with metabolically active cells.

(e) Make a labelled drawing of the TTC-stained sections.

Analysis

Part 1: External Structure

(f) Were you able to classify the samples by examining only the vein pattern of the leaves? Explain.

(g) What other leaf characteristics vary between monocots and dicots?

(h) Was it always easy to determine how the floral parts were grouped? Explain your answer.

Part 2: Internal Structure of Seeds

(i) Was the embryo visible in both the corn kernels and the bean seeds? Explain.

(j) How did cutting open the specimens help you to classify the plant?

Part 3: Internal Structure of Petioles

(k) Did the TBO stain affect all the tissue in the same way? Explain.

(l) Is celery a monocot or a dicot? Explain.

Part 4: Chemical Composition

(m) Compare the amount and location of starch in the monocot kernel and the dicot seed.

Part 5: Metabolic Activity

(n) Were dry or soaked seeds more metabolically active? How is this observation related to the energy needed for growth?

(o) Were all parts of each TTC-stained section equally active? Explain.

(p) Compare the results of TTC staining of the dicot seed and the monocot kernel.

Evaluation

(q) Based on your analysis, make a table that lists all the distinguishing characteristics of monocots and dicots that you observed.

(r) Some foods are high in starch, while others are high in fats or proteins. Based on your results, which type of seed appears to have the highest starch content?

(s) The cell walls of xylem are very thick and, in addition to cellulose, contain the chemical lignin, a very strong polymer. How do these characteristics help the plant to survive?

Synthesis

(t) People who live in Central America and Mexico consume many foods containing corn flour. Is their diet likely to be high in starch? Explain.

Figure 4
Testing plants

(u) Tests are conducted on plants for a variety of reasons (**Figure 4**). For each of the following, suggest reasons or situations in which testing of plant materials would be carried out.
  - (i) police forensics laboratory
  - (ii) pharmaceutical research
  - (iii) manufacture of wood-fibre products
  - (iv) monitoring of the effects of air pollution on plant health
  - (v) farming or food production

(v) For each of the situations in (u), indicate which plant characteristic would most likely be tested: structure, chemical composition, or metabolic activity. Justify your choice.

How Do Growing Plants Respond to Their Environment?

Inquiry Skills

○ Questioning	● Planning	● Analyzing
○ Hypothesizing	● Conducting	● Evaluating
● Predicting	● Recording	● Communicating

Three important environmental stimuli for plants are light, gravity, and touch. In this investigation, you will test the growth response of young seedlings to one or more of these stimuli.

Question

How do growing plants respond to environmental stimuli of light, gravity, and touch?

Prediction

(a) Predict how young seedlings will respond to light, gravity, and touch.

Materials

petri dish	2 cardboard boxes
4 presoaked corn or bean seeds	filter paper
4 young potted seedlings	cotton batting
paper towels	tape
	camera (optional)

Procedure

Part 1: Gravity

1. Obtain a petri dish, paper towels, and four presoaked corn kernels. Arrange the four corn kernels in the bottom of the petri dish as shown in **Figure 1**.

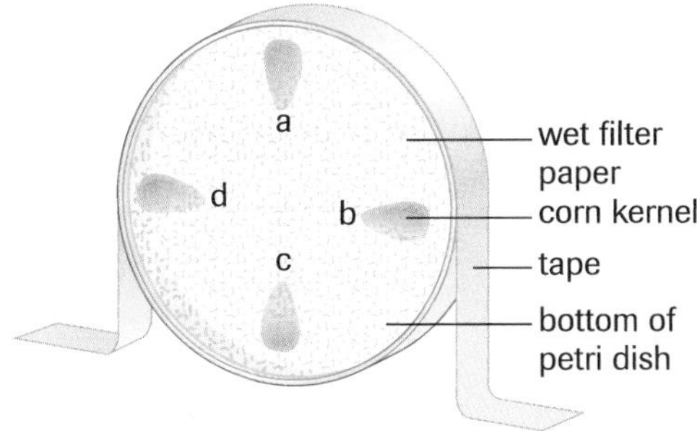

Figure 1

2. Cover the seeds with a piece of wet filter paper, being careful to keep the kernels in position.
3. Moisten the cotton batting with water. Fill the petri dish with enough of the moist cotton batting to hold the kernels in place.
4. Tape the petri dish shut and write your name on the lid. Stand the petri dish on its side and use tape to secure it in this position, as illustrated. Cover the petri dish to protect it from light.
5. Check the seedlings in the petri dish daily.

(b) Observe and record the direction of root and shoot growth.

Part 2: Light

6. Obtain four young potted plant seedlings. Label each plant pot with your name.
7. Obtain the cardboard boxes and cut a small window in the side of one box (**Figure 2**).
8. You will be using the two cardboard boxes turned upside down as growth chambers. Place the growth chambers in a well-lit part of the room, then put two potted seedlings under each.

Figure 2

(c) Observe and record the direction of shoot growth in both growth chambers daily.

Part 3: Touch

Experimental Design

(d) Design an experiment to investigate plants' response to touch. Your teacher will provide you with two different kinds of seedlings. Write your prediction, procedure steps, materials needed, and relevant safety cautions. Include a diagram of your experimental setup. Make sure your teacher approves your design before you begin.

(e) Conduct your experiment, being sure to observe and record your observations daily. You may wish to use a camera to record your plants' growth.

Analysis

Part 1: Gravity

(f) Why were the kernels positioned with their tips pointing in different directions?

(g) From which end of the corn kernels did the roots and the shoots grow?

(h) Describe how the growing seedlings responded to gravity.

Part 2: Light

(i) What was the purpose of using two plants in each chamber?

(j) Describe how light affected the growth of the young seedlings.

(k) Other than the bending of the stems, what differences were there between the plants grown in the two chambers?

Part 3: Touch

(l) How were the plants in your experiment given a touch stimulus?

(m) Did all the plant species respond to touch? Describe any differences you observed.

(n) Describe the response of the plants to touch.

Evaluation

(o) Summarize all of your findings in your notebook, using a table similar to **Table 1** below.

Table 1

Stimulus	Growth responses	
	Roots (+ or -)	*Shoots (+ or -)*
gravity		
light		
touch–species A		
touch–species B		

(p) Evaluate your prediction on the basis of your analysis.

(q) How do you know that the plants in Part 1 responded to gravity rather than moisture, light, or the direction in which they were planted?

(r) Did all the plant parts respond to environmental stimuli in the same way? Give specific examples to support your answer.

(s) Explain how your control in Part 3 showed you that the response was due to the touch stimulus.

(t) Identify any sources of error as you carried out your procedure in Part 3. How would you modify your procedure to reduce the effect of these sources of error if you had time to repeat your experiment?

Synthesis

(u) Which responses, if any, do you think are common to all plants? Explain.

(v) Why is it an advantage for a plant to be able to respond to touch?

(w) A plant producer wants to find species of wild vines to use as indoor house plants. Would you recommend that the producer look in forests or in grasslands? Explain.

(x) Would you expect roots to respond to light? How could you test your prediction?

(y) Most shoots respond to both gravity and light. Why is this an important adaptation for plants?

Extension

(z) Design an experiment to determine whether the growth response in **Figure 3** is a response to gravity or to light.

Figure 3

(aa) Test whether a spinning motion can mimic the effects of gravity. Tape petri dishes with germinating seeds to a turntable and observe their growth at different speeds. This experiment has direct applications for growing plants in a micro-gravity environment such as the International Space Station.

Control of Plant Development 4.15

In section 4.6, you learned that plant development involves both growth (increase in the number or size of cells) and differentiation (change in the form and/or function of the cells, tissues, and organs). Growth and differentiation take place throughout the life of the plant, as it proceeds through the various stages of its life cycle. How is plant development controlled? What factors affect the type of cell that will form from a zygote or meristem? Scientists do not yet fully understand the answers to these questions, but it is clear that in plants, development involves interplay between internal factors and external factors. This section will discuss some of those factors and how they interact.

Plant Growth Regulators

Plant growth regulators (plant hormones) are a group of chemicals that affect the rate of division, elongation, and differentiation of plant cells. They play a similar role to that of hormones in animals, and so are also known as plant hormones. However, animal hormones are produced by particular tissues in the body (for example, the pituitary gland), but plant growth regulators may be produced by any growing tissue of the plant. The type and amount of plant growth regulators produced and the sensitivity of a tissue to plant growth regulators vary according to the plant's stage in its life cycle and the environmental conditions in which it is growing. There are five well-characterized groups of plant growth regulators in most plants: auxins, gibberellins, cytokinins, abscisic acid, and ethylene. These groups are organized according to their chemical structure. As you will see, each group acts in a number of different ways.

plant growth regulator a chemical produced by plant cells that regulates plant growth and development; also called a plant hormone

DID YOU KNOW?

Auxins as Herbicides

The chemical herbicide 2-4-D is a synthetic auxin used to kill dicot weeds. It may kill weeds by stimulating rapid growth and exhausting the plant's food reserves.

Auxins

Auxins act primarily by promoting the elongation of cells. Different tissues respond differently to auxins. Auxins stimulate the growth and ripening of fruit but inhibit the growth of lateral buds. Auxins regulate many events in plant development; for example, they may stimulate cell division in the vascular cambium. Auxins also inhibit the dropping of fruit and leaves from the plant.

Synthetic auxins are chemicals with a structure similar to that of naturally occurring auxins. Synthetic auxins have a number of uses in plant production. Some synthetic auxins are used as herbicides (chemicals that control unwanted plants). How these chemicals kill weeds is not yet completely understood. Synthetic auxin herbicides kill only broad-leaved dicots, making them effective weed killers in grass (monocot) lawns.

Auxins can also be used to trigger the production of fruit from unpollinated flowers. For example, seedless tomatoes and cucumbers can be produced by applying synthetic auxins to unpollinated flowers. Auxins can also induce differentiation of roots (**Figure 1**). The mixtures and powders that are sold as rooting hormone contain synthetic auxins.

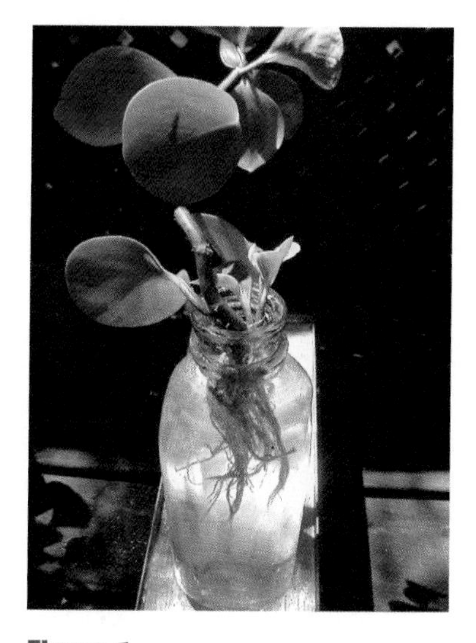

Figure 1
Auxins can induce cut shoots to form roots.

Gibberellins

Gibberellins promote both cell division and cell elongation. In some plants, they cause the stem to elongate just before the plant flowers. This process, known as bolting, produces a long stem that raises the flower to where it is easily reached by wind and pollinators. Many dwarf varieties of plants remain small because they do not produce normal levels of gibberellin. Gibberellins cause the fruit of many species to enlarge, and can play a role in the germination of dormant seeds. Gibberellins also have commercial uses. For example, grape growers often apply gibberellins to their crops to increase the size of the fruit.

Cytokinins

Cytokinins are plant growth regulators that stimulate cell division and leaf growth. The tissues of young fruit often contain relatively high concentrations of cytokinins. This group of plant growth regulators is also used commercially on flowers, vegetables, and fruits. For example, spraying cut flowers with cytokinins can extend their shelf life. Cytokinins are also used by the biotechnology industry in plant tissue culture.

Abscisic Acid

Recent evidence suggests that abscisic acid's main role is to coordinate stress responses in plants. Abscisic acid usually acts as an inhibitor of growth. For example, abscisic acid plays a role in inducing the dormancy of seeds and buds, to protect them from adverse environmental conditions. This plant growth regulator also induces the closing of stomata on plant leaves, which regulates the rate of transpiration.

Figure 2
Ethylene plays a role in regulating the timing of leaf loss in deciduous trees. Leaf loss increases the ability of these trees to survive harsh winter conditions.

Ethylene

Ethylene is a gas that is produced by a variety of plant tissues, especially within fruits. Ethylene plays a significant role in the timing of flower death, fruit ripening, and leaf and fruit loss (**Figure 2**). Many fruits start to produce ethylene just before they begin to convert stored starch into sugars, one of the final stages of fruit ripening. These sugars increase the sweetness of the fruit, so this process is very important to growers. Ethylene also is involved in the colour changes and tissue softening associated with fruit ripening. Fruit producers can control the rate at which their fruit ripens by controlling the levels of ethylene during storage. Ethylene levels are kept low when the fruit is being stored to prevent the fruits from ripening too quickly and spoiling (**Figure 3**). Ethylene may be applied externally to increase the rate of fruit ripening.

Figure 3
The ripening of fruits and vegetables sold in grocery stores may be delayed until just before they are to be sold.

Interactions with External Factors

Although they are stationary organisms, plants must be able to respond to changes in light, temperature, and moisture in some way, or they would be less likely to survive. Changes in environmental conditions usually trigger a change in the levels of plant growth regulators. This, in turn, causes a change in the plant tissues that enables the plant to respond in some way to the environmental change. Two examples of environmental responses mediated by plant growth regulators are tropisms and photoperiodism.

Figure 4
The tendrils of the sweet pea respond to touch by wrapping around whatever they are in contact with, allowing the plant to grow up off the ground.

Tropisms

Changes in the growth pattern or movement of plant tissues in response to directional environmental stimuli are called **tropisms**. For example, the tendrils of pea plants will grow around a supporting structure at the point of contact (**Figure 4**). The movement or growth of a plant in response to touch is called **thigmotropism**.

Most plants respond to environmental stimuli by changes in their growth patterns. It can take many days before the effects of these changes are observable. You have probably observed a plant that seems to be stretching toward the light. This appearance is the result of changes in the growth of the shoots in response to the stimulus of the light. When, as in this example, the movement or growth is toward the stimulus, the response is called a positive tropism. If the movement or growth is away from the stimulus, the response is called a negative tropism. Tropisms are important to the survival of the plant.

A change in the growth or movement of a plant in response to light is called **phototropism**. Plants exhibit phototropism when they are growing in an area that has uneven lighting. **Figure 5** shows the changes in the stem that cause a plant to bend toward the light. The cells on the side of the plant stem that is exposed to higher levels of light do not elongate as much as those on the other side. This uneven growth pushes the shoot in the direction of the light, which exposes more of the leaves. As a result, the plant can photosynthesize more and is more likely to survive. Phototropism also helps young plants to grow toward openings in the forest canopy.

tropism a change in the growth pattern or movement of a plant in response to an external stimulus

thigmotropism a change in the growth pattern or movement of a plant in response to touch

phototropism a change in the growth pattern or movement of a plant in response to light

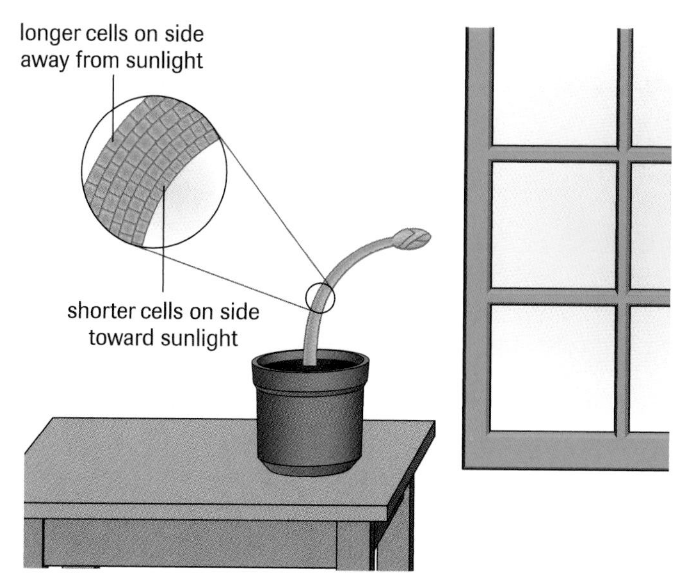

Figure 5
The cells that are away from a light source elongate more than those close to the light source, causing the stem to bend.

gravitropism (or **geotropism**) a change in the growth pattern or movement of a plant in response to gravity

Gravitropism, or geotropism, is a tropism in response to gravity. Why would a plant need to respond to gravity? Recall that the roots of a plant are usually underground, whereas the shoot (stems and leaves) is usually above ground. If plants did not detect and respond to gravity, shoots would be just as likely to grow down toward the ground as upward, and roots might grow up out of the ground. By detecting and responding to gravity, plants are able to orient themselves appropriately in their environment.

Photoperiodism

Plant development varies with the amount and quality of light. In regions of Earth with two or more seasons, plant growth and reproduction cycles are linked to these seasonal changes. For example, here in Canada many plants produce flowers in the spring; this provides them with the greatest amount of time for seed and fruit development and maturation before winter arrives. The leaves of deciduous trees fall in the autumn to protect the tree from the cold, dry conditions of winter. These changes are triggered by changes in the duration of natural daylight, or **photoperiod.** A plant that responds to changes in photoperiod is said to exhibit photoperiodism.

photoperiod the number of daylight hours

Different plants react differently to photoperiod length. Long-day plants flower and reproduce only when the days are lengthening—in Canada, during the spring. Short-day plants flower and reproduce only when the days are shortening, as in our summer and fall. **Figure 6** illustrates the relationship between photoperiod and plant activity here in the Northern Hemisphere.

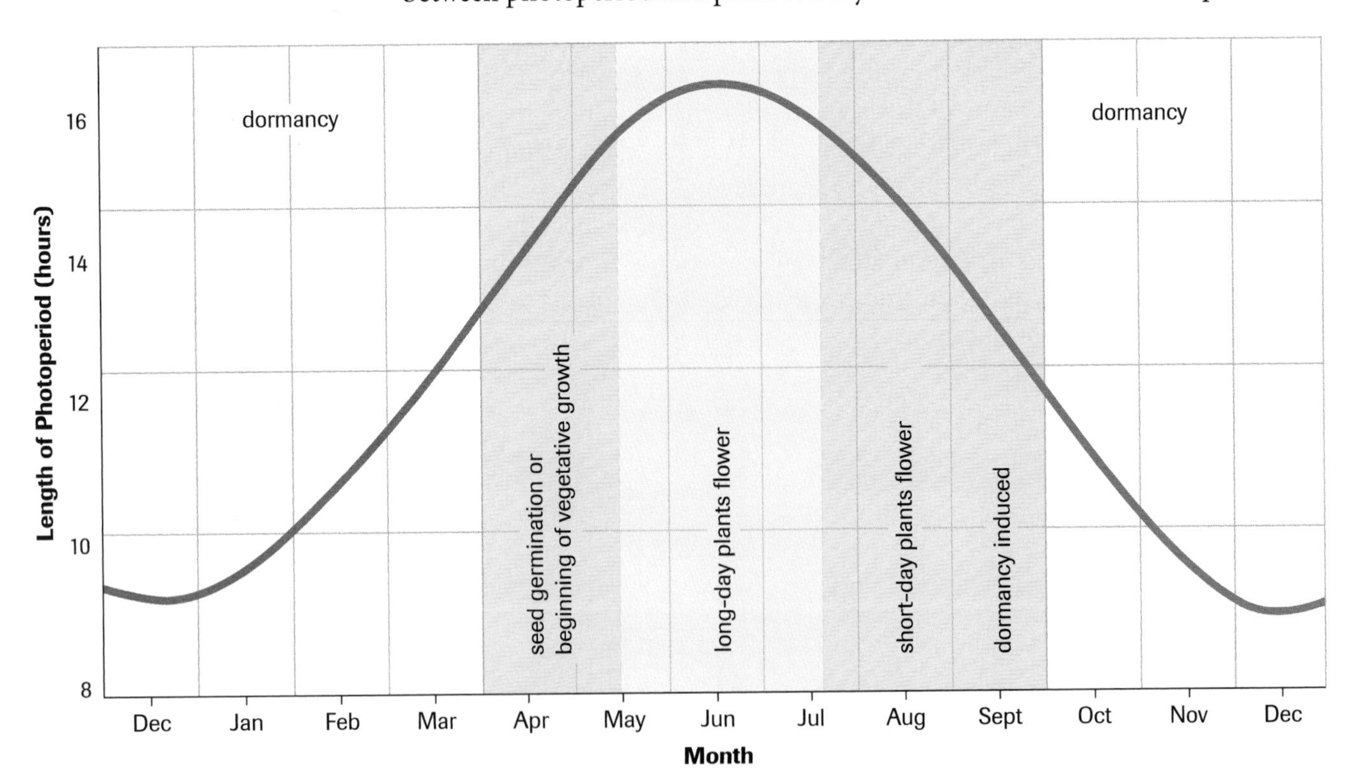

Figure 6
The relationship between the number of hours of daylight and plant growth and development. Note that both short-day and long-day seeds germinate at the same time, but flowering times differ. This graph reflects patterns in the Northern Hemisphere.

In artificial environments such as greenhouses, both the quantity and quality of lighting may be dramatically altered relative to the conditions in the natural environment. The control of lighting and other factors, such as temperature, humidity, and nutrient levels, permits greenhouse operators to provide whatever growing conditions their plants require whenever the operators wish. Careful regulation of these factors can stimulate plants such as the short-day poinsettia (**Figure 7**) to bloom at the precise time when they are most in demand.

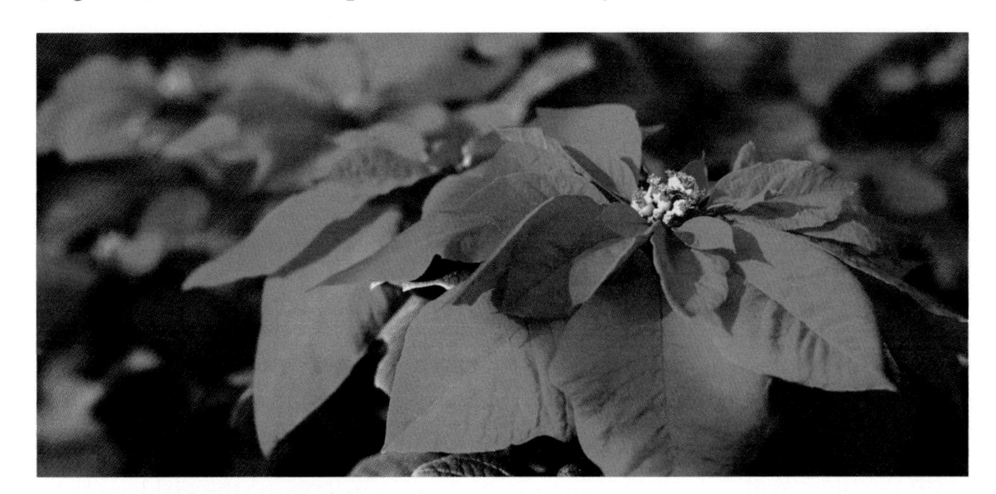

Figure 7
The precise timing of flowering is critical to the success of greenhouse operations.

Section 4.15 Questions

Understanding Concepts

1. In what way is control of growth and development in plants similar to that in animals?
2. Abscisic acid is a plant growth hormone that causes buds to become dormant. Would you expect the production of abscisic acid to be influenced by photoperiod? Explain.
3. Which plant tropisms are most common? Why?
4. Would you expect plants that normally live in the desert to exhibit more pronounced phototropism than forest-dwelling plants? Explain.
5. Explain how photoperiodism is critical to the survival and reproductive success of many plants.

Applying Inquiry Skills

6. Outline a simple experiment you could conduct to determine if a species was a short-day or a long-day plant.
7. The changing of leaf colour in the autumn is triggered mainly by the changing photoperiod. A tree is moved 1100 km north, from Kentucky, U.S., to central Ontario. Predict how the timing of leaf-colour change would compare when the tree is in Ontario to when it was in Kentucky.

Making Connections

8. Some plants, such as tomatoes, are day-neutral. The flowering period of day-neutral plants is not influenced by photoperiod. How might this be an advantage to greenhouse operators?
9. Auxins are produced in the apical tips of trees. One effect of auxins is to suppress the growth of buds farther down the branch. How might orchard growers or landscapers use this information when pruning trees or shrubs?
10. As fruits mature, they produce the plant growth regulator ethylene, which speeds ripening. When fruit must be shipped long distances, ethylene production can cause fresh fruit to spoil before it reaches its destination. Research this problem and find out what technologies are used to control ethylene production, and the effects of ethylene during shipping and storage of fresh fruits.

www.science.nelson.com

4.16 Investigation

Inquiry Skills

- ○ Questioning
- ● Hypothesizing
- ● Predicting
- ● Planning
- ● Conducting
- ● Recording
- ● Analyzing
- ● Evaluating
- ● Communicating

The Effect of Abscisic Acid on Transpiration

About 90% of all transpiration from leaves occurs through the stomata. The opening and closing of stomata are controlled by guard cells, which are influenced by light levels, the water content within the plant, and the plant growth regulator abscisic acid (ABA). Abscisic acid is known to initiate a number of stress responses in plants.

Question

How does ABA affect guard cells and the rate of transpiration?

Hypothesis/Prediction

(a) Formulate a hypothesis that relates ABA concentration to the function of guard cells.

(b) Predict whether providing ABA to the plant will increase or decrease the rate of transpiration from leaves.

Materials

ABA solution
3- to 7-day-old barley seedlings
single-edge razor blade
ruler
3 small vials or test tubes
electronic balance (accurate to 0.01 g)

Use the razor blade with extreme caution, cutting away from your hands.

Procedure

1. Label the 3 vials or test tubes as follows: "control," "leaf," and "leaf + ABA."
2. Add 3 mL of water to the vials labelled "control" and "leaf."
3. Add 3 mL of ABA solution to the vial labelled "leaf + ABA."
4. Using the razor blade, cut two 8-cm pieces from the upper tips of the leaves on the barley seedlings.
5. Place one leaf cutting in the vial labelled "leaf" and the other in the vial labelled "leaf + ABA."

(c) Measure and record the initial mass of each vial to the nearest 0.01 g.

6. Place the vials upright in a tray of shallow water under a light source.
7. After 24 h, remove the vials from the tray. Dry off any water on the outside of the vials.

(d) Measure and record the final mass of each vial to the nearest 0.01 g.

Analysis

(e) Calculate the change in mass for each vial. Record these values in your table.

(f) Was there a loss or gain of mass in each vial?

(g) Which vial had the least mass change? Which vial had the greatest mass change?

(h) Compare the change in mass of the vial labelled "leaf + ABA" with that of the vial labelled "leaf."

Evaluation

(i) Did the evidence from this investigation support your hypothesis? If not, suggest a new hypothesis that agrees with your findings.

(j) Evaluate your prediction on the basis of your analysis.

(k) What was the purpose of the control vial? Compare the change in mass of the control vial with the change in mass of the vials containing leaves.

(l) What caused the change in mass, if any, in each vial?

(m) Does ABA cause an increase or a decrease in the rate of transpiration?

Synthesis

(n) ABA concentration in guard cells increases at night and when the plant experiences water stress. Why would a plant benefit from increasing ABA concentrations under these conditions?

(o) The vial labelled "leaf + ABA" contained less than 0.0001 g of ABA. What does this suggest about the concentration and functioning of plant growth regulators?

Environmental Factors and Plant Growth 4.17

Plants, like all organisms, must obtain all they need from their environment. Each region on Earth has a unique range of environmental factors, including biotic factors such as the presence of diseases and competition, and abiotic factors such as levels of moisture, light, and nutrients, and range of temperature. The natural distribution of a plant species is therefore determined primarily by whether a region provides the environmental conditions that meet its particular needs (**Figure 1**). Other factors such as geographic barriers and human intervention can be equally important in determining where a particular plant species can grow successfully.

Even in a small area such as a local park or schoolyard, the distribution of the various plant species is determined by environmental factors, including light and soil conditions and exposure to stresses such as wind, fire, pollution, and competition, and by human intervention. **Table 1** details some factors that affect regional (large-scale) and local plant distribution.

Figure 1
Plants thrive in areas that meet all their needs, but will be absent in areas with an environment to which they are not adapted.

Table 1 Environmental Factors That Affect Plant Distribution

Regional factors	Local factors
annual minimum temperature average amount of precipitation and seasonal distribution length of the growing season geographic barriers such as oceans or mountain ranges	light levels (sunny or shady) soil quality snow cover competition pollinators diseases herbivores fire wind levels

Climate

Climate is the average conditions of temperature and precipitation of a region over time. Climate therefore determines many factors that affect plant growth, including the level of moisture, the amount and quality of light, and the range in temperature. Since plants are immobile, their natural distribution is strongly affected by climate. For example, maple, beech, and oak trees are the dominant species in southern Ontario, since they are well adapted to the moderate winters and relatively long, moist growing seasons of this region. These species are far less able to survive the longer, colder, drier winters of northern Ontario, and so forests in this region are dominated by coniferous species such as jack pine and black spruce. These species are well adapted to northern climate conditions.

DID YOU KNOW?

The Benefits of Snow
Mild winters can also be hard on plants. Snow acts as a protective blanket in winter, trapping moisture and moderating temperatures near and just below the soil surface. During mild winters, this snow cover may be missing, and low-growing plants can be damaged or killed by exposure to the cold, dry winter air.

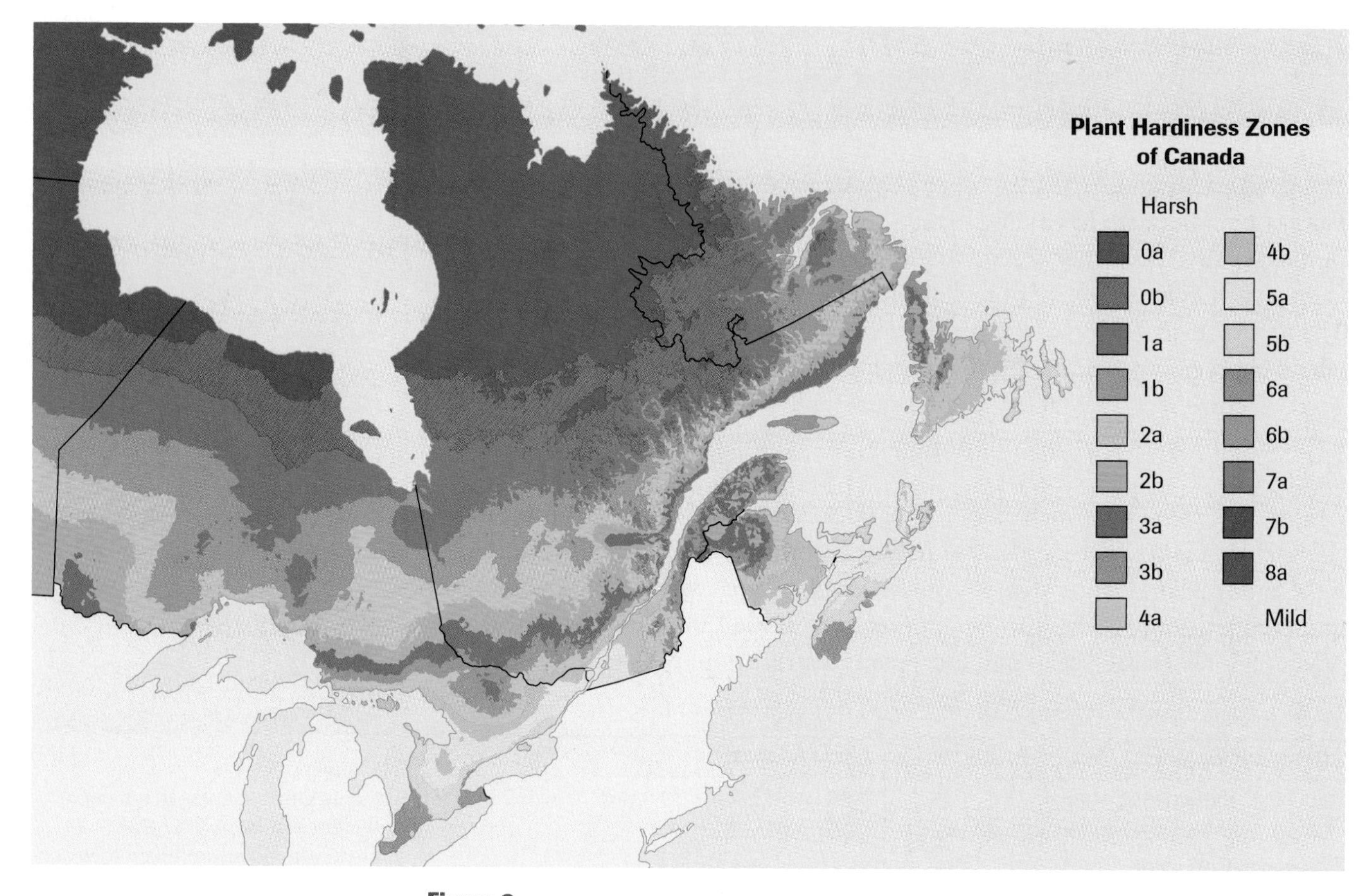

Figure 2
Hardiness zones in eastern Canada, from Natural Resources Canada

hardiness the ability of a plant to tolerate harsh climatic conditions, usually in reference to winter temperature extremes

hardiness zone a geographic region with a specific range of climatic conditions

The degree to which a plant can withstand harsh winter conditions is referred to as its **hardiness**. In 1960, scientists developed a scale from 0 to 11 to rank the hardiness of plant species. Each division is based on a range of climate conditions that can be found in various geographic zones in North America. **Figure 2** shows part of a recent map of **hardiness zones** from Natural Resources Canada.

Plant hardiness zones in Canada range from 0, with the most severe climate conditions, to 8. Hardiness zones 9, 10, and 11, with the mildest climate conditions, are found south of the Canadian border. The 12 main hardiness zones are divided further into subzones; an "a" subzone has a harsher climate than the corresponding "b" subzone.

Plants that are able to survive more severe climate conditions are given a low number on the hardiness scale, while plants that can survive only in mild climates are given a high number. Note that the map shows the geographic region where particular climate conditions exist, not the distribution of plant species. Plants often grow in regions outside the hardiness zone for which they are rated. For example, a plant rated at 4 on the hardiness scale could grow in any regions rated zone 4 or higher. It would be unlikely to survive in regions rated zone 3 or lower, however. **Table 2** gives the hardiness rating for some plant species found in Ontario.

Table 2 Hardiness Ratings of Some Ontario Plant Species

Plant	Hardiness rating
white spruce	1
red maple	3
trillium	4
magnolia	6

Professional and home gardeners use hardiness ratings to determine which plants are likely to survive in their region. It is important to keep in mind that hardiness ratings refer to the ability of a species to survive outdoors throughout the year. Human intervention allows many plant species to grow outside their hardiness range. For example, many decorative plants, such as impatiens, are not very hardy but are commonly grown outdoors during the spring and summer in regions outside their hardiness range. Such species cannot survive the winter climate in the region, however, and so must be replanted every year.

CAREER CONNECTION 

Landscapers and garden centre employees must carefully research and consider plant hardiness when making decisions about which plants to use in a particular location and when recommending plants to customers.

Soil Type

The ability of plants to survive in a region also depends on the characteristics of the soil. Soil is made up of particles of sand, clay, and silt, along with various amounts of humus. **Sand** consists of very small particles of rock. **Clay** is composed of small particles of minerals called silicates. **Silt** is a mixture of sand and clay particles that were carried by water and deposited elsewhere. **Humus** is partially decomposed material from living organisms. Humus is the main source of nutrients in soil, and helps soil to hold water. Soils are classified by the relative amounts of these materials (**Figure 3**).

sand very small pieces of rock

clay small particles of minerals

silt a mixture of sand and clay particles that is deposited by moving water

humus partially decomposed organic material

Type	Sandy soil	Clay soil	Loam soil
Composition	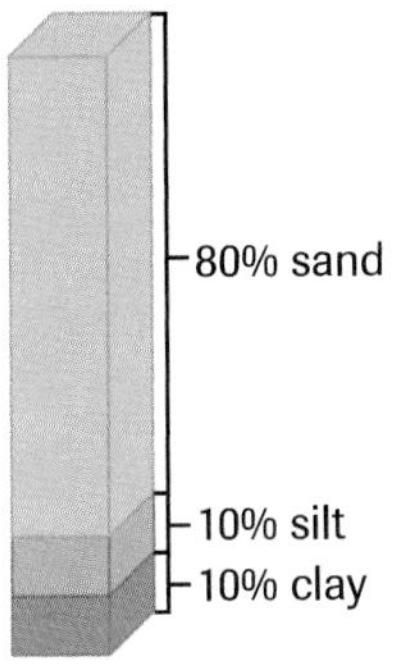 80% sand 10% silt 10% clay	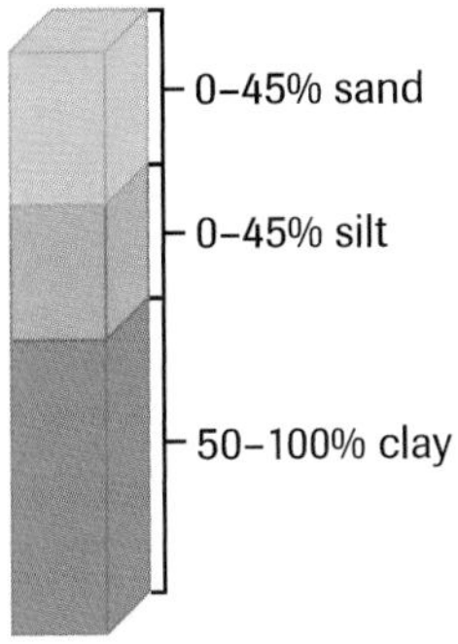 0–45% sand 0–45% silt 50–100% clay	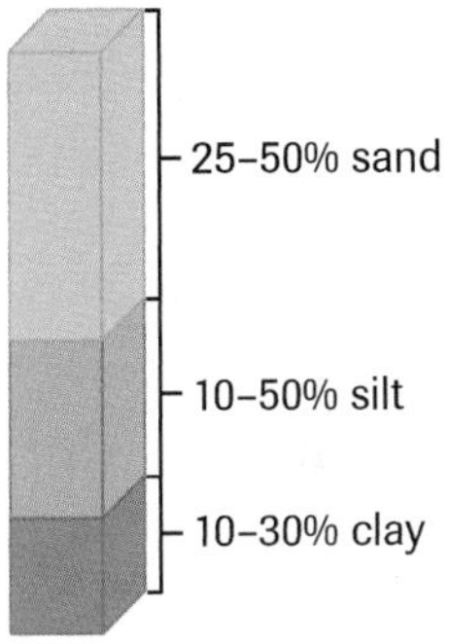 25–50% sand 10–50% silt 10–30% clay
Identification	• when moist soil is squeezed, forms a ball that quickly falls apart	• when moist soil is squeezed, forms a sticky, slimy ball	• when moist soil is squeezed, forms a ball that crumbles like moist cake
Characteristics	• soil drains very quickly • does not hold water well • nutrients wash away quickly • light soil	• soil does not drain well • holds water very well • dry soil takes up water very slowly • very heavy soil	• drains more slowly than sandy soil, but faster than clay • holds water relatively well, depending on amount of humus • medium-weight soil

Figure 3
The amount of humus can vary in all three soil types.

Soil characteristics can be changed dramatically by human action. Coarse sand can be added to clay soil, to lighten it and increase its drainage capacity. Humus may be added in the form of **compost** (partially decomposed material from living organisms) or **manure** (animal waste). Sometimes, gardeners make dramatic changes to the soil of an area. For example, people may buy commercial topsoil (loam) and layer it on existing soil, or even remove and replace the existing soil altogether.

compost a mixture of decomposing organic material, usually plant wastes such as food scraps, yard waste, or inedible plant materials

manure animal waste

Another important factor in soil quality is its pH level. Soil pH affects the chemical form of different soil nutrients, so that some may not be in a form that is usable by plants. Naturally occurring soils can have very different pH levels, from acidic to basic. Some plants are able to survive in soils with a wide range of pH levels, but most plants will grow best only in a limited pH range. Growers can usually identify which plants prefer acidic soils and which prefer alkaline soils, and often modify soil pH to suit particular plants. For example, crushed lime or ash may be added to raise soil pH, while materials such as peat moss or sulfur may be added to the soil to lower its pH value. Soil pH can be measured using simple, readily available test kits.

Nutrients

There are 17 nutrients that are essential for the healthy growth and reproduction of most vascular plants. The nutrients are usually obtained from the soil. Nine of these essential nutrients are **macronutrients**, which the plant needs to take up in relatively large quantities. The remaining eight nutrients are **micronutrients**, which are needed in much smaller amounts. Plant macronutrients include the elements carbon, oxygen, and hydrogen, which plants obtain primarily from carbon dioxide gas and water, and use in photosynthesis. These elements make up over 95% of the dry mass of plants. The remaining macronutrients and all eight micronutrients are obtained as dissolved ions, and are usually taken up by the plant roots. As you can see from the examples in **Table 3**, each nutrient performs vital functions in the plant. In their natural environment, plants must obtain adequate supplies of essential nutrients from the soil in which they grow. When a nutrient is in short supply, plants grow more slowly or produce fewer flowers, seeds, or fruits. When nutrient levels are very low, plants will show symptoms of nutrient deficiency.

macronutrient one of a group of nine nutrients required by plants in quantities greater than 1000 mg/kg of dry mass

micronutrient one of a group of eight nutrients required by plants in quantities less than 100 mg/kg of dry mass

Table 3 Selected Macronutrients

Macronutrient	Functions	Deficiency symptoms	Comments
nitrogen	constituent of all proteins, as well as nucleic acids and chlorophyll; very important for leaf growth	chlorosis: the yellowing of old leaves due to a reduction in chlorophyll	obtained from nitrate of ammonium ions; nitrogen gas cannot be used directly by plants
potassium	involved in water balance, including the operation of stomata; required for protein synthesis	retarded growth; chlorosis of older leaves	present in large amounts in most soils but often in insoluble form
phosphorus	component of ATP, nucleic acids, phospholipids, and some proteins; critical for cell division	lack of or poor seed and fruit development; leaves become dark and reddish; stunted growth	present mostly in fruit and seeds as well as meristematic tissue
calcium	constituent of cell walls; involved in membrane permeability	pronounced abnormalities; stunted growth, especially of roots; weakened condition	neutralizes harmful soil acids; presence can facilitate the uptake of phosphorus
magnesium	component of chlorophyll and molecules necessary for activity of some enzymes	chlorosis; sometimes reddening of leaves	enhances the uptake of phosphorus
sulfur	component of most proteins	stunted growth; yellowing of young leaves	usually plentiful in soil

Fertilizers

To produce the kind of growth they need, plant growers such as farmers must be able to recognize the symptoms of nutrient deficiency and provide appropriate nutrient supplements, or **fertilizers.** Fertilizers can be natural or synthetic. **Natural fertilizers** are obtained from natural sources and do not require chemical processing. Examples of natural fertilizers include manure (**Figure 4**) and compost (**Figure 5**).

fertilizer any minerals added to the soil, usually to replace those removed by planted crops

natural fertilizer fertilizer produced without human-directed chemical activities

Figure 4
Manure can be used on fields direct from livestock barns. Gardeners can purchase bagged, sterilized, composted manure to use in their gardens.

Figure 5
Home composting prevents the removal of valuable plant nutrients from an area.

Synthetic fertilizers are created by humans through chemical processes, and play a vital role in Canada's agricultural industry. Chemical fertilizers are primarily a mixture of nitrogen (N), phosphorus (P), and potassium (K), although they may contain other nutrients as well. Synthetic fertilizers are labelled with a three-number NPK code, indicating the percentages of nitrogen, phosphorus, and potassium they contain, always in that order (**Figure 6**). For example, an NPK code of 5-10-10 indicates that the fertilizer contains 5% nitrogen, 10% phosphorus, and 10% potassium. The remaining mixture is usually filler. The use of synthetic fertilizers increases the size and quality of crop harvests, but also adds to the cost of production.

synthetic fertilizer fertilizer produced by human-directed chemical processes

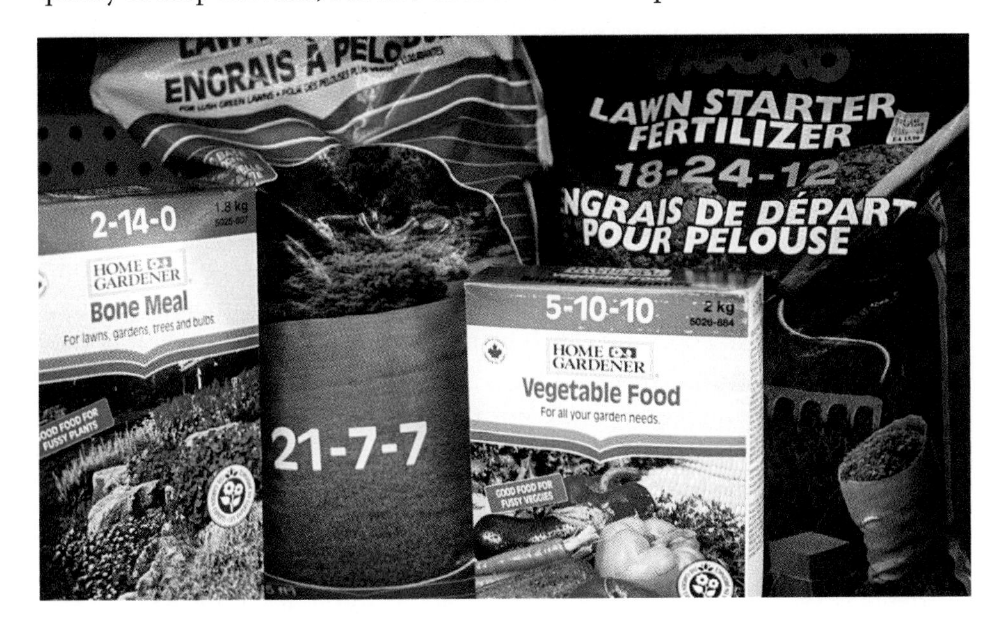

Figure 6
Synthetic fertilizers are labelled using an NPK code.

In a natural, balanced environment, any nutrients taken up by plants are replaced through natural processes such as the nitrogen cycle. Such cycles decompose the tissues that make up the plant body, close to where the plant was growing. The nutrients are therefore released in the area from which they were taken up. In agriculture, however, harvested plant material is removed from the area in which it grew. Over time, this practice can cause a steady depletion of soil nutrients. For example, corn cultivation can rapidly decrease the amount of available soil nitrogen, since high-protein corn plants take up large amounts of nitrogen as they grow. The nitrogen-rich corncobs are harvested and turned into foods for humans and livestock. Little, if any, of the nitrogen from the corncobs is returned to the fields where the corn was originally grown.

Figure 7
Municipal composting reduces the need for landfill space and produces valuable soil conditioner. Some programs use indoor digesters. The program shown here is outdoors and uses a large machine to regularly mix the yard waste with air.

Returning the nutrients in plant matter to the environment can keep soils healthy and minimize the need for synthetic fertilizers. Many municipalities have begun or are considering large-scale composting programs that will reduce the amount of plant waste that goes into landfills and will return the nutrients to the soil (**Figure 7**). However, a large amount of the nutrients in the plants we consume is released in our own wastes. Processing the sewage from large urban centres into sewage sludge, therefore, could generate a significant amount of fertilizer. If either large-scale composting or sewage is used as a source of fertilizer, however, we must ensure that any disease-causing pathogens are destroyed and that the levels of any other harmful contaminants are closely controlled.

Geographic Barriers

Few seeds and spores can cross over major geographic barriers such as mountain ranges and oceans, so such topographic features may limit natural plant distribution. As a result, often only a limited number of species will occur naturally in a region, even though many other plant species would be able to survive the environmental conditions. Some geographic barriers are so limiting that species rarely move outside these limits. An **endemic species** is found only within such geographic barriers, such as a species that is found only on one continent or on a large remote island. Throughout our history, humans have helped plants to overcome geographic barriers by transporting plants and growing them in new locations throughout the world. For example, tomatoes originated in South America but are now grown on every continent.

endemic species a species that does not naturally occur anywhere other than one geographic site on Earth

DID YOU KNOW?

Exotic Plants
The majority of food and ornamental plant varieties grown in Ontario are not native to Canada, but were imported as exotic species. This includes crops such as wheat, corn, soybeans, and potatoes. Most of these species are native to Eurasia. Corn and potatoes are native to the Americas.

Growing Plants for Human Use

People grow plants for a number of purposes. Plants may be grown for food, for the fibres they produce, to produce medicines, or for ornamentation. The first consideration in choosing which plants to grow is therefore whether they will provide what is needed. After that is decided, a grower must then find out about the environmental conditions the plants require and how these conditions will be provided. The most straightforward solution is to choose plants that will thrive in the existing environment. You might choose plant varieties according to their hardiness rating, then plant them in the soil type they prefer. If the nutrient requirements of those plants are also known, nutrients in the form of fertilizer might be added to the soil if needed.

Depending on why plants are to be grown, other variables may also be important. Choosing the appropriate plants requires some knowledge of plant growth and physiology. For ornamental gardens such as municipal parks or flowerbeds, the gardener might choose **annuals**, plants that live for only one growing season; **biennials**, plants that reproduce in their second year; or **perennials**, long-lived plants. The moisture and light conditions for each plant must also be considered. For example, most ferns thrive in moist, shaded areas, while most annuals prefer well-drained, sunny conditions. If plants are grown for their flowers, it is important to know about the timing of flowering. Does the plant produce flowers during the entire growing season, like most annuals, or does it have a much more limited flowering time, like most perennials? When selecting trees and shrubs for landscaping, growers must consider the amount of available space and the final size of the plant, as well as whether a deciduous or an evergreen variety would be most suitable.

Different factors must be considered when selecting a crop plant. For example, many fruits and vegetables require large amounts of water. How will adequate soil moisture be supplied? How do the nutritional and flavour characteristics of one variety compare with that of another? How long a growing season does the plant variety require? For example, watermelon plants can be grown in northern Ontario, but they will not have enough time between flowering and the first frosts of fall to produce mature edible fruit.

annual a plant that completes its life cycle in one year

biennial a plant that completes its life cycle in two years

perennial a plant that takes more than two years to complete its life cycle

Section 4.17 Questions

Understanding Concepts

1. What are the two main climatic factors that influence the distribution of plants?
2. What is the value of the hardiness-rating system to gardeners?
3. Can you grow white spruce in southern Ontario or magnolias in northern Ontario? Explain.
4. Peterborough is in hardiness zone 5a. Could someone living in Peterborough grow only those plants with a hardiness rating of 5a? Explain.
5. Many gardeners living in Ontario grow zone 10 flowering plants outdoors in the summer. How is this possible?
6. What are the three most important plant nutrients obtained as dissolved ions from the soil?
7. Explain why food plants diminish the supply of soil nutrients more quickly than nonfood plants.
8. Why is it critical to consider the length of the growing season when selecting food plants?
9. A gardener notices that leaves on many of her vegetable plants are turning yellow. She suspects that the plants have a nutrient deficiency. She has three bags of synthetic fertilizer, which have the following NPK ratings: 5-5-5, 10-0-10, and 0-15-15.
   (a) Which macronutrient deficiencies can produce these symptoms?
   (b) Which fertilizer is most likely to help? Which fertilizer is least likely to help? Explain.

Making Connections

10. Many people prefer to purchase organic fruits and vegetables, which are grown without the use of synthetic fertilizers or pesticides. What are the advantages and disadvantages of producing organic produce? How is this reflected in consumer demand and the cost of the food? Present your findings as a newspaper column that supports or refutes the stand that people should buy more organic produce.

www.science.nelson.com

Exploring

11. Determine the hardiness zones for your own home town and for each of the following cities: Vancouver, BC; St. Johns, NFLD; Edmonton, AB; Windsor, ON; and Timmins, ON. Use the plant hardiness zone map provided on the web site of Natural Resources Canada to identify each zone.

www.science.nelson.com

Planning a Vegetable Garden

Before planting, farmers and gardeners must choose from a wide selection of plant species and varieties (**Figure 1**). Each plant variety has particular characteristics that make it more or less suited to the conditions of a particular region and for particular market demands. In this activity, you will select three plant species for a vegetable garden. You will research the available varieties of these species and document their characteristics. You will then recommend one variety for each of the three plant species, and provide appropriate growing instructions for the varieties you recommend.

Materials

information sources on commercial plant varieties, such as nursery catalogues and seed packages

Procedure

1. Choose three plant species from the list below:
   - tomato
   - lettuce
   - sweet corn
   - bean
   - pumpkin
   - cucumber
   - carrot
   - squash
   - strawberry
   - pea
   - potato
2. Using electronic and paper resources such as catalogues, locate detailed information regarding the available varieties and their traits for the three plant species you chose.

www.science.nelson.com

3. For each of the plant species, compile information about three or more varieties in table format. You must include at least the following information in your tables:
   (i) the number of growing days required
   (ii) the qualities of each variety (e.g., colour, flavour, and size)
   (iii) the supplier and cost

(a) Complete a separate table for each species.

4. Research the growing conditions in your area. This information may be available from a local gardening centre.

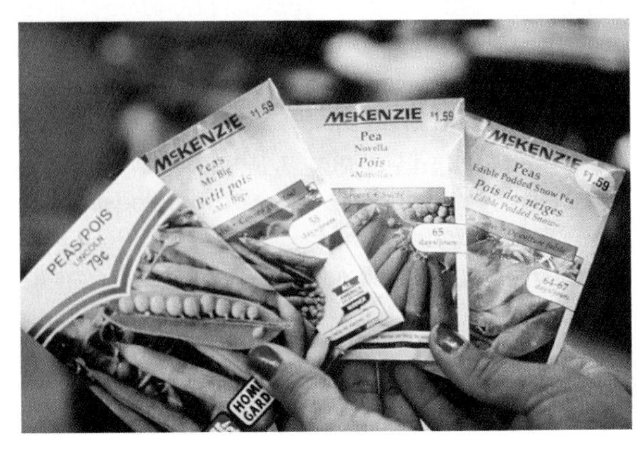

Figure 1
Seed packages of garden pea varieties

(b) How many growing days do you have in your region?

5. Using the information you have researched, select one variety of each species that you think would be the best choice for your region.

(c) Record your recommendations.

(d) Write a brief outline about how to grow each of your recommended plants.

Analysis

(e) Answer the following for each plant species:
   (i) Of the varieties you examined, which one required the longest growing season?
   (ii) Which characteristics are most important to consider when selecting a plant variety? Explain your reasoning.
   (iii) Did your recommended variety have all the best qualities, or did you have to give up one preferred quality in order to obtain another? Explain your answer.

Synthesis

(f) When choosing plant varieties to grow, what tradeoffs would people living in more northern climates most likely need to make?

Extension

(g) Order or purchase a sample of one of your recommended plant varieties. Plant, grow, and harvest your vegetable of choice, and then judge your success with a taste test!

Pest Management 4.19

Whether in natural or cultivated areas, humans compete with many other species for plant products. A host of fungi, insects, and other organisms feed on and in plants (**Figure 1**). At least one-third of the world's food production is lost to pests. In addition, the plants we use must compete for water, light, and nutrients with many unwanted plants (weeds).

Since we first began to cultivate plants, humans have looked for ways to eliminate or control these competitors. As agriculture grew to meet the demands of an increasing population, so did the need for effective pest control. Modern technology has given us a large variety of **pesticides** (**Figure 2**)—chemicals used to kill pests or unwanted **target species**. Pesticides are widely used in agriculture, forestry, and around homes and gardens. There are many different kinds, including insecticides (for killing insects), fungicides (for killing fungi), rodenticides (for killing rodents), and herbicides (for killing plants).

Figure 1
Pests can destroy the plants humans depend on; the devastation caused by gypsy moths on Canadian forests is shown here.

Figure 2
Reading the fine print on the pesticide containers is important. Does it harm pets? If so, what about the other mammals in your yard? Also check whether the pesticide might kill beneficial species of plants, animals, and insects.

The most commonly used pesticides are herbicides, fungicides, and insecticides. Some are **broad-spectrum pesticides**, killing a large variety of species, while others may be **selective pesticides**, killing a more limited range of species. Many crop fields are sprayed shortly after planting with a broad-spectrum herbicide, just before the planted crop emerges from the ground. This kills off any weed plants that have emerged, and gives the crop plants a head start toward becoming established before additional weeds begin growing. Insecticides are commonly used in and around homes, in the growing of fruits and vegetables, and in the forestry industry (**Figure 3**).

pesticide any chemical used to kill unwanted organisms

target species the particular pest species under consideration

broad-spectrum pesticide a pesticide used to kill a wide range of target and nontarget species

selective pesticide a pesticide that kills a more limited range of target and nontarget species; also called a narrow-spectrum pesticide

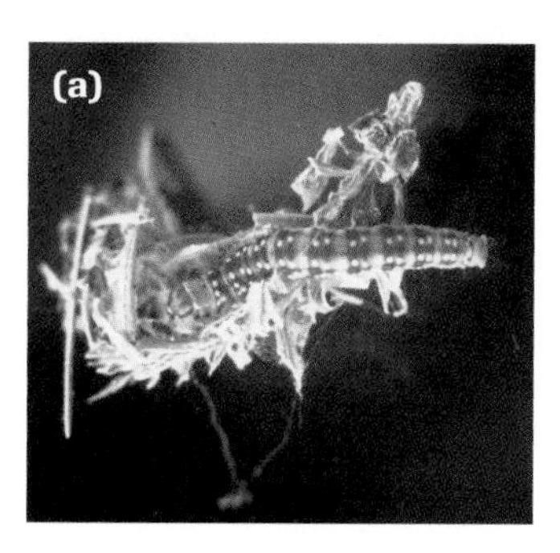

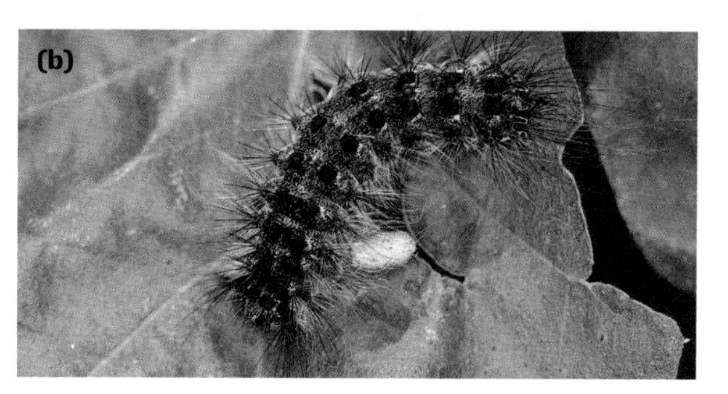

Figure 3
Two major forest insect pests are the larval stages of **(a)** the spruce budworm and **(b)** the gypsy moth.

Figure 4
In Canada, the peregrine falcon and other birds of prey have been victims of persistent chemicals, especially DDT. High levels of DDT-related residues interfered with calcium metabolism and prevented peregrine falcons from producing hard egg shells. The thin shells broke quickly before the chicks could develop. Those that managed to develop and hatch often had deformities that prevented them from thriving. As a direct result, the eastern race of the peregrine falcon is now extinct.

bioaccumulation the increase in concentration of a chemical contaminant as it is passed up through the food chain

Pesticides and Safety

All pesticides used in Canada are regulated under the federal *Pest Control Products (PCP) Act* and, in Ontario, under the *Pesticides Act.* These regulations are intended to protect human health and the natural environment through proper pesticide management. The legislation governs the sale, use, transportation, storage, and disposal of pesticides. The Government of Ontario maintains a database, the Pesticide Products Information System (PEPSIS), which lists the hundreds of pesticides approved for use in Ontario. **Table 1** lists the total mass of selected pesticides used in one year on Ontario crops. While these numbers may seem large, the total use of pesticides in Ontario is on the decline. For example, in 1998, total pesticide use had decreased by 36% over the 1983 total of 8127 tonnes.

Table 1 Pesticide Use on Ontario Cropland, 1998

Pesticide type	Mass used (tonnes of active ingredient)
herbicides	3919
insecticides	153
fungicides	649
all types	5214

Although pesticide use can dramatically increase food production at a relatively low cost in the short term, many people are concerned about the risks that pesticides may pose to the long-term health of humans and the environment. These concerns stem from a history of serious environmental problems that arose not long after the first widespread use of modern pesticides. The first synthetic pesticide to be used widely was DDT (dichlorodiphenyltrichlorethane), a low-cost, broad-spectrum insecticide. DDT use not only increased crop yields dramatically, it saved millions of human lives by controlling the mosquitoes responsible for spreading malaria, a potentially fatal disease. Unfortunately, these advantages were offset by the serious environmental impact of DDT.

DDT does not break down fully, so it persists in the environment, where it undergoes bioaccumulation. **Bioaccumulation** is the process by which the concentration of a compound increases through the food chain. DDT use caused significant declines in the populations of species at the top of the food chain, especially birds of prey such as ospreys, peregrine falcons, bald eagles, and pelicans (**Figure 4**). DDT was also linked to cancer and other human health concerns, and so is now banned in many countries, including Canada. Although newer pesticides are considered much safer than DDT, many people are concerned that the risks posed by continued widespread use of pesticides still outweigh the benefits.

Pesticide use can also seriously disrupt the natural balance of predators and prey within ecosystems by killing off large numbers of both target and nontarget species.

Alternative Methods of Pest Control

One important alternative to pesticides is the use of biological controls. All organisms, including pests, have natural predators, parasites, and diseases. Biological control reduces pest numbers by increasing the numbers of these natural controls. As a result, pesticide use can be reduced or eliminated. In greenhouses, for example, ladybugs (**Figure 5**) and parasitic wasps may be used to control other insects. Ladybugs are voracious predators of soft-bodied insects such as aphids, a common plant pest. Parasitic wasps lay their eggs on many species of insect larvae. When these eggs hatch, the wasp larvae emerge and then burrow into the pest larvae, which they consume. An increasingly popular type of biological control is **integrated pest management (IPM)**, which involves the careful monitoring of pest levels, the introduction of natural biological controls, and the use of synthetic pesticides only when necessary.

Figure 5
Ladybugs are harmless to plants but are an effective means of controlling insect pests.

integrated pest management (IPM) a pest-management strategy that uses a combination of careful monitoring, natural biological controls, and limited applications of synthetic pesticides

The most advanced and controversial technology being employed to control pests is genetic manipulation of plants. **Genetically modified organisms (GMOs)** are organisms that have been genetically altered by human intervention. Many GMOs contain useful genes from other organisms. Some GMO plants have genes that help them to resist pests. For example, Bt corn is corn that has been genetically modified by adding a gene from a soil bacterium. This gene enables the corn plant to manufacture a natural insecticide, Bt toxin, within its tissues. Any insect larvae that eat Bt corn are killed. Although Bt toxin is fatal to the larvae of moths and butterflies, it is considered safe for other species, including humans. When Bt corn is grown, there is no need to spray the crop with external insecticides. This saves money and prevents large quantities of insecticides from being released into the environment.

genetically modified organism (GMO) an organism in which the genetic material has been modified to suit human purposes, often by the introduction of genes from other species

Other examples of GMO crops grown in Canada include glyphosate-resistant canola and soybeans. Glyphosate is a broad-spectrum, synthetic, nonpersistent herbicide; glyphosate-resistant canola and soybean plants contain a gene for resistance to this herbicide. When farmers grow these GMOs, they can spray glyphosate whenever it is needed to reduce competition by weeds, at any time throughout the growing season. The reduction in weeds increases crop yields and can actually lead to a decrease in pesticide use.

DID YOU KNOW?

The Economics of Pest Control
In 1998, farmers in North America spent more than $500 million on insecticides for corn crops alone. It is widely accepted that use of Bt corn could dramatically lessen this use of pesticides.

While genetically modified plants are becoming increasingly popular within the agricultural community, many people object to the use of this technology. They argue that there are too many unknowns regarding GMOs. For example, the genes introduced into a cultivated plant may escape into wild plant populations with unforeseen environmental consequences. There have, in fact, been reports of GMOs crossing with wild populations and introducing genes into unintended species. There are also concerns about possible health risks from ingesting plants containing novel chemical compounds. Research is currently being carried out to address these concerns.

Section 4.19 Questions

Understanding Concepts

1. What are the major types of pesticides used in Ontario?
2. Is pesticide use on the increase or on the decline? Support your answer with data.
3. Distinguish between a broad-spectrum and a selective pesticide.
4. What aspects of pesticide use are governed by federal and provincial legislation?
5. Describe how persistent pesticides can lead to bioaccumulation and negative impacts on the environment.
6. Give two examples of biological controls and how they are used.
7. Describe how growing Bt corn can reduce the need for pesticide applications.
8. Growers of glyphosate-resistant canola still use pesticides in their fields. Why?
9. The common herbicide 2-4-D is a selective herbicide that kills dicots but not monocots. Would this herbicide be useful in lawn and golf-course maintenance? Explain.
10. Why might the use of broad-spectrum pesticides make it difficult to successfully incorporate biological control methods in the same field?

Applying Inquiry Skills

11. A pesticide technician is conducting tests to find the lowest concentration of a pesticide that can effectively control an insect pest. List at least three variables that must be controlled during the tests.

Making Connections

12. Although various insect predators and parasites can be purchased, they are expensive. Why might greenhouse operators be more willing to use these organisms for pest control than other plant growers?
13. Make a list of the ways that your family uses pesticides. For example, you may use pesticides to control mosquitoes, houseflies, fleas, dandelions, or other weeds. Outline the safety precautions your family should take when using pesticides for each use on your list.
14. Many people strongly believe that all genetically modified foods should be clearly labelled as such, so that consumers can choose whether or not to purchase them. Research the issue of labelling GMO foods. What are the advantages and disadvantages of labelling genetically modified foods? When your research is complete, write a letter to the editor of your local newspaper outlining your position on this issue.

www.science.nelson.com

Exploring

15. Many genetically modified crops are now available, including cotton, tomatoes, and potatoes. Research one of these plants and find out which of their characteristics have been genetically altered. Report your findings as a poster presentation.

www.science.nelson.com

Explore an Issue
Banning Pesticides in Urban Environments
4.20

Pesticides are used widely in Ontario in the agricultural and forestry industries, and along highways, railways, power lines, and gas-pipeline corridors. Most pesticide use in forestry and agriculture is for insect and weed management, in order to increase timber or crop yields. The application of herbicides along rights-of-way controls the growth of vegetation that poses a safety concern. In urban settings, homeowners, private contractors, and municipalities often use pesticides cosmetically, simply to control plants they find unattractive.

Lawn pesticide use banned!

1991—Town of Hudson, Quebec, passes bylaw banning the use of pesticides on lawns.

No more pesticide ban?

May 2002—Chemical industry fighting lawn-pesticide bans with increasing success.

New bylaw passed

August 2000—Halifax passes Bylaw P-800, the first major city in Canada to pass legislation aimed at reducing urban pesticide use.

Supreme Court upholds pesticide ban

June 2001—Canada's Supreme Court Upholds Hudson Ban on Pesticide Use—based on legal authority, not scientific concern over pesticides.

Communities say "No" to pesticides

December 2002—42 Ontario communities are considering laws to ban or restrict pesticide use on private property.

Figure 1
Pesticide use in urban areas can be a controversial issue.

As the headlines in **Figure 1** attest, people in recent years have become more concerned about widespread use of pesticides, especially in urban areas. They worry that excessive use of these chemicals poses a risk to both people and the environment. While many are willing to accept limited use of some pesticides in farming and forestry, they oppose the cosmetic use of pesticides on lawns, ornamental gardens, and parklands. They argue that the risks to the health of children and to the environment are too great a price to pay for making lawns and gardens weed free, especially in urban areas or near rural schools and daycares (**Figure 2**).

Supporters of pesticide use argue that these chemicals are important tools in the proper care of lawns and gardens. They point out that scientific evidence supports the claim that proper application of pesticides does not threaten our health or the environment. Many homeowners apply pesticides themselves to control weeds, or hire private landscape-maintenance companies to do so. Golf-course managers use herbicides to maintain the quality of their fairways and putting greens. Municipalities can save money by spraying herbicides to control weeds on parkland rather than paying staff to destroy them manually.

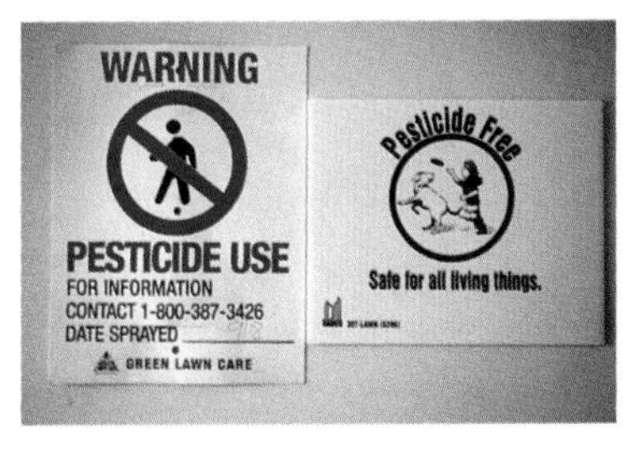

Figure 2
Many municipalities require posted warnings following the application of pesticides.

You may think that pesticide use should be more of a concern in rural, agricultural areas than in cities. Surprisingly, pesticide use per hectare in urban environments is actually twice that of rural farmland. People in urban areas are also more likely to be directly exposed to any applied pesticide, simply because more people are around. Many homeowners do not educate themselves about proper safety procedures, and even commercial licensed applicators do not always follow procedures correctly (**Table 1**).

Table 1 Ontario Ministry of the Environment Swat Team—2001 Pesticide Applicator Inspections Results

Districts	Number of inspections	POAs*
Hamilton	2	2
Toronto	41	25
Ottawa	2	2
Peterborough	2	2
Halton/Peel	9	5
York/Durham	15	5
Totals	71	41

* Number of tickets, summonses, or notices issued under the Provincial Offences Act (POA). These are issued when companies or individuals violate Ontario's environmental laws, such as by using improper signage on vehicles.

Understanding the Issue

1. Compare and contrast the reasons pesticides are used in urban areas with the reasons they are used in rural areas.
2. What is meant by "cosmetic use" of a pesticide?
3. What factors specific to urban areas increase the danger to people from pesticides?
4. Can the public be confident that all pesticide users are following proper regulations? Explain.

Take a Stand

Should pesticide use be banned?

Decision-Making Skills

○ Define the Issue ● Analyze the Issue ● Research
● Defend a Decision ● Identify Alternatives ● Evaluate

Statement:
The cosmetic use of pesticides in urban settings should be banned.

- In your group, research the issue and learn more about the use of pesticides in urban settings.
- Search for information in newspapers, periodicals, and CD-ROMs, and on the Internet.

www.science.nelson.com

(a) What are the most commonly used pesticides in cities?
(b) Do these pesticides pose a significant threat to the health of people and/or the environment?
(c) What are the alternatives to pesticide use?
(d) Can the continued use of pesticides be justified?
(e) Does your school and/or community have policies or bylaws concerning pesticide use on public and private land?
(f) Conduct a survey of your parents, neighbours, and fellow students. Prepare a set of questions to find out what they know about the science of pesticides, and whether they would support a ban or restrictions on pesticide use in the community.
(g) Make a list of the points and counterpoints that your group considered, then prepare a set of recommendations for your school or local town council or planning committee.
(h) How did your group reach a consensus on your recommendations? If you were to try this again, what would you do differently? Explain.

Case Study 4.21 IPM Technology

Integrated Pest Management (IPM) is a pest-control method that employs a combination of traditional and new techniques. IPM minimizes the use of chemical controls, thereby reducing possible health and environment risks and sometimes lowering costs. Some of the "tools" used for IPM include the following: crop-management practices that reduce the incidence of diseases, weeds, and insects; physical barriers and traps to protect crops from insect damage; pest-resistant plant varieties; and beneficial insects and bacteria.

Mate disruption is a promising new technique that can be used in IPM. Moths are among the most destructive orchard pests. Moths attract their mates using species-specific chemicals called **pheromones**. When they are ready to mate, female moths release minute quantities of pheromones into the air. The male moths are highly sensitive to pheromones of females of their species (**Figure 1**) and locate females by following these chemical scent trails. Scientists have identified and synthesized the pheromones of some moth species. Growers release the synthetic pheromones into the air as a decoy. Males follow the scent trails to dispensers containing the synthetic decoy pheromones, leaving females waiting for a mate. As a result, the reproductive rate of the moths is reduced.

Figure 1
The antennae of male moths are extremely sensitive and able to detect minute quantities of the sex pheromones released by the females.

pheromone a chemical emitted by an organism into the environment as a signal to another organism, usually of the same species

Grape production is big business in Ontario, which has more than 5000 ha of productive vineyards (**Figure 2**). The grape berry moth, *Endopiza viteana*, is the most destructive insect pest of Ontario grapes. In the Niagara region, *E. viteana* produces three generations of larvae per year. The first generation feeds on flowers or young berries, while the next two generations feed on the maturing fruit. Traditionally, grape growers controlled this pest with two to six applications of insecticides each year. From 1989 to 1993, mate-disruption methods were studied as an alternative to spraying for control of grape berry moths in vineyards.

Figure 2
Grapes are grown commercially in warmer regions of Ontario such as the Niagara region and Pelee Island.

The sex pheromone of the grape berry moth has been identified as the chemical 9-dodecenyl acetate. This compound was synthesized and made available for the study. Grape growers placed 1500 dispensers containing the synthetic pheromone in each of the 1.5-ha test plots. Control plots received traditional insecticide applications. **Table 1**, on the next page, summarizes the test results of the experiments on two farms where the trials were conducted.

Table 1 Percentage of Grape Clusters Infected with *E. viteana* at Harvest

Year	Farm 1		Farm 2	
	Pheromone	*Insecticide*	*Pheromone*	*Insecticide*
1989	45.8	20.6	4.8	16.9
1990	3.0	1.7	1.5	3.8
1991	4.5	3.3	8.4	19.1
1992	1.1	0.0	0.7	4.5
1993	0.4	0.2	7.8	11.9

Note: In 1989, dispensers contained 150 mg of active ingredients and were placed in the fields for the entire season. From 1990 to 1993, a different type of dispenser was used that contained only 66 mg of the active ingredients. These were used twice, with fresh dispensers replacing old ones after two months.

CAREER CONNECTION

Pest-management technicians are skilled, trained workers who are employed to control pests in many industries, including agriculture, forestry, catering, landscaping, and real estate management.

At present, only a small number of grape growers in the Niagara region are using this technology. In 1996, fewer than 200 of the 5400 ha of planted grapes were treated in this manner. The cost of traditional pesticide applications ranged from $45 to $210 per hectare for one growing season, while that of decoy mate-disruption methods ranged from $196 to $249 per ha, or about three times the cost, on average.

Case Study 4.21 Questions

Understanding Concepts

1. What combination of technologies is used in IPM?
2. (a) How do moths find their mates?
   (b) How is this information used to control their reproductive success?
3. What are the advantages of mate-disruption methods over pesticide applications?
4. In 1988, the Government of Ontario set a target of reducing the use of pesticides in the province by 50%. Does mate-disruption technology have the potential to contribute to this goal? Why?

Applying Inquiry Skills

5. (a) Using the data in **Table 1**, calculate the percent change in grape infestations of grape berry moths that occurred when the pest was controlled using pheromones, and when it was controlled using pesticides. Percent change is calculated as shown in the example below:

   Farm 1: percent change in 1989

   $$\% \text{ difference} = \frac{(45.8 - 20.6)}{20.6} \times 100\%$$

   = 122% increase in infestation rate with pheromones

   (b) Copy and complete **Table 2** in your notebook, using the values you calculated in (a).

Table 2 Percent Change in Infestation Rates with Pheromones versus Pesticides.

Year	Farm 1 (+/−)	Farm 2 (+/−)
1989	+ 122%	
1990		

   (c) Write a hypothesis to explain the differences in the results from Farm 1 and Farm 2.
   (d) There was a dramatic change in the infestation level in grapes after 1989. What factors might have contributed to this change?

Making Connections

6. Do you think consumers would pay more for food that was produced without the use of pesticides? Justify your response.
7. Apple growers can make use of IPM technology to control some insect pests in their orchards. Find out what species of moth is a common pest of apples in Ontario orchards. Analyze the costs and benefits of using mate-disruption methods to control this moth. Present your results as a poster or web page.

www.science.nelson.com

Working with Plants 4.22

As you have seen throughout this unit, humans use plants in many different ways. There are therefore many employment opportunities for those interested in working with plants. This section will briefly introduce a few of these.

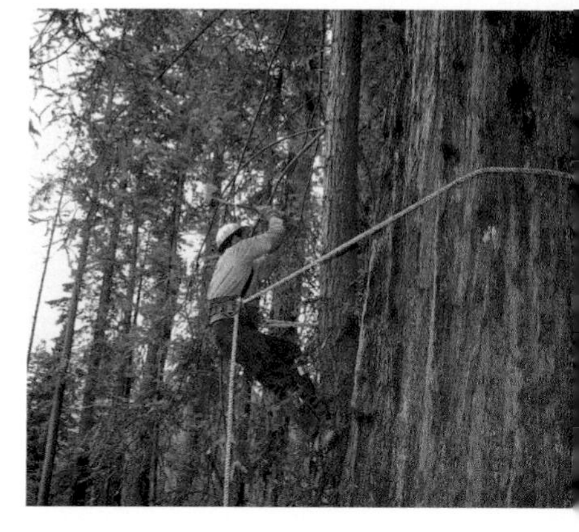

Figure 1
Professional tree climbers have the necessary training and skills to remove dead limbs and prune branches safely.

Arboriculture

How about a career as a professional tree climber (**Figure 1**)? Courses in tree climbing are often part of college programs in **arboriculture**, the cultivation of trees and shrubs. Students who study arboriculture learn how to raise and propagate trees and shrubs, and how to keep trees and shrubs healthy.

Trees are particularly valuable in urban centres, since they improve air quality by reducing the levels of carbon dioxide and other pollutants produced by human activities. They also provide green space, which provides a welcome change from the concrete and asphalt of cities. Urban trees must be well cared for to maintain their health. As trees age, they can become weak or unhealthy, or some of their limbs may die. These conditions can be dangerous, especially during windstorms when breakage can occur. **Arborists** (people who work in aboriculture) maintain trees by removing old, large trees or dead limbs, and pruning trees to grow away from power lines and other objects. Arborists also control pests and diseases on trees and shrubs.

arboriculture the activity of growing and caring for trees and shrubs

arborist a person who has expertise in the practical aspects of growing and maintaining trees and shrubs

silviculture the growing and management of forests and tree plantations

Forestry Technology

Forestry is one of Ontario's largest industries and a major employer. Graduates of programs in forestry technology (forestry technicians) are well-placed for jobs in this industry. Forestry technicians practise **silviculture**, the growing and management of forests and tree plantations. Forestry technicians must have a broad understanding of forest harvesting and management science and technology. They are responsible for measuring forest resources from observations in the field (**Figure 2**) and from aerial photographs.

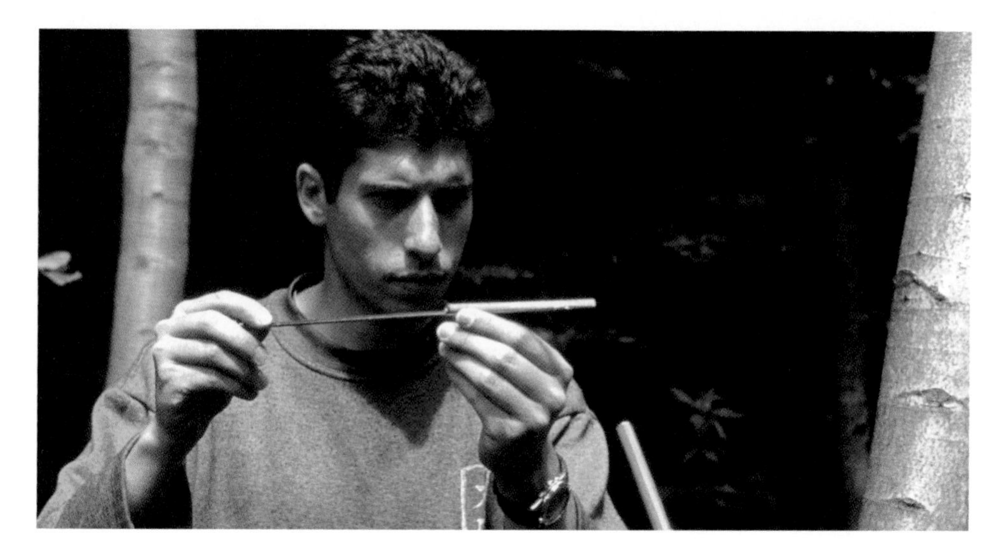

Figure 2
Forestry technicians may study core samples of the trees in an area to estimate the age of the trees and to monitor changes in the environmental conditions.

Forestry technicians are also knowledgeable about forest fires, including the ecological role of natural fires (**Figure 3**), firefighting methods, and the use of prescribed burns as a tool for forest management. Forestry technicians play a role in maintaining other wildlife species by using sustainable forestry practices.

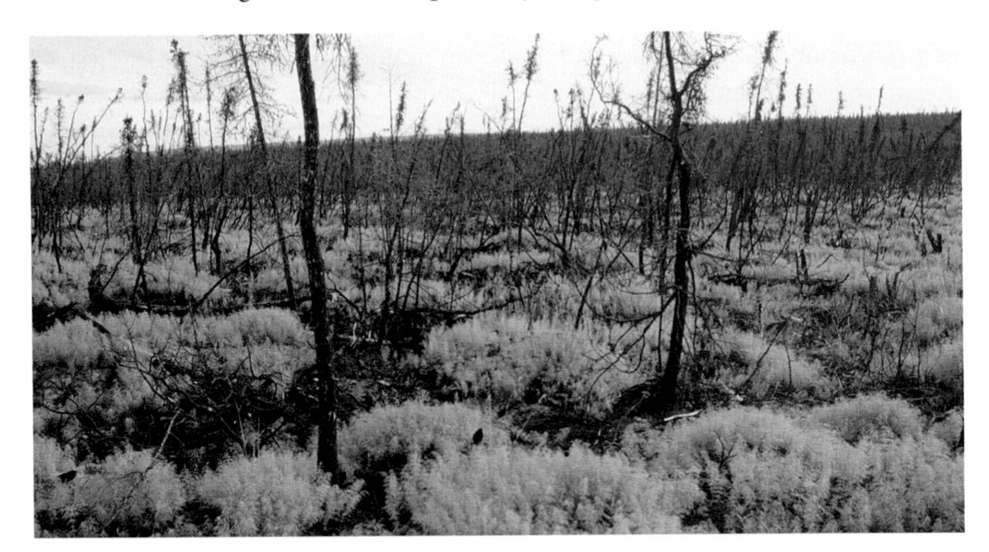

Figure 3
New life beginning after a large-scale forest fire

Horticulture

horticulture the cultivation of garden plants, including fruits, grasses, flowers, vegetables, and trees

Horticulture is the science and art of cultivating fruits, vegetables, flowers, and ornamental plants. Practitioners of this discipline (horticulturalists) work in nurseries, greenhouses, and garden centres, and for landscaping firms. Horticulturists require a thorough understanding of the environmental needs and life cycles of a wide variety of plant species. They may work on anything from small, home garden projects to large public displays such as the Butchart Gardens in Victoria, B.C. (**Figure 4**). Water gardens are also becoming popular and an area of specialization for horticulturists (**Figure 5**).

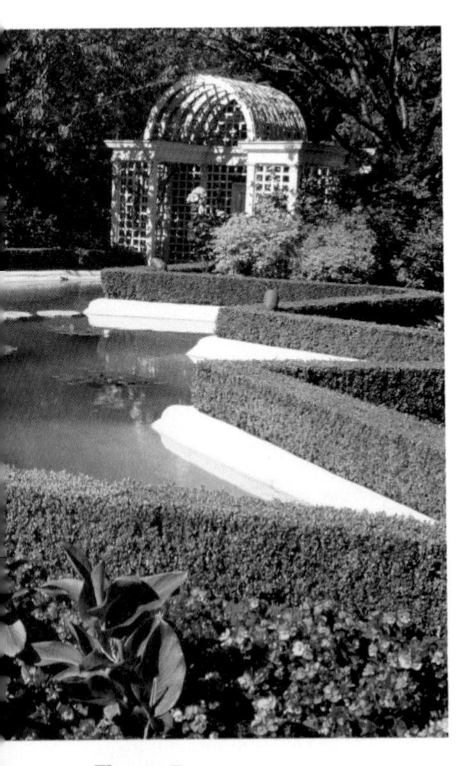

Figure 5
A large water feature at Butchart Gardens

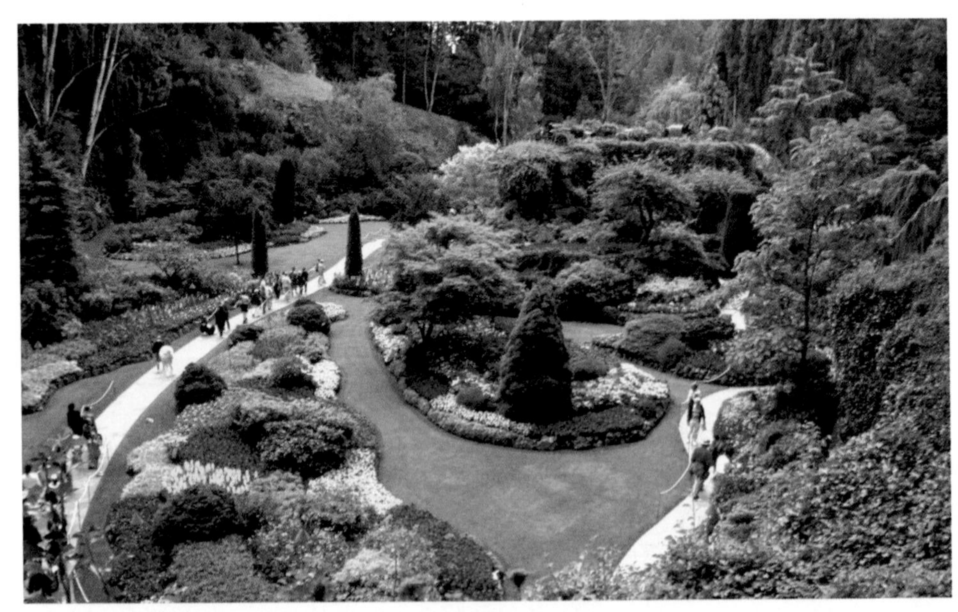

Figure 4
Each year more than 1 million annuals, comprising 700 different varieties, are planted throughout the large displays of the Butchart Gardens.

The design and construction of ornamental gardens is a major task. Many gardens are now designed using computer-assisted-drawing (CAD) software. This technology enables designers to provide their clients with a realistic image of the finished garden before work begins (**Figure 6**). The technology can even show how the appearance of the plants will change in winter, spring, summer, and fall.

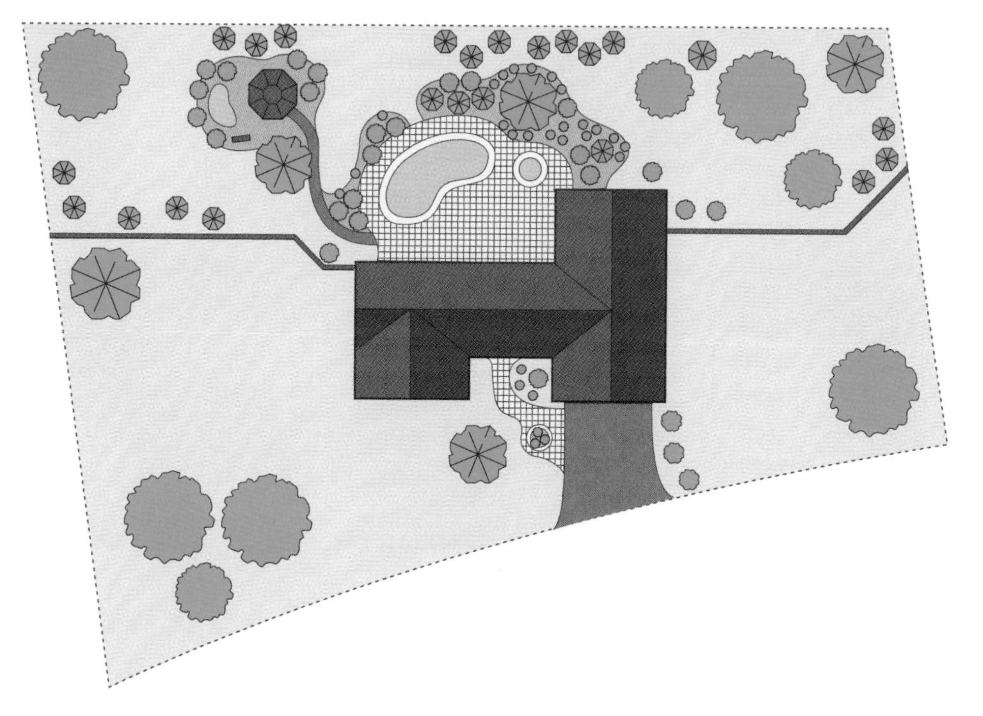

Figure 6
Software enables designers to produce accurate layouts of gardens.

Hydroponics

Growing plants without soil is now common practice in homes and commercial greenhouses. In **hydroponics**, the soil is replaced by a sterile solution of aerated water and essential plant nutrients, or sterile sand and the nutrient solution (**Figure 7**). Growing plants hydroponically can reduce or eliminate pathogens and pests that normally reside in soils.

hydroponics a technique for growing plants in solution rather than in soil

Successful hydroponics operations require precise regulation of nutrient concentrations that are tailored to the changing needs of the growing plants. Hydroponics technicians must therefore be knowledgeable about the environmental factors that affect the growth and development of the plants with which they work. They must also be able to maintain and operate the specialized equipment in a hydroponics operation.

Figure 7
Hydroponically grown lettuce has a much larger root system than lettuce grown in soil.

Turf Management

Expertise in growing and maintaining lawns, or **turf management**, is highly sought after. Golf courses, sports fields, large public parks, and home and estate lawns all require care to remain in good condition (**Figure 8**, on the next page). The financial success of commercial operations such as golf courses and soccer stadiums depends heavily on their turf-management practices.

turf management the planting, growing, and maintaining of grass lawns and fields

Maintaining high-quality lawns requires careful attention to drainage, irrigation, fertilizer application, mowing, and pest management. Turf managers have training in plant and soil science, irrigation systems, and pest control.

DID YOU KNOW ?

The Grass of Greens
Most putting greens are planted with creeping bent grass. This unusual variety of grass likes cool temperatures and must be mowed and watered every 1 to 3 days. Construction of a single putting green can cost more than $100 000.

Figure 8
Golf courses are expensive to design and construct. Expertise in turf management is essential for their success.

Section 4.22 Questions

Understanding Concepts

1. Make a list of plant uses that correspond to each of the specialties discussed in this section.
2. What factors make large trees of great interest and concern in urban settings?
3. How does the study of arboriculture compare with the study of silviculture?
4. In which areas discussed in this section are irrigation and soil drainage of greatest concern? Explain your choices.
5. Which areas discussed in this section involve management of large monocultures?

Making Connections

6. For occupations in each of the specialties of forestry, horticulture, hydroponics, and turf management, state which of the following areas of knowledge would be required. Explain your reasoning.
   (a) plant diversity
   (b) natural ecosystems or ecology
   (c) tree anatomy and physiology
   (d) irrigation systems
   (e) monocot physiology
   (f) aesthetics
   (g) planting zones
7. Various colleges in Ontario offer training programs for the specialties discussed in this section. Choose one such program and conduct research to answer the following questions:
   (a) What are the entrance requirements for the program?
   (b) How many years or semesters of study are required?
   (c) What types of courses are taken during the program? List at least six courses.
   (d) What career opportunities are available for graduates of the program?

www.science.nelson.com

Case Study 4.23

Aquatic Plants and Phytoremediation

Many plant species are aquatic, living in ponds, rivers, marshes, and other wet ecosystems. **Emergent plants** are vascular aquatic plants that grow in shallow water, with their roots below the water surface and their leaves held in the air above. Common examples of emergents include cattails and arrowheads (**Figure 1**). These plants are often found growing along shorelines and in marsh environments. These environments usually provide a steady supply of water and plenty of sunlight, so cattail marshes and other wetlands often grow very rapidly, removing large amounts of nutrients from the soil. Emergent plants are therefore of great use in new technologies aimed at reducing nutrient runoffs from agricultural, mining, and industrial operations, and for the treatment of sewage waste water. This technology is called **phytoremediation**, which is the use of plants to clean up polluted soil and water.

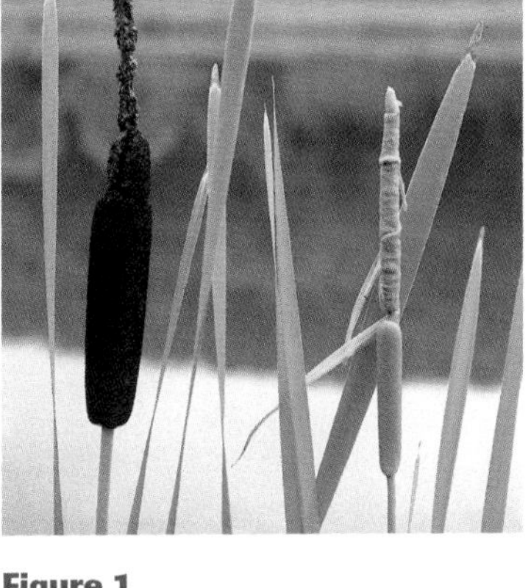

Figure 1
Cattails are common wetland plants.

emergent plant an aquatic plant that grows in shallow water with most of its shoots above the water

phytoremediation the use of plants to repair pollution-damaged ecosystems

For example, constructed marshes are sometimes created and planted to cleanse the water draining from abandoned mines. This water can have high levels of toxins such as heavy metals, which are a health risk to humans and other organisms. Cleaning up heavy-metal contamination by conventional methods can be very costly. However, some species of aquatic plants can take up heavy metals through their roots and not be harmed. The heavy metals then accumulate in the plant tissues, removing them from the immediate environment. Plants may accumulate concentrations of these metals high enough to make them worth harvesting as a mineral resource.

Aquatic plant species can also be used to treat sewage waste water, an approach with both environmental and financial benefits. Conventional treatment plants produce effluents that can contain toxic chlorinated compounds, as well as air pollutants and solid-waste sewage sludge. Conventional water-treatment facilities are also expensive to operate. Although using artificial marshes to treat waste water offers some benefits over conventional methods, it has drawbacks as well (see **Table 1**). This technology is still under development and so will likely be improved by further research.

DID YOU KNOW?

Indoor Wetlands
Some buildings, such as the Ontario Science Centre and the head office of the Body Shop in Toronto, have indoor constructed wetlands. These wetlands are used to treat waste water and improve indoor air quality.

Table 1 Using Constructed Wetlands to Treat Waste Water

Advantages	Disadvantages
Cost: Constructed wetlands can be considerably less expensive to construct and operate than conventional facilities.	Susceptibility to disease: Since they are living systems, wetlands can be damaged by diseases and cannot be quickly repaired.
Aesthetics: Wetlands are more appealing to the eye and nose than conventional water-treatment plants.	Water level: Winter conditions, spring runoff, and heavy rains can interfere with the functioning of wetlands.
Habitat creation: Wetlands construction creates biologically diverse habitats, rather than destroying them for treatment-plant construction.	Phosphorus: Constructed wetlands are poor at phosphorus uptake and removal; phosphorus can be a serious pollutant.
Added value: Constructed wetlands can also be used as recreational areas.	Life expectancy: Constructed wetlands last about 20 years, whereas conventional systems last 40 years.

DID YOU KNOW?

Remediation of Chernobyl

To remove radioactive wastes from pond water contaminated by the Chernobyl nuclear reactor accident, scientists floated sunflower plants on rafts with their roots dangling in the water. The water was cleaned at a cost of only a few dollars per 1000 L, much less than the cost of "advanced" technologies.

Restoring Cootes Paradise

Cootes Paradise is a 250-ha wetland located at the western end of Hamilton Harbour, on Lake Ontario (**Figure 2**). During the last century, human activity in the area, including the introduction of carp (an exotic fish species), resulted in serious degradation of the ecology of Cootes Paradise. By 1987, the harbour was severely polluted. It was placed on a list of 43 areas of concern in the Great Lakes basin, and a remedial action plan was put into place to begin rehabilitating the harbour. Cootes Paradise is now owned and managed as part of the Royal Botanical Gardens.

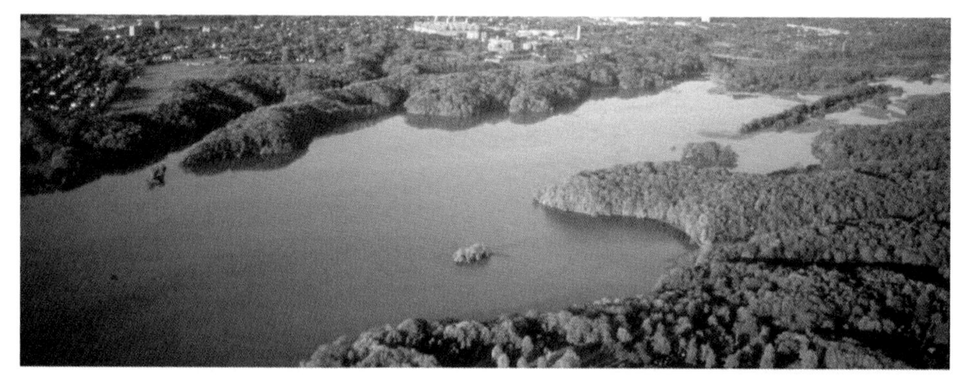

Figure 2
Cootes Paradise is a valuable wetland located at the extreme western end of Lake Ontario.

The main threats to the Cootes Paradise habitat were pollution from the overflow of sewers and the Dundas Sewage Treatment Plant, and destruction caused by carp (**Figure 3**). Carp were thriving at the expense of native fish species. They were also damaging the aquatic ecosystem by uprooting plants and reducing water clarity. To rehabilitate Cootes Paradise, polluting overflows into the wetland were dramatically reduced and a carp barrier/fishway was constructed across a canal connecting the marsh to the rest of the harbour. The Cootes Paradise Fishway prevents large carp from entering the marsh, while allowing native fish species to pass. Workers at the Royal Botanical Gardens also conducted experiments in aquatic plant establishment and developed the Aquatic Plant Nursery, which now provides 70 different species of local, native aquatic plants used in replanting the wetland (**Figure 4**).

Figure 3
During spring mating and spawning in shallow water, carp stir up sediments and make the water turbid. The turbidity prevents light from reaching submerged aquatic plants. Carp also feed in the shallow bottom sediments and damage and dislodge the roots of emergent plants.

Figure 4
Workers at the Aquatic Plant Nursery at the Royal Botanical Gardens grow local plants used in many rehabilitation programs.

Project Paradise, the Royal Botanical Garden's program to restore the habitat of Cootes Paradise, is the only one of its kind in North America. The project involves many initiatives, including the following:

- Providing local schools with seeds, potting materials, and growing instructions to produce 10 000 native aquatic plants.
- Planting more than 4 ha of habitat with emergent plants in the late 1990s. Species included cattail, arrowhead, and sweetflag (**Figure 5**).
- Installing floating cattail rafts to encourage nesting by black terns.
- Creating several small islands and improving spawning beds to enhance bird and fish habitat.
- Building the Cootes Paradise Fishway, which reduced carp numbers from 70 000 to less than 5000 in its first year of operation.
- Constructing an innovative barrier from discarded Christmas trees, which protects aquatic vegetation from damage by erosion and carp.

Figure 5
Aquatic plants such as these arrowheads were planted in Cootes Paradise to re-establish the native plants and to remove pollutants.

The remediation program is ongoing, but initial results are very promising. Water clarity has improved dramatically, vegetation is flourishing, and many species that had been absent for decades are returning to the wetland. Birds such as the least bittern, marsh wren, and Virginia rail have returned and are nesting. The grey treefrog—last seen in 1950—has also returned to Cootes Paradise. Perhaps most promising is the rapid return of underwater or **submergent plants** that had been entirely absent prior to 1997.

submergent plant an aquatic plant that grows entirely underwater

Case Study 4.23 Questions

Understanding Concepts

1. Define and give examples of phytoremediation.
2. What characteristic of marshes might account for their high biodiversity and productivity?
3. What factors were most responsible for damaging Cootes Paradise?
4. How do carp threaten aquatic vegetation?
5. There are advantages and disadvantages to using constructed wetlands for sewage waste-water treatment. Describe how the following might influence the choice between a constructed wetland and a traditional treatment facility:
   (a) serious financial concerns of the community
   (b) an interest in preserving biodiversity in a sensitive habitat
   (c) severe winter conditions in northern Ontario
   (d) local annual spring flooding problems
   (e) a growing concern about chlorine pollution in the local watershed
6. Many aquatic plant species used in the Cootes Paradise rehabilitation program grow throughout southern Ontario. Why do you think the Royal Botanical Gardens preferred to use the seeds from local plants for their Aquatic Plant Nursery?
7. List and describe the methods used to rehabilitate Cootes Paradise.
8. Compare the biodiversity of Cootes Paradise before and after the rehabilitation program was begun.

Applying Inquiry Skills

9. A key factor in the survival of submergent species is water clarity. Obtain a small sample of fine soil and mix it vigorously in a test tube of water. How does mixing affect the amount of light that penetrates the water? How long does it take for the soil to settle and the water to clear?

Making Connections

10. The common carp, *Cyprinus carpio*, was intentionally introduced into Ontario in 1880 and was first reported in Cootes Paradise in 1908. Research this fish species. Find out why it was brought to Ontario.

www.science.nelson.com

11. The Cootes Paradise rehabilitation program is dependent on many volunteers to help with the planting of aquatic species and the removal of invasive purple loosestrife. What factors would influence your decision to participate in a similar volunteer program in your region?

Unit 4 SUMMARY

Key Understandings

4.1 Plants and Biodiversity

- plants play the most critical role in supporting Earth's biodiversity
- many plants and plant communities are threatened, including some in Ontario

4.2 Classes of Plants: Nonseed Plants

- four main divisions of plants dominate terrestrial ecosystems: mosses, ferns, gymnosperms, and angiosperms
- plant reproductive cycles include an alternation of generations
- mosses and ferns reproduce sexually using free-swimming sperm

4.3 Classes of Plants: Seed Plants

- gymnosperms and angiosperms have advanced vascular systems and produce pollen and seeds
- angiosperms are the largest division of plants
- angiosperms produce flowers for sexual reproduction and fruits for dispersion
- angiosperms can be subdivided into monocots and dicots

4.4 Activity: Collecting and Classifying Plants

- plant specimens are collected, examined, and classified using identification keys and or field guides

4.5 Vascular Plant Structure

- the plant body is composed of a below-ground root and an above-ground shoot system
- there are four plant tissues: meristematic tissue, dermal tissue, ground tissue, and vascular tissue
- mitotic division in plant bodies is restricted to meristems
- dermal tissue covers the exterior surfaces of plants, and serves defensive and protective functions
- ground tissues perform most of the storage and metabolic activities of plants
- vascular tissue is composed of xylem tissue and phloem tissue; its role is to transport materials within the plant

4.6 Plant Development

- growth is an increase in the size of an organism, and differentiation is a change in its form or function
- primary growth includes the growth in the length of roots and shoots and occurs in apical meristem regions
- secondary growth increases the diameter of roots and shoots and results from the activity of lateral meristem regions
- differentiation takes place from the cells of the apical meristems and the lateral meristems

4.7 Investigation: Oxygen Consumption by Pea Seeds

- the oxygen consumption of germinating plant seeds is investigated using a respirometer

4.8 Roots and Stems

- plants have extensive root systems that provide support, take up water and minerals, and store food
- stems support the leaves and reproductive organs, transport materials between the roots and shoots, and store food
- wood is composed of xylem tissue, and woody stems are a valuable source for pulp-and-paper and lumber products

4.9 Tech Connect: Engineering Wood Products

- advances in technology are applied to the manufacture of a variety of large-dimension wood products such as plywood and oriented strand board

4.10 Leaves

- the structure of leaves enhances their ability to perform photosynthesis while preventing excess water loss
- gas exchange and transpiration are regulated by the opening and closing of stomata
- leaves are the source of many useful products, including medicines

4.11 Activity: Separating Compounds in Plant Leaves

- paper chromatography is used to separate and identify leaf pigments

4.12 Seeds and Fruits

- angiosperms produce seeds and fruits

- angiosperms include two main groups of plants, the monocots and the dicots
- monocots are nonwoody and include the grasses, which provide the major source of food for humans and livestock
- dicots include all the flowering trees and many flowering herbaceous plants; most commercially grown fruits and vegetables are dicots
- the successful pollination of crops is of vital economic and ecological concern

4.13 Activity: Monocots and Dicots

- samples of monocots and dicots are sectioned and stained; their internal and external structures are compared, and their metabolic activity and chemical composition are examined

4.14 Investigation: How Do Growing Plants Respond to Their Environment?

- the influence of gravity, light, and touch on plant growth are investigated

4.15 Control of Plant Development

- plant growth factors regulate the growth and development of plants, and include auxins, gibberellins, cytokinins, abscisic acid, and ethylene
- many plant growth factors are used commercially to influence flowering and fruit production
- tropisms are changes in the growth pattern or movement of plants in response to environmental conditions, and are mediated by plant growth factors
- photoperiodism is an example of how plants respond to environmental factors such as light and temperature

4.16 Investigation: The Effect of Abscisic Acid on Transpiration

- the influence of abscisic acid on guard cells and transpiration rates is investigated

4.17 Environmental Factors and Plant Growth

- plant distribution is influenced by large-scale and small-scale factors
- plant hardiness zones are used to relate the geographic range of climate conditions to the hardiness of plant species
- plants require different soil types and pH levels
- plants need different levels of mineral nutrients, including nitrogen, phosphorus, and potassium
- sources of these nutrients can be either natural (manure and compost) or synthetic (commercial fertilizers)
- geographic barriers restrict the distribution of plants in nature, but can be overcome by human intervention

4.18 Activity: Planning a Vegetable Garden

- a range of variables are considered in selecting varieties of vegetables for a home garden

4.19 Pest Management

- pesticides are used extensively to protect desirable plants from insects and fungi, and reduce competition from weeds
- pesticide use raises serious safety concerns; alternatives are being actively researched and employed

4.20 Explore an Issue: Banning Pesticides in Urban Environments

- discussing the cosmetic use of pesticides in urban settings

4.21 Case Study: IPM Technology

- new technologies such as mate disruption have the potential to reduce the need for pesticides

4.22 Working with Plants

- arboriculture, forestry, horticulture, hydroponics, and turf management are all areas of specialization that offer employment opportunities in working with plants
- working with plants requires a wide range of knowledge, skills, and training

4.23 Case Study: Aquatic Plants and Phytoremediation

- the remediation of Cootes Paradise by replanting native aquatic plants is explored as an example of phytoremediation, a technology that uses plants to clean up polluted soil and water

Key Terms

4.1
biodiversity
monoculture
invasive species

4.2
seed
moss
vascular system
fern
alternation of generations
sporophyte
gametophyte
bryophyte
rhizoid
frond
rhizome

4.3
gymnosperm
angiosperm
woody
conifer
evergreen
pollen grain
herbaceous plant
nectar
pollinate
dormant
monocot
cotyledon
dicot
endosperm
seed coat
stamen
anther
pistil
style
stigma
ovary
fruit

4.5
shoot
root
terminal bud
axillary bud
leaf blade
leaf petiole
root tip
primary root
secondary root
meristematic tissue
growth
differentiation
dermal tissue
epidermis
ground tissue
vascular tissue
xylem
phloem
sieve-tube element
sieve tube
companion cell

4.6
apical meristem
lateral meristem
primary growth
secondary growth

4.8
taproot
fibrous root
root cap
vascular cambium
bark
annual ring

4.10
vein
node
internode
simple leaf
compound leaf
mesophyll cell
palisade mesophyll
spongy mesophyll
stoma
transpiration
guard cell

4.15
plant growth regulator
tropism
thigmotropism
phototropism
gravitropism
photoperiod

4.17
hardiness
hardiness zone
sand
clay
silt
humus
compost
manure
macronutrient
micronutrient
fertilizer
natural fertilizer
synthetic fertilizer
endemic species
annual
biennial
perennial

4.19
pesticide
target species
broad-spectrum pesticide
selective pesticide
bioaccumulation
integrated pest management (IPM)
genetically modified organism (GMO)

4.21
pheromone

4.22
arboriculture
arborist
silviculture
horticulture
hydroponics
turf management

4.23
emergent plant
phytoremediation
submergent plant

▸ MAKE a summary

This unit has emphasized three distinct aspects of plant science:

- plant diversity, anatomy, and physiology
- factors affecting the growth and development of plants
- uses and value of plants to humans and the natural environment

In your notebook, complete the following tables:

(a) Summarize plant structure and function.

Major plant part	Structures and tissues	Functions and adaptations

(b) Summarize the factors that affect the growth and development of plants.

Internal factors	Effect on plant	Technology used to influence factor

External factors	Effect on plant	Technology used to influence factor

(c) Summarize the values and uses of plants in nature and in meeting human needs.

Plant value in nature	Human uses and needs for plants

Unit 4 Plant Structure and Physiology

PERFORMANCE TASK

Criteria

Process

- Use laboratory equipment properly and safely.
- Document procedures used to answer each inquiry.
- Record observations in an organized and effective manner.
- Find and gather relevant information.

Product

- Compile, organize, and present research data using a variety of formats.
- Include properly labelled drawings or digital images of specimens.
- Address each customer inquiry thoroughly and justify your recommendations.
- Demonstrate an ability to apply knowledge of plant biology to specific problems.
- Use both scientific terminology and the English language effectively.

Your Growing Concerns!

Plants are extremely valuable organisms that provide humans with a great many products. Experts in the agriculture and forestry industries produce some of these products on large scale, such as fibres, wood products, and foods. However, homeowners are often responsible for tending plants used in other ways—in lawns, flowerbeds, and vegetable gardens (**Figures 1** and **2**). Most homeowners have no special expertise in plant science, and so have many questions about the selection and care of plants. They obtain answers to these questions by doing their own research or by seeking the advice of trained professionals. Before investing much time, effort, and money in digging new gardens and purchasing plants, homeowners often consult the experts at their local garden centre or nursery.

In this activity, you will play the role of an employee at a new garden centre called ***Your Growing Concerns!*** As an employee trained in plant science, you have been assigned to an expert group whose duty is to answer customer inquiries. Your customers will have a variety of questions and concerns. Your group will be asked to

- identify plant specimens and plant parts
- help choose appropriate plants for a particular garden situation
- recommend control methods for a particular pest

This activity will require you to apply the knowledge and laboratory skills you have developed in this unit, as well as to conduct research using electronic and print resources.

Questions

(a) The members of your expert group will be given a number of questions representing customer inquiries. These inquiries may or may not be accompanied by a living or preserved plant specimen. The following is a sample customer inquiry:

Your Growing Concerns!

Customer Inquiry: *I am interested in planting some spring flowering bulbs in my flower garden. I have heard that daffodils, crocuses, and tulips all grow well in my region. I want to plant some bulbs in a shade garden and others in an open sunny location. Which should I choose for each location? Also, please advise me on the height and colours that might be available.*

Cindy B., New Liskeard, ON

Materials

As you proceed through this activity, list the materials you use. Record any reference sources for the information you include in your recommendations.

Procedure

1. Obtain a set of customer-inquiry cards from your teacher.
2. Once you have received living and preserved materials to identify or classify, examine the specimens using a hand lens or dissecting microscope. Make cross sections of the specimens, if necessary.
3. Make drawings or take photographs to document the appearance of the specimens.

(b) Document the characteristics of each specimen and, if necessary, determine whether it is a monocot or a dicot.

4. Research and gather any information needed to respond to all customer inquiries. Take care to note your sources of information.

 www.science.nelson.com

Analysis

(c) Answer each inquiry by making recommendations based on your research and assessment of the information you have gathered. Include explanations and justifications with each response. The customers should have a clear understanding of how you arrived at your recommendations.

(i) If you are recommending a particular plant variety or varieties to a customer, prepare a "Growing Tips" card similar to the one below. A sample template for a vegetable garden plant is given on the left of the card, and a sample template for a flowering plant on the right.

Your Growing Concerns! Growing Tips Information Card

Vegetable Plant: Variety Name	Flowering Plant: Variety Name
Characteristics (if they apply): colour, size, flavour, hardiness, growing days required, watering needs	Characteristics (if they apply): colour, blooming time, height, annual or perennial, hardiness
How to plant them: when to plant, how deep, etc.	Where to plant them: sun/shade, moist or dry conditions, special considerations
Cost and supplier	Cost and supplier

(ii) If you are answering a customer question about pest control, obtain relevant safety information for any pesticides. Recommend methods that are not only effective but also the least likely to pose a hazard to the health of the customer or his or her local environment.

Figure 1
The proper selection of plants helps to ensure healthy and beautiful gardens.

Figure 2
Knowledge of the particular needs of each plant is very important. This plant, a rhododendron, requires shade and very acidic soil to grow well.

Understanding Concepts

1. Make a table in your notebook with headings for the four main groups of terrestrial plants, and then place the following terms under the appropriate heading(s).
   (a) produce pollen
   (b) gametophyte is free living
   (c) male sex cells are mobile sperm
   (d) no vascular tissue
   (e) reproduce using seeds
   (f) sporophyte is small and nonphotosynthetic
   (g) produce flowers and fruits
   (h) are all woody trees
2. For each characteristic below, explain how it enhances a plant's ability to survive and reproduce in a terrestrial environment.
   (a) waxy cuticle
   (b) vascular tissue
   (c) woody tissue
   (d) pollen
   (e) seeds
   (f) flowers
   (g) fruit
3. Make a table with headings for each major plant part. Under each heading, list foods that are obtained from that plant part.
4. As in question 3, create a table with headings for each major plant part. Under each heading, list *nonfood* products obtained from these same plant parts.
5. In this unit, you collected and classified a number of plants that grow in your local environment. State the plant characteristics that were most useful in this process.
6. Angiosperms can be subdivided into monocots and dicots. Compare and contrast the physical characteristics of these two plant subgroups.
7. List the four main types of plant tissues and describe the key functions of each.
8. Draw a fully labelled diagram of a respirometer. How does the design of the respirometer ensure that only oxygen-gas consumption is measured?
9. Describe the vascular system of plants. Include brief descriptions of the characteristics and functions of xylem and phloem.
10. State the most common function of each of the following seed structures:
    (a) seed coat
    (b) endosperm
    (c) cotyledon
    (d) embryo
11. Explain how each of the following is related to pollination:
    (a) production of nectar
    (b) large, colourful flowers
    (c) cones
    (d) increased apple production in orchards
    (e) releasing bumblebees in greenhouses
    (f) attempts to produce seedless cucumbers
12. State the primary roles in plants of each of the following plant growth regulators:
    (a) auxin
    (b) ethylene
    (c) cytokinin
    (d) gibberellin
13. For each plant growth regulator in question 12, describe one way in which plant growers or users use the compound for a technological application.
14. For each of the following plant structures, give two examples of an adaptation that is found in some species.
    (a) roots
    (b) leaves
    (c) stems
15. Using specific examples, explain how both regional and local factors influence the distribution of plants.
16. What is a plant hardiness zone? Could a zone 3 plant survive in a zone 4 climate? Explain.
17. Compare and contrast natural and synthetic fertilizers. Include examples of each.
18. What tropism(s) could have caused the growth pattern in **Figure 1**? Explain.

Figure 1

19. Copy and complete **Table 1** in your notebook:

Table 1

Nutrient	Main function(s) within the plant
nitrogen	
potassium	
phosphorus	
calcium	
magnesium	
sulfur	

20. Clearly explain how the harvesting of crops can result in a rapid loss of nutrients from the soil. Include an example.
21. Determine the mass of each kind of nutrient in a 10-kg bag of 15-5-10 commercial fertilizer.
22. Why is information about hardiness zones and hardiness ratings not useful or necessary when choosing annual flowering plants?
23. Give two examples each of situations in which one might use the following pesticides:
    (a) insecticides
    (b) fungicides
24. Name three uses of herbicides other than weed control in agriculture and home gardens.
25. What is "cosmetic use" of a pesticide? Why might people be opposed to such a use?
26. Greenhouse operators grow plants in entirely artificial environments. Briefly outline the ways in which a greenhouse operator might control the following:
    (a) pests
    (b) pollination
    (c) plant nutrients
27. The grape berry moth can be controlled using mate disruption. Explain this method of pest control by answering the following questions:
    (a) What is a pheromone? How and when is it produced?
    (b) What is 9-dodecenyl acetate?
    (c) How is this compound released in vineyards?
    (d) How does the release of the compound lower the numbers of moth larvae infecting grapes?
28. Compare native species and non-native species growing in the same area with regard to the following characteristics:
    (a) number of diseases and parasites
    (b) number of natural herbivores
    (c) reaction of local herbivores
    (d) chances of being invasive
29. In your notebook, draw a table similar to **Table 2** and list five specific characteristics that are used for plant identification and classification. Provide examples of plants with and without each characteristic. A sample answer is given below.

Table 2

Characteristic used to identify and classify plants	Plants with (+) and without (−) the characteristic
1. compound leaves	ash (+), maple (−)
2.	

30. Define each of the following terms, and suggest an occupation for which training in this area would be helpful or appropriate. Use a dictionary to find the meaning of any new terms.
    (a) arboriculture
    (b) horticulture
    (c) agronomy
    (d) silviculture
    (e) viticulture
    (f) hydroponics
    (g) turf management
31. Define each of the following terms and explain how each of them can benefit the plant.
    (a) gravitropism
    (b) phototropism
    (c) thigmotropism
    (d) photoperiod
32. What factors are most responsible for the great amount of biological diversity of tropical rain forests? What human activities are the biggest threat to this diversity?
33. During what months of the year would you expect long-day plants to bloom? Why?

Applying Inquiry Skills

34. Identify the structures shown in the following photomicrographs (**Figure 2**):

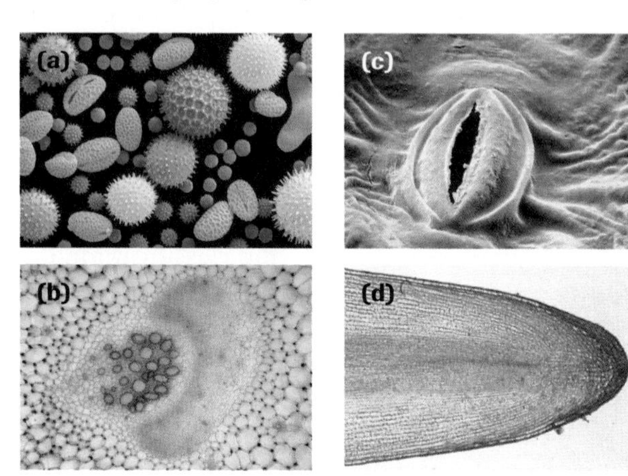

Figure 2

35. Based on the soil test results given in **Table 3**, what actions would you take to make the soil more suitable for the growing of azaleas, which prefer fertile, acidic soil?

Table 3

Test	Result
nitrogen	low
phosphorus	medium
potassium	high
water retention	poor
acidity (pH)	7.8 (basic)

36. Based on the results of your chromatography experiment, do you think scientists could use chromatography to separate chemicals other than plant pigments? Explain.

37. Interpret the test results given in **Table 4**. Describe the probable type of tissue, the metabolic activity, and the chemical nature of the samples.

Table 4

Tissue	Stain	Result
A	TTC	red
	toluidine blue O (TBO)	pink-purple
B	TTC	colourless
	Lugol's iodine	purple-black
C	toluidine blue O (TBO)	blue-green
	Lugol's iodine	yellow

38. Some plants have leaves that are deep red or purple. Based on your knowledge of photosynthesis, do you predict that these plants lack chlorophyll? Obtain a sample of a non-green leaf (such as from a coleus). Apply your skills in chromatography to conduct an experiment and test your prediction.

39. Henna is a reddish-brown pigment extracted from leaves of the henna plant. People sometimes use henna to dye their hair. List the steps you could use to separate this chemical from the leaves.

Making Connections

40. One challenge to growing plants in a space station arises from plant tropisms. Which plant tropism is likely to cause the most problems? How might engineers design a space-station greenhouse to overcome these problems?

41. Photoperiod is one factor that must be considered in order to successfully grow plants in a space station. Research the length of daylight for the International Space Station. What problems might arise, and how could they be overcome?

www.science.nelson.com

42. Potatoes can produce a toxic protein, called solanin, in photosynthesizing tissues. When potato tubers are left exposed to light for long periods of time, they turn green. Based on your knowledge of plant pigments, what is likely the cause of the green colour? Is there a safety concern about eating green potatoes?

43. Spiders produce a large number of different toxins, some of which are selectively toxic to other insects. Plant biologists have suggested that plants could be genetically engineered to produce these spider toxins. Ideally, the plants would be toxic to all insect pests and safe for humans and livestock. How do you think consumers would react to such plants? What are the potential risks and benefits of developing such plants?

44. In your notebook, copy and complete **Table 5.** Conduct research to find out the plant source of each product, the country of origin of the plant, and the uses of the product.

 www.science.nelson.com

Table 5

Plant product	Plant and country of origin	Properties and uses of product
tannin		
cotton fibres		
vegetable oil		
vanilla		
quinine		
gutta-percha		
saffron		

45. The world's most important foods and products are derived from plant crops. Conduct research to answer the following questions:
    (a) What is cassava? In what parts of the world is it a major food crop?
    (b) Mangoes and bananas are among the world's most important fleshy fruits. Which countries are major exporters of these foods?
    (c) What countries are the leading exporters of coffee beans?
    (d) From what plant do we obtain chocolate? What countries are major growers of this crop?

 www.science.nelson.com

46. Select one of the following plant products: coffee, chocolate, corn oil, cotton, table sugar, or natural rubber. Conduct research to determine what technologies are necessary for processing the natural plant material. Use a flow chart or other graphic organizer to show the steps followed to produce the final consumer product.

 www.science.nelson.com

47. In the May 2002 Parks and Recreation Ontario Update, the following advice was provided to municipal officials:

"When environmental lobbying takes place in your local municipality, and landscaping/lawn care firms inevitably line up against them, don't get caught in the crossfire! Based on the experiences of cities where this has occurred, outright lies, exaggerations, and fear-mongering tactics may be part of the dialogue, along with blatant misinformation campaigns."

    (a) What does this suggest about the care needed to evaluate and judge opinions and information?
    (b) Provide examples from your own research in this unit that illustrate bias in the reporting or interpretation of information.
    (c) What strategies can you use to detect possible bias?

48. Biological pest control is becoming increasingly popular. Many garden centres now sell ladybugs and praying mantis to home gardeners. Research these two insects.
    (a) What types of insect pests is each able to control?
    (b) Consult local garden centres in your area to see if these insects are available and, if so, at what cost?
    (c) Would you consider using these predators to help control pests in your garden? Why or why not?

 www.science.nelson.com

49. A variety of plant species that have been introduced into Ontario have become invasive. Choose and research one such species and answer the following:
    (a) When was the species introduced?
    (b) Where did the species originate?
    (c) What problems, if any, has this species caused?
    (d) What actions, if any, are being taken to control the spread of this species?

 www.science.nelson.com

unit 5

Environmental Science

At one time, many people thought that humans had the right to be dominant over all other living things and over Earth itself. Although this way of thinking may have changed, in many ways human behaviour has not. We use natural resources without regard for the fact that they are in limited supply; we pollute the very land, air, and water that we and other organisms have to live in; and we do this because we are either unable or unwilling to see and think beyond our own life span.

As a species, we have few if any predators; we have learned to eliminate or control most diseases; and we have chosen not to limit our numbers, even though we have the ability to do so. Every environment has a limit to its ability to support populations of living things. At the current population growth rate and the current rate of resource consumption, we are fast approaching the limit of Earth's ability to support the human population.

What do we need to do to ensure that future generations will have the resources they need to survive? The first step is to understand the current situation—where we are now; the second step is to understand where we are headed and, if necessary, plan to change the future. In this unit, you will further your understanding of what you already know about the natural environment and develop an understanding of what is required to sustain it.

Overall Expectations

In this unit, you will be able to

- demonstrate an understanding of factors that influence the sustainability of the natural environment and evaluate their importance;
- analyze how various factors influence the relationships between organisms and the natural environment;
- explain why it is important to be aware of the impact of human activities on the natural environment.

Unit 5 Environmental Science

ARE YOU READY?

Prerequisites

Concepts

- biotic and abiotic factors
- food chains and food webs
- process of bioaccumulation
- trophic levels
- biogeochemical cycles
- resource limits of an ecosystem

Skills

- investigate factors that affect ecological systems and the consequences of changes in these factors
- select and use appropriate vocabulary to communicate scientific ideas
- use graphs to compile, organize, and interpret data
- analyze issues that affect the sustainability of ecosystems
- present informed opinions about environmental issues

Knowledge and Understanding

1. Use **Figure 1** to identify the following:
   (a) two biotic and two abiotic factors
   (b) a producer, a consumer, and a decomposer
   (c) three different food chains of at least three organisms each

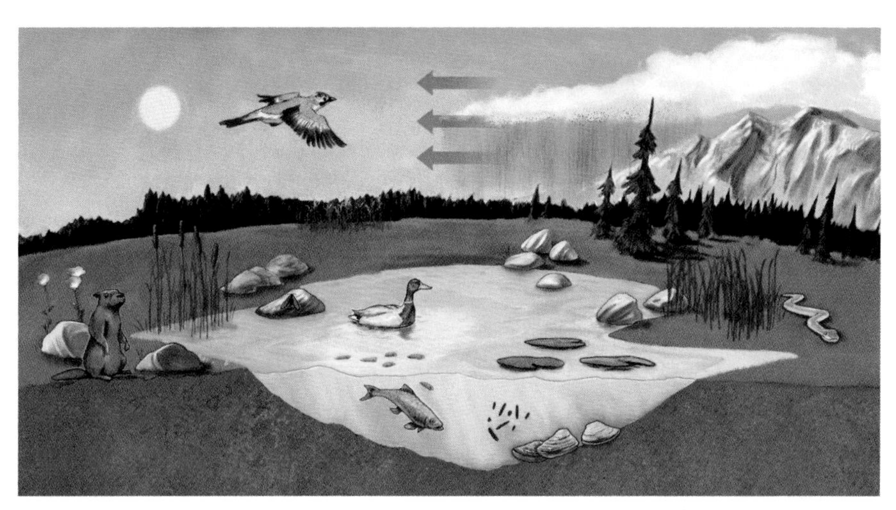

Figure 1
An ecosystem

2. In your notebook, sketch **Figure 2** or write labels to represent the organisms. Draw arrows to complete a food web.

Figure 2

3. (a) Define the term *bioaccumulation.*
   (b) Using an example, explain how bioaccumulation can lead to a serious environmental problem.
4. Explain what happens to the population numbers and the available energy as you go up the trophic levels in an ecosystem.
5. Describe how water is cycled within ecosystems.

Inquiry and Communication

6. A field study of temperatures throughout the day in a forest produced the observations in **Table 1**.
   (a) Identify the variables being measured in this study.
   (b) Propose a hypothesis to explain the variation in temperature among the locations.
   (c) Use these data to produce a line graph illustrating the observations.

Table 1 Temperature Readings (°C) Taken over a 24-h Period in a Forest

Location	Time of day								
	12 midnight	3 A.M.	6 A.M.	9 A.M.	12 noon	3 P.M.	6 P.M.	9 P.M.	12 midnight
air	17.5	17.0	17.5	18.5	19.5	21.5	21.0	19.0	18.0
litter	17.0	17.0	16.5	17.0	18.5	19.5	20.0	19.0	18.0
topsoil	16.5	16.5	16.0	15.5	16.0	16.5	17.5	18.0	17.5
subsoil	14.0	14.0	13.5	13.5	13.5	13.0	13.5	14.0	14.0

7. Use **Table 1** or the graph you constructed in question 6 to answer the following questions:
   (a) Which location in the forest showed the greatest variation in temperature?
   (b) Which location in the forest showed the least variation in temperature?
   (c) Suggest ways that the temperature variations might affect the organisms living in the forest.
8. (a) Match the photographs shown in **Figure 3** with the following biomes:
   tundra, boreal forest, grassland, and *temperate deciduous forest.*
   (b) What evidence from the photographs did you use to classify the biomes?

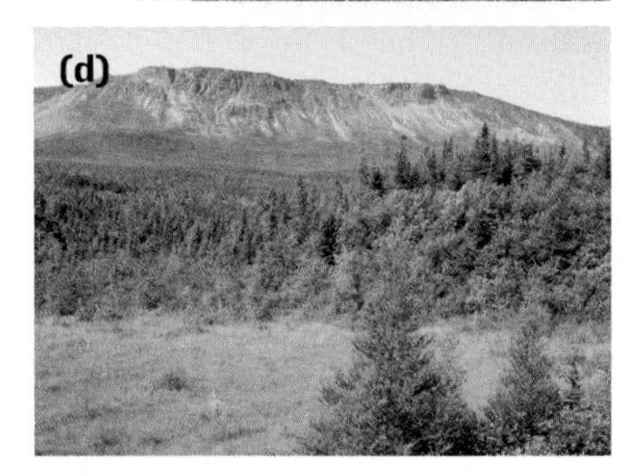

Figure 3
Biomes

9. The graph in **Figure 4** shows the growth in the world human population over the last 50 years.
   (a) Describe the trend in population growth.
   (b) Is the rate of population growth increasing, decreasing, or staying the same? Explain your answer.
   (c) Based on the graph, predict human population growth for the next 30 years. Do you think it is very likely that the population will actually reach that number? Explain.
   (d) At the current population growth rate, in approximately what year will the world population reach 12 billion?

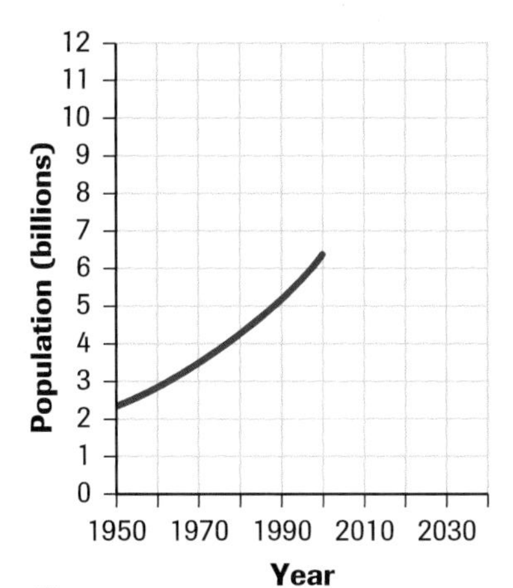

Figure 4
World population, 1950–2000

Making Connections

10. Attitudes toward the environment have changed over time. Using specific examples, describe evidence that Canadians' attitudes have changed over time. Is there still a need for changes in attitude? Explain your answer.

GETTING STARTED

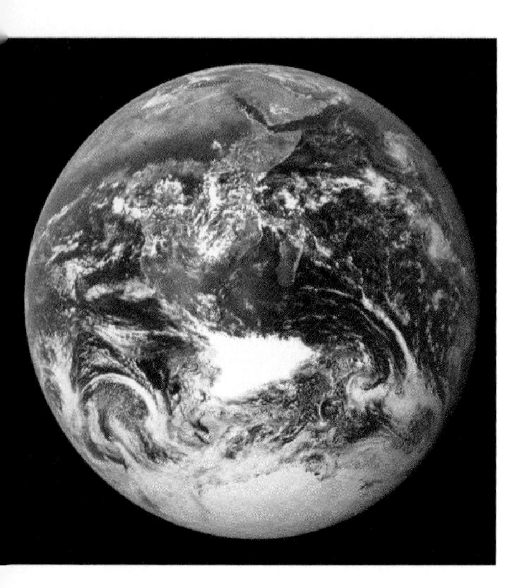

Figure 1
Earth

"We do not inherit the Earth from our grandparents; we borrow it from our grandchildren."

Lester Russell Brown

Space travel has given us a different perspective of Earth. It has enabled us to view our home from a distance. Prior to this development, most people could never imagine a picture like that in **Figure 1**. What has this view of Earth done for us?

Only when we can see the whole Earth from a distance do we realize that everything on Earth is connected to everything else. There are no real boundaries. The atmosphere that envelopes Earth is continuous and free to move and flow. The oceans that give the planet its name, "The Blue Planet," are a continuous body of water, even though we talk about the oceans as if they are separate bodies of water.

This distant view of Earth shows us only the big picture—a beautiful blue planet of water and land surrounded by a thin envelope of air. It does not allow us to see the countless continuous interactions among the living and nonliving components that take place in this system. We do not see the insect feeding on the leaves of a tree, or the frog feeding on the insect, or the bird eating the frog (**Figure 2**), or any other details of a complex web of activity that keeps the system running. Neither can we see the impacts of human activity—the treeless hills; the smog hanging over cities; and the polluted rivers, lakes, and oceans—or the efforts of humans to prevent species extinction or to preserve the natural environment. Is the big picture really as perfect as it appears?

Figure 2
Interactions among biotic and abiotic components

REFLECT on your learning

1. Explain the following statement: "An ecosystem is constantly changing, yet it remains the same."
2. Describe a typical food web in your region. Be sure to include producers, consumers, and decomposers.
3. Why should we be concerned with air and water pollution that is happening on the side of Earth opposite to where we live?
4. Do you think that Earth can withstand the current negative impacts of human activity? What evidence supports your opinion?

In the 1970s, James Lovelock, a scientist and inventor, proposed a new version of a very old idea, suggesting that Earth functions like a living organism and is not simply an environment where living things exist. The basis of his Gaia theory is that Earth is a self-sustaining system that adapts and maintains a balance in response to changing conditions. To support his theory, he uses certain characteristics of Earth: for example, the amount of oxygen in the atmosphere has been stable for 200 million years and the temperature has been relatively steady for billions of years, even though the sun's energy has increased significantly during that time.

Lovelock's Gaia theory—named after the Greek goddess Gaia (**Figure 3**), who was believed to have formed the living world from chaos—is a controversial but useful way of looking at Earth. It may give us some hope for the future. If the amount of oxygen in the atmosphere has remained stable for the past 200 million years, then it should continue to remain stable. The Gaia theory, however, may give us a false sense of security. People may assume that Earth will maintain a balance no matter what human activity may do to it. This is probably not a wise assumption. Has Earth been able to maintain a balance simply because, up to 100 years ago, the impact of human activity had been insignificant? Can Earth continue to maintain a balance given the assault of human activities over the past 100 years? What effect will the current rate of increase in human population and the accompanying increase in the use of resources have on Earth? Can we afford to take a chance?

Figure 3
The Greek goddess Gaia

▶ *TRY THIS activity* **Doubling Time**

Starting with one drop of water, determine how many doublings are required to make 1 L of water.

(a) Predict how many doublings will be required to make 1 L of water.

Materials: eyedropper, graduated cylinders (10 mL, 50 mL, 100 mL), plastic container (> 1 L)

1. Start by dropping one drop of water in the 10-mL graduated cylinder. Count this as doubling time 0.
2. Add one drop and record as doubling time 1; add 2 drops and record as doubling time 2; add 4 drops and record as doubling time 3; and so on. Also record the total number of drops and determine how many drops are in 1 mL of water.
3. To save time, at an appropriate point you will discontinue using the eyedropper and use the graduated cylinders to measure the amount of water for doubling.

(b) How many doublings did it take to make 1 L of water? How accurate was your prediction?

(c) Estimate how many drops are in 1 L of water.

(d) If each doubling time required 1 h, how many days would it take to reach 1 L?

(e) Imagine that the drops of water were fish in an aquarium. What do you think would happen to the population of fish and the water? Do you think that the population could keep on doubling? Explain.

5.1 Studying the Environment

Compared with the physical sciences (e.g., physics, chemistry), ecology is a relatively new science that has emerged only in the 20th century. The word **ecology** was coined by a German scientist, Ernst Haeckel, in 1866 to describe the relationships of living things to each other and to their abiotic environment.

ecology the study of the interactions of living organisms with one another and with their nonliving environment of matter and energy

Although ecology was not formally recognized as a science until the 1920s, people have been studying the environment forever. Early humans depended on their observations of the environment for their survival. Even in more recent times, the early inhabitants and settlers of Canada depended on their observations of nature for their livelihood and for their survival. Hunters, fishers, and farmers developed an understanding of the natural environment from their own observations. This understanding was not based on scientific research as we know it today, but it was just as valid and important.

TRY THIS activity **The Black Box**

1. Your teacher will give you a sealed black box. Your task is to attempt to find out everything you can about the contents of the box without opening it.

(a) Describe the process that you follow in investigating the black box.

(b) Record *all* of your observations.

(c) What inferences can you make and what conclusions can you draw from your observations?

2. Open the box only when you are instructed to do so by your teacher.

(d) Was there anything surprising about the contents of the black box? Explain.

(e) Try to explain why you did not discover everything about the contents of the box.

How Do We Know What We Know?

We learn about the world around us by making careful observations and analyzing these observations. This procedure is the basis of all science. Scientific investigation involves asking appropriate questions, using an organized method for making observations, and carefully analyzing the information gathered through this process to draw conclusions and make predictions. This process has led to the development of scientific laws, principles, and theories that other scientists use as the basis of further study.

We can't always make direct observations from which we can draw conclusions. We have to make indirect observations and measurements that provide the basis for inferences. An **inference** is a conclusion based on logical reasoning. If we observe that flowers planted in soil with added compost grow better than flowers planted in regular soil, we can infer that the compost has caused the flowers to grow better. However, inferences need to be verified before they can be accepted as fact. That is where science comes in.

inference a logical conclusion that is based on observations or prior knowledge, but that has not been scientifically tested

Misunderstandings about Science

The first misunderstanding about science is that it is always done using the **scientific method**. This method is generally described as a series of steps that scientists follow in their research. The steps include asking a testable question, developing a hypothesis or possible answer to the question, designing and carrying out a controlled experiment, collecting and analyzing data, drawing a conclusion based on the data, and comparing the conclusion with the hypothesis to determine whether or not the results support the hypothesis. While variations of this method are very common in science, it should be recognized that valid and valuable science can be conducted without carrying out a controlled experiment.

In environmental science, when an entire ecosystem or biome is your laboratory, a controlled experiment is often not possible. For example, it is impossible to carry out a controlled experiment to test the hypothesis that an increase in water temperature causes a disruption in the reproduction of certain species of lake trout. To address this kind of hypothesis, scientists have to rely on data obtained through an observational study. **Observational studies** are used to study natural phenomena where it is impractical to carry out a controlled experiment or where it is important not to disturb or disrupt the natural order. Studying the reproductive behaviour of lake trout in captivity would probably reveal little valuable information about the reproductive behaviour of lake trout in their natural environment.

A second misunderstanding about science is that scientific investigations must produce quantitative data (**Figure 1**). While there are lots of **quantitative** data produced during environmental investigations (e.g., measurements of temperature or oxygen levels, population counts, etc.), valuable evidence can also be collected in **qualitative** form, that is, descriptions of observations made using the senses (e.g., colours, sounds, smells, behaviours, etc.) (**Figure 2**). Environmental research probably relies more on qualitative data than do other areas of scientific research. Qualitative evidence can be recorded using words or pictures.

scientific method the systematic method used in scientific investigations of the natural world, which includes designing and carrying out controlled experiments to test hypotheses

observational study a scientific study in which the scientist attempts to gain scientific information by observing natural phenomena without interfering with the normal order of the environment

quantitative referring to quantities or numerical data

qualitative referring to qualities or non-numerical data

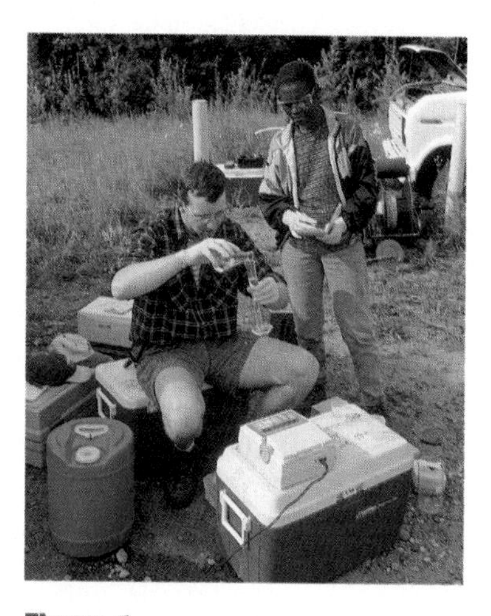

Figure 1
A technician measures the oxygen content of stream water.

Environmental Science versus Environmental Decision Making

The information we gather from studying the environment is of little value unless we use it for a specific purpose. Usually, we use the information to make decisions about environmental issues or problems. In order to make informed decisions we need valid and reliable information—scientific facts. Without this scientific information, our decisions can be based only on personal opinion, beliefs, or emotions. For example, a sound decision on whether or not a proposed hydroelectric project should proceed can be made only after all the relevant scientific data have been gathered and analyzed and all perspectives from all stakeholders have been considered. Environmental science is the method by which this information can be gathered. Environmental decision making is the process used to analyze all of the relevant information to help make a decision on the issue (**Figure 3**, on the next page). One of the tools that is commonly used in this process is a risk–benefit analysis. Information about

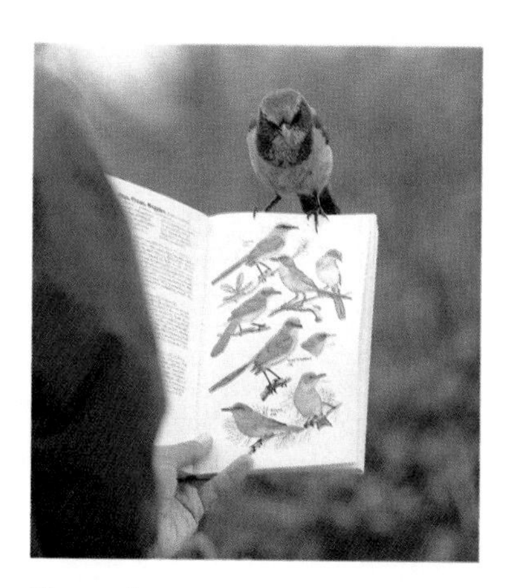

Figure 2
Scientists often rely on qualitative observations of the natural world.

Figure 3
Generating electricity from water or wind sources causes different environmental effects. Scientific knowledge of these effects is important in making informed decisions.

the potential negative impacts on, or the risks to, the environment, wildlife, and different groups of people is weighed against the potential positive economic and social benefits of the development. A risk–benefit analysis alone cannot resolve the issue or provide a definite solution to the problem, but it is a valuable method of looking at both sides of the issue.

Table 1 summarizes the steps in the decision-making process. A more complete description can be found in Appendix A2.

Table 1 The Environmental Decision-Making Process

Steps in the decision-making process	
1. Define the issue	Explain the issue as clearly as possible and identify the stakeholders involved.
2. Identify alternatives or positions	Examine the issue from a variety of perspectives and think of as many alternative solutions as possible.
3. Research the issue	Develop a plan to find reliable and relevant sources of information. Gather information from many different sources.
4. Analyze the issue	Determine the relevance and significance of the information gathered, and conduct a risk–benefit analysis to evaluate each of the possible solutions. Consider ecological, economic, social, and cultural effects.
5. Make and defend a decision	Weighing all the evidence, take a position on the issue. Use the supporting information to justify your decision.
6. Evaluate the process	Carefully examine the thinking that led to the individual or group decision.

Section 5.1 Questions

Understanding Concepts

1. How is an inference different from a prediction?
2. Differentiate between quantitative data and qualitative data. Is one more important than the other? Explain.
3. An environmental scientist is studying the impact of a gas pipeline on the migration and feeding habits of caribou. Give three examples of quantitative data and three examples of qualitative data in this investigation.
4. Explain the difference between environmental science and environmental decision making. Use an example to illustrate each.
5. In a controlled experiment, the investigator changes one variable (the independent variable) to determine how it affects another variable (the dependent variable). How is this different from an observational study?

Applying Inquiry Skills

6. You are a wildlife biologist assigned to study the populations of songbirds in a forest. Which type of scientific investigation would be most appropriate to answer the question "What are the feeding habits of the forest songbirds?" Explain your answer.
7. Locate an article about a controversial environmental issue. Analyze the article using the following questions to guide your analysis:
   (a) What is the issue?
   (b) What perspective(s) is (are) represented in the article?
   (c) Who is the author? Does he or she have the credentials to write on this issue? What biases might the author have?
   (d) Does the author take a position on the issue? If so, what is his or her position?
   (e) Does the author use valid scientific evidence to support his or her position, or is the article based on opinion?

Making Connections

8. Use the risk–benefit analysis described in Appendix A2 to evaluate the following statement related to an environmental concern. (*Note:* In this scenario, you are permitted to make up scientific and other information in order to complete your analysis.) What assumptions must you make to complete the analysis?

   Statement: "The construction of a golf course is planned for an area that includes a small section of ecologically sensitive sand dunes."

Levels of Organization

Starting with the smallest subatomic particles, the organization of matter on Earth becomes increasingly complex. **Figure 1** represents the organization of matter from subatomic particles to the universe. The study of ecology is concerned mainly with the levels ranging from the organism to the biosphere.

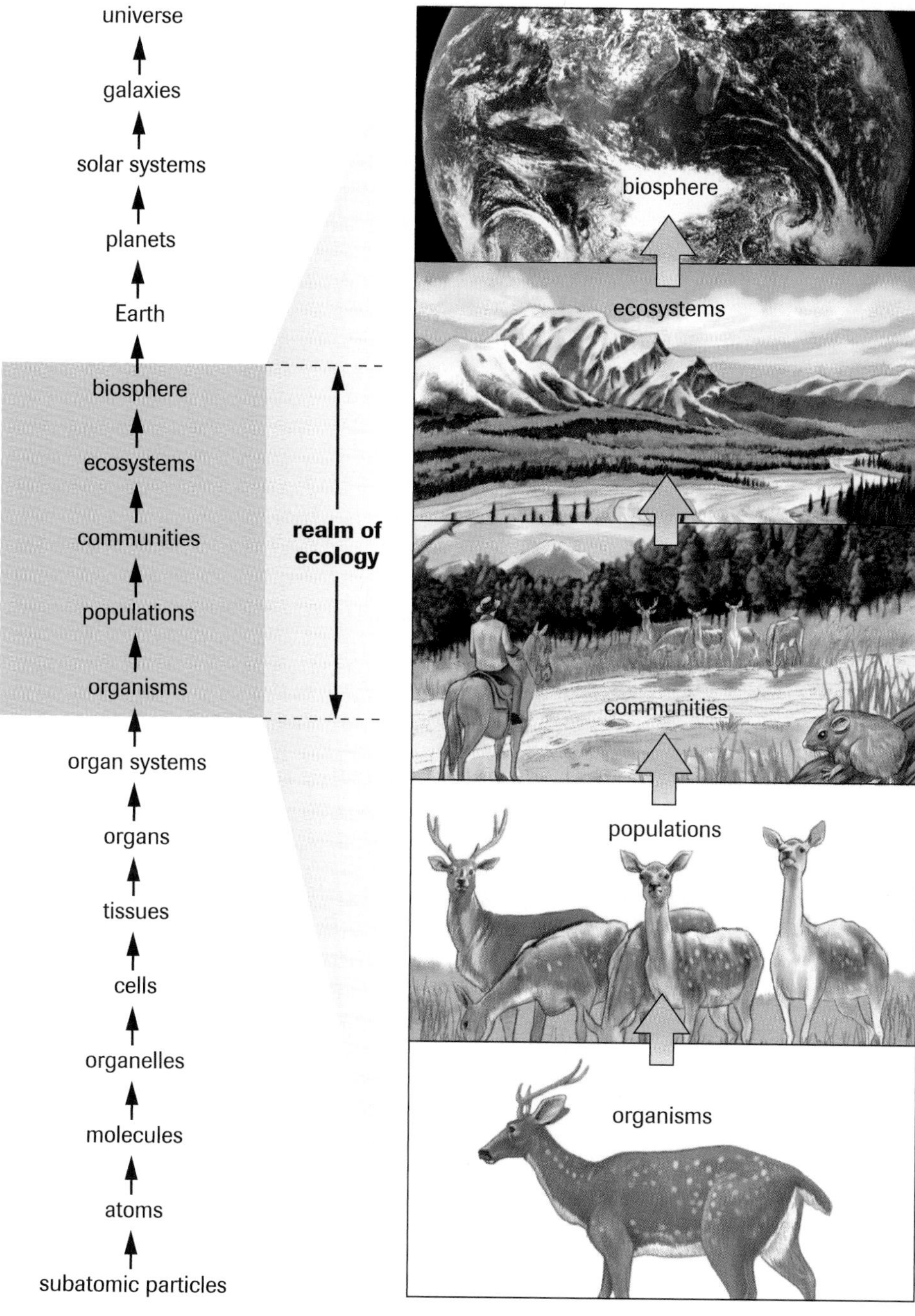

Figure 1
Levels of organization of matter

The Biosphere and Biomes

biosphere that part of Earth inhabited by plants and animals (all communities of living things on Earth)

biome a large area of Earth characterized by distinctive climate and soil conditions with a biological community adapted to those conditions

Starting with the big picture at the most complex level, the **biosphere** represents the entire living system on Earth. The biosphere is normally divided into two parts—terrestrial (land) and aquatic (water). It is not practical to study the terrestrial or the aquatic part of the biosphere as a single large system. The terrestrial part of the biosphere is usually divided into **biomes**, which are sections of the biosphere that have similar characteristics as defined by climate, soil conditions, and living organisms. There are many different classification systems for biomes, but we will use the following: tundra, boreal forest (or taiga), temperate deciduous forest, grassland, tropical rain forest, and desert (**Figure 2**). The aquatic part of the biosphere includes the oceans, freshwater environments, and wetlands.

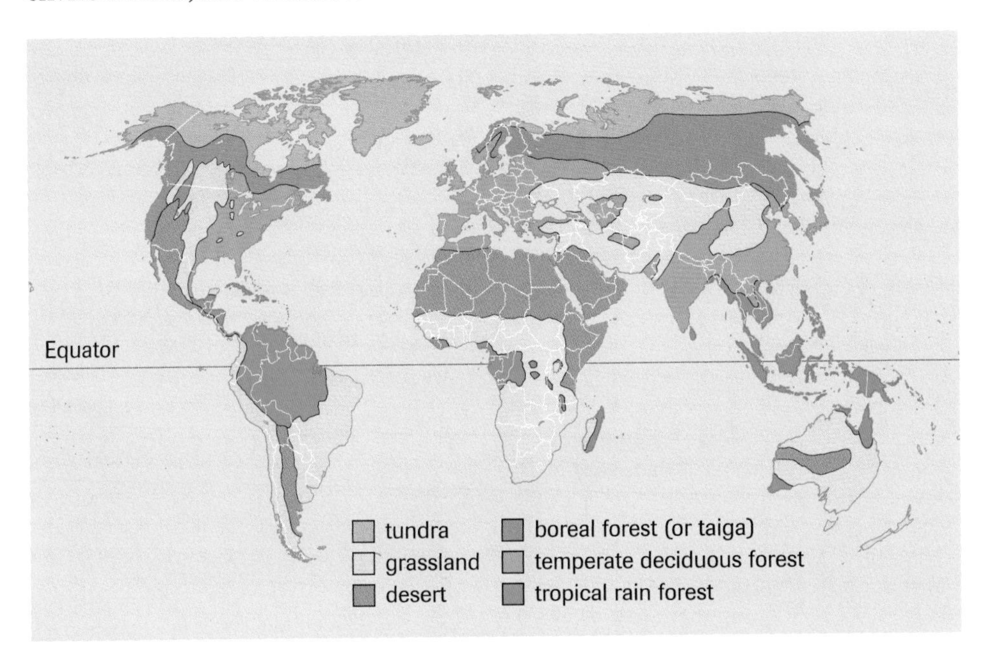

Figure 2
Earth's major biomes are defined based mainly on differences in climate, soil, and vegetation.

The main factor that characterizes a biome is its climate, which is defined mainly by long-term patterns of temperature and precipitation. Obviously, a biome's geographic location on Earth determines its climate. More northerly regions receive less energy from the sun than regions near the equator, and the average temperatures are therefore different. Regions on the windward side of mountains will receive more precipitation than those on the leeward side—the side protected from the wind. So it is these two characteristics, temperature and precipitation, that largely determine what organisms can live in a specific location on Earth.

The surface of Earth is not divided into continuous (contiguous) sections. For example, the desert biome as a whole comprises a number of separate desert regions scattered around Earth (see **Figure 2**). However, the conditions that describe these regions (climate, soil conditions, and living organisms) are similar in all deserts. Boreal forests are found around the world in the northern latitudes, and they are similar whether they are found in North America or in Europe.

Ecological Roles in an Ecosystem

Starting at the other end of the ecological organization, we begin with the individual organism. Seldom do we see a single organism in any natural environment. We generally see a group of organisms of the same species living together in a geographic area. Such a group of organisms is called a **population**, and the area where it lives is referred to as its **habitat**. And we hardly ever see a single population in any area. There are usually a number of different populations interacting with one another in an area. This group of populations in a geographic area makes up a biological **community**. All of the living organisms in an ecosystem are referred to as the **biotic** components, and the nonliving matter and energy (e.g., air, soil, water, wind, and solar energy) are referred to as the **abiotic** components. The members of the biological community (the biotic components) interact with one another and with the nonliving environment (the abiotic components), making up what we call an **ecosystem**.

population a group of individuals of the same species living in a particular geographic area at the same time

habitat the place or type of environment where an organism or a population of organisms lives

community populations of all species living in a particular area at a particular time

biotic referring to the living organisms and their products in an ecosystem

abiotic referring to the nonliving components of an ecosystem

ecosystem a community of different species interacting with one another and with the abiotic environment

The term *ecosystem* is a general term used to describe the interactions of the biotic and abiotic components of the environment. It is not usually described in terms of its physical space or its complexity. An ecosystem can be as small as the space in a rotting log on the forest floor or as big as an ocean. You could even consider Earth to be one big ecosystem. An ecosystem can have many different populations or only a few, a complex food web or a simple food chain. The term is simply a convenient way of referring to the interactions of the biotic and abiotic components of the environment. The context in which the term is used will provide an indication of the size of the physical space under consideration. We often use an additional descriptive term, for example, backyard ecosystem, ocean ecosystem, or forest ecosystem, to help define the size of the space to which we are referring.

The living organisms of any ecosystem fill different roles. The producers or autotrophs are the organisms that produce food for themselves and other organisms by using the solar energy from sunlight to convert simple inorganic substances (water and carbon dioxide) into more complex chemicals (carbohydrates) through photosynthesis. On land, green plants are the most important producers. All members of the kingdom Plantae are autotrophs. In aquatic environments, algae (kingdom Protista) and certain types of bacteria (kingdom Eubacteria) provide food through photosynthesis. A special group of eubacteria are able to produce food by using the energy in inorganic materials to convert raw materials into usable nutrients. These organisms are called **chemotrophs** and are able to produce food without the energy from sunlight.

chemotroph a producer that converts the energy found in organic chemical compounds into other forms of energy without the use of sunlight

All organisms that are not producers are consumers, or heterotrophs. All members of the kingdoms Animalia, Fungi, and Archaebacteria and some members of the kingdoms Eubacteria and Protista are heterotrophs. There are several types of consumers, all of which depend directly or indirectly on producers for their food energy and materials. Since they all depend on producers, they obtain their energy indirectly from the sun. **Table 1**, on the next page, summarizes the types of consumers and provides common examples of each type.

Table 1 Consumers and Their Roles in Ecosystems

Type of consumer	Ecological role	Example organisms
herbivores or primary consumers	feed directly on green plants	zooplankton, insects, cows, sheep, deer, moose
carnivores or secondary consumers	feed indirectly on green plants by eating herbivores	cats, dogs, bobcats, wolves, certain snakes, whales
tertiary consumers, including:	feed on secondary consumers	hawks, eagles, sharks, pike, humans
omnivores	eat both plants and animals	black bears, pigs, raccoons, humans
scavengers	feed on dead organisms	vultures, hyenas, crayfish
detritus feeders	feed on fragments of dead or decaying tissues or organic wastes	earthworms, dung beetles, maggots, shrimp
decomposers	live on or within their food source, breaking down complex molecules in detritus into simpler compounds; play a major role in the recycling of nutrients in the environment	fungi, bacteria

Types of Species in Ecosystems

Aside from classification according to their feeding requirements, the different species in an ecosystem can be classified by another method. This classification is based on the role of the species and how they interact with one another and their environment. There are four types of species that may be found in any ecosystem: endemic (native), exotic (immigrant), indicator, and keystone.

Endemic species are those that are native to an area. In other words, they normally live and thrive in that particular area. For example, moose are an endemic species in the boreal forest of Ontario. As far as can be determined, moose have always lived there. **Exotic species** are those that have moved into an area or have been introduced by humans, intentionally or accidentally (**Figure 3(a)**). Even though there is a large thriving population, moose are an exotic species on the island portion of the province of Newfoundland and Labrador (**Figure 3(b)**). This current population of more than 120 000 grew from six moose that were introduced to the island, two in 1878 and four more in 1904. Exotic species may be beneficial to humans but may also be harmful to endemic species because they compete for food and space.

exotic species a species that has migrated to or has been introduced by humans to a geographic area

(a)

(b)

Figure 3
(a) The zebra mussel, an exotic species in the Great Lakes
(b) Moose are endemic to Ontario but exotic to Newfoundland and Labrador.

All species in an ecosystem fall into one of the two categories above. However, there are two additional categories that ecologists use to describe the role of a species in the ecosystem.

Indicator species are so named because the success or failure of these species indicates what is happening to an ecosystem. Frogs are a good indicator species because they are sensitive to a variety of environmental changes, such as increased ultraviolet radiation, chemical pollutants, and habitat loss resulting from development (**Figure 4(a)**). In addition, since they spend part of their life on land and part in water, frogs provide information about both the terrestrial and the aquatic habitats. They are an indicator of the overall health of the ecosystem.

indicator species a species sensitive to small changes in environmental conditions. Monitoring the health of its population is used to indirectly monitor the health of the environment in which it lives.

All species play an important role in their ecosystem. However, some species are so critical that their ecosystem would be dramatically affected if the species itself were affected. The extinction of a **keystone species** would almost certainly result in the extinction of other species, and the entire ecosystem would suffer. The prairie dog is an example of a keystone species because so many other species depend directly or indirectly on it for their survival. The sea otter is another example of a keystone species because it feeds on sea urchins, thereby preserving the kelp beds that are home to many other species (**Figure 4(b)**). Since humans can affect their environments more than any other species, they are also considered a keystone species.

keystone species a species that is critically important to the balance and survival of the particular ecosystem in which it lives

(a)

(b)

Figure 4
(a) The success or failure of frogs provides an indication of the health of an ecosystem.
(b) Sea otters are key to maintaining the balance in the kelp bed ecosystem.

Section 5.2 Questions

Understanding Concepts

1. Distinguish between autotrophs and heterotrophs.
2. Distinguish among a population, a community, and an ecosystem.
3. What two factors determine whether a particular part of Earth's surface will be tundra or tropical forest? Explain.
4. Explain why amphibians are good indicator species. Suggest reasons why their populations might decline.
5. Use examples from your geographic area to explain what is meant by the term *exotic species*.
6. Explain why humans are considered a keystone species.
7. Name and define four different types of consumers and give two examples of each.

Applying Inquiry Skills

8. Design a scientific investigation to find out whether or not a local species is a keystone species.
9. Which of the species in **Figure 5** do you think is an indicator species? Explain why you think so.

Making Connections

10. Conduct research using print and electronic resources to create a brief description of the four major biomes found in Canada. In your description, include information about the geography, climate, vegetation, and wildlife that are typical of each biome.

www.science.nelson.com

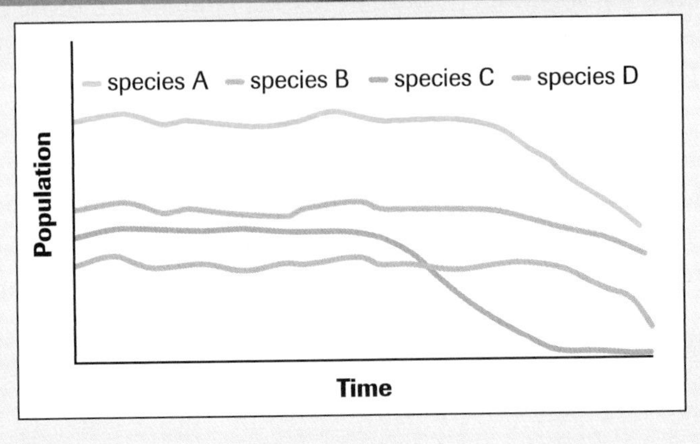

Figure 5

11. A biologist researching tropical species claims that it is not important to protect the tundra because most of the world's biodiversity is in the tropical rain forest. Explain your reaction to this claim. Consider ecological, environmental, social, and cultural perspectives.
12. "Because we are the most intelligent living things, humans have the right to control all other species of living things on the planet." Write a paragraph outlining your reaction to this statement. Consider, ecological, environmental, social, and cultural perspectives.
13. Conduct research about an exotic species in your area. Explain the effects of its introduction on the environment. Write a brief report outlining your findings.

www.science.nelson.com

Biomes and Ecozones of Canada

The major biomes of Canada (tundra, taiga, temperate deciduous forest, and grassland) and its aquatic ecosystems are too large and diverse to be studied and described scientifically so that the description is of practical use. Ecozones are used to accurately describe much smaller geographic regions. In Canada, 15 terrestrial and 5 marine ecozones have been identified (**Figure 1**). In addition to the biophysical characteristics that define a biome (e.g., climate, soil, landforms, vegetation, wildlife, and water), human activities are sometimes used to define an ecozone. For example, agriculture is an important defining characteristic of the prairie ecozone. **Table 1**, on the next page, summarizes the boreal shield ecozone.

Even the ecozones are fairly large geographic areas, however, so each ecozone has been further subdivided into a number of smaller ecoregions. The data gathered on the ecozones and ecoregions are intended to be used as the basis for ecological planning and sustainable resource management.

Procedure

1. From the map in **Figure 1**, select an ecozone that interests you (e.g., somewhere in Canada you have visited, or somewhere a friend or relative lives).
2. Use print and electronic resources (and your personal experience, if appropriate) to research the ecozone of your choice.

3. Using the summary of the boreal shield ecozone in **Table 1** (on the next page) as a model, describe your chosen ecozone.

Analysis

(a) What did you learn about the ecozone that surprised you? Why did it surprise you?

(b) Based on your research, can you conclude that all locations within an ecozone are very much alike? Explain.

(c) Under which category of information do you think there is the greatest variability among ecozones?

(d) Under which category of information do you think there is the greatest variability within an ecozone?

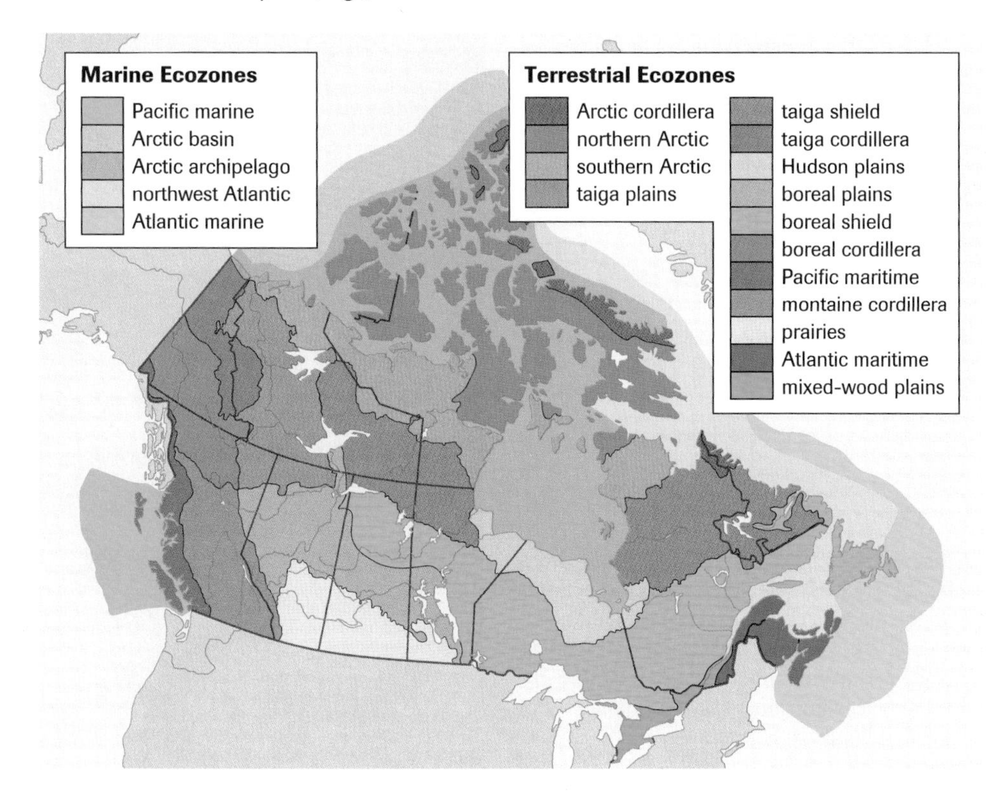

Figure 1
The terrestrial and marine ecozones of Canada

CAREER CONNECTION

An ecosystem management technician analyzes the big picture of human activity in ecosystems and attempts to manage that activity in a way that benefits both humans and the environment. Research the skills and training needed for such a role.

www.science.nelson.com

Evaluation

(e) Do you think that the ecozone classification system for describing ecosystems is a valuable one? Explain its advantages and disadvantages.

Synthesis

(f) The climate and the soil of an ecozone determine what types of plants can be grown in an area. For your chosen ecozone, indicate what types of crops a home gardener might be able to grow, and describe the requirements of some of the more common crops.

Table 1 The Boreal Shield Ecozone

The boreal shield ecozone is Canada's largest ecozone, ranging from Newfoundland and Labrador in the east to northern Saskatchewan in the west. While much of this ecozone is settled, most of it remains a wilderness area. The boreal shield ecozone has been subdivided into 30 ecoregions.	
Climate Most of this ecozone has a continental climate–long cold winters and short warm summers. The climate in the eastern part of the ecozone is moderated somewhat by the influence of the Atlantic Ocean, and, in the central region of Ontario, by the Great Lakes. This moderating effect also increases the level of precipitation and tends to lower the average temperature in summer and raise the average temperature in winter.	
Vegetation This ecozone is over 80% forested, with the dominant trees being spruce, fir, and tamarack. In the southern parts of the zone, there are more deciduous trees, such as birch, aspen, and poplar. The other 20% of the zone is mostly rocky, barren areas covered by lichens and shrubs.	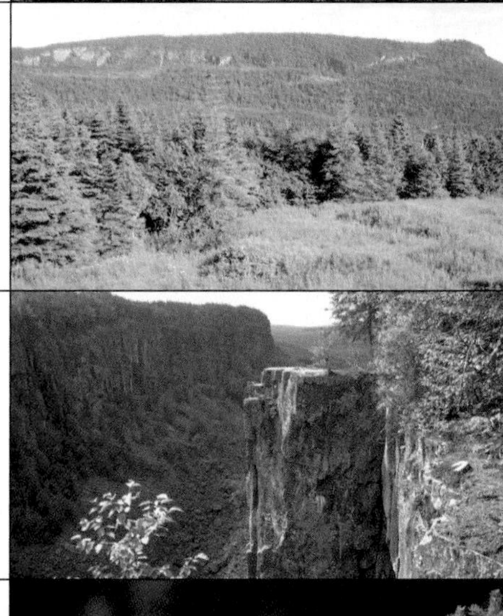
Landforms and soils The landscape, which was carved by glaciers, is neither mountainous nor flat but a mixture of rolling hills, valleys, and wetlands. Bare bedrock at the surface is common. The soils and surface materials are mainly glacial deposits. There are numerous small- to medium-sized lakes, a feature that is also characteristic of an area carved by glaciers. In the wetland areas, especially in northwestern Ontario and Newfoundland, peatlands and bogs are common. This area also contains the drainage basins of many large river systems, including the St. Lawrence and the Nelson/Churchill Rivers.	
 **Wildlife** Since much of the ecozone is wilderness, there is a wide variety of wildlife. Common mammals include caribou, deer, moose, black bear, wolf, fox, lynx, skunk, raccoon, and chipmunk. Most common among the many species of birds are geese, ducks, loons, owls, blue jays, and grosbeaks.	
Human activities Most human activities are connected with the development of the rich natural resources of the ecozone. Mining, forestry, hydroelectric generation, recreation and tourist activities, along with hunting, fishing, and trapping, are the main commercial activities. Although agriculture does occur throughout the ecozone, it is not a major activity because in most areas the climate and soil conditions are not ideal. The main urban centres include Flin Flon in Manitoba; Sault Ste. Marie, Sudbury, Timmins, and Thunder Bay in Ontario; Chicoutimi and Rouyn-Noranda in Quebec; and St. John's in Newfoundland and Labrador.	

The Flow of Energy and Matter in Ecosystems 5.4

Most of the energy entering an ecosystem is in the form of solar energy. A very small portion of it (less than 1%) is used by producers in the process of photosynthesis. In the process of photosynthesis, solar energy is converted into chemical energy and is stored in molecules such as glucose. Plants themselves use most of this chemical energy to fuel their metabolic processes, but some of it is passed on to herbivores, who consume plants as food. The herbivores, in turn, convert the chemical energy into usable energy as they move about, keep themselves warm, and carry out other processes such as respiration, repair and maintenance, and reproduction. Some of this usable energy is converted into heat energy and is lost to the environment. The portion of chemical energy that is stored in the herbivore's body is transferred to the carnivore when the carnivore eats the herbivore. This whole process of the transfer of energy through the ecosystem is described as **energy flow**. It is a one-directional process: once the energy is used by an organism it is not available to be used again by another organism.

energy flow the one-way transfer of energy within an ecosystem

The transfer of energy from one organism to the next in an ecosystem is called a food chain (**Figure 1**). Since many organisms have more than one source of food, the relationship is rarely that of a simple food chain. In most ecosystems, there is a great diversity of organisms, and hundreds of energy pathways are possible. This combination or network of many food chains is referred to as a food web (**Figure 2**, on the next page). Each energy transfer level in a food chain or food web is referred to as a **trophic level**. So, producers are at the first trophic level, primary consumers at the second trophic level, and so on. At each level, some of the available energy is used by the organisms and much is lost as heat energy.

trophic level a step in the transfer of matter and energy along a food chain or through a food web. Each organism is assigned to a trophic level depending on its primary source of food.

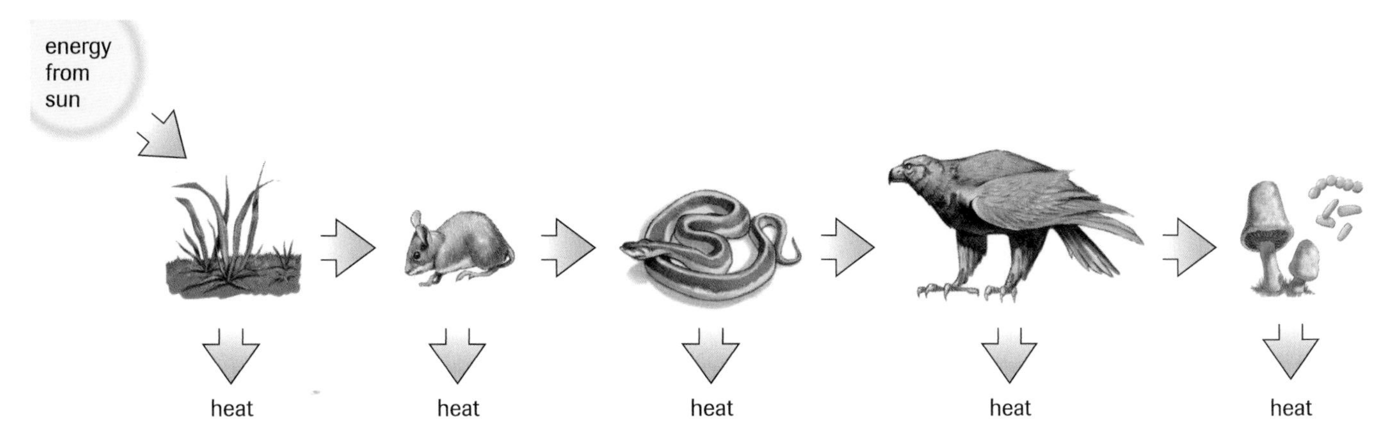

Figure 1
One-way energy flow through a food chain

Energy Pyramids

Unfortunately, the transfer of energy from one trophic level to the next is not very efficient. The amount of energy lost as heat energy with each transfer is much greater than the amount that is available for use by the next organism. Only about 10% of the energy is transferred from one trophic level to the next, with 90% being lost to the environment as heat.

Figure 2
A simplified food web showing a number of energy paths through the ecosystem

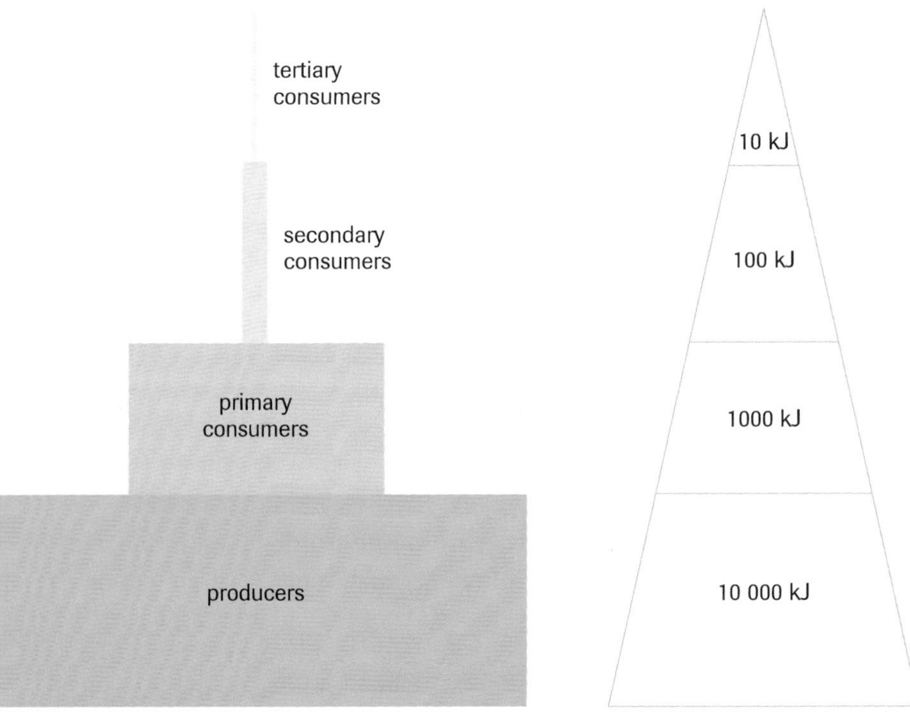

Figure 3
An energy pyramid showing the loss of usable energy at each trophic level. Energy is measured in joules (J).

This loss of energy can be represented graphically using an **energy pyramid** (**Figure 3**). It becomes obvious that, since there is a fixed total amount of energy, more usable energy is available at the lower levels of the food chain. Organisms at the highest trophic level receive a small fraction of the energy that is available at the lowest level in the food chain.

energy pyramid a model that illustrates energy flow from producers at the beginning of food chains to consumers farther along

Energy pyramids illustrate that humans can obtain more energy by consuming food directly from producers—grains, vegetables, and fruits—than by feeding these producers to livestock and then eating livestock products (**Figure 4**). The evidence suggests that eating organisms early in food chains or energy pyramids can sustain larger populations in countries with insufficient food production. For example, in 2000, China harvested 462.5 million tonnes of various grains, an amount that was sufficient to ensure that the entire population could be fed. In the past, the people of China consumed fish in addition to grains, but very few meat products. More recently, many Chinese have begun to include more meat products in their diet, a change that has led to challenges in food production. Vegetarianism may be a partial solution to the worldwide shortage of food!

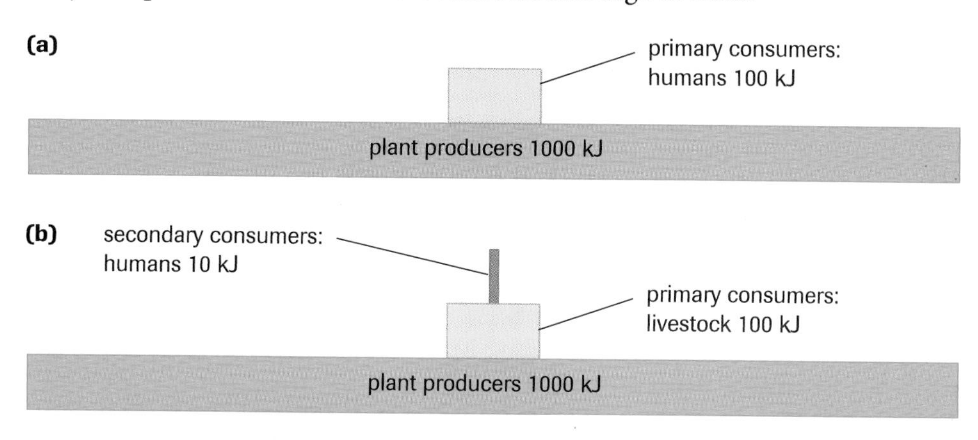

Figure 4
These energy pyramids compare the relative efficiency of energy transfer to humans from **(a)** the primary trophic level and **(b)** a secondary trophic level.

▸ *TRY THIS activity* *Calculate Your Energy Intake*

You can relate the foods you eat to the amount of land required to produce those foods.

(a) In a table, record the amount of everything you eat in 24 h. Also indicate the type of food (animal or plant product).

1. Use a kilojoule counter, or record kilojoule values from food packaging, to calculate the number of kilojoules you consumed from plant products and from animal products.
2. Multiply each of your results by 365 to get an average annual intake of kilojoules.
3. Divide the annual average intake of kilojoules from plant products by 8350 kJ/m^2 to determine the amount of land needed to produce the plant products you consume.
4. Divide the annual average intake of kilojoules from animal products by 835 kJ/m^2 to determine the amount of land needed to produce the animal products you consume.

(b) Which food products required less land?

(c) Compare your results with those of your classmates. Are there significant differences among classmates? Analyze the observations and suggest an explanation for these differences.

(d) Is it more efficient to eat plant or animal products? Explain.

(e) How might you and your classmates change your diets to reduce the amount of land required to produce your food?

(f) How might your results compare with those for people living in other parts of the world?

Pyramids of Numbers

The more trophic levels there are, the greater the total loss of energy; this explains why it is unusual to have more than four trophic levels in an ecosystem. If a fifth trophic level existed, the amount of available energy would support only a very small number of organisms at that level. For example, the population of eagles in an ecosystem is much smaller than the populations of primary and secondary consumers (e.g., rabbits, fish, etc.) that they feed on. The top consumer in the energy pyramid is therefore more at risk when its ecosystem becomes disrupted. Such a small population may even be wiped out if its food supply is reduced.

The reduction in available energy as you move up the trophic levels means that fewer organisms can be supported. In many ecosystems, the **pyramid of numbers** parallels the energy pyramid. As you go up the trophic levels, the number of organisms that can be supported decreases. For example, in the grassland ecosystem, there are a large number of producers (grass plants) and a much smaller number of primary consumers (herbivores) (**Figure 5(a)**). In some ecosystems, the typical pyramid shape may be modified at the lower trophic levels. For example, the forest ecosystem has a smaller number of producers (trees) and a much larger number of primary consumers (insects) (**Figure 5(b)**).

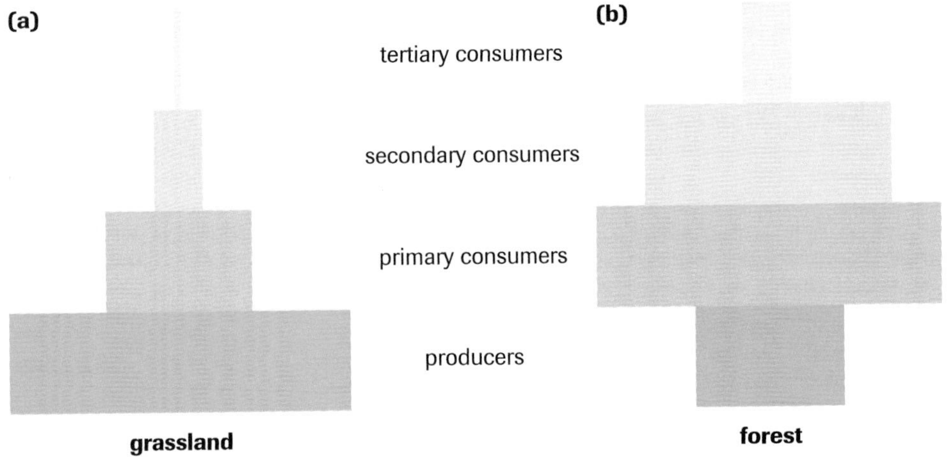

pyramid of numbers a model that illustrates the number of organisms of a particular type that can be supported at each trophic level

(a)

(b)

tertiary consumers

secondary consumers

primary consumers

producers

grassland

forest

Figure 5
Pyramid of numbers for **(a)** a grassland ecosystem and **(b)** a forest ecosystem

The Productivity of Ecosystems

The productivity of an ecosystem is determined by how much solar energy can be captured by photosynthesis and converted into usable chemical energy in a given period of time. The more energy captured by plants, the more productive the ecosystem; this means that more plant and other living material (biomass) is produced. Some geographic locations and some types of ecosystems are more productive than others. The most productive ecosystems on Earth are the shallow waters around the coasts of continents and the tropical rain forests because these regions have a good supply of sunlight, water, and nutrients (**Figure 6**). The least productive regions are the deserts, because of their extremely low precipitation and high temperatures, and the deep oceans, because of their lack of sunlight. **Table 1** shows the productivity of different ecosystems in terms of their production of biomass.

Figure 6
Estuaries are among the most productive ecosystems on Earth.

Table 1 Productivity of Ecosystems

Ecosystem	Average productivity (g/m²/year)
algal beds and coral reefs	2500
tropical rain forest	2200
swamp and marsh	2000
estuaries	1500
temperate deciduous forest	1200
boreal forest	800
temperate grassland	600
lakes and streams	250
tundra	140
open ocean	125
desert	90
extreme desert	3

Section 5.4 Questions

Understanding Concepts

1. Why is energy flow in an ecosystem described as one-way?
2. Explain what is meant by trophic levels in an ecosystem.
3. Transfer of energy in an ecosystem is very inefficient. Use an energy pyramid to explain this statement.
4. Most ecosystems have three, four, or five (rare), trophic levels. Why don't we see ecosystems with ten trophic levels?
5. Distinguish between an energy pyramid and a pyramid of numbers.
6. A pyramid of numbers is not always pyramid shaped. Explain.
7. Why are coral-reef ecosystems more productive than deserts? Consider the conditions present in each area.
8. What does an energy pyramid tell us about the amount of energy available to top carnivores?
9. A food chain consists of grain, cattle, and human populations. If the amount of energy in the harvested grain is 730 kJ, how much energy is available for the human population?
10. What would happen to an ecosystem if all of the producers were eliminated? all the decomposers?

Applying Inquiry Skills

11. **Table 2** provides estimates of the average primary productivity of different habitats in kJ/m²/year. In your notebook, complete the table, comparing the value of food energy that could be supplied for human consumption if we were to feed on the plants directly or if we were to feed on animals.

Table 2

Type of ecosystem	Tropical rain forest	Grassland	Open ocean
primary productivity (kJ/m²/year)	38.0	9.20	5.00
energy available to primary consumers (kJ/m²/year)			
energy available to secondary consumers (kJ/m²/year)			

Making Connections

12. You have a choice of a hamburger meal or a garden salad, either of which will provide you with 3000 kJ of energy. How will your decision affect the energy required from the ecosystem?

5.5 Biogeochemical Cycles

biogeochemical cycle the process by which matter cycles from living organisms to the nonliving, physical environment and back again (e.g., the carbon cycle, the nitrogen cycle, the water cycle, and the phosphorus cycle)

Even on cloudy days there is continuous input of energy to Earth from the sun. However, Earth is essentially a closed system when it comes to matter. If organisms are to survive, grow, and reproduce, they must have a continuous supply of matter in the form of nutrients. These nutrients get transferred from one organism to another (through food chains and webs) and from living organisms to the abiotic environment and back again. This cycling of matter involves interactions among living things, Earth, and chemicals. It is referred to as the **biogeochemical cycle**. While all materials are cycled, there are five main cycles that ensure the continuity of life on Earth because they are the main ingredients of living matter—carbon, oxygen, water, nitrogen, and phosphorus.

DID YOU KNOW?

Chemical Cycling

Carbon, oxygen, nitrogen, and water can exist in gaseous form, so the cycling of these chemicals can occur around the world and over long periods of time. Carbon dioxide molecules that are exhaled by you could be consumed by a plant in South America.

The Carbon and Oxygen Cycles

The cycling of carbon through the biotic and abiotic environment is largely due to the use of carbon dioxide by plants in photosynthesis. Because plants give off oxygen during photosynthesis, we often look at the oxygen cycle in conjunction with the carbon cycle. Plants use water from their immediate environment and energy from the sun to convert gaseous carbon dioxide into complex organic compounds such as glucose. In the same process, the oxygen atoms are removed from the carbon dioxide and released into the atmosphere as oxygen. This

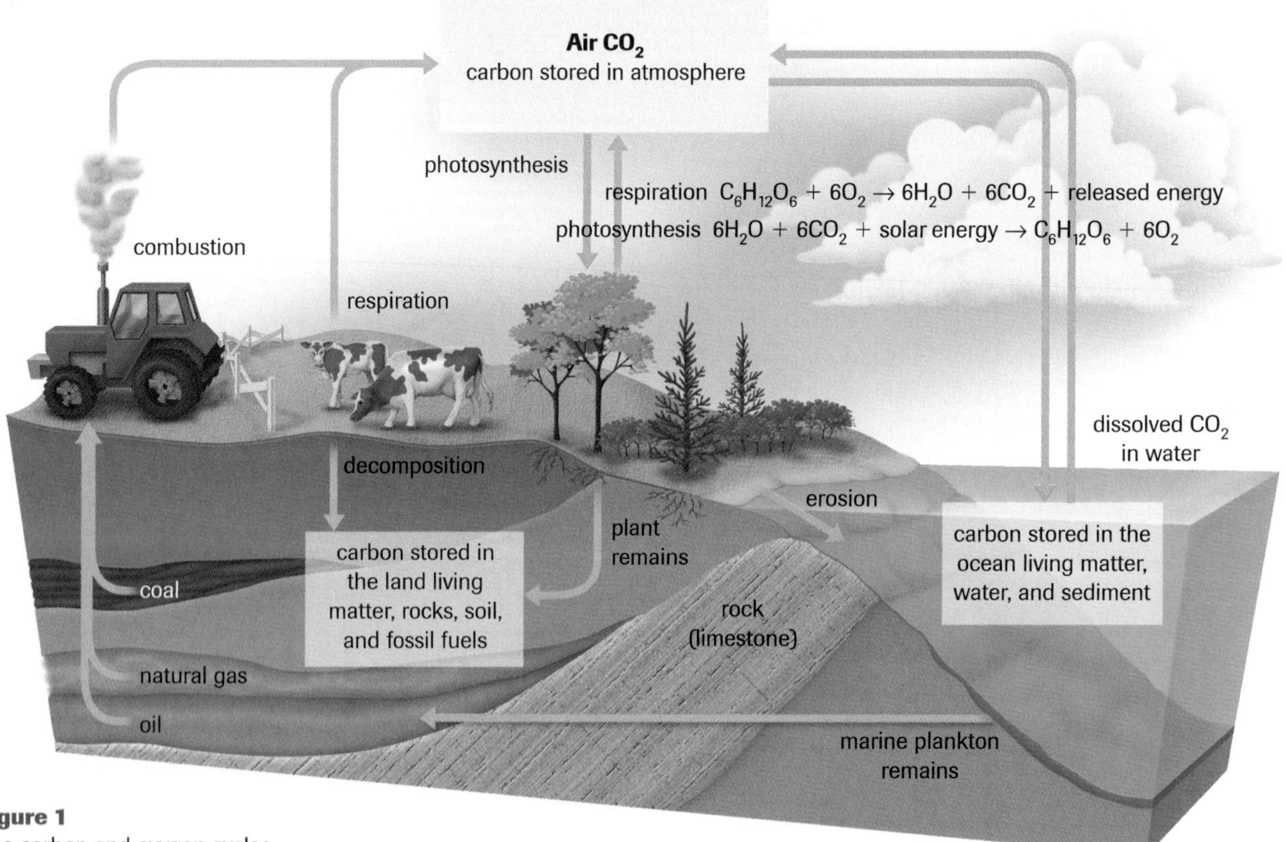

Figure 1
The carbon and oxygen cycles

process is reversed by consumers. They use oxygen from the atmosphere and the organic compounds produced by plants and, through cellular respiration, return the carbon dioxide to the atmosphere (**Figure 1**).

Some carbon is not released back to the atmosphere for a long period of time. Some of the carbon-containing organic molecules produced by plants are trapped for hundreds of years—either in the wood of living trees or in the dead organic material of Earth. Some carbon that was part of living organisms has been trapped for millions of years—underground in fossil fuels, or in ocean sediments as carbonates.

The Water or Hydrologic Cycle

Table 1 summarizes the processes that drive the water (or hydrologic) cycle. The water cycle is powered by two sources—solar energy from the sun and gravity. Solar energy causes the evaporation of water that gets it into the atmosphere. When water vapour condenses to a liquid, the force of gravity brings it back to Earth, and gravity also causes water to run to the lowest points on Earth. **Figure 2** shows a simplified model of the water cycle.

Table 1 Processes That Drive the Water Cycle

Process	Description	Energy source/production
evaporation	the conversion of liquid water to water vapour	solar energy
transpiration	the loss of water from the leaves of plants to the atmosphere	solar energy
condensation	the conversion of water vapour to liquid water	release of heat energy to the atmosphere
precipitation	the descent of water from the atmosphere to Earth in various forms: rain, sleet, snow, or hail	gravity
infiltration	the movement of water into the soil	gravity
percolation	the flow of water through the soil into storage areas (aquifers) in permeable rock	gravity
runoff	the flow of water over the surface of the land into streams, rivers, lakes, and oceans	gravity

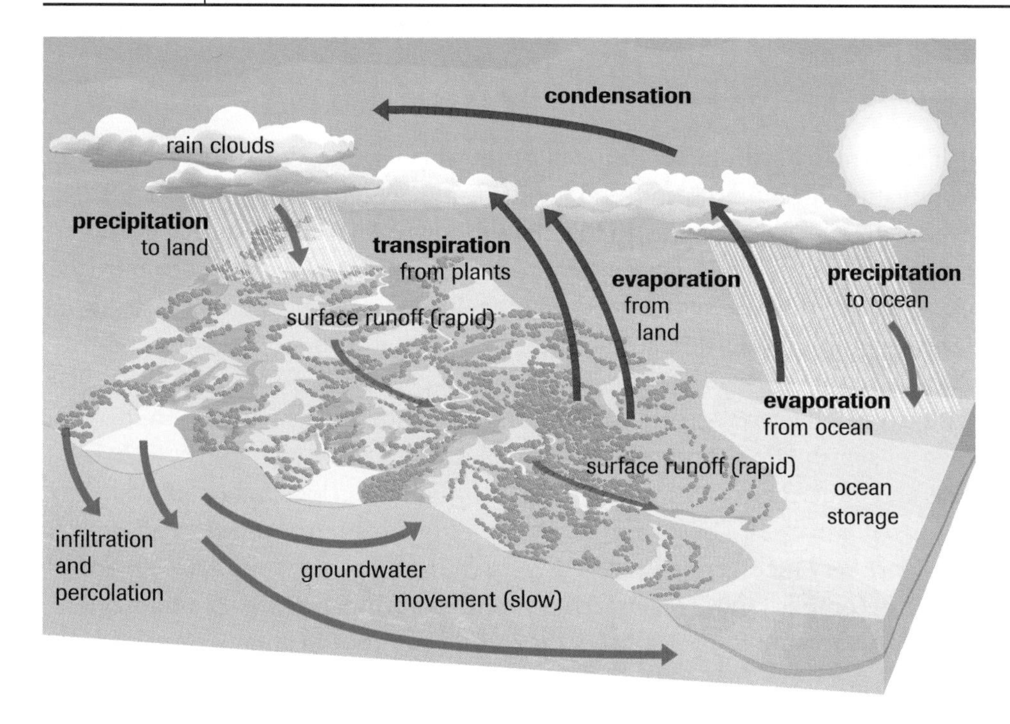

Figure 2
The water cycle

The Nitrogen Cycle

The role of bacteria in ecosystems is clearly demonstrated in the nitrogen cycle. Without bacteria, there would be no nitrogen cycle; without nitrogen, life is not possible. The atmosphere is about 78% nitrogen gas (N_2), but this nitrogen is not soluble in water and cannot be used directly by plants. It first must be converted to a usable form that is soluble in water and can be absorbed into plants.

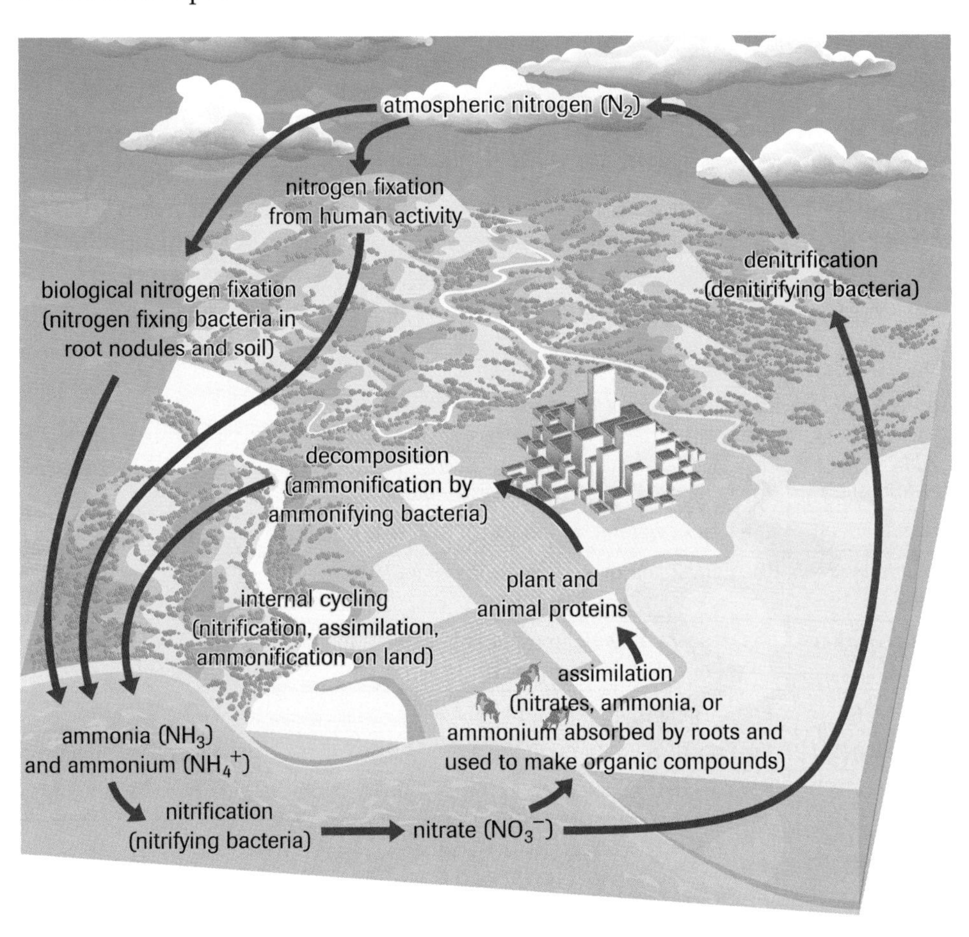

Figure 3
The nitrogen cycle

nitrogen fixation the conversion of nitrogen gas from the atmosphere to ammonia by nitrogen-fixing bacteria

nitrogenase an enzyme used by nitrogen-fixing bacteria in anaerobic conditions to break down atmospheric nitrogen gas

nitrification the conversion of ammonia or ammonium ions to nitrates by nitrifying bacteria in the soil

There are five steps in the nitrogen cycle—nitrogen fixation, nitrification, assimilation, ammonification, and denitrification (**Figure 3**). All steps except assimilation involve bacteria. In the first step, **nitrogen fixation**, special nitrogen-fixing bacteria use an enzyme, **nitrogenase**, to break apart the atoms in nitrogen gas (N_2). The nitrogen atoms are then combined with hydrogen molecules to form ammonia (NH_3) and ammonium (NH_4^+). Nitrogenase functions only in the absence of oxygen, so the bacteria must keep the enzyme away from contact with oxygen. Bacteria accomplish this by living under layers of slime on the roots of some plant species or inside nodules on the roots of legumes such as peas and beans. Nitrogen can also be fixed by lightning, by combustion, and by industrial processes. The industrial process of making nitrogenous fertilizers requires a very high temperature and pressure and lots of energy. Bacteria are able to accomplish the same process without the high temperature and pressure.

Soil bacteria are responsible for the second step, **nitrification**. Two types of bacteria, *Nitrosomonas* and *Nitrococcus*, change the ammonia and ammonium to nitrite ions (NO_2^-), and a third type, *Nitrobacter*, then converts the nitrite to nitrate ions (NO_3^-).

In the third step, **assimilation**, ammonia, ammonium ions, or nitrate ions are absorbed by plants and used in the construction of proteins and chlorophyll. When animals eat the plants, they use the nitrogen compounds from plants in the construction of DNA, amino acids, and proteins.

assimilation the uptake of ammonia, ammonium ions, or nitrates by plants and their conversion into organic molecules containing nitrogen

As organisms produce wastes or die, the nitrogen compounds go back to the ecosystem. However, they again have to be changed in order to be used once more. Decomposer bacteria convert the nitrogen-rich wastes into ammonia in a process called **ammonification.**

ammonification the part of the nitrogen cycle during which nitrogen compounds are converted to ammonia by ammonifying bacteria in the soil

The fifth and final step of the cycle, **denitrification**, is the opposite of nitrification. Other specialized bacteria convert the nitrates to gaseous nitrogen (N_2) and nitrous oxide (N_2O), which are released into the atmosphere and the cycle starts again. The cycle is a continuous process, so at any point in time some nitrogen can be found in every compound in each step of the cycle.

denitrification the part of the nitrogen cycle during which nitrates are converted by denitrifying bacteria into nitrogen gas and released into the atmosphere

The Phosphorus Cycle

Since phosphorus does not exist in the gaseous state, the phosphorus cycle takes place on the land and in the oceans. Unlike the carbon and water cycles, the phosphorus cycle is very slow. Over the short term (hundreds of years: human time scale), phosphorus flows from the land to the oceans; over the long term (millions of years: geologic time scale), phosphorus cycles back from the oceans to the land. **Figure 4** illustrates the phosphorus cycle covering both short- and long-term processes.

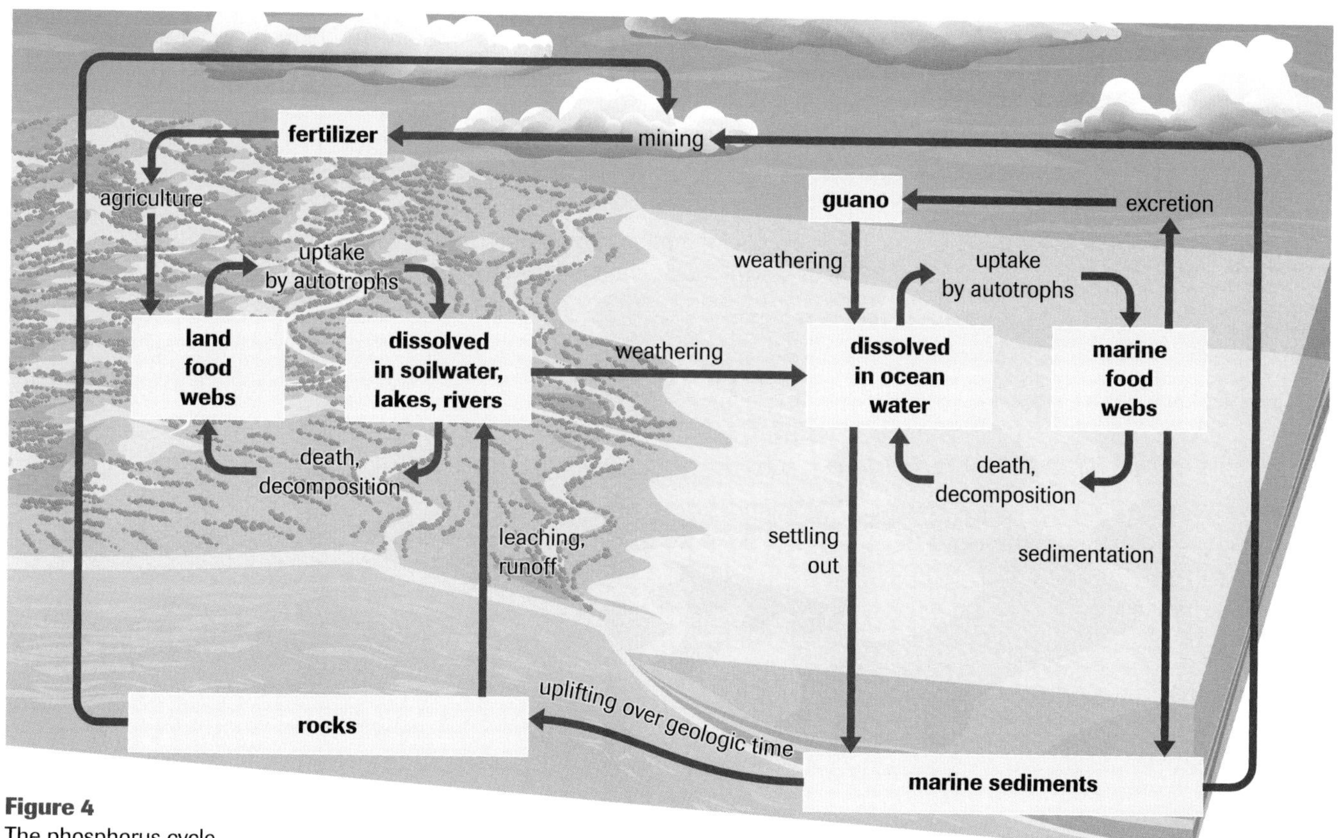

Figure 4
The phosphorus cycle

DID YOU KNOW?

Guano

Bird, bat, and seal droppings are called guano. Guano contains high levels of phosphates and nitrates and is a very valuable fertilizer. On some bird islands off the coast of Peru where there is very little rainfall, guano accumulates to a depth of almost 50 m. There is a profitable guano industry where this fertilizer is mined.

Rocks and sediments containing phosphorus are the primary source of phosphorus in the environment. Phosphate ions (PO_4^{3-}) are absorbed through the roots of plants and used in the construction of many biological molecules and compounds. Phosphorus is then moved through the ecosystem in the same way as other nutrients. Plants get their phosphorus from the soil; consumers in the second trophic level get their phosphorus by eating plants; and so on, up the food chain.

Section 5.5 Questions

Understanding Concepts

1. Explain the meaning of the word *biogeochemical.*
2. Use the following terms to describe the water cycle: *evaporation, transpiration, condensation, precipitation, infiltration, percolation,* and *runoff.*
3. Explain the carbon cycle, including the following terms in your description: *photosynthesis, cellular respiration, combustion, erosion,* and *carbon storage.*
4. Name and briefly describe the steps in the nitrogen cycle.
5. "Some of the carbon atoms in your body were once part of a plant that was eaten by a dinosaur." Could this statement be true? Explain your answer.
6. We generally associate decomposers with death. How do decomposers contribute to life?
7. Farmers add fertilizers containing nitrogen and phosphorus to their crops. Why don't they add carbon as well?

Making Connections

8. Use the Internet and other resources to research the seabird- and bat-guano industry: the status of the industry, locations where it is being pursued, and the current price of guano fertilizers. Evaluate the industry in terms of its economic and environmental benefits and costs. Prepare a brief report outlining your findings.

www.science.nelson.com

Reflecting

9. "The cycling of matter is important but not essential for the continuation of life on Earth." State and justify your opinion about this statement.

Interactions: Competition and Predation

As you learned earlier, populations do not live in isolation. They interact with each other and with other species in their community. These interactions have been studied by scientists, and patterns have been observed and described. Consumers, for example, cannot survive without the species that are their food supply. Even some producers cannot survive without other species. For example, some species of flowering plants depend on insects to pollinate them so that they can reproduce. In a community or ecosystem, there are many interactions occurring all the time. Some of the interactions are obvious (e.g., the dependence of one species on another for food), while others are not revealed without detailed study (e.g., how the flowering plants on a forest floor might affect the trees of the forest).

Competition

Within a community, many species live in the same general habitat. Each species, however, has adapted to a particular niche in order to survive. The **niche** is the organism's role in the community—what it eats, what eats it, how it reproduces, its preferred temperature range, and other factors that describe how it lives. For example, a fox shares a habitat with other carnivores, such as the wolf and the lynx, but even though it relies on some of the same species for food, it occupies a different niche and uses different strategies for survival. The fox experiences **interspecific competition** (*inter* meaning "between," and *specific* referring to species). It survives, however, because it has adapted to hunting at night, thereby reducing its direct competition with other carnivores. The variety of shorebirds (e.g., herons, sandpipers, plovers, oystercatchers, and turnstones) has also adapted to competition with other species on sandy beaches and mud flats. Different species feed at different distances on either side of the water line, at different tide heights, and at different depths in the sand or mud.

niche an organism's role in its habitat, including how it interacts with the biotic and abiotic resources in its environment

interspecific competition competition between individuals of different species for an essential common resource that is in limited supply

The members of every species also experience **intraspecific competition** (*intra* meaning "within") because the members of each species compete with one another for the same food supply.

intraspecific competition competition among individuals of the same species for resources in their habitat

Interspecific competition is most significant and noticeable when the niches of two different species overlap. The more the niches overlap, the greater the competition. Over the long term, species that compete with one another for similar resources have evolved mechanisms to reduce the effect of the competition. One such mechanism is called **resource partitioning**. For example, resource partitioning occurs between hawks and owls. Although they hunt the same species, hawks hunt by day and owls by night. **Figure 1**, on the next page, shows the resource partitioning among different species of warblers in a spruce tree.

resource partitioning avoidance of, or reduction in, competition for resources by individuals of different species occupying overlapping niches

We generally consider competition to be a negative phenomenon. If you own a business and another person comes to your neighbourhood and establishes the same sort of business, then competition is not good for you. You now have to share your resource. This is also true in nature, but the relationship is not

Figure 1
Resource partitioning. Each species of warbler spends most of its feeding time in a different portion of the tree.

necessarily a simple, direct interaction. In some cases, competition may be helpful. For example, competition may eliminate the weakest members of a population so that only the strongest survive to reproduce. So, in the long term, the population may grow stronger and be better able to compete with other species.

Predation

predation an ecological relationship in which a member of one species (predator) catches, kills, and eats a member of another species (prey)

Predation is an example of interspecific competition in which one species—the predator—feeds directly on another species—the prey. In this relationship, the individual prey is harmed while the predator benefits. In the longer term, the population of prey may actually benefit because predators often kill off the sick, weak, and old members of the population. Although this will reduce the prey population, it will also increase the food supply available for each of the remaining members of the population and will ensure that only the strongest members of the population survive to reproduce.

Humans often have an emotional reaction to the predator–prey relationship. When we see a lynx attempting to catch and eat a hare, we tend to sympathize with the hare (**Figure 2**). We have to realize that the lynx is doing what its

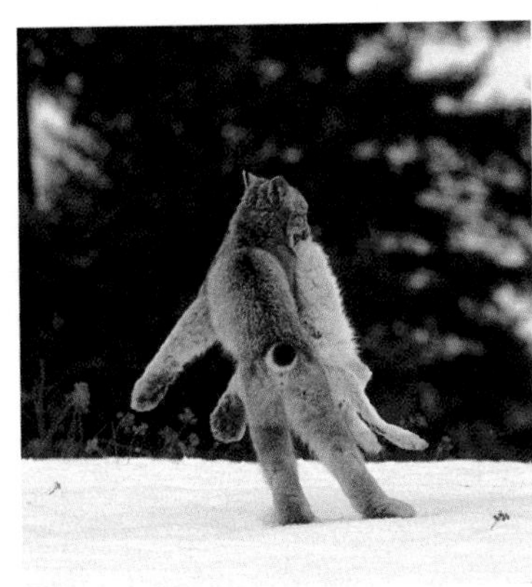

Figure 2
The lynx is a natural predator of the hare.

instincts tell it to do—feed itself and its young. At the same time, it is ensuring that the hare population does not grow out of control, an important role in maintaining an ecological balance.

Predator and Prey Strategies

Both predator and prey have developed strategies for survival—predators for getting food and prey for avoiding becoming food. Predators have two main strategies for capturing their prey—pursuit and ambush. Some predators, such as hawks or lynx, simply pursue and catch their prey by being able to move faster than their prey. Others, such as wolves, cooperate as a group to hunt and kill their prey. Other predators have developed characteristics that enable them to ambush their prey. For example, some fish have colouration that makes them nearly invisible against the stream or lake bottom. Camouflage is a successful ambush strategy because predators simply have to wait until their prey comes to them (**Figure 3(a)**).

Prey have also developed strategies for survival. Camouflage (or protective colouration in the case of prey) works just as well for prey as it does for predators. An organism that looks like the background of its environment is difficult to see and, therefore, hard for predators to detect (**Figure 3(b)**).

Some organisms, such as poisonous frogs, have developed chemical defences. After an unpleasant experience or two, predatory snakes learn very quickly to leave these frogs alone. Skunks have a gland from which they can spray a bitter-tasting and foul-smelling chemical. Sometimes, an organism that is not able to defend itself from predators gains protection because it resembles another species that is dangerous. This defence, known as mimicry, protects the hawk moth caterpillar because it can change its appearance to resemble a snake (**Figure 3(c)**). Still other prey gain protection by living with a large number of others of their species in herds (caribou), flocks (starlings), or schools (herring).

Figure 3
Strategies for survival:
(a) A stick insect simply waits for its prey to come close enough to be grabbed.
(b) This flatfish avoids detection by assuming the colour of the ocean bottom.
(c) The hawk moth caterpillar mimics the appearance of a snake.

Section 5.6 Questions

Understanding Concepts

1. Explain the concept of resource partitioning and give an example.
2. Distinguish between interspecific and intraspecific competition.
3. Describe three strategies that prey use to avoid being eaten by predators. Give an example of each.
4. What are the two main ways that predators capture their prey? Give an example of each.
5. For prey species, there is safety in numbers. This strategy improves the probability of survival of caribou in a herd because a wolf will not likely attack a large herd of caribou. In what other way might living in a herd be an advantage for an individual caribou? (*Hint:* Think of being lost in a crowd.)

Applying Inquiry Skills

6. Briefly describe an experiment that you might conduct to determine if two species of butterfly that resemble each other are an example of mimicry.

Reflecting

7. (a) Describe how you feel when you see a cat catch a bird or a mouse.
   (b) Why do you feel that way?
   (c) What do you think you should do when your pet cat catches a bird or a mouse? Explain.

5.7 Activity

Observing Competition between Species

In experiments conducted in 1934, Russian ecologist G.F. Gause tested the population-growth theory—that two species with similar requirements could not live together in the same community. Gause predicted that, when two species competed for the same resource, one species would consume most of the resource and cause the other species to become extinct. His experiments led him to the conclusion that, if resources are limited, no two species can remain in competition for exactly the same niche indefinitely. This became known as Gause's principle, or the principle of competitive exclusion.

Question

What happens to populations when they compete for resources in the same ecological niche?

Observations

In the first experiment, *Paramecium aurelia* and *Paramecium caudatum* were cultured separately and then cultured together. The results are shown in the graphs in **Figure 1**. A separate experiment was then conducted where *Paramecium bursaria* and *Paramecium caudatum* were cultured together. The results are shown in **Figure 2**.

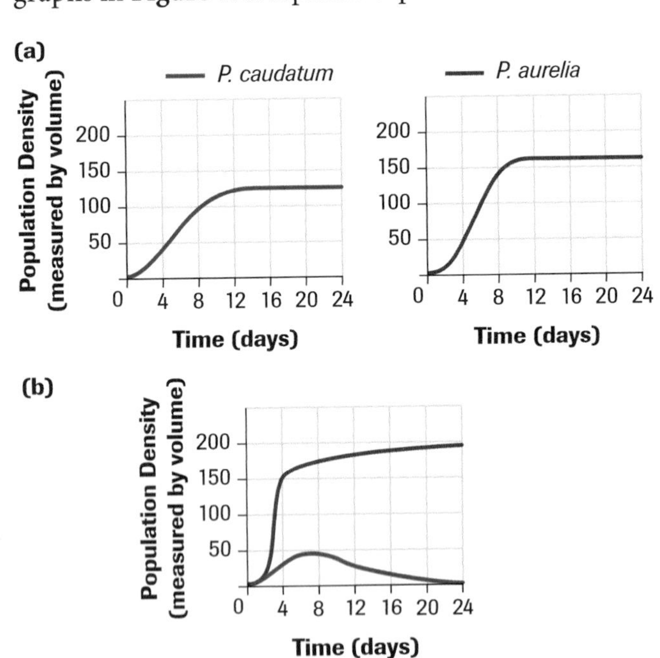

Figure 1
Gause's results for the first experiment: **(a)** *P. caudatum* and *P. aurelia* cultured individually, and **(b)** *P. caudatum* and *P. aurelia* cultured together

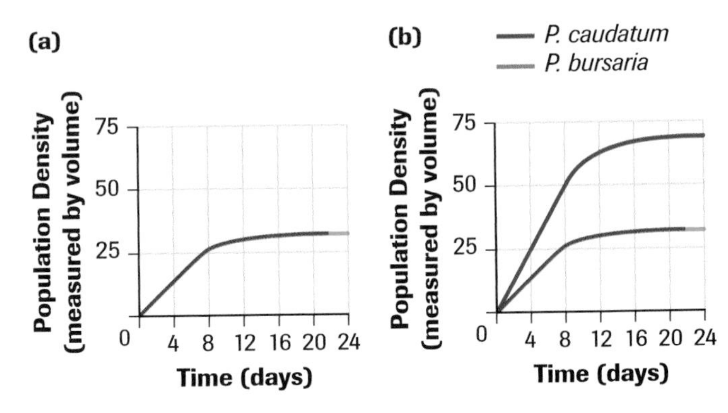

Figure 2
Gause's results for the second experiment: **(a)** *P. bursaria* cultured individually, and **(b)** *P. bursaria* and *P. caudatum* cultured together

Analysis

(a) Compare the population densities of the paramecium populations in both experiments.

(b) In each of the paired experiments, do species grown in combination reach the same density as species grown independently?

(c) What are the limits in population density achieved by each of the species when
  (i) grown independently?
  (ii) combined with another species?

Evaluation

(d) Did the results of the first experiment support Gause's prediction? Explain.

(e) Did the results of the second experiment support Gause's prediction? Explain.

(f) "The ability of *P. aurelia* to grow faster in the presence of *P. caudatum* is an example of exploitative competition." Explain what this statement means.

Synthesis

(g) Propose an explanation for why both species of paramecium survived in the second experiment.

Interactions: Symbiotic Relationships

Symbiosis refers to a relationship in which individuals of two different species live in close, usually physical, contact. At least one of the two species benefits from the relationship. As you learned in Unit 2, there are three main types of symbiotic relationships—mutualism, commensalism, and parasitism. Mutualism exists when both species in the relationship benefit and neither is harmed. Commensalism occurs when one organism benefits and the other neither benefits nor is harmed. Parasitism occurs when one organism benefits at the expense of another organism's well-being.

Mutualism

Even though it may appear that two species are cooperating to benefit each other, mutualism is not cooperation. It is simply two species that take advantage of each other without harming each other. Lichens are a classic example of mutualism (**Figure 1**). Lichens are a combination of algae and fungi living together. The algae are photosynthetic and produce food for themselves and the fungi. The fungi hold moisture and mineral nutrients for themselves and the algae, and provide a suitable habitat for the algae. Lichens constitute a microecosystem because they create an environment where other organisms such as insects can also live. Lichens grow very slowly and can live for very long periods of time—hundreds, if not thousands, of years. Much is still unknown about the relationship between the algae and fungi in lichens, and, because of this, it is almost impossible to grow lichens in an artificial (not natural) environment.

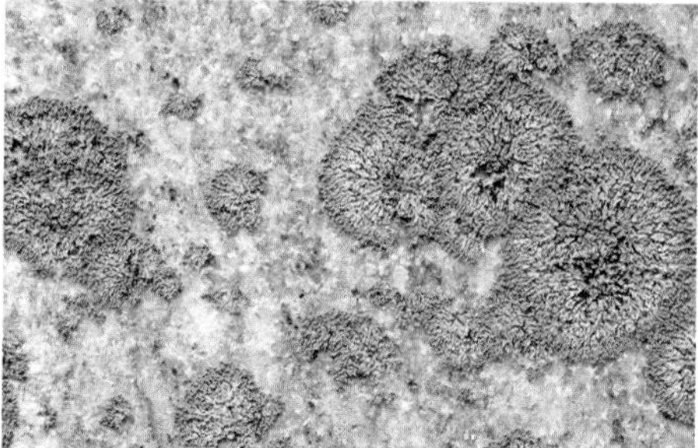

Figure 1
Lichens found in Canada.

Another example of mutualism is the relationship between mycorrhizal fungi and the roots of many terrestrial plants. The fine threads of the fungus form a network among the soil particles and the roots of plants and help the plants absorb nutrients such as phosphorus (**Figure 2**). In return, the fungi obtain nutrients such as sugars that the plant has made during photosynthesis. Mycorrhizal fungi are so beneficial to some plants that the plants' seeds will not germinate in the fungi's absence. The seeds of wild orchids, for example, will not germinate without the fungal relationship, and some species require the relationship throughout their life.

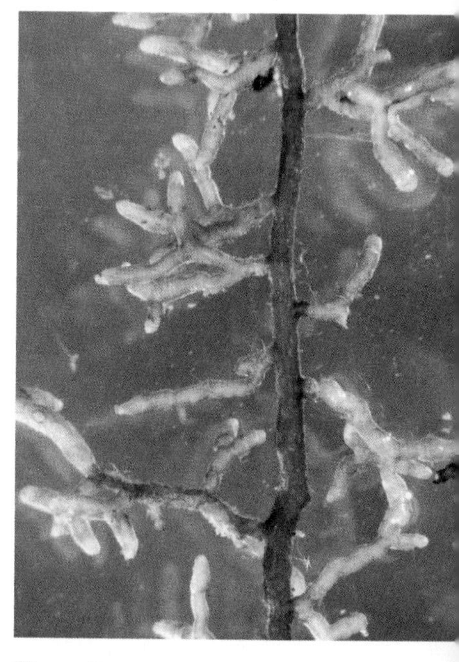

Figure 2
Network of mycorrhizal fungi around the roots of a plant

Commensalism

In commensalism, it is often difficult to determine if the unaffected species is, in fact, being harmed or benefited. A common example of commensalism is the relationship between remora and sharks (**Figure 3(a)**). The remora have suction disks on the back of their heads with which they attach themselves to sharks. When the sharks travel, the remora get a free ride—they do not use energy for swimming. They also benefit by being able to feed on small scraps of food left from the sharks' prey.

Another example of commensalism is shown in **Figure 3(b)**. The clown fish lives among the tentacles of the anemone. The tentacles of the anemone have nematocysts, or stinging cells, that it uses to paralyze and capture prey. The clown fish is not stung by these cells, so its living space offers it protection from other fish and leftovers from the anemone's meal.

(a)

(b)

Figure 3
The remora **(a)** gets a free ride and access to new food sources. The clown fish **(b)** gets protection and food.

Parasitism

One of the basic principles of survival in nature is to use as little energy as possible in obtaining food. That is why parasitism is so common. Parasites live and feed on the most nutritious environments on Earth—the bodies of other living organisms. The host organism provides a continuous, easy supply of nutrients for the parasite. Their dependence on their host is so great that parasites cannot complete their life cycle without the host. Biologists estimate that as many as 25% of all animal species may be parasites.

In most cases, parasites do not cause significant harm to their hosts. Otherwise, they would be harming the very environment that they depend on for survival. However, some parasites are responsible for serious diseases in humans (e.g., malaria and African sleeping sickness) and may cause the death of the host.

endoparasite a parasite that lives and feeds within the host's body

ectoparasite a parasite that lives and feeds on the outside surface of the host

Endoparasites are organisms that live inside the body of their hosts. Tapeworms live in the digestive systems of different organisms (including humans) and use the digested nutrients as a food supply. Many single-celled protozoans live in the bloodstream.

Ectoparasites are organisms that live and feed on the outside surface of their host. The most common examples are parasitic insects such as lice, fleas, and ticks. Human head lice are a common parasite that, contrary to popular opinion, are not the result of poor personal hygiene (**Figure 4**). They are an equal-opportunity parasite and will inhabit any head. These tiny insects (2–3 mm) cannot fly or jump, so they can be transmitted from head to head only by direct contact or by sharing hats, combs, and brushes.

Figure 4
Human head louse

Head lice get their nutrients by feeding directly on the blood of the host, and they cannot survive for more than a day or so without feeding. The lice themselves do little harm to their human host, so the most serious problem they cause is annoyance and the effort required to get rid of them. Head lice are host-specific; that is, they will live only on a human head and will not survive on other animals.

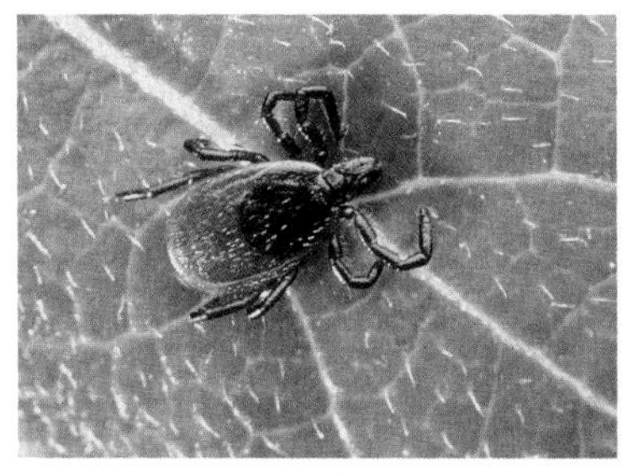

(a)

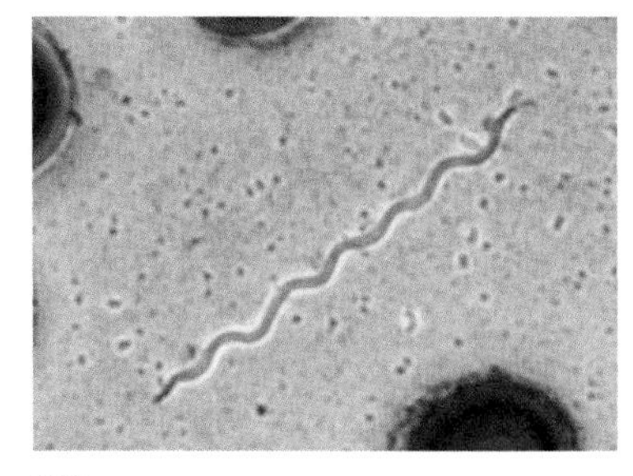

(b)

Figure 5
(a) Black-legged, or deer, tick
(b) Bacterium, carried by the ticks, that causes Lyme disease

Ticks are also blood-feeding parasites and are found on many mammals, birds, and reptiles. An important difference between lice and ticks is that ticks can transmit bacteria, viruses, and protozoa that cause serious diseases in humans and other animals. For example, Lyme disease is caused by a bacterium, *Borrelia burgdorferi*, that is transmitted by ticks found on mice (the main source of the bacterium), white-tailed deer, and other warm-blooded animals (**Figure 5**). Lyme disease can cause severe conditions and can be difficult to diagnose because of its similarity to other diseases.

An example even closer to home is found in the mouths of most humans. An amoeba known as *Entamoeba gingivalis* lives in and on the teeth and gums of most people. It was once thought that the amoeba was a harmless scavenger feeding on bits of food, but it is now known to feed on bacteria and red and white blood cells (**Figure 6**). White blood cells are an important line of defence against infection, so the amoeba is now considered a parasite. The only way this protozoan can be transmitted from one person to another is through mouth-to-mouth contact.

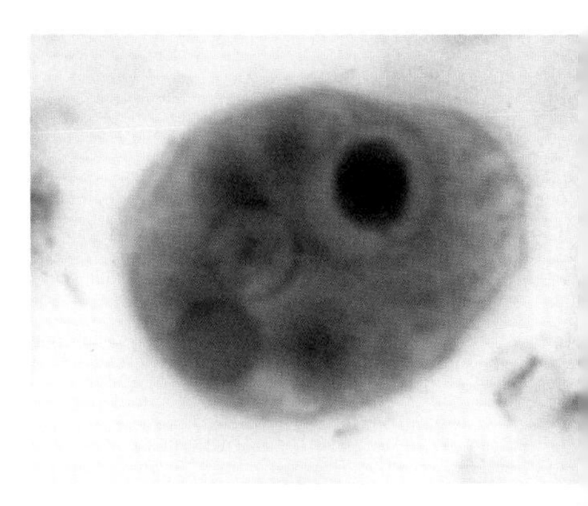

Figure 6
This amoeba has ingested a red blood cell.

Section 5.8 Questions

Understanding Concepts

1. Define mutualism, commensalism, and parasitism and give an example of each type of relationship. What do these relationships have in common?
2. Distinguish between endoparasites and ectoparasites.
3. Why don't parasites generally harm their hosts?
4. Why is parasitism so common in nature?

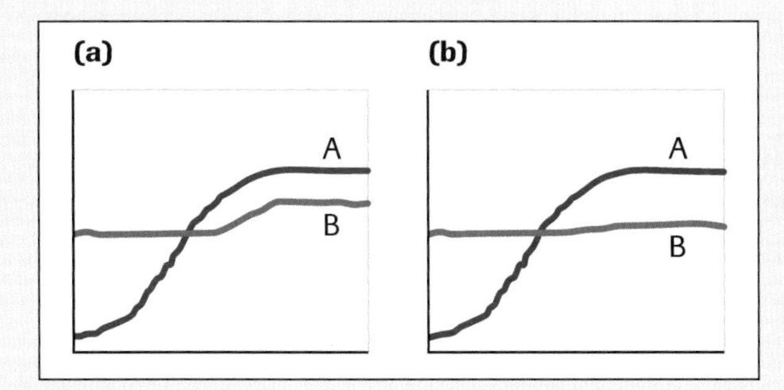

Figure 7

Applying Inquiry Skills

5. The graphs in **Figure 7** represent the growth of two populations of different species in an ecosystem. In a group, discuss what type of symbiotic relationship each of these graphs could represent. Explain your reasoning.

Making Connections

6. Use the Internet and other resources to research Lyme disease. Present your findings as an informative pamphlet.
   (a) What are the symptoms of and treatments for it?
   (b) What is its status in your province?
   (c) Is Lyme disease a serious health concern? Justify your answer.
   (d) What measures can you take to help protect yourself from Lyme disease?

www.science.nelson.com

5.9 How Populations Change

Populations of organisms are always changing. Some populations, such as the Vancouver Island marmot, are declining to the point where extinction is a possibility (**Figure 1(a)**). Currently in Canada there are more than 300 species that are threatened with extinction. Other populations, such as the greater snow goose, have grown to the level where they are threatening their habitat (and that of other species) and their own survival is at stake (**Figure 1(b)**). The greater snow goose population has grown from 50 000 in 1965 to more than 1 million currently.

(a)

(b)

Figure 1
(a) The Vancouver Island marmot population is threatened by low numbers.
(b) The greater snow goose is threatened by overpopulation.

Changes in population numbers and in patterns of distribution of individuals can have direct effects on the local ecosystem and may affect the well-being of other species within the community. Can the extinction of a species be avoided? What are the consequences of rapid population growth and overpopulation? Wildlife technicians use special methods to monitor and measure changes in populations. They gather the data necessary to predict future trends in the growth of populations. This information can be used to determine the health of individual species and entire ecosystems, to develop policies and plans of action to save species from extinction, and to address the impacts of rapidly growing populations.

CAREER CONNECTION

Wildlife technicians conduct biological surveys using various types of equipment to collect samples (e.g., by trapping animals for tagging) and to record observations and measurements on the samples.

Population Characteristics

All ecosystems have limited biotic and abiotic resources at any given time. Biotic resources, such as prey, vary in availability. Some abiotic resources, such as space and light, vary little, while others, such as temperature and water, vary greatly. There is, therefore, a limit to the number of individuals that an environment can support at any given time. The **carrying capacity** of an ecosystem is the maximum number of organisms that can be sustained by available resources over a given period of time. Since environmental conditions vary, carrying capacity also varies over time.

carrying capacity the maximum number of organisms that can be supported by available resources over a given period of time

Another limit to population growth is the organism's reproductive ability or its **biotic potential**. Even under ideal conditions with unlimited resources, every species will reach a maximum rate of reproduction.

biotic potential the maximum rate at which a population can increase under ideal conditions

Populations are generally described using four main characteristics—size, density, dispersion, and age distribution. **Population size** simply refers to the total number of individuals within the habitat. It changes over time and is limited by the carrying capacity of the ecosystem.

population size the number of individuals of a species occupying a given area or volume at a given time

Population density (D) refers to the number of organisms living in a geographic area. It is calculated by dividing the total numbers counted (N) by the space (S) occupied by the population.

population density the number of individuals of the same species that occur per unit area or volume

$$D = \frac{N}{S}$$

SAMPLE problem 1

If 480 moose live in a 600-ha region of Algonquin Park, what is their population density (D)?

Solution

$$D = \frac{N}{S}$$

$$D = \frac{480 \text{ moose}}{600 \text{ ha}}$$

$$D = 0.8 \text{ moose/ha}$$

The population density will vary according to the species and the habitat. As shown in **Table 1**, small organisms usually have higher population densities than larger organisms.

Table 1 Typical Densities of Different Populations

Population	Typical density
jack pine	380/ha
field mice	250/ha
moose	0.8/ha
soil arthropods	500 000/m^2

Population density can be deceiving because of unused or unsuitable space within a habitat. For example, the moose in Algonquin Park do not use open lake water. Therefore, if the 600 ha of total habitat included 70 ha of open lake water, only 530 ha are available to the moose. The **crude density** of the moose population is 0.8/ha, but the actual or **ecological density** is 480 moose/530 ha, or 0.9 moose/ha.

crude density population density measured in terms of the number of organisms of the same species within the total area of the entire habitat

ecological density population density measured in terms of the number of individuals of the same species per unit area or volume actually used by the individuals

The biotic and abiotic conditions vary from place to place in a particular area. For this reason, the individuals are dispersed or distributed throughout the range in different ways. Biologists have identified three main dispersion patterns among wild populations: clumped, uniform, and random. Most populations show patchy or **clumped dispersion**, in which organisms are densely grouped in areas of the habitat with favourable conditions for survival. Clumped dispersion may also be the result of social behaviour, such as fish swimming in large schools to gain protection from predators (**Figure 2(a)**, on the next page). In contrast to clumped dispersion, organisms may show **uniform dispersion**, in which individuals are evenly distributed throughout the habitat. This pattern may result from competition among individuals that set up territories for feeding, breeding, or nesting. Penguins nesting on South Georgia Island in the South Atlantic Ocean show a nearly uniform dispersion pattern (**Figure 2(b)**, on the next page). Individuals show **random dispersion** when they are not influenced by other individuals and when the habitat is virtually the same throughout. Some species of trees in tropical rain forests show random dispersion, although this pattern is rare among wild populations (**Figure 2(c)**, on the next page).

clumped dispersion a distribution pattern in which individuals in a population are more concentrated in certain parts of a habitat

uniform dispersion a distribution pattern in which individuals are equally spaced throughout a habitat

random dispersion a type of population distribution in which individuals live throughout a habitat in an unpredictable and patternless manner

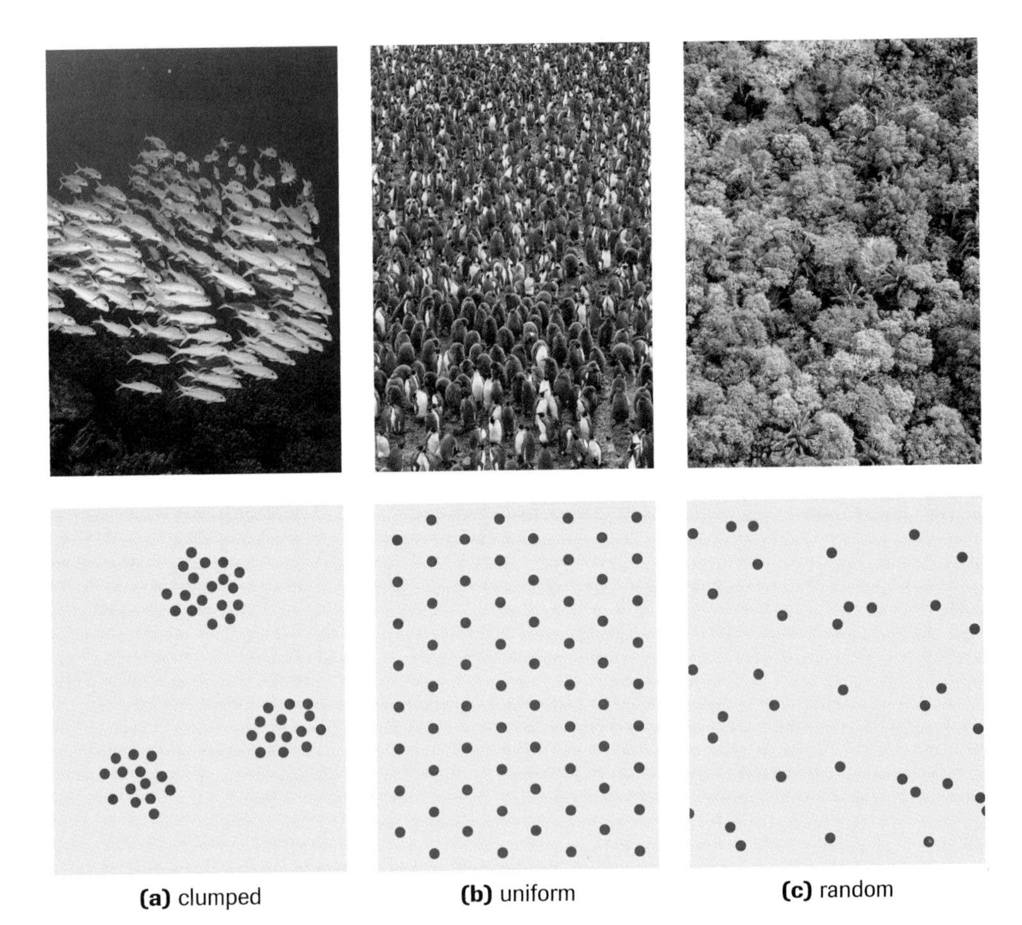

Figure 2
Populations generally show one of three patterns of dispersion:
(a) Yellow goatfish are often found clumped in schools.
(b) Nesting King penguins show a uniform pattern.
(c) In tropical rain forests, trees of the same species can be randomly dispersed.

Population Dynamics

population dynamics changes in population characteristics determined by natality, mortality, immigration, and emigration

Populations are always changing. Depending on the species and on environmental conditions, populations experience natural hourly, daily, seasonal, and annual fluctuations in numbers. Changes in population characteristics are known as **population dynamics**. Births (natality), deaths (mortality), immigration (the number that move into an existing population), and emigration (the number that move away from an existing population) can be used to determine the growth rate of a population in a given period of time. The population growth rate of any given population is calculated mathematically using the following formula:

$$\text{population growth rate} = \frac{[(\text{births} + \text{immigration}) - (\text{deaths} + \text{emigration})]}{\text{initial population}} \times 100\%$$

$$\text{population growth rate} = \frac{[(b + i) - (d + e)]}{n} \times 100\%$$

The growth rate is expressed as a percentage. This formula is useful for biologists to determine the changes in population size over time. If the number of individuals that were born and migrated into the population is higher than the number of individuals that died and emigrated, the population will have positive growth, increasing in size. Conversely, if the decrease from deaths and emigration is greater than the increase from births and immigration, the population will experience negative growth, decreasing in size. If the gains from births and immigration equal the losses from deaths and emigration, the population is said to have zero growth and will remain constant.

If migration into and out of a population occurs, the population is called an **open population**. The human population in Canada is an example of an open population. In some populations, such as the population of trout in a lake, migration into and out of the population does not occur. This is referred to as a **closed population**, a population in which only births and deaths determine population growth.

open population a population in which changes in number and density are determined by births, deaths, immigration, and emigration

closed population a population in which changes in number and density are determined by natality and mortality alone

Patterns of Population Growth

When birth rates and death rates remain constant, populations grow at a fixed rate in a fixed time interval. Such a fixed rate of growth can be expressed as a ratio or percentage per unit of time, such as 1.05 or 5% per year. In other words, the population will be 1.05 times as large as it was in the previous year, or we say it has increased by 5%. Populations whose growth rate is constant are said to experience **geometric growth**. The growth rate can be determined by comparing the population size in one year with the population size at the same time in the previous year. In species that show this pattern of growth, the population grows rapidly during the breeding season and then declines throughout the remainder of the year until the next breeding season begins, when it increases again. But the overall trend is a continuous increase from year to year. The geometric growth rate, symbolized by the Greek letter *lambda* (λ), can be calculated by the following formula:

geometric growth a pattern of population growth in which organisms reproduce at fixed intervals at a constant rate

$$\lambda = \frac{N_{(t+1)}}{N_{(t)}}$$

where N is the population size, t is the previous year, and $(t + 1)$ is the current year.

SAMPLE problem 2

The white-tailed deer population in an area was determined to be 2000 individuals. At the same time the previous year, the population in the same area was 1600. What is the deer's geometric growth rate?

Solution

We know the geometric growth rate can be calculated using the formula

$$\lambda = \frac{N_{(t+1)}}{N_{(t)}}$$

where

$$\lambda = \frac{2000}{1600}$$

$$\lambda = 1.25$$

The geometric growth rate is 1.25 or 25% per year.

Population growth can be illustrated by graphing the change in population size over time. **Figure 3**, on the next page, shows the change in the white-tailed deer population over a period of ten years if its constant geometric growth rate is 1.25 or 25%. The data are shown in **Table 2**, on the next page.

Table 2 Growth of a White-Tailed Deer Population

Time (years)	Population size
0	1 600
1	2 000
2	2 500
3	3 125
4	3 906
5	4 884
6	6 105
7	7 631
8	9 539
9	11 924
10	14 905

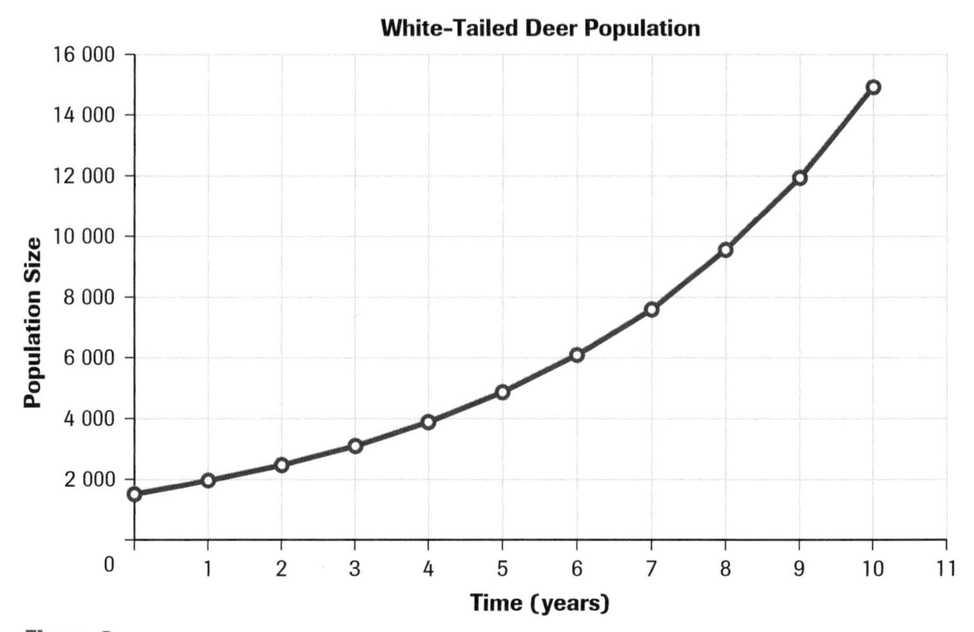

Figure 3
A graphical representation of the geometric growth of a white-tailed deer population, in which the population grows at a constant rate of 1.25 or 25% yearly

exponential growth a pattern of population growth in which organisms reproduce continuously at a constant rate

DID YOU KNOW ?

E. coli

Under ideal conditions, a single *E. coli* bacterium can reproduce by binary fission every 12 min. After 12 min there would be two bacteria, after 24 min there would be four cells, and so on. If this continued unchecked for the next 24 h, there would be enough *E. coli* cells to cover the surface of Earth to a depth of more than 1 m!

Some species, those that have no specific breeding season, reproduce on a continuous rather than an intermittent or irregular basis. **Exponential growth** describes a pattern in which populations grow continuously at a fixed rate in a fixed time interval. If a population has unlimited resources, its growth may be exponential. The graph in **Figure 4** shows the typical J-shaped curve of exponential growth. In this case, an original yeast-cell population under ideal conditions doubles every 24 h. The data are shown in **Table 3**. As the increasing slope of the graph indicates, the amount of growth gets bigger as the population gets bigger.

Table 3 Yeast-Cell Population Growth

Time (days)	Population size
0	2 500
1	5 000
2	10 000
3	20 000
4	40 000
5	80 000
6	160 000
7	320 000
8	640 000
9	1 280 000

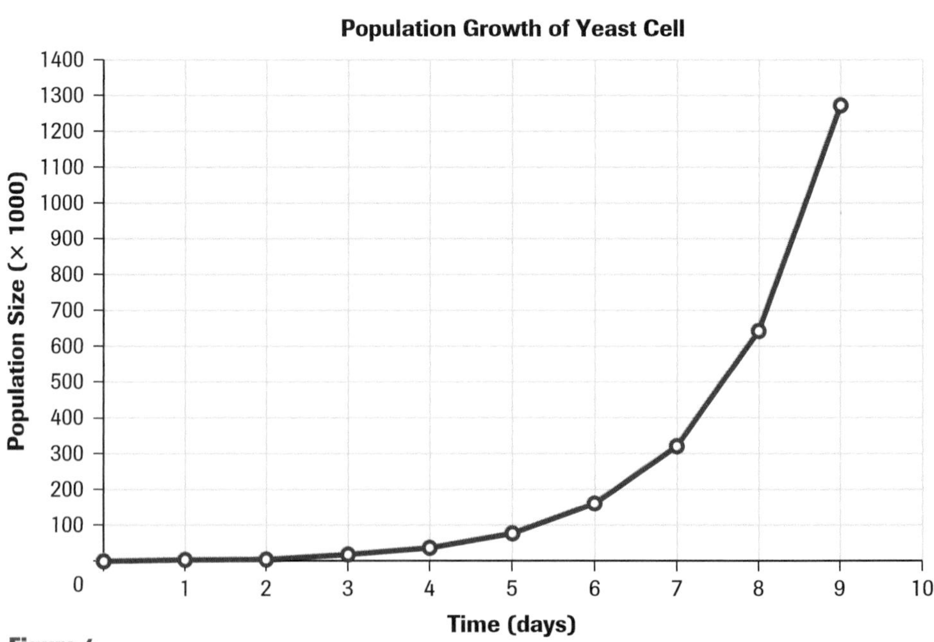

Figure 4
In exponential growth, the larger the population gets, the faster it grows.

Logistic growth can be seen in a population of fur seals on St. Paul Island, Alaska. In 1911, fur seal hunting was banned on St. Paul Island, because the seal population had become extremely low. Because their numbers were so severely depressed, the seals had many unused resources to support the recovering population. The population began to grow rapidly until it stabilized around its carrying capacity, as shown by the graph in **Figure 5**.

logistic growth a model of population growth describing growth that levels off as the size of the population approaches the carrying capacity of the environment

When a population's growth is controlled by its environment, the graph pattern is an S shape. The S-shaped graph of logistic growth shows that, at the beginning, the pattern of population growth is similar to geometric and exponential growth. It starts slowly and increases more rapidly as time goes on. As time passes, however, the population experiences **environmental resistance** (e.g., limited food, space, and other resources), and the population growth slows down—reproduction decreases and the number of deaths increases. The population levels off at the carrying capacity of the environment. At this point, the population is in **dynamic equilibrium** because the number of births equals the number of deaths, resulting in no increase in population size.

environmental resistance any factors that limit a population's ability to reach its biotic potential when it nears or exceeds the environment's carrying capacity

dynamic equilibrium (population) the condition of a population in which the birth rate equals the death rate and there is no net change in population size

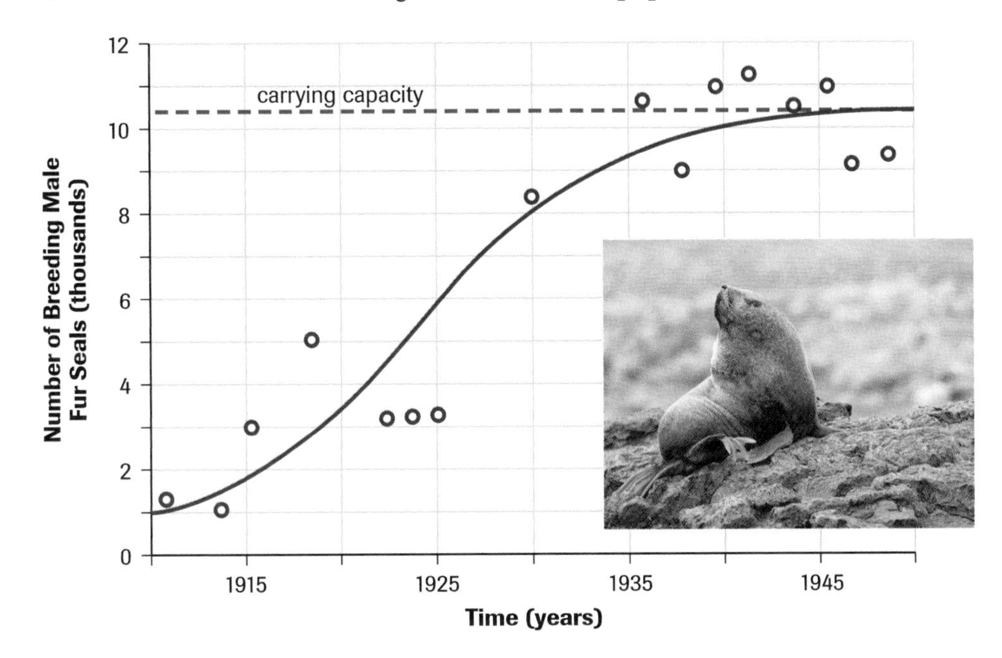

Figure 5
A population that grows logistically results in an S-shaped curve.

Population Fluctuations

Even after a population stabilizes around the carrying capacity of the ecosystem, it continues to change. Biologists have identified four types of population changes—stable, irruptive, irregular, and cyclic—as illustrated in **Figure 6**, on the next page. *Stable* populations show the least fluctuation, with only slight variations above and below the carrying capacity. The populations of species whose habitats are fairly constant are generally very stable. A temporary increase in the carrying capacity because of increased food supply, however, might cause a normally stable population to explode or *irrupt*. This sudden increase in population is followed by a just-as-sudden crash of the population to a more stable level. For largely unknown reasons, some species show *irregular* population changes. They experience large fluctuations above and below the carrying capacity with no observable pattern in the changes.

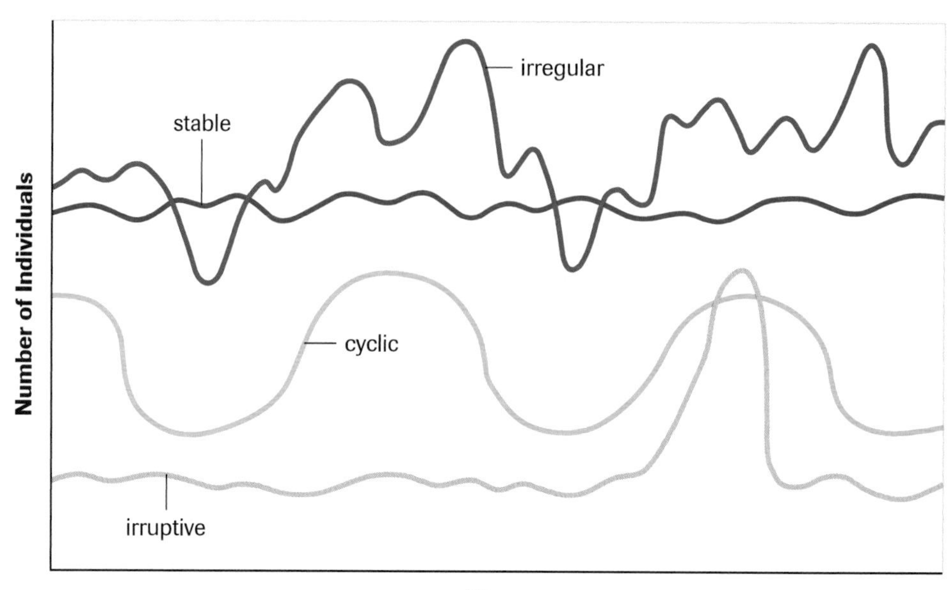

Figure 6
Types of population changes

Many predator–prey relationships cause the populations of both species to become *cyclic* (**Figure 7**). For example, lynx populations tend to increase when the population of snowshoe hares is high and decrease as the hare population decreases. In this scenario, the predator population changes follow slightly behind the prey population changes. When the hare population is low, the lynx have little food and their population declines. A reduction in the lynx population allows the hare population to recover and increase. The lynx population does not increase again until the lynx begin to reproduce and survive. Both hare and lynx populations grow until the increase in the lynx population causes the hare population to decline. As the lynx population increases, more hares are eaten. The resulting lowered hare population leads to starvation among lynx, slowing the lynx population's growth rate. Although it appears that these two populations regulate each other, it is now believed that other factors, such as food supply for the hares, also contribute to the cyclic nature of its population.

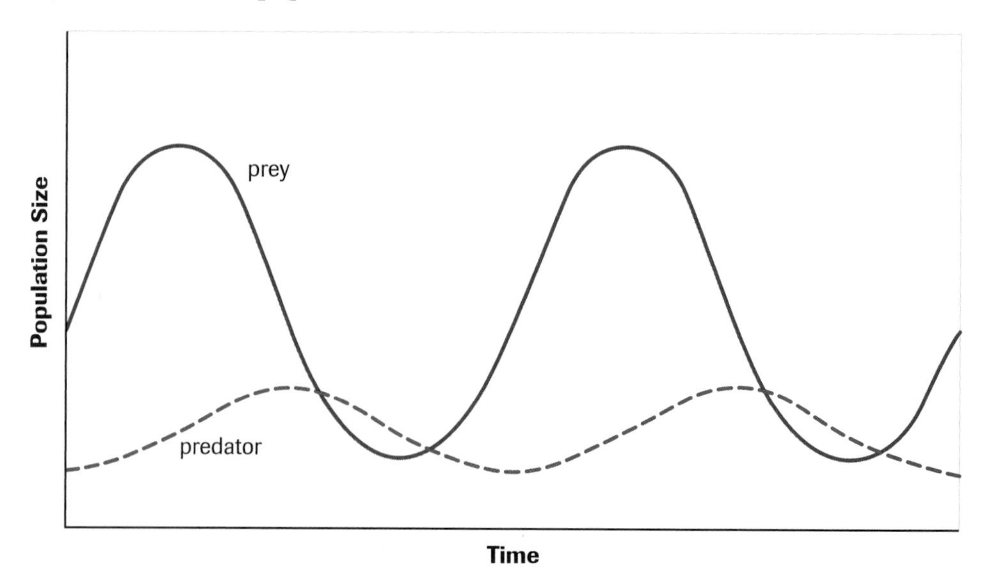

Figure 7
A model of predator–prey population cycles

Section 5.9 Questions

Understanding Concepts

1. Describe the concepts of carrying capacity and biotic potential and explain their roles in population dynamics.
2. Explain, using an example, the difference between crude density and ecological density.
3. Define the terms *natality, mortality, emigration,* and *immigration* and explain how they affect populations.
4. (a) Calculate the crude density of a population of painted turtles if 90 turtles are counted in a 450-ha park.
   (b) Calculate the ecological density of the population if 150 ha of the park is unsuitable habitat.
5. Give an example to explain the difference between an open population and a closed population.
6. Explain why it would be a big mistake to eliminate a major predator from a community.
7. The growth rate for a population of 90 field mice in six months is 400%. If the number of births was 340, the number of deaths was 45, and there was no emigration, calculate the number of mice that migrated into the field.
8. A nesting colony of gannets on Île Bonaventure exhibits geometric growth. During the year, an initial population of 50 000 birds had 32 000 births and 29 000 deaths.
   (a) Calculate the geometric growth rate.
   (b) Estimate the population size after 2 years and after 10 years.
9. Explain the differences among stable, irruptive, irregular, and cyclic population changes.

Applying Inquiry Skills

10. Six ground finches began nesting on an island in 1990. Biologists monitored numbers in this population for years, compiling their data as shown in **Table 4**.
    (a) Graph the changes in this population over the 10-year period.
    (b) Estimate the carrying capacity of the island. Label this value on your graph.

Table 4 Ground Finch Population 1990–1999

Year	Population size
1990	6
1991	18
1992	35
1993	58
1994	170
1995	280
1996	477
1997	359
1998	296
1999	283

11. Sketch two graphs representing two populations of the same species of bacteria. Population A has unlimited resources. Population B is limited by available space.

Making Connections

12. To try to limit effects from the exponential growth of populations of zebra mussels, scientists are proposing such strategies as using ultrasonic vibrations, adding chlorine or other chemicals to the water to kill the mussels, and manual removal.
    (a) Research three potential action plans to deal with effects from zebra-mussel growth.

www.science.nelson.com

    (b) Summarize your findings in a PMI chart. You should include social, scientific and/or technological, and environmental criteria for each plan.
13. Use the Internet and other resources to research the requirements for a career as a wildlife technician.

www.science.nelson.com

5.10 Why Populations Change

Figure 1
Competition for resources among zebra mussels will eventually limit their population growth.

Zebra mussels were unknown in the Great Lakes until the mid-1980s, when they were likely introduced through bilge (filthy waters) discharged from ships from the Caspian Sea, where the mussels are an endemic (prevalent) species. The mussel population expanded very rapidly throughout the Great Lakes and had a significant impact on other species. In recent years, researchers have observed a dramatic decline in the populations of zebra mussels in areas where the populations exploded a few years earlier (**Figure 1**). Why would a population thrive and then experience a serious downturn?

There are two types of factors that cause population sizes to change—those that are related to the density of the population (density-dependent factors), and those that are not related to the density of the population (density-independent factors).

Density-Dependent Factors

density-dependent factor a factor that influences population regulation, having a greater impact as population density increases or decreases

A **density-dependent factor** that limits population growth is one that intensifies as the population increases in size. These factors include competition, predation, disease, and others. Intraspecific competition occurs when individuals of a population of the same species rely on the same resources for survival. As a population grows, individuals may migrate to another region. If migration is not possible, then the population density increases and there are more individuals competing for the same resources. So the growth rate slows down. Consider a population of bears in a forested area that has a carrying capacity of 85 bears. When the population grows larger than the carrying capacity, there is intense competition among the bears for the remaining resources. Stronger bears that are able to obtain food will survive, while weaker bears will starve or risk death by moving from the area to seek another habitat with adequate resources. As competition for food increases, the amount of food per individual often decreases. This decrease in nutrition results in a decrease in an individual's growth and reproductive success. **Figure 2** shows that as the total population increases, the average number of offspring decreases.

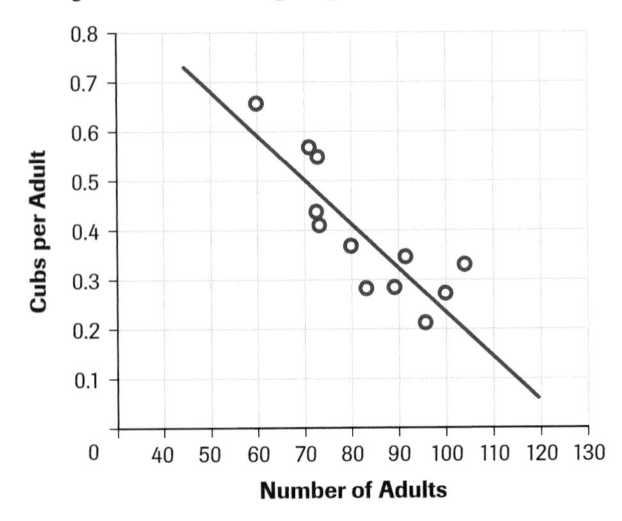

Figure 2
The average number of grizzly bear cubs decreases as the total population increases.

Predation is another major density-dependent factor that limits population growth. Obviously, a predator has an effect on the population of a prey, but it may have a more pronounced effect if the predator prefers that prey because they are plentiful or are easy to catch. For example, among the fish that make up their total diet, sharks will have a more significant effect on the populations of fish prey that are slower or that do not have places to hide.

Disease can also be a significant density-dependent factor that limits population size. In areas with dense or overcrowded populations, pathogens are able to pass from host to host more easily than in areas where the population is spread out. The overcrowding of farm animals can lead to the spread of disease, such as foot-and-mouth disease in cattle and encephalitis (brain swelling) in poultry (**Figure 3**).

Allee effect a density-dependent phenomenon that occurs when a population fails to reproduce enough to offset mortality once the population density is very low; such populations usually do not survive

minimum viable population size the smallest number of individuals needed for a population to continue for a given period of time

Figure 3
A risk of increased population density is that it can allow disease to spread more rapidly, resulting in a reduction in population size.

While most density-dependent factors act on high-density populations, there are factors that reduce population growth rates at low densities. Scientist Warder Allee discovered that all plant and animal populations suffer a decrease in their growth rates as they reach small sizes or very low densities. This is known as the **Allee effect**. For example, if a population is too small, it may become difficult for individuals to find mates. Some threatened species, especially those with low reproductive success, may be susceptible to the Allee effect. This might lead to their extinction.

All species must have a minimum number in order to survive. The **minimum viable population size** is the smallest number of individuals that allows the population to sustain itself. The minimum viable population size consists of enough individuals to allow the population to cope with variations in natality and mortality, as well as environmental changes or disasters. Scientists use this number to determine when a population should be considered to be at risk. In 1941, biologists were concerned that the whooping crane would become extinct, as the population worldwide had decreased to 23 birds; 21 living in the wild and 2 kept in captivity (**Figure 4**). The population of whooping cranes had fallen well below the predicted minimum viable population size, but human intervention saved the cranes from extinction. An ambitious breeding program and legal protection of the whooping crane in its winter and summer habitats have restored the wild population to around 300 birds. Even with this improvement, the species is still considered endangered.

Figure 4
Minimum viable population size is only a prediction. The whooping crane did not become extinct, even though its population size fell well below the minimum viable population size.

Figure 5
Swainson's hawks are decreasing in number as a result of pesticide use.

Density-Independent Factors

Factors such as human intervention, extreme weather changes, or other conditions not related to population density are known as **density-independent factors.** For example, some organisms may not breed in extreme temperatures. With a reduced birth rate, the population declines. With the return of moderate temperatures, reproductive success improves and the population grows again.

Pesticide use is an example of a density-independent factor in the form of human intervention. Biomagnification of pesticides in the food chain has led to a drastic decrease in the population of Swainson's hawks throughout North America (**Figure 5**). These birds migrate between the Canadian prairies and South America. It was discovered that during their stay in Argentina they feed on insects, mostly grasshoppers, in fields that have been heavily sprayed with highly toxic pesticides. These pesticides, which are banned in North America, are used to control the grasshopper population that threatens human food crops.

Limiting factors prevent populations from achieving their biotic potential. These limiting factors may be shortages of environmental resources such as light, space, water, or nutrients, and will determine the carrying capacity of the environment. Of all the resources that individuals or populations require for growth, the resource in shortest supply is called the **limiting factor**, and it determines how much the individual or the population can grow. For example, a plant requires nitrogen, carbon dioxide, and sunlight for it to be able to grow and reproduce (**Figure 6**). If it uses up all of the available nitrogen, it can no longer grow, even though sunlight and carbon dioxide are still plentiful. In this case, nitrogen is the limiting factor.

density-independent factor a factor that influences population regulation regardless of population density

limiting factor any essential resource that is in short supply or unavailable

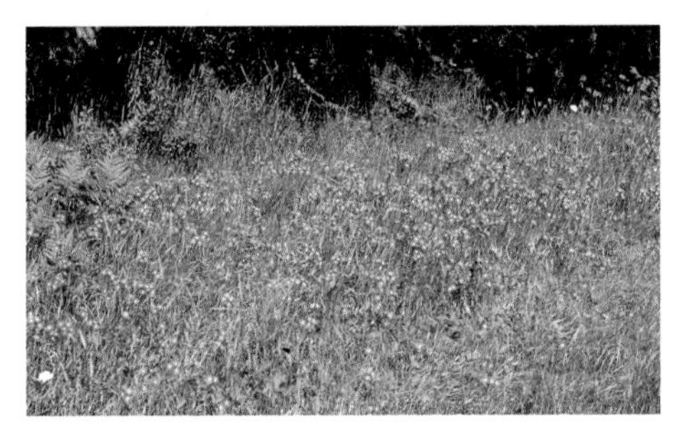

(a)

(b)

Figure 6
The orange hawkweed plants in **(a)** are flourishing in the open field, while space is the limiting factor for those in **(b)**.

Survival

The factors outlined above, and other factors, work together to determine the survival rates of populations. There are three general patterns of survival among populations, and these can be represented graphically by survivorship curves (**Figure 7**). For species that produce many offspring, the probability of death is much greater early in the life of the individuals. For example, female lobsters hatch thousands of free-swimming larvae, only a few of which survive to become adult lobsters. Their chance of survival is very low early in life but increases with age. This is referred to as *early loss.* The probability of survival of other animals, such as humans and elephants, decreases with age. In other words, most deaths occur later in life. This pattern is called *late loss.* In *constant loss* survivorship, such as that shown by songbirds, the probability of survival does not change with age, and there is a fairly constant death rate at all ages.

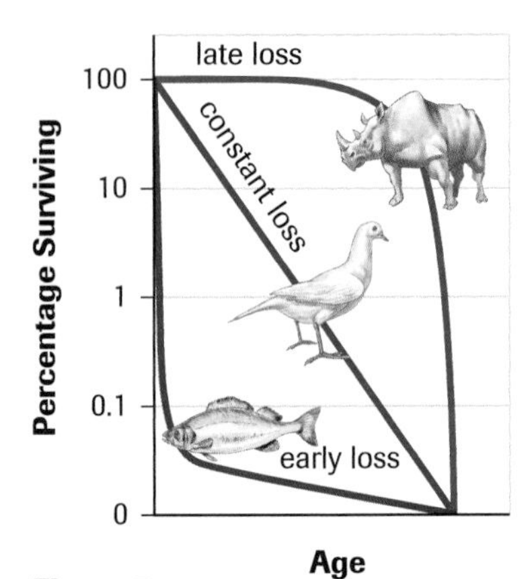

Figure 7
These survivorship curves show the generalized patterns of survival of various species at different ages.

Minor increases and decreases in the size of populations pose no serious risks to the survival of a species. More drastic population growth or decline can result in adverse changes to the habitat or even to the population itself. Too many births or too much immigration may result in the population overshooting the carrying capacity of its habitat. When this happens, the number of deaths starts to increase and the number of births starts to decline, resulting in a reduction in population size. Very high rates of mortality in a population may threaten a species with extinction. Scientists study fluctuations in population size and density to understand natural patterns and to try to predict the ability of an organism to withstand serious impacts such as natural disasters or destructive human activities.

Section 5.10 Questions

Understanding Concepts

1. Explain, with examples, the difference between density-dependent and density-independent factors.
2. What is the minimum viable population size?
3. Classify the following scenarios as density dependent or density independent:
   (a) A forest fire destroys a great deal of habitat in Algonquin Park.
   (b) Many aquatic organisms die as a result of adverse weather conditions during the breeding season.
   (c) A young, aggressive hawk invades the geographic range of established hawks, driving weaker birds from the range.
4. Identify one density-dependent factor and one density-independent factor that were not discussed in this section. Explain how they might affect the growth of a population.
5. Explain early loss, late loss, and constant loss and give an example of a species that shows each type of survivorship.
6. Why do biotic factors that affect a population tend to be density-dependent while abiotic factors tend to be density-independent?

Applying Inquiry Skills

7. Study the graph in **Figure 8** depicting a chickadee population. The graph shows population density versus brood size (the number of eggs to be hatched at one time).
   (a) Is this a case of density-dependent or density-independent regulation?
   (b) Sketch a corresponding graph to illustrate population density versus food supply. Explain the reasoning behind the shape of your graph.

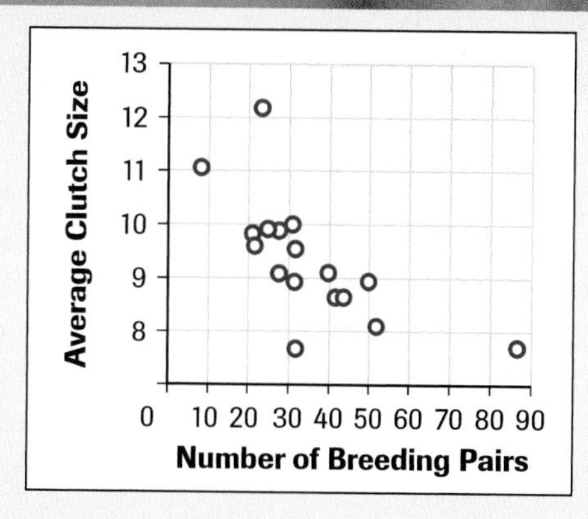

Figure 8

8. Describe the pattern of survival shown by the survivorship graph in **Figure 9**. Explain how this pattern might be observed in the human population in a developing country.

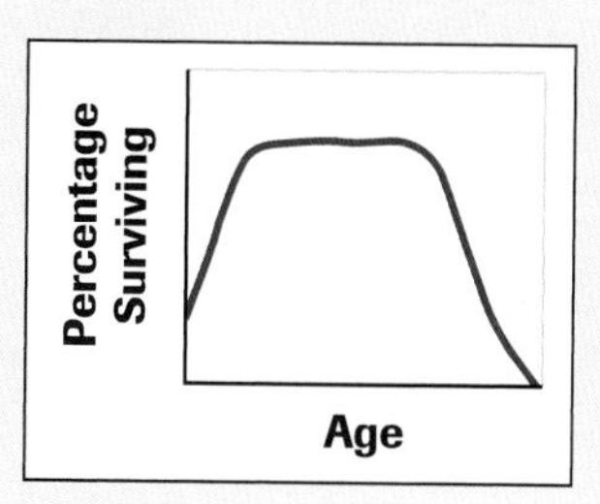

Figure 9

Making Connections

9. The moose population on Isle Royale in Lake Michigan was established from a few moose that crossed over the ice to the island. There were no natural predators so the population thrived for years. Explain what will happen when the population exceeds the carrying capacity of the island. What management practices could have been implemented to prevent this from happening?

Tech Connect

5.11 Satellite Telemetry

Researchers have a whole new set of tools to help them learn about natural ecosystems. Among these new tools is satellite telemetry. It involves sending a signal from a transmitter on Earth to a satellite orbiting Earth. This signal is relayed back to a computer at a receiving station on Earth. The tool is especially useful for studying the movements of migrating animals such as leatherback sea turtles. Although the turtles nest in tropical areas, they spend most of their time in cooler waters such as the North Atlantic. **Figure 1** shows the range of the leatherback turtle in Canada.

Leatherback turtles get their name from the leatherlike texture of their carapace or shell. Since the carapace is somewhat flexible and not hard like those of other turtles, the glue that is normally used to attach a transmitter on hard-shell turtles will not stick. Technologists, however, have devised a new method of attaching the transmitter. The leatherback turtles are fitted with a backpack that contains a personal computer the size of a compact disc player, and a transmitter (**Figure 2**). The unit is referred to as the platform transmitting terminal (PTT). The computer is equipped with a sensor so that it "knows" when the transmitter is above water. When the turtle surfaces, the PTT sends a signal that is picked up by an orbiting satellite overhead. This information is used to pinpoint the location of the turtle at the time the signal was received.

While this sounds simple, it is not without problems. The satellite that receives the signals circles Earth about every 100 min. This means that it can pick up the signal from any single transmitter for a very short period—10 to 15 min. In addition, it takes 3 to 5 min for the satellite to zero in on the transmitter and determine its location. So the transmitter must be on the surface for that period of time while a satellite is passing overhead. In some ways it is a "hit or miss" connection.

Although the backpack may be an annoyance, the turtle is not harmed in any way. The system is designed to avoid any long-term interference with the turtle's normal lifestyle; when the power supply is exhausted (after 8 to 10 months), the PTT shuts down and the unit falls off the turtle.

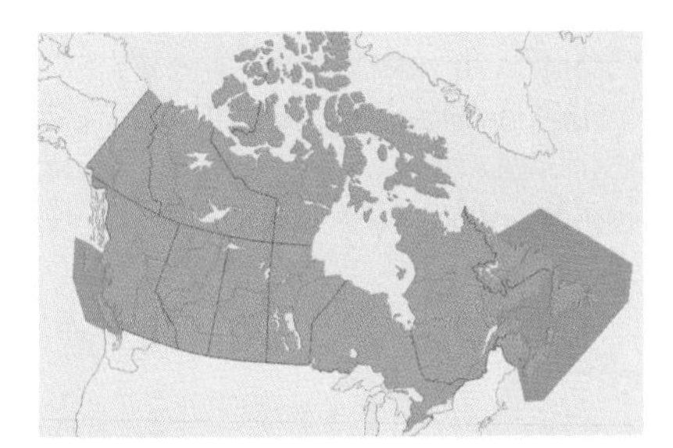

Figure 1
The range of the leatherback turtle in Canada

Figure 2
The backpack (PTT) contains a small computer and a transmitter.

DID YOU KNOW?

Leatherback turtles

- They are the world's largest reptiles.
- They are the most endangered species of marine turtle.
- They eat only jellyfish.
- They have the largest migration range of any reptile.
- They swim even when sleeping and can travel 45 to 65 km/day.
- They have been known to swim more than 12 000 km a year.
- They can dive as deep as 1 km.

Tech Connect 5.11 Questions

Understanding Concepts

1. Briefly describe how satellite telemetry works.
2. Explain why satellite telemetry is a useful tool for scientists studying the leatherback turtle.
3. Explain why it might be uncommon to receive a signal from a tagged leatherback turtle every day.

Making Connections

4. The PTT can be equipped to gather other data. What other data do you think would be useful to a scientist studying the leatherback turtle?
5. Use the Internet to research other social or environmental applications of satellite telemetry.

www.science.nelson.com

6. In a small group, discuss the costs and benefits of conducting this type of research. What is the information obtained from this research used for? Who benefits? Is it of any benefit to the turtles? Is it of any benefit to humans? Should taxpayers pay for such research, or should it be funded by the private sector? Do the benefits justify the costs? On a priority scale of 1 to 10 (10 is highest), where does this type of research rank?

Estimating Population Size 5.12

Accurate estimates of population size and density can be extremely valuable. In the forestry industry, staff must be able to determine the population density and size of valuable tree species. This information is essential in determining allowable harvest rates while still maintaining healthy and sustainable populations.

In very rare cases, an exact count of the total number of individuals in a population can be made. However, populations are dynamic, and their numbers and geographic locations change over time, generally making a precise count impossible. Instead, wildlife technicians count a sample of the population at a particular time and then estimate the size of the total population. In some cases, an estimate or index of population size can be made using indirect indicators, such as the number of fecal droppings, the number of tracks, or the number of nests or burrows (**Figure 1**).

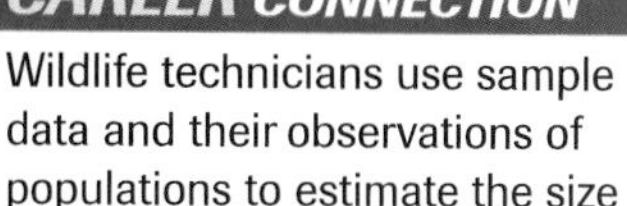

Wildlife technicians use sample data and their observations of populations to estimate the size and density of populations.

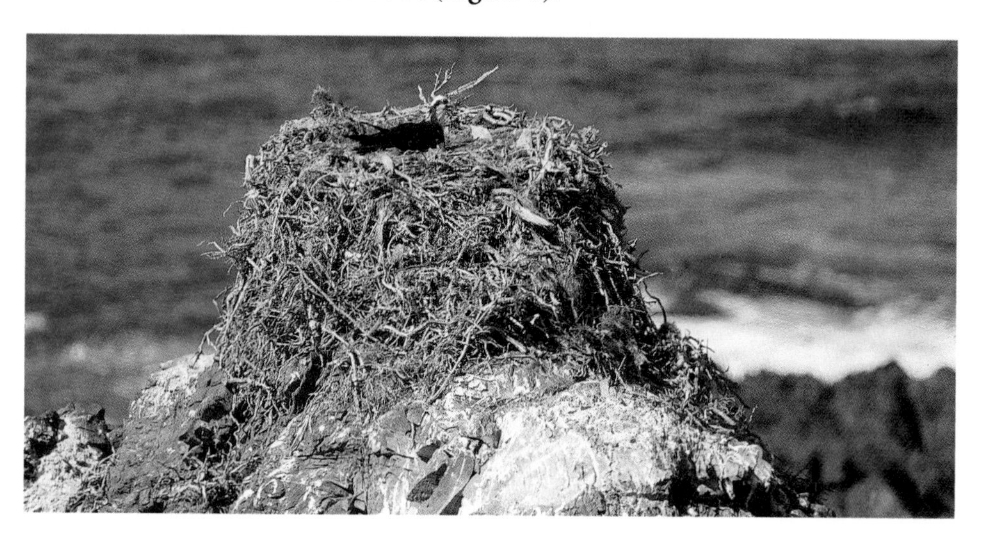

Figure 1
The number of osprey nests can be a useful indicator when estimating population size.

Mark–Recapture Sampling

A common sampling technique for estimating the size and density of mobile animal populations, such as fish, is the **mark–recapture method**, in which a sample of animals is captured, marked in some way, and then released. The marked individuals will spread randomly throughout the population. Using the mark–recapture method, often with the assistance of commercial and sport fishers, fisheries technologists are able to make good estimates of population sizes and densities on which to base management programs, such as restocking programs.

mark–recapture method a sampling technique for estimating population size and density by comparing the proportion of marked and unmarked animals captured in a given area; sometimes called capture–recapture

The techniques for capturing and marking individual organisms must be carefully planned so that the chances of each individual being caught are equal. Marking techniques must not harm the organism or restrict its normal activities and must ensure that the marks remain clearly visible for the duration of the study (**Figure 2**). Marks must also not alter the chances of recapture. The accuracy of mark–recapture as a sampling method for estimating population numbers and/or density depends on the following assumptions:

Figure 2
A tag attached to the dorsal fin of a salmon.

- Every organism in a population has an equal chance of being captured.
- During the time period between the initial marking and the subsequent recapture, the proportion of marked to unmarked animals remains constant.
- The population size does not increase or decrease during the sampling study.

Under ideal conditions, marked and unmarked individuals of the population are captured at random and, during the period of the sampling, no new individuals enter the population, no marked animal dies, and none of the marked animals leaves the population.

In using a sample to estimate the size of the total population, the wildlife technicians assume that the ratio of marked fish to the size of the entire population is equal to the ratio of marked recaptures to the size of a sample. The following formula can be used to calculate the total population size from a sample:

$$\frac{\text{total marked } (M)}{\text{total population } (N)} = \frac{\text{number of recaptures } (m)}{\text{size of second sample } (n)}$$

or $\frac{M}{N} = \frac{m}{n}$, which can be rewritten as $N = \frac{M \times n}{m}$.

Consider a fish population of unknown size from which 26 individuals are randomly captured, marked, and released, as shown in **Figure 3(a)–(c)**. Assume that the released individuals move randomly through the population, as illustrated in **Figure 3(d)**. If a second sample of 21 individuals is captured sometime later in which 3 individuals are found marked, as shown in **Figure 3(e)**, we may use these values to estimate the population size.

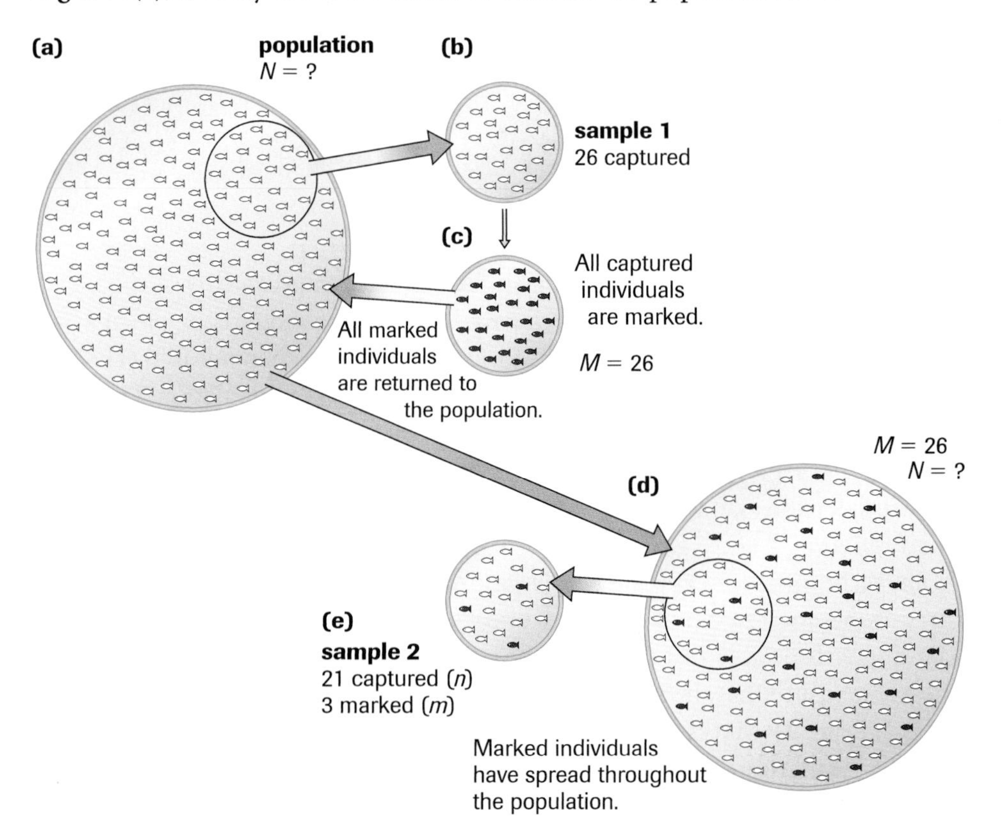

Figure 3
Mark–recapture sampling of a fish population

We assume that the ratio of marked individuals to the size of the total population equals the ratio of marked individuals recaptured in the second sample to the size of the second sample. If we let N represent the population size, then, using the formula above, we can estimate the population size to be 182 (**Figure 4**).

$$\frac{M}{N} = \frac{m}{n}$$

$$N = \frac{M \times n}{m}$$

$$N = \frac{26 \times 21}{3}$$

$$N = 182$$

Figure 4
Calculating the population size using the mark–recapture method

Quadrat Sampling

A common sampling technique for stationary or small organisms, such as plants or insects, is to isolate a defined area and count samples of one or more populations of organisms within that area. A large area, such as a tract of grassland or forest, can be sampled using small selected plots in which a sampling frame, or **quadrat**, is placed (**Figure 5**). The number of individuals of one or more species can be counted within each quadrat. Population size and density can be estimated through calculations based on counts within a representative area or all of the selected areas. Average estimates of population size and density for the entire area can then be determined based on these calculations. Quadrat sampling is most effective for stationary species, such as the populations of different tree species in a forest. For the most valid and reliable information, the sampling should be random and should make up about 10% of the total area being studied.

quadrat a sampling frame used for estimating population size; frames can be real or virtual

Figure 5
Quadrats can be used to estimate the population of different species in a large tract of grassland.

▶ *SAMPLE* problem

A student wants to estimate the population size and density of birch trees in a woodlot measuring 100 m by 100 m. On a diagram of the woodlot, she splits the area into a 10 by 10 grid and randomly selects 10 sample quadrats–(0,0), (1,4) (2,6), (3,9), (4,3), (5,5), (6,1), (7,8), (8,2), and (9,7) (**Figure 6**). These areas of the woodlot are marked off, and the birch trees are counted. She finds 15, 25, 40, 20, 10, 30, 25, 5, 15, and 35 trees in the ten quadrats. Estimate the density and size of the birch-tree population in the woodlot.

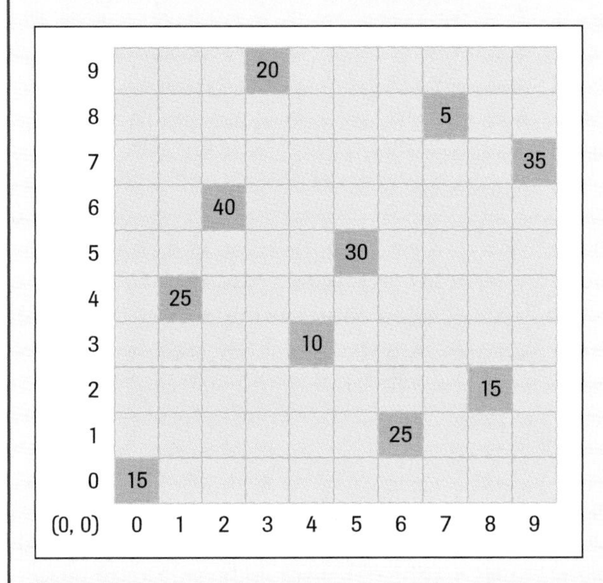

Figure 6
Grid representing a 100 m by 100 m woodlot, showing ten sample quadrats

Solution

An estimate of the population density of the birch trees may be obtained by calculating the average density of the sampled quadrats. Each sample quadrat has an area of 100 m^2.

$$\text{average sample density} = \frac{\text{total number of individuals}}{\text{total sample area}}$$

$$\text{average sample density} = \frac{15 + 25 + 40 + 20 + 10 + 30 + 25 + 5 + 15 + 35 \text{ birch trees}}{10 \times 100 \text{ m}^2}$$

$$\text{average sample density} = \frac{220 \text{ birch trees}}{1000 \text{ m}^2}$$

$$\text{average sample density} = 0.22 \text{ birch trees/m}^2$$

Assuming that the sample area is representative of the total study area, the estimated population density is 0.22 trees/m^2.

An estimate of the population size may be obtained by multiplying the estimated population density by the total area of the study site.

$$\text{estimated population size} = (\text{estimated population density}) \times (\text{total area of study site})$$

$$\text{estimated population size} = (0.22 \text{ trees/m}^2) \times (10\,000 \text{ m}^2)$$

$$\text{estimated population size} = 2200 \text{ trees}$$

Section 5.12 Questions

Understanding Concepts

1. What conditions must be met and what assumptions must be made in carrying out a sampling procedure?
2. Using the mark–recapture method, wildlife researchers surveyed an area of wetlands where 80 wood ducks were captured in traps, marked with permanent metal bands, and then released. Two weeks later, 110 wood ducks were captured. Of the wood ducks recaptured, 12 were marked. Estimate the total size of the wood-duck population in the survey area.
3. In a river in British Columbia, 430 sockeye salmon were captured and marked on the fin with a uniquely numbered T-bar anchor tag. Two weeks later, a total of 154 sockeye salmon were recaptured, and 15 had the tags on their fin.
   (a) Estimate the sockeye salmon population in this river during the study.
   (b) Identify conditions that must be met in this study to obtain a valid estimate.
4. To estimate the size of the slug population on a golf course, biology students randomly selected five 1-m^2 quadrats in a site 10 m by 10 m. The numbers of slugs in each quadrat were 4, 8, 9, 5, and 1. Estimate the density and size of the slug population in this study site.

Applying Inquiry Skills

5. For a biology project, students decided to investigate the effects of pollution on a population of grasshoppers. The students investigated a population of grasshoppers in the 2-ha school yard and in a 1-ha field near a local oil refinery. Estimates of population size were used as indicators of survival. Data were collected using the mark–recapture method in which a tiny drop of coloured nail polish was used to mark captured grasshoppers.
   (a) From the data in **Tables 1** and **2**, calculate the grasshoppers' population size and density for each area.
   (b) What conclusions can be drawn from the data?

Table 1 Mark–Recapture of Grasshoppers in School Yard

number of grasshoppers marked	280
number of grasshoppers captured in a second sample	130
number of marked grasshoppers recaptured	8

Table 2 Mark–Recapture of Grasshoppers Near Oil Refinery

number of grasshoppers marked	150
number of grasshoppers captured in a second sample	70
number of marked grasshoppers recaptured	26

6. A homeowner decided to estimate the density and size of the population of cinch bugs in her lawn. Her lawn measured 15 m by 25 m. She decided to sample her lawn and count the cinch bugs in the sample areas. She made a wire-frame quadrat measuring 0.5 m by 0.5 m and sampled five areas at random. In these five sample areas she captured and counted 75, 120, 67, 48, and 94 insects.
   (a) Draw a scale drawing of the lawn and randomly assign and indicate the sample areas.
   (b) Determine the average population density of cinch bugs in her lawn.
   (c) Estimate the size of the cinch bug population in her lawn.
   (d) What assumptions must she make in estimating the density and size of the population?
   (e) What would you recommend to make her estimates more reliable?
7. Prairie dogs live in grassland habitats where they form large burrowing colonies. Describe the method you would use to estimate the population size, and give reasons for your choice.

Making Connections

8. Another common technique used for sampling and estimating population size is the transect method. Use the Internet and other resources to find information about this method and write a brief report.

www.science.nelson.com

9. In a group, brainstorm and discuss challenges that biologists experience in estimating population characteristics for wild populations of the following:
   (a) whales that migrate along the western coasts of North and South America
   (b) algae that live in a water body receiving excess fertilizers in runoff from cropland
   (c) caribou that inhabit an Arctic tundra environment
   (d) amphibians that live in a marsh
   (e) insects in a field

Estimating Population Size

Inquiry Skills
○ Questioning ○ Planning ● Analyzing
○ Hypothesizing ● Conducting ● Evaluating
● Predicting ● Recording ● Communicating

Ecologists need to count organisms for many different reasons. They may want to count cells on a microscope slide to test whether or not food-processing plants are meeting health standards, to count trees in a forest to assess the health of an ecosystem, or to count salamanders in a stream to assess the impact of pollution. For many organisms, it is virtually impossible to actually count each individual in the population.

In this investigation, a bag of polystyrene "peanuts" represents a population of animals in a study area. You will use the mark–recapture method described in section 5.12 to estimate the number of peanuts in a bag, or the total population of organisms in the study area.

Question

How can you estimate the size of a population?

Prediction

(a) Predict the total number of organisms in your simulated population. Record your prediction.

Materials

foamed polystyrene "peanuts" in a plastic bag
black marker
note paper and pencil

Procedure

1. Create a data table to record your observations.
2. Obtain a plastic bag of foamed polystyrene peanuts from your teacher.
3. Randomly capture 40 to 50 peanuts from the bag.
4. Mark each captured peanut with black ink and return them all to the bag. Shake and stir the contents of the bag thoroughly to mix the peanuts.

(b) Record in your data table the number you have captured and marked.

5. Without looking into the bag, recapture a sample of similar size.

(c) Record the total number of recaptured peanuts and the number of those that are marked.

6. After recording, return all the peanuts to the bag. Shake the bag to mix the peanuts.
7. Obtain data for two additional recaptures by repeating steps 5 and 6. Be sure to record your results as you did in (c).

Analysis

(d) From each of your three sets of recapture data, calculate a population estimate (N) for the peanuts in the bag by using the following formula:

$$N = \frac{\text{total marked} \times \text{total recaptured in each sample}}{\text{total number of recaptures that were marked}}$$

Calculate an average population estimate from your three answers.

Evaluation

(e) What is the purpose of mixing the peanuts thoroughly after you returned the samples to the bag? What would this represent in a real population?

(f) Why is it important not to look in the bag when sampling the population?

(g) Compare your three values for the peanut population in your bag: your original prediction (a); your average estimate (d); and the actual population size as revealed by your teacher. How close was your prediction to the actual population? How close was your average estimate to the actual population?

(h) List three factors that could have reduced the accuracy of your total estimate.

(i) Change one of the factors you listed in (h) and perform two more counts using the mark–recapture method. Analyze the new data and discuss how this factor influenced your population estimates.

(j) If migration occurred in this population, how would this affect the reliability of your estimate?

Synthesis

(k) Suggest a modification of this investigation to show how this method could be used to monitor changes in a population over time.

5.14 Investigation

Plant and Animal Survey

Inquiry Skills

- ○ Questioning
- ○ Hypothesizing
- ● Predicting
- ● Planning
- ● Conducting
- ● Recording
- ● Analyzing
- ● Evaluating
- ● Communicating

A 50-ha area outside your town has been identified as the site of a proposed golf course. One of the town's municipal bylaws requires the developer to conduct an independent environmental impact assessment (EIA).

As an environmental management technician, you have been hired as a member of the team that will conduct the EIA. The assessment has been divided into a number of different components, and your task is to conduct a survey of the plants and animals in the proposed development area. This survey is necessary to understand the potential impacts of the development on the wildlife of the area.

In this investigation, you will visit the proposed area and conduct a qualitative survey to identify the plants and animals that are common to the area. The survey is not intended to be a detailed description of every type of organism in the area or an accurate count of each species. For practical reasons, you will survey only the trees and vertebrate animals in the area. In some cases, it may not be possible to observe directly animals that are common to the area. In such cases, use indirect evidence (tracks, nests, pellets, etc.) to determine if they are present in the ecosystem.

Question

What plants and animals are found in the proposed development area?

Prediction

With other members of your team, use your personal experience and knowledge of the area to discuss what organisms might be found in the area.

(a) Make a prediction about what trees and vertebrates will be common in the proposed development area.

Experimental Design

(b) Design an observational study to enable you to complete the plant and animal survey of the proposed development area. Include the following in your design:

- a map of the proposed development area
- a detailed description of the proposed procedure, including the sampling method(s)
- a list of all required materials and safety precautions
- method(s) of recording observations

Procedure

1. Submit the proposed design, including a list of materials and the safety precautions to your teacher for approval.
2. Carry out the procedure.
3. Using appropriate resources, classify the most common plants and animals in the region. (It is not essential to identify the family, genus, and species names of each organism. Use the common names of organisms for identification purposes.)
4. Write a report (see Appendix A4). Include in your report ways to improve your design.

Analysis

(c) What are the dominant plants and animals in the proposed development area?

Evaluation

(d) Did your survey confirm the kinds of plants and animals that you expected to find? Explain.

(e) Describe any problems or difficulties in carrying out the procedure.

(f) If you were to repeat this experiment, what would you do differently to improve the survey?

Synthesis

(g) Write a report using the data collected in this study to prepare a section of the EIA that includes the following information:

(i) any plant or animal species that would be considered threatened/endangered in this area

(ii) potential factors that might influence the population size of any of the species

(iii) three preventive measures that might reduce the impact on these species

(h) Research the educational requirements for an environmental management technician.

www.science.nelson.com

5.15 Human Population Growth

The world's human population shows clumped dispersion, and its growth is exponential. It is clumped because populations have tended to settle and survive in locations that offer the best habitat—a sufficient food supply and enough space. The exponential pattern of growth demonstrates that, while the rate may be constant, the amount of increase grows rapidly. Exponential growth is like compound interest on your bank account—you gain interest on the principal amount and on the interest previously earned. **Figure 1** shows that although the growth rate has been decreasing over the last 50 years, the total size of the population continues to increase rapidly.

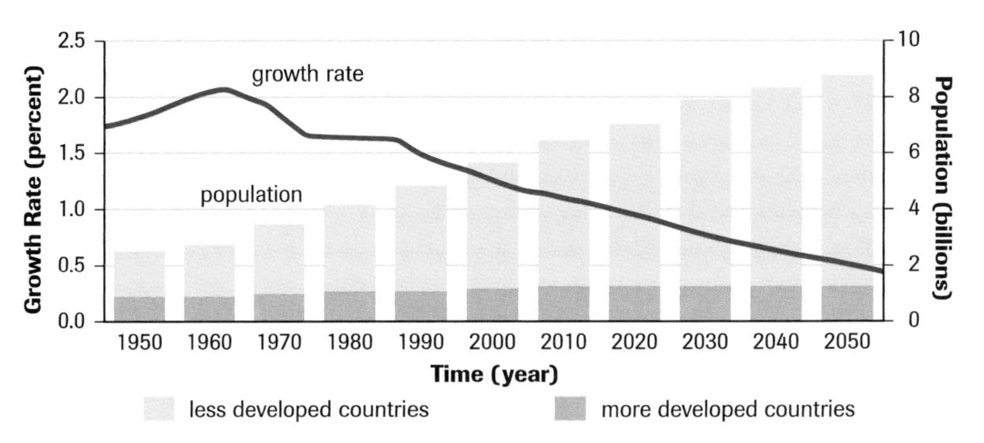

Figure 1
The annual growth rate has been steadily declining since its peak in the early 1960s. However, the world population has been increasing steadily.

Since emigration and immigration aren't factors in the change of Earth's human population, the net increase or decrease in the population is really the difference between the number of births and the number of deaths. The growth rate for human populations is usually expressed as a percentage. The growth rate is the difference between the numbers of births and deaths as a percentage of 1000 people. The annual rate of population growth is determined by the following equation:

$$\text{annual population growth} = \frac{(\text{births} - \text{deaths})}{1000} \times 100\%$$

If the number of births equals the number of deaths, the annual change rate is zero—the total population does not change. If the number of deaths exceeds the number of births, the change rate is a negative number—the total population decreases.

The annual world population growth rate in 2002 was 1.3%. This is down from about 2.2% at the peak growth rate in the early 1960s. However, during the same period of time (since the early 1960s), the world population has doubled from 3 billion to more than 6 billion people. With 6.2 billion people in the world, and a growth rate of 1.3%, the population will increase by about 81 million each year (22 918 per day, 9247 per hour, 154 per minute, or 2.6 per second). The populations of some regions of the world are increasing more than others. **Figure 2** shows the annual rates of population change for the various areas of the world in 2001.

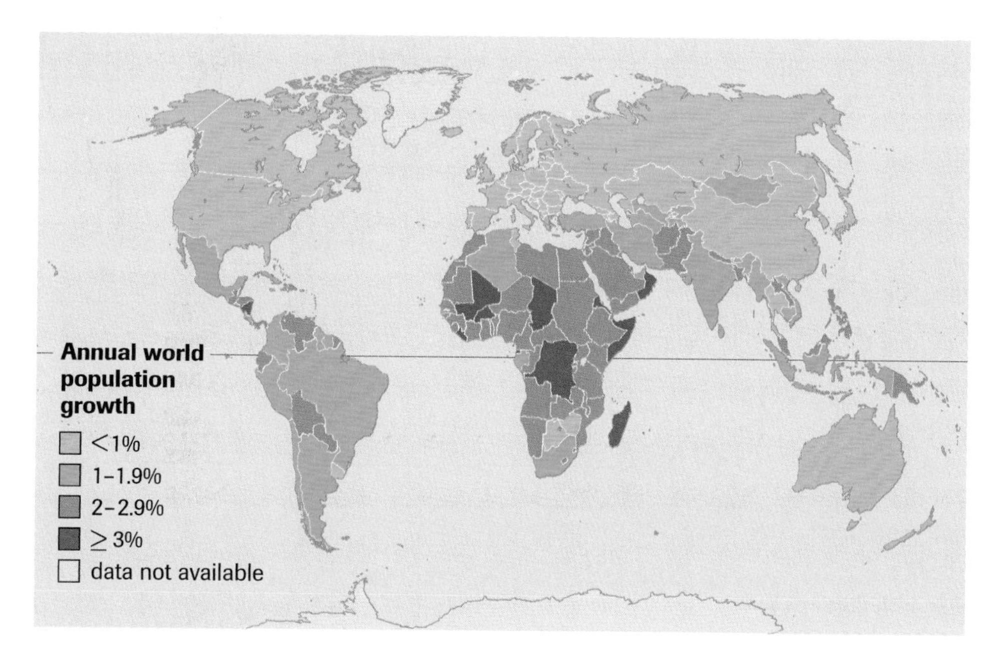

Figure 2
Average annual rate of population increase in different regions of the world

Doubling Time

One of the ways of describing the growth of a population is to calculate the **doubling time.** This is simply the time it takes for a population to double at a given growth rate. This prediction is based on several assumptions. For example, in predicting the size of a population in the future, we assume a certain growth rate. We also assume that no major catastrophe such as a world war or worldwide epidemic will suddenly reduce the population. It has been calculated that a population growing at an annual rate of 1% would double in approximately 70 years. If the rate increased to 2%, the doubling time would be 35 years. So a general rule is that the doubling time is equal to 70 divided by the annual growth rate. **Figure 3** shows the graphs of populations growing at different rates over the same time period.

doubling time the time it takes for a population to double in size at a given rate of increase; approximately equal to 70 divided by the annual growth rate

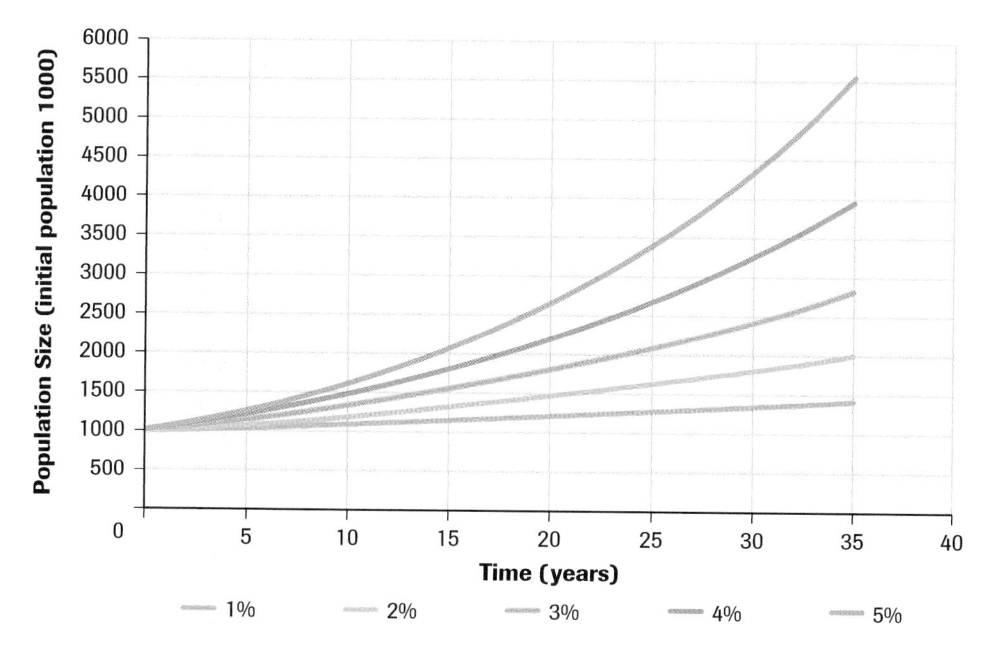

Figure 3
Populations growing at different rates

Section 5.15 Questions

Understanding Concepts

1. Explain how a population can increase even though its growth rate is decreasing. Create an example to illustrate your explanation.
2. Why aren't immigration and emigration included in the formula for determining the global human population growth rate?
3. Explain what is meant by the term *doubling time*. What assumptions do we make when we predict future population sizes using doubling time?
4. In 1999, the world population reached 6 billion people. The annual growth rate at that time was 1.3%. At this rate, in what year will the world's population reach 12 billion people?
5. Can a country's population double indefinitely? Why or why not?

Applying Inquiry Skills

6. Using a spreadsheet program to do your calculations, determine in what year the world population will reach
   - 8 billion, and
   - 10 billion.

   (Use the current growth rate of 1.3%.)
7. Birth rate and death rate are expressed as the number of births and deaths for every 1000 people in a given year. **Table 1** shows population data for selected countries. (*Note:* For the following questions, assume that immigration and emigration are not significant factors.)
   (a) Calculate the annual percentage of population growth for the countries listed.
   (b) Explain how birth and death rates determine the rate of natural population growth.
   (c) Identify countries that you consider to have a large annual increase in growth. Describe two factors that may account for these large numbers.
   (d) Why do some countries, such as the United Kingdom and Germany, experience little or negative growth?
   (e) In 2002, the population of Angola was 12.7 million. In what year will its population be double the current level?
   (f) Russia's population was 143.5 million in 2002. What will its population be in 2007?
8. Using an atlas and **Figure 2** (page 409), identify the countries that have the highest population growth rates.

Table 1 Sample Population Data for 2002 for Selected Countries

Country	Birth rate (per 1000 people)	Death rate (per 1000 people)
Angola	48	20
Australia	13	7
Canada	11	7
Egypt	27	7
Estonia	9	14
Ethopia	40	15
Germany	9	10
Greece	10	10
Jamaica	20	5
Nigeria	41	14
Pakistan	30	9
Palestinian Territory	40	4
Paraguay	31	5
Russia	9	16
United Kingdom	11	10

Making Connections

9. Some countries are experiencing a higher population growth than others. Which factor, high birth rates or low death rates, do you think is probably more significant? Use the Internet and other resources to research the factors that influence population growth rates.

www.science.nelson.com

10. What factors do you think account for the variation in birth and death rates among countries? Explain how each factor affects birth or death rates.
11. **Figure 1** (page 408) shows that the world's population continues to increase, although at a slower rate.
    (a) Suggest possible reasons that might explain why the world's population continues to increase.
    (b) We assume that Earth has a finite carrying capacity. What is likely to happen to the world's population when this carrying capacity is reached?
    (c) If the trend in the growth rate continues, the world's population growth will eventually reach zero. Do you think that Earth can support the world's population until zero population growth is reached? Explain your answer.

Case Study
Effect of Population Growth on the Environment 5.16

Doubling a population will obviously affect the environment, but the degree of the effect depends on the size of the population. A population that doubles from 10 000 to 20 000 will have less effect on the environment than a population that doubles from one to two million people. The impact on the environment also depends on the time over which the population doubles. A population with an annual growth rate of 5% will double in 14 years, whereas a population with a growth rate of 0.5% will double in 140 years. The impact of a slowly growing population on the environment may be just as significant in the long term but will not be as dramatic as the impact of a short-term change.

The richest 20% of the world's population consume 86% of the available resources. The poorest 20% of the world's people consume less than 2% of the available resources. In developing countries, environmental problems are often the result of poor people trying to meet their basic needs—for food, water, shelter, and fuel. In the developed countries, environmental problems such as air and water pollution are usually the result of industrial activity to support high levels of consumption.

Atmospheric CO_2 and Global Warming

Emissions of carbon dioxide (CO_2) resulting from fossil-fuel burning are almost certain to be the dominant influence on the trends in atmospheric CO_2 concentration in the 21st century. Carbon dioxide, along with other gases such as methane and water vapour, is a major contributor to the phenomenon known as the **greenhouse effect**. These gases allow the wavelengths of visible light and some ultraviolet radiation from the sun to pass through to Earth's surface. When the light is absorbed by the surface, much of it is radiated back out as infrared radiation (heat) and is then absorbed by these gases. The atmosphere, with the greenhouse gases, acts as a thermal heat blanket keeping Earth's temperature within an acceptable range (**Figure 1**, on the next page). This range of temperatures can change, however, in response to changes in the concentrations of the greenhouse gases. When the concentrations increase, additional global warming is likely to occur.

greenhouse effect a result of certain atmospheric gases, such as CO_2, water vapour, and methane, trapping heat in the atmosphere by letting visible sunlight penetrate to Earth's surface while absorbing most wavelengths of infrared radiation that radiate from Earth's surface

An increasing population using increasing amounts of fossil fuels that release billions of tonnes of carbon dioxide each year is adding to the concentrations of greenhouse gases in the atmosphere (**Figure 2**, on the next page). It is obvious that the industrialized countries are the greatest contributors of carbon dioxide from the burning of fossil fuels (**Figure 3**, on the next page). It is difficult to predict accurately what the impact of increasing greenhouse gas concentrations will be, but most climate specialists agree that greenhouse gases are a contributing cause of increasing air temperatures. The Intergovernmental Panel on Climate Change (IPCC), a network of about 2000 of the world's leading climate scientists from 70 countries, concluded in their 2001 report that "There is new and stronger evidence that most of the warming observed over the last

Figure 1
Greenhouse gases—such as water vapour, carbon dioxide, methane, and nitrous oxide—absorb heat radiation, preventing it from escaping back into space. These gases have the same effect as the glass that prevents infrared radiation from leaving a greenhouse.

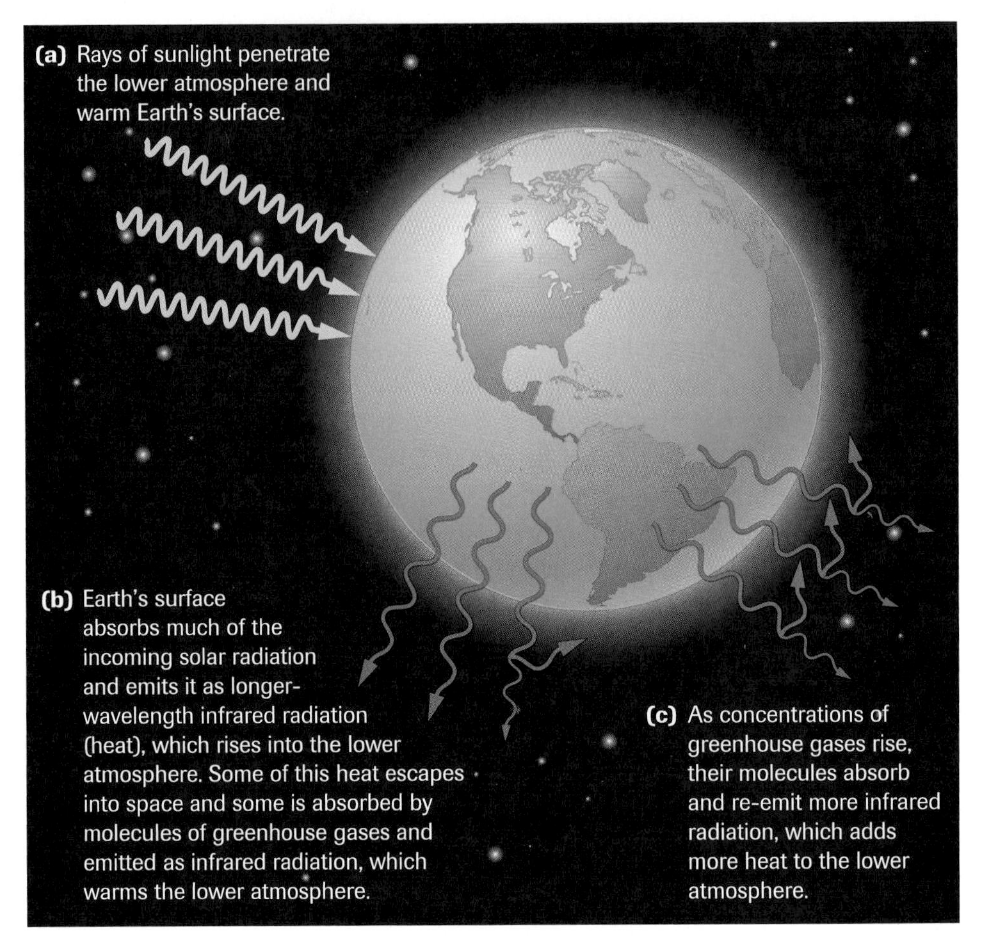

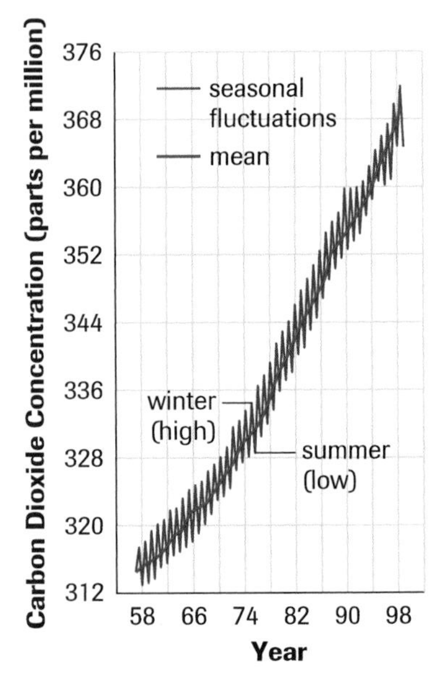

Figure 2
Since measurements of atmospheric CO_2 concentrations began in 1958, the average levels have been rising steadily.

50 years is attributable to human activities." They also predicted with 90% to 95% confidence that Earth's average temperature will increase by 1.4°C to 5.8°C between 2000 and 2100.

During the cold Canadian winters, we may jokingly welcome global warming. However, the possible harmful effects of global warming on the environment and on humans are numerous and far reaching (**Table 1**). This is no joking matter!

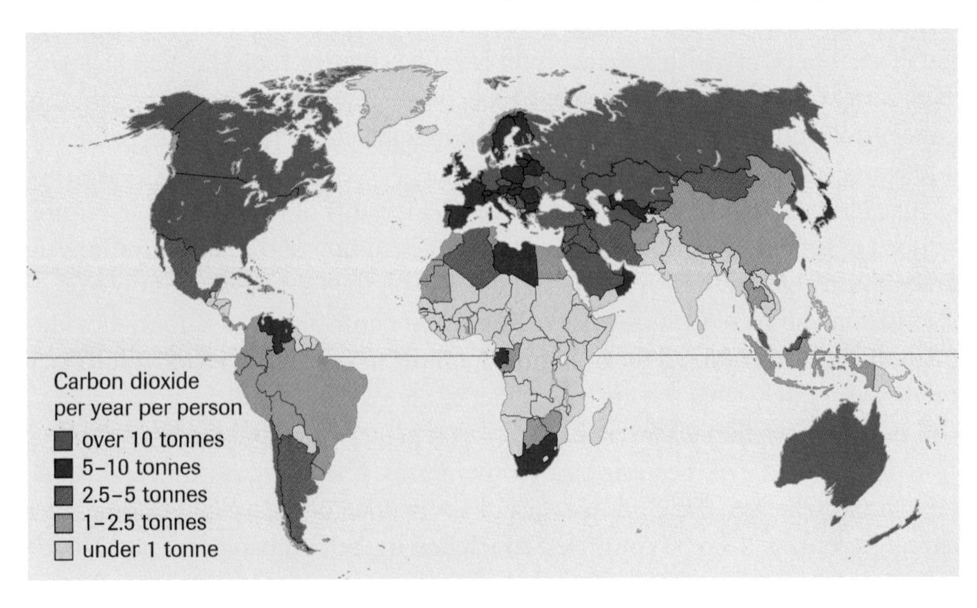

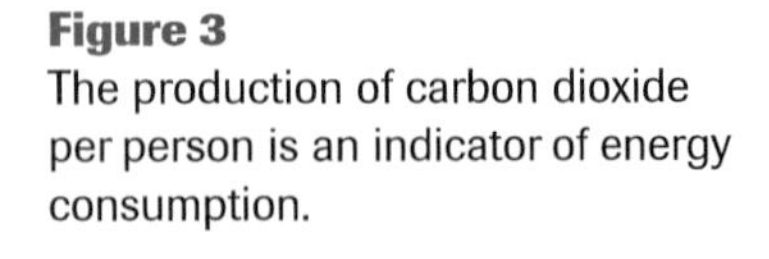
Figure 3
The production of carbon dioxide per person is an indicator of energy consumption.

Table 1 Possible Effects of Global Warming

Areas affected	Examples of possible effects
weather extremes	• prolonged heat waves and droughts • increased flooding • more intense hurricanes, typhoons, tornadoes, and violent storms
water resources	• change in water supply • decreased water quality • increased drought or flooding
biodiversity	• extinction of some plant and animal species • loss of habitat • disruption of aquatic life
forests	• change in forest composition and locations • disappearance of some forests • increased fires from drying • loss of wildlife habitat and species
sea level and coastal areas	• melting of polar ice caps and rising sea levels • flooding of low-lying islands and coastal cities • flooding of coastal estuaries, wetlands, and coral reefs • beach erosion • disruption of coastal fisheries • contamination of coastal aquifers (groundwater storage)
agriculture	• shifts in food-growing areas • changes in crop yields • increased irrigation demands • increased pests, crop diseases, and weeds in warmer areas
human populations	• increased deaths • more environmental refugees, increased migration
human health	• increased deaths from heat and disease • disruption of food and water supplies • spread of tropical diseases to temperate areas • increased respiratory disease • increased water pollution from coastal floods

The increase in greenhouse gases and its contribution to global warming is recognized as a worldwide problem. In December 1997, delegates from 161 countries gathered in Kyoto, Japan, to work out a new treaty to address the global warming problem. The treaty, which has become known as the Kyoto Protocol, or the Kyoto Accord, would result in a reduction of greenhouse gases to about 5.2% below 1990 levels by 2012. This treaty has not been fully ratified (approved), although some countries and major industrial organizations have made a commitment to address the problem. Five years after the treaty was signed, only 55 of the 84 countries that originally signed the treaty have officially committed themselves to follow its recommendations. The United Kingdom is committed to a 12.5% reduction in CO_2 emissions by 2010, and a royal commission has recommended a 60% cut by 2050. The Canadian parliament ratified the treaty in December 2002, but the debate over its implementation has become a controversial political issue. The United States, however, has stated that it will definitely not participate in the treaty, even though it is the world's leading contributor of atmospheric CO_2 (**Figure 4**, on the next page). However, 42 additional countries have agreed to accept and follow the recommendations, even though they were not involved in the conference that created the treaty.

DID YOU KNOW?

Growth and CO_2

Some plants, such as wheat and rice, show an increased growth rate when exposed to increased levels of CO_2. Corn, however, is not responsive to additional atmospheric CO_2.

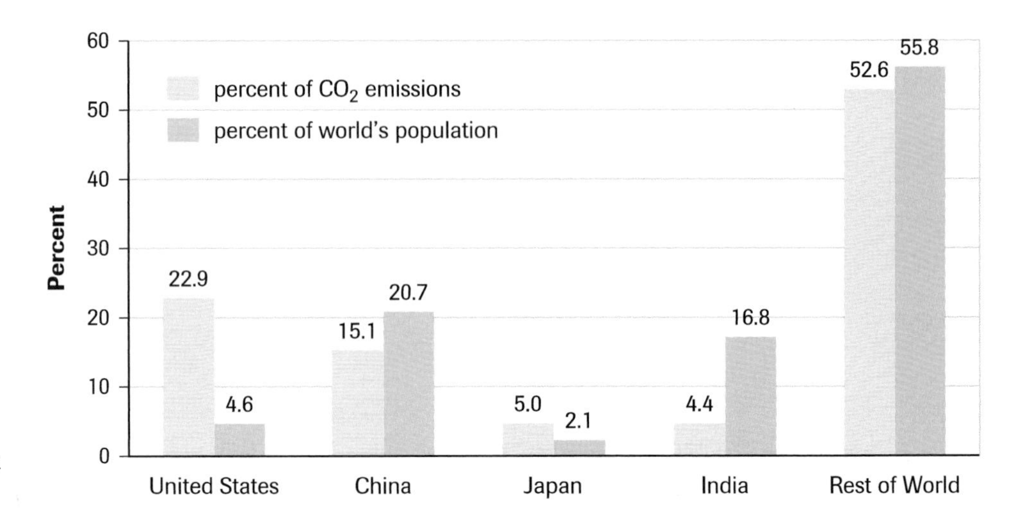

Figure 4
Four countries with 44.2% of the world's population contribute almost half of the global CO_2 emissions.

Deforestation for Food Production

As the world population grows, the need for increased food production also grows. The challenge is to provide the food where it is needed. It could be argued that Earth can sustain the current population of over 6 billion; the real problem is in distribution of the available food. But what happens as the population continues to grow? By the year 2020, Earth's human population is expected to exceed 8 billion. Will the food production capacity of Earth continue to grow?

The graph in **Figure 5** shows three possible outcomes for future populations. One possibility is that the population will continue to grow. This will require technological advances that will increase the ability of Earth to produce the food and other resources needed for survival. It will also require other changes, such as the willingness of people living in the developed world to change their lifestyles and consume less. A second possible future is that the population will stabilize because it will reach the carrying capacity of Earth. This possibility will also require technological advances and changes in lifestyle. The third

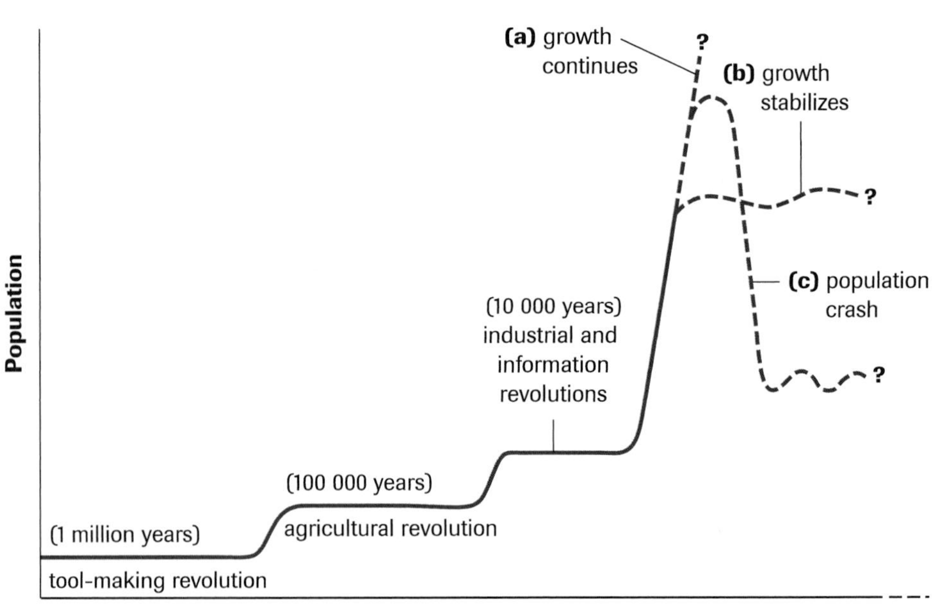

Figure 5
Three possible future outcomes:
(a) continued population growth
(b) stable population
(c) population crash

possible future is a population crash. This is a likely scenario, because the evidence to date suggests that technology will not be able to keep up with requirements and people will not be willing to significantly change their lifestyles. Paul Ehrlich, a biology professor at Stanford University and well-known environmentalist, has continually warned that humanity is living in unsustainable ways that stretch the ecological limits of the planet's ability to support human life.

In an effort to increase food production, people are clearing forested lands to convert them into farmland. Growing populations, mostly in equatorial regions, are cutting down large areas of forest for new farmland and settlements. On the surface, this might seem acceptable since other types of plants are replacing the trees. However, these forests help purify the air; act as giant sponges regulating the flow of water; influence regional and local climates; provide numerous habitats for wildlife species; and support food webs, energy flow, and nutrient cycling. The direct consequences of deforestation include the following:

- increased global warming resulting from increased levels of CO_2
- increased soil erosion and other types of land problems
- increased incidence of flooding caused by runoff
- loss of habitat for many species and resulting loss of biodiversity
- increased rate of species extinction

Deforestation contributes to the level of carbon dioxide in the atmosphere in two ways: first, the removal of trees means that fewer plants are using the carbon dioxide from the air; and second, the burning of forests releases the carbon that was stored in the tree trunks into the atmosphere as CO_2 (**Figure 6**). Almost 70% of deforested land is converted to agricultural land. However, there are other causes of deforestation, including commercial logging, mining, and flooding for hydroelectric projects (**Figure 7**). It is extremely difficult to determine accurately how much forest is being cut each year, but scientists estimate that between 50 000 and 170 000 km^2 (5–17 million ha) of global tropical forest are lost each year.

Figure 6
Clearing land for agriculture often involves burning the forest.

Figure 7
Much of the logging in tropical forests produces lumber for export.

At the global level, the net loss in forest area during the 1990s was an estimated 94 million ha (**Table 2**). This was the combined effect of a deforestation rate of 14.6 million ha per year and a rate of forest increase of 5.2 million ha per year. Deforestation of tropical forests is almost 1% per year.

Table 2 Change in Forested Land by Region, 1990–2000

Region	Total land area (x 10^6 ha)	Total forest 1990 (x 10^6 ha)	Total forest 2000 (x 10^6 ha)	Change 1990–2000 (x 10^6 ha)	% of land forested in 1990	% of land forested in 2000
Africa	2 963.3	702.5	649.9	–52.6	23.7	21.9
Asia and the Pacific	3 463.2	734.0	726.3	–7.7	21.2	21.0
West Asia	372.4	3.6	3.7	0.1	1.0	1.0
Europe	2 359.4	1 042.0	1 051.3	9.3	44.2	44.6
Latin America and the Caribbean	2 017.8	1 011.0	964.4	–46.6	50.1	47.8
North America	1 838.0	466.7	470.1	3.4	25.4	25.6
World	13 014.1	3 959.8	3 865.7	–94.1	30.4	29.7

Source: FAO Forestry Paper 140, Global Forest Resources Assessment 2000

▸ *TRY THIS* activity — *If the World Were an Apple!*

In this activity, an apple represents the planet Earth. You will develop an appreciation for the limited amount of land that is available to support all of the people and many of the other animals on Earth.

1. Cut an apple into quarters. Set aside three of the quarters. These three quarters represent the oceans. The oceans are not habitable by people but do provide some resources to help feed us.

CAUTION: Knives are sharp. Always cut away from yourself and your classmates.

2. Cut the remaining quarter in half. You now have two one-eighth ($^1/_8$) sections. Set aside one of the $^1/_8$ pieces. This piece represents mountainous areas, deserts, swamps, and other areas where people can't live.
3. Slice the remaining $^1/_8$ piece into four equal sections. Each section represents one-thirty-secondth ($^1/_{32}$). Three of these $^1/_{32}$ sections represent areas that are not appropriate for growing food.
4. Carefully peel the remaining $^1/_{32}$ section of Earth. Try not to remove any apple with the peel. This peel represents the soil, a very thin layer of Earth's crust in which all of our food is grown.

(a) In a group, discuss what can happen to the very thin layer of soil on Earth. Is it increasing? decreasing? Explain.

(b) Where else does food come from besides the soil on farmlands?

(c) Think about your own diet. Does most of your food come from producers or consumers? What impact does this have on the total available food energy?

Case Study 5.16 Questions

Understanding Concepts

1. Explain how adding CO_2 to the atmosphere causes global warming.
2. India has nearly four times the population of the United States. Compare the CO_2 emissions of these two countries.
3. Explain why deforestation has a twofold effect when it comes to the increasing level of CO_2 in the atmosphere.
4. Describe the impact of increased worldwide meat production on the total available food supply.

Applying Inquiry Skills

5. Use **Table 2** (page 416) to answer the following questions:
   (a) Which region of the world experienced the greatest change in forested area between 1990 and 2000? the smallest change?
   (b) Which region had the greatest percentage of its forested area deforested between 1990 and 2000? the least?
   (c) Which, if any, regions increased their forested area during this decade?
   (d) What percentage of the world's forest was lost during this period?
   (e) Suggest a possible explanation for your answers to (a) and (c) above.
6. From the map in **Figure 3** (page 412), identify five countries that are among the highest producers of CO_2 and five countries that are among the lowest producers. What observation can you make about the economic status of these countries?
7. **Figure 2** (page 412) shows the fluctuations in CO_2 concentrations from winter to summer in the Northern Hemisphere. Explain these fluctuations.

Making Connections

8. Research to determine the current status of the Kyoto Accord. In addition to the United States, what prominent countries have not ratified the treaty? Why have these countries not ratified the treaty?

www.science.nelson.com

9. Which of the three possible future outcomes for the world population do you think is most likely? Explain your reasoning. Propose a course of action that you think will lead to the most desirable outcome.
10. A simple solution for the problem of deforestation is for governments to prohibit the cutting of trees.
    (a) Is this simple solution a realistic one? Explain.
    (b) You are invited to an international conference on the problem of deforestation. In preparation for the conference, draft a proposal that offers a solution to the problem.
11. Tropical rain forests have been described as nature's pharmacy. Why do you think they are described this way? Research to find out the kinds of medicinal products that come from tropical forests. Prepare a report on one such product, including information about the source plant, the status of the source, its medicinal use, and its economic value.

www.science.nelson.com

5.17 Explore an Issue
Agriculture and the Environment

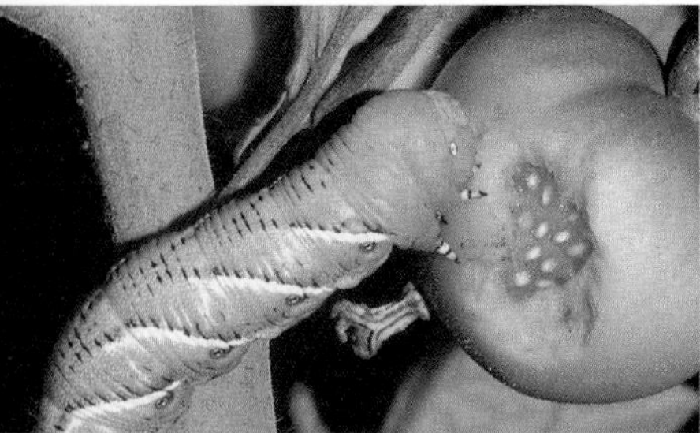

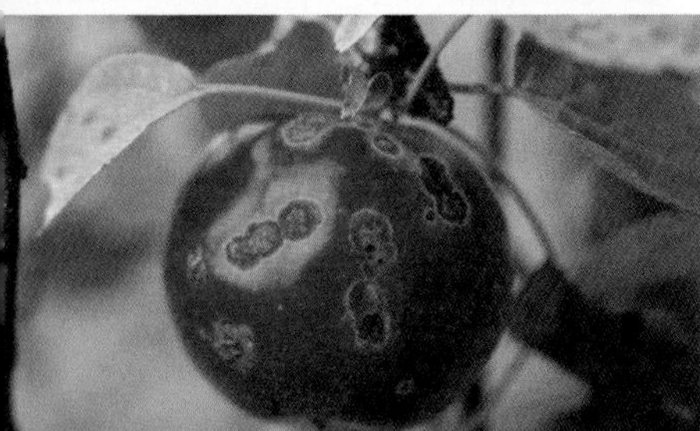

Figure 1
A variety of pests compete for the food intended for human consumption.

Your pet dog has just had a litter of puppies. You have decided to keep all of the puppies, but you have also decided that you will not increase the amount of dog food that you buy. You also allow some of the puppies to eat more than their share of the available food. The impact of your decisions on this small population of animals should be obvious. The food supply will not be sufficient to support all of the individuals; some are likely to die.

You would never make such an irresponsible decision for your pets, but we are collectively making this decision in regard to the world's population. There is a limited supply of food, but the population continues to increase. Furthermore, the available food is unevenly shared among the world's population.

In the past 50 to 60 years, there have been a number of technological advances that have resulted in a dramatic increase in global food production. Improvements in equipment (e.g., farm machinery and high-tech fishing boats and gear), chemicals (e.g., chemical fertilizers and pesticides), and genetic engineering (e.g., high-yield and disease-resistant crop varieties) have been effective in increasing the amount of food produced. However, the environment has paid a price.

Chemical Pesticides

Food producers often compete with a number of pests such as insects, weeds, rodents, and fungi that can consume or ruin a significant portion of the food intended for human consumption (**Figure 1**). To help control these pests, we have developed a variety of chemicals or pesticides that will kill them. Worldwide, about 2.3 million tonnes of pesticides—insecticides, herbicides, rodenticides, and fungicides—are used each year.

Pesticide use has a number of environmental consequences. For example, most pesticides are not target-specific; that is, they kill organisms other than the pest that they are intended to kill. Many pesticides also contain chemicals that are toxic to humans, causing problems with the nervous system and the immune system and increasing the risk of various forms of cancer. **Table 1** outlines the advantages and disadvantages of using chemical pesticides to control pests.

Table 1 Advantages and Disadvantages of Chemical Pesticides

Advantages	Disadvantages
• save human lives from diseases carried by pests • increase food supplies and lower food-production costs • increase profit for farmers • are more efficient (faster and better) than natural alternatives • save many humans from starvation • new pesticides are safer and more effective in smaller doses	• prompt insects and other pests to develop resistance • kill predators that help control pest populations • do not stay in the area where they are applied • harm wildlife • pose risks to human health • kill non-pest species in the same habitat

Chemical Fertilizers

Agriculture disrupts the normal cycling of matter. When plant material is removed from farmland to be distributed as food for animals and humans, the nutrients are removed from the cycle and do not decay to release the nutrients back to the soil. As farming continues, the natural nutrients of soils become depleted, and organic or chemical fertilizers are used to replace mineral nutrients. Chemical fertilizers are those that are manufactured from petrochemicals, natural rock, and synthetic chemicals. **Table 2** outlines the advantages and disadvantages of using chemical fertilizers.

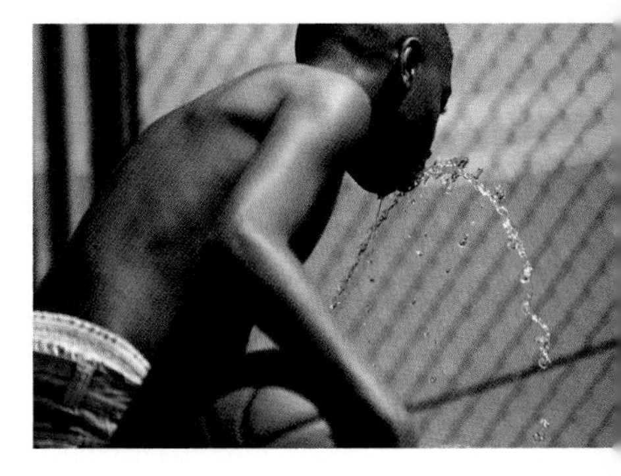

Figure 2
In certain locations, the water that we consume every day may be affected by agricultural activities.

Table 2 Advantages and Disadvantages of Chemical Fertilizers

Advantages	Disadvantages
• increase food production by about 40% • lower production costs by about 30% • increase profit for farmers • are easy to transport, store, and apply • make nutrients immediately available to plants because of their solubility	• don't add organic material to the soil • supply only 3 of more than 20 nutrients required by plants • cause water pollution in nearby water bodies. • require large amounts of energy for their production • release nitrous oxide, a greenhouse gas • make nutrients available in the soil for only short periods of time

The main problem associated with the use or overuse of chemical fertilizers is the contamination of water (**Figure 2**). The nitrogen from fertilizers is converted by bacteria in the soil into nitrates. These nitrates are leached into the groundwater and could end up in the drinking-water supply. High nitrate levels in drinking water are considered dangerous to human health. The phosphorus from fertilizers runs into water bodies and causes increased growth of algae. When the algae die, their decomposition removes oxygen from the water, causing problems for fish and other organisms (**Figure 3**).

Figure 3
The decomposition of algae in lakes consumes the oxygen needed by fish and other organisms.

Sustainable Agriculture

Chemical fertilizers and various chemical pesticides are effective in increasing worldwide food production and reducing its costs. However, both types of substances have serious environmental impacts. The challenge for the agriculture industry is to meet the increasing demand for food while at the same time respecting the consumer's expectation that the environment will not be harmed. This is the challenge of sustainable agriculture.

sustainable agriculture method of growing crops and raising livestock based on organic fertilizers, soil conservation, water conservation, biological pest control, and minimal use of nonrenewable fossil-fuel energy

Sustainable agriculture includes such practices as the use of organic fertilizers (e.g., animal manure and compost), crop rotation, and natural or biological pest control. These practices may not be as effective or efficient as the chemical alternatives but they have little detrimental effect on the environment. How do we find the balance?

Understanding the Issue

1. What are the advantages of more environmentally friendly methods of fertilizing crops and controlling pests over the chemical approach?
2. How do chemical fertilizers produce water pollution?
3. Explain the links between the use of chemical fertilizers and human health.
4. What do we mean when we say that a pesticide is not target-specific?
5. Explain what is meant by the term *sustainable agriculture.*
6. Why is sustainable agriculture important for the future?

Take a Stand

Decision-Making Skills

- Define the Issue
- Analyze the Issue
- Research
- Defend a Decision
- Identify Alternatives
- Evaluate

Can we produce the food required to feed the world's increasing population using sustainable agricultural practices?

The current level of food production would not be possible without the control of competing pests and the use of fertilizers to increase yields. There are essentially two alternatives: the use of chemical fertilizers and pesticides or the use of organic fertilizers and biological pest-control methods. Which approach to fertilizers and pest control will solve the problem? Both methods have advantages and disadvantages.

To feed the increasing world population without destroying habitat, the agriculture industry is challenged to increase the amount of food either by increasing the yield or by increasing the amount of land used to grow crops. In order to increase the yield, farmers will have to use fertilizers and control pests.

Statement: Sustainable agricultural practices can produce enough food to feed the world's increasing population.

- In your group, research the issue to learn more about sustainable agricultural practices.
- Search for information in newspapers, periodicals, and CD-ROMs and on the Internet.

www.science.nelson.com

(a) Identify the issue presented by the question. Explain why the issue is controversial.

(b) Who are the stakeholders in this issue? Identify and explain the different perspectives that are represented by the stakeholders.

(c) Complete a risk–benefit analysis based on the information from your research.

(d) Take a stand on the issue. In an audiovisual presentation, defend your position.

(e) How did your group reach a decision on which position to defend? Were there things you could have done differently? Explain.

5.18 Remote Sensing

Tech Connect

If you could look at Earth from outside the atmosphere, you would see dozens of observation satellites orbiting at different altitudes, generally ranging from 500 to 1000 km above Earth's surface. Imagine being able to distinguish wheat from corn using such a satellite!

Remote sensing is a general term that refers to the process of obtaining information about something from a distance. Although we usually think of satellites when we talk about remote sensing, it can be done from aircraft as well.

Radar remote sensing involves sending a signal from a satellite to the surface of Earth. Part of this signal is reflected back to the satellite and is then transmitted to a receiving station somewhere on Earth (**Figure 1**).

Agriculture is just one of the many applications of remote sensing. Even within the agriculture area there are many different uses for remote-sensing information, including the following:

- classifying crop types
- assessing crop conditions (water stress, nitrogen status, insect damage, and weather damage)
- estimating crop yield
- mapping soil characteristics
- mapping soil-management practices
- monitoring farming practices

Figure 2 shows an image produced by scanning a field with a temperature sensor. The blue areas are the coolest areas, and the yellow and orange areas are the warmest. Since the evaporation of water reduces the plant's surface temperature, it can be concluded that the cooler areas have sufficient water. When water is not in sufficient supply, there is little or no evaporation from the surface of the plant leaves, and the surface temperature increases, indicating the need for water.

Prior to this technology, farmers had to depend on their own ground observations of their fields to decide when to irrigate or apply fertilizers or pesticides and to determine how much water, fertilizer, or pesticide to apply. Using remote sensing, farmers can get more accurate and cost-effective information, and can respond quickly to the requirements of their crops.

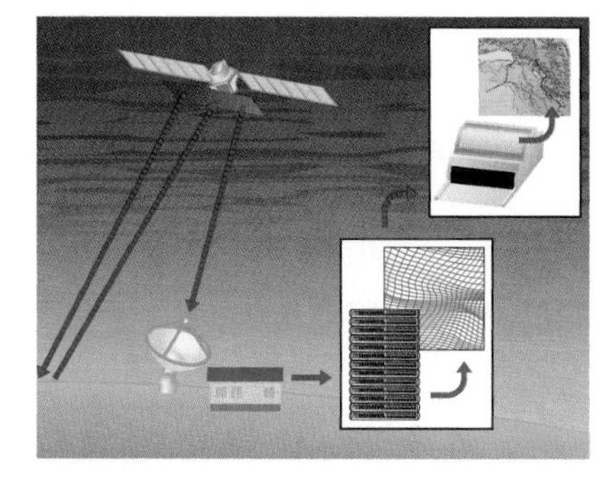

Figure 1
Satellites in orbit around Earth send signals to Earth, and receive reflected signals back. These signals are relayed to a receiver on Earth and transformed into digital satellite images and photographs.

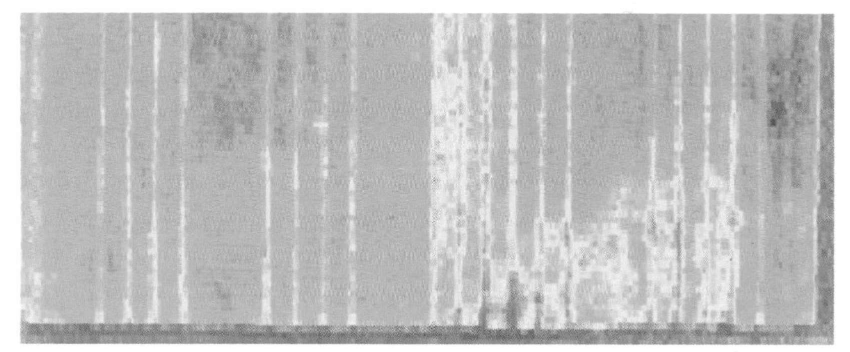

Figure 2
If the field above were irrigated uniformly, some areas of the field would receive more water than the plants need, while other areas would not receive enough. Irrigating only those areas requiring water reduces water use without reducing crop yield.

Tech Connect 5.18 Questions

Understanding Concepts

1. Briefly explain the concept of remote sensing.
2. How is remote sensing used in the agriculture industry? How has it helped agriculture practices?

Making Connections

3. The information from observation satellites must be recorded and analyzed. Research to find out the educational requirements and the job description of a remote-sensing technician.

www.science.nelson.com

4. In a small group, use the model in Appendix A2 to complete a risk–benefit analysis of this technology. Consider the perspectives of a government agriculture manager, a large-scale commercial farmer, a taxpayer, an environmentalist, a politician, and any others who may have an interest in this technology.

5.19 Lifestyles and Sustainability

Sustainability

sustainability the ability of an ecosystem to maintain ecological processes, functions, biodiversity, and productivity over time

Sustainability is a relatively new term that has come into common use since we began to recognize that Earth's resources are not limitless. It refers to the ability of a society to satisfy the basic needs of its people without depleting or degrading its natural resources to the point where future generations of humans and other species will not be able to meet their basic needs. Sustainability is like living off the interest income from your bank account. If you continue to use up the principal (the capital) you will eventually be bankrupt. Earth's resources are the capital, and the interest is the amount that can be replenished or replaced as it is used. If we use Earth's resources at a rate faster than they can be replenished, the capital will start to diminish or shrink. As Mathis Wackernagel, program director of the nonprofit sustainability organization Redefining Progress, says, "Like any responsible business that keeps track of spending and income to protect financial assets, we need ... to protect our natural assets. And if we don't ... we will prepare for ecological bankruptcy."

Humans are a highly intelligent and successful species that has been able to take full advantage of Earth's habitats and resources. This success has placed many stresses on the biosphere, but we do have the intelligence to recognize these stresses and make changes. Making such changes requires more than just recognition; it requires a desire and a willingness to change the behaviours that have led to the environmental problems.

There are several factors that will affect our ability to maintain sustainable societies. One factor is population growth. If population growth continues, Earth's carrying capacity will be reached very quickly. Another factor that affects the ability of Earth to support the population is the overconsumption of Earth's mineral, water, and living resources. The unequal distribution of the world's wealth and political instability are additional factors that affect the ability of societies to sustain themselves. Individual levels of consumption are increasing, especially in North America and other developed areas. Even if the world population stabilizes, it will still be difficult for Earth to provide the resources that are being demanded. These two factors together create a situation that will stretch the ecological limits of the planet's ability to support human life.

The evidence of unsustainable practices is plentiful (**Figure 1**). However, recognition of the problems and a willingness to do something about it have brought about some changes to more sustainable practices (**Figure 2**).

(a)

(b)

Figure 1
Evidence of unsustainable activity exists wherever humans live.
(a) Thousands of tonnes of mine tailings (waste products from the mining and/or processing of ore) severely polluted this stream in northern Ontario.
(b) Extensive use of water for irrigation drained this large body of water near the Aral Sea in Central Asia.

(a)

(b)

(c)

Figure 2
Sustainable practices such as conserving water **(a)**, recycling **(b)**, and planting trees **(c)** are becoming more common.

What Can Be Done?

Over the past few decades, we have seen a change in the decision-making models of governments, organizations, and industries. Traditionally, economic, social, and environmental decisions were made independently (**Figure 3(a)**). For example, developments that provided jobs and contributed to society's overall economic wealth were undertaken without consideration for their impacts on the environment. Current thinking is that the economy exists within society and society exists in the physical environment. Each of these areas affects and is affected by the others, so a decision in one area can't be made without considering the contributions of and the impacts on the other areas (**Figure 3(b)**). Today, proposals for most new economic developments in a society must include an environmental-impact assessment.

In 1992, the U.S. National Academy of Sciences and the Royal Society of London issued a joint report that contained the following warning:

> *If current predictions of population growth prove accurate and patterns of human activity on the planet remain unchanged, science and technology may not be able to prevent either irreversible degradation of the environment or continued poverty for much of the world.... Sustainable development can be achieved, but only if irreversible degradation of the environment can be halted in time.*

To make the changes recommended in this report and to address the environmental challenges we face requires a shift in thinking on a number of fronts:

- from cleaning up pollution to preventing pollution
- from disposing of waste to preventing and reducing waste
- from protecting species to protecting habitats
- from degrading environments to restoring environments
- from increasing resource use to less wasteful resource use
- from population growth to population stabilization

Humans need to take responsibility for their actions, whether on a local scale or on a global scale. The wealthiest 20% of people on Earth are presently enjoying a much higher quality of life than the remaining 80%, while imposing a far greater share of the stresses on the planet's life-support systems. Individuals can do little about the decisions of governments or major corporations; they do, however, have control of their personal lifestyle decisions.

(a)

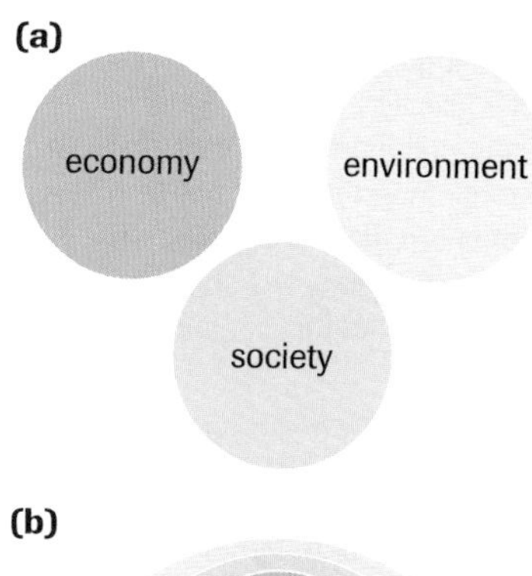

(b)

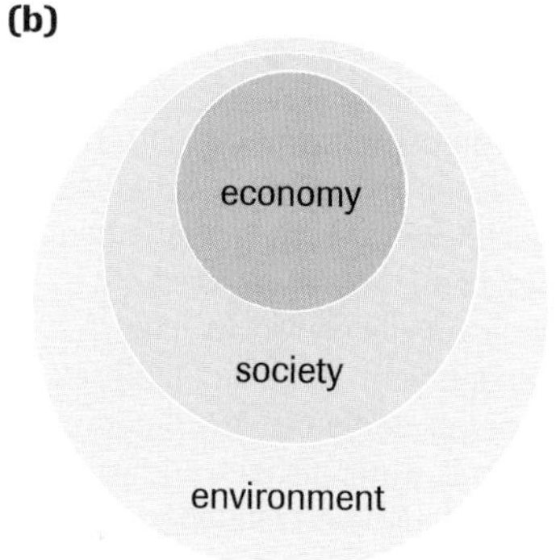

Figure 3
Decision making must take social, economic, and environmental factors into account in order to ensure sustainable solutions **(a)**. Social and economic decisions must be made in the context of the physical environment **(b)**.

DID YOU KNOW?

The Aluminum Can Crunch

Each year North Americans use more than 100 billion aluminum pop cans—enough to go around Earth 120 times! Large quantities of electric power are needed to make these cans. Some of this power is produced from huge hydroelectric dams that flood large portions of tropical rain forest and other habitats. The recycling of aluminum cans produces 95% less air pollution and 97% less water pollution and uses 95% less energy than the refining of new aluminum metal. Recycling one aluminum can saves enough energy to run a TV for three hours!

CAREER CONNECTION

The *green* job market is a fast growing segment of the economy. Environmental career opportunities are available in a number of fields: sustainable forestry management; parks and recreation; air- and water-quality control; waste management; recycling; land-use planning; ecological restoration; and soil, water, fishery, and wildlife conservation and management.

Every human being contributes directly or indirectly to the environmental problems that we face today. Solving these problems may seem to be a monumental or hopeless task. However, if every individual contributes to the problem, then every individual can contribute to the solution. The following guidelines may help you move toward more responsible environmental actions:

- Don't be trapped by hopelessness or blinded by faith in science and technology.
- Recognize that no one can do everything. Focus on things that you can do something about.
- Don't feel guilty or afraid, and don't use fear and guilt to try to motivate other people to get involved.
- Remember that there is no single correct or best solution. There are many solutions, all of which work to some degree.
- Think beyond your own lifetime to that of your children and grandchildren.

Table 1 summarizes the areas where changes need to be made and offers some examples of individual actions that can make a difference.

Table 1 Personal Actions toward Sustainable Living

Category	Examples of individual actions
Preserve biodiversity and protect the soil.	• Plant trees. • Reduce your use of paper and wood products. • Compost.
Promote sustainable agriculture and reduce pesticide use.	• Waste less food. • Eat less meat and more grains, vegetables, and fruits. • Use pesticides only when absolutely necessary and in the smallest possible quantities. • Don't worry about not having the perfect lawn.
Use natural chemicals or methods to control common insect pests and weeds.	• Raise the cutting level of the lawn mower to control lawn weeds. • Make your own flypaper using honey on yellow-coloured paper. • Sprinkle boric acid in dark, warm places to kill cockroaches.
Save energy and reduce outdoor air pollution.	• Reduce fossil-fuel use. Join a car pool or use mass transit. • Set your heating thermostat a little lower or your air-conditioner thermostat a little higher. • Plant trees to help absorb CO_2. • Purchase only energy-efficient vehicles and appliances.
Reduce exposure to indoor air pollutants.	• Use wood or linoleum flooring rather than carpet. • Remove your shoes before entering your house. • Don't use aerosol-spray products.
Save water.	• Reduce the amount of water used for flushing the toilet. • Turn off faucets while brushing teeth or shaving. • Sweep driveways and walks rather than hosing them off. • Wash only full loads of clothes. • Take shorter showers.
Reduce water pollution.	• Don't pour pesticides, paints, solvents, or other harmful chemicals down the drain or onto the ground. • Use organic fertilizers rather than chemical fertilizers. • Use biodegradable detergents and soaps.

Table 1 (continued)

Category	Examples of individual actions
Reduce solid waste and hazardous waste.	• Carry your lunch in reusable containers rather than wrapping it in disposable plastic or aluminum foil. • Carry groceries in reusable bags or a basket. • Reuse, recycle, and compost.

Section 5.19 Questions

Understanding Concepts

1. Use an example to explain the concept of sustainability.
2. Describe two factors that determine whether or not we exceed the carrying capacity of Earth.
3. Why is a shift in our thinking required in order to ensure a sustainable future? In what areas must our thinking change?
4. Why is it not wise to be "blinded by faith in science and technology" when it comes to solving the world's environmental problems?

Making Connections

5. A lumber company has applied for a 40-year lease of 150 000 ha of forested land from the provincial government. Their business plan calls for an annual harvest of 20 000 ha. The environmental component of the plan suggests that they will replant 10 000 ha in each of the first five years, 15 000 ha in each of the next five years, and 20 000 ha annually after that. The particular species of tree requires 40 years to grow to harvestable size. You are a member of the team that will evaluate the proposal.
   (a) Describe the status of the leased land at the end of the 40 years if the plan is followed.
   (b) Does this proposal respect the principles of sustainability? Explain.
   (c) What will be your recommendation on this application? Provide a rationale for your recommendation.
6. Brainstorm with a group to expand the list of personal actions presented in **Table 1**. These should be actions that you are able to do. Which of these are you willing to do?
7. Select one of the career areas mentioned in the Career Connection on page 424. Use the Internet and other resources to find the educational requirements and the job prospects in that area.

GO www.science.nelson.com

8. Do you think we have paid attention to the warnings issued in the report by the U.S. National Academy of Sciences and the Royal Society of London? Explain why or why not.

Reflecting

9. Some people believe that Earth will never run out of resources because science and technology will be able to replace depleted resources or substitute other materials for them. Do you agree or disagree with this line of thinking? Explain your answer, remembering to consider the effects that some technologies can have on the environment and sustainability.

Unit 5 SUMMARY

Key Understandings

5.1 Studying the Environment
- science is not necessarily limited to use of the scientific method and reliance on quantitative data
- environmental science provides the basis for environmental decision making

5.2 Ecological Roles
- different organisms—autotrophs, heterotrophs, and chemotrophs—have different ecological roles in ecosystems
- most ecosystems have different types of species: endemic, exotic, indicator, and keystone

5.3 Activity: Biomes and Ecozones of Canada
- biomes are divided into smaller geographic areas called ecozones to facilitate more accurate description

5.4 The Flow of Energy and Matter in Ecosystems
- energy pyramids show the flow of energy from one trophic level to the next
- pyramids of numbers show the numbers of organisms at each trophic level

5.5 Biogeochemical Cycles
- the carbon and oxygen cycles are often linked because of the relationship between photosynthesis and respiration
- the water cycle is driven by solar energy and gravity
- the nitrogen cycle: nitrogen fixation, nitrification, assimilation, ammonification, and denitrification
- phosphorus is a long-term cycle because phosphorus does not exist as a gas

5.6 Interactions: Competition and Predation
- species interact with one another by competing for food, space, and other resources in the ecosystem
- individuals can compete with other individuals of the same species (intraspecific) or with members of different species (interspecific)
- both predators and prey use strategies to increase their chances of survival

5.7 Activity: Observing Competition between Species
- analysis of data from experiments with competing species in the same environment

5.8 Interactions: Symbiotic Relationships
- when species interact, the relationship can be mutualistic, commensalistic, or parasitistic

5.9 How Populations Change
- population size is usually determined by the carrying capacity of the environment and its biotic potential
- most populations are distributed in a clumped dispersion pattern
- populations can grow geometrically, exponentially, or logistically
- exponential population growth can continue to occur only with unlimited resources
- natural populations generally grow to reach a dynamic equilibrium

5.10 Why Populations Change
- some factors that limit population growth depend on the density of the population, while other factors are unrelated to the density
- populations require a minimum size in order to survive

5.11 Tech Connect: Satellite Telemetry
- satellite transmitters are used to track animals that migrate or move over long distances

5.12 Estimating Population Size
- mark–recapture and quadrat sampling are two methods used to estimate population size

5.13 Investigation: Estimating Population Size
- estimating the size of a simulated population

5.14 Investigation: Plant and Animal Survey
- a qualitative survey and classification of plants and animals in the local area

5.15 Human Population Growth
- the world-population growth rate is the net annual increase per 1000 people, expressed as a percentage
- the doubling time of a population is approximately equal to 70 divided by the annual growth rate

5.16 Case Study: Effect of Population Growth on the Environment
- the world's resources are unevenly shared
- the addition of CO_2 to the atmosphere contributes to the greenhouse effect and global warming

- the developed countries contribute more to global warming than developing countries
- most deforestation around the world occurs because of increased demand for farmland

5.17 Explore an Issue: Agriculture and the Environment

- chemical pesticides have increased crop yields and reduced diseases but have also caused human-health and environmental problems
- chemical fertilizers have increased food production, but overuse has contaminated water supplies
- the goal of sustainable agriculture is to meet the demands for food without serious environmental impacts

5.18 Tech Connect: Remote Sensing

- uses of satellite technology include classifying crop types, assessing crop conditions, and estimating yields

5.19 Lifestyles and Sustainability

- at current population growth and consumption rates, Earth's carrying capacity will soon be exceeded
- individuals can contribute to global sustainablility by adjusting their lifestyle

Key Terms

5.1
ecology
inference
scientific method
observational study
quantitative
qualitative

5.2
biosphere
biome
population
habitat
community
biotic
abiotic
ecosystem
chemotroph
exotic species
indicator species
keystone species

5.4
energy flow
trophic level
energy pyramid
pyramid of numbers

5.5
biogeochemical cycle
nitrogen fixation
nitrogenase
nitrification
assimilation
ammonification
denitrification

5.6
niche
interspecific competition
intraspecific competition
resource partitioning
predation

5.8
endoparasite
ectoparasite

5.9
carrying capacity
biotic potential
population size
population density
crude density
ecological density
clumped dispersion
uniform dispersion
random dispersion
population dynamics
open population
closed population
geometric growth
exponential growth
logistic growth
environmental resistance
dynamic equilibrium

5.10
density-dependent factor
Allee effect
minimum viable population size
density-independent factor
limiting factor

5.12
mark–recapture method
quadrat

5.15
doubling time

5.16
greenhouse effect

5.17
sustainable agriculture

5.19
sustainability

▸ *MAKE a summary*

Select a species of plant or animal from your local area and consider its role and functions as an individual, as a part of a population, and as a part of an ecological community. Describe what life would be like for an individual of this species in terms of the following:

- what ecozone it is in and whether or not it is a common species in this ecozone
- its role in the ecosystem (what it eats and what eats it)
- current population status and potential changes as a result of interspecific interactions
- factors (density-dependent and density-independent) that might affect the species
- any other population dynamics learned in this unit
- the impact of human activities on it or its habitat, including what can be done or has been done to help ensure a sustainable population

PERFORMANCE TASK

Criteria

Process

- Use a variety of sources, including print and electronic research tools.
- Conduct a thorough search of available reference materials related to sustainability.
- Compile and summarize the information.
- Select appropriate methods for the analysis and presentation of your findings.
- Evaluate the quality of data and information obtained.

Product

- Compile, organize, and present the data using a variety of formats, including tables, graphs, diagrams, and graphic organizers.
- Demonstrate an understanding of the relationships among the concepts taught in this unit.
- Express and present your findings in a creative and convincing manner.
- Use both scientific terminology and the English language effectively.

Ecological Footprints

The concept of ecological footprints helps us visualize and compare ecological effects of different human lifestyles. Many factors such as food consumed, energy used, water required, and waste generated must be considered to assess an ecological footprint.

It is estimated that if the amount of productive land on Earth were divided equally among every human alive in 2002, there would be 1.9 ha available for each human. This is referred to as the Earth's biocapacity. This number does not take into account the space needed by other species. If we estimate that other species need about 12% of the available space, then the average biocapacity drops to 1.7 ha per person.

A recent report prepared by Mathis Wackernagel, program director of the nonprofit sustainability organization Redefining Progress, presents the ecological footprint of 146 nations. The report indicates that an average of 2.3 ha are required to support each person on Earth. The global average was significantly lower than that of industrialized countries such as Canada, where 8.8 ha were required per person. If all the people in the world had an ecological footprint as large as that of each Canadian, the global population would need another four Earths to sustain itself!

If the world's population reaches 10 billion (at the current growth rate, in a little over 30 years) and the total amount of productive land does not increase, there will be only 1.1 ha of biocapacity per person, less than what people use today in countries such as Iraq, Indonesia, Peru, and Nigeria.

Questions

(i) Why might various countries have very different ecological footprints?

(ii) Do all people in the same country have the same ecological footprint?

(iii) Can humans change their ecological footprints?

Part 1: Comparing Ecological Footprints

1. Study **Figure 1(a)**, which shows the average per capita ecological footprint for the country, and **Figure 1(b)**, which shows the total ecological footprint for each country.

Analysis

(a) With two or three other students, list possible reasons for the differences among the per capita footprints of the countries illustrated in **Figure 1**.

(b) Rank the countries from the largest to the smallest per capita footprint. Then rank the countries from the largest to the smallest total footprint. Explain why the ranking of the countries' total footprints does not simply repeat the ranking of the per capita footprints.

Synthesis

(c) Why are the total footprints for Canada and the United Kingdom very small compared with that for the United States while their per capita footprints are only a little smaller?

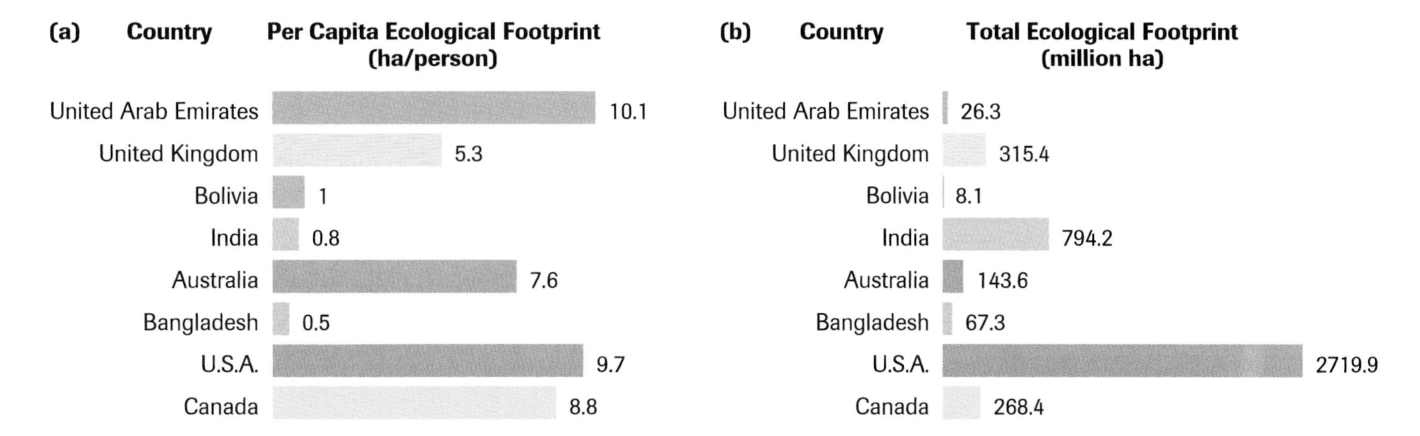

Figure 1
A country's total ecological footprint **(b)** is calculated by multiplying the per capita footprint **(a)** by the country's total population.

(d) Why is the total footprint for India very large compared with that for the United Arab Emirates while its per capita footprint is much smaller?

(e) Eric Krause, an environmental policy official with the city of Toronto, has noted that "it would be better for the environment if 10% of the population reduced its ecological footprint by 90%, than if 90% of the population reduced its ecological footprint by 10%." Do you think he was referring to any 10% of the population or to 10% from specific countries? Explain.

Part 2: Calculating Your Home's Ecological Footprint

2. Your teacher will provide you with a printed copy or a spreadsheet of an ecological footprint calculator, or you can use an online ecological footprint calculator.

www.science.nelson.com

Analysis

(f) Use the calculator to determine your own or your family's ecological footprint. If Earth's per capita biocapacity is 1.7 ha, calculate the number of Earths we would need if everyone in the world had the same lifestyle as you or your family.

(g) Compare your footprint with those of your classmates. What factors could account for any differences? Calculate the class average.

Evaluation

(h) Do you think the questionnaire is an accurate and useful indicator of your real ecological footprint? If not, how would you change it?

Part 3: Preparing a Presentation

3. Using the Internet and print resources, research the concept of the ecological footprint and the factors that influence it.

www.science.nelson.com

(i) Prepare an audio-visual presentation that explains the concept of the ecological footprint and outlines the components that are used in the calculation. Include in your presentation a list of simple, achievable suggestions for reducing one's ecological footprint.

Understanding Concepts

1. What is the main energy source for all biological activities on Earth?
2. What are trophic levels? Explain how they are related to an ecological pyramid of energy.
3. Using the diagram in **Figure 1**, calculate the amount of energy transferred from one level to another if the top consumer's energy is 25 J. Name the areas of the pyramid based on the organism's trophic level. Assume a transfer of 10% of energy at each level.

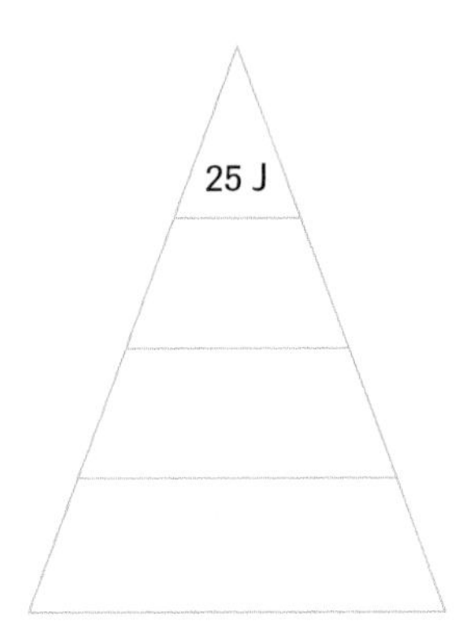

Figure 1

4. Analyze the energy pyramid in **Figure 2**. How could this food chain be adjusted to provide more food energy for a dense and growing human population?

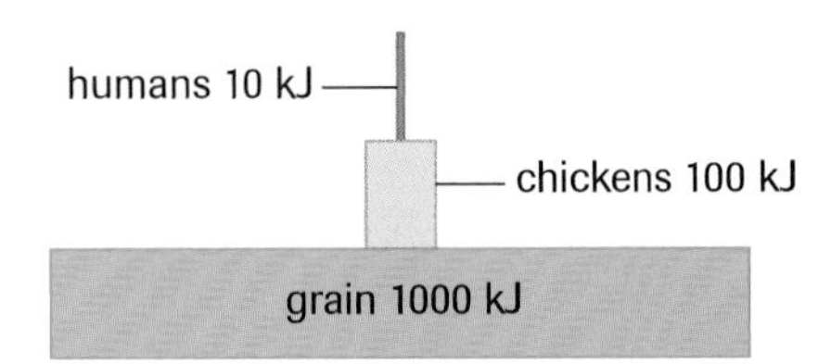

Figure 2

5. Calculate the density of a population of southern flying squirrels if 1020 squirrels were counted in a 68-ha area.
6. Describe three patterns of population dispersion and give an example of a population that shows each pattern.
7. Explain, using an example, the difference between crude density and ecological density.
8. Describe the concept of carrying capacity and explain its role in population dynamics.
9. What is the Allee effect? Give an example of a population that is affected by this density-dependent limiting factor.
10. Using examples, describe what might happen if an important prey species were eliminated from a community.
11. Identify similarities and differences between predators and parasites.
12. Name and describe the mechanism by which certain gases trap heat within Earth's atmosphere.
13. Explain, using an example for each, the differences between interspecific and intraspecific competition.
14. Classify each of the following as density-dependent or density-independent factors:
    (a) fertilizer runoff causing increased algae population in a pond
    (b) the impact of predators on a population of snow geese
    (c) the spread of an infectious disease in a wild population
    (d) the effects of habitat loss on wild populations
15. In a group, discuss an example of each type of symbiotic relationship in which humans are involved.
16. Parasites are often classified according to where they live. What two terms are involved? Explain each.
17. Does the human population show a uniform, clumped, or random dispersion pattern? Explain.
18. What do energy pyramids show about the suitability of plants versus animals for meeting human food demands?
19. If the human-population growth rate dropped to 0.56%, what would be the doubling time for the world population?
20. According to the 2001 census, the population of Canada had reached 30 007 094 people.
    (a) What is the population density of Canadians if Canada's land area is 9 976 000 km^2?
    (b) Using our population as an example, explain why the ecological density of a species is usually greater than its crude density.

21. In 2000, the population of Ontario was 11 697 600. Calculate the growth rate of the Ontario population between July 21, 2000, and June 30, 2001, given the following data from Statistics Canada:
    - 130 672 births
    - 87 565 deaths
    - 149 868 immigrants
    - 32 156 emigrants
22. Using **Figure 3** below, answer the following questions:
    (a) What are the top carnivores in this food web?
    (b) What are the producers in this food web?
    (c) If the rat population became extinct, what organism(s) would lose a food source?
    (d) How many organisms depend on salmon as a food supply?
    (e) List the primary consumers and the secondary consumers.

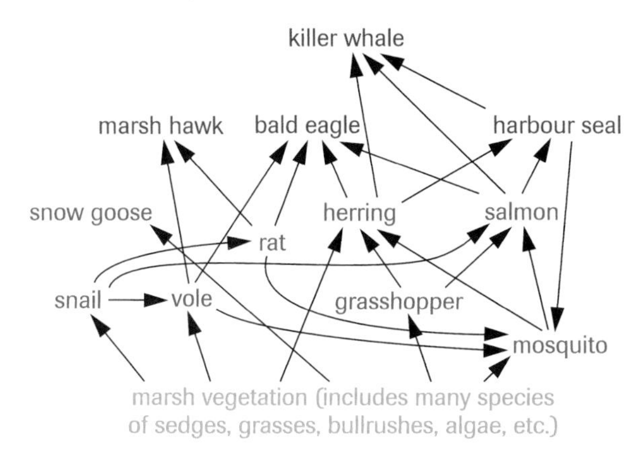

Figure 3

23. *Natality, mortality, immigration,* and *emigration* are all terms related to any population.
    (a) Write a paragraph that defines the terms and explains how each process affects the provincial populations and the total population of Canada.
    (b) Is the population of Canada an open population or a closed population? Which of the terms would not apply to a closed population?
24. Explain what happens once a population reaches a dynamic equilibrium.
25. In some cases, a small population may have a greater effect on the local and global environment than a large population. Explain this statement using specific examples of a human population's pattern of consumption.
26. Assess the impacts of agriculture on the natural environment.

Applying Inquiry Skills

27. A fisheries technician studied a population of trout in a 120-ha lake in northern Ontario. She found that the fish lived only in scattered reed beds that represented only 15% of the total area of the lake. The data from her mark–recapture sampling are summarized in **Table 1**.

Table 1 Mark–Recapture Analysis of Trout

number of trout marked	185
number of trout captured in second sample	208
number of marked trout recaptured	13

    (a) How many trout would you estimate to be in the study area?
    (b) What is the density of the trout population per hectare?
    (c) What is the dispersion pattern of the fish?
    (d) Considering the dispersion pattern, how might you present a different interpretation of the population density?
28. Meadow voles, sometimes referred to as field mice, are extremely common small rodents that breed actively throughout the year, with females capable of having young at less than two months of age. The data in **Table 2**, on the next page, represent the growth over time of a population of voles living in a small grassland.
    (a) Make a graph of the data.
    (b) Based on your graph, is the meadow vole population showing logistic, geometric, or exponential growth? Explain.
    (c) From your graph, estimate the size of the vole population for the next three months.
    (d) What conditions are necessary for a population to grow this way?
    (e) Do you think the population will continue to increase indefinitely? Explain.

Table 2 Monthly Growth for a Population of Meadow Voles

Month	Meadow vole population
January	3 920
February	5 488
March	7 683
April	10 756
May	15 058
June	21 081
July	29 513
August	41 318
September	57 845
October	80 983
November	113 376

29. Students sampled aquatic insect larvae living on a small section of river bottom measuring 2 m by 0.8 m. They found approximately 5000 black fly larvae in their sample.
    (a) What was the population density of this species?
    (b) Estimate the number of black fly larvae living in a similar habitat of river bottom measuring 4 m by 10 m.

30. Study the graphs of two different populations (**Figure 4**): a population of bacteria growing in a laboratory and a population of owls living in a forest. Which graph represents which population? Explain your reasoning.

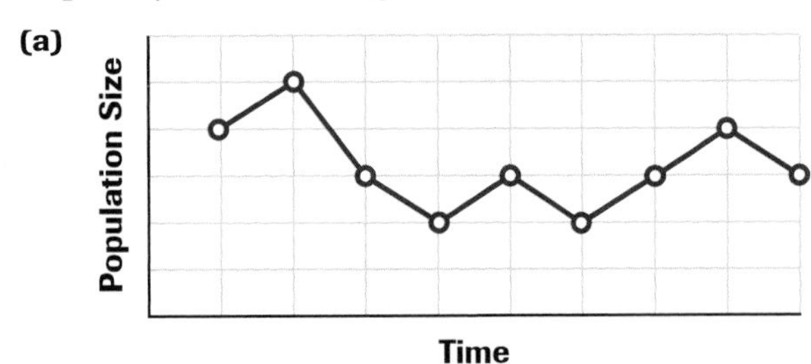

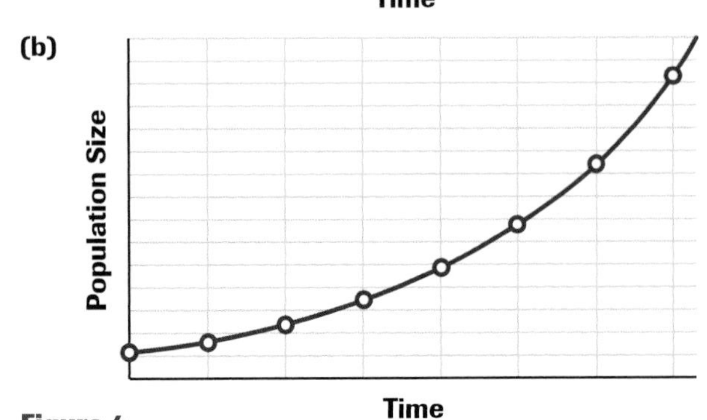

Figure 4

31. During the past 15 years, white-tailed deer populations in Canada have significantly affected woody plants in forest ecosystems. To estimate the white-tailed deer population in three tracts of forest in southern Ontario, investigators captured and tagged 175 white-tailed deer.
    (a) After three weeks, 135 deer were recaptured and 21 had bands. Calculate the population size.
    (b) If each tract of forest was 125 ha, calculate the crude population density.
    (c) Of the total area of the forest, 94 ha consist of small lakes and ponds and rocky cliffs. Calculate the ecological density of the deer population.

32. Explain why the crests and troughs of the graphs in **Figure 5** do not coincide.

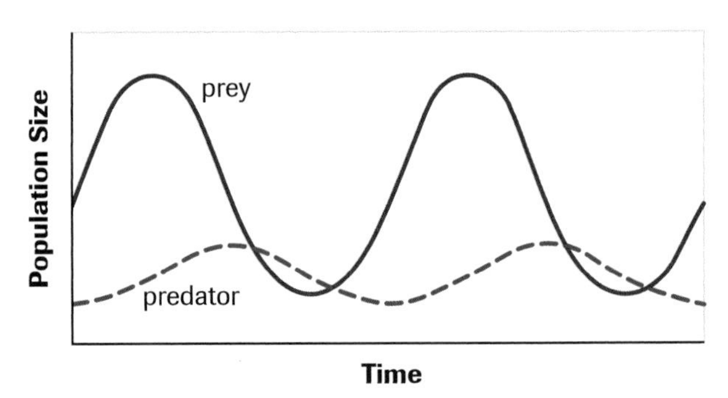

Figure 5

Making Connections

33. The Nova Scotia Leatherback Turtle Working Group (LTWG) is a research and conservation project committed to the recovery of the endangered leatherback turtle. Since 1995, group members have been tagging turtles in the North Atlantic Ocean of eastern Canada.

    Use the Internet to explore the science of tagging turtles. What methods are used to tag leatherback turtles? How does the research play a role in learning about the migration routes of the leatherback turtles? Gather satellite data on any species tagged with radio transmitters, and plot their movements on a map. What information can be obtained from these data regarding migration and other behaviours?

www.science.nelson.com

34. Describe the impact of increased worldwide meat production on agriculture, water resources, and waste management.

35. Research the sea lamprey, an exotic species that has had a great impact on the fish communities of the Great Lakes.
    (a) Describe the niche of the sea lamprey.
    (b) Find out how sea lampreys may have entered the Great Lakes ecosystem.
    (c) Identify the interspecific interactions of the sea lamprey and its effects on the Great Lakes.
    (d) Describe some economic setbacks faced by Canadian fisheries as a result of sea lamprey invasion, and outline any control efforts.

www.science.nelson.com

36. With a small group of your classmates, discuss the following questions:
    (a) Do you think it would be possible, theoretically, for humans to survive in a world that consisted of only water, shelter, oxygen, and grain?
    (b) Do you think humans would want to live in such a world?

37. Discuss what aspects of your lifestyle would need to be examined in order to determine your personal ecological footprint.

38. How reasonable is it for humans to continue to dispose of millions of tonnes of garbage each day in landfills and incinerators around the world? Research the advantages and disadvantages of this approach.

www.science.nelson.com

39. Of the following global environmental issues, select the one that you think is the most important:
    - world population growth
    - pollution
    - food production
    - food distribution
    - overconsumption
    - global warming

    Prepare a letter to the federal minister of the environment outlining the issue and suggesting what Canada could or should do to help solve the problem.

40. Vegans are individuals who consume plants only. They claim that a larger number of people could be fed if populations switched to a vegan diet. Their opponents argue that there is not enough productive land to grow enough plant products to sustain the world's population.
    (a) With two or three other students, discuss arguments and counterarguments for each position.
    (b) Create a PMI chart to summarize your thinking. (See Appendix A5 for a description of a PMI chart.)

41. Canadians produce the most garbage per person of any society in the world. In 1996, almost 9 million tonnes of municipal solid wastes were generated in Ontario alone, 80% of which was deposited in landfills. How does this influence the ecological footprint of Canada's 30 million people relative to the populations of less industrialized countries?

42. The mayor of a large city claims that she will convert her city into a sustainable ecosystem. How realistic is this goal? Explain your answer.

43. The Canadian International Development Agency (CIDA) is a government-funded organization whose mandate is to help developing countries achieve sustainable economic and social development. The organization is involved in many projects around the world to support better food production. Research one CIDA-funded project that invests in food-production technologies. Compile a report on your research to provide the following information:
    (a) What specific problem(s) does the project address?
    (b) What role does CIDA have in the project?
    (c) Describe specific details of the project.
    (d) Evaluate and discuss the potential of the project to improve the long-term sustainability of the population and the natural environment.

www.science.nelson.com

Appendices

A1 Scientific Inquiry

Planning an Investigation

In our attempts to further our understanding of the natural world, we encounter questions, mysteries, or events that are not easily explained. We can use controlled experiments or observational studies to help us look for answers or explanations. The methods used in scientific inquiry depend, to a large degree, on the purpose of the inquiry.

Controlled Experiments

Controlled experiments are performed when the purpose of the inquiry is to create or test a scientific concept. In a controlled experiment, an independent variable is purposefully and steadily changed to determine its effect on a second, dependent variable. All other variables are controlled or kept constant.

The common components for controlled experiments are outlined in the flow chart below. *Even though the sequence is presented as linear, there are normally many cycles through the steps during an actual experiment.*

	Stating the Purpose	Asking the Question	Hypothesizing/ Predicting	Designing the Investigation
Process Description	**Choose a topic that interests you. Determine whether you are going to carry out a given procedure or develop a new experimental design. Indicate your decision in a statement of the purpose.**	**Your Question forms the basis for your investigation. Controlled experiments are about relationships, so the Question could be about the effects on variable A when variable B is changed. The Question could be about what causes the change in variable A. In this case, you might speculate about possible variables and determine which variable causes the change.**	**A hypothesis is a tentative explanation. You must be able to test your hypothesis, which can range in certainty from an educated guess to a concept that is widely accepted in the scientific community. A prediction is based on a hypothesis or a more established scientific explanation, such as a theory. In the prediction, you state what outcome you expect from your experiment.**	**The design of a controlled experiment identifies how you plan to manipulate the independent variable, measure the response of the dependent variable, and control all the other variables.**
Example: Diffusion and Osmosis	The purpose of this investigation is to examine the conditions under which water moves into and out of cells.	Under what conditions does water move into and out of cells?	If a model cell is placed in a hypotonic solution (having lower concentrations of solute than the cell), water will move into the cell. If a cell is placed in a hypertonic solution (having higher concentrations of solute than the cell), water will move out of the cell.	Three model cells are constructed from dialysis tubing. Two are filled with distilled water and one with a starch suspension. One of the cells with distilled water is placed in a beaker of starch suspension. The other two cells are placed in beakers of distilled water. Iodine is added to the distilled water in the beakers. (Iodine is used as an indicator to detect the presence of starch.) The mass of each cell is recorded and other observations are made at the beginning and after 10 min and 20 min.

A

Table 1 Observations of Model Cells

Model cell	Initial mass (g)	Mass after 10 min (g)	Mass after 20 min (g)	Other observations
cell A—dialysis tube with distilled water in beaker of distilled water				
cell B—dialysis tube with starch suspension in beaker of distilled water				
cell C—dialysis tube with distilled water in beaker of starch suspension				

There are many ways to gather and record observations during an investigation. It is helpful to plan ahead and think about what data you will need and how best to record them. This helps to clarify your thinking about the Question posed at the beginning, the variables, the number of trials, the procedure, and your skills. It will also help you organize your evidence for easier analysis later.

Gathering, recording, and organizing observations

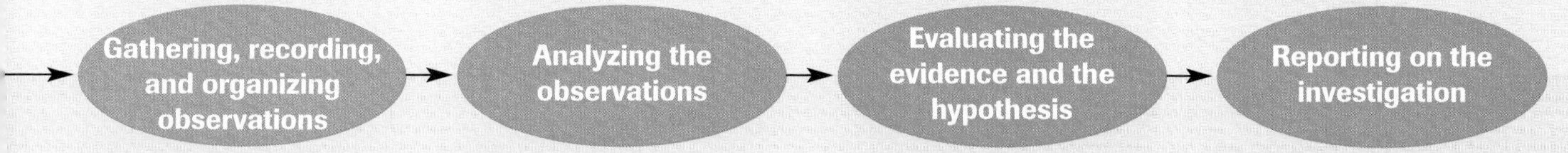

The measurements and observations will be recorded in a table like **Table 1**.

After thoroughly analyzing your observations, you may have sufficient and appropriate evidence to answer the Question posed at the beginning of the investigation.

Analyzing the observations

The results will be analyzed to determine if there were any changes in mass and if there were any other observable changes. Changes in mass and/or colour should enable us to determine what is happening with the cells.

At this stage of the investigation, you will evaluate the processes that you followed to plan and perform the investigation. Evaluating the processes includes reviewing the design and the procedure. You will also evaluate the outcome of the investigation, which involves assessing the evidence—whether it supports the hypothesis or not—and the hypothesis itself.

Evaluating the evidence and the hypothesis

To evaluate this investigation, we have to ask ourselves several questions. Is our model cell an appropriate model? Do the results allow us to determine how water moves into and out of cells? Does the evidence gathered support our hypothesis? Are there possible sources of error that may invalidate the evidence?

In preparing your report, your aim should be to describe your design and procedure accurately, and to report your observations accurately and honestly.

Reporting on the investigation

For the format of a typical lab report, see Appendix A4.

Observational Studies

Often the purpose of inquiry is simply to study a natural phenomenon with the intention of gaining scientifically significant information to answer a question. Observational studies involve observing a subject or phenomenon in an unobtrusive or unstructured manner, often with no specific hypothesis. A hypothesis to describe or explain the observations may, however, be generated after repeated observations, and modified as new information is collected over time.

The flow chart below summarizes the stages and processes of scientific inquiry through observational studies. *Even though the sequence is presented as linear, there are normally many cycles through the steps during an actual study.*

	Stating the Purpose	Asking the Question	Hypothesizing/ Predicting	Designing the Investigation
Process Description	**Choose a topic that interests you. Determine whether you are going to replicate or revise a previous study, or create a new one. Indicate your decision in a statement of the purpose.**	**In planning an observational study, it is important to pose a general question about the natural world. You may or may not follow the Question with the creation of a hypothesis.**	**A hypothesis is a tentative explanation. In an observational study, a hypothesis can be formed after observations have been made and information gathered on a topic. A hypothesis may be created in the analysis.**	**The design of an observational study describes how you will make observations relevant to the Question.**
Example: Local Vegetation	The purpose of our investigation is to conduct an inventory of the plants in our local area to determine the most common trees and to determine if the vegetation in our local area is typical of that found in the boreal forest biome.	What are the most common species of trees in our local region? Is the vegetation in our region representative of the vegetation in the boreal forest biome?	There is considerable variation in the vegetation found throughout any biome. However, one would expect that certain species would be more common than others and would be found throughout the biome. Since we are geographically located in what is defined as the boreal forest biome, we would expect that the most common trees in this area would be those that define the vegetation of the boreal forest (i.e., spruce, fir, and pine).	We will conduct an inventory of trees in 10 sample areas of our local region. Each sample area will be 10 000 m^2 (100 m $\times$ 100 m). All species of trees will be identified (using the common and scientific names) and counted in each sample area. The sample areas will be selected by placing a scaled grid over a map of the region and then randomly selecting 10 cells of the grid.

A

Table 2 Inventory of Trees in Sample Areas of Local Region

	Sample area										
Tree species (common and scientific name)	**1**	**2**	**3**	**4**	**5**	**6**	**7**	**8**	**9**	**10**	**Total**

There are many ways to gather and record observations during an investigation. During your observational study, you should quantify your observations where possible. All observations should be objective and unambiguous. Consider ways to organize your information for easier analysis.

Gathering, recording, and organizing observations

The data will be recorded in a table like **Table 2**.

After thoroughly analyzing your observations, you may have sufficient and appropriate evidence to answer the Question posed at the beginning of the investigation. You may also have enough observations and information to form a hypothesis.

Analyzing the observations

The data from the table will be plotted on a bar graph to show the frequency distribution of the various species. This will make it easier to identify the most common species.

At this stage of the investigation, you will evaluate the processes used to plan and perform the investigation. Evaluating the processes includes evaluating the materials, the design, the procedure, and your skills. The results of most such investigations will suggest further studies, perhaps correlational studies or controlled experiments to explore tentative hypotheses you may have developed.

Evaluating the evidence and the hypothesis

In the evaluation of this investigation, we must first decide whether our sample areas collectively represent the forested areas of our local region. We can then compare the frequencies of the different species in the local area with the frequencies of those species in the typical boreal forest. The results of this investigation might lead us to speculate about the reasons (e.g., climate, topography, and soil) for any differences in vegetation between our local area and the typical boreal forest.

In preparing your report, your aim should be to describe your design and procedure accurately, and to report your observations accurately and honestly.

Reporting on the investigation

For the format of a typical lab report, see Appendix A4.

A2 Decision Making

Modern life is filled with environmental and social issues that have scientific and technological dimensions. An issue is defined as a problem that has at least two possible solutions rather than a single answer. There can be many positions, generally determined by the values that an individual or a society holds, on a single issue. Which solution is "best" is a matter of opinion; ideally, the solution that is put into practice is the one that is most appropriate for society as a whole.

The common elements of the decision-making process are outlined in the graphic below.

Even though the sequence is presented as linear, you may go through several cycles before deciding you are ready to defend a decision.

Defining the issue

Process Description

The first step in understanding an issue is to explain why it is an issue, describe the problems associated with the issue, and identify the individuals or groups, called stakeholders, involved in the issue. You could brainstorm the following questions to research the issue: Who? What? Where? When? Why? How? Develop background information on the issue by clarifying facts and concepts, and identifying relevant attributes, features, or characteristics of the problem.

Example: Genetically Modified Foods

Genetically modified (GM) foods are foods that have been altered by the insertion of genes from a different species. Crops can be genetically modified to grow quickly, to be resistant to diseases and pests, or to be efficient at absorbing nutrients from the soil.

There is growing public debate about such foods. Have they been tested enough to assure us that there will be no long-term effects from eating these foods? What will be the impact of genetically engineered species on natural species?

The issue is the overall safety of foods produced from genetically engineered organisms. In this debate, there are two positions: you either support genetic engineering in agriculture or you do not support it. **Table 1**, on page 442, lists the potential groups or stakeholders who may be positively or negatively affected by the issue. Develop your knowledge of the background information on the issue by clarifying facts and identifying features.

Identifying alternatives/positions

Process Description

Examine the issue and think of as many alternative solutions as you can. At this point, it does not matter if the solutions seem unrealistic. To analyze the alternatives, you should examine the issue from a variety of perspectives. Stakeholders may bring different viewpoints to an issue and these may influence their position on the issue. Brainstorm or hypothesize how various stakeholders would feel about your alternatives.

Example: Genetically Modified Foods

One possible solution for people concerned about GM food is to ban its production. Since farmers who grow GM foods would lose income, it might be necessary to offer subsidies to these farmers.

Think about how various stakeholders might feel about the alternatives. What would be the perspective of a consumer? a local politician? a farmer? a geneticist? an economist? a nutritionist? **Table 2**, on page 442, lists the possible perspectives on an issue.

Remember that one person could have more than one perspective, or two people looking at an issue from the same perspective might disagree. For example, geneticists might disagree about the impact of genetically engineered species on natural species.

Researching the issue

Process Description

Formulate a research question that helps to limit, narrow, or define the issue. Then develop a plan to find reliable and relevant sources of information. Outline the stages of your information search: gathering, sorting, evaluating, selecting, and integrating relevant information. You may consider using a flow chart, concept map, or other graphic organizer to outline the stages of your information search. Gather information from many sources, including newspapers, magazines, scientific journals, the Internet, and the library.

Example: Genetically Modified Foods

A possible research question: What are the advantages and disadvantages of genetically modified foods?

A

There are five steps that must be completed to effectively analyze the issue:

1. **Establish criteria for determining the relevance and significance of the data you have gathered.**
2. **Evaluate the sources of information.**
3. **Identify and determine what assumptions have been made. Challenge unsupported evidence.**
4. **Determine any relationships associated with the issue.**
5. **Evaluate the alternative solutions, possibly by conducting a risk–benefit analysis.**

Analyzing the issue

The relationships related to the issue include the following: GM food has arisen from the need to produce more and better food for the world's population. A consequence may be that consumers will refuse to buy GM products, thereby increasing the demand for traditionally produced food and increasing prices accordingly. Another consequence is that public funds are being spent to support research in genetically modified food crops.

Table 3, on page 443, shows a risk–benefit analysis of producing GM foods.

After analyzing your information, you can answer your research question and take an informed position on the issue. You should be able to defend your solution in an appropriate format—debate, class discussion, speech, position paper, multimedia presentation (e.g., computer slide show), brochure, poster, or video.

Your position on the issue must be justified using supporting information that you have researched. You should be able to defend your position to people with different perspectives. Ask yourself the following questions:

- **Do I have supporting evidence from a variety of sources?**
- **Can I state my position clearly?**
- **Can I show why this issue is relevant and important to society?**
- **Do I have solid arguments (with solid evidence) supporting my position?**
- **Have I considered arguments against my position, and identified their faults?**
- **Have I analyzed the strong and weak points of each perspective?**

Defending the decision

By reviewing the research and performing a risk–benefit analysis on the information, we conclude that the benefits of using GM foods are greater than the risks. The production of GM foods will solve an immediate problem of world hunger by producing more food faster, by producing food with a higher nutritional content, and by reducing the need for chemical use in food production. The use of GM foods provides an immediate solution to world hunger, which we consider to be a major benefit for their production. There are, however, calculated risks involved in full-scale GM food production. We may produce new organisms whose long-term effects on an ecosystem cannot be known ahead of time. However, our evidence indicates that the probability of serious damage is minimal when compared with the overall benefits that GM foods will bring to society.

The final phase of decision making includes evaluating the decision itself and the process used to reach the decision. After you have made a decision, carefully examine the thinking that led to your decision.

Some questions to guide your evaluation:

- **What was my initial perspective on the issue? How has my perspective changed since I first began to explore the issue?**
- **How did we make our decision? What process did we use? What steps did we follow?**
- **In what ways does our decision resolve the issue?**
- **What are the likely short- and long-term effects of the decision?**
- **To what extent am I satisfied with the final decision?**
- **What reasons would I give to explain our decision?**
- **If we had to make this decision again, what would I do differently?**

Evaluating the process

This issue came to my attention through the many newspaper articles I had read that warned of a growing danger to people and the environment with the production and sale of GM foods. After researching the issue, I found that the vast majority of recent scientific reports conclude that, while there are some long-term unknowns, the production and consumption of GM products are as safe as those of non-GM foods. In fact, some studies indicate that the production of some organically grown foods could be even more dangerous. While every effort was made to obtain the most current, scientific information on GM foods and their effects on humans, other species, and the environment, the number and quality of studies focusing on the long-term environmental impact of GM foods is relatively low. These studies need to be carried out. By examining such research, we may be able to make better-informed decisions regarding the use of these foods.

Table 1 Stakeholders in the GM Food Debate

Stakeholders	Viewpoint
scientists	GM foods may end world hunger.
farmers	GM crops require less pesticides and herbicides.
doctors	GM foods can be created to solve health problems. For example, children whose major diet staple is rice often have a vitamin A deficiency. Rice can be genetically engineered to produce vitamin A.
environmentalists	Genes from GM crops may spread to create superweeds and superbugs.
politicians	GM foods can end world hunger.
health critics	Genetic alterations may produce substances that are poisonous or that trigger allergies and disease.

Table 2 Perspectives on an Issue

Perspective	Focus of the perspective
cultural	customs and practices of a particular group
ecological	interactions among organisms and their natural habitat
economic	the production, distribution, and consumption of wealth
educational	the effects on learning
emotional	feelings and emotions
environmental	the effects on physical surroundings
esthetic	artistic, tasteful, beautiful
moral/ethical	what is good/bad, right/wrong
legal	the rights and responsibilities of individuals and groups
spiritual	the effects on personal beliefs
political	the effects on the aims of a political group or party
scientific	logical or research information based
social	the effects on human relationships, the community, or society
technological	machines and industrial processes

A

A Risk–Benefit Analysis Model

Risk–benefit analysis is a tool used to organize and analyze information gathered in research. A thorough analysis of the risks and benefits associated with each alternative solution can help you decide on the best alternative.

- Research as many aspects of the proposal as possible. Look at it from different perspectives.
- Collect as much evidence as you can, including reasonable projections of likely outcomes if the proposal is adopted.
- Classify each potential result as being either a benefit or a risk.
- Quantify the size of the potential benefit or risk (perhaps as a dollar figure, as a number of lives affected, or on a scale of 1 to 5).
- Estimate the probability (percentage) of that event occurring.
- By multiplying the size of a benefit (or risk) by the probability of its happening, you can calculate a probability value for each potential result.
- Total the probability values of all the potential risks and all the potential benefits.
- Compare the sums to help you decide whether to accept the proposed action.

Table 3 shows a risk–benefit analysis of the issue. Note that there could be other possible results. Although you should try to be objective in your assessment, the beliefs of the person making the risk–benefit analysis will have an effect on the final sums. The possible outcomes considered for analysis, the assessment of the relative importance of a cost or benefit, and the probability of the cost or benefit actually arising will vary according to who does the analysis. For example, would you agree completely with the values placed in the "Cost" and "Benefit" columns of the analysis in **Table 3**?

Table 3 Risk-Benefit Analysis of Producing Foods from Genetically Modified Organisms

Risks				**Benefits**			
Possible result	**Cost of result (scale of 1 to 5)**	**Probability of result occurring (%)**	**Cost × probability**	**Possible result**	**Benefit of result (scale of 1 to 5)**	**Probability of result occurring (%)**	**Benefit × probability**
GM foods increase human health risks.	very serious 5	research is inconclusive (50%)	250	Food supplies increase and food becomes cheaper.	great 5	very likely (90%)	450
GM crops are more competitive and eliminate natural species.	serious 4	likely (80%)	320	GM foods have higher nutritional value than their natural counter-parts.	great 5	likely (75%)	375
GM species negatively affect other species in the food chain.	serious 4	somewhat likely (60%)	240	GM crops require less chemical fertilizers.	high 4	somewhat likely (60%)	240
Total risk value			**810**	**Total benefit value**			**1065**

A3 Technological Problem Solving

There is a difference between science and technology. The goal of science is to understand the natural world. The goal of technological problem solving is to develop or revise a product or a process in response to a human need. The product or process must fulfill its function but, in contrast with scientific problem solving, it is not essential to understand why or how it works.

Technological solutions are evaluated based on such criteria as simplicity, reliability, efficiency, cost, and ecological and political consequences.

Even though the sequence presented in the graphic below is linear, there are normally many cycles through the steps in any problem-solving attempt.

Process Description

Defining the problem

This process involves recognizing and identifying the need for a technological solution. You need to state clearly the question(s) that you want to investigate to solve the problem and the criteria you will use as guidelines and to evaluate your solution. In any design, some criteria may be more important than others. For example, if a tool measures accurately and is economical, but is not safe, then it is clearly unacceptable.

Identifying possible solutions

Use your prior knowledge, experience, and creativity to propose possible solutions.

During brainstorming, the goal is to generate many ideas without judging them. They can be evaluated and accepted or rejected later.

To visualize the possible solutions, it is helpful to draw sketches. Sketches are often better than verbal descriptions in communicating an idea.

Planning

Planning is the heart of the entire process. Your plan will outline your processes, identify potential sources of information and materials, define your resource parameters, and establish evaluation criteria.

Seven types of resources are generally used in developing technological solutions to problems—people, information, materials, tools, energy, capital, and time.

Example: Aquarium Activity and Fish Death

Defining the problem

You are conducting an investigation to determine how temperature affects respiration rate in mammals. You need an instrument to measure the consumption of oxygen. A respirometer measures oxygen consumption during respiration. Respirometers are commercially available but may be inaccessible because of cost or other limitations. Your challenge then is to design a respirometer that will enable you to measure oxygen consumption with a degree of accuracy sufficient to answer your scientific question.

Identifying possible solutions

Generate as many ideas as possible about the functioning of a respirometer and about potential designs.

Planning

In this problem, you may need, and be restricted by, the following resources:
People—yourself, your partner(s), and your biology teacher
Information—You already understand the biochemical process of respiration. You need to understand how a respirometer works. What chemical reactions occur? What variable is measured as an indicator of oxygen consumption? In what units do you measure oxygen consumption?
Materials—Your choice of materials is limited by the proposed design of your respirometer, cost, availability, and time. You are restricted to materials that can be obtained in school or at home. Materials to consider include a large tray, beaker or flask, rubber stopper, rubber tubing, glass tubing, pipette, graduated cylinder, plastic bottles, carbon dioxide (CO_2) absorbent such as potassium hydroxide (KOH) solution or pellets, plasticene, or silicone caulking.
Tools—The tools you will need will depend on your proposed design. For this problem, you are likely to require only a knife and scissors. Your design should not require any specialized tools or machines.
Energy—The only energy requirement for this problem is your own energy.
Capital—The capital resources must be minimal; otherwise, it may be more efficient to purchase a commercial model.
Time—Because of the time limit on the scientific investigation, there is an even shorter time limit on the construction of a respirometer that will be used to collect data in the investigation. You should be able to construct your respirometer within 30 min.
The following are design criteria:

- large enough to contain a small mammal (e.g., mouse or gerbil)
- accurate to at least +/− 5 mL/min
- must incorporate a thermometer
- can be reused without significant maintenance
- safe for animal subject

A

In this phase, you will construct and test your prototype using trial and error. Try to change only one variable at a time. Use failures to inform the decisions you make before your next trial. You may also complete a cost–benefit analysis on the prototype.

To help you decide on the best solution, you can rate each potential solution on each of the design criteria using a five-point rating scale, with 1 being poor, 2 fair, 3 good, 4 very good, and 5 excellent. You can then compare your proposed solutions by totalling the scores.

Once you have made the choice among the possible solutions, you need to produce and test a prototype. While making the prototype, you may need to experiment with the characteristics of different components. A model, on a smaller scale, might help you decide whether the product will be functional. The test of your prototype should answer three basic questions:

- **Does the prototype solve the problem?**
- **Does it satisfy the design criteria?**
- **Are there any unanticipated problems with the design?**

If these questions cannot be answered satisfactorily, you may have to modify the design or select another potential solution.

In presenting your solution, you will communicate your solution, identify potential applications, and put your solution to use.

Once the prototype has been produced and tested, the best presentation of the solution is a demonstration of its use—a test under actual conditions. This demonstration can also serve as a further test of the design. Any feedback should be considered for future redesign. Remember that no solution should be considered the absolute final solution.

Evaluation is not restricted to the final step. However, it is important to evaluate the final product using the criteria established earlier, and to evaluate the processes used while arriving at the solution. Consider the following questions:

- **To what degree does the final product meet the design criteria?**
- **Did you have to make any compromises in the design? If so, are there ways to minimize the effects of the compromises?**
- **Did you exceed any of the resource parameters?**
- **Are there other possible solutions that deserve future consideration?**
- **How did your group work as a team?**

→ **Constructing/testing solutions** → **Presenting the preferred solution** → **Evaluating the solution and process**

Table 1 illustrates a rating for two different respirometer designs. Note that although Design 1 came out with the highest rating, there is one factor (subject safety) that suggests that we should go with Design 2. This is what is referred to as a tradeoff. We have to compromise on criteria such as size and accuracy in order to ensure that the apparatus is safe for the subject specimen. By reviewing or evaluating the product and the process so far, we may be able to modify Design 2 to optimize its performance in meeting the other criteria.

Table 1 Rating for Respirometer Designs

Design criterion	Design 1	Design 2
size	4	3
accuracy	4	3
thermometer	5	3
ease of reuse	4	3
subject safety	2	5
total score	**19**	**17**

1—poor, 2—fair, 3—good, 4—very good, 5—excellent

Design 1 scored higher than Design 2; however, we feel that the safety of the subject is a very important factor that must be addressed in our design. While our best effort went into designing and creating a respirometer, a comparison between our instrument and a commercial model is recommended. By comparing the results between our instrument and a commercial model, an estimation of our instrument's accuracy can be determined. As a final step, our design could be improved, thus justifying the creation and use of our instrument over the purchase and use of a commercial model.

A4 Lab Reports

When carrying out investigations, it is important that scientists keep records of their plans and results, and share their findings. In order to have their investigations repeated (replicated) and accepted by the scientific community, scientists generally share their work by publishing papers in which they provide details of their design, materials, procedure, evidence, analysis, and evaluation.

Lab reports are prepared after an investigation is completed. To ensure that you can accurately describe the investigation, it is important to keep thorough and accurate records of your activities as you carry out the investigation.

Investigators use a similar format in their final reports or lab books, although the headings and order may vary. Your lab book or report should reflect the type of scientific inquiry that you used in the investigation and should be based on the following headings, as appropriate. (See **Figure 1**, on pages 448–450, for a sample lab report.)

Title

At the beginning of your report, write the section number and title of your investigation. In this course, the title is usually given, but if you are designing your own investigation, create a title that suggests what the investigation is about. Include the date the investigation was conducted and the names of all lab partners (if you worked as a team).

Purpose

State the purpose of the investigation. Why are you doing this investigation?

Question

This is the Question that you attempted to answer in the investigation. If it is appropriate to do so, state the Question in terms of independent and dependent variables.

Hypothesis/Prediction

Based on your reasoning or on a concept that you have studied, formulate an explanation of what should happen (a hypothesis). From your hypothesis you may make a prediction, a statement of what you expect to observe, before carrying out the investigation. Depending on the nature of your investigation, you may or may not have a hypothesis or a prediction.

Experimental Design

This is a brief general overview (one to three sentences) of what was done. If your investigation involved independent, dependent, and controlled variables, list them. Identify any control or control group that was used in the investigation.

Materials

This is a detailed list of all materials used, including sizes and quantities where appropriate. Be sure to include safety equipment such as goggles, laboratory apron, gloves, and tongs, where needed. Draw a diagram to show any complicated set-up of apparatus.

Procedure

Describe, in detailed, numbered steps, the procedure you followed to carry out your investigation. Include steps to clean up and dispose of waste.

Observations

This includes all qualitative and quantitative observations you made. Be as precise as possible when describing quantitative observations, include any unexpected observations, and present your information in a form that is easily understood. If you have only a few observations, this could be a list; for controlled experiments and for many observations, a table would be more appropriate.

Analysis

Interpret your observations and present the evidence in the form of tables, graphs, or illustrations, each with a title. Include any calculations, the results of which can be shown in a table. Make statements about any patterns or trends you observed. Conclude the analysis with a statement based only on the evidence you have gathered, answering the question that initiated the investigation.

Evaluation

The evaluation is your judgment about the quality of evidence obtained and about the validity of the prediction and hypothesis (if present). This section can be divided into two parts—evaluation of the investigation and evaluation of the prediction (and hypothesis). The following questions and suggestions should help you in each part of the process.

Evaluation of the Investigation

- Did the design enable you to answer the question?
- As far as you know, is the design the best available or are there flaws that could be corrected?
- Were the steps in the investigation in the correct order and adequate to gather sufficient evidence?
- What steps, if done incorrectly, could have significantly affected the results?
- What improvements could be made to the procedure?

Sum up your conclusions about the procedure in a statement that begins like this: "The procedure is judged to be adequate/inadequate because..."

- What specialized skills (e.g., measuring) might have an effect on the results?
- Was the evidence from repeated trials reasonably similar?
- Can the measurements be made more precise?

Sum up your conclusions about the required skills in a statement that begins like this: "The skills are judged to be adequate/inadequate because..."

- What are the sources of uncertainty and error in my investigation?
- Based on any uncertainties and errors you have identified, do you have enough confidence in your results to proceed with the evaluation of the prediction and hypothesis?

State your confidence level in a statement like this: "Based on my evaluation of the investigation, I am certain/I am moderately certain/I am very certain of my results."

Evaluation of the Prediction (and Hypothesis)

- Does the predicted answer clearly agree with the answer in your analysis?
- Can any difference be accounted for by the sources of uncertainty or error listed earlier in the evaluation?

Sum up your evaluation of the prediction in a statement that begins like this: "The prediction is judged to be verified/inconclusive/falsified because..."

- Is the hypothesis supported by the evidence?
- Is there a need to revise the hypothesis or to replace it with a new hypothesis?

If the prediction was verified, the hypothesis behind it is supported. If the results were inconclusive or the prediction is falsified, then the hypothesis is questionable. Sum up your evaluation of the hypothesis in a statement that begins like this: "The hypothesis being tested is judged to be acceptable/unacceptable because..."

Investigation 2.5—Movement of Water Into and Out of Cells
April 15, 2015

By Barry La Drew and Eileen Jong

Purpose
The purpose of this investigation is to examine the conditions under which water moves into and out of cells, using dialysis tubing as a model of a cell membrane.

Question
Under what conditions does water move into and out of cells?

Hypothesis/Prediction
If a model cell is placed in a hypotonic solution (having lower concentrations of solute than the cell), water will move into the cell. If a model cell is placed in a hypertonic solution (having higher concentrations of solute than the cell), water will move out of the cell.

Experimental Design
Three model cells are constructed from dialysis tubing. Two are filled with distilled water and one with starch suspension. One of the cells with distilled water is placed in a beaker of starch suspension. The other two cells are placed in beakers of distilled water. Iodine is added to the distilled water in the beakers. (Iodine is used as an indicator to detect the presence of starch.) The mass of each cell is recorded and other observations are made at the beginning and after 10 min and 20 min.

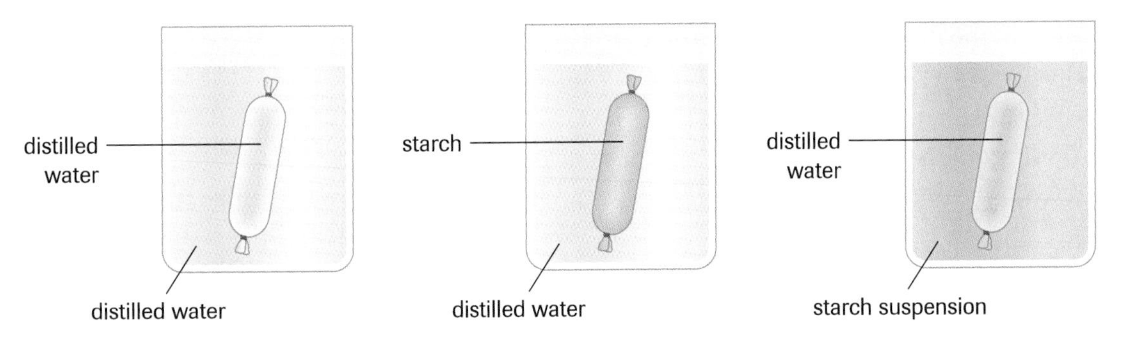

Materials
laboratory apron
medicine dropper
funnel
scissors
100-mL graduated cylinder
three 250-mL beakers
triple-beam balance
gloves
dialysis tubing
paper towels
distilled water
4% starch suspension
iodine

Figure 1
Sample lab report

A

Procedure

1. A laboratory apron and gloves were put on.
2. Three strips of dialysis tubing (each about 25 cm long) were cut. The strips were then soaked in a beaker of tap water for approximately 2 min.
3. To find the opening, one end of the dialysis tube was rubbed between our fingers. A knot was tied near the other end of the tube. These steps were repeated for the other tubes.
4. Using a graduated cylinder, 15 mL of 4% starch suspension was measured. The suspension was poured through a funnel into the open end of one of the dialysis tubes.
5. The second and third dialysis tubes were filled with 15 mL of distilled water. In one tube, 20 drops of iodine were added.
6. A knot was tied in the open end of each tube to close it.
7. The outside of all the tubes was rinsed with distilled water to remove any fluids that may have leaked out during the tying process.
8. Excess water was gently blotted from the dialysis tubes and the mass of each tube was measured.
9. An observation table similar to **Table 1** was constructed and the measurements recorded.
10. The dialysis tube with the starch suspension was placed into a beaker containing 100 mL of distilled water. The tube with only distilled water was placed in a second beaker containing 100 mL of distilled water. The tube with distilled water and iodine was placed in a beaker containing 100 mL of starch suspension.
11. Twenty drops of iodine were added to each of the beakers containing distilled water. All three tubes were observed closely for any colour change at the beginning.
12. After 10 min, the dialysis tubes were removed from the beakers. Any excess liquid from the tubes was gently blotted off and the mass of each tube measured and recorded.
13. The dialysis tubes were then returned to the appropriate beakers.
14. After another 10 min, the tubes were removed again and blotted dry, and the mass of each was measured and recorded.
15. The contents of the tubes were poured down the sink and the sink was rinsed. The tubes were placed in the regular garbage disposal.

Observations

Table 1 Observation Table

Model cell	Initial mass (g)	Mass after 10 min (g)	Mass after 20 min (g)	Other observations
dialysis tube with distilled water in beaker of distilled water	17.5	17.5	17.5	no observable change in the solution inside the tube
dialysis tube with starch suspension in beaker of distilled water	24.7	25.3	28.4	suspension inside the dialysis tube turned dark
dialysis tube with distilled water in beaker of starch suspension	18.5	17.8	15.7	starch suspension in the beaker turned dark

Analysis

- The tube with the distilled water in the beaker containing distilled water showed no change in mass.
- The tube with the starch suspension increased in mass during each 10-min period. The solution in this tube turned dark.
- The tube with the distilled water in the beaker with the starch suspension decreased in mass during each 10-min period. The starch suspension in the beaker turned dark.

The mass of the cell with the starch suspension inside increased because water moved into the cell. The suspension in this tube turned dark, which suggests that iodine also must have moved into the tube along with the water molecules.

The mass of the cell with the starch suspension outside decreased because water moved out of the cell. The suspension outside the cell turned dark, which suggests that iodine also must have moved outside the tube along with the water molecules.

The tube with the distilled water acted as a control. It had the same solution inside as outside. There was no observable change in it.

From our observations, we can conclude that water will move out of a cell in a hypertonic solution and into a cell in a hypotonic solution.

Evaluation

The design is adequate because the observations allowed us to reach a conclusion about the conditions under which water moves into and out of cells. The procedure is easy to follow and does not require any special skills.

Water or starch suspension remaining on the tubes after they were taken out of the beakers may have affected the measurement of the mass of the tubes. However, this should not make a difference since this error would have affected the mass of all tubes. Errors in measurement would probably not have accounted for the increase in the mass of the tube with the starch or the decrease in the mass of the tube with distilled water. So, while there may have been errors in measurement, they should not have affected the quality of the evidence.

The evidence supports the prediction. To verify our conclusion, we could repeat the investigation or do a similar investigation with different solutions.

Since we were working with model cells, we cannot be sure that this would happen with real cells.

A5 Graphic Organizers

Graphic organizers such as those outlined in this section can help you to understand a topic, and assist you in forming a clear, concise answer.

PMI Chart

A PMI chart is used to examine both sides of an issue. Positive aspects of a topic or issue are recorded in the P (plus) column. Negative aspects are recorded in the M (minus) column. Interesting or controversial questions are recorded in the I (interesting) column (**Table 1**).

Table 1 A PMI Chart

P	M	I

Table 2 A KWL Chart

K	W	L

KWL Chart

A KWL chart can help you identify prior knowledge and experience, decide what new information you want to learn about, and reflect on your learning. Before you begin a new concept, lesson, or unit, list what you know about a topic in the K column and what you want to know in the W column. After studying the new topic, list what you learned in the L column (**Table 2**).

Venn Diagram

A Venn diagram is used to show similarities and differences in two or more concepts. Write all similarities between the concepts in the overlapping section of the circles and all unique traits of each concept in the nonoverlapping parts of the appropriate circles (**Figure 1**).

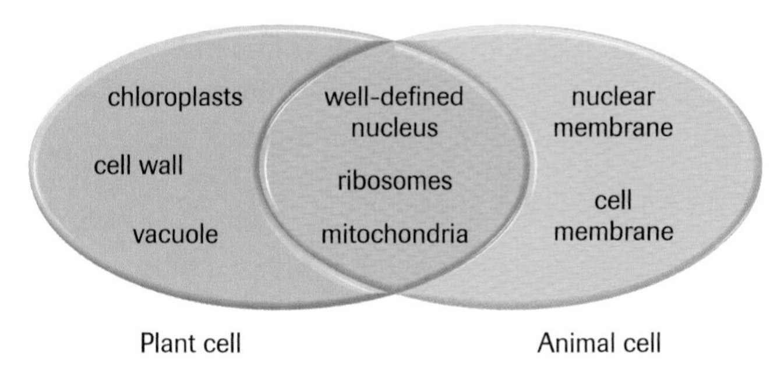

Figure 1
Venn diagram: plant and animal cells

Fishbone Diagram

A fishbone diagram is used to identify separate causes and effects. In the head of the fish, identify the effect, topic, or result. At the end of each major bone, identify the major subtopics or categories. On the minor bones that attach to each major bone, add details about the subtopics or possible causes of each effect or result (**Figure 2**).

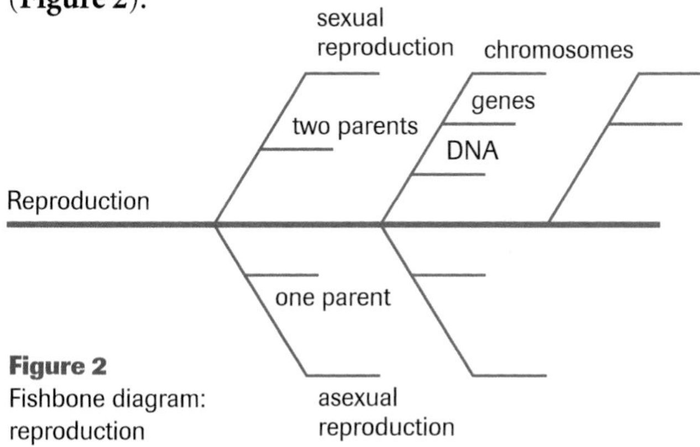

Figure 2
Fishbone diagram: reproduction

The Concept Map

Concept maps are used to show connections between ideas and concepts, using words or visuals. Put the central idea in the middle of a sheet of paper. Organize the ideas most closely related to each other around the centre. Draw arrows between the ideas that are related. On each arrow, write a short description of how the terms are related to each other (**Figure 3**).

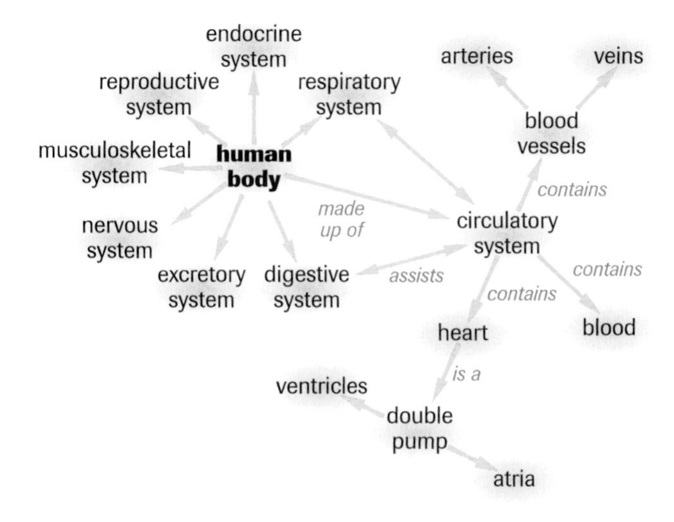

Figure 3
Concept map: human body

A6 Math Skills

Scientific Notation

It is difficult to work with very large or very small numbers when they are written in common decimal notation. Usually it is possible to accommodate such numbers by changing the SI prefix so that the number falls between 0.1 and 1000; for example, 237 000 000 mm can be expressed as 237 km and 0.000 000 895 kg can be expressed as 0.895 mg. However, this prefix change is not always possible, either because an appropriate prefix does not exist or because it is essential to use a particular unit of measurement. In these cases, the best method of dealing with very large and very small numbers is to write them using scientific notation. Scientific notation expresses a number by writing it in the form $a \times 10^n$, where $1 < |a| < 10$ and the digits in the coefficient a are all significant. **Table 1** shows situations where scientific notation would be used.

Table 1 Examples of Scientific Notation

Expression	Common decimal notation	Scientific notation
124.5 million kilometres	124 500 000 km	1.245×10^8 km
6 billion nanometres	0.000 000 006 nm	6×10^{-9} nm
5 gigabytes	5 000 000 000 bytes	5×10^9 bytes

To multiply numbers in scientific notation, multiply the coefficients and add the exponents; the answer is expressed in scientific notation. Note that when writing a number in scientific notation, the coefficient should be between 1 and 10 and should be rounded to the same certainty (number of significant digits) as the measurement with the least certainty (fewest number of significant digits). Look at the following examples:

$$(4.73 \times 10^5 \text{ m})(5.82 \times 10^7 \text{ m}) = 27.5 \times 10^{12} \text{ m}^2 = 2.75 \times 10^{13} \text{ m}^2$$

$$(3.9 \times 10^4 \text{ N})(5.3 \times 10^{-3} \text{ m}) = 21 \times 10^1 \text{ N}\bullet\text{m} = 2.1 \times 10^2 \text{ N}\bullet\text{m}$$

On many calculators, scientific notation is entered using a special key, labelled EXP or EE. This key includes "× 10" from the scientific notation; you need to enter only the exponent. For example, to enter

7.5×10^4 press 7.5 EXP 4

3.6×10^{-3} press 3.6 EXP +/−3

Uncertainty in Measurements

There are two types of quantities that are used in science: exact values and measurements. Exact values include defined quantities (1 m = 100 cm) and counted values (5 cars in a parking lot). Measurements, however, are not exact because there is some uncertainty or error associated with every measurement.

There are two types of measurement error. **Random error** results when an estimate is made to obtain the last significant figure for any measurement. The size of the random error is determined by the precision of the measuring instrument. For example, when measuring length, it is necessary to estimate between the marks on the measuring tape. If these marks are 1 cm apart, the random error will be greater and the precision will be less than if the marks are 1 mm apart.

Systematic error is associated with an inherent problem with the measuring system, such as the presence of an interfering substance, incorrect calibration, or room conditions. For example, if the balance is not zeroed at the beginning, all measurements will have a systematic error; if using a metre stick that has been worn slightly, all measurements will contain an error.

The precision of measurements depends on the gradations of the measuring device. **Precision** is the place value of the last measurable digit. For example, a measurement of 12.74 cm is more precise than a measurement of 12.7 cm because the first value was measured to hundredths of a centimetre, whereas the latter was measured to tenths of a centimetre.

When adding or subtracting measurements of different precision, the answer is rounded to the same precision as the least precise measurement. For example, using a calculator, add

11.7 cm + 3.29 cm + 0.542 cm = 15.532 cm

The answer must be rounded to 15.5 cm because the first measurement limits the precision to a tenth of a centimetre.

No matter how precise a measurement is, it still may not be accurate. **Accuracy** refers to how close a

(a)

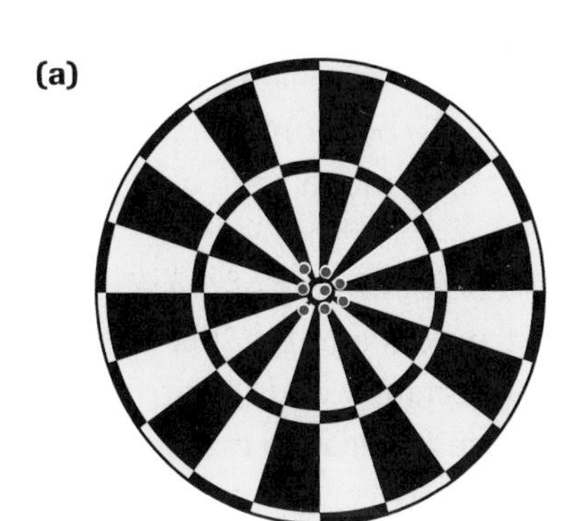

(b)

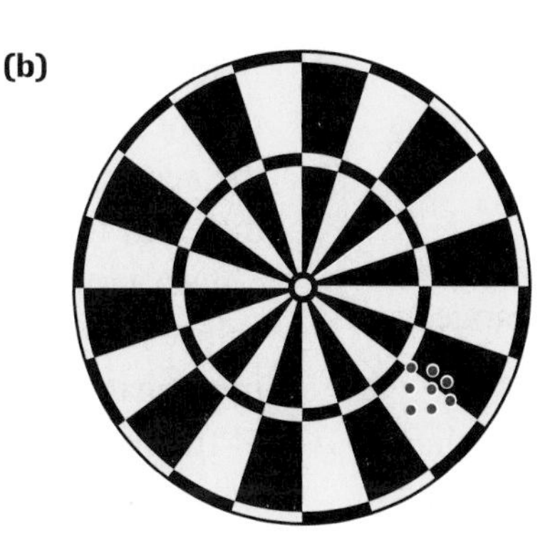

(c)

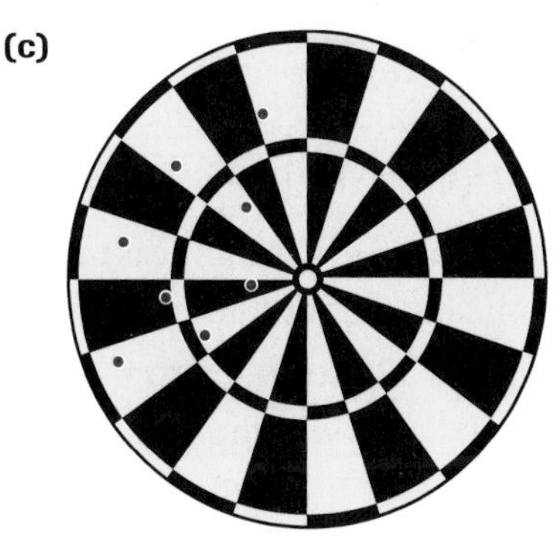

Figure 1
The positions of the darts in each of these figures are analogous to measured or calculated results in a laboratory setting. The results in **(a)** are precise and accurate, in **(b)** they are precise but not accurate, and in **(c)** they are neither precise nor accurate.

value is to its true value. The comparison of the two values can be expressed as a percentage difference. The percentage difference is calculated as follows:

$$\% \text{ difference} = \frac{|\text{experimental value} - \text{predicted value}|}{\text{predicted value}} \times 100\%$$

Figure 1 shows an analogy between precision and accuracy, and the positions of darts thrown at a dartboard.

How certain you are about a measurement depends on two factors: the precision of the instrument used and the size of the measured quantity. More precise instruments give more certain values. For example, a mass measurement of 13 g is less precise than a measurement of 12.76 g; you are more certain about the second measurement than the first. Certainty also depends on the measurement. For example, consider the measurements 0.4 cm and 15.9 cm; both have the same precision. However, if the measuring instrument is precise to ± 0.1 cm, the first measurement is 0.4 ± 0.1 cm (0.3 cm or 0.5 cm) for an error of 25%, whereas the second measurement could be 15.9 ± 0.1 cm (15.8 cm or 16.0 cm) for an error of 0.6%. For both factors—the precision of the instrument used and the value of the measured quantity—the more digits there are in a measurement, the more certain you are about the measurement.

Significant Digits

The certainty of any measurement is communicated by the number of significant digits in the measurement. In a measured or calculated value, significant digits are the digits that are certain plus one estimated (uncertain) digit. Significant digits include all digits correctly reported from a measurement.

Follow these rules to decide if a digit is significant:

1. If a decimal point is present, zeros to the left of the first non-zero digit (leading zeros) are not significant.
2. If a decimal point is not present, zeros to the right of the last non-zero digit (trailing zeros) are not significant.
3. All other digits are significant.
4. When a measurement is written in scientific notation, all digits in the coefficient are significant.
5. Counted and defined values have infinite significant digits.

Table 2 shows some examples of significant digits.

Table 2 Certainty in Significant Digits

Measurement	Number of significant digits
32.07 m	4
0.0041 g	2
5×10^5 kg	1
6400 s	2
100 people	3

An answer obtained by multiplying and/or dividing measurements is rounded to the same number of significant digits as the measurement with the fewest number of significant digits. For example, we could use a calculator to solve the following equation:

$$77.8 \text{ km/h} \times 0.8967 \text{ h} = 69.76326 \text{ km}$$

However, the certainty of the answer is limited to three significant digits, so the answer is rounded up to 69.8 km.

Rounding Off

The following rules should be used when rounding answers to calculations.

1. When the first digit discarded is less than five, the last digit retained should not be changed.
   3.141 326 rounded to 4 digits is 3.141
2. When the first digit discarded is greater than five, or if it is a five followed by at least one digit other than zero, the last digit retained is increased by 1 unit.
   2.213 724 rounded to 4 digits is 2.214
   4.168 501 rounded to 4 digits is 4.169
3. When the first digit discarded is five followed by only zeros, the last digit retained is increased by 1 if it is odd, but not changed if it is even.
   2.35 rounded to 2 digits is 2.4
   2.45 rounded to 2 digits is 2.4
   −6.35 rounded to 2 digits is −6.4

Measuring and Estimating

Many people believe that all measurements are reliable (consistent over many trials), precise (to as many decimal places as possible), and accurate (representing the actual value). But there are many things that can go wrong when measuring.

- There may be limitations that make the instrument or its use unreliable (inconsistent).
- The investigator may make a mistake or fail to follow the correct techniques when reading the measurement to the available precision (number of decimal places).
- The instrument may be faulty or inaccurate; a similar instrument may give different readings.

For example, when measuring the temperature of a liquid, it is important to keep the thermometer at the proper depth and the bulb of the thermometer away from the bottom and sides of the container. If you sit a thermometer with its bulb at the bottom of a liquid-filled container, you will be measuring the temperature of the bottom of the container and not the temperature of the liquid. There are similar concerns with other measurements.

To be sure that you have measured correctly, you should repeat your measurements at least three times. If your measurements appear to be reliable, calculate the mean and use that value. To be more certain about the accuracy, repeat the measurements with a different instrument.

Every measurement is a best estimate of the actual value. The measuring instrument and the skill of the investigator determine the certainty and the precision of the measurement. The usual rule is to make a measurement that estimates between the smallest divisions on the scale of the instrument.

Graphs

There are many types of graphs that you can use to organize your data. You need to identify which type of graph is best for your data before you begin graphing. Three of the most useful kinds are bar graphs, circle (pie) graphs, and point-and-line graphs.

Bar Graphs

When at least one of the variables is qualitative, use a bar graph to organize your data (**Figure 2**). For example, a bar graph would be a good way to present the data collected from a study of the number of plants (quantitative) and the type of plants found (qualitative) in a local nursery. In this graph, each bar stands for a different category, in this case a type of plant.

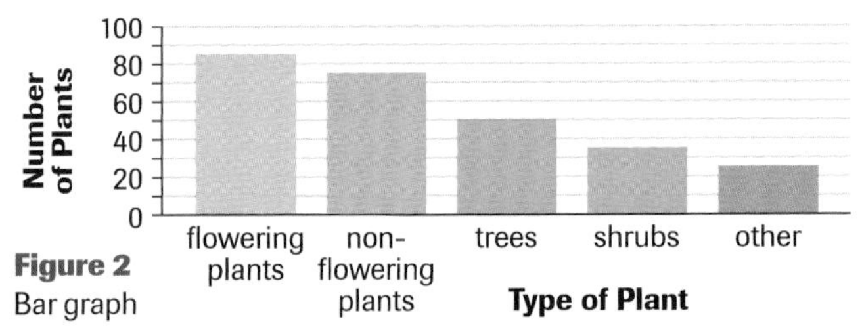

Figure 2
Bar graph

Circle Graphs

Circle graphs and bar graphs are used to present similar types of data. A circle graph is used if the quantitative variable can be changed to a percentage of a total quantity (**Figure 3**). For example, if you surveyed a local nursery to determine the types of plants found and the number of each, you could make a circle graph. Each piece in the graph stands for a different category (e.g., the type of plant). The size of each piece is determined by the percentage of the total that belongs in each category.

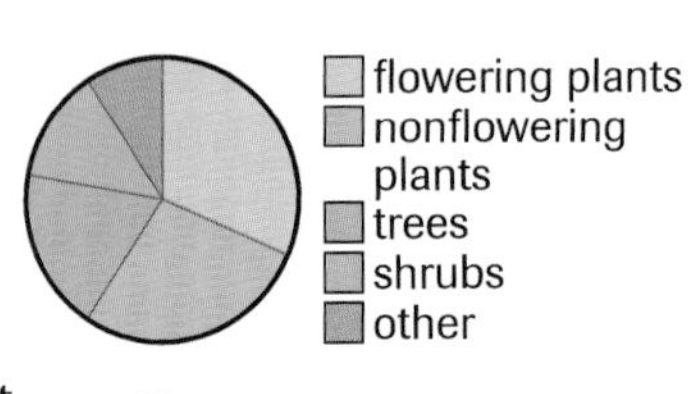

Figure 3
Circle graph

Point-and-Line Graphs

When both variables are quantitative, use a point-and-line graph. For example, we can use the guidelines below and the data in **Table 3** to construct the point-and-line graph shown in **Figure 4**.

Table 3 Number of Brine Shrimp Eggs Hatched in Salt Solutions of Various Concentrations

Day	2% salt	4% salt	6% salt	8% salt
1	0	0	0	1
2	0	11	2	3
3	0	14	8	5
4	2	20	17	8
5	5	37	25	15
6	6	51	37	31

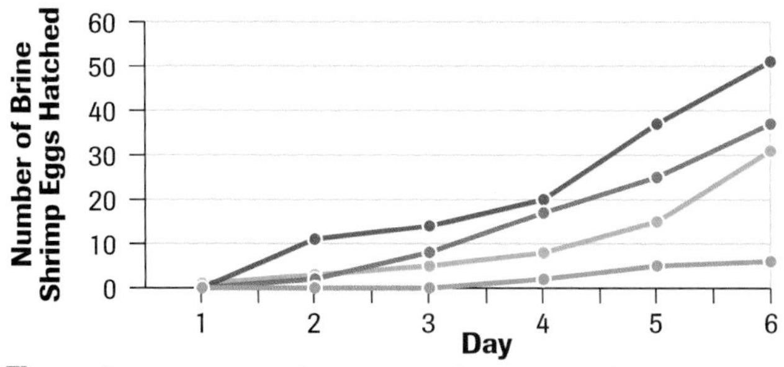

Figure 4 Point-and-line graph

1. Use graph paper and construct your graph on a grid. The horizontal edge on the bottom of this grid is the *x*-axis and the vertical edge on the left is the *y*-axis. Remember that a larger graph is easier to interpret.
2. Decide which variable goes on which axis and label each axis, including the units of measurement. The independent variable is generally plotted along the *x*-axis and the dependent variable along the *y*-axis. The exception to this is when you plot a variable against time: regardless of which is the independent or dependent variable, always plot time on the *x*-axis. This convention ensures that the slope of the graph always represents a rate.
3. Title your graph. The title should be a concise description of the data contained in the graph.
4. Determine the range of values for each variable. The range is the difference between the largest and smallest values. Graphs often include a little extra length on each axis, to make them appear less cramped.
5. Choose a scale for each axis. This will depend on how much space you have and the range of values for each axis. Each line on the grid usually increases steadily in value by a convenient number, such as 1, 2, 5, 10, or 50.
6. Plot the points. Start with the first pair of values, which may or may not be at the origin of the graph.
7. After all the points are plotted, draw a line through the points to show the relationship between the variables, if possible. Not all points may lie exactly on a line; small errors in each measurement may have occurred and moved the points away from the perfect line. Draw a line that comes closest to most of the points. This is called the line of best fit—a smooth line that passes through or between the points so that there are about the same number of points on each side of the line. The line of best fit may be straight or curved.
8. If you are plotting more than one set of data on one graph, use different colours or symbols to indicate the different sets, and include a legend (**Figure 4**).

Spreadsheets

A spreadsheet is a useful tool for creating different graph types such as column, bar, line, and pie to display data.

The following steps show how the data for the plants in a local nursery can be used to create a column (or bar) graph.

Step 1

Enter the data in the spreadsheet.

Step 2

Select the graph type that will display the data in the most appropriate form.

Step 3

Select the data from the spreadsheet that will be plotted on the graph.

Step 4

Enter the remaining graph information, such as the graph title and axes titles.

Step 5

Indicate where the graph is to be located in the spreadsheet. It can be included on the same sheet as the data or on a new sheet.

A7 Research Skills

General Research

There is an incredible amount of information available to us, from many different sources. We now have the ability to access more information than at any other time in history. Information is simply information, however. Before you can make effective use of it, you must know how to gather information efficiently and how to assess its credibility.

Collecting Information

- Before you begin your research, list the most important words associated with your research, so you can search for appropriate topics.
- Brainstorm a list of possible resources. Consider all the sources of information available to you. Rank the list, starting with the most useful resource.
- Search out and collect information from a variety of resources.
- Ask yourself, "Do I understand what this resource is telling me?"
- Check when the resource was published. Is it up to date?
- Consider the source of the information. From what perspective is it written? Is it likely to be biased?
- Keep organized notes or files while doing your research.
- Keep a complete list of the resources you used, so you can quickly find the source again if you need to, and so you can make a bibliography when writing your lab report.
- Review your notes. After your research, you may want to alter your original position or hypothesis, or research in a slightly different direction.

Assessing the Credibility of Information Sources

Understanding and evaluating the work of others is an important part of research. Think about how many messages, opinions, and pieces of information you hear and see every day. When you do research, you may access information from the Internet, textbooks, magazines, chat lines, television, radio, and through many other forms of communication. Is all of this information "correct"? Are all of these information sources "reliable"? How do we know what to believe and what not to believe? Often science, or the appearance of science, is used to convince us that claims are true. Sometimes this method of reporting is used to encourage us to buy something or just to catch our interest. Even serious stories on scientific work are sometimes difficult to interpret.

To analyze information, you have to use your mind effectively and critically. When you encounter a "scientific" report in the media, analyze the report carefully, and see if you can identify the following:

- the type of investigation that is being reported
- the dependent and independent variables in any reported investigation
- the strengths and weaknesses in the design of the investigation

PERCS

A useful framework for evaluating the credibility of information gathered from various sources is PERCS (**Table 1**). This framework, developed at Central Park East Secondary School in New York City, NY, uses a series of questions to critically assess information and arguments concerning an issue. These questions can help you to evaluate the information you collect.

Table 1 The PERCS Checklist

Perspective	From whose viewpoint are we seeing or reading or hearing? From what angle or perspective?
Evidence	How do we know what we know? What's the evidence and how reliable is it?
Relevance	So what? What does it matter? What does it all mean? Who cares?
Connections	How are things, events, or people connected to one another? What is the cause and what is the effect? How do they "fit" together?
Supposition	What if...? Could things be otherwise? What are or were the alternatives? Suppose things were different.

A

Internet Research

The world wide web's accessibility and ease of use has led to an amazing increase in the amount of information available to us. However, as a research tool, the Web lacks the quality assurance that editors provide with print publications. In other words, anybody can post just about anything on the Web, without any proof of authenticity.

There is a huge variety of information on the Internet (facts, opinions, stories, interpretations, and statistics) created for many purposes (to inform, to persuade, to sell, to present a viewpoint, and to create or change an attitude or belief). For each of these various kinds and purposes, information exists on many levels of quality or reliability: from very good to very bad with every shade in between. Given this variability, it is crucial that you critically evaluate the material you find.

Search Strategy

Before you begin searching the Internet, you should think about the information you are searching for. What is your topic? What are the key concepts in your Question? What words would best describe your subject? Try to be as precise as possible. Are there other ways that you can express these key concepts? When you have answered these questions, you will have a list of search terms to start with. Be willing to add to and subtract from your list as you evaluate what you have found to see if it is relevant and useful. (Later we will explore how to evaluate sources.)

Table 2 shows three primary ways of searching the Internet for documents and web pages.

Table 2 Ways of Searching the Internet

Search engine	Meta search engine	Subject gateway (or directory)	E-mail, discussion lists, databases
Searches using keywords that describe the subject you are looking for.	Enables you to search across many search engines at once.	Provides an organized list of web pages, divided into subject areas. Some gateways are general and cover material on many subjects.	Puts you in touch with individuals who are interested in your research topic.
AltaVista Canada ca.altavista.com	MetaCrawler www.metacrawler.com	About www.about.com	
Lycos Canada www.lycos.ca	Search.com www.search.com	Looksmart www.looksmart.com	
Google www.google.ca	Ask Jeeves www.ask.com	Yahoo www.ca.yahoo.com	
Go.com infoseek.go.com	Dogpile www.dogpile.com	Librarians' Index to the Internet lii.org	
HotBot www.hotbot.com	Highway 61 www.highway61.com	Infomine infomine.ucr.edu	
Webcrawler www.webcrawler.com		WWW Virtual Library vlib.org/overview.html	

Search Engines

Search engines are automated programs that create an index of web pages in the Internet. When you use a search engine, you are not actually searching the Internet; you are searching through this index. The larger search engines currently index more than 100 million web pages each, but no search engine indexes every single web page. Most search engines look through, and index, all the words in the web pages they find. When you type some words in a search engine, it searches through its index to find web pages that contain as many of those words as it can. Search engines try to rank the pages so that the most relevant results are shown first. Most search engines provide online help/search tips. Always look at these to find tips for better searching.

Use a search engine to look for a specific web site or web page, information about a specific company or organization, or a specific subject. A very general subject search will probably produce too many results to be useful.

There are several reasons why you might not find what you are looking for. Although search engines update their indexes frequently, it can take several months for the engine to crawl the Web to find new web pages. If the page you are looking for is new, it might not be indexed yet.

You will probably have a favourite search engine, but if you do not find what you are looking for, try using another. Many web pages are indexed in only a couple of search engines. If you are not having much success, you could try using a meta search engine.

Meta Search Engines

Meta search engines allow you to search across several search engines at the same time. They take your search and run it in several search-engine indexes. They will only return the top few results (usually between six and ten) from each search engine.

Use a meta search engine to look for a specific web site or web page, information about a specific company or organization, or a specific subject. If you do not find what you are looking for, try searching individual search engines. For general subject searching, it might be better to use a subject gateway.

Subject Gateways

Subject gateways are directories of web resources on particular subjects. Usually the gateways are compiled by subject experts and may be published by professional societies, universities, or libraries. They contain particularly useful and important resources, possibly including databases and electronic journals. Use a subject gateway when you want resources in a particular subject area. Although subject gateways contain far fewer resources than search engines, the resources have usually been selected and evaluated by subject experts, so they are often useful places to start.

E-mail, Discussion Lists, and Databases

Although there are directories (or lists) of e-mail addresses on the Internet, which can be searched free of charge, none of them is comprehensive. Just because you do not find a person's e-mail address in the directories does not necessarily mean that the address doesn't exist. If you know the organization for which someone works, try its home page to see if an internal directory is available.

E-mail discussion lists are an extremely useful resource for researchers. They enable you to communicate with, and ask questions of, other people who are interested in similar topics. When you send an e-mail message to an e-mail discussion list, it is forwarded to all the people who are members of that particular list. Many lists maintain archives of past discussions.

There are different types of discussion lists. An open list is open to anyone, and the messages are forwarded automatically without human intervention. A closed list is available to only certain people, such as the employees of a company. A moderated list is monitored by a person, often known as the list owner, who decides whether or not to forward the messages to other members.

There are many free databases of reference material on the Internet. A useful directory to them is www.isleuth.com, which allows you to search multiple databases simultaneously for maximum results.

Search Results

Once you have done a search, you will be confronted with a list of web pages, and a number (often very high)

A

of "matches" for your search. This can seem daunting at first, but do not be put off. The most relevant pages should appear at the top of the list. There is often some information to help you decide which pages to look at in detail. You can always refine your search to reduce the number of "matches" you receive. There are several ways of refining or improving your search.

- Use more search terms to get fewer, more relevant records.
- Use fewer search terms to get more records.
- Search for phrases (words next to each other in the order you specified) by enclosing search terms in quotation marks (e.g., "robert menzies").
- Choose search engines that allow you to refine your search results (e.g., AltaVista).
- Limit your searches to Canadian sites by using local search engines, or limit your searches to sites with .ca at the end of their domain name.
- Use Boolean operators: + (an essential term) and —(a term that should be excluded).

Every page on the Web has a unique address or URL (Universal Resource Locator). Looking at the URL can help you decide whether or not a page will be useful. The URL sometimes tells you the name of the organization hosting the site or can give a clue that you are viewing a personal page (often indicated by a ~ symbol in the URL). Some organizations are likely to provide more reliable information than others. The address includes a domain name, which also contains clues to the organization hosting the web page (**Table 3**). For example, the URL weatheroffice.ec.gc.ca/canada_e.html is a page showing a weather map of Canada. "ec.gc.ca" is the domain name for Environment Canada—probably a fairly reliable source.

Table 3 Some Organization Codes

com or co	commercial
edu or ac	educational
org	nonprofit organizations
net	networking providers
mil	military
gov	government
int	international organizations

Evaluating Your Sources

Anyone with access to a server can put material on the Web; there are almost no controls on what people choose to write and publish. It is your job as a researcher to evaluate what you find in order to determine if it suits your needs. As a result, web pages should be viewed with even more caution than most print material.

Use the following questions to determine the quality of a web resource. The greater number of questions answered "yes," the more likely that the source is of high quality.

Authority: Is it clear who is sponsoring the creation and maintenance of the page? Does the site seem to be permanent, or part of a permanent organization? Is there information available describing the purpose of the sponsoring organization? Is there a way of verifying the legitimacy of the page's sponsor? For instance, is a phone number or address available to contact the organization for more information? Is it clear who developed and wrote the material? Are that person's qualifications for writing on this topic stated?

Accuracy: Are the sources for factual information given so they can be verified? Is it clear who has the responsibility for the accuracy of the information presented? If statistical data are presented in graphs or charts, are they labelled clearly?

Objectivity: Is the page, and the information included, provided as a public service? Does it present a balance of views? If there is advertising on the page, is it clearly separated from the informational content?

Currency: Are there dates on the page to indicate when the page was written, first placed online, and last revised or edited? Are there any other indications that the material is updated frequently? If the information is published in print in different editions, is it clear what edition the page is from? If the material is from a work that is out of print, has an effort been made to update the material?

Coverage: Is there an indication that the page has been completed and is not still under construction? If there is a print equivalent to the web page, is there clear indication of whether the entire work or only a portion of it is available on the Web?

A8 Scientific Investigation Skills

Care and Use of the Microscope

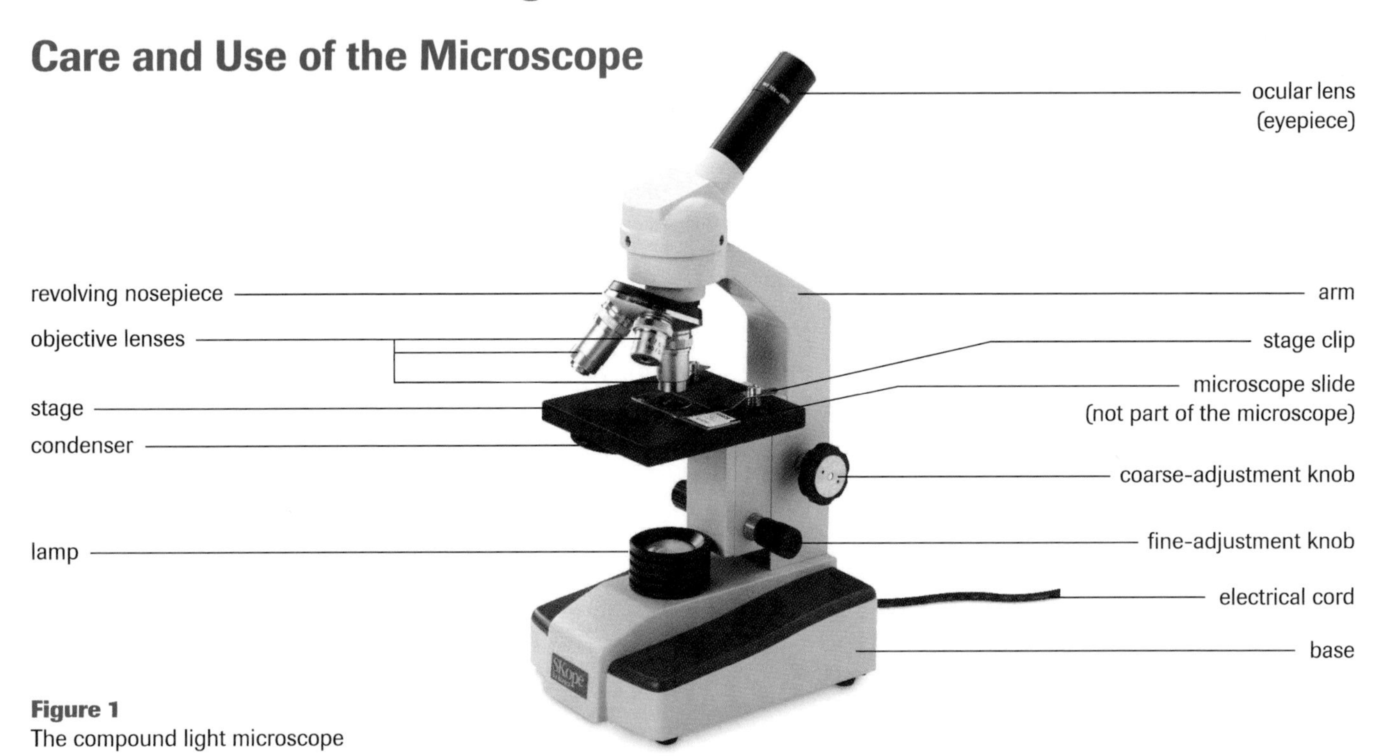

Figure 1
The compound light microscope

The microscope (**Figure 1**) is a useful tool in making observations and collecting data during scientific investigations in biology.

The eyepiece lens usually magnifies 10×. This information is printed on the side of the eyepiece. The microscopes that you will likely use have three objective lenses: low (magnifies 4×), medium (magnifies 10×), and high (magnifies 40×). To determine the total magnification, multiply the magnification of the eyepiece lens by the magnification of the objective lens. For example, the magnification obtained using the medium-power objective lens is 10× multiplied by 10×, or 100×. In other words, a specimen viewed under medium power will appear 100× larger than it actually is.

Use of the Compound Light Microscope

1. Obtain a microscope from the storage area. Grasp the arm with one hand and use the other to support the base of the microscope.
2. If the microscope has a built-in light supply, plug it in. Place the cord so that it will not be hooked accidentally.
3. Rotate the revolving nosepiece until the shortest (low-power) objective lens clicks into place.
4. Place a prepared slide of the letter *f* on the stage and centre it. Hold the slide in place with the stage clips.
5. Turn the coarse-adjustment knob away from you to lower the lens down as far as possible. Watch from the side to ensure that the lens does not contact the slide.
6. Keeping both eyes open, look through the eyepiece and turn the coarse-adjustment knob toward you until the object comes into view. Use the fine-adjustment knob to focus the image.
7. Adjust the condenser to control the amount of light and fix the contrast. If the microscope has a mirror in the base, adjust it to receive the appropriate amount of light.
8. Compare the orientation of the letter *f* as viewed on the slide with the orientation as viewed through the eyepiece.
9. Compare the movement of the object on the slide with the movement of the object as viewed through the eyepiece.

10. With the image in focus and centred in the field of view, rotate the nosepiece until the next longer objective lens clicks into place.
11. Use the fine-adjustment knob to refocus the image, if necessary.
12. Readjust the condenser or the mirror to regulate the amount of light. The higher the lens power, the more light that is necessary.
13. Repeat the previous three steps with the high-power lens.
14. View other prepared slides to practise locating objects within the field of view and focusing the microscope.
15. When you have finished viewing a specimen, always rotate the nosepiece so that the shortest (low-power) lens is centred before making any adjustments with the focusing knobs.
16. Remove and clean the slide and cover slip and return them to their appropriate location.
17. Return the microscope to the storage area.

Oil Immersion

Many microscopes are equipped with a fourth objective lens of 100× magnification, or the highest power lens can be replaced with a special oil-immersion lens. The oil-immersion objective is used to reduce the reflection of light from the slide and to bend the light rays to focus on the specimen. This provides more light and allows the observer to see the specimen in more detail. Follow these guidelines for using the oil-immersion lens.

1. Adjust the light and focus the specimen with the high-power (40×) objective lens, making sure that the specimen is in the centre of the field of view.
2. Rotate the nosepiece so that it is halfway between the 40× and the oil-immersion 100× lenses.
3. Place a drop of immersion oil on the slide at the point that was directly under the lens.
4. Rotate the 100× lens into position so that the tip of the lens contacts the drop of oil.
5. Use only the fine-adjustment knob to refocus the image. If you use the coarse-adjustment knob to focus, you may contact the slide and break it or damage the lens.
6. Using the condenser and lighting control (if available), adjust the lighting as necessary to produce the clearest possible image.
7. When you are finished, rotate the nosepiece so that it is halfway between the 100× and 40× lenses. This will ensure that the 40× lens does not contact the oil.
8. Remove the slide and clean the oil from the slide with lens paper moistened with ether or another solvent.
9. Clean the tip of the oil-immersion lens with lens paper moistened with ether. Do not rub the lens. Rather, pull the lens paper across the surface of the lens to lift the oil off the lens and onto the paper.

The Stereomicroscope or Dissecting Microscope

Unlike the compound light microscope, which uses transmitted light to view very thin, transparent objects, the stereomicroscope uses incident light to view objects that are thicker and opaque. The stereomicroscope, or dissecting microscope (**Figure 2**), has advantages over the compound light microscope for some types of observations. The two eyepieces produce a stereoscopic effect, so we are able to see an object in three dimensions. It is used to look at objects that are too small to be seen clearly with the unaided eye but too large to be viewed properly with a compound light microscope. For example, the stereomicroscope may be used to observe small live specimens that are too large to fit under a cover slip, or to observe bacterial cultures on a streak plate. Because it is used for viewing relatively larger objects, the stereomicroscope generally has lower magnification (usually 6× to 40×) than a compound light microscope.

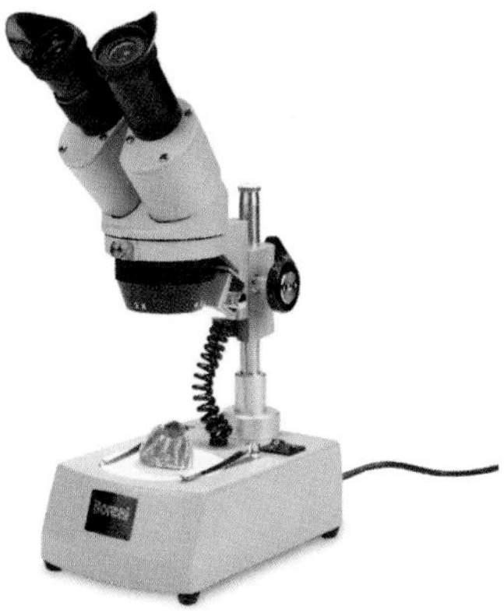

Figure 2
The stereomicroscope, or dissecting microscope

Investigating Depth of Field

The depth of field is the amount of an image that is in sharp focus when it is viewed under a microscope.

1. Cut two pieces of thread of different colours.
2. Make a temporary dry mount by placing one thread over the other in the form of an "X" in the centre of a microscope slide. Cover the threads with a cover slip.
3. Place the slide on the microscope stage and turn on the light.
4. Position the low-power objective lens close to, but not touching, the slide.
5. View the crossed threads through the ocular lens (eyepiece). Slowly rotate the coarse-adjustment knob until the threads come into focus.
6. Rotate the nosepiece to the medium-power objective lens. Focus on the upper thread by using the fine-adjustment knob. You will probably notice that you cannot focus on the lower thread at the same time. The depth of the object that is in focus at any one time represents the depth of field.
7. Repeat step 6 for the high-power objective lens. The stronger the magnification, the shallower the depth of field.

Determining Field of View

It is often necessary to measure the size of objects viewed through the microscope. The field of view is the circle of light seen while looking through the eyepiece. Once the size of the field of view is determined, it is possible to estimate the size of a specimen by comparing it with the size of the field of view.

1. With the low-power objective lens in place, lay a microscope slide on the stage and place a transparent millimetre ruler on the slide.
2. Viewing the microscope stage from the side, position the millimetre marks on the ruler immediately below the objective lens.
3. Looking through the eyepiece, focus the marks on the ruler using the coarse- and fine-adjustment knobs.
4. Move the ruler so that one of the millimetre marks is at the edge of the field of view.
5. Use the following equations to calculate the diameter of the medium- and high-power fields of view. (*Note*: Magnification is shortened to mag. in the equations.)

   For medium power:

   $$\text{diameter} = \text{total mag.}_{\text{low power}} \times \frac{\text{diameter}_{\text{low power}}}{\text{total mag.}_{\text{medium power}}}$$

   For high power:

   $$\text{diameter} = \text{total mag.}_{\text{low power}} \times \frac{\text{diameter}_{\text{low power}}}{\text{total mag.}_{\text{high power}}}$$

6. Using a table similar to **Table 1**, determine the magnification of each of the lenses on your microscope, and record the diameter of the field of view.

Table 1

Lens	Magnification	Eyepiece magnification	Total magnification	Diameter (mm)
low		10×		
medium		10×		
high		10×		

Estimating Size

- **Figure 3** shows the edge of a ruler under low power with the markings 1 mm apart. The diameter of the field of view is estimated to be 1.3 mm.

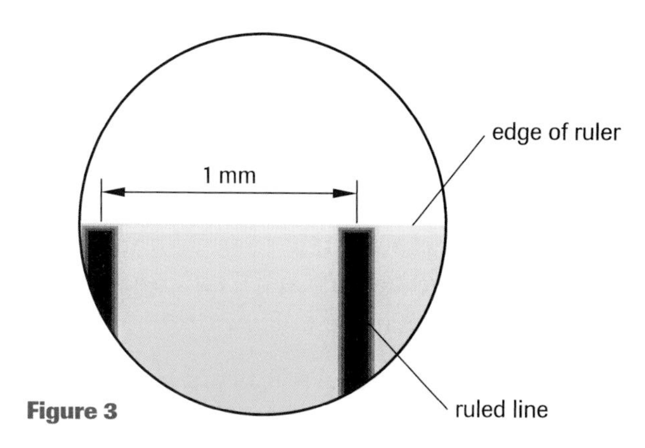

Figure 3

- Estimate the length and width of the specimen by comparing it to the diameter of the field of view.
- **Figure 4** shows a skin cell viewed under high power. One might estimate it to be 0.2 mm wide.

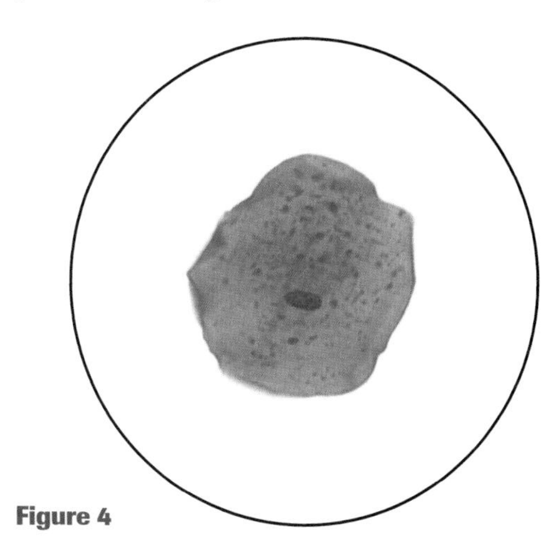

Figure 4

Preparing a Wet-Mount Slide

Specimens of all types can be mounted in a fluid medium before they are examined. Water is the most convenient, but 30% glycerol can also be used. Glycerol helps make the material more transparent, and it does not dry out as fast as water does. A cover slip is used so that there are no reflecting surfaces.

1. Clean a slide and cover slip, holding them by the edges so that you do not leave fingerprints on them. Lay them on a clean, dry surface.
2. Place a drop of water or glycerol in the centre of the slide.
3. Transfer the specimen into the water or glycerol.
4. Hold the edges of the cover slip between thumb and forefinger. Place the cover slip in an almost vertical position on the slide so that the cover slip just touches the edge of the drop of water or glycerol (**Figure 5**).

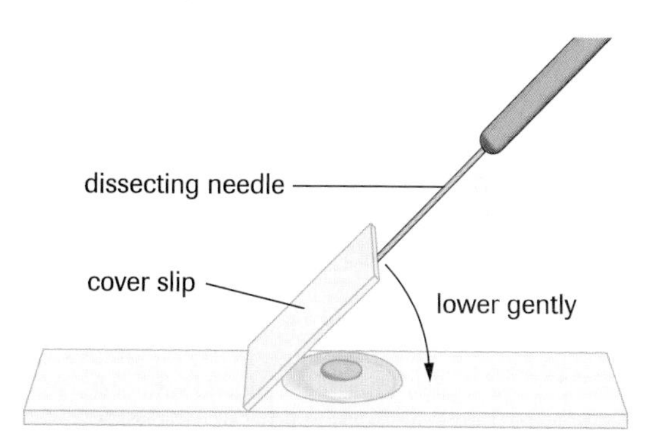

Figure 5
Preparing a wet mount

5. The water or glycerol will spread along the edge of the cover slip. Supporting the cover slip with the point of a needle, gently lower the cover slip onto the slide.
6. Use blotting paper to dry off any excess water or glycerol around the cover slip. Dry any water or glycerol that accidentally drops onto the stage of the microscope or onto the objective lenses.

Staining Techniques

Staining procedures are used to increase the contrast between cells of various tissues and the mounting medium, and to make the features more visible. Biological stains may be temporary or permanent. In this course, we are not concerned with creating permanent slides, so temporary staining is adequate.

Stains react differently with different components of cells; they may affect the cell walls, they may stain the cytoplasm, or they may be used as a test for a specific substance (e.g., iodine for starch).

There are many stains available, but those listed in **Table 2** are more common.

Table 2

Stain	Features and uses
Lugol's (iodine) solution	• especially good for plant cells (e.g., onion cells) • turns blue-black in presence of starch • cellulose walls remain unstained
Sudan IV indicator	• stains all fatty substances orange-pink or red • fresh sections should be left in stain for 20 min • wash with 50% alcohol to remove excess stain • fats seen as droplets within cells
toluidine blue O (TBO)	• water-soluble double stain • lignified cells stain green, greenish blue, or bright blue • xylem elements stain a greenish colour • non-lignified cell walls stain pinkish colour • nuclei become visible and chloroplasts keep their distinctive green colour • specimens should be left in stain for about 30 s, rinsed with water, and then dry mounted
methylene blue	• smears are fixed with heat and allowed to cool • stain with the methylene blue solution 2–7 min • wash slide and blot dry • intracellular granules stain a ruby-red to black colour • remainder of the cell staining a pale blue colour
triphenyltetrazolium chloride (TTC)	• used to identify living or dead tissue • mix stain with equal volume of sample and allow to stand at room temperature overnight (preferably in the dark) • living cells stain red
crystal violet, potassium iodide, safranin	See Gram staining.

In most cases, the staining technique is simple. After a wet-mount slide has been prepared, a drop of the stain is placed beside and under one corner of the cover slip. A piece of paper towel or filter paper touched to the water on the opposite side of the cover slip will draw the stain through the entire specimen in a few seconds. The stained specimen may then be observed.

Gram Staining

The Gram stain technique, named after pioneering Danish microbiologist Hans Christian Gram, is used to help identify unknown species of bacteria. Examination of a Gram-stained smear of material from an infection site can provide important information quickly. This information, which is possible within minutes, may be critical in initiating treatment in serious cases such as gangrene, pneumonia, or meningitis, where it would not be wise to wait for culture results, which could take one or two days.

The staining procedure is outlined below. Staining of Gram positive (coccus) and Gram negative (bacillus) is summarized in **Figure 6**.

- Air dry the bacterial smear and heat fix by passing the slide through a flame several times.
- Stain the slide with crystal (gentian) violet. Wash with tap water after 10 s.
- Stain again with potassium iodide for 10 s. Wash with tap water.
- Wash the slide with alcohol. (Some bacteria will remain stained while others will lose their colouring.)
- Stain again with safranin. Gram positive bacteria will retain the original purple crystal violet stain. Gram negative bacteria will assume the pink colouring of the safranin.

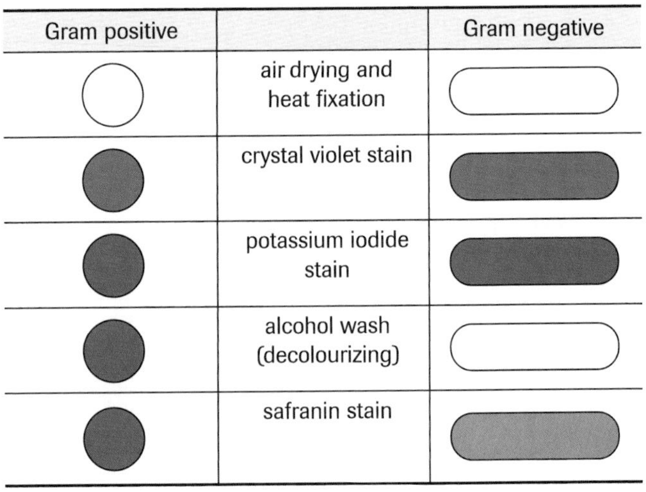

Gram positive		Gram negative
	air drying and heat fixation	
	crystal violet stain	
	potassium iodide stain	
	alcohol wash (decolourizing)	
	safranin stain	

Figure 6
Gram staining

The Stethoscope and Sphygmomanometer

Medical personnel often have to rely on external observations in an attempt to determine what is going on inside the human body. These observations are non-invasive procedures and are very important in helping medical personnel decide what further tests are required to diagnose health problems. Two of the most common instruments used to make these observations are the stethoscope (**Figure 7**) and the sphygmomanometer (**Figure 8**).

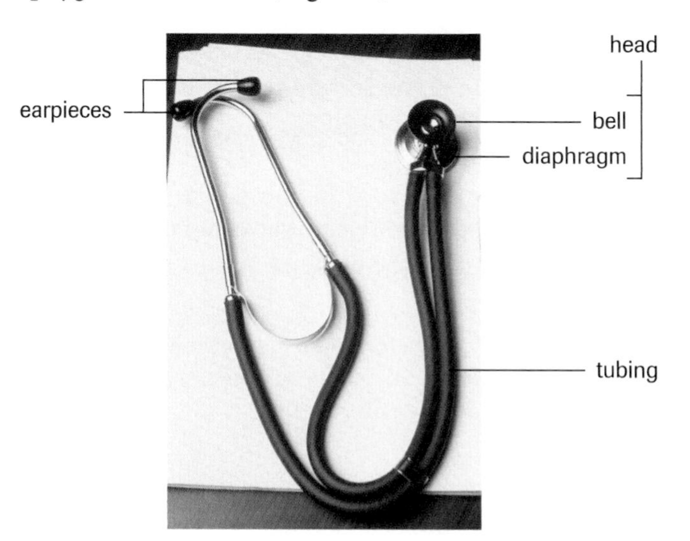

Figure 7
The stethoscope

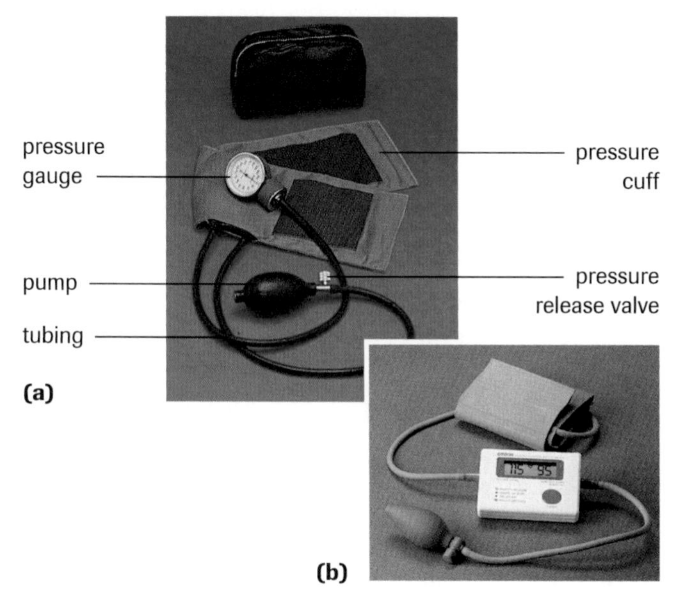

Figure 8
Two of the many types of sphygmomanometer: **(a)** analog display and **(b)** digital display

Using the Stethoscope

The stethoscope is used to listen to sounds inside the body. These sounds are generated by the heart, lungs, and intestinal tract. The stethoscope is also used along with the sphygmomanometer to measure blood pressure. Not all stethoscopes have a bell, but most stethoscopes have a bell and a diaphragm—the bell is used to listen to low-pitched sounds and the diaphragm for high-pitched sounds.

Heart

Listen to the heart sounds by placing the earpieces in your ears, and the diaphragm or bell of the stethoscope on the chest in the general area of the heart. The heart is located on the left side around the 4th to the 6th rib almost directly under the breast. The normal heart sounds are the regular *lubb-dubb* that represents one beat of the heart. The sounds are produced by the closing of the valves. The *lubb* sound is caused when the atrioventricular valves shut. The *dubb* sound is caused by the closing of the semilunar valves.

Certain conditions will produce other heart sounds as well. A quiet *whoosh* after the *lubb-dubb* sound is known as a murmur. This may be caused by incomplete opening or closing of the valves.

Respiratory System

Place the stethoscope on your chest and breathe in and out deeply and slowly. Different sounds can be heard in different locations. The stethoscope may also be placed on the back to listen to the lungs. For normal lungs, you should hear only the rush of air into and out of the lungs. Abnormal lung sounds include crackles and wheezes. If the lung rubs on the chest wall, there may be friction rubs. Crackles sound like a sheet of paper being crushed. They indicate that there is fluid in the lungs. Wheezes are high-pitched whistling noises. Friction rubs are squeaky sounds that you might expect when two objects are rubbed together.

Abdomen

The stethoscope can be used to listen to sounds produced by the digestive system. Place the stethoscope over the upper abdomen to hear the normal gurgling sounds of the stomach and over the lower abdomen to listen to the intestines. Gurgling sounds are normal. A doctor would determine if the digestive system is overactive or underactive.

Using the Sphygmomanometer

The sphygmomanometer is an instrument that gives an indirect measurement of blood pressure. It consists of a gauge (manometer) to register pressure, an inflatable compression bag with a strong cloth cover (cuff), an inflating device (inflation bulb or pump), and a means of deflating (pressure release valve).

Use the sphygmomanometer and stethoscope, and the procedure outlined below, to measure blood pressure:

1. The subject should be relaxed and comfortably seated with the arm extended straight at heart level.
2. Place the centre of the pressure cuff over the brachial artery in the upper arm just above the elbow (**Figure 9**). The cuff should be wrapped snugly around the arm but should not be too tight.

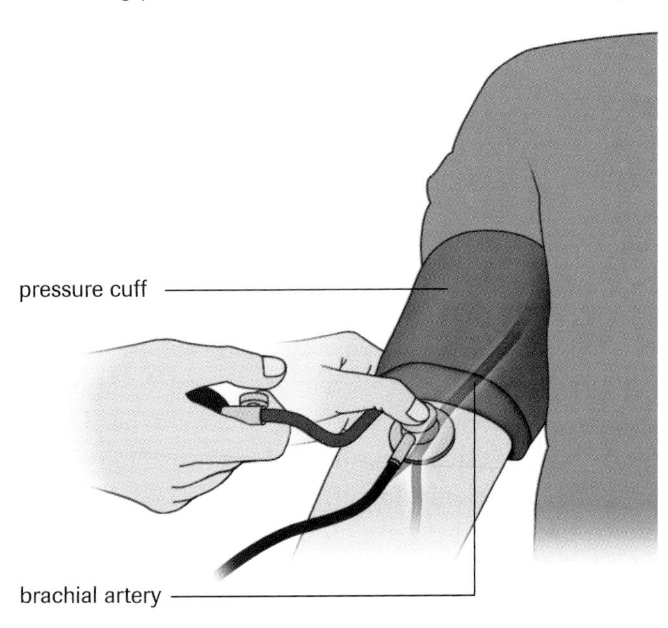

Figure 9
Proper placement of the stethoscope and sphygmomanometer

3. Place the head of the stethoscope over the brachial artery just below the elbow (**Figure 10**). Ensure that it does not touch the pressure cuff, as this will produce noise that might interfere with your listening.

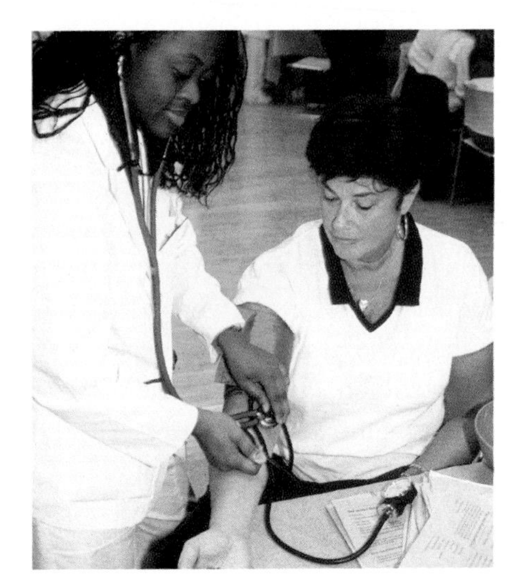

Figure 10
Using the stethoscope and sphygmomanometer to measure blood pressure

4. Close the pressure release valve and use the pump to inflate the pressure cuff to 180 mm Hg.

Do not inflate the pressure cuff beyond 180 mm Hg, and do not leave the pressure on for longer than one minute.

5. Carefully open the pressure release valve so that the pressure drops at 2–3 mm Hg per second.
6. As the pressure drops, listen for clear tapping sounds. The pressure at which you can hear clear sounds for two consecutive beats is the systolic pressure.
7. Continue slowly releasing the pressure. At some point the repetitive sounds will disappear. The pressure at this point is the diastolic pressure.
8. Once both pressures have been noted, quickly release the pressure by opening the pressure release valve completely.
9. Record both measurements to the nearest 2 mm Hg.
10. It is often recommended to verify the reading by taking a second reading a few minutes after the first.

Do not try to interpret what you consider to be abnormal observations. Only qualified medical personnel should interpret observations to make diagnoses.

A

Microbiological Culturing Techniques

Microorganisms are usually cultured on a medium that provides a physical habitat and all of the nutrients required by the organisms (**Figure 11**). The medium may be liquid, solid, or semi-solid. Liquid media are usually in the form of broth. Solid media contain agar as a solidifying agent. Semi-solid media also contain agar, but in insufficient quantities to fully solidify.

Agar is a carbohydrate that is extracted from red seaweed and is used as a gelling agent. Agar itself provides mainly the physical habitat for microorganism growth. To promote proper growth, however, nutrients must be added to the agar. Nutrient agar, the most commonly used medium for microorganism cultures, contains agar (the gelling agent), beef extract (provides carbohydrates, nitrogen, vitamins, and salts), and peptone (a protein derivative that helps control pH). The amount of nutrient agar (23 g) required to make one litre of medium consists of 15 g of agar, 3 g of beef extract, and 5 g of peptone. A simple and inexpensive nutrient agar mixture can be made by adding 15 g of plain agar and 2 beef boullion cubes to one litre of distilled water. This medium will support the growth of many microorganisms and is adequate for senior high school applications.

Agar is a desirable compound as a culture medium because of its physical properties. It dissolves at temperatures near the boiling point of water (100°C) and once it is dissolved, solidifies at room temperature (<40°C). This allows it to be poured at temperatures that will not kill most cells and to be incubated at a range of temperatures (up to 40°C) without melting.

The streak plate (a petri dish containing solidified nutrient agar) (**Figure 11**) is commonly used for the growth and identification of bacterial cells. A sample of the microorganism is spread onto the surface of the agar (see inoculation procedure below), and the colonies of cells will grow only on the surface and will be easy to observe.

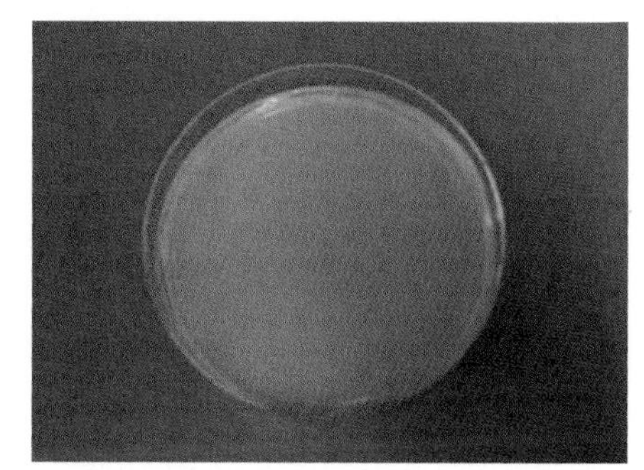

Figure 11
An agar streak plate

Preparing Agar Plates

The following procedure will assist you in making nutrient medium and preparing plates for microorganism culture:

1. Observe necessary safety precautions: wear safety glasses, tie back long hair, and never leave a flame unattended.
2. Obtain the required equipment and materials. Sizes and amounts will depend on the quantity of nutrient medium being prepared.
   - sterile petri dishes (15 × 100 mm standard)
   - Erlenmeyer flask
   - 2 beakers (larger than the Erlenmeyer flask)
   - glass stirring rod
   - thermometer
   - distilled water
   - nutrient agar
   - tongs

   Note: Do NOT remove the petri dish lids until you are ready to pour the prepared nutrient agar.
3. Measure quantities of distilled water and nutrient agar according to **Table 3**.

Table 3

Yield (15 x 100 mm petri dishes)	Nutrient agar (g)	Distilled water (mL)
50 plates	23.0	1000
25 plates	11.5	500
15 plates	6.9	300
10 plates	4.6	200

4. Sterilize (wash and dry) the Erlenmeyer flask, glass stirring rod, and thermometer.
5. Fill a large beaker about half full with tap water.
6. Gently boil the water and keep it boiling. This is your hot water bath.
7. Add the distilled water and nutrient agar to the Erlenmeyer flask and stir to mix.
8. Holding the neck of the Erlenmeyer flask with tongs, carefully lower it into the hot water bath.
9. Stir the mixture constantly while heating until the agar is fully dissolved. It will look clear yellow, with no powder floating. Boil the mixture for 1 min. Remove from the hot water bath.
10. Using tongs, carefully remove the flask and place it on a heat-resistant pad. Continue stirring the agar solution. Have your partner prepare a cold water bath by filling the second large beaker about one-third full of room-temperature tap water.
11. Carefully lower the Erlenmeyer flask into the cold water bath, holding its neck with tongs. Place a sterile lab thermometer in the mixture and monitor the temperature. Stir constantly as the agar cools to between 45°C and 50°C. Do not let it cool further because it will start to solidify.
12. One at a time, lift the lid of each petri dish on a 45° angle and pour enough agar into each dish to cover the bottom—about 3–6 mm deep. Replace the lid immediately. You should work quickly so that the agar doesn't solidify in the Erlenmeyer flask.
13. Swirl each dish very gently to spread the agar evenly. Leave the agar plates untouched until the agar is fully cooled and solidified. The agar medium is now ready for storage or use.
14. To store the agar plates, stack them upside down in the refrigerator. The purpose of turning the plates upside down is to prevent condensation from forming and dripping down onto the agar surface when the plates are removed from the refrigerator.
15. Wash all glassware, the lab bench, and your hands with soap and warm water.

Inoculation

1. If the agar plates have been refrigerated, allow them to warm to room temperature before using them.
2. Collect bacteria from a single source using a clean swab.
3. Open the petri dish lid just enough to allow the swab inside while inoculating the medium to help prevent contamination from air-borne particles. Do not allow the swab to touch the petri dish.
4. Using the swab, make a pattern of inoculation lines (parallel lines, tic-tac-toe, zig-zag, your initials, etc.) on the agar surface. This will help demonstrate that the growing culture is the result of your inoculation and not the result of some sort of air-borne or accidental contamination. Do not penetrate the surface of the agar. See examples of zig-zag patterns in **Figure 12** and a culture pattern in **Figure 13**.

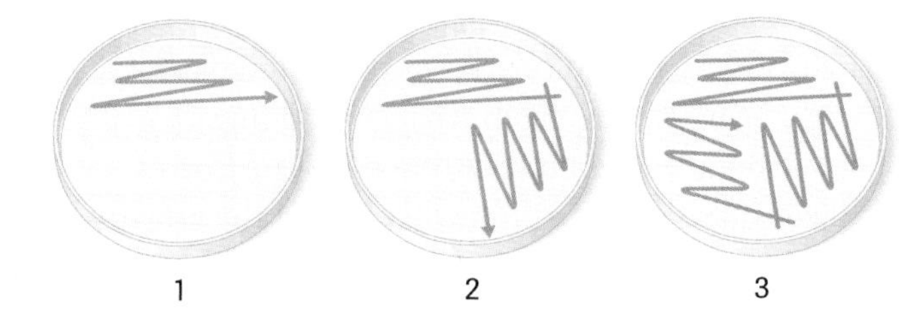

Figure 12
Zig-zag inoculation pattern. Note that only one stroke of the second pattern overlaps the first area, and only one stroke of the third pattern overlaps the second area.

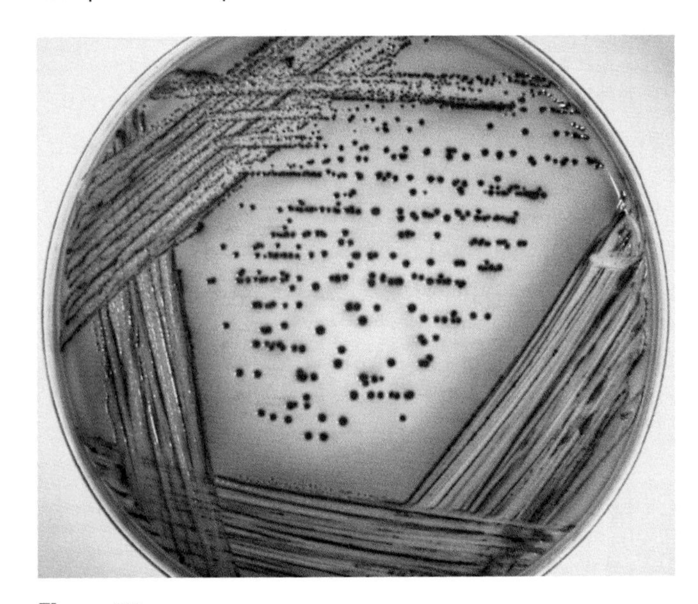

Figure 13
Microorganism culture patterns on an agar streak plate

5. Close the lid immediately after inoculating the medium.
6. Tape the petri dish lid closed in the form of an "X." Label it with the location from which your sample was taken, the date, and other required information. Write small so as not to obscure the viewing area through the lid.
7. Keep at least one plate as a control. This will alert you to possible prior contamination of the agar plates.
8. To incubate, turn the plates upside down and put them in a warm, dark place. The ideal temperature for incubation is 32°C. Bacterial growth should start to become visible in about 2–3 days.

Never incubate plates at 37°C because it could support the growth of pathogens.

9. Wash your hands thoroughly with soap and water.

Since many microorganisms are single-celled, their growth is not defined in terms of cell size but in terms of the increase in the number of cells as a result of cell division. It is much easier to measure the increase in cell number than it is to measure the increases in cell size. The changes in the size of the microorganism colonies on the surface of the agar plate indicate the growth of the number of cells. Colonies are described and identified using their physical characteristics—size, form, margin, texture, opacity, and colour. **Figure 14** illustrates the typical margins, colony forms, and elevations that are used to describe microorganism colonies. (**Figure 15** shows actual bacterial colonies of various forms.)

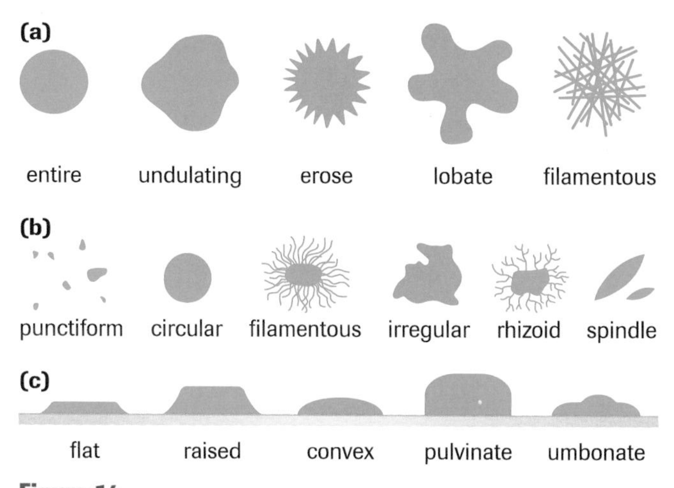

Figure 14
Margins **(a)**, colony forms **(b)**, and elevation types **(c)** of bacterial colonies

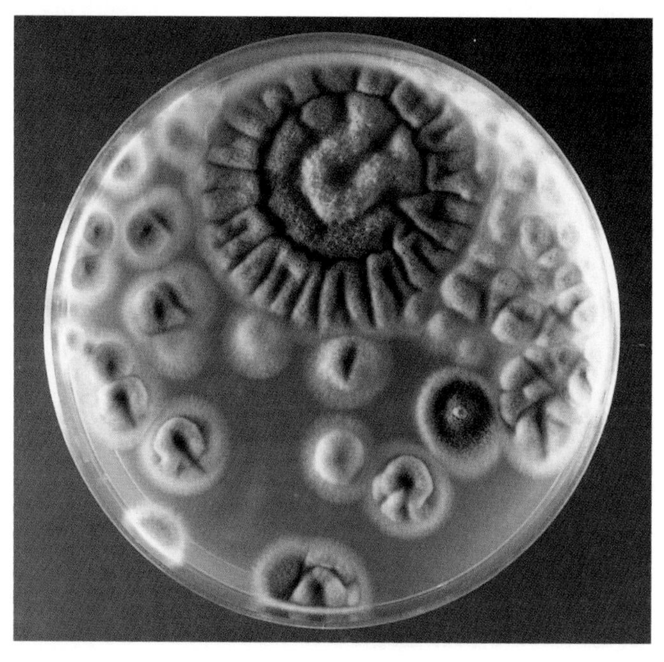

Figure 15
Bacterial colonies growing on a plate of nutrient agar

Constructing and Using a Dichotomous Key

A dichotomous key is a tool that is widely used in the biological sciences for easily and quickly identifying an unknown organism. Dichotomous means "divided into two parts," so the key presents a series of choices between two statements based on specific characteristics of the organism (**Figure 16**). Selecting the correct statement in each choice of the series will always lead to the identification of the organism.

(a)

1a.	Bean is round.	**garbanzo bean**
1b.	Bean is elliptical or oblong.	Go to 2.
2a.	Bean is white.	**white northern bean**
2b.	Bean has dark pigments.	Go to 3.
3a.	Bean has mottled or spotty pigmentation (colouring).	**pinto bean**
3b.	Bean is evenly pigmented (coloured).	Go to 4.
4a.	Bean is black.	**black bean**
4b.	Bean is reddish-brown.	**kidney bean**

(b)

1a. Bean is elliptical or oblong. Go to 2a.
  2a. Bean has dark pigments (colours). Go to 3a.
    3a. Bean colour is solid. Go to 4a.
    3b. Bean colour is mottled or spotty. **pinto bean**
      4a. Bean is black. **black bean**
      4b. Bean is reddish-brown. **kidney bean**
  2b. Bean is white. **white northern bean**
1b. Bean is round. **garbanzo bean**

Figure 16
Two formats of a dichotomous key for the identification of beans:
(a) Two choices presented together. Choices are easy to see and compare.
(b) Choices are grouped by relationships. Choices are separated but relationships are easy to see.

Tips for Constructing a Key

Constructing a dichotomous key will be easier if you use the following guidelines:

- Start with the most general characteristics and progress to more detailed characteristics.
- Depending on the format you are using, leave a space between each pair of statements (**Figure 16(a)**) or indent each pair (**Figure 16(b)**) to make the key easier to read.
- Where possible, use measurements rather than vague descriptors such as "large" and "small."
- Use measurements that are constant or have a small range rather than those that are highly variable.
- Use characteristics that are constant and generally available to the user rather than seasonal characteristics or those seen only under certain conditions.
- Attempt to write the descriptions as positive rather than negative statements; for example, "something is..." rather than "something is not..."
- Start the description with the part that is being described; for example, "leaves are red" rather than "red leaves are present."
- Start each choice of the pair of statements with the same word, if possible.
- Start each new pair of statements with different words, if possible.

Tips for Using a Key

When using a dichotomous key, keep the following guidelines in mind:

- Read both choices in a pair of statements carefully. The first description may seem logical, but the second may be more appropriate.
- If you aren't certain of the meaning of the terms used in the descriptions, check them in a glossary, dictionary, or other reference book.
- Since two organisms of the same species are seldom identical, use more than one sample if possible.
- If you aren't sure of which choice to make, follow both choices. After a couple of pairs of statements, it may become obvious that one division doesn't fit your sample.
- When a measurement is provided in the statement, do not estimate or guess. Use a calibrated scale to measure your sample.
- Keep notes indicating the sequence of identification that you followed. This will allow you to check your process and identify any mistakes made along the way.
- Double-check your identification. Find a description of the organism and see if your sample matches the description.

Biological Drawings

Biological drawings are an essential part of recording your observations, both microscopic and macroscopic. These drawings are used to communicate, so it is very important that they are as accurate as possible, clear, well-labelled, and easy to understand. You will be required to make drawings of external features of whole specimens, parts of specimens, dissections, and prepared slides.

Preparation

- Use plain, white paper and a sharp, hard pencil (2H or 4H). Sharpen your pencil often to ensure clear, fine lines.
- Plan your drawing to fit on the page. Ensure that it is large enough to show the details.
- Leave space for labelling to the right of the drawing.

Drawing Tips

- Draw only what you see. Do not copy diagrams from books or draw what you think you should see.
- Don't sketch. Use firm, clear lines. When recording microscope observations, use only the outline of structures for low-power observations; show details in high-power observations (**Figure 17**).

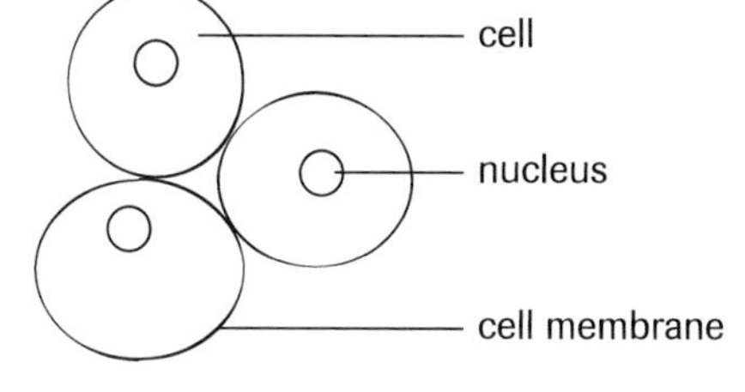

Figure 17
Clear, solid lines showing outlines of cells

- If important details are too small to be shown clearly, they should be shown in an enlarged drawing on the side.
- Don't use colouring or shading. To show darker areas, stippling (dots) may be used. Draw double lines to indicate thick structures (**Figure 18**).

Figure 18
Use stippling to show darker areas and double lines to show thick structures.

- If a specimen shows repetitive structures, it is sufficient to draw only a representative section in detail (**Figure 19**).
- Don't draw on both sides of the paper.

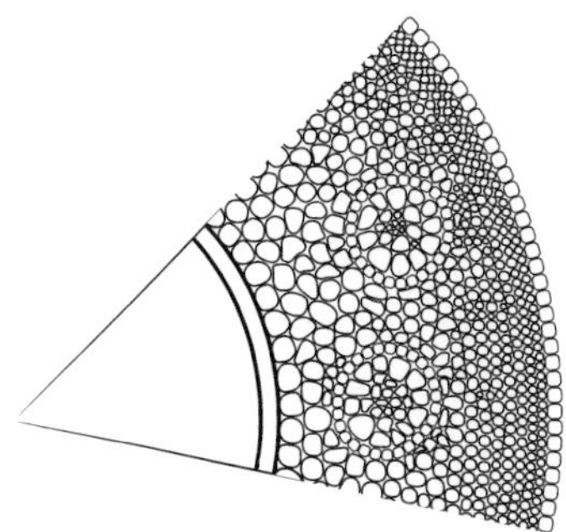

Figure 19
A representative drawing of a stem cross section.

Labelling

- All important structures should be labelled clearly. Always use the singular form for the labels (e.g., chloroplast, not chloroplasts).
- Connect labels to the appropriate parts using only horizontal lines if possible (**Figure 20**).
- Don't label too close to the drawing, and never write on the drawing itself. It is preferable to list your labels in an even column down the right side (**Figure 21**).
- For drawings of microscope observations, indicate the magnification.
- Title the drawing. Use the name of the specimen and any other information that will help identify the drawing. Underline the title.

Figure 20
Connect your labels with horizontal lines.

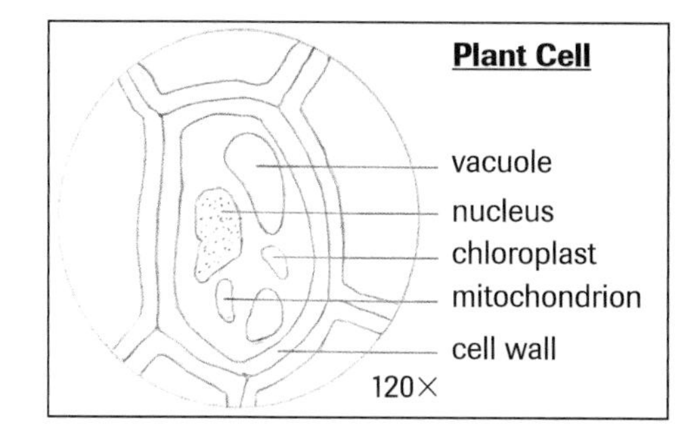

Figure 21
Neat labelling makes your drawing clear.

B1 Safety Conventions and Symbols

Although every effort is undertaken to make the science experience a safe one, there are some risks associated with certain scientific investigations. These risks are generally associated with the materials and equipment used, and the disregard of safety instructions that accompany investigations. However, there may also be risks associated with the location of the investigation, whether in the science laboratory, at home, or outdoors. Most of these risks pose no more danger than one would normally experience in everyday life. With an awareness of the possible hazards, knowledge of the rules, appropriate behaviour, and a little common sense, these risks can be practically eliminated.

Remember, you share the responsibility not only for your own safety, but also for the safety of those around you. Always alert the teacher in case of an accident.

In this text, chemicals, equipment, and procedures that are hazardous are highlighted in red and are preceded by .

WHMIS Symbols and HHPS

The Workplace Hazardous Materials Information System (WHMIS) provides workers and students with complete and accurate information regarding hazardous products. All chemical products supplied to schools, businesses, and industries must contain standardized labels and be accompanied by Material Safety Data Sheets (MSDS) providing detailed information about the product. Clear and standardized labelling is an important component of WHMIS (**Table 1**). These labels must be present on the product's original container or be added to other containers if the product is transferred.

The Canadian Hazardous Products Act requires manufacturers of consumer products containing chemicals to include a symbol specifying both the nature of the primary hazard and the degree of this hazard. In addition, any secondary hazards, first aid treatment, storage, and disposal requirements must be noted. Household Hazardous Product Symbols (HHPS) are used to show the hazard and the degree of the hazard by the type of border surrounding the illustration (**Figure 1**).

	Corrosive
	This material can burn your skin and eyes. If you swallow it, it will damage your throat and stomach.
	Flammable
	This product or the gas (or vapour) from it can catch fire quickly. Keep this product away from heat, flames, and sparks.
	Explosive
	Container will explode if it is heated or if a hole is punched in it. Metal or plastic can fly out and hurt your eyes and other parts of your body.
	Poison
	If you swallow or lick this product, you could become very sick or die. Some products with this symbol on the label can hurt you even if you breathe (or inhale) them.

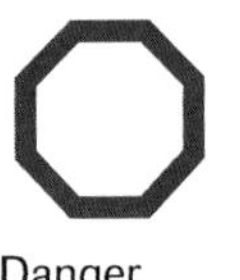

Danger

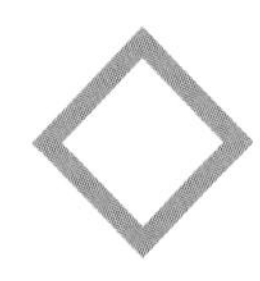

Warning

Caution

Figure 1
Hazardous household product symbols

B

Table 1 The Workplace Hazardous Materials Information System (WHMIS)

Class and type of compounds	WHMIS symbol	Risks	Precautions
Class A: *Compressed Gas* Material that is normally gaseous and kept in a pressurized container		• could explode due to pressure • could explode if heated or dropped • possible hazard from both the force of explosion and the release of contents	• Ensure container is always secured. • Store in designated areas. • Do not drop or allow to fall.
Class B: *Flammable and Combustible Materials* Materials that will continue to burn after being exposed to a flame or other ignition source		• may ignite spontaneously • may release flammable products if allowed to degrade or when exposed to water	• Store in properly designated areas. • Work in well-ventilated areas. • Avoid heating. • Avoid sparks and flames. • Ensure that electrical sources are safe.
Class C: *Oxidizing Materials* Materials that can cause other materials to burn or support combustion		• can cause skin or eye burns • increase fire and explosion hazards • may cause combustibles to explode or react violently	• Store away from combustibles. • Wear body, hand, face, and eye protection. • Store in proper container that will not rust or oxidize.
Class D: *Toxic Materials, Immediate and Severe* Poisons and potentially fatal materials that cause immediate and severe harm		• may be fatal if ingested or inhaled • may be absorbed through the skin • have a toxic effect even in small volumes	• Avoid breathing dust or vapours. • Avoid contact with skin or eyes. • Wear protective clothing, and face and eye protection. • Work in well-ventilated areas and wear breathing protection.
Class D: *Toxic Materials, Long-Term Concealed* Materials that have a harmful effect after repeated exposures or over a long period		• may cause death or permanent injury • may cause birth defects or sterility • may cause cancer • may cause allergies	• Wear appropriate personal protection. • Work in a well-ventilated area. • Store in appropriate designated areas. • Avoid direct contact. • Use hand, body, face, and eye protection. • Ensure respiratory and body protection is appropriate for the specific hazard.
Class D: *Biohazardous Infectious Materials* Infectious agents or a biological toxin causing a serious disease or death		• may cause anaphylactic shock • includes viruses, yeasts, moulds, bacteria, and parasites that affect humans • includes fluids containing toxic products • includes cellular components	• Special training is required to handle these materials. • Work in designated biological areas with appropriate engineering controls. • Avoid forming aerosols. • Avoid breathing vapours. • Avoid contamination of people and/or area. • Store in special designated areas.
Class E: *Corrosive Materials* Materials that react with metals and living tissue		• eye and skin irritation on exposure • severe burns/tissue damage on longer exposure • lung damage if inhaled • may cause blindness if contacts eyes • environmental damage from fumes	• Wear body, hand, face, and eye protection. • Use breathing apparatus. • Ensure protective equipment is appropriate. • Work in a well-ventilated area. • Avoid all direct body contact. • Use appropriate storage containers and ensure proper non-venting closures.
Class F: *Dangerously Reactive Materials* Materials that may have unexpected reactions		• may react with water • may be chemically unstable • may explode if exposed to shock or heat • may release toxic or flammable vapours • may vigorously polymerize • may burn unexpectedly	• Handle with care, avoiding vibration, shocks, and sudden temperature changes. • Store in appropriate containers. • Ensure storage containers are sealed. • Store and work in designated areas.

B2 Safety in the Laboratory

General Safety Rules

Safety in the laboratory is an attitude and a habit more than it is a set of rules. It is easier to prevent accidents than to deal with the consequences of an accident. Most of the following rules are common sense:

- Do not enter a laboratory unless a teacher or other supervisor is present, or you have permission to do so.
- Familiarize yourself with your school's safety regulations.
- Make your teacher aware of any allergies and other health problems you may have.
- Wear eye protection, a laboratory apron or coat, and gloves when appropriate.
- Wear closed shoes (not sandals) when working in the laboratory.
- Place your books and bags away from the work area. Keep your work area clear of all materials except those that you will use in the investigation.
- Do not chew gum, eat, or drink in the laboratory. Food should not be stored in refrigerators in the laboratory.
- Know the location of MSDS information, exits, and all safety equipment, such as the fire blanket, fire extinguisher, and eyewash station.
- Avoid sudden or rapid motion in the laboratory that may interfere with someone carrying or working with chemicals or using sharp instruments.
- Never engage in horseplay or practical jokes in the laboratory.
- Ask for assistance when you are not sure how to do a procedural step.
- Never attempt any unauthorized experiments.
- Never work in a crowded area or alone in the laboratory.
- Always wash your hands with soap and water before and after you leave the laboratory. Definitely wash your hands before you touch any food.
- Use stands, clamps, and holders to secure any potentially dangerous or fragile equipment that could be tipped over.
- Do not taste any substance in a laboratory.
- Never smell chemicals unless specifically instructed to do so by the teacher. Do not inhale the vapours, or gas, directly from the container. Take a deep breath to fill your lungs with air, then waft or fan the vapours toward your nose.
- Clean up all spills, even water spills, immediately.
- If you are using a microscope with a mirror, never direct the mirror to sunlight. The concentrated reflected light could hurt your eyes badly.
- Do not forget safety procedures when you leave the laboratory. Accidents can also occur outdoors, at home, and at work.

Eye and Face Safety

- Always wear approved eye protection in a laboratory, no matter how simple or safe the task appears to be. Keep the safety goggles over your eyes, not on top of your head. For certain experiments, full face protection may be necessary.
- If you must wear contact lenses in the laboratory, be extra careful; whether or not you wear contact lenses, do not touch your eyes without first washing your hands. If you do wear contact lenses, make sure that your teacher is aware of it. Carry your lens case and a pair of glasses with you.
- Do not stare directly at any bright source of light (e.g., a burning magnesium ribbon, lasers, the Sun). You will not feel any pain if your retina is being damaged by intense radiation. You cannot rely on the sensation of pain to protect you.
- Never look directly into the opening of flasks or test tubes.

Handling Glassware Safely

- Never use glassware that is cracked or chipped. Give such glassware to your teacher or dispose of it as directed. Do not put the damaged item back into circulation.

- Never pick up broken glassware with your fingers. Use a broom and dustpan.
- Do not put broken glassware into garbage containers. Dispose of glass fragments in special containers marked "Broken Glass."
- Heat glassware only if it is approved for heating. Check with your teacher before heating any glassware.
- Be very careful when cleaning glassware. There is an increased risk of breakage from dropping when glassware is wet and slippery.
- If you need to insert glass tubing or a thermometer into a rubber stopper, get a cork borer of a suitable size. Insert the borer into the hole of the rubber stopper, starting from the small end of the stopper. Once the borer is pushed all the way through the hole, insert the tubing or thermometer through the borer. Ease the borer out of the hole, leaving the tubing or thermometer inside. To remove the tubing or thermometer from the stopper, push the borer from the small end through the stopper until it shows at the other end. Ease the tubing or thermometer out of the borer.
- Protect your hands with heavy gloves or several layers of cloth before inserting glass into rubber stoppers.

Using Sharp Instruments Safely

- Make sure your instruments are sharp. Surprisingly, one of the main causes of accidents with cutting instruments is using a dull instrument. Dull cutting instruments require more pressure than sharp instruments and are, therefore, much more likely to slip.
- Always transport a scalpel in a dissection case or box. Never carry the scalpel from one area of the laboratory to another with an exposed blade.
- Select the appropriate instrument for the task. Never use a knife when scissors would work best.
- Always cut away from yourself and others.

Fire Safety

- Immediately inform your teacher of any fires. Very small fires in a container may be extinguished by covering the container with a wet paper towel or a ceramic square to cut off the supply of air. Alternatively, sand may be used to smother small fires. A bucket of sand with a scoop should be available in the laboratory.
- If anyone's clothes or hair catch fire, tell the person to drop to the floor and roll. Then use a fire blanket to help smother the flames. Never wrap the blanket around a person on fire; the chimney effect will burn the lungs. For larger fires, immediately evacuate the area. Call the office or sound the fire alarm if close by. Do not try to extinguish larger fires. Your prime concern is to save lives. As you leave the classroom, make sure that the windows and doors are closed.
- If you use a fire extinguisher, direct the extinguisher at the base of the fire and use a sweeping motion, moving the extinguisher nozzle back and forth across the front of the fire's base. Different extinguishers are effective for different classes of fires. The fire classes are outlined below. Fire extinguishers in the laboratory are 2A10BC. They extinguish classes A, B, and C fires.
- Class A fires involve ordinary combustible materials that leave coals or ashes, such as wood, paper, or cloth. Use water or dry chemical extinguishers on class A fires.
- Class B fires involve flammable liquids such as gasoline or solvents. Carbon dioxide or dry chemical extinguishers are effective on class B fires.
- Class C fires involve live electrical equipment, such as appliances, photocopiers, computers, or laboratory electrical apparatus. Carbon dioxide or dry chemical extinguishers are recommended for class C fires. Do not use water on live electrical devices as this can result in severe electrical shock.
- Class D fires involve burning metals, such as sodium, potassium, magnesium, or aluminum. Sand, salt, or graphite can be used to put out class D fires. Do not use water on a metal fire as this can cause a violent reaction.
- Class E fires involve a radioactive substance. These require special consideration at each site.

Heat Safety

- Keep a clear workplace when performing experiments with heat.
- Make sure that heating equipment, such as the burner, hot plate, or electric heater, is secure on the bench and clamped in place when necessary.
- Do not use a laboratory burner near wooden shelves, flammable liquids, or any other item that is combustible.
- Take care that the heat developed by the heat source does not cause any material close by to get hot enough to burst into flame. Do not allow overheating if you are performing an experiment in a closed area. For example, if you are using a light source in a large cardboard box, be sure you have enough holes at the top of the box and on the sides to dissipate heat.
- Before using a laboratory burner, make sure that long hair is always tied back. Do not wear loose clothing (wide, long sleeves should be tied back or rolled up).
- Always assume that hot plates and electric heaters are hot and use protective gloves when handling.
- Do not touch a light source that has been on for some time. It may be hot and cause burns.
- In a laboratory where burners or hot plates are being used, never pick up a glass object without first checking the temperature by lightly and quickly touching the item, or by placing your hand near but not touching it. Glass items that have been heated stay hot for a long time, even if they do not appear to be hot. Metal items such as ring stands and hot plates can also cause burns; take care when touching them.
- Never look down the barrel of a laboratory burner.
- Always pick up a burner by the base, never by the barrel.
- Never leave a lighted burner unattended.
- Any metal powder can be explosive. Do not put these in a flame.
- When heating a test tube over a laboratory burner, use a test-tube holder and a spurt cap. Holding the test tube at an angle, with the open end pointed away from you and others, gently move the test tube back and forth through the flame.
- To heat a beaker, put it on the hot plate and secure with a ring support attached to a utility stand. (A wire gauze under the beaker is optional.)
- Remember to include a cooling time in your experiment plan; do not put away hot equipment.

To use a burner:

- Tie back long hair and tie back or roll up wide, long sleeves.
- Secure the burner to a stand using a metal clamp.
- Check that the rubber hose is properly connected to the gas valve.
- Close the air vents on the burner. Use a sparker to light the burner.
- Open the air vents just enough to get a blue flame.
- Control the size of the flame using the gas valve.

Electrical Safety

- Water or wet hands should never be used near electrical equipment such as a hotplate, a light source, or a microscope.
- Do not use the equipment if the cord is frayed or if the third pin on the plug is missing. If the teacher allows this, then make sure the equipment has a double-insulated cord.
- Do not operate electrical equipment near running water or a large container of water.
- Check the condition of electrical equipment. Do not use if wires or plugs are damaged.
- If using a light source, check that the wires of the light fixture are not frayed, and that the bulb socket is in good shape and well secured to a stand.
- Make sure that electrical cords are not placed where someone could trip over them.
- When unplugging equipment, remove the plug gently from the socket. Do not pull on the cord.

Handling Chemicals Safely

Many chemicals are hazardous to some degree. When using chemicals, operate under the following principles:

- Never underestimate the risks associated with chemicals. Assume that any unknown chemicals are hazardous.
- Use a less hazardous chemical wherever possible.
- Reduce exposure to chemicals as much as possible. Avoid direct skin contact, if possible.
- Ensure that there is adequate ventilation when using chemicals.

The following guidelines do not address every possible situation but, used with common sense, are appropriate for situations in the high school laboratory.

- Obtain an MSDS for each chemical and consult the MSDS before you use the chemical.
- Know the emergency procedures for the building, the department, and the chemicals being used.
- Wear a laboratory coat and/or other protective clothing (e.g., apron, gloves), as well as appropriate eye protection at all times in areas where chemicals are used or stored.
- Never use the contents from a bottle that has no label or has an illegible label. Give any containers with illegible labels to your teacher. When leaving chemicals in containers, ensure that the containers are labelled. Always double-check the label—once, when you pick it up and a second time when you are about to use it.
- Carry chemicals carefully using two hands, one around the container and one underneath.
- Always pour from the side opposite the label on a reagent bottle; your hands and the label are protected as previous drips are always on the side of the bottle opposite the label.
- Do not let the chemicals touch your skin. Use a laboratory scoop or spatula for handling solids.
- Pour chemicals carefully (down the side of the receiving container or down a stirring rod) to ensure that they do not splash.
- Always pour volatile chemicals in a fume hood or in a well-ventilated area.
- Never pipet or start a siphon by mouth. Always use a pipet suction device (such as a bulb or a pump).
- If you spill a chemical, use a chemical-spill kit to clean up.
- Return chemicals to their proper storage place according to your teacher's instructions.
- Do not return surplus chemicals to stock bottles. Dispose of excess chemicals in an appropriate manner as instructed by your teacher.
- Clean up your work area, the fume hood, and any other area where chemicals were used.
- Wash your hands immediately after handling chemicals and before and after leaving the laboratory, even if you wore gloves. Definitely wash your hands before you touch any food.

Handling Microorganisms Safely

- If you have a cut or abrasion, take appropriate protective measures to avoid contamination with microorganisms.
- Never put anything in your mouth.
- Use sterilized equipment and aseptic procedures.
- For the aseptic transfer of cultures, wipe the laboratory work surfaces with a disinfectant solution before and after an activity (10% bleach solution or 3% Lysol or Dettol solution).
- When culturing, grow bacteria and fungi on solids (agar) rather than liquids (broth) to avoid spills and aerosol formation. Choose substances such as nutrient agar that do not favour the growth of pathogens. Disposable petri dishes should be used.
- Do not grow cultures of spores collected from telephones, door knobs, washrooms, etc., as they could be contaminated with pathogens.
- Do not grow soil bacteria because of the possibility of culturing tetanus-causing organisms.
- Never culture microorganisms using material from waste containers of polluted water.
- Cultures should be grown at room temperature or in the range of 25°C to 32°C.
- When storing cultures, do not stack petri dishes more than three high on incubator shelves. Tall stacks are a potential hazard if they fall over when the incubator is opened.

- Clean up any spills of microorganism cultures immediately.
- Do not leave cultures for long periods before exposure.
- Dispose of anything that begins to produce an odour.
- Wash hands thoroughly with soap and water after working with any cultures.
- For additional information, refer to Appendix A8, Microbiological Culturing Techniques.

Handling Animals and Plants Safely

- Do not perform any investigation on any animal that might cause suffering or pain, or that might pose a health hazard to you or anyone else in the school.
- Animals that live in the classroom should be treated with care and respect, and be kept in a clean, healthy environment.
- Ensure that your teacher is aware of any plant or animal allergies that you may have.
- Never bring a plant, animal, or other organism to school without receiving prior permission from the teacher.
- Keep cages and tanks clean—both for your health and the health of the organism. Most jurisdictions recommend no live mammals or birds in the laboratory. Reptiles often carry *Salmonella.*
- Wear gloves and wash your hands before and after feeding or handling an animal, touching materials from the animal's cage or tank, or handling bacterial cultures.
- Wild or sick animals should never be brought into the laboratory. Dead animals, wild or tame, that have died from unknown causes should also not be brought into the laboratory.
- Preserved specimens should be removed from the preservative with gloves or tongs, and rinsed thoroughly in running water.
- Before going on field trips, become familiar with any dangerous plants and animals that may be common in the area (e.g., stinging nettles and poisonous plants).

Waste Disposal

Waste disposal at school, at home, and at work is a societal issue. To protect the environment, federal and provincial governments have regulations to control wastes, especially chemical wastes. For example, the WHMIS program applies to controlled products that are being handled. Most laboratory waste can be washed down the drain or, if it is in solid form, placed in ordinary garbage containers. However, some waste must be treated more carefully. It is your responsibility to follow procedures and to dispose of waste in the safest possible manner according to your teacher's instructions.

Flammable Substances

Flammable liquids should not be washed down the drain. Special fire-resistant containers are used to store flammable liquid waste. Waste solids that pose a fire hazard should be stored in fireproof containers. Care must be taken not to allow flammable waste to come into contact with any sparks, flames, other ignition sources, or oxidizing materials. The method of disposal depends on the nature of the substance.

Corrosive Solutions

Solutions that are corrosive but not toxic, such as acids, bases, and oxidizing agents, should be disposed of in a container provided by the teacher, preferably kept on the teacher's desk. Do not pour corrosive solutions down the drain.

Heavy-Metal Solutions

Heavy-metal compounds (e.g., lead, mercury, and cadmium compounds) should not be poured down the drain. These substances are cumulative poisons and should be kept out of the environment. A special container should be kept in the laboratory for heavy-metal solutions. Pour any heavy-metal waste into this container. Remember that paper towels used to wipe up solutions of heavy metals, as well as filter papers with heavy-metal compounds embedded in them, should be treated as solid toxic waste.

Toxic Substances

Solutions of toxic substances, such as oxalic acid, should not be poured down the drain; they should be disposed of in the same manner as heavy-metal solutions, but in a separate container.

Microorganisms

For disposal, place petri dishes in a proper disposal bag labelled with a WHMIS biohazard symbol, and autoclave at 140 kPa for 20 min before discarding in the regular garbage. If an autoclave is not available, call the district Health and Safety Office for a pickup of biohazardous materials. An alternative procedure is to tape the petri dish closed completely and put it in a secondary container before disposal with regular waste or incineration.

Organic Material

Remains of plants and animals can generally be disposed of in school garbage containers. Before disposal, organic material should be rinsed thoroughly to rid it of any excess preservative.

First Aid

The following guidelines apply in case of an injury, such as a burn, cut, chemical spill, ingestion, inhalation, or splash in the eyes.

- Always inform your teacher immediately of any injury.
- Know the location of the first-aid kit, fire blanket, eyewash station, and shower, and be familiar with the contents and operation of them.
- If the injury is a minor cut or abrasion, wash the area thoroughly. Using a compress, apply pressure to the cut to stop the bleeding. When bleeding has stopped, replace the compress with a sterile bandage. If the cut is serious, apply pressure and seek medical attention immediately.
- If the injury is the result of chemicals, drench the affected area with a continuous flow of water for 15 min. Clothing should be removed as necessary. Retrieve the Material Safety Data Sheet (MSDS) for the chemical; this sheet provides information about the first-aid requirements for the chemical.
- If you get a solution in your eye, quickly use the eyewash or nearest running water. Continue to rinse the eye with water for at least 15 min. This is a very long time—have someone time you. Unless you have a plumbed eyewash system, you will also need assistance in refilling the eyewash container. Have another student inform your teacher of the accident. The injured eye should be examined by a doctor.
- If you have ingested or inhaled a hazardous substance, inform your teacher immediately. The MSDS provides information about the first-aid requirements for the substance. Contact the Poison Control Centre in your area.
- If the injury is from a burn, immediately immerse the affected area in cold water or run cold water gently over the burned area. This will reduce the temperature and prevent further tissue damage.
- In case of electric shock, unplug the appliance and do not touch it or the victim. Inform your teacher immediately.
- If a classmate's injury has rendered him or her unconscious, notify the teacher immediately. The teacher will perform CPR, if necessary. Do not administer CPR unless under specific instructions from the teacher. You can assist by keeping the person warm and by reassuring him or her once conscious.

B

B3 Field Trip Safety

Science investigations often take place in locations other than the school classroom or laboratory. Any investigation conducted outside the school building is considered a field trip. Field trips present different hazards than those encountered in the school laboratory. It is very important to know these hazards. Use common sense and the following guidelines to reduce or eliminate the risk of accidents:

- Wear footwear, clothing, and sunscreen (if necessary) that is appropriate for the season and the weather conditions. Depending on the activity, you should also wear safety gear such as goggles, gloves, or PFDs (personal floatation devices).
- Follow all instructions regarding safety from your teacher.
- Behave in an orderly manner. Running and rough play can cause injuries.
- Do not wander off alone. Stay with your group and in visual contact with your teacher. If you become separated from the group and are lost, stay where you are and call out frequently to let others know where you are.
- Do not enter dangerous areas. These include steep cliffs, loose gravel that may slide, swamps, riverbanks, bodies of water, or wave-washed seashores (**Figure 1**).
- If you get caught in a thunderstorm with lightning, avoid any exposed or open areas such as hilltops. Never stand near tall objects such as trees or power poles.
- Never handle or feed wild animals, even young ones. Use binoculars or a small telescope to observe animals from a safe distance. Do not harass or annoy the animals. If an animal shows signs that it feels threatened or disturbed, sit quietly or move slowly away from it.
- Never taste any berries or leaves from wild plants. Some plants are poisonous (**Figure 2**). Some people may be allergic to certain plants that are harmless to most people. If you are allergic to any plants, notify your teacher before going on the trip.
- If you are allergic to bee or other insect stings, inform your teacher and ensure that the appropriate medication is available.
- Do not disturb the natural environment or collect any plant or animal specimens more than is absolutely necessary. Do not leave litter or garbage behind. In fact, you should try to leave the area cleaner than you found it.
- Always wash your hands thoroughly with soap and water when you return from a field trip.

Figure 1
Avoid slopes with loose gravel.

Figure 2
Poison ivy will cause serious skin irritation.

Appendix C REFERENCE

C1 Numerical Prefixes and Units

Throughout *Nelson Biology 11: College Preparation* and in this reference section, we have attempted to be consistent in the presentation and usage of units. As far as possible, *Nelson Biology 11: College Preparation* uses the International System of Units (SI). However, some other units have been included because of their practical importance, wide usage, or use in specialized fields. For example, Health Canada and the medical profession continue to use millimetres of mercury (mm Hg) as the units for measurement of blood pressure, although the Metric Practice Guide indicates that this unit is not to be used with the SI.

The most recent *Canadian Metric Practice Guide* (CAN/CSA-Z234.1-89) was published in 1989 and reaffirmed in 1995 by the Canadian Standards Association.

Numerical Prefixes

Prefix	Power	Symbol
deca-	10^{1}	da
hecto-	10^{2}	h
kilo-	10^{3}	k*
mega-	10^{6}	M*
giga-	10^{9}	G*
tera-	10^{12}	T
peta-	10^{15}	P
exa-	10^{18}	E
deci-	10^{-1}	d
centi-	10^{-2}	c*
milli-	10^{-3}	m*
micro-	10^{-6}	m*
nano-	10^{-9}	n*
pico-	10^{-12}	p
femto-	10^{-15}	f
atto-	10^{-18}	a

* commonly used

Common Multiples

Multiple	Prefix
0.5	hemi-
1	mono-
1.5	sesqui-
2	bi-, di-
2.5	hemipenta-
3	tri-
4	tetra-
5	penta-
6	hexa-
7	hepta-
8	octa-
9	nona-
10	deca-

Some Examples of Prefix Use

0.0034 m = 3.4×10^{-3} m = 3.4 **milli**metres or 3.4 mm
1530 L = 1.53×10^{3} L = 1.53 **kilo**litres or 1.53 kL

C2 Greek and Latin Prefixes and Suffixes

Greek and Latin Prefixes

Prefix	Meaning
a-	not, without
ab-	away from
abd-	led away
acro-	end, tip
adip-	fat
aer-, aero-	air
agg-	to clump
agro-	land
alb-	white
allo-	other
angio-	blood vessel
amphi-	around, both
amyl-	starch
an-	without
ana-	up
andro-	man
ant-, anti-	opposite
anth-	flower
archae-, archaeo-	ancient
archi-	primitive
arter-, artero-, arterio-	artery
aur-	gold
aut-, auto-	self
azo-	nitrogen
baro-	weight (pressure)
bi-	twice
bio-	life
blast-, blasto-	sprout (budding)
carcin-	cancer
card-, cardio-, cardia-	heart
chlor-, chloro-	green
chondr-	cartilage
chrom-, chromo-	colour
co-	with
coronary	heart
cut-	skin
cya-, cyan-	blue
cyt-, cyto-	cell
dendr-, dendri-, dendro-	tree
dent-, denti-	tooth
derm-	skin
di-	two
dors-	back
dys-	difficulty
ec-, ecto-	outside
em-	inside
en-	in
end-, endo-	within
epi-	at, on, over
equi-	equal
erythro-	red
eu-	true
ex-, exo-	away, out
flag-	whip

Prefix	Meaning
gamet-, gamo-	marriage, united
gastr-	stomach
genit-	reproductive organs
geo-	Earth
glyc-	sweet
halo-	salt
haplo-	single
hem-, hema-, hemato-	blood
hemi-	half
hepat-, hepa-	liver
hepta-	seven
hetero-	different
histo-	web
holo-	whole
homeo-	the same
hydro-	water
hyper-	above
hypo-	below
infra-	under
inter-	between
intra-	inside of, within
intro-	inward
iso-	equal
karyo-	nut, kernel
lact-, lacti-, lacto-	milk
leuc-, leuco-	white
lip-, lipo-	fat
lymph-, lympho-	clear water
lys-, lyso-	break up
macro-	large
mamm-	breast
meg-, mega-	great
melan-	black
meningo-	membrane
mes-, meso-	middle
meta-	after, transition
micr-, micro-	small
mono-	one
morpho-	form, shape
muc-, muco-	slime
multi-	many
myco-	mushroom
myo-	muscle
myxo-	slime, mucous
nas-	nose
necro-	corpse
neo-	new
nephr-	kidney
neur-, neuro-	nerve
noct-	night
odont-, odonto-	tooth
oligo-	few
oo-	egg
orni-	bird

Prefix	Meaning
oss-, osseo-, osteo-	bone
ovi-	egg
pale-, paleo-	ancient
para-	beside, near
patho-	disease
peri-	around
petro-	rock
phag-, phago-	eat
pharmaco-	drug
phono-	sound
photo-	light
phylo-	tribe
phyto-	plant
pneum-	air
pod-	foot
poly-	many
pro-	forward
pseud-, pseudo-	false
pyr-, pryo-	fire
radio-	ray
ren-	kidney
rhizo-	root
sacchar-, saccharo-	sugar
sapr-,sapro-	rotten
septi-	seven
soma-	body
spermato-	seed

Prefix	Meaning
sphero-	ball (shape)
sporo-	seed
squam-	scale
sub-	beneath
super-, supra-	above
staphylo-	bunch of
strepto-	twisted
sym-, syn-	with, together
telo-	end
tens-	strain, pressure
therm-, thermo-	temperature, heat
thromb-, thrombo-	clot
tox-	poison
trans-	across
trich-	hair
ultra-	beyond
uro-	tail, urine
vas-, vaso-	vessel
ven-, veno-	vein
vita-	life
vitro-	glass
vivi-	alive
xanth-, xantho-	yellow
xer-, xero-	dry
xyl-	wood
zoo-	animal
zygo-	yoke

C

Greek and Latin Suffixes

Suffix	Meaning
-aceous	like
-ase	enzyme
-blast	budding
-cide	kill
-crin	secrete
-cut	skin
-cyst	bladder
-cyte	cell
-emia	blood
-gen	born, agent
-genesis	formation
-geny	production of
-gram	written
-graph, -graphy	to write
-gynous	woman
-ism	process, system
-itis	inflammation
-ium	structure, tissue
-logy	the study of
-lysis	loosening
-lyt	dissolvable
-mere	share
-metry	measure
-mycin	mushroom
-mnesia	memory
-nomy	related to

Suffix	Meaning
-oid	like
-ol	alcohol
-ole	oil
-oma	tumour
-osis	a condition
-otic	pertaining to
-ous	full of, abounding in
-pathy	suffering
-ped	foot
-phage	eat
-phile	loving
-phyll	leaf
-phyte	plant
-plasty	surgical repair
-pod	foot
-sis	a condition
-some	body
-stas, -stasis	halt
-stat	to stand, stabilize
-tone, -tonic	strength
-troph	nourishment
-ty	state of
-vorous	eat
-yl	wood
-zoa	animal
-zyme	ferment

C3 Human Systems

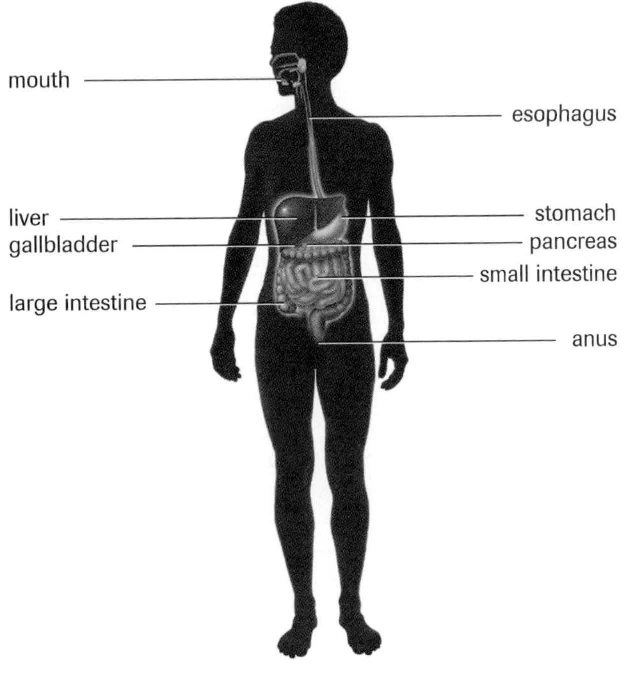

Digestive System

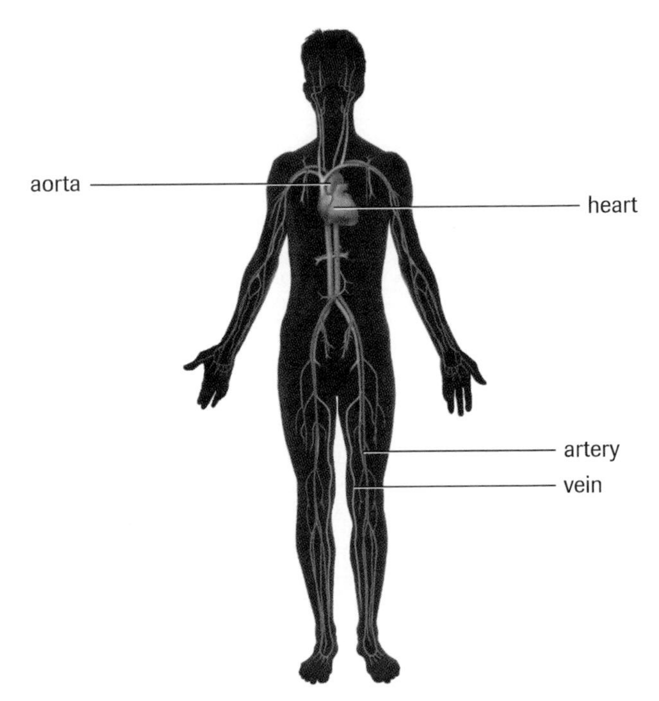

Circulatory System

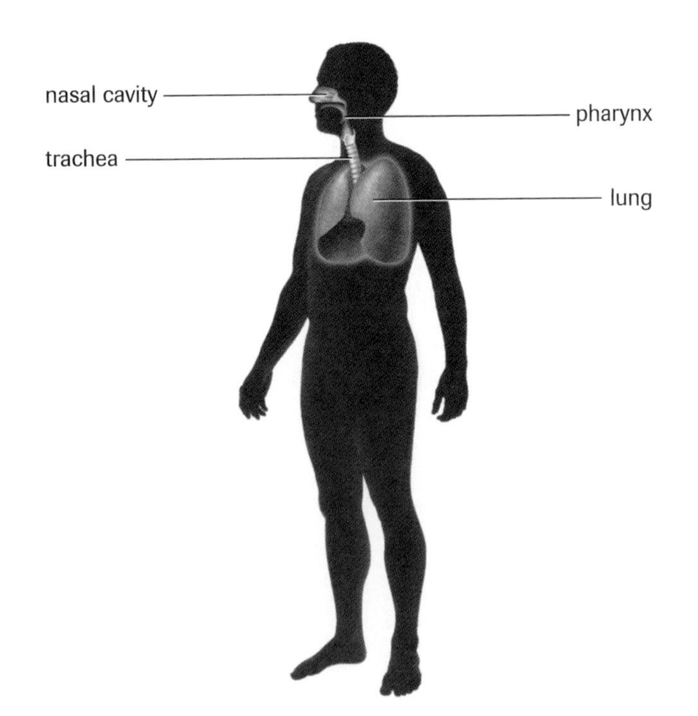

Respiratory System

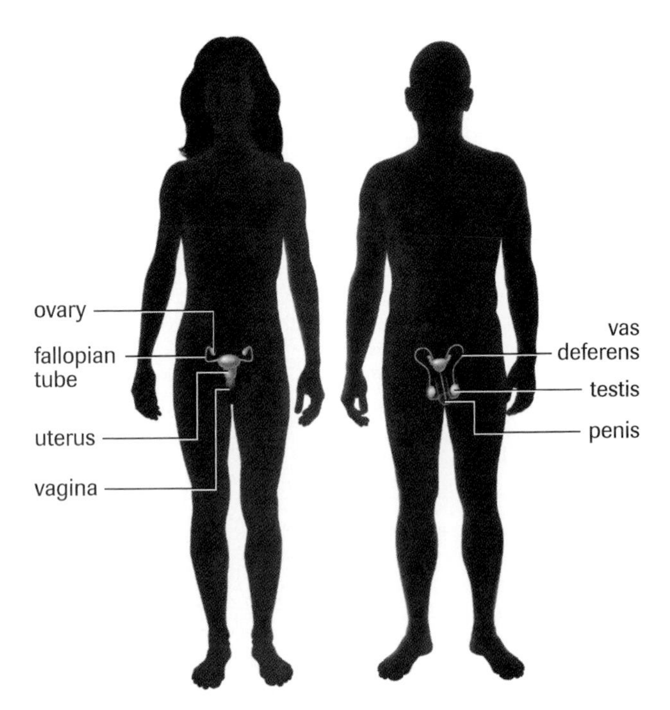

Reproductive System

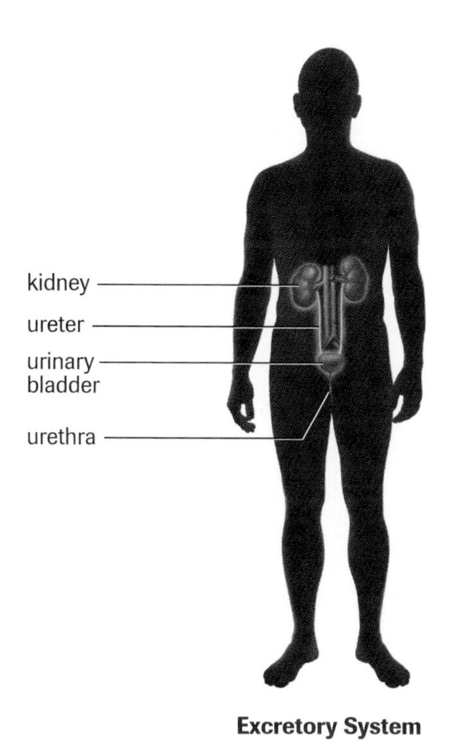

Excretory System

skull
humerus
ribs
ulna
pelvis
radius
femur
knee joint
fibula
tibia
skeletal muscle
tendon
ligament

Locomotion: Skeletal and Muscular Systems

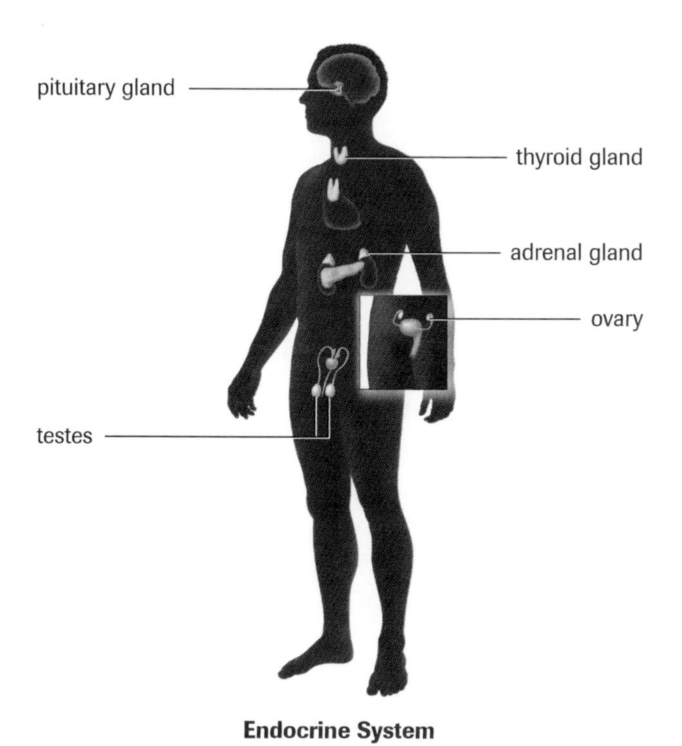

Endocrine System

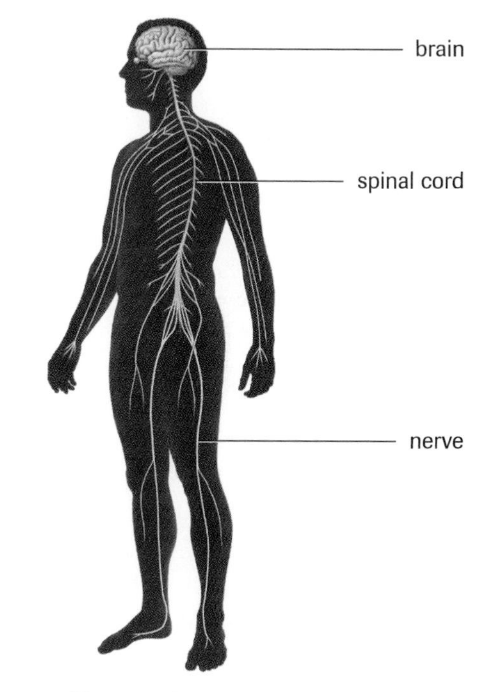

Nervous System

Appendix D ANSWERS

This section includes numerical and short answers to Section and Unit Review questions. The short answers provided are not necessarily complete answers to the questions.

Section 1.1, p. 9

2. invention of the microscope; microorganisms can be seen only with a microscope
3. no; viruses cannot perform some of the basic functions of life (i.e., reproduction) on their own
4. magnification involves making the image larger than the object; resolution is the ability to distinguish between objects that are very close together
7. $249.95; the magnification and the resolution are stated on the label

Section 1.2, p. 15

2. a large water-filled vacuole presses up against the cell membrane and the cell wall creating turgor pressure
3. nucleoplasm is the substance inside the nucleus; cytoplasm includes the cytosol and organelles between the nuclear envelope and the cell membrane
4. muscle cells use more energy than skin cells
5. (a) rough endoplasmic reticulum contains ribosomes; smooth endoplasmic reticulum does not
   (b) rough endoplasmic reticulum is a site of protein synthesis; smooth endoplasmic reticulum is thought to be a site of lipid synthesis
6. lipid and protein molecules of the phospholipid bilayer are in constant motion, and move in to fill the hole produced by the needle when it's removed
7. apoptosis is sometimes called cell suicide; lysosomes break open and digest the cell
8. A=outer membrane, B=crista, C=inner membrane, D= matrix

Section 1.4, p. 24

1. (a) 2 carbon; 4 hydrogen; 2 oxygen
   (b) 2 hydrogen; 1 sulfur; 4 oxygen
   (c) 6 carbon; 12 hydrogen; 6 oxygen
2. (a) ionic bond
3. hydrogen bonds between water molecules hold the molecules together
4. (a) hydrogen bonds
   (b) molecule B is nonpolar; it cannot form strong intermolecular bonds with water so it is excluded
5. (a) acidic
   (b) basic; they will neutralize stomach acid

Section 1.5, p. 35

1. very light element; can form four stable covalent bonds
2. carboxyl, sulfhydryl, phosphate
3. monosaccharide is a simple sugar (i.e., glucose); disaccharide is a double sugar (i.e., maltose)
4. diet rich in saturated fats and cholesterol has been linked to heart disease and stroke
5. amino acid that cells cannot make from simpler substances; must be obtained from diet
6. may cause important enzymes to denature and lose their function
7. polypeptide is a chain of amino acids; protein is a chain of amino acids that has taken on a three-dimensional shape and a specific function
8. a change in the shape of an enzyme that reduces its activity
9. (a) DNA
   (b) nucleus (also mitochondria and chloroplasts)
   (c) RNA; transports genetic information out of the nucleus
10. A always bonds to T, and C always bonds to G
11. nucleotide sequence in DNA determines the amino acid sequence in protein
13. to protect against DNA-destroying enzymes in the cytoplasm
15. (a) albumin denatures
16. (a) contains cellulose that draws water from the intestinal cells and keeps solid wastes moist
    (b) prunes, mineral oil
17. (a) vegetable oil is reacted with hydrogen
    (b) to make it look like butter
18. Curly hair can be straightened, and straight hair can be curled
19. (a) by placing in boiling water for a brief period of time
    (b) blanching prevents the browning reaction
20. (a) candles, lipstick

Section 1.8, p. 42

1. functional food products claim to provide health effects in addition to the nutritional value of the food
2. *Ginkgo biloba* "aids memory"; St. John's wort "balances personality"; *Echinacea* "relieves cold symptoms"
3. many believe that advertisers make unsubstantiated claims on food labels
4. heart disease, osteoporosis, and high blood pressure
5. some functional foods may produce effects that aggravate an already existing condition in a patient

Section 1.9, p. 43

1. blood cells, yeast cells
2. helps the technician keep track of counted and uncounted cells
4. Coulter Counter counts cells more quickly than a technician with a hemacytometer can;

a Coulter Counter cannot be used to count very small cells such as bacteria

Section 1.11, p. 50

1. (a) an undifferentiated cell that has the potential to become any other type of cell
   (b) stem cells may be used to produce transplant tissues
2. types of genes that are active (turned on) determine the identity of a cell
3. (a) a type of skin cell
   (b) they may be transformed into a number of other cell types such as bone and cartilage
   (c) Adult bodies have a relatively small number of stem cells, but a very large number of fibroblasts. Fibroblasts are easy to obtain because they are in easily accessible skin.
5. (a) used to replace bone lost due to gum disease and cartilage lost due to osteoarthritis

Section 1.12, p. 55

1. (a) reduces the activation energy of the reaction it catalyzes
   (b) heat may denature vital enzymes and reduce their functions
4. (a) low pH causes the enzyme to reduce its activity because it is denatured
   (b) pH = 7
5. (a) convert starch into glucose in the sweetener industry; also used to clean blood and grass stains from fabrics

Section 1.14, p. 64

1. (b) glucose in liver cells
2. salt may form a hypertonic solution with steak juices, drawing water out of steak tissue by osmosis, producing a dry steak
3. fertilizer may form a hypertonic solution in the soil that may draw water out of plant cells by osmosis, causing plant cells to die
4. fish egg would take on water by osmosis and may swell and burst
5. (a) rate of diffusion increases
   (b) rate of diffusion decreases
6. (a) yes; concentration of oxygen is higher on one side than the other
   (b) yes; there is a higher concentration of non-diffusable solutes (starch) on one side of the membrane
7. at 5%, osmosis will not occur; 20% would cause osmosis out of the blood cells
8. salt dissolves in saliva, forming a hypertonic solution that draws water out of skin by osmosis
9. (a) the surrounding solution must be replaced with clean, fresh water
   (b) blood plasma will not be cleaned; wastes will not diffuse out of blood
10. (a) filter paper allows some tea solutes through to the water in the cup, but holds back small particles
    (b) water surrounding the tea bag becomes more coloured, then at some point it remains the same colour
11. spray fresh water on the vegetables periodically

Section 1.16, p. 71

1. (b) sodium ions
4. (a) phagocytosis
   (b) A = pseudopod
   B = solid particle
   C = phagocytotic vesicle
   (c) may fuse with the phagocytotic vesicle and digest its contents
5. cell membrane would decrease in size
6. hormones, enzymes

Section 1.17, p. 76

1. plants, plantlike protists, cyanobacteria
3. (a) chlorophyll
   (b) light-absorbing molecules in plants (other than chlorophyll); they transfer light energy to chlorophyll
   (c) yellow xanthophylls and orange carotenoids
4. (a) chloroplast
   (b) A = granum
   B = stroma
   C = inner membrane
   D = outer membrane
   (c) within thylakoid membranes
5. (a) $6CO_2 + 6H_2O \rightarrow C_6H_{12}O_6 + 6O_2$
   (b) oxygen from the air; oxygen from water
   (c) released to the atmosphere by plants

Section 1.18, p. 82

1. photosynthesis does not produce enough ATP to power all of the cellular processes of life
2. (a) aerobic respiration uses oxygen; anaerobic respiration does not use oxygen
3. protein synthesis, cell division, phagocytosis
4. (a) $C_6H_{12}O_6 \rightarrow 2C_3H_6O_3$ (glucose) (pyruvate)
   (b) cytoplasm
   (c) mitochondria
5. (a) 2 ATP molecules
   (b) 36 ATP molecules
6. (a) the action of lactic acid on nerve cells
7. 2 ATP molecules
9. winemaking, bread making

Unit 1 Review, pp. 88–91

1. (a) cells are the basic structural and functional units of life
   (b) all cells arise from the division of other cells
   (c) all living things are composed of one or more cells
2. (b) 150×
7. (a) amyloplasts store starch; chromoplasts store coloured pigment molecules
   (b) amyloplast—Figure 1, chromoplast—Figures 2 and 3
8. (a) sperm cells contain a lot of mitochondria; fat cells contain few mitochondria
9. embedded in thylakoid membranes in chloroplasts
10. (a) a different set of genes are active in skin cells than in nerve cells
    (b) stem cells are not specialized; they have the potential to become almost any type of specialized cell
11. an atom is composed of a single atom, but a molecule contains two oxygen atoms attached by a double covalent bond

D

12. hydrogen bond, ionic bond, covalent bond
13. (a) H_2S and HBr are polar.
    (b) CH_4 is hydrophobic; H_2S and HBr are hydrophilic.
14. (a) Water dissolves more solutes than any other solvent.
    (b) Polar solvents dissolve polar (and ionic) solutes; nonpolar solvents dissolve nonpolar solutes. Oil does not dissolve in water; alcohol dissolves in water.
15. (a) a solution that may neutralize small amounts of acids and bases
    (b) approximately equal concentrations of an acid and a base
    (c) carbonic acid-bicarbonate buffer
16. plant—build cell walls; humans—prevent constipation
17. (a) oil triglycerides are unsaturated or polyunsaturated; butter triglycerides are saturated
    (b) are hydrogenated to convert them to margarine
    (c) both are hydrophobic; steroids are ring molecules, while fatty acids are linear molecules
18. plants use wax to create waterproof coatings; animals such as bees use waxes to build honeycombs
19. (a) protein
    (b) amino, carboxyl
20. all are composed of proteins
21. (a) identity of a protein depends on the number and the sequence of amino acids in its structure
    (b) sequence of nucleotides in DNA determines the sequence of amino acids in proteins
22. (a) creates crevices that act as active sites where substrates attach
    (b) they become denatured; denatured enzymes lose activity and cannot catalyze reactions effectively
    (c) denatures the proteins in a steak, making the steak easier to chew
23. (a) DNA stores genetic information; RNA takes information from the nucleus to ribosomes, where it is used to produce proteins
    (b) messenger RNA; transports the genetic information in DNA from the nucleus to the ribosomes in the cytoplasm
24. (a) transfers energy to energy-requiring reactions in a cell
    (b) it will die
26. (a) A = simple diffusion, B = facilitated diffusion, C = active transport
    (b) a transmembrane carrier protein
    (c) method C uses ATP, and may transport solutes against a concentration gradient
27. (a) when the solute concentration is uniform throughout
    (b) particles move in all directions at the same rate
    (c) no, active transport uses energy to pump solute particles against a concentration gradient, and may cause all particles to go one way
28. protein carriers of active transport require the use of ATP; facilitated diffusion carriers do not use ATP
29. (a) A is hypotonic, B is isotonic, C is hypertonic
    (b) they will burst because water will continue to enter the cells by osmosis
31. enters a mitochondrion and is processed through the reactions of oxidative respiration
32. (a) glycolysis
    (b) 2 ATP molecules
33. muscle cells during strenuous exercise
36. approximately 0.1 mm
43. biodegradable surgical stitches and some types of contact lenses

Section 2.1, p. 101

1. allow scientists around the world to be exact when they are identifying/categorizing the same organism
2. a method of naming organisms by using two Latin names: the genus and the species; both are italicized
4. all bacteria were classified in Kingdom Monera; the division of bacteria into Archaebacteria and Eubacteria resulted in the creation of a six-kingdom system
5. kingdom → phylum → class → order → family → genus → species

Section 2.3, p. 107

1. most scientists do not consider viruses to be living organisms; only living organisms are classified in taxonomy systems
4. micro is 10^{-6}; nano is 10^{-9}

Section 2.4, p. 112

1. most archaebacteria thrive under extreme conditions that other organisms could not tolerate, e.g., without oxygen, in extremely hot environments, in extremely salty environments
2. digestion of sewage and oil spills; alternative fuel source; cancer research; production of bioplastics; production of enzymes for food processing, perfume manufacture, pharmaceuticals, and enzymes used in molecular biology
4. in the absence of oxygen, without air circulation, e.g., in the soil, inside the body, inside a sealed container (e.g., home canning)
5. genetic material is exchanged through the pilus; the new organisms are genetically different from their parents
7. (a) streptococci
   (b) bacillus (c) coccus

Section 2.6, p. 115

1. transported and stored without refrigeration; bacteria destroyed; unopened container has a shelf life of 9–12 months; no additives or preservatives added; nutritional value is identical to fresh milk

3. aseptic packaging is a safe way of transporting and storing a perishable product
5. any group that wants to consume milk over a period of time without access to refrigeration

Section 2.7, p. 117

1. recyclers in the human intestine, releasing digestive enzymes and breaking down organic matter; probiotics; makes vitamin K, which humans require to be healthy
3. as organisms metabolize, heat is released; if temperature of compost relates directly to bacterial activity, the bacteria were most active on day 3
4. as bacteria metabolize, enzymes are released; in nature these enzymes promote decay of organic matter; in industry, these enzymes are harvested to produce consumer products

Section 2.8, p. 119

1. vomiting, cramps, bloody diarrhea, fever, death in some cases
2. ground beef, processed meats, sprouts, leafy green produce, unpasteurized milk and juice, contact with cattle; drinking inadequately chlorinated water or swimming in contaminated lakes

Section 2.9, p. 123

1. external (i.e., skin, sweat, mucus, cilia, tears); lymphatic system (i.e., tonsils, adenoids, lymph nodes, thymus, spleen, lymphocytes, macrophages); third line of defence (i.e., antigens, memory cells, plasma cells, antibodies)
2. through a special circulatory system of vessels and nodes
4. moisture droplets in the air, dust particles, direct contact; fecal contamination, animal bites or wounds

Section 2.10, p. 124

1. orally or by injection; genetically engineered into foods that are then eaten (e.g., banana with hepatitis vaccine, potato with cholera vaccine); can also be absorbed through the skin or nasal passages

Section 2.12, p. 127

1. e.g., penicillin, streptomycin, tetracycline, ciprofloxacin
4. is related to the greater exposure of microorganisms to antibiotics; mutations occur with each exposure, and, before long, a large population of resistance bacteria exists

Section 2.13, p. 133

1. chloroplasts; other identifying features: 1 or 2 flagella; eyespot; pellicle; central nucleus; contractile and food vacuoles; and mitochondria
2. *Euglena* is a single-celled organism; algae are multicellular organisms. Algae have no tissue structure, unlike plants, which are highly specialized into tissues.
7. Amoeba; pseudopods move toward, flow around, and engulf food particles; cell membrane pinches off to form a vacuole around the food, and, inside, enzymes digest the food
11. require at least one host, often two, to complete their life cycles; in some cases, the parasite reproduces asexually in one host and sexually in the other
12. 2 147 483 648 new organisms, assuming all survived and reproduced

Section 2.15, p. 140

1. fungi are heterotrophs (consumers) and lack the structures and ability to produce food
2. often have many nuclei per cell; have few or no storage molecules; have no roots; often have chitin in their cell walls; are heterotrophs; do not reproduce by seed
7. both are fruticose lichen
8. commercially made bread contains preservatives and mould retardants to extend shelf life

Section 2.17, p. 144

3. reduces the diversity of microorganisms directly, as these chemicals kill organisms in the soil; nutrient cycling is reduced and plant growth is affected; changes are seen in the food chain

Section 2.18, p. 148

5. does not produce the side effects associated with insulin produced from pigs or cows; this technology is expensive and is not yet universally available

Unit 2 Review, pp. 154–157

2. (a) Eubacteria
   (b) Protista (c) Protista
   (d) other
   (e) Archaebacteria
   (f) Eubacteria
   (g) Protista (h) Fungi
4. invading virus takes over control of the host cell and directs it to expend energy making viral particles; normal functioning of the host cell stops
6. alternate fuel sources, acetone, bioplastics, stonewashed jeans, enzymes used in food processing (dairy products, vinegar), perfume manufacture, antibiotics and other pharmaceuticals
14. eukaryotic; many cell organelles; cell walls; most are anchored in soil or other substrate; non-motile; reproduction can be asexual, sexual, or both
16. nutritionally (as a dietary supplement), agriculturally (in a process called effective microorganisms technology), industrially (in the production of ethanol), and medically (in gene therapy)
18. diphtheria (3); whooping cough (4); oral polio (1); measles (1); mumps (1); smallpox (5); tetanus (3); injected polio (2)
20. smallpox (N); measles (N); anthrax (Y); *E. coli* food poisoning (Y); *Streptococcus* throat infection (Y); rabies (N); malaria (Y); viral throat infection (N)
22. (a) kills microorganisms in the soil and decreases the natural coating of microorganisms on leaf surfaces

D

27. (a) protist A is *Euglena*, protist B is *Paramecium*, and protist C is *Amoeba*
    (b) protist A—a producer; protists B and C—consumers or parasites
28. (b) pH 5.5
    (c) moisture level, light level, type of bread used, container, location, handling
    (d) growth at all pH levels
29. (a) lactic bacteria, yeasts, moulds, and proteolytic bacteria
    (b) organisms that break protein down into simpler, soluble substances using a hydrolytic process
    (e) 4.4

Section 3.1, p. 166

2. related functions (physiology) or structures (anatomy)
3. pancreas (digestive and endocrine); liver (digestive and excretory); blood vessels (circulatory and respiratory)

Section 3.2, p. 167

1. ligament repair; gallstone removal; kidney stone removal; hernia repairs; spleen, appendix, and uterus removal; correction of spinal injuries; bladder, eye, and colon procedures; correction of birth defects; heart procedures

Section 3.3, pp. 176–177

3. cold temperatures cause muscles to twitch (shivering); a decrease in blood flow to body surfaces; an increase in heart rate
4. sensory—carry impulses from receptors such as the eye and ear to the central nervous system (brain and/or spinal cord); motor—carry impulses from the central nervous system to muscles/glands
5. impulses would reach the spinal cord (and possibly the brain), but not the muscle cells
6. (a) a compound released by one type of cell that travels through the bloodstream
   (b) hormonal response is slower and longer-lasting than a nervous response

Section 3.4, p. 182

1. stimulants increase heart rate and breathing rate; depressants decrease heart rate and breathing rate
3. short term—slower heart rate, slower breathing rate, increased urine output; long term—cirrhosis
4. nerve cells in the brain adjust to long-term exposure to a drug by producing fewer neurotransmitter receptor molecules, causing neurons to become less sensitive to the drug

Section 3.5, p. 188

1. break food down into nutrients that are then absorbed and transported by the circulatory system; glucose, amino acids, fatty acids, and glycerol
2. mouth (teeth begin mechanical digestion, salivary glands contribute enzymes); pharynx; esophagus; stomach; small intestine; (pancreas, liver, gallbladder contribute digestive substances); large intestine; rectum, anus
4. amylase—breaks down carbohydrates into maltose sugars

   pepsin—begins the digestion of protein to long-chain polypeptides

   trypsin—converts long-chain peptides into short-chain peptides

   lipase—breaks down fats into fatty acids and glycerol

Section 3.8, p. 194

1. Body Mass Index; Advantage: easy calculation, more standardized way of assessing ideal body weight than the use of height-weight charts; Disadvantage: does not take into account the proportion of muscle mass to fat mass
2. drain on Canada's Health Care system; coronary artery disease, stroke, osteoarthritis, cancers, diabetes, respiratory diseases, gastrointestinal abnormalities, and psychosocial difficulties

Section 3.9, p. 202

1. brings oxygen and nutrients to cells; takes wastes away from cells; relays chemical messages throughout the body; helps maintain acceptable levels of fluids; permits the transport of immune cells throughout the body
2. circulatory includes the heart, blood vessels, and the blood; cardiovascular system includes only the heart and blood vessels
3. 2 876 083 200 times in a lifetime

Section 3.10, p. 207

1. blood supply to the heart muscles is blocked
9. smoking, physical inactivity, high blood cholesterol, high blood pressure, obesity, poor eating habits, stress, diabetes, and genetics

Section 3.12, p. 214

1. broken down in the liver in a process called deamination
4. (a) D—the kidney
   (b) E—the ureter
   (c) C—the renal artery
   (d) F—the urinary bladder

Section 3.15, pp. 223–224

1. gas exchange; obtaining and supplying oxygen to the blood from the environment and releasing waste carbon dioxide from the blood into the environment
2. respiratory system exchanges gases directly with the blood of the circulatory system

Section 3.17, p. 229

4. all diseases of the respiratory system and all air pollutants
5. individual who smokes 12 cigarettes a day, 135 mg of tar per day
7. (a) 3980 mL, 350 mL, 1220 mL

Section 3.18, p. 239

1. to help reduce heat loss
3. sperm cells are most efficiently produced at temperatures slightly below body temperature
4. seminiferous tubules, male

   in females: passageway for urine

   in males: passageway for urine and semen

   vagina, female

   female, site of fertilization

6. if not fertilized, a human egg cell travels through a fallopian tube and the uterus, and is discharged through the vagina
9. brings the circulatory systems of the mother and the baby close together; also produces hormones that prevent menstruation in the mother
10. a microscopic examination of cells scraped from the surface of the cervix; used for the early detection of abnormal cells in the cervix

Section 3.19, p. 250

1. protection for organs; production of blood cells; storage of minerals; support for muscles
2. skeleton contains strong bones that form a foundation for thinner, weaker bones
4. ossification is the process by which cartilage is replaced by compact bone
5. so that the shape of the head may change during childbirth
7. smooth muscle contracts automatically while skeletal muscle contractions are under conscious control
8. the bones to which they are attached would not move
9. simple—a break in a bone that does not protrude through the skin; compound—a piece of the broken bone protrudes through the skin
12. The child could have a greenstick fracture of the tibia.

Unit 3 Review, pp. 262–265

4. protective mucus layer in the stomach prevents the acid from digesting the stomach lining
5. folds in the wall of the small intestine, villi and microvilli, increase the surface area, allowing for better absorption of nutrients
7. A: adrenal gland; adrenaline

   B: testes; testosterone

   C: thyroid gland; thyroxine
10. (a) autonomic nervous system
    (b) somatic nervous system
    (c) autonomic nervous system
    (d) somatic nervous system
12. Seminal fluid is a useful substance that leaves the body; urine is a waste substance. Secretion is the release of useful substances by cells of the body; excretion is the release of wastes.
13. atria—thin-walled chambers of the heart that receive blood from veins; ventricles—muscular, thick-walled chambers of the heart that deliver blood to the arteries
14. false; the pulmonary arteries, which carry blood to the lungs, have deoxygenated blood
15. high levels of cholesterol in the blood, smoking, diabetes mellitus, high blood pressure, lack of regular exercise, rapid weight gain or loss, and genetic factors (congenital problems)
20. (b), (c), (e), (f), (a), (d)
24. all are living, moist, large, and have a rich blood supply
33. cartilage at the ends of bones at the joints, and slippery synovial fluid in synovial sacs help prevent wear and tear
34. (a) tendons attach muscles to bones
    (b) ligaments attach bones to each other at joints
35. (a) A
    (b) B
    (c) C
    (d) D
    (e) antagonistic muscles; when one muscle contracts, the other muscle relaxes
39. (a) patient D

Section 4.1, p. 274

1. each plant species supports an average of 50 different animal species; the loss of even a single plant species can significantly reduce the biodiversity of an ecosystem
3. habitat destruction
4. it will have few or no natural predators, parasites, or pathogens

Section 4.2, p. 279

1. (b), (e), (f) & (a), (c), (d)
2. absence of specialized vascular tissue for support and water transport
3. their male gametes are sperm and require a thin film of water to reach the female gametes
4. to lower soil pH and help retain moisture
5. yes; a large volume of water would dissolve away the acid-causing chemicals in the moss

Section 4.3, p. 284

1. they are adaptations to survival and sexual reproduction in the dry conditions of terrestrial environments
2. specialized flowers for pollination and fruits to protect and assist in seed dispersal
3. wood, paper, varnishes, turpentine, disinfectants, fuel
4. monocot and dicots; monocot seeds contain a single cotyledon; dicot seeds contain two cotyledons
5. seed coat—protection from desiccation

   endosperm—food storage

   cotyledon—food storage (may become first photosynthetic seed leaves)

   embryo—the immature plant
6. they are wind pollinated

Section 4.5, p. 289

1. prevents water loss, keeps out pathogens, and may possess physical features and/or chemical toxins to deter herbivores
4. growth and differentiation
5. phloem; this is living cellular material and would be much more nutritious than just the cell walls of xylem tissue
6. it probably serves a protective function
7. ground tissue; this tissue stores nutrients

Section 4.6, p. 292

1. meristems found at the tips of roots and shoots; cause increase in length of plant
2. primary growth is all growth of apical meristems and first-year growth of lateral meristems; secondary growth is all growth from

D

lateral meristems after the first year

3. provides added strength
4. lateral meristems
5. apical meristems; lateral meristems

Section 4.8, p. 299

1. roots absorb water and minerals from soil, and support and anchor the plant; stems provide support for the shoot and link roots to leaves and reproductive structures; roots absorb water and minerals almost exclusively
2. storage; carrots (roots), potatoes (stems)
4. preventing erosion; fibrous, since these hold soil in place
5. they would likely be eaten
7. (a) approximately 11 years old
   (b) heartwood
   (c) sapwood

Section 4.9, p. 300

1. its size
3. flooring, roofing, panelling, furniture

Section 4.10, p. 304

1. photosynthesis
2. (a) prevents water loss
   (b) photosynthesis
   (c) gas exchange
   (d) gas exchange and photosynthesis
   (e) regulates stomatal openings
3. water and carbon dioxide; oxygen gas and carbohydrates
4. chemical toxins, surface hairs, cactus spines

Section 4.12, p. 309

1. monocots (specifically grasses)
2. dicots
3. tomatoes, squash, cucumbers produced from flowers and contain seeds
4. seeds store protein, carbohydrates and fats, and fibre, and so are very nutritious
6. fruits protect the seed and aid in its dispersal, often by attracting and providing food to herbivores, which then carry away the seeds
7. most angiosperms will not produce fruit until their flowers have been pollinated

Section 4.15, p. 319

1. both are regulated by hormones

Section 4.17, p. 327

1. temperature and precipitation
6. nitrogen, phosphorus, potassium
7. humans use plants that are high in nutrient content as food crops
8. they must have ample time to complete their reproductive life cycle and produce fruit and seeds
9. (a) nitrogen, potassium
   (b) 10-0-10, 0-15-15

Section 4.19, p. 332

1. herbicides, insecticides, and fungicides
3. broad spectrum pesticides kill a much greater variety of species than selective pesticides

Section 4.21, p. 336

1. traditional techniques such as pesticides and crop management, as well as biological controls
2. (a) using pheromones

Section 4.22, p. 340

2. high ornamental value, can't be replaced, and are a potential safety hazard if unhealthy (falling limbs, etc.)
5. forestry, hydroponics, turf management

Section 4.23, p. 343

1. using plants to repair pollutant-damaged environments; Cootes Marsh and Chernobyl
2. abundant water and nutrient supply
3. pollution from sewer overflows and introduced carp

Unit 4 Review, pp. 350–353

1. mosses—(b), (c), (d), (f); ferns—(b), (c); gymnosperms—(a), (e), (h); angiosperms—(a), (e), (g)
2. (a) prevents water loss
   (b) internal transport system and support
   (c) support
   (d) permits sexual reproduction without water
   (e) protects the plant embryo from desiccation
   (f) enhances sexual reproduction
   (g) protects seeds and aids in their dispersal
7. meristematic tissue—cell division for growth; gives rise to other tissue and cell types

   dermal tissue—produces a waxy cuticle; has microscopic hairs on root dermal cells to increase water and mineral absorption; protects the plant

   ground tissue—photosynthesis; stores nutrients

   vascular tissue—transports water, minerals, sugars, and other nutrients
10. (a) protection from desiccation
    (b) food storage/supply for growing embryo
    (c) food storage/supply for growing embryo; first photosynthetic organ
    (d) the embryo is a young plant
12. (a) cell elongation
    (b) flower death and fruit ripening, leaf and fruit drop
    (c) cell division and leaf growth
    (d) cell division and elongation
21. 1.5 kg nitrogen, 0.5 kg phosphorus, and 1.0 kg potassium compounds
34. (a) pollen
    (b) vascular bundle
    (c) stoma
    (d) root tip

Section 5.1, p. 362

3. quantitative data—number of caribou, length of pipeline, speed of migration; qualitative data—feeding habits, health of caribou, types of food

Section 5.2, p. 368

1. autotrophs—make their own food; heterotrophs—cannot make their own food, so they feed on autotrophs
3. temperature and precipitation
4. amphibians live in both terrestrial and aquatic habitats and are affected by changes in either

9. species C is probably an indicator species because it responded to changes before the other species

Section 5.4, p. 375

9. 7.3 kJ
11. primary consumers—3.80, 0.92, 0.50; secondary consumers—0.380, 0.092, 0.050

Section 5.6, p. 383

4. ambush—e.g., angler fish; pursuit—e.g., eagle

Section 5.8, p. 387

5. (a) mutualism—both populations flourish, B increases as A increases
   (b) commensalism—A flourishes but B is not affected

Section 5.9, p. 395

4. (a) 0.2 turtles/ha or 1 turtle/5 ha
   (b) 0.3 turtles/ha or 1 turtle/3.33 ha
7. 65 mice immigrated
8. (a) 1.06, or 6%
   (b) 56 180 after 2 years; 89 542 after 10 years
11. Population A Population B

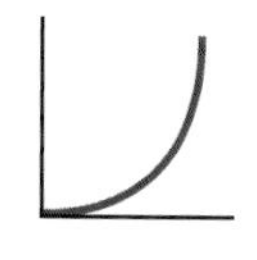

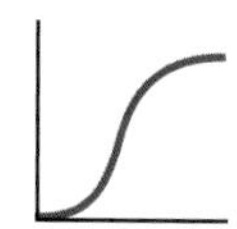

Section 5.10, p. 399

3. (a) density independent
   (b) density independent
   (c) density dependent

7. (a) density dependent
   (b)

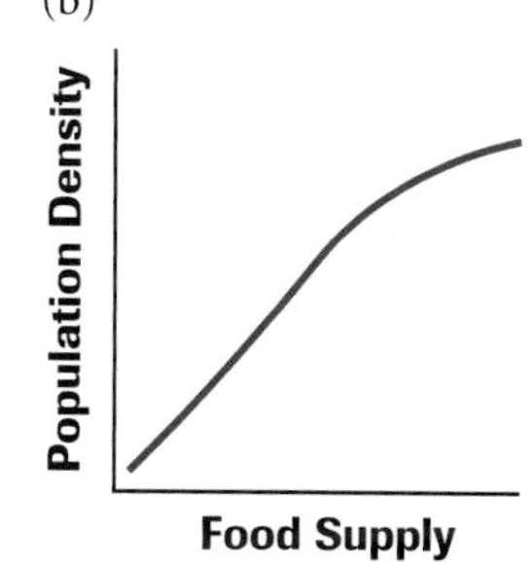

Section 5.12, p. 405

2. 733
3. (a) 4415
4. population density = 0.54 slugs/m^2

   population size = 54 slugs
5. (a) School yard: population size = 4550 grasshoppers; density = 2275 grasshoppers/ha
   Near oil refinery: population size = 404 grasshoppers; density = 404 grasshoppers/ha
6. (b) 323.2 cinch bugs/m^2
   (c) 121 200 cinch bugs

Section 5.15, p. 410

4. 2053
5. no; resources (food, water, space) will be limited
6. 8 billion in 2022

   10 billion in 2039

Section 5.16, p. 417

2. CO_2 emissions of the United States is more than 5 times greater than that of India
5. (a) Africa; North America
   (b) Latin America and the Caribbean; West Asia
   (c) West Asia, Europe, and North America
   (d) 0.7%

Section 5.19, p. 425

2. population growth and overconsumption
5. (a) All of the original forest will be cut after 8 years. Only 95 000 ha will have been reforested in that time, none of which will be of harvestable size.
   (b) no; they are proposing to harvest more than they replant

Unit 5 Review, pp. 430–433

1. the sun
3. first trophic level—25 000 J

   second trophic level—2500 J

   third trophic level—250 J
4. humans could feed directly from the producer, or first trophic, level
5. 15 squirrels/ha
14. (a) density independent
    (b) density dependent
    (c) density dependent
    (d) density independent
16. endoparasites—inside the body of the host; ectoparasites—on the outside of the host's body
17. clumped—humans live where conditions are most favourable
19. 125 years
20. (a) 3 persons/km^2
    (b) some areas, e.g., Rocky Mountains, Great Lakes, and far North are not suitable human habitats
21. population growth rate = 1.4%
22. (a) killer whale, bald eagle, marsh hawk
    (b) marsh vegetation
    (c) marsh hawk, bald eagle
    (d) three—harbour seal, killer whale, and bald eagle
    (e) primary consumers—snail, vole, snow goose, herring, grasshopper, and mosquito; secondary consumers—rat, herring, salmon, marsh hawk, bald eagle, harbour seal, and killer whale
24. increases will equal decreases—births and immigration will offset deaths and emigration
27. (a) 2960
    (b) 24.7 trout/ha
    (c) clumped (in reed beds)
    (d) ecological population density will be higher because only 15% of the lake is suitable habitat; ecological population density = 164.4 trout/ha
29. (a) 3125 larvae/m^2
    (b) 125 000 larvae
30. graph (a): owls living in forest

    graph (b): bacteria in a laboratory
31. (a) 1125 deer
    (b) 3 deer/ha
    (c) 4 deer/ha

Glossary

A

abiotic referring to the nonliving components of an ecosystem

accessory pigment a multicoloured pigment in chloroplast membranes that assists chlorophyll in absorbing light energy

acrosome a vesicle in a sperm cell containing enzymes that allow the sperm to penetrate an egg cell

activation energy the energy required for a chemical reaction to occur

active immunity lasting protection against an invading antigen through the manufacture of antibodies

active site the location where the substrate binds to an enzyme

active transport the movement of substances through a membrane against a concentration gradient using membrane-bound carrier proteins and energy from ATP

addiction a compulsive need for a harmful substance. Addiction is characterized by tolerance for the substance and withdrawal symptoms.

adenosine triphosphate (ATP) a compound used as a source of chemical energy in cells

adrenaline a hormone secreted by the adrenal glands that increases heart rate, increases blood pressure, causes perspiration, and increases muscle tension

aerobic requiring oxygen for respiration

aerobic cellular respiration a form of cellular respiration that uses oxygen and produces large amounts of ATP

albumen the protein-rich nutrient surrounding the yolk in a bird egg

Allee effect a density-dependent phenomenon that occurs when a population fails to reproduce enough to offset mortality once the population density is very low; such populations usually do not survive

alternation of generations a life cycle that alternates between sporophyte individuals and gametophyte individuals

alveolus one of the tiny air sacs within the lungs in which gas exchange occurs between the air and the blood

amino acid a protein subunit

ammonification the part of the nitrogen cycle during which nitrogen compounds are converted to ammonia by ammonifying bacteria in the soil

amylase an enzyme that breaks down complex carbohydrates

anaerobic conducting respiration processes in the absence of oxygen

anaerobic cellular respiration a form of cellular respiration that does not use oxygen and produces small amounts of ATP

angioplasty a procedure in which a blocked artery is opened by inflating a small balloon at the end of a catheter

angiosperm a member of the kingdom Plantae that has a vascular system and produces enclosed seeds within flowers

annual a plant that completes its life cycle in one year

annual ring the amount of xylem produced in one year visible in the cross section of a woody stem. The number of annual rings indicates the age of a woody plant.

antagonistic muscles the skeletal muscles that are arranged in pairs and that work against each other to make a joint move

anther an organ in a flower that produces pollen grains

antibiotic a chemical produced synthetically or by microorganisms that inhibits the growth of or destroys certain other microorganisms

antibody a protein that inactivates a foreign substance in the body by binding to its surface

antigen any substance that causes the formation of antibodies

antiseptic a chemical that destroys or impedes the growth of disease-causing organisms without harming body cells

aorta the largest artery in the body leading directly from the heart and carrying oxygenated blood to the tissues of the body

apical meristem a meristem located at any growing tip of shoots or roots, which contributes to increases in the length of plant tissues

apoptosis cell suicide carried out by lysosomes

arboriculture the activity of growing and caring for trees and shrubs

arborist a person who has expertise in the practical aspects of growing and maintaining trees and shrubs

Archaebacteria in a six-kingdom system, a group of prokaryotic microorganisms distinct from Eubacteria that possess an unusual cell-wall structure and that thrive in harsh environments such as salt lakes and thermal vents

arteriosclerosis a group of disorders that cause the walls of blood vessels to thicken, harden, and lose their elasticity

assimilation the uptake of ammonia, ammonium ions, or nitrates by plants and their conversion into organic molecules containing nitrogen

atherosclerosis a degeneration of the arteries caused by the accumulation of plaque along the inner wall

atrioventricular (AV) node a small mass of tissue through which impulses from the sinoatrial node are passed to the ventricles

atrium a thin-walled chamber of the heart that receives blood from veins

autonomic nervous system motor neurons that function without conscious control

autotroph an organism (such as a plant) that obtains energy directly from light

axillary bud a bud that will form a side branch from a stem

axon the part of a neuron that carries impulses away from the cell body

B

bacteriophage a category of viruses that infect and destroy bacterial cells

bark the outer layers of older stems, branches, and trunks, consisting of all tissues from the vascular cambium outward

biennial a plant that completes its life cycle in two years

bile salts the components of bile that break down large fat globules

binomial nomenclature a method of naming organisms by using two names—the genus name and the species name. Scientific names are italicized.

bioaccumulation the increase in concentration of a chemical contaminant as it is passed up through the food chain

biodiversity a measure of the number and abundance of different species living in a particular ecosystem

biogeochemical cycle the process by which matter cycles from living organisms to the nonliving, physical environment and back again (e.g., the carbon cycle, the nitrogen cycle, the water cycle, and the phosphorus cycle)

biome a large area of Earth characterized by distinctive climate and soil conditions with a biological community adapted to those conditions

bioremediation a process that uses biological mechanisms to destroy, transform, or immobilize environmental contaminants

biosphere that part of Earth inhabited by plants and animals (all communities of living things on Earth)

biotechnology the use of living things in industrial or manufacturing applications

biotic referring to the living organisms and their products in an ecosystem

biotic potential the maximum rate at which a population can increase under ideal conditions

blood pressure the pressure exerted on the walls of the arteries when the ventricles of the heart contract. Blood pressure is measured in millimetres of mercury, or mm Hg.

Body Mass Index (BMI) a calculation obtained by dividing an individual's weight in kilograms by his or her height in metres squared

bone fracture a broken bone

Bowman's capsule the cuplike structure that surrounds the glomerulus

broad-spectrum pesticide a pesticide used to kill a wide range of target and nontarget species

bronchiole one of the small air passages located within and throughout each lung

bronchus one of the air passages from the trachea that goes to the right or left lung

bryophyte a member of a group of nonvascular plants that includes species of mosses, liverworts, and hornworts

buffer a solution containing chemicals that can neutralize small amounts of acid and base

bulk transport the movement of large quantities of materials into or out of a cell

C

Calvin cycle reactions of photosynthesis in which carbon dioxide molecules are used to produce carbohydrates such as glucose

capsid the protective protein coat of viruses

carbohydrate a molecule containing carbon, hydrogen, and oxygen, primarily used by living organisms as a source of energy

cardiac catheterization the insertion of a catheter into the coronary artery through an incision in an artery in the groin

cardiac muscle the involuntary muscle of the heart

cardiovascular system part of the circulatory system comprising the heart and blood vessels

carrier an infected individual who carries disease but does not become infected herself/himself

carrier protein a transmembrane protein that facilitates the diffusion of certain substances through a membrane

carrying capacity the maximum number of organisms that can be supported by available resources over a given period of time

cartilage the semi-solid, flexible connective tissue

catalyst a chemical that speeds up chemical reactions

cellular respiration a set of reactions partially occurring in mitochondria that produces compounds that cells use as a source of energy

cellulose glucose polymer found in plant cell walls

central nervous system the brain and spinal cord

cervix the opening of the cervix into the vagina

chemotroph a producer that converts the energy found in organic chemical compounds into other forms of energy without the use of sunlight

chitin a modified form of cellulose that forms the tough exoskeleton of insects, fungi, and crustaceans

chlorophyll the green pigment that begins the process of photosynthesis. It is found in the thylakoid membranes of plant cell chloroplasts.

chloroplast a cell organelle found only in plant cells. It contains chlorophyll and carries out the reactions of photosynthesis.

chondrocyte a cartilage cell

chyme a mixture of partly digested food, water, and gastric juice

circulatory system the body system consisting of the heart, blood vessels, and blood that is responsible for the transportation of materials around the body and plays a major role in the regulation of body temperature

clay small particles of minerals

cloaca in female birds, the external organ containing the oviduct opening; in male birds, the external organ containing the vas deferens opening

closed population a population in which changes in number and density are determined by natality and mortality alone

clumped dispersion a distribution pattern in which individuals in a population are more concentrated in certain parts of a habitat

collagen the connective protein fibres found in bone matrix

colon the largest segment of the large intestine, where water reabsorption occurs

colony a visible growth of microorganisms containing thousands of cells

commensalism a symbiotic relationship in which one organism benefits and the other organism is unaffected. It is categorized as a +/0 relationship.

community populations of all species living in a particular area at a particular time

compact bone dense bone

companion cell a phloem cell that contains a nucleus and other cell organelles, and that controls the cellular activity of sieve-tube elements

compost a mixture of decomposing organic material, usually plant wastes such as food scraps, yard waste, or inedible plant materials

compound leaf a leaf that is divided into two or more leaflets

concentration gradient a difference in concentration between two areas

conifer a plant that reproduces sexually using specialized male and female cones

coronary bypass surgery a surgical procedure in which a vein graft is used to shunt the blood supply around the blocked area in a coronary artery

corpus luteum a yellowish mass of cells in an ovary after ovulation that secretes the hormone progesterone

cotyledon a leaflike structure of the embryo of a seed plant

covalent bond a bond formed by the sharing of electrons between atoms in a molecule

crenation clumping of cells (usually red blood cells) that have become scallop-shaped when placed in a hypertonic solution

crista an infolding of the inner membrane of mitochondria

crude density population density measured in terms of the number of organisms of the same species within the total area of the entire habitat

cyst a cell that has a hardened protective covering on top of the cytoplasmic membrane

cytosol the aqueous solution inside a cell in which cell organelles are suspended

D

deamination the removal of the amino group from an organic compound

denaturation a change in the three-dimensional shape of a protein caused by high temperatures or harsh acidic, basic, or salty environments

dendrite the part of a neuron that carries impulses toward the cell body

denitrification the part of the nitrogen cycle during which nitrates are converted by denitrifying bacteria into nitrogen gas and released into the atmosphere

density-dependent factor a factor that influences population regulation, having a greater impact as population density increases or decreases

density-independent factor a factor that influences population regulation regardless of population density

depressant a drug that slows down the action of the central nervous system, often causing a decrease in heart and breathing rates

dermal tissue cells specialized for covering the outer surfaces of leaves, stems and roots

detoxify to remove the effects of a poison

diagnosis the process of determining the type of illness a person may have, and possibly the cause of the illness

diaphragm a sheet of muscle that separates the organs of the chest cavity from those of the abdominal cavity

diastole relaxation (dilation) of the heart, during which the cavities of the heart fill with blood

dichotomous key a two-part key used to identify living things. *Di-* means "two."

dicot an angiosperm that produces seeds containing embryos with two cotyledons

differentiation a change in the form of a cell, tissue, or organ to allow it to carry out a particular function

dikaryotic describes cells that contain two haploid nuclei, each of which came from a separate parent

disaccharide a sugar molecule composed of two monosaccharides linked by a covalent bond

disinfectant a strong chemical used to destroy or impede the growth of disease-causing organisms

dormant a condition marked by no or very slow growth that is overcome only by specific environmental conditions

doubling time the time it takes for a population to double in size at a given rate of increase; approximately equal to 70 divided by the annual growth rate

duodenum the first segment of the small intestine

dynamic equilibrium a state of balance where particles move in all directions at equal rates

dynamic equilibrium (population) the condition of a population in which the birth rate equals the death rate and there is no net change in population size

E

ecological density population density measured in terms of the number of individuals of the same species per unit area or volume actually used by the individuals

ecology the study of the interactions of living organisms with one another and with their nonliving environment of matter and energy

ecosystem a community of different species interacting with one another and with the abiotic environment

ectoparasite a parasite that lives and feeds on the outside surface of the host

ectoplasm the thin, semirigid (gelled) layer of the cytoplasm under the cytoplasmic membrane

electrocardiogram the graphical record produced by an electrocardiograph

electrocardiograph an instrument that detects the electrical signals of the heart

electrolyte a solution composed of positive and/or negative ions dissolved in water

emergent plant an aquatic plant that grows in shallow water with most of its shoots above the water

endemic species a species that does not naturally occur anywhere other than one geographic site on Earth

endocrine system the organ system that regulates internal environmental conditions by secreting hormones into the bloodstream

endocytosis a form of bulk transport used to bring large amounts of material into the cell from the extracellular fluid

endoparasite a parasite that lives and feeds within the host's body

endoplasm the fluid part of the cytoplasm that fills the inside of the cell. The endoplasm is responsible for an amoeba's shape as it moves.

endosperm a tissue in angiosperm seeds that stores food

endospore a dormant cell of bacilli bacteria that contains genetic material encapsulated by a thick, resistant cell wall. This form of cell develops when environmental conditions become unfavourable.

endotherm an organism that maintains a near-constant body temperature. All birds and mammals are endotherms.

energy flow the one-way transfer of energy within an ecosystem

energy pyramid a model that illustrates energy flow from producers at the beginning of food chains to consumers farther along

environmental resistance any factors that limit a population's ability to reach its biotic potential when it nears or exceeds the environment's carrying capacity

enzyme a protein that speeds up chemical reactions

enzyme-substrate complex an enzyme with its substrate attached to the active site

epidermis the outermost surface layer of living cells

epiglottis the strcture that covers the opening of the trachea during swallowing

ethanol fermentation a form of anaerobic respiration in which glucose is converted into ethanol, ATP, and carbon dioxide gas

Eubacteria in a six-kingdom system, a group of prokaryotic microorganisms that possess a peptidoglycan cell wall and reproduce by binary fission

eukaryotic a type of cell that has a true nucleus, surrounded by a nuclear envelope

evergreen a plant that retains its leaves throughout the year, such as pines and spruces

exocytosis the movement of large amounts of material out of a cell by means of secretory vesicles

exotic species a species that has migrated to or has been introduced by humans to a geographic area

expiration the action of breathing out, or exhaling, air

exponential growth a pattern of population growth in which organisms reproduce continuously at a constant rate

extensor the muscle that must contract to straighten a joint

external intercostal muscles muscles between the ribs that raise the rib cage, increasing volume and reducing air pressure within the chest

extracellular fluid the watery environment outside a cell

F

facilitated diffusion the diffusion of solutes through a membrane assisted by proteins

fern a member of the kingdom Plantae that has a vascular system and does not produce seeds

fertilizer any minerals added to the soil, usually to replace those removed by planted crops

fibroblast an unspecialized skin cell

fibrous root a root system in which large numbers of relatively equal-sized roots arise from the base of the plant immediately below the ground

flexor the muscle that must contract to bend a joint

flow phase the phase of the menstrual cycle in which menstruation occurs

follicle-stimulating hormone (FSH) the hormone secreted by the pituitary gland at puberty that stimulates the development of egg cells in the ovary

follicular phase the phase of the menstrual cycle in which follicles develop within the ovary

frond the leaf portion of a fern

fruit a structure that develops from the ovary of a pollinated angiosperm, in which the seed develops

functional group a cluster of atoms that gives compounds specific chemical properties

Fungi a group of organisms formerly regarded as simple plants lacking chlorophyll, but now classified in a separate kingdom

G

gametophyte the form of a plant that reproduces sexually

genetically modified organism (GMO) an organism in which the genetic material has been modified to suit human purposes, often by the introduction of genes from other species

geometric growth a pattern of population growth in which organisms reproduce at fixed intervals at a constant rate

germinate to grow or sprout; refers specifically to the embryo inside a plant seed

gestation period the normal period of time from conception to birth

gland an organ that secretes hormones and other useful substances

glomerulus a capillary bed that is the site of fluid filtration

glycolysis a series of reactions occurring in the cytoplasm that breaks a glucose molecule into two pyruvate molecules and forms two molecules of ATP

goblet cell a specialized cell that produces mucus

gonad the organ of males and females that produces and stores gametes

granum a stack of thylakoids found in chloroplasts

gravitropism a change in the growth pattern or movement of a plant in response to gravity

greenhouse effect a result of certain atmospheric gases, such as CO_2, water vapour, and methane, trapping heat in the atmosphere by letting visible sunlight penetrate to Earth's surface while absorbing most wavelengths of infrared radiation that radiate from Earth's surface

greenstick fracture a fracture, commonly found in children, in which the bone does not break all the way through

ground tissue the internal nonvascular tissues of a plant. They perform many functions, such as photosynthesis and food storage.

growth the increase in volume or dry weight of an organism

guard cell one of a pair of cells around each stoma in the epidermis of a leaf or stem that regulates the opening and closing of the stoma

gymnosperm a member of the kingdom Plantae that has a vascular system and produces naked seeds

H

habitat the place or type of environment where an organism or a population of organisms lives

haploid nucleus, cell, or organism with one set of chromosomes. The haploid number is designated as n.

hardiness the ability of a plant to tolerate harsh climatic conditions, usually in reference to winter temperature extremes

hardiness zone a geographic region with a specific range of climatic conditions

hemolysis swelling and bursting of red blood cells placed in a hypotonic solution

herbaceous plant a nonwoody plant, such as grasses and tulips

heterotroph an organism (such as an animal) that obtains energy by eating other organisms

homeostasis the maintenance of a healthy balance of all chemical reactions in an organism

homologous structures body parts that are similar in structure but differ in function

hormone a compound released by one type of cell that has an effect on other cells of the body

hormone-releasing factor a hormone released by the hypothalamus of the brain that controls the secretion of various hormones of the anterior lobe of the pituitary gland

horticulture the cultivation of garden plants, including fruits, grasses, flowers, vegetables, and trees

host range the limited number of host species, tissues, or cells that a virus or other parasite can infect

humus partially decomposed organic material

hydrogen bond the force of attraction between highly polar molecules containing combinations of oxygen, hydrogen, and nitrogen atoms

hydrophilic a substance that dissolves in water because it is polar

hydrophobic a substance that does not dissolve in water because it is nonpolar

hydroponics a technique for growing plants in solution rather than in soil

hypertonic solution a solution that has a higher solute concentration than some other solution

hypotonic solution a solution that has a lower solute concentration than some other solution

I

immovable joint bones that are attached so tightly that little or no movement is possible

implantation the attachment of an embryo to the endometrium

impulse an electrical signal that travels through neurons

indicator species a species sensitive to small changes in environmental conditions. Monitoring the health of its population is used to indirectly monitor the health of the environment in which it lives.

induced-fit model a model of enzyme activity that describes an enzyme as a protein molecule that changes shape to better accommodate the substrate

inference a logical conclusion that is based on observations or prior knowledge, but that has not been scientifically tested

insertion the place where a muscle is attached to a moving bone

inspiration the action of breathing in, or inhaling, air

integrated pest management (IPM) a pest-management strategy that uses a combination of careful monitoring, natural biological controls, and limited applications of synthetic pesticides

internode the space between two successive nodes on the same stem

interspecific competition competition between individuals of different species for an essential common resource that is in limited supply

intraspecific competition competition among individuals of the same species for resources in their habitat

invasive species an introduced species that out-competes native species in an ecosystem

ionic bond the attraction between positive and negative ions in an ionic compound

isotonic solution a solution of equal solute concentrations

J

joint any place in a skeleton where two or more bones meet

K

keystone species a species that is critically important to the balance and survival of the particular ecosystem in which it lives

L

lactate (lactic acid) fermentation a form of anaerobic respiration in which glucose is converted into lactate and ATP

lacteal a small vessel that provides the products of fat digestion access to your circulatory system

larynx the enlarged portion of the trachea containing a pair of vocal cords

lateral meristem a meristem located within a stem or root that contributes to increases in the diameter of plant tissues

leaf blade the wide portion of a leaf

leaf petiole a stalk that supports the leaf and attaches it to a stem

lichen a combination of a green alga or cyanobacterium and a fungus growing together in a symbiotic relationship

ligament a strong, stretchable band of connective tissue that holds the bones of movable joints together

light reactions of photosynthesis reactions of photosynthesis in which light energy is absorbed by chlorophyll and transferred to ATP

limiting factor any essential resource that is in short supply or unavailable

lipase a fat-digesting enzyme

lipid a large molecule composed of carbon, oxygen, and hydrogen, with a higher proportion of hydrogen atoms than carbohydrates

locomotion the ability to move from place to place under one's own power

logistic growth a model of population growth describing growth that levels off as the size of the population approaches the carrying capacity of the environment

luteal phase the phase of the menstrual cycle in which the uterus lining (the endometrium) is prepared to receive a fertilized egg

luteinizing hormone (LH) the hormone secreted by the anterior pituitary gland that stimulates the development of egg cells in the ovary, the monthly release of mature egg cells, and the development and maintenance of a yellowish mass called the corpus luteum in the ovary after fertilization

lymphocyte a white blood cell (plasma and memory cells) that can make antibodies when stimulated by antigens

lysis the destruction or bursting open of a cell. An example is when an invading virus replicates in a bacterium and many viruses are released.

lysogeny the dormant state of a virus

M

macromolecule a very large organic molecule

macronutrient one of a group of nine nutrients required by plants in quantities greater than 1000 mg/kg of dry mass

macrophage a white blood cell that engulfs and destroys pathogens

magnification the ability of lenses to enlarge the image of a specimen

maltase a carbohydrate-digesting enzyme

manure animal waste

mark-recapture method a sampling technique for estimating population size and density by comparing the proportion of marked and unmarked animals captured in a given area; sometimes called capture-recapture

matrix the protein-rich fluid that fills the interior space of a mitochondrion

meristematic tissue plant tissue composed of cells that are able to divide repeatedly by mitosis. This tissue gives rise to all other plant tissue and cell types.

mesophyll cell a photosynthetic ground tissue cell

micronutrient one of a group of eight nutrients required by plants in quantities less than 100 mg/kg of dry mass

microvilli microscopic fingerlike outward projections of the cell membrane

mineral an element (such as copper, iron, calcium, phosphorus, etc.) required by the body, often in trace amounts. Minerals are inorganic materials.

minimum viable population size the smallest number of individuals needed for a population to continue for a given period of time

mitochondrial DNA the DNA in mitochondria

Monera in a five-kingdom system, organisms that lack a true nucleus

monocot an angiosperm that produces seeds containing embryos with one cotyledon

monoculture a habitat dominated by a single plant species

monomer the individual component of a polymer

monosaccharide the simplest type of sugar molecule

moss a member of the kingdom Plantae that lacks a vascular system

motor neuron a nerve cell that carries an impulse from the brain or spinal cord to a muscle or gland

movable joint a joint that allows bones to move

mucus a lubricating substance, secreted by glands in the stomach walls, that protects the stomach

musculoskeletal system the organ system composed of the skeleton (skeletal system) and muscles (muscular system)

mutualism a symbiotic relationship in which both organisms benefit. Since neither is harmed, it is categorized as a +/+ relationship.

mycelium a collective term for the branching filaments that make up the part of a fungus not involved in sexual reproduction

mycorrhiza a symbiotic relationship between the hyphae of certain fungi and the roots of some plants

myelin sheath the fatty white tissue that covers some axons

myocardial infarction (*myo*, referring to muscle, *cardial*, referring to the heart, and *infarct*, meaning "death") death of heart muscle tissue due to lack of oxygen and other nutrients; commonly known as a heart attack

N

narcotic a psychoactive drug that relieves pain and makes you sleepy

natural fertilizer fertilizer produced without human-directed chemical activities

nectar a sweet liquid produced by flowers that attracts insects or other animals

negative feedback a process that restores conditions to an original state

nephron a functional unit of the kidneys

nerve a bundle of neurons

nerve cell or **neuron** a specialized cell that uses electrical signals to communicate with other cells

nervous system the control system that enables animals to detect a stimulus and coordinate a response

neurotransmitter a chemical that transmits nerve-cell impulses from one nerve cell to another

niche an organism's role in its habitat, including how it interacts with the biotic and abiotic resources in its environment

nitrification the conversion of ammonia or ammonium ions to nitrates by nitrifying bacteria in the soil

nitrogenase an enzyme used by nitrogen-fixing bacteria in anaerobic conditions to break down atmospheric nitrogen gas

nitrogen fixation the conversion of nitrogen gas from the atmosphere to ammonia by nitrogen-fixing bacteria

node the attachment site of a leaf to a stem

nonpolar covalent bond a covalent bond in which charges are not separated because electrons are equally shared

normal level or **reference level** an average value for measurement of a body structure or a function

normal range or **reference range** the average ranges of values for measurements of body structures or functions above or below which a person is suspected of having an illness

nuclear envelope the double membrane surrounding the nucleus

nucleoplasm the fluid material inside the nucleus

O

observational study a scientific study in which the scientist attempts to gain scientific information by observing natural phenomena without interfering with the normal order of the environment

open population a population in which changes in number and density are determined by births, deaths, immigration, and emigration

organ a structure composed of different tissues specialized to carry out a specific function

organ system a group of organs that have related functions

origin the place where a muscle is attached to a stationary bone

osmosis the net movement of water across a selectively permeable membrane from an area of high concentration to an area of low concentration

ossification the process in which cartilage is replaced by compact bone

osteocyte a bone cell

ovary an organ that contains egg cells

oxidative respiration a series of reactions occurring in the mitochondrion that uses oxygen to convert pyruvate into carbon dioxide, water, and ATP

oxygen debt the extra oxygen required to clear the body of lactate that accumulates in muscle cells during vigorous exercise

P

palisade mesophyll a layer of mesophyll cells at right angles to the leaf surface

Pap test the microscopic examination of cells scraped from the surface of the cervix to detect the presence of abnormal cells

parasitism a symbiotic relationship in which one organism (the parasite) benefits at the expense of another organism (the host), which is often harmed but usually not killed. It is categorized as a +/− relationship.

parasympathetic nerve a nerve that returns the body to a normal state after a stressful situation

passive immunity temporary protection against a particular disease by the direct introduction of antibodies

pathogen a disease-causing microorganism

pepsin a protein-digesting enzyme produced by the stomach

peptidase an enzyme that breaks peptides into amino acids

peptide bond a bond between amino acids in a protein molecule

perennial a plant that takes more than two years to complete its life cycle

periosteum the tough outer covering of bones

peripheral nervous system all nerves except those in the brain and spinal cord that relay information between the central nervous system and other parts of the body

peristalsis rhythmic, wavelike contractions of smooth muscles that move food along the gastrointestinal tract

pesticide any chemical used to kill unwanted organisms

phagocytosis the bulk transport of solids into the cell

phagocytotic vesicle a vesicle that is formed within a cell's cytoplasm when cells engulf solid particles by phagocytosis

pheromone a chemical emitted by an organism into the environment as a signal to another organism, usually of the same species

phloem vascular tissue that transports sugars and other solutes through the plant

phospholipid a fat molecule in a cell membrane

photoperiod the number of daylight hours

photosynthesis a set of chemical reactions that converts carbon dioxide and water into molecules that plant cells use for food

phototropism a change in the growth pattern or movement of a plant in response to light

phylogeny the history of the evolution of a species or a group of species

phytoremediation the use of plants to repair pollution-damaged ecosystems

pinocytosis the bulk transport of (liquid) extracellular fluid into the cell

pinocytotic vesicle a vesicle that is formed within a cell's cytoplasm when cells engulf extracellular fluid by pinocytosis

pistil the female reproductive part of a flower, consisting of the stigma, style, and ovary

pituitary gland a gland at the base of the brain that secretes more types of hormones than any other endocrine gland

plant growth regulator a chemical produced by plant cells that regulates plant growth and development; also called a plant hormone

plaque a sticky yellow substance, composed of fatty deposits, calcium, and cellular debris, that builds up on the inside of arteries

plasmid a small piece of genetic material

plastid an organelle in plant cells that stores useful materials or carries out vital functions such as photosynthesis

pleural membrane a thin, fluid-filled membrane that surrounds the outer surface of the lungs and lines the inner wall of the chest cavity

pneumonia a lung infection caused by a virus, bacterium, or fungus that leads to the accumulation of fluid within the lung

polar covalent bond a covalent bond in which positive and negative charges are separated because of unequally shared electrons

pollen grain a specialized plant structure from which male sex cells are released

pollinate to transfer pollen from one plant to another

polymer a large molecule composed of long chains of repeating subunits

polypeptide a short chain of amino acids

polysaccharide a complex carbohydrate composed of long chains of monosaccharides

population a group of individuals of the same species living in a particular geographic area at the same time

population density the number of individuals of the same species that occur per unit area or volume

population dynamics changes in population characteristics determined by natality, mortality, immigration, and emigration

population size the number of individuals of a species occupying a given area or volume at a given time

predation an ecological relationship in which a member of one species (predator) catches, kills, and eats a member of another species (prey)

primary growth all growth of apical meristems and first-year growth of lateral meristems

primary root the central, often largest, portion of the root

progesterone the hormone secreted by cells of the corpus luteum that stimulates the cells of the lining of the uterus to prepare for a possible pregnancy

prokaryotic a type of cell that does not have its chromosomes surrounded by a nuclear envelope

protein an unbranched polymer of amino acids

protein synthesis the production of proteins

Protista a kingdom originally proposed for all unicellular organisms. More recently, multicellular algae have been added to the kingdom.

pseudopod a fingerlike projection of the cell membrane that surrounds and encloses solid particles in the extracellular fluid that are brought into a cell by phagocytosis

psychoactive drug a drug that affects the nervous system and often results in changes to behaviour

pulmonary circuit the system of blood vessels that carries deoxygenated blood to the lungs and oxygenated blood back to the heart

pulse change in the diameter of the arteries that can be felt on the body's surface following heart contractions

pyramid of numbers a model that illustrates the number of organisms of a particular type that can be supported at each trophic level

Q

quadrat a sampling frame used for estimating population size; frames can be real or virtual

qualitative referring to qualities or non-numerical data

quantitative referring to quantities or numerical data

R

random dispersion type of population distribution in which individuals live throughout a habitat in an unpredictable and patternless manner

recombinant DNA fragments of DNA from two or more different organisms

renal artery a blood vessel that carries waste-laden blood to the kidneys

renal vein a blood vessel that carries purified blood from the kidneys to the body

reproductive cloning the production of fully formed cloned organisms

reproductive system the organ system in males and females that performs the functions of sexual reproduction

resolution the ability to show objects that are close together as separate and distinct

resource partitioning avoidance of, or reduction in, competition for resources by individuals of different species occupying overlapping niches

respiration all processes involved in the exchange of oxygen and carbon dioxide between cells and the environment

restriction enzyme an enzyme that attacks a specific sequence of nitrogenous bases on a DNA molecule to create fragments of DNA with unpaired bases

rhizoid a simple rootlike filament used to absorb water and anchor mosses to surfaces

rhizome an underground stem

root the portion of the plant that typically grows downward and absorbs water and minerals from the soil

root cap a loose mass of cells forming a protective cap covering the apical meristem of root tips

root tip the end of a growing root

rough endoplasmic reticulum (RER) endoplasmic reticulum with attached ribosomes

S

sand very small pieces of rock

saprophyte an organism that obtains nutrients from dead or nonliving organic matter

saturated triglyceride a triglyceride containing fatty acids with only single covalent bonds between carbon atoms

scientific method the systematic method used in scientific investigations of the natural world, which includes designing and carrying out controlled experiments to test hypotheses

secondary growth all growth from lateral meristems occurring after the first year

secondary root a branching root arising from the primary root

secretory vesicle a membrane sac containing substances that are transported out of a cell

seed a specialized plant structure containing an embryonic plant and food supply

seed coat a protective layer covering a seed

selectively permeable allows only certain substances to pass through it

selective pesticide a pesticide that kills a more limited range of target and nontarget species; also called a narrow-spectrum pesticide

seminal fluid a nutrient-rich fluid secreted by several glands of the male reproductive system

sensory neuron a nerve cell that carries an impulse from a receptor to the brain or spinal cord

septum a wall of muscle that separates the right heart pump from the left

shell gland the gland in the bird oviduct that forms a hard shell around the egg

shoot typically, the above-ground portions, such as the stem and leaves, of vascular plants

sieve tube a long tube formed from many individual sieve-tube elements joined end to end

sieve-tube element a cell type in the phloem that lacks nuclei, mitochondria, or vacuoles, and has large pores through the cell walls at either end

silt a mixture of sand and clay particles that is deposited by moving water

silviculture the growing and management of forests and tree plantations

simple diffusion the movement of particles from an area of higher concentration to an area of lower concentration until particle concentration is equal throughout

simple leaf a leaf that is not divided into leaflets

sinoatrial (SA) node a small mass of tissue in the right atrium that originates the impulses stimulating the heartbeat

skeletal muscle the voluntary muscle that makes the bones of the skeleton move

skull the curved, fused bones that form a hard protective covering for the brain

smooth endoplasmic reticulum (SER) endoplasmic reticulum without ribosomes

smooth muscle the involuntary muscle found in the lining of many internal organs

sodium-potassium pump an active-transport mechanism that pumps sodium and potassium ions into and out of a cell

somatic nervous system motor neurons that are under conscious control

species a group of organisms that resemble one another physically, behaviourally, or genetically and that can interbreed under natural conditions to produce fertile offspring

sphincter a constrictor muscle that surrounds a tubelike structure

sphygmomanometer a device used to measure blood pressure

spongy bone a porous bone

spongy mesophyll loosely arranged mesophyll cells within the leaf

sporangium a reproductive structure in which spores are produced

spore a reproductive cell that can produce a new organism without fertilization. Fungal spores have thick, resistant outer coverings to protect them.

sporophyte the form of a plant that reproduces asexually

stamen the male reproductive part of a flower, consisting of an anther and filament

stem cell an unspecialized cell that may transform into a specialized cell

stent a small metallic mesh cylinder that is inserted into a blocked artery to allow blood to flow freely and to prevent the recurrence of the blockage

stigma the sticky surface at the tip of the style of a flower

stimulant a drug that speeds up the action of the central nervous system, often causing an increase in heart and breathing rates

stoma a small opening in the surface of a leaf controlled by a pair of guard cells

stroma the protein-rich fluid that fills the space between thylakoid membranes and the inner membrane of a chloroplast

style the stalk on the female reproductive tissue of a flower

submergent plant an aquatic plant that grows entirely underwater

substrate the reactant that an enzyme acts on when it catalyzes a chemical reaction

sustainability the ability of an ecosystem to maintain ecological processes, functions, biodiversity, and productivity over time

sustainable agriculture method of growing crops and raising livestock based on organic fertilizers, soil conservation, water conservation, biological pest control, and minimal use of nonrenewable fossil-fuel energy

suture the immovable joints between bones of the skull

symbiotic relationship a relationship between two organisms in which both partners benefit from the interaction

sympathetic nerve a nerve that prepares the body for danger

synapse the connection between the terminal knob of a neuron's axon and a dendrite of an adjacent neuron

synovial fluid lubricating fluid between the bones of a joint

synthetic fertilizer fertilizer produced by human-directed chemical processes

systemic circuit the system of blood vessels that carries oxygenated blood to the tissues of the body and deoxygenated blood back to the heart

systole contraction of the ventricles, during which blood is pushed out of the heart

T

taproot a root system with a large central root and many smaller side branches

target species the particular pest species under consideration

taxon a category used to classify organisms

tendon a tough, largely inflexible strip of connective tissue that attaches muscle to bone

terminal bud the growing tip of a shoot

terminal knob the part of a neuron that attaches it to another cell

therapeutic cloning the production of cloned embryos for the purpose of obtaining stem cells

thigmotropism a change in the growth pattern or movement of a plant in response to touch

thylakoid a membrane-bound compartment found in chloroplasts

tissue a group of cells that works together to perform a specialized task

toxin a poison produced in the body of a living organism. It is not harmful to the organism itself but only to other organisms.

trachea the windpipe through which air passes from the pharynx toward the lungs

transmembrane protein a protein molecule in a membrane that spans the thickness of the phospholipid bilayer

transpiration water loss from a plant surface by the process of evaporation

triglyceride a lipid molecule composed of glycerol and three fatty acids

trophic level a step in the transfer of matter and energy along a food chain or through a food web. Each organism is assigned to a trophic level depending on its primary source of food.

tropism a change in the growth pattern or movement of a plant in response to an external stimulus

trypsin a protein-digesting enzyme

turf management the planting, growing, and maintaining of grass lawns and fields

turgor pressure pressure, caused by water in the cytoplasm and vacuoles, that helps keep the cell membrane pressed firmly against the cell wall of a plant cell

U

uniform dispersion a distribution pattern in which individuals are equally spaced throughout a habitat

unsaturated triglyceride a triglyceride containing fatty acids with double bonds between carbon atoms

urea nitrogen waste formed from two molecules of ammonia and one molecule of carbon dioxide

ureter one of the tubes that conducts urine from the kidneys to the bladder

urethra the tube that carries urine from the bladder to the exterior of the body

urine the solution of dissolved wastes produced by the kidneys and stored in the bladder

V

vaccine a suspension prepared from dead or weakened viral or bacterial cells

vascular cambium a lateral meristem that is responsible for creating secondary xylem and phloem tissue

vascular system a system of specialized cells or tissues that transport water and dissolved materials between cells in the body of an organism

vascular tissue tissue composed of cells specialized in transporting water and other substances among cells

vegetative any growth or development that does not involve sexual reproduction

vein a vascular bundle that forms part of the conducting and support tissue of a leaf

ventricle a muscular, thick-walled chamber of the heart that delivers blood to the arteries

vertebral column a stack of short, hollow, cylindrical bones that provides a protective covering for the spinal cord

vesicle a membranous sac containing proteins and lipids that may be formed by the endoplasmic reticulum and Golgi apparatus

villi small fingerlike projections that extend into the small intestine, increasing surface area for absorption

virus a microscopic particle capable of reproducing only within living cells

vitamin an organic molecule needed in trace amounts for normal growth and metabolic processes

W

woody describes a plant in which some parts are composed of wood

xylem vascular tissue that transports water and minerals

Index

C

Index

D

E

Index

F

G

H

Index

I

J

K

L

M

N

O

P

Index

Q

R

S

U

V

W

X

Y

Z

Photo Credits

Unit Opening Photos: Unit 1: iv, viii ©SPL/Photo Researchers; Unit 2: iv, 92 PhotoLink/Photodisc/Gettyimages; (inset) ©Andrew Syred/Science Photo Library; Unit 3: v, 158 ©Lester V. Bergman/CORBIS/MAGMA; Unit 4: vi, 266 ©Inge Spence/Visuals Unlimited; (inset top) ©Richard Choy/Peter Arnold; (inset bottom) ©Barbara Magnuson/Visuals Unlimited; Unit 5: vi, 354 ©J.K. Keener/firstlight.ca.

Unit 1: 2 Jeremy Jones; 3 (Fig 4) courtesy Boreal; (Fig 5) Doug Fraser; 4 (Fig 1) ©SIU/Visuals Unlimited; (Fig 2) Chase Jarvis/PhotoDisc/Gettyimages; 5 (Fig 3) ©M. Kalab/Visuals Unlimited; (Fig 4) Julie Greener; 6 (Fig 1) Courtesy Sally Corporation; (Fig 2) Photodisc/Gettyimages; 7 (Fig 3) ©Kathy Talaro/Visuals Unlimited; (Fig 4) courtesy Boreal; 8 (Fig 5) ©Stanley L. Flegler/Visuals Unlimited; (Fig 6) London School of Hygiene & Tropical Medicine/Science Photo Library; (Fig 7) ©Brad Mogen/Visuals Unlimited; (Fig 8) ©Dr. Stanley Flegler/Visuals Unlimited; 9 ©Harold Fisher/Visuals Unlimited; 11 (Fig 5) ©Don W. Fawcett/Visuals Unlimited; 12 (Fig 6) ©CNRI/Science Photo Library; (Fig 7) ©K.G. Murti/Visuals Unlimited; 13 (Fig 9) ©Dr. Gopal Murti/Science Photo Library; 14 (Fig 10) W.P. Wergin/E. H. Newcomb/University of Wisconsin/Biological Photo Service; 22 (Fig 8) ©Dennis Drenner/Visuals Unlimited; (Fig 9) ©Science VU/Visuals Unlimited; 24 Julie Greener; 27 (a) ©Dave B. Fleetham/Visuals Unlimited; (b) ©Phil A. Dodson/Photo Researchers; 28 ©G.W. Willis, M.D./Visuals Unlimited; 30 (Fig 12a) Emma Lee/Life File/Photodisc/Gettyimages; (b) Scott Camazine/Photo Researchers; (Fig 15, left to right) C-Squared Studios/Photodisc/Gettyimages; ©P. Motta/Photo Researchers; ©John D. Cunningham/Visuals Unlimited; 40 (Fig 1) ©Dick Hemingway; (Fig 2) Julie Greener; 42 Julie Greener; 43 ©Science Photo Library; 50 (a) ©SIU/Visuals Unlimited; (b) ©Mediscan/Visuals Unlimited; 52 Courtesy Thomas Steitz/ Yale University; 54 Julie Greener; 59 (Fig 1) ©Mediscan/Visuals Unlimited; (Fig 2) © Bill Beatty/Visuals Unlimited; 63 ©Dr. David Phillips/Visuals Unlimited; 67 (a) ©Jeff Greenberg/Visuals Unlimited; (b) ©David M. Phillips/Visuals Unlimited; (c) Michael W. Clayton, Dept. of Botany, University of Wisconsin; 70 (Fig 7) ©Dr. David Phillips/Visuals Unlimited; (Fig 8) ©M. Schliwa/ Visuals Unlimited; 72 (a) © Inge Spence/Visuals Unlimited; (b) ©J. Richardson/Visuals Unlimited; 73 (a) ©Robert Calentine/Visuals Unlimited; (c) ©George Chapman/Visuals Unlimited; 76 Visuals Unlimited; (a, inset) ©Jerome Wexler/Visuals Unlimited; 79 ©D. Fawcett/Visuals Unlimited; 80 (Fig 3, left to right) ©Gary Randall/ Visuals Unlimited; Photodisc/Gettyimages; Royalty-Free/CORBIS/ MAGMA; (Fig 5) CP Archive/AP Photo/Amy Sancetta; 88 (Fig 1 & 3) Julie Greener; (Fig 2) ©Jerome Wexler/Visuals Unlimited; 89 (Fig 6, left to right) Photodisc/Gettyimages; ©John Gerlach/Visuals Unlimited; ©Dick Hemingway.

Unit 2: 97 ©Lee D. Simon/Photo Researchers; 103 (Fig 10, left to right) ©David M. Phillips/Visuals Unlimited; ©Dr. Hans Reichenbach; ©Biophoto Associates/Photo Researchers; ©K.G. Murti/Visuals Unlimited; ©T.E. Adams/Visuals Unlimited; 105 ©Alfred Pasieka/Science Photo Library; 106 (Fig 5) Science VU/Visuals Unlimited; (Fig 6) ©K.G. Murti/Visuals Unlimited; 108 (b) ©USDA/Visuals Unlimited; 109 (a) ©M. Kalab/Visuals Unlimited; (b) ©T.J. Beveridge/Visuals Unlimited; (c) ©Annie Griffiths Belt/CORBIS/MAGMA; (Fig 3) ©Fred Hossler/Visuals Unlimited; 110 ©Dr. Stephen Hogg, School of Dental Sciences, University of Newcastle, UK; 111 ©Dr. Hans Reichenbach; 112 ©George P. Chapman/Visuals Unlimited; 114 ©Royalty-Free/CORBIS/MAGMA; 116 ©Loring Nies, Purdue School of Civil Engineers; 117 ©UFCSIM/Visuals Unlimited; 118 (Fig 1) ©Arthur M. Stegelman/Visuals Unlimited; (Fig 2) CP Photo/Kevin Frayer; 123 Photo Researchers; 125 ©Barry Cohen; 127 Grambo Photography/firstlight.ca; 128 ©Biophoto Associates/Photo Researchers; 129 (a) ©R. Kessel-G. Shih/Visuals Unlimited; (b) ©Cabisco/Visuals Unlimited; 130 (Fig 5) ©Cabisco/Visuals Unlimited; (Fig 6a) ©Eric Grave/Science Photo Library; (b) ©CNRI/Science Photo Library; (c) Roy Ficken/Memorial University; (d) S. Flegert/Visuals Unlimited; 132 (Fig 10a) ©R. Calentine/Visuals Unlimited; (b) Onderstepoort Veterinary Research Institute; 136 (Fig 1 left, top to bottom) ©Janice Palmer; (Fig 1 right, top to bottom) ©Jana R. Jurar/Visuals Unlimited; ©Kevin & Betty Collins/Visuals Unlimited; ©Walter H. Hodge/Peter Arnold; ©Janice Palmer; 137 (Fig 3, left to right) © Dr. Tony Brain/Science Photo Library; ©Dr. Jeremy Burgess/Science Photo Library; © J.L. Lepore/Photo Researchers; © J. Forsdyke/Science Photo Library; © CNRI/Science Photo Library; (Fig 4) ©G.A. Maclean/Oxford Scientific Films; 138 (Fig 6) ©Lee D. Simon/Photo Researchers; (Fig 7) ©Visuals Unlimited; 139 ©Janice Palmer; 140 ©Janice Palmer; 143 ©Cabisco/Visuals Unlimited; 146 ©Dr. Gopal Murti/Visuals Unlimited; 153 courtesy Prof. Thomas A. Zitter, Cornell University, Ithaca, N.Y; 154 (Fig 1) ©Eric Grave/Science Photo Library; (Fig 3) ©Cabisco/Visuals Unlimited; 155 ©R.F. Ashley/Visuals Unlimited.

Unit 3: 163 ©Reuters New Media/CORBIS; 166 (a) ©CNRI/Science Photo Library; (b) ©Ohio-Nuclear Corporation/Science Photo Library; (c) ©CNRI/Science Photo Library; (d) ©Mihau Kulyk/Science Photo Library; 167 ©Dr. R.F.R. Slehiller/Science Photo Library; 181 (left) ©Doug Martin/Photo Researchers; (right) ©Will & Deni McIntyre/Photo Researchers; 187 ©Manfred Kage/Science Photo Library; 189 courtesy Warren Bishop, M.D., University of Iowa; 192 Meredith White-McMahon; 193 Jeff Malone/PhotoDisc/ Gettyimages; 194 (Fig 3 top, left to right) Bettmann/CORBIS/MAGMA; Hulton-Deutsch Collection/CORBIS/MAGMA; Bettmann/CORBIS/MAGMA; (bottom left) ©Mitchell Gerber/CORBIS/MAGMA; (bottom right) Archivo Inconografisco SA/CORBIS/MAGMA; 199 (Fig 6) ©Cabisco/Visuals Unlimited; (Fig 7) ©Cardi-Thoracic Centre/ Freeman Hospital/Science Photo Library; 200 (top) ©James A. Sullivan, www.cellsalive.com; (centre left) ©Don Fawcett/Visuals Unlimited; (centre right) ©David M. Phillips/Visuals Unlimited; (a) James A. Sullivan, www.cellsalive.com; (b, c) Michael Rolf Meuller; 203 (a) Royalty-Free/CORBIS/MAGMA; (b) ©Visuals Unlimited; 206 (a) ©Cabisco/Visuals Unlimited; (b) ©Alfred Pasieka/Visuals Unlimited; 207 ©Z&B Barran/STONE; 208 ©Lester Lefkowitz/CORBIS/MAGMA; 209 ©Sheila Terry/Science Photo Library; 210 ©Cabisco/Visuals Unlimited; 213 ©David M. Phillips/Visuals Unlimited; 216 Julie Greener; 218 ©Paul Seheult, Eye Ubiquitous/CORBIS/MAGMA; 222 (Fig 10) Eyewire/Gettyimages; (Fig 9, left) ©Didrick Johnck/CORBIS SYGMA/MAGMA; (right) ©Dave. B. Fleetham/Visuals Unlimited; 223 Doug Fraser; 226 (top) ©Simko/Visuals Unlimited; (bottom) © Bernhard Wittich/Visuals Unlimited; 227 ©Mediscan/Visuals Unlimited; 228 (a) Photo Researchers, Inc.; (b) ©Mediscan/Visuals Unlimited; 230 (Fig 1) CP Photo/Moncton Times and Transcript/Greg Agnew; (Fig 2) ©David M. Phillips/Visuals Unlimited; 240 CP Photo/Hamilton Spectator-John Rennison; 242 (Fig 3) Franmaur Enterprises, Inc.; (Fig 4) ©L. Bassett/Visuals Unlimited; 243 ©Photo Researchers; 248 (Fig 14, left to right) ©Dr. Michael K. McLennan/Diagnostic Radiologist, Markham Stouffville Hospital, Markham, Ontario; ©Science Photo Library; ©Mediscan/Visuals Unlimited; 250 ©Photo Researchers; 251 ©L. Bassett/Visuals Unlimited.

Unit 4: 270 Doug Fraser; 271 ©Janice Palmer; 272 (Fig 1) Doug Fraser; (Fig 2) ©G & C Merker/Visuals Unlimited; 273 (Table 1, left, top to bottom) ©David Sieren/Visuals Unlimited; ©Bill Beatty/Visuals Unlimited; ©David Matherly/Visuals Unlimited; (Table 1, right, top to bottom) ©Robert L. Johnson, Cornell University, www.forestryimages.org; ©Inge Spence/Visuals Unlimited; ©David Sieren/Visuals Unlimited; (Fig 3) ©Janice Palmer; 275 (a) ©Janice Palmer; (b) ©Joe McDonald/Visuals Unlimited; 277 ©Janice Palmer; 278 (Fig 5) ©John G. Roberts/First Light; (Fig 6) Doug Fraser; 279 J. H. Robinson/Photo Researchers; 280 (Fig 1a) ©Gary C. Will/Visuals Unlimited; (b) ©Ned Therrien/ Visuals Unlimited; (Fig 2a) ©Janice Palmer; (b) Doug Fraser; 281 (Fig 3a) ©Rob & Ann Simpson/Visuals Unlimited; (b) Doug Fraser; 282 Julie Greener; 287 Julie Greener; 296 (Fig 3b) Doug Fraser; (Fig 4) Mike Sumner, University of Manitoba; 299 ©Janice Palmer; 300 Julie Greener; 303 ©Dick Hemingway; 304 ©Ryan McVay/ Photodisc/Gettyimages; 306 Doug Fraser; 307 ©Janice Palmer; 308 (Fig 2) ©Inge Spence/Visuals Unlimited; (Fig 3) © Visuals Unlimited; 309 (Fig 4) ©Inge Spence/Visuals Unlimited; (Fig 5) Doug Fraser; 311 (Fig 3) Mike Sumner, University of Manitoba; 312 ©Keith Dannemiller/CORBIS SABA/MAGMA; 314 ©Jerome Wexler/Photo Researchers; 315 ©Barry Cohen; 316 (Fig 2) ©Pat Anderson/Visuals Unlimited; (Fig 3) ©Barry Cohen; 317 ©D.W. Stock/Visuals Unlimited; 319 ©Dick Hemingway; 321 ©Adam Jones/Visuals Unlimited; 325 ©Janice Palmer; 326 ©Janice Palmer; 328 ©Janice Palmer; 329 (Fig 1) ©Rob Simpson/Visuals Unlimited; (Fig 2) ©Janice Palmer; (Fig 3a) ©David McComb, USDA Forest Service, www.forestryimages.org; (b) ©Rob & Ann Simpson/Visuals Unlimited; 330 ©Joe McDonald/Visuals Unlimited; 331 ©Kevin & Betty Collins/Visuals Unlimited; 333 ©Janice Palmer; 335 (Fig 1) ©Gilbert Twiest/Visuals Unlimited; (Fig 2) L. Diane Lackie/Visuals Unlimited; 337 (Fig 1) ©Inge Spence/Visuals Unlimited; (Fig 2) Philippe Psaila/Science Photo Library; 338 (Fig 3) ©Steve McCutcheon/Visuals Unlimited; (Fig 4) ©William F. Smith/Lone Pine Photo; (Fig 5) ©Clarence W. Norris/Lone Pine Photo; 339 (Fig 7) ©Janice Palmer; 340 ©Clarence W. Norris/Lone Pine Photo; 341 ©Bill Beatty/Visuals Unlimited; 342 (Fig 2 & 4) courtesy Royal Botanical Gardens, www.rbg.ca; (Fig 3) ©Gary Meszaras/Visuals Unlimited; 343 ©Steve Callahan/Visuals Unlimited; 349 (Fig 1) ©Dick Hemingway; (Fig 2) ©Adam Jones/Visuals Unlimited; 350 ©Jerome Wexler/Photo Researchers; 352 (a) ©RMF/Visuals Unlimited; (b) ©Mike Clayton, Dept. of Botany, University of Wisconsin; (c) ©Dr. Stanley Flegler/Visuals Unlimited; (d) London Scientific Films/Oxford Scientific Films.

Unit 5: 357 ©V. Last/Geographical Visual Aids; 358 (Fig 1) NASA; (Fig 2, left to right) Paul Beard/Photodisc/Gettyimages; ©Joe McDonald/CORBIS/MAGMA; Jeremy Woodhouse/Photodisc/ Gettyimages; 359 "The Millennial Gaia" by Oberon Zell. 7 1/2" Polystone statue. The Mythic Images Collection: http://www.mythicimages.com/gaia.html. 1-888-MYTHIC-1; 361 (Fig 1) ©H.G. Rose/Visuals Unlimited; (Fig 2) ©Joe McDonald/Visuals Unlimited; 362 (top) ©V. Last/Geographical Visual Aids; (bottom) Glenn Allison/Photodisc/Gettyimages; 366 (top to bottom) ©Frank Awbrey/Visuals Unlimited; ©Royalty-Free/CORBIS/MAGMA; Photodisc/Gettyimages; Don Farrell/ Photodisc/Gettyimages; Digital Vision/Gettyimages; Ryan McVay/Photodisc/Gettyimages; Ian Cartwright/Photodisc/ Gettyimages; 367 (Fig 3a) ©Ann & Rob Simpson/Visuals Unlimited; (Fig 3b) Eyewire/Gettyimages; (Fig 4a) Digital Vision/Gettyimages; (Fig 4b) PhotoLink/Photodisc/ Gettyimages; 370 (top to bottom) Alex L. Fradkin/Photodisc/ Gettyimages; ©V. Last/Geographical Visual Aids; ©V. Last/ Geographical Visual Aids; James Gritz/Photodisc/Gettyimages; ©V. Last/Geographical Visual Aids; 374 ©Martin G. Miller/Visuals Unlimited; 382 (Fig 1, left to right) ©Rob & Ann Simpson/Visuals Unlimited; ©Rob Simpson/Visuals Unlimited; ©Rob & Ann Simpson/Visuals Unlimited; ©Steve Maslowski/Visuals Unlimited; ©Rob Simpson/Visuals Unlimited; (Fig 2) Tom Ulrich/Visuals Unlimited; 383 (a) Greg Neise/Visuals Unlimited; (b) ©Gerald & Buff Corsi/Visuals Unlimited; (c) ©Stephen J. Krasemann/Photo Researchers; 385 (Fig 1) V. Last/Geographical Visual Aids; (Fig 2) ©Dan Richter/Visuals Unlimited; 386 (Fig 3a) ©John D. Cunningham/Visuals Unlimited; (b) James Gritz/ Photodisc/ Gettyimages; (Fig 4) Science Photo Library; 387 (Fig 5a) ©Science VU/ARS/Visuals Unlimited; (b) ©Mike Abbey/Visuals Unlimited; (Fig 6) Prof. Steve J. Upton, Kansas State University; 388 (a) Dr. Andrew Bryant/Marmot Recovery Foundation; (b) ©Arthur Morris/Visuals Unlimited; 390 (Fig 2, left to right) © Photo Researchers; ©Fritz Polking/Visuals Unlimited; ©Will & Deni McIntyre/Photo Researchers; 393 ©Arthur Morris/Visuals Unlimited; 396 ©Gary Meszaros/Visuals Unlimited; 397 (Fig 3) ©Inge Spence/ Visuals Unlimited; (Fig 4) ©Ken Lucas/Visuals Unlimited; 398 (Fig 5) ©John Gerlach/Visuals Unlimited; (Fig 6) ©Janice Palmer; 400 ©Nova Scotia Leatherback Turtle Working Group; 401 (Fig 1) ©Tim Hauf/Visuals Unlimited; (Fig 2) ©Gunter Marx Photography/ CORBIS/MAGMA; 403 Bayard H. Brattstrom/Visuals Unlimited; 415 (Fig 6 top) Digital Vision/ Gettyimages; (Fig 6) ©V. Last/Geographical Visual Aids; (Fig 7) ©D. Cavagnaro/Visuals Unlimited; 418 (Fig 1 top to bottom) ©David Wrobel/Visuals Unlimited; ©Gerald & Buff Corsi/Visuals Unlimited; ©Gerard Fuehrur/Visuals Unlimited; ©R.Calentine/Visuals Unlimited; ©McCutcheon/Visuals Unlimited; 419 (Fig 2) Photodisc/ Gettyimages; (Fig 3) ©Dick Thomas/Visuals Unlimited; 421 Thomas R. Clarke, U.S. Water Conservation Laboratory, U.S.D.A. Agricultural Research Service; 422 (a) Doug Fraser; (b) ©A. J. Cunningham/ Visuals Unlimited; 423 (a) courtesy DPI Industries; (b) ©Janice Palmer; (c) ©Geoff Manasse/Photodisc/ Gettyimages.

Appendices: vii, 434 Royalty-Free/CORBIS/MAGMA; 460 courtesy Boreal; 461 courtesy Boreal; 465 (Fig 7) ©SIU/Visuals Unlimited; (Fig 8) courtesy Boreal; 466 ©Jeff Greenberg/Visuals Unlimited; 467 Meredith White-McMahon; 468 ©G.W. Willis, M.D./Visuals Unlimited; 469 ©A.M. Siegelman/Visuals Unlimited; 480 (Fig 1) ©V. Last/Geographical Visual Aids; (Fig 2) Ken Samuelsen/Photodisc/Gettyimages.